SQUID Sensors:
Fundamentals, Fabrication and Applications

NATO ASI Series

Advanced Science Institutes Series

A Series presenting the results of activities sponsored by the NATO Science Committee, which aims at the dissemination of advanced scientific and technological knowledge, with a view to strengthening links between scientific communities.

The Series is published by an international board of publishers in conjunction with the NATO Scientific Affairs Division

A Life Sciences	Plenum Publishing Corporation
B Physics	London and New York
C Mathematical and Physical Sciences	Kluwer Academic Publishers
D Behavioural and Social Sciences	Dordrecht, Boston and London
E Applied Sciences	
F Computer and Systems Sciences	Springer-Verlag
G Ecological Sciences	Berlin, Heidelberg, New York, London,
H Cell Biology	Paris and Tokyo
I Global Environmental Change	

PARTNERSHIP SUB-SERIES

1. Disarmament Technologies	Kluwer Academic Publishers
2. Environment	Springer-Verlag / Kluwer Academic Publishers
3. High Technology	Kluwer Academic Publishers
4. Science and Technology Policy	Kluwer Academic Publishers
5. Computer Networking	Kluwer Academic Publishers

The Partnership Sub-Series incorporates activities undertaken in collaboration with NATO's Cooperation Partners, the countries of the CIS and Central and Eastern Europe, in Priority Areas of concern to those countries.

NATO-PCO-DATA BASE

The electronic index to the NATO ASI Series provides full bibliographical references (with keywords and/or abstracts) to more than 50000 contributions from international scientists published in all sections of the NATO ASI Series.
Access to the NATO-PCO-DATA BASE is possible in two ways:

– via online FILE 128 (NATO-PCO-DATA BASE) hosted by ESRIN,
Via Galileo Galilei, I-00044 Frascati, Italy.

– via CD-ROM "NATO-PCO-DATA BASE" with user-friendly retrieval software in English, French and German (© WTV GmbH and DATAWARE Technologies Inc. 1989).

The CD-ROM can be ordered through any member of the Board of Publishers or through NATO-PCO, Overijse, Belgium.

Series E: Applied Sciences - Vol. 329

SQUID Sensors: Fundamentals, Fabrication and Applications

edited by

Harold Weinstock

Air Force Office of Scientific Research,
Washington, DC, U.S.A.

Springer-Science+Business Media, B.V.

Proceedings of the NATO Advanced Study Institute on
SQUID Sensors: Fundamentals, Fabrication and Applications
Acquafredda di Maratea, Italy
June 18–30, 1995

A C.I.P. Catalogue record for this book is available from the Library of Congress

ISBN 978-94-010-6393-7   ISBN 978-94-011-5674-5 (eBook)
DOI 10.1007/978-94-011-5674-5

Printed on acid-free paper

---

to
Linda, Steve *and* Allen

TABLE OF CONTENTS

PREFACE   ix

ACKNOWLEDGMENTS   xi

INTRODUCTION   xiii

1. SQUID FUNDAMENTALS   1
   J. Clarke

2. ADVANCED SQUID READ-OUT ELECTRONICS   63
   D. Drung

3. SQUID GRADIOMETERS IN REAL ENVIRONMENTS   117
   J. Vrba

4. DC SQUIDS: DESIGN, OPTIMIZATION   179
   AND PRACTICAL APPLICATIONS
   R. Cantor

5. FABRICATION OF HIGH-TEMPERATURE   235
   SQUID MAGNETOMETERS
   A. I. Braginski

6. PULSE TUBE REFRIGERATORS: A COOLING OPTION   289
   FOR HIGH-T_c SQUIDS
   C. Heiden

7. HIGH-RESOLUTION MAGNETIC IMAGING: 307
   CELLULAR ACTION CURRENTS
   AND OTHER APPLICATIONS
   J. P. Wikswo, Jr.

8. THE VOLUME CONDUCTOR PROBLEM IN BIOMAGNETISM   361
   M. J. Peters, S. P. van den Broek and F. Zanow

9. MAGNETOCARDIOGRAPHY, AN INTRODUCTION   395
   S. N. Erné and J. Lehmann

viii

10. MAGNETOCARDIOGRAPHIC AND ELECTROCARDIOGRAPHIC   413
    MAPPING STUDIES
        G. Stroink, M. J. R. Lamothe and M. J. Gardner

11. NEUROMAGNETISM AND ITS CLINICAL APPLICATIONS   445
        G. L. Romani, C. Del Gratta and V. Pizzella

12. INTEGRATING COMPETING TECHNOLOGIES WITH MEG   491
        M. C. Gilardi, G. Rizzo, G. Lucignani and F. Fazio

13. SUPERCONDUCTING MAGNETIC GRADIOMETERS   517
    FOR MOBILE APPLICATIONS WITH AN EMPHASIS
    ON ORDNANCE DETECTION
        T. R. Clem, G. J. KeKelis, J. D. Lathrop,
        D. J. Overway and W. M. Wynn

14. SUPERCONDUCTING ACCELEROMETERS,   569
    GRAVITATIONAL-WAVE TRANSDUCERS,
    AND GRAVITY GRADIOMETERS
        H. J. Paik

15. THE USE OF SQUIDS FOR NONDESTRUCTIVE EVALUATION   599
        G. B. Donaldson, A. Cochran and D. McA. McKirdy

16. THE MAGNETIC INVERSE PROBLEM FOR NDE   629
        J. P. Wikswo, Jr.

INDEX   697

GLOSSARY OF ABBREVIATIONS AND ACRONYMS   701

PREFACE

This book will be of value to anyone who wishes to consider the use of SQUID-based magnetic sensing for any one of a number of practical applications. The focus here is to examine in detail how SQUID technology is used and how the results of the measurements obtained can be interpreted to provide useful information in a variety of real-world applications. The concentration is on those areas that have received the most attention, namely biomagnetism and nondestructive evaluation, but the topics chosen include as well, geophysics, underwater ordnance detection, accelerometry and a few somewhat more exotic applications. To provide a reasonable perspective, an attempt has been made to consider competing technologies for most applications, and in some cases to consider how SQUID-based technology may be integrated with other technologies to provide an optimum total-system configuration.

It is also the intention of the editor, that this book will be of major value to those scientists and engineers who will be required to build both the essential components and complete cryogenic SQUID systems which will be utilized in the various applications presented. Thus, there is a comprehensive review of the principles of SQUID operation, and a detailed exposition on the fabrication of high-temperature-superconducting (HTS) SQUIDs. Although the market is currently dominated by low-temperature-superconducting (LTS) SQUIDs, it is reasonably certain that in the near future HTS SQUIDs will take over in most situations. Hence, the well-developed LTS SQUID fabrication technology, which is documented elsewhere [1], has been omitted here in favor of a detailed exposition of the evolving HTS SQUID technology. Information also is provided on the latest methods for improved noise reduction and signal enhancement, and there are complete descriptions of some of the most sophisticated multi-sensor systems on the market today.

Is there another book that provides similar information? The answer is a resounding no! A 1982 NATO Advanced Study Institute (ASI) [2] was the last one (prior to the 1995 NATO ASI upon which these proceedings are based) to deal exclusively with SQUID applications, but it dealt only with biomagnetism, and it took place at a time when only single-sensor SQUID systems were available. It is clearly time for an update to that ground-breaking NATO ASI. There is, however, another fine book which must be mentioned and which is very complementary to the current one. At first glance, it might appear that the book entitled *Principles and Applications of Superconducting Quantum Interference Devices*, edited by Antonio Barone [1], covers the same ground. However, close examination will show that in the area of applications, the only major overlap is in neuromagnetism, with the same chapter author in this subject, Gian-Luca Romani, for both books. There is overlap, as well, on the principles of rf and dc SQUIDs and on the fabrication of HTS SQUIDs. Since the earlier book is now five years old, there is good reason to re-examine the subject of HTS SQUID fabrication, particularly because this current volume actually examines the application of HTS SQUIDs to the real world, something not considered in the earlier volume. As has already been mentioned above, the earlier book provides an excellent exposition of the practical fabrication of LTS SQUIDs. It is unique also in the coverage of fundamental physics experiments in a chapter authored by Blas Cabrera. No attempt has been made in the current volume to include a discussion of Cabrera's ingenious original studies, primarily because the focus has been placed here on applications that extend beyond the research laboratory and which potentially can result in a major market for the manufacturers of SQUID-based technology.

Washington, DC   Harold Weinstock
August 1996

References

1.   Takada, S. and Koyanagi, M. (1992) Fabrication of integrated dc-SQUID with refractory tunnel junctions, in A. Barone (ed.), *Principles and Applications of Superconducting Quantum Inteference Devices*, World Scientific Publishing Co., Singapore, pp. 151-186.
2.   Williamson, S.J., Romani, G.L., Kaufman, L. and Modena, I. (eds.) (1983), *Biomagnetism: An Interdisciplinary Approach*, Plenum Press, New York.

ACKNOWLEDGMENTS

This book compromises the official proceedings of the NATO Advanced Study Institute (ASI) held 18-30 June 1995 at the Hotel Villa del Mare in Acquafredda di Maratea, Italy. Since without generous financial (and other non-monetary forms of) support from the Scientific and Environmental Affairs Division of NATO, the Institute would not have been possible, I first must thank this organization for making the Institute and these proceedings a reality. Its continued support of four ASIs which I have proposed and directed in the field of superconductivity since 1988 is greatly appreciated. Additional financial support has been provided by my employer, the Air Force Office of Scientific Research (AFOSR), which has contributed funds to provide travel support to US researchers through a grant to the National Science Foundation. I am indebted, as well, to the National Science Foundation for contributing some of its own funds to augment this travel support. I must thank AFOSR additionally for allowing me to work on activities related to the ASI during the course of my normal duties and to use some of its resources for the benefit of the ASI. The indulgence and enthusiastic support of my AFOSR supervisor, Dr. Horst Wittmann, is especially acknowledged, as is the support of Dr. Helmut Hellwig, who was Director of AFOSR until January of this year.

Special thanks go to Mrs. Sandra Ronayne, a Management Assistant at AFOSR, for aiding in all administrative aspects of the ASI. Without her assistance it would have been far more difficult to handle the large number of details and unexpected problems that inevitably arise in directing such an Institute. Prior to and during the ASI I received considerable aide from the very efficient staff of the Villa del Mare. The meeting and lodging facilities, the food and general ambiance, all contributed to a successful Institute. I especially wish to thank Mrs. Maria Armiento of the Villa del Mare for all her assistance.

I must reserve my greatest thanks for the 15 distinguished scientists who were the ASI lecturers and who (in some cases together with their collaborators) subsequently became the authors of the chapters that follow. It is not easy to devote 2 full weeks to presenting lectures and workshops to approximately 80 young researchers from 19 countries. The preparation for the ASI, the considerable effort that goes into writing an extensive manuscript and the problems of dealing with the peccadilloes of this fussy editor, add considerably to the work they must carry out on behalf of the ASI. For all of this, I am very much in their debt. I wish to thank also the aforementioned ASI participants for showing by their enthusiastic attendence that SQUID sensors and their applications are subjects of significant interest to them, thus providing additional satisfaction and incentive to all the lecturers.

For a little over 5 weeks in the spring of 1995 I again had the opportunity to engage directly in SQUID-based NDE research at INSA de Lyon in France. This not only gave me a chance to increase my knowledge base in SQUID science and technology, but I additionally was able to engage in considerable planning for both the ASI and this proceedings. I wish to thank INSA de Lyon for affording me this opportunity, and I especially wish to thank Professor Pierre-Louis Vuillermoz for once again being a perfect host and a stimulating colleague.

Finally, I wish to thank my wife and sons for putting up once again with my passion for directing NATO ASIs and for editing the subsequent proceedings.

Washington, DC   Harold Weinstock
August 1996

INTRODUCTION

The application of SQUID (Superconducting QUantum Interference Device) magnetic sensing has been around for over 25 years. This field has come a long way from December 31, 1969 when a SQUID was operated in a magnetically-shielded room at the Francis Bitter National Magnet Laboratory, MIT to produce a magnetocardiogram [1]. This work was done by Edgar A. Edelsack (Office of Naval Research), who had suggested this study, James E. Zimmerman (National Bureau of Standards - now NIST), a co-inventor of the rf SQUID, and David Cohen (MIT), who had constructed the shielded room and had been carrying out non-SQUID magnetocardiographic studies, following the initial work of this type in 1963 by G. M. Baule and R. McFee [2]. The bulk-lead, point-contact rf SQUID magnetometer used in that pioneering study is generations removed from today's sophisticated commercial and laboratory systems with gradiometer-coil configurations, thin-film technology and sophisticated auxiliary electronic circuitry providing noise reduction and increased sensitivity, subjects which will be covered in detail within the main body of this volume.

A major step in the progression of SQUID magnetic sensing was the introduction of gradiometer coils by the group of Samuel J. Williamson and Lloyd Kaufman at New York University in the mid 1970's [3]. Located 12 floors above a steady flow of New York subway trains and initially without the benefit of a shielded room, they provided courage to many others who worked in a less than ideal environment.

Up until the early 1980's almost all SQUID magnetic sensing was done using a single sensor per cryogenic system, regardless of the application being addressed, e.g., biomagnetic or geophysical, and even as late as 1985 the largest multisensor system in operation had only 7 sensors to map the magnetic field simultaneously within an area of a few square centimeters. For those with only a single sensor, it was required to take data for several minutes at each of tens of grid points on a human head, all the while hoping that the signal related to the brain being studied had not changed during the hours it took to take the entire set of measurements and relying on *supposedly* precise determination of each grid point.

At a workshop held at UCLA in September 1984 (when only one 5-sensor system was in active use) for the purpose of discussing progress in the interpretation of neuromagnetic signals, I asked the question: "How many sensors do you really need to obtain a reliable mapping of the magnetic field of the human brain in order to localize some particular brain activity or function?" There was no immediate response to this question, so to be provocative I said: "Should it be 100?" This was met with great laughter because of the apparent absurdity of the suggestion. One attendee, Jack Beatty of UCLA, added to the sense of frivolity by stating that every night before going to bed he "prayed for a 100-channel system." It is perhaps worth recalling that a commercial single-channel system sold for $30,000 to $40,000 in those days and that the first seven-channel system (with four additional SQUIDs providing local-field reference information) cost approximately $500,000. Consider the situation today, little more than a decade later, for which there are a handful of companies offering systems with on the order of 100 or more channels, and there is one project recently completed in Japan, at the ad-hoc Superconductor Sensor Laboratory outside Tokyo, in which there is a 256-channel system. Perhaps more importantly, the use of SQUID neuromagnetometry has reached the stage of acceptance whereby within the past three years (in the United States) insurance companies have provided reimbursement for the cost of diagnostic testing on patients scheduled for neurosurgery. The reader will find a comprehensive update of this subject in the chapter authored by Gian-Luca Romani.

The emergence of multi-sensor systems also has had the effect of reviving serious interest in SQUID-based magnetocardiography as a potential non-invasive diagnostic tool applied to the discovery and localization of arrythmias. Additionally, the monitoring of the foetal heart using a SQUID appears to be emerging as a serious medical application, particularly with the use of high-temperature-superconductor (HTS) SQUID magnetometry. The chapters written by Sergio Erné and Gerhard Stroink specifically address these areas.

Outside of the biomedical world, the largest potential market for SQUID sensors is in the area of nondestructive evaluation (NDE), known also as nondestructive testing (NDT) and nondestructive inspection (NDI). I have had the opportunity to be among the first to apply SQUID magnetometry to this applications area, having done some crude experiments at the Naval Research Laboratory in late 1982 and into the first half of 1983. In the early stages of this work I was visited by Gordon Donaldson, who indicated that he was about to start some SQUID NDE studies in his laboratory at the University of Strathclyde on steel that is used in off-shore oil platforms. Both studies resulted in the publication of the first SQUID NDE papers in the proceedings of *SQUID '85* [4, 5], while another version of my work was published as well in an NDE publication [6].

Gordon Donaldson and his colleagues in this volume address the significant expansion of research in SQUID-based NDE in the years since the above-mentioned studies over a decade ago. Probably the greatest near-term potential for application to NDE is the use of SQUIDs to detect the magnetic fields created by eddy currents induced in the bodies of aging aircraft. Currently, there is not a single NDE technology which satisfactorily addresses the need of both the world's military and commercial aviation communities. Studies carried out principally over the past seven years by John Wikswo's group at Vanderbilt University indicate that the added sensitivity of a SQUID at low frequencies - down to 1 Hz if necessary - may make SQUID-based eddy-current sensing the technology of choice in the inspection of aircraft for subsurface cracks and corrosion. John Wikswo addresses, in one of the chapters herein, the question of analyzing eddy-current-generated magnetic field plots, so that these data may be transformed into current-density plots which may reveal unseen cracks and corrosion. Especially exciting is his group's ability to use a phase-sensitive-detection method of measurement and analysis to probe eddy currents as a function of depth, something that is impossible to do with conventional eddy-current techniques.

Many other applications are covered in this book, most dealing with areas in the real world in which the SQUID sensors must compete with other technologies, and it is not yet entirely certain whether any of these applications will yield a significant market for SQUID-systems manufacturers. I feel compelled to echo here a much-quoted comment I made at a 1990 NATO Advanced Study Institute on applications of superconductivity: **Never use a SQUID when a simpler, less-expensive technology will do the job. Use a SQUID only when nothing else (less expensive) will satisfy your needs.**

One rather specialized application which appears to satisfy the above advice in favor of the SQUID is that of ordnance detection, specifically the detection of underwater mines in shallow harbors, with a possible extension to the detection of land mines. Until recently research in this area was classified, but the chapter by Ted Clem on this topic represents one of the first detailed explanations of work that has been done for the past several years at the Naval Surface Warfare Center.

The applications presented in this book are not all inclusive. Most notably, no attention has been paid to the area of paleomagnetism, nor to the use of SQUID magnetometry in other areas of fundamental research, such as in the search for magnetic monopoles. The emphasis is clearly on areas of application beyond the research laboratory, although the chapter by Ho Jung Paik in part describes some unique work on SQUID-based, gravitational-wave transducers. Although only now in the planning stage, I further believe there is a role for HTS SQUID-based gradiometers in the detection of underground chemical and nuclear explosions, and in detecting precursors to major earthquakes several days prior to eruption.

While this book focuses strongly on SQUID applications, it is designed to provide a solid introduction to SQUID fundamentals, to fabrication of (primarily HTS) SQUIDs, to signal analysis and interpretation, and to the construction of the auxiliary electronics and total systems required to make the applications possible. It is not meant to cover SQUID applications that are far outside the realm of what is commonly construed as magnetometry, e.g., it does not include SQUIDs as circuit elements in digital logic or memory, as in single-flux-quantum (SFQ) circuitry [7].

This introduction has been formulated to provide the reader with a historical perspective for SQUID applications and to provide some rationale for the inclusion of the topics which appear, both those which relate to the principles and fabrication, and those which relate specifically to applications and analysis. In the course of this discussion I have mentioned only some of the chapters and their authors, while omitting others. In doing this I have not meant to imply that some of these contributions are more important. Clearly, the importance of any particular section of this book will relate to the individual needs and interests of each reader. Similarly, my discussion of historical events is meant only to be illustrative and is not complete. To all my hard-working authors and to all of the important contributors to this exciting scientific field, I offer my deep-felt thanks and my apologies for any inadvertent sins of omission.

It is the hope of both the authors of the chapters of this book, all of whom have presented some original research in the contributions they have made, and of myself, that our readers will find the information provided of value in improving the capabilities of SQUID-based, magnetic-sensing systems and in extending the use of these systems to both existing, known applications areas and in extending this fascinating technology to as yet undeveloped, unknown applications areas.

Washington, DC   Harold Weinstock
August 1996

References

1.   Cohen, D., Edelsack, E.A. and Zimmerman, J.E.(1970) Manetocardiograms taken inside a shielded room with a superconducting point contact magnetometer, *Appl. Phys. Lett.* **16**, 278-280.

2.   Baule, G.M. and McFee, R. (1963) Detection of the magnetic field of the heart, *Am. Heart J.* **66**, 95-96.

3.   Brenner, D., Williamson, S.J. and Kaufman, L. (1975) Visually evoked magnetic fields of the human brain, *Science* **190**, 1339-1340.

4.   Weinstock, H. and Nisenoff, M. (1985) Nondestructive evaluation of metallic structures using a SQUID magnetometer, in H.D. Hahlbohm and H. Lubbig (eds.), *SQUID '85 - Superconducting Quantum Interderence Devices and their Applications*, Walter de Gruyter & Co., Berlin, pp. 853-858.

5.   Bain, R.J.B., Donaldson, G.B., Evanson, S. and Hayward, G. (1985) SQUID gradiometric detection of flaws in ferromagnetic structures, *ibid*, pp. 841-846.

6.   Weinstock, H. and M. Nisenoff (1986) Defect detection with a SQUID magnetometer, in D.O. Thompson and D.E. Chimenti (eds.), *Review of Progress in Quantitative Nondestructive Evaluation, Vol. 5A*, Plenum Pulishing Corporation, New York, pp. 699-704.

7.   Likharev, K.K. (1993) Rapid single-flux-quantum logic, in H. Weinstock and R.W. Ralston (eds.) *The New Superconducting Electronics*, Kluwer Academic Publishers, Dordrecht, pp. 423-452.

SQUID FUNDAMENTALS

JOHN CLARKE
Department of Physics, University of California
and
Materials Sciences Division, Lawrence Berkeley National Laboratory
Berkeley, California 94720

ABSTRACT. DC Superconducting QUantum Interference Devices (SQUIDs) incorporating two resistively shunted tunnel junctions are routinely fabricated from thin films of low-transition-temperature (T_c) superconductors. An integrated superconducting input coil couples the SQUID to the signal source. Typical dc SQUIDs operating at 4.2K have a magnetic flux noise of $10^{-6}\Phi_0$ Hz$^{-1/2}$ corresponding to a noise energy of 10^{-32} JHz^{-1} at frequencies f above the l/f noise knee, which may be below 1Hz ($\Phi_0 \equiv$ h/2e is the flux quantum). Recently, the performance of thin-film rf SQUIDs, which involve a single junction, has improved significantly, and the sensitivity of a device operated at 3 GHz approaches that of dc SQUIDs. In the last two years, there have been dramatic improvements in the performance of both dc and rf SQUIDs made from high-T_c thin films, and noise energies of about 10^{-30} JHz^{-1} and magnetic field noise levels below 10fTHz$^{-1/2}$ at frequencies down to a few Hz have been achieved at 77K. Multilayer thin-film flux transformers are now available. Instruments based on low-T_c SQUIDs include magnetometers, magnetic gradiometers, voltmeters, susceptometers, amplifiers, and displacement sensors; their applications vary from neuromagnetism and magnetotelluric sounding to the detection of gravity waves and magnetic resonance.

1. Introduction

Superconducting QUantum Interference Devices (SQUIDs) are the most sensitive detectors of magnetic flux currently available. They are amazingly versatile, being able to measure any physical quantity that can be converted to a flux, for example, magnetic field, magnetic field gradient, current, voltage, displacement, and magnetic susceptibility. As a result, the applications of SQUIDs are wide ranging, from the detection of tiny magnetic fields produced by the human brain and the measurement of fluctuating geomagnetic fields in remote areas to the detection of gravity waves and the observation of spin noise in an ensemble of magnetic nuclei.

SQUIDs combine two physical phenomena, flux quantization, the fact that the flux Φ in a closed superconducting loop is quantized [1] in units of the flux quantum $\Phi_0 \equiv$ h/2e $\cong$ 2.07 x 10^{-15} Wb, and Josephson tunneling [2]. There are two kinds of SQUIDs. The first [3], the dc SQUID, consists of two Josephson junctions connected in parallel in a superconducting loop, and is so named because it can be operated with a steady current bias. The second [4,5], the rf SQUID, involves a single Josephson junction interrupting the current flow around a superconducting loop, and is operated with a radiofrequency flux bias. In both cases, the output from the SQUID is periodic with period Φ_0 in the magnetic flux applied to the loop. One generally is able to detect an output signal corresponding to a flux change of much less than one flux quantum.

1

H. Weinstock (ed.), SQUID Sensors: Fundamentals, Fabrication and Applications, 1–62.
© *1996 Kluwer Academic Publishers.*

In this chapter I give an overview of the current state of the SQUID art. I cannot hope to describe all of the SQUIDs that have been made or, even less, all of the applications in which they have been successfully used. I begin, in Sec. 2, with a brief review of the resistively-shunted Josephson junction, with particular emphasis on the effects of noise. Section 3 contains a description of the dc SQUID: how these devices are made and operated, and the limitations imposed by noise. Section 4 contains a similar description of the properties of rf SQUIDs. In Sec. 5, I describe a selection of instruments based on SQUIDs and mention some of their applications. Section 6 contains a discussion of the present state of the art of high temperature SQUIDs, and Sec. 7 some concluding remarks.

Sections 2-5 are identical to those in the chapter I wrote for an earlier NATO Advanced Study Institute [6], apart from Sec. 3.6 which I have partly rewritten. However, I have substantially rewritten and updated Sec. 6 on high-T_C SQUIDs and magnetometers, the performance of which has improved tremendously in the intervening three years. I have also rewritten Sec. 7.

Some topics I touch on briefly are dealt with in much greater detail by the authors of other chapters in these proceedings. The chapter by Dietmar Drung provides a comprehensive review of alternative readout schemes for dc SQUIDs, and that by Alex Braginski gives an overview of high-T_C thin film fabrication and describes rf SQUIDs and their applications in some detail. I have said nothing about nondestructive evaluation with SQUIDs, which is covered by Gordon Donaldson and John Wikswo, or about gravity gradiometers, which are discussed by Ho Jung Paik. Finally, I have omitted the vast subject of biomagnetism, which is surveyed in great detail by the remaining authors.

2. The Resistively-Shunted Junction

A Josephson junction [2] consists of two superconductors separated by a thin insulating barrier. Cooper pairs of electrons are able to tunnel through the barrier, maintaining phase coherence in the process. The applied current, I, controls the difference $\delta = \phi_1 - \phi_2$ between the phases of the two superconductors according to the current-phase relation

$$I = I_0 \sin \delta, \tag{2.1}$$

where I_0 is the critical current, that is, the maximum supercurrent the junction can sustain. When the current is increased from zero, initially there is no voltage across the junction, but for $I > I_0$ a voltage V appears, and δ evolves with time according to the voltage-frequency relation

$$\dot{\delta} = 2eV / \hbar = 2\pi V/\Phi_0 . \tag{2.2}$$

A high quality Josephson tunnel junction has a hysteretic current-voltage (I - V) characteristic. As the current is increased from zero, the voltage switches abruptly to a nonzero value when I exceeds I_0, but returns to zero only when I is reduced to a value much less than I_0. This hysteresis must be eliminated for SQUIDs operated in the conventional manner, and one does so by shunting the junction with an external shunt resistance. The "resistively shunted junction" (RSJ) model [7, 8] is shown in Fig.1(a). The junction has a critical current I_0 and is in parallel with its self-capacitance C and its shunt resistance R, which has a current noise source I_N (t) associated with it. The equation of motion is

$$C\dot{V} + I_0 \sin \delta + V/R = I + I_N (t). \tag{2.3}$$

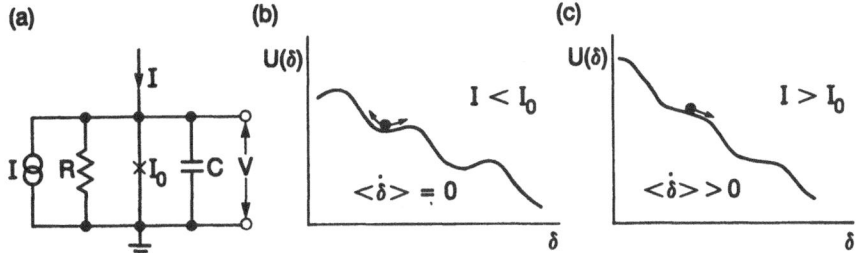

Figure 1. (a) The resistively-shunted Josephson junction; (b) and (c) show the tilted washboard model for $I < I_0$ and $I > I_0$.

Neglecting the noise term for the moment and setting $V = \hbar\dot{\delta}/2e$, we obtain

$$\frac{\hbar C}{2e}\ddot{\delta} + \frac{\hbar}{2eR}\dot{\delta} = I - I_0 \sin\delta = -\frac{2e}{\hbar}\frac{\partial U}{\partial\delta} , \tag{2.4}$$

where

$$U = -\frac{\Phi_0}{2\pi}(I\delta + I_0 \cos\delta). \tag{2.5}$$

One obtains considerable insight into the dynamics of the junction by realizing that Eq. (2.4) also describes the motion of a ball moving on the "tilted washboard" potential U. The term involving C represents the mass of the particle, the 1/R term represents the damping of the motion, and the average "tilt" of the washboard is proportional to -I. For values of $I < I_0$, the particle is confined to one of the potential wells [Fig. 1(b)], where it oscillates back and forth at the plasma frequency [2] $\omega_p = (2\pi I_0/\Phi_0 C)^{1/2}[1- (I/I_0)^2]^{1/4}$. In this state $<\dot{\delta}>$ and hence the average voltage across the junction are zero ($<>$ represents a time average). As the current is increased to I_0, the tilt increases, and when I exceeds I_0, the particle rolls down the washboard; in this state $<\dot{\delta}>$ is nonzero, and a voltage appears across the junction [Fig.1(c)]. As the current is increased further, $<\dot{\delta}>$ increases, as does V. For the nonhysteretic case, as soon as I is reduced below I_0 the particle becomes trapped in one of the wells, and V returns to zero. In this, the overdamped case, we require [7, 8]

$$\beta_c \equiv (2\pi I_0 R/\Phi_0)RC = \omega_J RC \lesssim 1; \tag{2.6}$$

$\omega_J / 2\pi$ is the Josephson frequency corresponding to the voltage $I_0 R$.

We introduce the effects of noise by restoring the noise term in Eq. (2.4) to obtain the Langevin equation

$$\frac{\hbar C}{2e}\ddot{\delta} + \frac{\hbar}{2eR}\dot{\delta} + I_0 \sin\delta = I + I_N(t). \tag{2.7}$$

In the thermal noise limit, the spectral density of $I_N(t)$ is given by the Nyquist formula

$$S_I(f) = 4k_B T/R, \tag{2.8}$$

where f is the frequency. It is evident that $I_N(t)$ causes the tilt in the washboard to fluctuate with time. This fluctuation has two effects on the junction. First, when I is less than I_0, from time to time fluctuations cause the total current $I + I_N(t)$ to exceed I_0, enabling the particle to roll out of one potential minimum into the next. For the underdamped junction, this process produces a series of voltage pulses randomly spaced in time. Thus, the time average of the voltage is nonzero even though $I < I_0$, and the I - V characteristic is "noise-rounded" at low voltages. [9] Because this thermal activation process reduces the observed value of the critical current, there is a minimum value of I_0 for which the two sides of the junction remain coupled together. This condition may be written as

$$I_0\Phi_0/2\pi \gtrsim 5k_BT, \qquad (2.9)$$

where $I_0\Phi_0 / 2\pi$ is the coupling energy of the junction [2] and the factor of 5 is the result of a computer simulation [10]. For T = 4.2K, we find $I_0 \gtrsim 0.9\mu A$.

The second consequence of thermal fluctuations is voltage noise. In the limit $\beta_c \ll 1$ and for $I > I_0$, the spectral density of this noise at a measurement frequency f_m that we assume to be much less than the Josephson frequency f_J is given by [11,12]

$$S_V(f_m) = \left[1+\frac{1}{2}\left(\frac{I_0}{I}\right)^2\right] \frac{4k_BTR_d^2}{R} \cdot \quad \left\{ \begin{array}{c} \beta_c \ll 1 \\ I > I_0 \\ f_m \ll f_J \end{array} \right\} \qquad (2.10)$$

The first term on the right-hand side of Eq. (2.10) represents the Nyquist noise current generated at the measurement frequency f_m flowing through the dynamic resistance $R_d \equiv dV/dI$ to produce a voltage noise - see Fig. 2. The second term, $(1/2)(I_0 / I)^2 (4k_BT/R) R_d^2$, represents Nyquist noise generated at frequencies $f_J \pm f_m$ mixed down to the

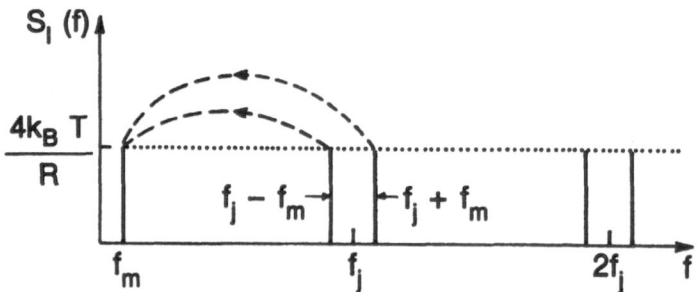

Figure 2. Schematic representation for the noise terms in Eq.(2.10). The Nyquist noise generated in the resistor at frequency f_m contributes directly at f_m; that generated at $f_J \pm f_m$ is mixed down to f_m.

measurement frequency by the Josephson oscillations and the inherent nonlinearity of the junction. The factor $(1/2)(I_0 / I)^2$ is the mixing coefficient, and it vanishes for sufficiently large bias currents. The mixing coefficients for the Nyquist noise generated near harmonics of the Josephson frequencies $2f_J$, $3f_J$, ... are negligible in the limit $f_m / f_J \ll 1$.

At sufficiently high bias current, the Josephson frequency f_J exceeds k_BT/h, and quantum corrections [13] to Eq. (2.10) become important, provided the term $(1/2)(I_0 / I)^2$ is not too small. The requirement for observing significant quantum corrections is $eI_0 R / k_B T \gg 1$. The spectral density of the voltage noise becomes

$$S_V(f_m) = \left[\frac{4k_BT}{R} + \frac{2eV}{R}\left(\frac{I_0}{I}\right)^2 \coth\left(\frac{eV}{k_BT}\right)\right]R_d^2, \qquad \begin{Bmatrix} \beta_c \ll 1 \\ I > I_0 \\ f_m \ll f_J \end{Bmatrix} \qquad (2.11)$$

where we have assumed that $hf_m / k_B T \ll 1$, so that the first term on the right-hand side of Eq. (2.11) remains in the thermal limit. In the limit $T \to 0$, the second term, $(2eV/R)(I_0/I)^2 R_d^2$, represents noise mixed down from zero point fluctuations near the Josephson frequency.

This concludes our review of the RSJ, and we now turn our attention to the dc SQUID.

3. The dc SQUID

3.1. A FIRST LOOK

The essence of the dc SQUID [3] is shown in Fig. 3(a). Two junctions are connected in parallel on a superconducting loop of inductance L. Each junction is resistively shunted to eliminate hysteresis on the I -V characteristics, which are shown in Fig. 3(b) for $\Phi = n\Phi_0$ and $(n + 1/2)\Phi_0$, where Φ is the external flux applied to the loop and n is an integer. If we bias the SQUID with a constant current ($> 2 I_0$), the voltage across the SQUID oscillates with period Φ_0 as we steadily increase Φ, as indicated in Fig. 3(c). The SQUID is generally operated on the steep part of the V - Φ curve where the transfer coefficient, $V_\Phi \equiv |(\partial V/\partial \Phi)_I|$, is a maximum. Thus, the SQUID produces an output voltage in response to a small input flux $\delta\Phi$ ($\ll \Phi_0$), and is effectively a flux-to-voltage transducer.

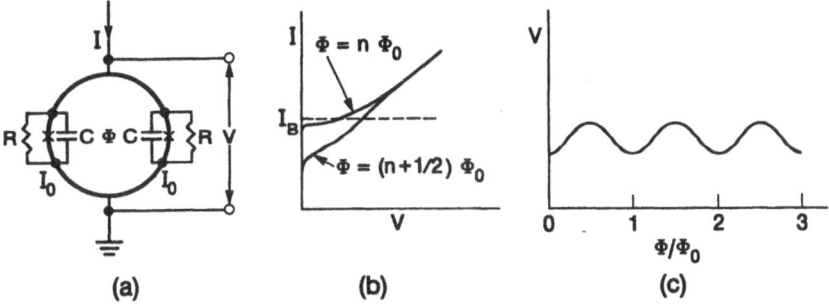

Figure 3. (a) The dc SQUID; (b) I-V characteristics; (c) V vs. Φ/Φ_0 at constant bias current I.

Before we give a detailed description of the signal and noise properties of the SQUID, it may be helpful to give a simplified description that, although not rigorous, gives some insight into the operation of the device. We assume the two junctions are identical and arranged symmetrically on the loop. We further assume, for simplicity, that the bias current is swept from zero to a value above the critical current of the two junctions at a frequency much higher than $d\Phi/\Phi_0 dt$. In the absence of any applied flux (or with $\Phi = n \Phi_0$), there is no current circulating around the loop and the bias current divides equally between the two junctions. The measured critical current is $2I_0$ (if we ignore noise rounding). If we apply a magnetic flux, Φ, the flux in the loop will be quantized and will generate a current $J = -\Phi / L$, where we have neglected the effects of the two junctions [Figs. 4(a) and (b)]. The circulating current adds to the bias current flowing through junction 1 in Fig. 4(a) and subtracts from that flowing through junction 2. In this naive picture, the critical current of junction 1 is reached when $I/2 + J = I_0$, at which point the current flowing through junction 2 is $I_0 - 2J$. Thus, the SQUID switches to the voltage state when $I = 2I_0 - 2J$. As Φ is increased to $\Phi_0 / 2$, J increases to $\Phi_0 / 2L$ [Fig. 4(b)], and the critical current falls to $2I_0 - \Phi_0 / L$ [Fig. 4(c)]. As the flux is increased beyond $\Phi_0 / 2$, however, the SQUID makes a transition from the flux state $n = 0$ to $n = 1$, and J changes sign [Fig. 4(b)]. As we increase Φ to Φ_0, J is reduced to zero and the critical current is restored to its maximum value $I_m = 2I_0$ [Fig. 4(c)].

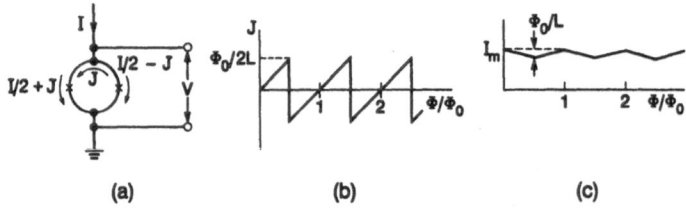

Figure 4. Simplistic view of the dc SQUID: (a) a magnetic flux Φ generates a circulating current J that is periodic in Φ as shown in (b); as a result (c), the maximum supercurrent I_m is also periodic in Φ.

In this way the critical current oscillates as a function of Φ.

Continuing with our simplified model, we see that the voltage change across the SQUID (at the peak of the current sweep) as we change Φ from 0 to $\Phi_0 / 2$ is $\Delta V = (\Phi_0 / L) R/2$, where R/2 is the parallel resistance of the two shunts. Hence, $V_\Phi = \Delta V / (\Phi_0 / 2) = R / L$.

We also can estimate the equivalent flux noise of the SQUID. If the noise voltage across the SQUID is $V_N(t)$ with a spectral density $S_V(f)$, the corresponding flux noise referred to the SQUID loop is just

$$S_\Phi(f) = S_V(f) / V_\Phi^2 . \tag{3.1}$$

A convenient way of characterizing the flux noise is in terms of the noise energy per unit bandwidth,

$$\varepsilon(f) = S_\Phi(f) / 2L. \tag{3.2}$$

If we assume that the noise in the SQUID is just the Nyquist noise in the shunt resistors with spectral density $4k_B T (R/2)$, we find $\varepsilon(f) = k_B TL/R$. Although these results are not quantitatively correct, they do give the correct scaling with the various parameters. For example, we see that to lower $\varepsilon(f)$ we should reduce T and L while using the largest possible value of R subject to the I - V characteristic remaining nonhysteretic.

Exact results for the signal and noise can be obtained only from computer simulation. The results show that the plots of the circulating supercurrent and the critical current vs. Φ become smoothed. Furthermore, the noise voltage is higher than Nyquist noise because of mixed-down noise; unfortunately the magnitude of this noise cannot be obtained analytically.

One final remark is appropriate at this point. To observe quantum interference effects, we require the modulation depth of the critical current, Φ_0 / L, to be much greater than the root mean square noise current in the loop, $< I_N^2 >^{1/2} = (k_B T / L)^{1/2}$. We shall return to this issue in Sec. 6 in the context of high-T_c SQUIDs.

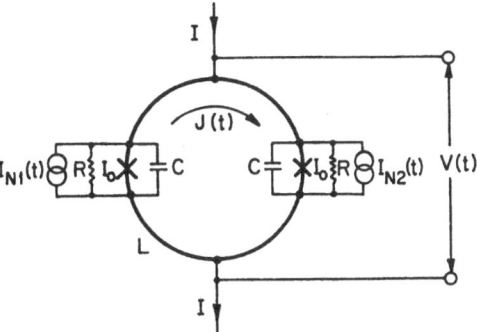

Figure 5. Model of dc SQUID showing noise sources associated with the shunt resistors.

3.2. THERMAL NOISE IN THE SQUID : THEORY

A model for noise calculations is shown in Fig. 5. This figure shows two independent Nyquist noise currents, $I_{N1}(t)$ and $I_{N2}(t)$, associated with the two shunt resistors. The phase differences across the junctions, $\delta_1(t)$ and $\delta_2(t)$, obey the following equations: [14-16]

$$V = \frac{\hbar}{4e}\left(\dot\delta_1 + \dot\delta_2 \right) , \tag{3.3}$$

$$J = \frac{\Phi_0}{2\pi L}\left(\delta_1 - \delta_2 - \frac{2\pi\Phi}{\Phi_0} \right) , \tag{3.4}$$

$$\frac{\hbar C}{2e}\ddot\delta_1 + \frac{\hbar}{2eR}\dot\delta_1 = \frac{I}{2} - J - I_0 \sin\delta_1 + I_{N1}, \tag{3.5}$$

and

$$\frac{\hbar C}{2e}\ddot\delta_2 + \frac{\hbar}{2eR}\dot\delta_2 = \frac{I}{2} + J - I_0 \sin\delta_2 + I_{N2} . \tag{3.6}$$

Equation (3.3) relates the voltage to the average rate of change of phase; Eq. (3.4) relates the current in the loop, J, to $\delta_1 - \delta_2$ and to Φ; and Eqs. (3.5) and (3.6) are Langevin equations coupled via J. These equations have been solved numerically for a limited range of values of the noise parameter $\Gamma = 2\pi k_B T/I_0\Phi_0$, reduced inductance $\beta = 2 LI_0 / \Phi_0$ and hysteresis parameter β_c. For typical SQUIDs in the ^{4}He temperature range, $\Gamma = 0.05$. One computes the time-averaged voltage V vs. Φ, and hence finds V_Φ, which, for a given value of Φ, peaks smoothly as a function of bias current. The transfer function exhibits a shallow maximum around $(2n + 1) \Phi_0 / 4$. One computes the noise voltage for a given value of Φ as a function of I, and finds that the spectral density is white at frequencies much less than the Josephson frequency. For each value of Φ, the noise voltage peaks smoothly at the value of I where V_Φ is a maximum. From these simulations, one finds that the noise energy has a minimum when $\beta \approx 1$. For $\beta = 1$, $\Gamma = 0.05$, $\Phi = (2n + 1) \Phi_0 / 4$ and for the value of I at which V_Φ is a maximum, the results can be summarized as follows:

$$V_\Phi \approx R/L, \tag{3.7}$$

$$S_V(f) \approx 16k_BTR, \qquad (3.8)$$

and

$$\varepsilon(f) \approx 9k_BTL/R. \qquad (3.9)$$

We see that our rough estimate of V_Φ in Sec. 3.1 was rather accurate, but that the assumption that the noise spectral density was given by the Nyquist result underestimated the computed value by a factor of about 8.

It is often convenient to eliminate R from Eq. (3.9) using the expression $R = (\beta_C \Phi_0 / 2\pi I_0 C)^{1/2}$. We find

$$\varepsilon(f) \approx 16 \, k_BT(LC/\beta_C)^{1/2}. \quad (\beta_C \lesssim 1) \qquad (3.10)$$

Equation (3.10) gives a clear prescription for improving the resolution: one should reduce T, L and C. A large number of SQUIDs with a wide range of parameters have been tested and found to have white noise energies generally in good agreement with the predicted values. It is common practice to quote the noise energy of SQUIDs in units of $\hbar$ ($\approx 10^{-34}$J sec = 10^{-34}JHz^{-1}).

In closing this discussion, we emphasize that although $\varepsilon(f)$ is a useful parameter for characterizing the resolution of SQUIDs with different inductances, it is not a complete specification because it does not account fully for the effects of current noise in the SQUID loop. We defer a discussion of this point to Sec. 5.4.

3.3. PRACTICAL DC SQUIDS

Modern dc SQUIDs are invariably made from thin films with the aid of either photolithography or electron-beam lithography. A major concern in the design is the need to couple an input coil inductively to the SQUID with rather high efficiency. This problem was elegantly solved by Ketchen and Jaycox, [17, 18] who introduced the idea of depositing a spiral input coil on a SQUID in a square washer configuration. The coil is separated from the SQUID with an insulating layer (Fig. 6). Figure 7 shows a typical example [19] of one of these designs, involving trilayer Nb/Al-Al$_2$O$_3$/Nb junctions [20]. The SQUIDs are made in batches of 400 on oxidized, 100 mm-diameter silicon wafers. After the Nb base electrode and a thin Al layer have been sputtered, the Al is oxidized in a reduced pressure of O$_2$ and the Nb counterelectrode is then deposited. The entire trilayer is formed without removing the wafers from the controlled atmosphere of the sputter system. One defines the junction areas by anodizing a small ring of the counterelectrode, and the base electrode is etched to form the SQUID washer. In subsequent operations one adds the Nb layer that forms the input and flux modulation coils and makes the connection to the counterelectrode, the shunt resistors (Mo or Pb), and the final Nb layer that makes the connection to the innermost turn of the input coil. The insulation between each layer is SiO$_2$, formed by plasma-enhanced chemical vapor deposition (PECVD). All patterning is performed with reactive ion etching. The estimated SQUID inductance is $L = 0.29$ nH and the measured shunt resistance $R = 4.0 \, \Omega$.

Design guidelines for square washer SQUIDs have been given by Jaycox and Ketchen [18], who showed that a square washer (with no slit) with inner and outer edges d and w has an inductance L (loop) $= 1.25\mu_0 d$ in the limit $w \gg d$. They gave the following expressions for the inductances of the SQUID, L, and of the spiral coil, L_i, and for the mutual inductance, M_i, and coupling coefficient, α^2, between the spiral coil and the SQUID:

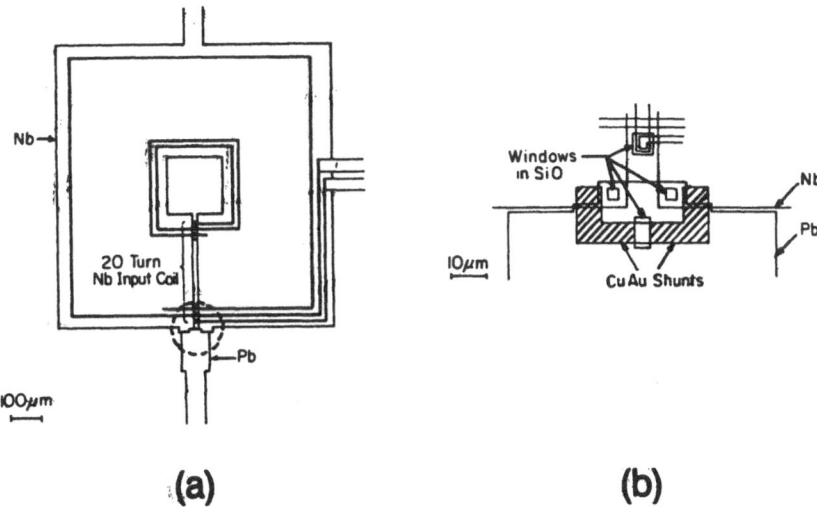

Figure 6. (a) Configuration of planar dc SQUID with overlaid spiral input coil; (b) expanded view of junctions and shunts.

$$L = L \text{ (loop)} + L_j, \tag{3.11}$$

$$L_i = n^2 (L-L_j) + L_s, \tag{3.12}$$

$$M_i = n(L-L_j) \tag{3.13}$$

and

$$\alpha^2 = (1-L_j / L) / [1+L_s / n^2(L-L_j)]. \tag{3.14}$$

Here, L_j is the parasitic inductance associated with the junctions (and possibly with the slit), n is the number of turns of the input coil and L_s is the stripline inductance of this coil, which is generally much smaller than L_i for n ≳ 20. Measured parameters are generally in good agreement with these predictions.

References [21-28] are a selection of papers describing SQUIDs fabricated on the basis of the Ketchen-Jaycox design. Some of the devices involve edge junctions in which the counter-electrode is a strip making a tunneling contact to the base electrode only at the edge. This technique enables one to make junctions with a small area and thus a small self-capacitance without resorting to electron-beam lithography. However, stray capacitances are often critically important. As has been emphasized by a number of authors, parasitic capacitance between the square washer and the input coil can produce resonances that, in turn, induce structure on the I-V characteristics and give rise to excess noise. Knuutila et al. [27] success-fully damped the resonances in the input coil by terminating the stripline with a matched resistor. In an alternative scheme, Foglietti et al. [25] introduced additional damping across the two junctions in series by depositing a thin-film resistor across the slit in the square washer. Other approaches have been devised to reduce the parasitic capacitance. For example, Muhlfelder et al. [26] coupled the SQUID to the signal source via an intermediate superconducting transformer, thereby reducing the number of turns on the SQUID

(a)

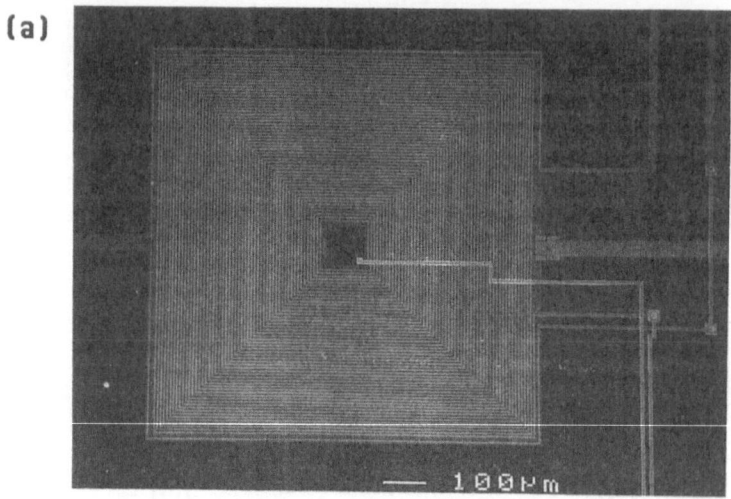

(b)

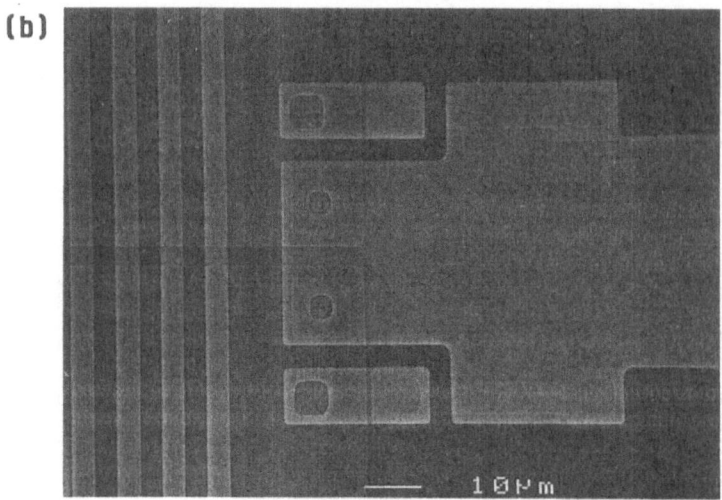

Figure 7. (a) Electron micrograph of planar dc SQUID with 53-turn input coil; the modulation and feedback coil consists of a single-turn outside the input coil. (b) Electron micrograph showing junctions on either side of the slit in the square washer. (Courtesy A. Barfknecht, M. S. Colclough and A. de la Cruz, Conductus, Inc. [19].)

washer; Carelli and Foglietti [29] fabricated "fractional turn SQUIDs" with many loops in parallel that are coupled to a thin-film input coil surrounding them.

Most SQUIDs are patterned with conventional photolithographic techniques which yield linewidths of 2-3μm. Recently, however, Ketchen et al. [30] fabricated devices in which the input coils had linewidths of 0.5μm. The inductances of these designs were in good agreement with predicted values. Ultimately, one expects a new generation of SQUIDs with submicron features and a corresponding reduction in the noise energy.

3.4 FLUX-LOCKED LOOP

In most, although not all, practical applications one uses the SQUID in a feedback circuit as a null detector of magnetic flux [31]. One applies a modulating flux to the SQUID with a peak-to-peak amplitude $\Phi_0/2$ and a frequency f_m usually between 100 and 500kHz, as indicated in Fig. 8. If the quasistatic flux in the SQUID is exactly $n\Phi_0$ the resulting voltage is a rectified version of the input signal, that is, it contains only the frequency $2f_m$ [Fig. 8(a)]. If this voltage is sent through a lock-in detector referenced to the fundamental frequency f_m, the output will be zero. On the other hand, if the quasistatic flux is $(n + 1/4)\Phi_0$, the voltage across the SQUID is at frequency f_m [Fig. 8(b)], and the output from the lock-in will be a maximum. Thus, as one increases the flux from $n\Phi_0$ to $(n + 1/4)\Phi_0$, the output from the lock-in will increase steadily; if one reduces the flux from $n\Phi_0$ to $(n - 1/4)\Phi_0$, the output will increase in the negative direction [Fig. 8(c)].

The alternating voltage across the SQUID is coupled to a low-noise preamplifier, usually at room temperature, via either a cooled transformer [32] or a cooled LC series-resonant circuit [31]. The first of these options presents an impedance N^2R_d to the preamplifier, and

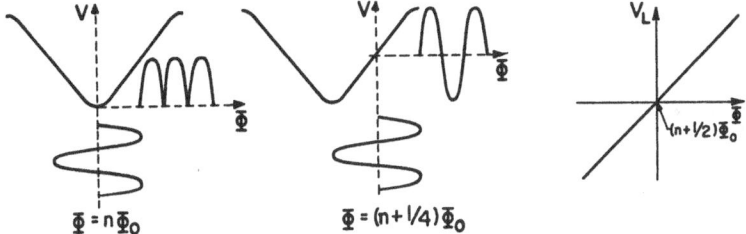

Figure 8 Flux modulation scheme showing voltage across the SQUID for (a) $\Phi = n\Phi_0$ and (b) $\Phi = (n+1/4)\Phi_0$. The output V_L from the lock-in detector vs. Φ is shown in (c).

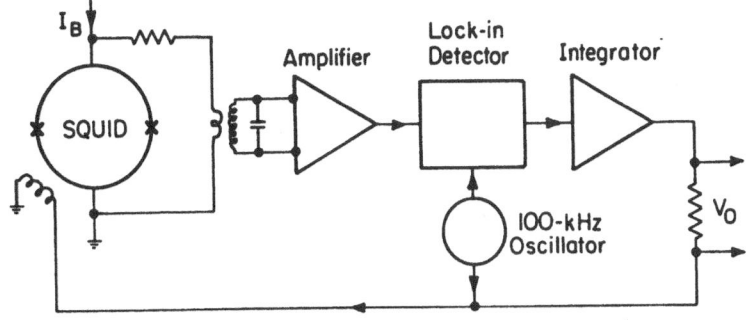

Figure 9. Modulation and feedback circuit for the dc SQUID.

the second an impedance Q^2R_d, where R_d is the dynamic resistance of the SQUID at the bias point, N is the turns ratio of the transformer, and Q is the quality factor of the tank circuit. The value of N or Q is chosen to optimize the noise temperature of the preamplifier; with careful design, the noise from the amplifier can be appreciably less than that from the SQUID at 4.2 K.

Figure 9 shows a typical flux-locked loop in which the SQUID is coupled to the preamplifier via a cooled transformer. An oscillator applies a modulating flux to the SQUID. After amplification, the signal from the SQUID is synchronously detected and sent through an integrating circuit. The smoothed output is connected to the modulation and feedback coil via a large series resistor R_f. Thus, if one applies a flux $\delta\Phi$ to the SQUID, the feedback circuit will generate an opposing flux $-\delta\Phi$, and a voltage proportional to $\delta\Phi$ appears across R_f. This technique enables one to measure changes in flux ranging from much less than a single flux quantum to many flux quanta. The use of a modulating flux eliminates 1/f noise and drift in the bias current and preamplifier. Using a modulation frequency of 500kHz, a double transformer between the SQUID and the preamplifier, and a two-pole integrator, Wellstood *et al.* [28] achieved a dynamic range of $\pm 2 \times 10^7$ Hz$^{1/2}$ for signal frequencies up to 6 kHz, a frequency response from 0 to 70kHz (±3 dB), and a maximum slew rate of 3×10^6 Φ_0 sec^{-1}.

3.5 THERMAL NOISE IN THE DC SQUID : EXPERIMENT

One determines the spectral density of the equivalent flux noise in the SQUID by connecting a spectrum analyzer to the output of the flux-locked loop. A representative power spectrum [33] is shown in Fig. 10: above a 1/f noise region, the noise is white at frequencies up to the roll-off of the feedback circuit. In this particular example, with L = 200pH and R = 8Ω, the measured flux noise was $S_\Phi^{1/2} = (1.9 \pm 0.1) \times 10^{-6}\Phi_0Hz^{-1/2}$, in reasonable agreement with the predictions of Eqs. (3.7) and (3.8). The corresponding flux-noise energy was 4×10^{-32} JHz$^1 \approx 400$ $\hbar$. Many groups have achieved noise energies that are comparable or, with lower values of L or C, somewhat better.

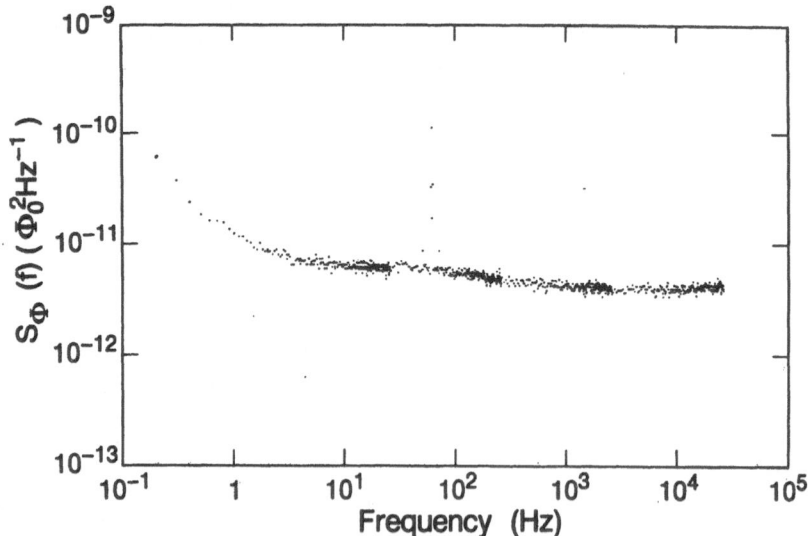

Figure 10. Spectral density of equivalent flux noise for dc SQUID: L = 0.2nH, R = 8Ω, and T = 4.2K. (Courtesy F.C. Wellstood [33])

To investigate the temperature dependence of their SQUIDs, Wellstood *et al.* [34] cooled them in a dilution refrigerator to temperatures below 1K , and used a second dc SQUID as a preamplifier. They found that the noise energy scaled accurately with T at temperatures down to about 150mK, below which the noise energy became nearly constant. This saturation was traced to heating in the resistive shunts, which prevented them from cooling much below 150mK. This heating is actually a hot-electron effect: the bottleneck in the cooling process is the rate at which the electrons can transfer energy to the phonons which, in turn, transfer energy to the substrate [35, 36]. The temperature of the shunts was lowered by connecting each of them to a CuAu "cooling fin" of large volume. The hot electrons diffuse into the fins where they rapidly transfer energy to other electrons. Since the "reaction volume" is now greatly increased, the numbers of electrons and phonons interacting are also increased, and the electron gas is cooled more effectively. In this way, the effective electron temperature was reduced to about 50mK when the SQUID was at a bath temperature of 20mK, with a concomitant reduction in ε to about $5\hbar$. Subsequently, Ketchen *et al.* [37] have achieved a noise energy of about $3\hbar$ at 0.3 K in a SQUID with L = 100 pH and C = 0.14 pF.

3.6 1/f NOISE IN DC SQUIDS

The white noise in dc SQUIDs is well understood. However, some applications of SQUIDs, for example neuromagnetism, require good resolution at frequencies down to 0.1 Hz or less, and the level of the 1/f or "flicker" noise becomes very important.

There are at least two separate sources of 1/f noise in the dc SQUID [38]. The first arises from 1/f fluctuations in the critical current of the Josephson junctions, and the mechanism for this process is reasonably well understood [39]. In the process of tunneling through the barrier, an electron becomes trapped on a defect in the barrier and is subsequently released. While the trap is occupied, there is a local change in the height of the tunnel barrier and hence in the critical current density of that region. As a result, the presence of a single trap causes the critical current of the junction to switch randomly back and forth between two values, producing a random telegraph signal. If the mean time between pulses is τ, the spectral density of this process is a Lorentzian,

$$S(f) \propto \frac{\tau}{1+(2\pi f\tau)^2} ,$$ (3.15)

namely white at low frequencies and falling off as $1/f^2$ at frequencies above $1/2\pi\tau$. In many cases, the trapping process is thermally activated, and τ is of the form

$$\tau = \tau_0 \exp (E/k_BT),$$ (3.16)

where τ_0 is a constant and E is the barrier height.

In general, there may be several traps in the junction, each with its own characteristic time τ_i. One can superimpose the trapping processes, assuming them to be statistically independent, to obtain a spectral density [40]

$$S(f) \propto \int dE\ D(E)\left[\frac{\tau_0\exp(E/k_BT)}{1+(2\pi f\tau_0)^2\exp(2E/k_BT)}\right] ,$$ (3.17)

where D(E) is the distribution of activation energies. The term in square brackets is a strongly peaked function of E, centered at $\tilde{E} \equiv k_BT\ln(1/2\pi f\tau_0)$, with a width $\sim k_BT$. Thus, at a given temperature, only traps with energies within a range k_BT of $\tilde{E}$ contribute significantly to the noise. If one now assumes D(E) is broad with respect to $k_B T$, one can take $D(\tilde{E})$ outside the integral, and carry out the integral to obtain

$$S(f,T) \propto \frac{k_B T}{f} D(\tilde{E}). \qquad (3.18)$$

In fact, one obtains a 1/f-like spectrum from just a few traps.

The magnitude of the 1/f noise in the critical current depends strongly on the quality of the junction as measured by the current leakage at voltages below $(\Delta_1 + \Delta_2)/e$, where Δ_1 and Δ_2 are the energy gaps of the two superconductors. Traps in the barrier enable electrons to tunnel in this voltage range, a process producing both leakage current and 1/f noise. Thus, for a given technology, junctions with low subgap leakage currents will have low 1/f noise. Figure 11 shows an example of a Nb-Al$_2$O$_3$-Nb junction with a single trap [41]. The junction was resistively shunted and voltage biased at typically 1.5 µV; the noise currents were measured with a SQUID. At 4.2 K [Fig.11(a)], the noise is approximately Lorentzian; the switching process producing the noise is shown in the inset. Figure 11(b) shows that at 1.5 K the noise is substantially reduced as the trap freezes out. By measuring the temperature dependence of the random telegraph signal, Savo et al. [41] found that τ obeyed Eq. (3.16) with $\tau_0 = 10$ s and E = 1.8 meV. Furthermore, τ was exponentially distributed, as expected,

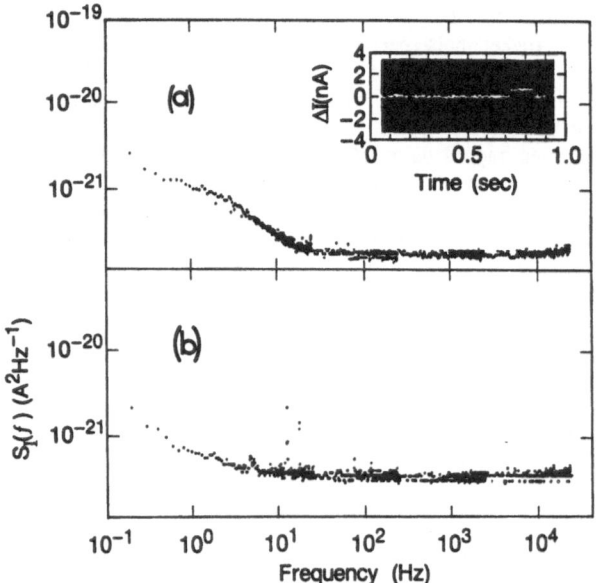

Figure 11. Spectral density of fluctuations in the critical current of a single Nb-Al$_2$O$_3$-Nb tunnel junction at (a) 4.2K and (b) 1.5K. Inset in (a) shows fluctuations vs. time (from ref. 41).

with an average value of 107 ms at 4.2 K.

The second source of 1/f noise in SQUIDs arises from the motion of flux lines trapped in the body of the SQUID [38]. This mechanism manifests itself as a flux noise; for all practical purposes the noise source behaves as if an external flux noise were applied to the SQUID. Thus, the spectral density of the 1/f flux noise scales as V_Φ^2, and, in particular, vanishes at $\Phi = (n \pm 1/2)\Phi_0$ where $V_\Phi = 0$. By contrast, critical current noise is still present when $V_\Phi = 0$, although its magnitude does depend on the applied flux.

The level of 1/f flux noise appears to depend strongly on the microstructure of the thin films. For example, SQUIDs fabricated at Berkeley with Nb loops sputtered under a particular set of conditions show 1/f flux noise levels of typically [38] $10^{-10}\ \Phi_0^2\ Hz^{-1}$ at 1 Hz. On the other hand, SQUIDs with Pb loops in exactly the same geometry exhibited a 1/f noise level of about $2 \times 10^{-12}\ \Phi_0^2 Hz^{-1}$ at 1 Hz, arising from critical current fluctuations. Tesche et al. [42] reported a 1/f noise level in Nb-based SQUIDs of about $3 \times 10^{-13}\ \Phi_0^2 Hz^{-1}$, while Foglietti et al. [43] found a critical current 1/f noise corresponding to $2 \times 10^{-12}\ \Phi_0^2\ Hz^{-1}$, also in Nb-based devices. Thus, we conclude that the quality of the Nb films plays a significant role in the level of 1/f flux noise. It is of considerable fundamental and practical interest to understand the mechanism in detail.

There is an important practical difference between the two sources of 1/f noise: critical current noise can be reduced by a suitable modulation scheme, whereas flux noise cannot. To understand how to reduce critical current 1/f noise, we first note that at constant current bias the spectral density of the 1/f voltage noise across the SQUID can be written in the approximate form

$$S_V(f) \approx \frac{1}{2}\left[(\partial V/\partial I_0)^2 + L^2 V_\Phi^2\right] S_{I_0}(f). \qquad (3.19)$$

In Eq. (3.19), we have assumed that each junction has the same level of critical current noise, with a spectral density $S_{I_0}(f)$. The first term on the right is the "in-phase mode", in which each of the two junctions produces a fluctuation of the same polarity. This noise is eliminated (ideally) by the conventional flux modulation scheme described in Sec. 3.4, provided the modulation frequency is much higher than the 1/f noise frequency. The second term on the right of Eq. (3.19) is the "out-of phase" mode in which the two fluctuations are of opposite polarity and, roughly speaking, result in a current around the SQUID loop. This term appears, therefore, as a flux noise, vanishing for $V_\Phi = 0$, but is not reduced by the usual flux modulation scheme. Fortunately, there are schemes by which this second term, as well as the first, can be reduced. These include the bias reversal scheme introduced by Koch et al. [38], which is similar to that available on the dc SQUID system manufactured by BTi [44]. An alternative scheme, second harmonic detection (SHAD), was developed by Foglietti et al. [43].

We briefly describe the scheme of Koch et al. [38]; the principle is illustrated in Fig. 12 and the practical implementation in Fig. 13. The SQUID is flux-modulated with a 100 kHz square wave of peak-to-peak amplitude $\Phi_m = \Phi_0/2$. Synchronously with the modulation the bias current I through the SQUID is reversed, typically at a frequency $f_r = 3.125$ kHz. The resistance bridge shown in Fig. 13 minimizes the 3.125 kHz switching transients across the transformer. Simultaneously with the bias reversal a flux $\Phi/2$ is applied to the SQUID. In Figs. 12(a) and (b) we see that the bias reversal changes the sign of the voltage across the SQUID while the flux shift ensures that the sign of the flux-to-voltage transfer function remains the same. The transformer coupling the SQUID to the preamplifier is often tuned at the modulation frequency with a Q of about 3, so that any 100 kHz signals at the secondary are approximately sinusoidal.

We assume that the SQUID is operated in the usual flux-locked loop, with the output from the lock-in detector integrated and fed back to the SQUID (Fig. 13). Thus, the 100 kHz signal across the SQUID consists of just the tiny error signal. Suppose now that we apply a small external flux $\delta\Phi$ to the SQUID at a frequency well below f_r. The V-Φ curves are shifted as in Figs. 12(a), and the 100 kHz flux modulation switches the SQUID between the points 1 and 2 for positive bias and 3 and 4 for negative bias. As a function of time, the voltage V across the SQUID is as shown in Fig. 12, and the signal across the tuned transformer V_t is at the fundamental frequency. When this signal is mixed with the reference

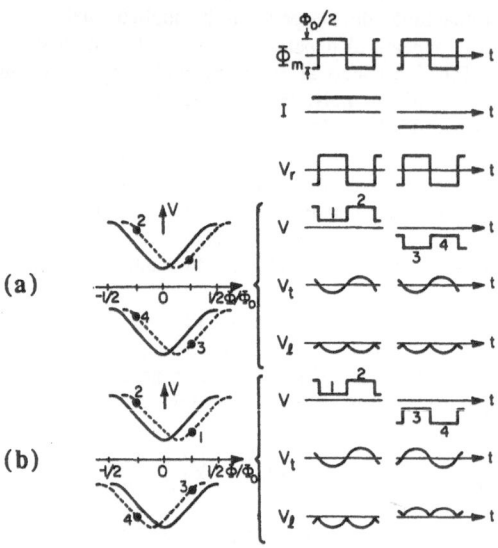

Figure 12. Principle of bias reversal scheme to reduce l/f noise due to out-of-phase critical current fluctuations. The left-hand column shows the V-Φ curves (solid lines), and the dashed lines indicate the effect of (a) an external flux change δΦ and (b) a flux change generated by out-of-phase critical current fluctuations. The right-hand column shows, as a function of time t, (top to bottom) the flux modulation Φ_m, the bias current I, and the reference voltage V_r used to lock-in detect the signal from the SQUID; the next three rows are for an external flux change δΦ, and show the voltage V across the SQUID, the voltage V_t across the secondary of the tuned transformer and the output V_ℓ of the lock-in detector; the last three rows show the same voltages for an out-of-phase critical current fluctuation.

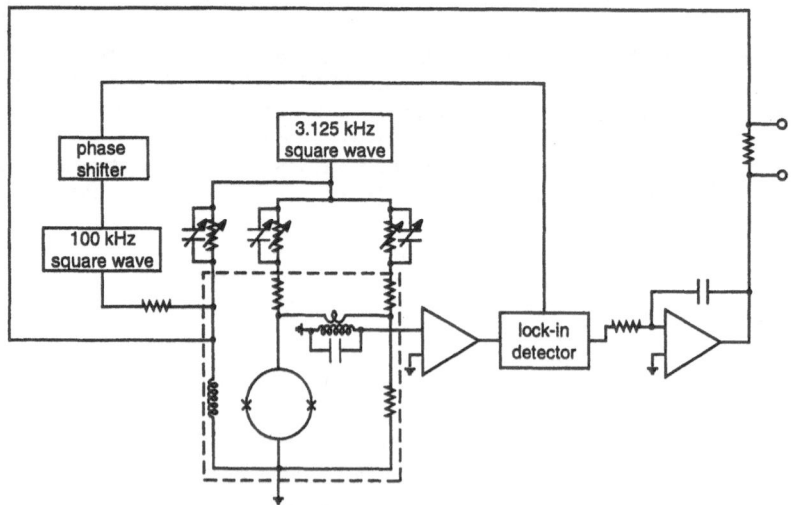

Figure 13. Schematic for flux-locked loop with bias current reversal. Cryogenic components are enclosed in the dashed box.

voltage V_r , the output from the lock-in detector V_ℓ will consist of a series of negative-going peaks for both polarities of the bias current. The average of this output produces a negative signal proportional to $\delta\Phi$ which is then used to cancel the flux applied to the SQUID. Thus, in the presence of bias reversal and flux shift, the SQUID responds to an applied flux in the usual way.

We consider now the effects of 1/f noise on the critical currents. The in-phase mode is eliminated by the 100kHz flux modulation as in Sec. 3.4. Suppose, instead, we have an out-of-phase critical current fluctuation at a frequency below f_r. Because the flux generated by this fluctuation *changes sign* when the bias current is reversed, the V-Φ curves are displaced in opposite directions. As a result, the voltage across the SQUID undergoes a phase change of π when the bias current is reversed, as shown in Fig. 12. Consequently, the voltage at the output of the lock-in due to the out-of-phase critical current fluctuation changes sign each time the bias current is reversed, and the time average of the signal over periods much longer than 1/f_r is zero. Thus, the 1/f noise due to both in-phase and out-of-phase critical current fluctuations is eliminated by this scheme.

As we shall see in Sec. 6, the 1/f noise in the critical current of high-T_c SQUIDs is severe, and it is essential to use a suitable noise reduction scheme for low-frequency applications. As stressed before, no bias reversal scheme can remove the 1/f noise due to the motion of flux.

3.7 ALTERNATIVE READ-OUT SCHEMES

Although the flux modulation method described in Sec. 3.4 has been used successfully for many years, a number of alternate schemes recently have been developed. These efforts have been motivated, at least in part, by the need to simplify the electronics required for the multichannel systems used in neuromagnetism - see Sec. 5.1. Fujimaki and co-workers [45] and Drung and co-workers [46] have devised schemes in which the output from the SQUID is sensed digitally and fed back as an analog signal to the SQUID to flux-lock the loop. Fujimaki *et al.* [45] used Josephson digital circuitry to integrate their feedback system on the same chip as the SQUID so that the flux-locked signal was available directly from the cryostat. The system of Drung and co-workers, however, is currently the more sensitive, with a flux resolution of about $10^{-6}\,\Phi_0Hz^{-1/2}$ in a 50 pH SQUID. These workers also were able to reduce the 1/f noise in the system using a modified version of the modulation scheme of Foglietti *et al.* [43]. Although they need further development, cryogenic digital feedback schemes offer several advantages: they are compact, produce a digitized output for transmission to room temperature, offer wide flux-locked bandwidths, and need not add any noise to the intrinsic noise of the SQUID.

In a quite different approach, Mück and Heiden [47] have operated a dc SQUID with hysteretic junctions in a relaxation oscillator. The oscillation frequency depends on the flux in the SQUID, reaching a maximum at $(n+1/2)\Phi_0$ and a minimum at $n\Phi_0$. A typical frequency modulation is 100kHz at an operating frequency of 10 MHz. This technique produces large voltages across the SQUID so that no matching network to the room temperature electronics is required. The room temperature electronics is simple and compact, and the resolution at 4.2 K is about $10^{-5}\Phi_0Hz^{-1/2}$ with an inductance estimated to be about 80pH.

More recently, Drung [48] introduced the concept of additional positive feedback (APF). In this technique, the SQUID is shunted with a small resistor in series with an inductance that is magnetically coupled to the SQUID. When the flux applied to the SQUID is changed, there is a re-distribution of the current between the SQUID and the parallel shunt, and the inductance links an additional flux to the SQUID. This feedback is positive or negative, depending on the sign of V_Φ. The voltage across the shunted SQUID is connected directly to a low-noise preamplifier. After additional stages of amplification and integration, the signal is fed back to another coil coupled to the SQUID to flux-lock the SQUID in the usual way; the phase of the feedback is chosen to ensure that the additional feedback is positive. APF enhances V_Φ by an order of magnitude or more, boosting its value to the point at which the preamplifier noise is comparable with or less than the intrinsic SQUID noise. This technique eliminates the need for a coupling network between the SQUID and the preamplifier, and allows for a rather simple, direct-coupled feedback circuit.

In yet another novel approach, Seppä [49] has used adaptive noise cancellation to reduce the preamplifier noise, again allowing one to couple the SQUID directly to the room temperature preamplifier without sacrificing noise performance. In this mode of operation, the SQUID is voltage biased and coupled to an inductor connected across it. The voltage noise of the preamplifier induces a current noise in the inductor and thus a flux noise in the SQUID. The phase of the feedback is such that the voltage noise generated across the SQUID cancels the preamplifier voltage noise.

4   The rf SQUID

4.1 PRINCIPLES OF OPERATION

The rf SQUID has continued to be widely used because of its long-standing commercial availability, but for many years had seen very little development. Recently, however, there has been an upsurge of interest in rf SQUIDs, spurred in part by the advent of high-T_c superconductivity. After a rather brief account of the principles and noise limitations, following rather closely descriptions in earlier reviews [50, 51], I shall describe one of these recent advances.

The rf SQUID [4, 5] shown in Fig. 14 consists of a superconducting loop of inductance L interrupted by a single Josephson junction with critical current I_0 and a nonhysteretic current-voltage characteristic. Flux quantization [1] imposes the constraint

$$\delta + 2\pi\Phi_T/\Phi_0 = 2\pi n \qquad (4.1)$$

on the total flux Φ_T threading the loop. The phase difference δ across the junction determines the supercurrent

$$I_S = -I_0 \sin(2\pi\Phi_T/\Phi_0) \qquad (4.2)$$

flowing in the ring. A quasistatic external flux Φ thus gives rise to a total flux

$$\Phi_T = \Phi - LI_0 \sin(2\pi\Phi_T/\Phi_0). \qquad (4.3)$$

The variation of Φ_T with Φ is sketched in Fig. 15(a) for the typical value $LI_0 = 1.25\ \Phi_0$. The regions with positive slope are stable, whereas those with negative slope are not. A "linearized" version of Fig. 15(a) showing the path traced out by Φ and Φ_T is shown in Fig. 15(b). Suppose we slowly increase Φ from zero. The total flux Φ_T increases less rapidly than Φ because the response flux $-LI_S$ opposes Φ. When I_S reaches I_0, at an applied flux Φ_C and a total flux Φ_{TC}, the junction switches momentarily into a nonzero voltage state and the SQUID jumps from the k = 0 to the k = 1 quantum state. If we subsequently reduce Φ from a value just above Φ_C, the SQUID remains in the k = 1 state until $\Phi = \Phi_0 - \Phi_C$, at which point I_S again exceeds the critical current and the SQUID returns to the k = 0 state. In the same way, if we lower Φ to below $-\Phi_C$ and then increase it, a second hysteresis loop will be traced. We note that this hysteresis occurs provided $LI_0 > \Phi_0/2\pi$; most practical SQUIDs are operated in this regime. For $LI_0 \approx \Phi_0$, the energy ΔE dissipated when one takes the flux around a single hysteresis loop is its area divided by L :

$$\Delta E \approx I_0\Phi_0. \qquad (4.4)$$

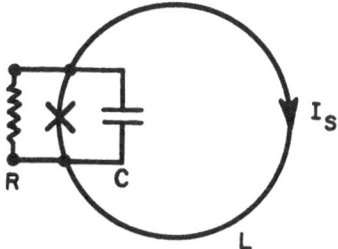

Figure 14. The rf SQUID.

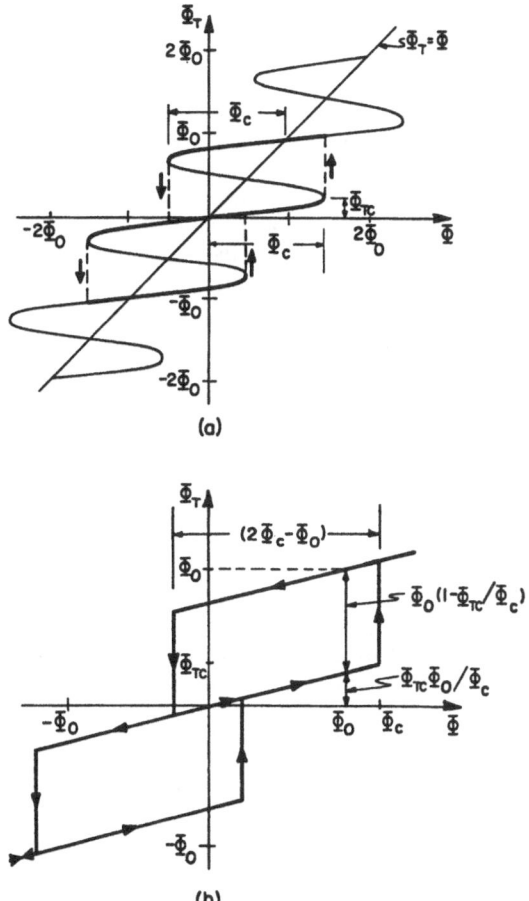

(a)

(b)

Figure 15. The rf SQUID: (a) total flux Φ_T vs. Φ for $LI_0 = 1.25\ \Phi_0$; (b) values of Φ_T as Φ is quasistatically increased and then decreased.

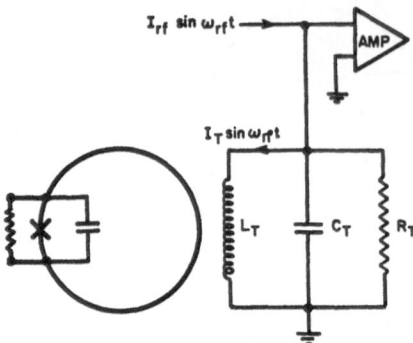

Figure 16. The rf SQUID inductively coupled to a resonant tank circuit.

We now consider the radio frequency operation of the device. The SQUID is inductively coupled to the coil of an LC-resonant circuit with a quality factor $Q = R_T/\omega_{rf}L_T$ via a mutual inductance $M = K(LL_T)^{1/2}$ – see Fig. 16. Here, L_T, C_T, and R_T are the inductance, capacitance and (effective) parallel resistance of the tank circuit, and $\omega_{rf}/2\pi$ is its resonant frequency, typically 20 or 30 MHz. The tank circuit is excited at its resonant frequency by a current I_{rf} $\sin\omega_{rf}t$, which generates a current of amplitude $I_T = QI_{rf}$ in the inductor. The voltage V_T across the tank circuit is amplified with a preamplifier having a high input impedance. First, consider the case $\Phi = 0$. As we increase I_{rf} from zero, the peak rf flux applied to the loop is $MI_T = QMI_{rf}$, and V_T increases linearly with I_{rf}. The peak flux becomes equal to Φ_C when $I_T = \Phi_C/M$ or $I_{rf} = \Phi_C/MQ$, at A in Fig. 17. The corresponding peak rf voltage across the tank circuit is

$$V_T^{(n)} = \omega_{rf}L_T\Phi_C/M, \tag{4.5}$$

where the superscript (n) indicates $\Phi = n\Phi_0$, in this case with n=0. At this point the SQUID makes a transition to <u>either</u> the k = +1 state or the k = -1 state. As the SQUID traverses the hysteresis loop, energy ΔE is extracted from the tank circuit. Because of this loss, the peak flux on the next half cycle is less than Φ_C, and no transition occurs. The tank circuit takes many cycles to recover sufficient energy to induce a further transition, which may be into either the k = +1 or -1 states. If we now increase I_{rf}, transitions are induced at the same values of I_T and V_T but, because energy is supplied at a higher rate, the stored energy builds up more rapidly after each energy loss ΔE, and transitions occur more frequently. In the absence of thermal fluctuations (Sec. 4.2), the "step" AB in Fig. 17 is at constant voltage. At B, a transition is induced on each positive and negative rf peak, and a further increase in I_{rf} produces the "riser" BC. At C, transitions from the k = ±1 to the k = ±2 states occur, and a second step begins. As we continue to increase I_{rf}, we observe a series of steps and risers.

If we now apply an external flux $\Phi = \Phi_0/2$, the hysteresis loops in Fig. 15(b) are shifted by $\Phi_0/2$. Thus, a transition occurs on the positive peak of the rf cycle at a flux $(\Phi_C - \Phi_0/2)$, whereas on the negative peak the required flux is $-(\Phi_C + \Phi_0/2)$. As a result, as we increase I_{rf} from zero, we observe the first step at D in Fig. 17 at

$$V_T^{(n+1/2)} = \omega_{rf}L_T(\Phi_C-\Phi_0/2)/M. \tag{4.6}$$

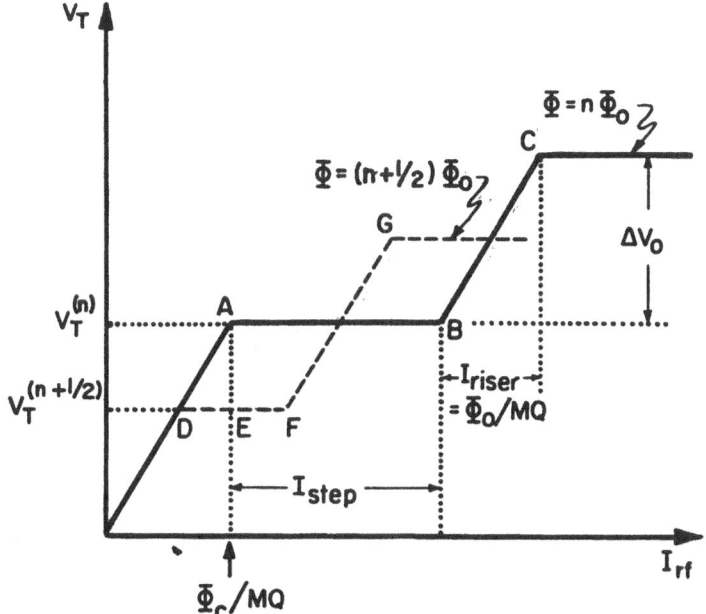

Figure 17. V_T vs. I_{rf} in the absence of thermal noise for $\Phi = n\Phi_0$, $(n+1/2)\Phi_0$.

As we increase I_{rf} from D to F, the SQUID traverses only one hysteresis loop, corresponding to the $k = 0$ to $k = +1$ transition at $(\Phi_c - \Phi_0/2)$. A further increase in I_{rf} produces the riser FG, and at G, transitions begin at a peak rf flux $-(\Phi_C + \Phi_0/2)$. In this way, we observe a series of steps and risers for $\Phi = \Phi_0/2$, interlocking those for $\Phi = 0$ (Fig. 17). As we increase Φ from zero, the voltage at which the first step appears will drop to a minimum (D) at $\Phi_0/2$ and rise to its maximum value (A) at $\Phi = \Phi_0$. The change in V_T as we increase Φ from 0 to $\Phi_0/2$, found by subtracting Eq. (4.6) from Eq.(4.5), is $\omega_{rf}L_T\Phi_0/2M$. Thus, for a small change in flux near $\Phi = \Phi_0/4$, we find the transfer function

$$V_\Phi = \omega_{rf}L_T/M. \tag{4.7}$$

At first sight, Eq.(4.7) suggests that we can make V_Φ arbitrarily large by reducing K sufficiently. However, we obviously cannot make K so small that the SQUID has no influence on the tank circuit, and we need to establish a lower bound on K. To operate the SQUID, we must be able to choose a value of I_{rf} that intercepts the first step for all values of Φ : this requirement is satisfied if the point F in Fig. 17 lies to the right of E, that is, if DF exceeds DE. We can calculate DF by noting that the power dissipation in the SQUID is zero at D and $\Delta E(\omega_{rf}/2\pi) \approx I_0\Phi_0\omega_{rf}/2\pi$ at F. Thus, $(I_{rf}^{(F)} - I_{rf}^{(D)})v_T^{(n+1/2)}/2 = I_0\Phi_0\omega_{rf}/2\pi$ (I_{rf} and v_T are peak, rather than rms values). Furthermore, we easily can see that $I_{rf}^{(E)} - I_{rf}^{(D)} = \Phi_0/2MQ$. Assuming $LI_0 \approx \Phi_0$ and using Eq.(4.5), we find that the requirement $I_{rf}^{(E)} > I_{rf}^{(D)}$ can be written in the form

$$K^2Q \gtrsim \pi/4. \tag{4.8}$$

If we set $K \approx Q^{-1/2}$, Eq.(4.7) becomes

$$V_\Phi \approx \omega_{rf}(QL_T/L)^{1/2}. \tag{4.9}$$

To operate the SQUID, one adjusts I_{rf} so that the SQUID remains biased on the first step - see Fig. 17 - for all values of Φ. The rf voltage across the tank circuit is amplified and demodulated to produce a signal that is periodic in Φ. A modulating flux, typically at 100kHz and with a peak-to-peak amplitude of $\Phi_0/2$, is also applied to the SQUID, just as in the case of the dc SQUID. The voltage produced by this modulation is lock-in detected, integrated, and fed back as a current into the modulation coil to flux-lock the SQUID.

4.2 THEORY OF NOISE IN THE RF SQUID

A detailed theory has been developed for noise in the rf SQUID [52-60]; in contrast to the case for the dc SQUID, noise contributions from the tank circuit and preamplifier are also important. We begin by discussing the intrinsic noise in the SQUID. In the previous section we assumed that transitions from the $k = 0$ to the $k = 1$ state occurred precisely at $\Phi = \Phi_c$. In

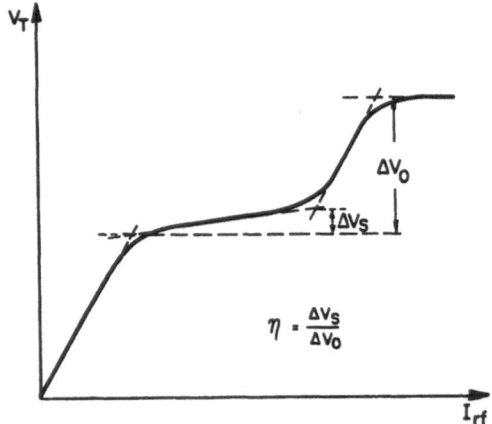

Figure 18. V_T vs. I_{rf} showing the effects of thermal noise.

fact, thermal activation causes the transition to occur stochastically, at lower values of flux. Kurkijärvi [52] calculated the distribution of values of Φ at which the transitions occur; experimental results [61] are in good agreement with his predictions. When the SQUID is driven with an rf flux, the fluctuations in the value of flux at which transitions occur have two consequences. First, noise is introduced on the peak voltage V_T, giving an equivalent intrinsic flux noise spectral density [53, 57]

$$S_\Phi^{(i)} \approx \frac{(LI_0)^2}{\omega_{rf}} \left[\frac{2\pi k_B T}{I_0 \Phi_0} \right]^{4/3}. \tag{4.10}$$

Second, the noise causes the steps to tilt (Fig. 18), as we easily can see by considering the case for $\Phi = 0$. In the presence of thermal fluctuations the transition from the $k = 0$ to the $k = 1$ state (for example) has a certain probability of occurring at any given value of the total flux

$\Phi + \Phi_{rf}$. Just to the right of A in Fig. 17 this transition occurs at the peak of the rf flux once in many rf cycles. Thus, the probability of the transition occurring in any one cycle is small. On the other hand, at B a transition must occur at each positive and negative peak of the rf flux, with unity probability. To increase the transition probability, the peak value of the rf flux and hence V_T must increase as I_{rf} is increased from A to B. Jackel and Buhrman [54], introduced the slope parameter η defined in Fig. 18, and showed that it was related to $S_\Phi^{(i)}$ by the relation

$$\eta^2 \approx S_\Phi^{(i)} \omega_{rf}/\pi\Phi_0^2 , \qquad (4.11)$$

provided η was not too large. This relation is verified experimentally.

The noise temperature T_a of typical rf amplifiers operated at room temperature is substantially higher than that of amplifiers operated at a few hundred kilohertz, and is therefore not negligible for rf SQUIDs operated at liquid ^{4}He temperatures. Furthermore, part of the coaxial line connecting the tank circuit to the preamplifier is at room temperature. Since the capacitances of the line and the amplifier are a substantial fraction of the capacitance of the tank circuit, part of the resistance damping the tank circuit is well above the bath temperature. As a result, there is an additional contribution to the noise which we combine with the preamplifier noise to produce an effective noise temperature T_a^{eff}. The noise energy contributed by these extrinsic sources can be shown to be [54, 58] $2\pi\eta k_B T_a^{eff}/\omega_{rf}$. Combining this contribution with the intrinsic noise, one finds

$$\varepsilon \approx \frac{1}{\omega_{rf}} \left(\frac{\eta\pi^2\Phi_0^2}{2L} + 2\pi\eta k_B T_a^{eff} \right) . \qquad (4.12)$$

Equation (4.12) shows that ε scales as $1/\omega_{rf}$, but one should bear in mind that T_a tends to increase with ω_{rf}. Nonetheless, as we shall see shortly, improvements in performance have been achieved by operating the SQUID at frequencies much higher than the usual 20 or 30 MHz.

4.3 PRACTICAL rf SQUIDs

Although generally less sensitive than the dc SQUID, the rf SQUID is entirely adequate for a wide range of applications. It is has been commercially available since the early 1970's, notably from BTi (formerly SHE). Figure 19 shows a cut-away drawing of the BTi rf SQUID [44], which has a toroidal configuration machined from Nb. One way to understand this geometry is to imagine rotating the SQUID in Fig. 14 through 360° about a line running through the junction from top to bottom of the page. This procedure produces a toroidal cavity connected at its center by the junction. If one places a toroidal coil in this cavity, a current in the coil produces a flux that is tightly coupled to the SQUID. In Fig. 19, there are actually two such cavities, one containing the tank circuit-modulation-feedback coil and the other the input coil. This separation eliminates cross-talk between the two coils. Leads to the two coils are brought out via screw-terminals. The junction is made from thin films of Nb. This device is self-shielding against external magnetic field fluctuations, and has proven to be reliable and convenient to use. In particular, the Nb input terminals enable one to connect different input circuits in a straightforward way. A typical device has a white noise energy of 5×10^{-29} JHz^{-1}, with a l/f noise energy of perhaps 10^{-28} JHz^{-1} at 0.1Hz.

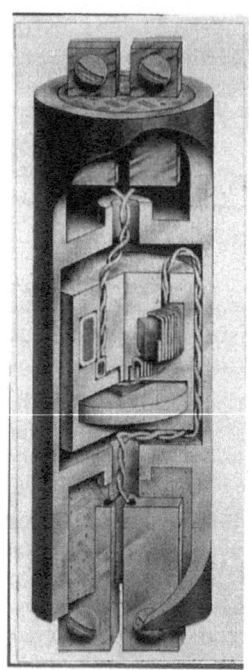

Figure 19. Cut-away drawing of toroidal SQUID (courtesy D. Paulsen, BTi, Inc.).

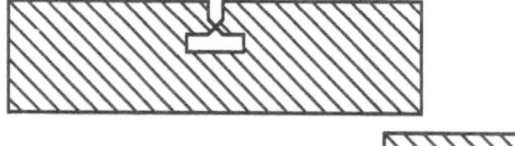

Figure 20. Microwave rf SQUID consisting of Nb film microstrip resonator coupled to microstrip input and output line (from ref. [65]).

As we saw in Eq. (4.12), one can improve the noise energy by operating the rf SQUID at high frequencies [62, 63]. One can also reduce T_a^{eff} by cooling the preamplifier [62, 64], thereby reducing T_a and reducing the temperature of the tank circuit to that of the helium bath. Recently, Mück and co-workers [65] have taken advantage of both of these techniques to make substantial improvements in the performance of rf SQUIDs.

To operate a SQUID at 3GHz, Mück used an ingenious microstrip configuration, shown in Fig. 20. The SQUID consisted of a 100-nm-thick niobium film, with a hole varying from 50 x 50μm^2 to 200 x 200μm^2, deposited on a 1-mm-thick sapphire substrate. The junction was either a microbridge or a Nb-Al$_2$O$_3$-Nb tunnel junction with a resistive shunt to reduce β_c to about 0.8. The sapphire substrate was mounted on a printed circuit board, the copper cladding serving as a ground plane for the microstrip resonator. The microwaves were coupled in via a second microstrip (Fig. 20) and the SQUID signal was coupled out via the same microstrip; a directional coupler separated the incoming bias from the outgoing signal. The Q of the resonator was typically 500 to 1000, and K was chosen to make $K^2Q \approx 1$. For SQUIDs with a tunnel junction, a peak-to-peak signal as high as 80μV was obtained, corresponding to

$V_\Phi = 160\mu V/\Phi_0$. Using a room temperature amplifier with an rms noise of 0.5 nVHz$^{-1/2}$, Mück obtained a flux noise of $3 \times 10^{-6}\Phi_0$ Hz$^{-1/2}$ in a flux-locked loop at frequencies down to 0.1 Hz. The best noise energy achieved was about 2×10^{-31} JHz^{-1}. SQUIDs were also fabricated with integrated Nb coils deposited on them in the same way as for dc SQUIDs.

To improve the performance further, Mück [65] used a cooled GaAs HEMT (High Electron Mobility Transistor) as a preamplifier, achieving a noise energy of about 3×10^{-32} JHz^{-1} in a flux-locked loop at 1Hz. This is a remarkable improvement in the performance of the rf SQUID, to a value that is comparable with that of typical thin-film dc SQUIDs.

5. SQUID-Based Instruments.

Both dc and rf SQUIDs are used as sensors in a far-ranging assortment of instruments. I here briefly discuss some of them: my selection is far from exhaustive, but does include the more commonly used instruments.

Each instrument involves a circuit attached to the input coil of the SQUID. We should recognize from the outset that, in general, the presence of the input circuit influences both the signal and noise properties of the SQUID while the SQUID, in turn, reflects a complex impedance into the input. Because the SQUID is a nonlinear device a full description of the interactions is complicated, and we shall not go into the details here. However, one aspect of this interaction, first pointed out by Zimmerman [66], is easy to understand. Suppose we connect a superconducting pick-up loop of inductance L_p to the input coil of inductance L_i to form a magnetometer, as shown in Fig. 21(a). It is easy to show that the SQUID inductance L is reduced to the value

$$L' = L\left[1 - \alpha^2 L_i / (L_i + L_p)\right], \tag{5.1}$$

where α^2 is the coupling coefficient between L and L_i. We have neglected any stray inductance in the leads connecting L_i and L_p, and any stray capacitance. The reduction in L tends to increase the transfer coefficient of both the dc SQUID [Eq.(3.7)] and the rf SQUID [Eq.(4.9)]. In most cases, the reduction of L and the change in the noise properties will be detectable, but they will not have a major effect on the results presented here.

5.1 MAGNETOMETERS AND GRADIOMETERS

One of the simplest instruments is the magnetometer [Fig. 21(a)]. A pick-up loop is connected across the input coil to make a superconducting flux transformer. The SQUID and input coil are generally enclosed in a superconducting shield. If one applies a magnetic flux, $\delta\Phi^{(p)}$, to the pick-up loop, flux quantization requires that

$$\delta\Phi^{(p)} + (L_i + L_p)J_s = 0, \tag{5.2}$$

where J_s is the supercurrent induced in the transformer. We have neglected the effects of the SQUID on the input circuit. The flux coupled into the SQUID, which we assume to be in a flux-locked loop, is $\delta\Phi = M_i | J_s | = M_i\delta\Phi^{(p)} / (L_i + L_p)$. We find the minimum detectable value of $\delta\Phi^{(p)}$ by equating $\delta\Phi$ with the equivalent flux noise of the SQUID. Defining $S_\Phi^{(p)}$ as the spectral density of the equivalent flux noise referred to the pick-up loop, we find

$$S_\Phi^{(p)} = \frac{(L_p+L_i)^2}{M_i^2} S_\Phi. \tag{5.3}$$

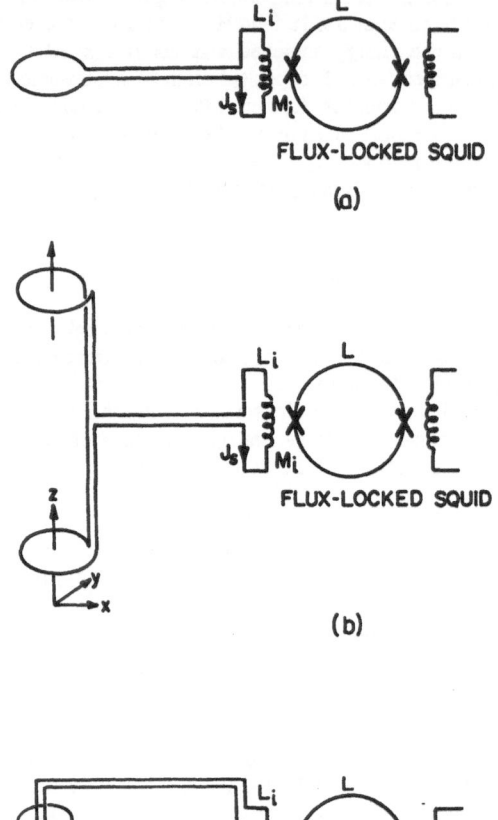

(a)

(b)

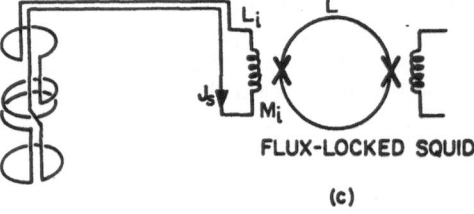

(c)

Figure 21. Superconducting flux transformers:(a) magnetometer,
(b) first-derivative gradiometer, (c) second-derivative gradiometer.

Introducing the equivalent noise energy referred to the pick-up loop, we obtain

$$\frac{S_\Phi^{(p)}}{2L_p} = \frac{(L_p+L_i)^2}{L_i L_p} \frac{S_\Phi}{2\alpha^2 L} . \tag{5.4}$$

We observe that Eq.(5.4) has the minimum value

$$S_\Phi^{(p)} / 2L_p = 4\varepsilon(f) / \alpha^2 \tag{5.5}$$

when $L_i = L_p$. Thus, a fraction $\alpha^2/4$ of the energy in the pick-up loop is transferred to the SQUID. In this derivation we have neglected noise currents in the input circuit arising from noise in the SQUID, the fact that the input circuit reduces the SQUID inductance, and any possible coupling between the feedback coil of the SQUID and the input circuit. Having obtained the flux resolution for $L_i = L_p$, we can immediately write down the corresponding magnetic field resolution $B_N^{(p)} = (S_\Phi^{(p)})^{1/2}/\pi r_p^2$, where r_p is the radius of the pick-up loop:

$$B_N^{(p)} = 2\sqrt{2} L_p^{1/2} \varepsilon^{1/2}/\pi r_p^2 \alpha . \tag{5.6}$$

For a loop made from wire of radius r_0, one finds [67] $L_p = \mu_0 r_p[\ln(8r_p/r_0) - 2]$, where $\mu_0 = 4\pi \times 10^{-7}$ henry/meter; for a reasonable range of values of r_p/r_0 we can set $L_p \approx 5\mu_0 r_p$. Thus, we obtain $B_N^{(p)} \approx 2(\mu_0\varepsilon)^{1/2}/\alpha r_p^{3/2}$. This indicates that one can, in principle, improve the magnetic field resolution indefinitely by increasing r_p, keeping $L_i = L_p$. Of course, in practice, the size of the cryostat will impose an upper limit on r_p. If we take $\varepsilon = 10^{-28}$ JHz^{-1} (a somewhat conservative value for an rf SQUID), $\alpha = 1$, and $r_p = 25$ mm, we find $B_N^{(p)} \approx 5 \times 10^{-15}$ tesla $Hz^{-1/2} = 5 \times 10^{-11}$ gauss $Hz^{-1/2}$. This is a much higher sensitivity than that achieved by any nonsuperconducting magnetometer.

Magnetometers have usually involved flux transformers made of Nb wire. For example, one can make the rf SQUID in Fig. 19 into a magnetometer merely by connecting a loop of Nb wire to its input terminals. In the case of the thin-film dc SQUID, one can make an integrated magnetometer by fabricating a Nb loop across the spiral input coil. In this way, Wellstood et al. [28] achieved a magnetic field white noise of 5×10^{-15} tesla $Hz^{-1/2}$ using a pick-up loop with a diameter of a few millimeters. One application of magnetometers is in geophysics [68] – see Sec. 5.7.

An important variation of the flux transformer is the gradiometer. Figure 21(b) shows an axial gradiometer that measures $\partial B_z/\partial z$. The two pick-up loops are wound in opposition and balanced so that a uniform field B_z links zero net flux to the transformer. A gradient $\partial B_z/\partial z$, on the other hand, does induce a net flux and thus generates an output from the flux-locked SQUID. Figure 21(c) shows a second-order gradiometer that measures $\partial^2 B_z/\partial z^2$; Fig. 22(a) is a photograph of a practical version. Thin-film gradiometers based on dc SQUIDs were made as long ago [32] as 1978, and a variety of devices [27,69-74] have been reported since then. Most thin film gradiometers have been planar, and therefore measure an off-diagonal gradient, for example, $\partial B_z/\partial x$ or $\partial^2 B_z/\partial x\partial y$. A representative device is shown in Fig. 22(b) [72]. However, Hoenig et al. [74] made a 37-channel biomagnetic system involving thin film first-derivative axial gradiometers. The pick-up loops are deposited on flexible printed circuit board which is then folded and bonded onto a supporting block. The leads from each gradiometer are bonded to the input coil of a thin-film dc SQUID.

The most important application of the gradiometer thus far is in neuromagnetism [75], notably to detect weak magnetic signals emanating from the human brain. The gradiometer discriminates strongly against distant noise sources, which have a small gradient, in favor of locally generated signals. One can thus use a second-order gradiometer in an unshielded environment, although the present trend is toward using first-order gradiometers in a shielded room of aluminum and mu-metal that greatly attenuates the ambient magnetic noise. In this application, axial gradiometers of the type shown in Fig. 21(a) actually sense magnetic field, rather than the gradient, because the distance from the signal source to the pick-up loop is less than the baseline of the gradiometer. The magnetic field sensitivity referred to one pick-up loop is typically $10fTHz^{-1/2}$. Considerable progress in multichannel instruments has been made in recent years, and one with 122 gradiometers has recently been marketed [76].

(a)

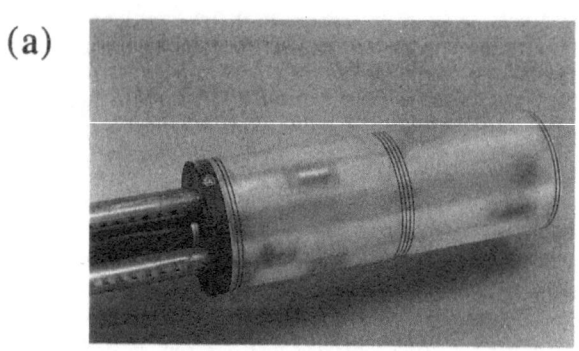

(b)

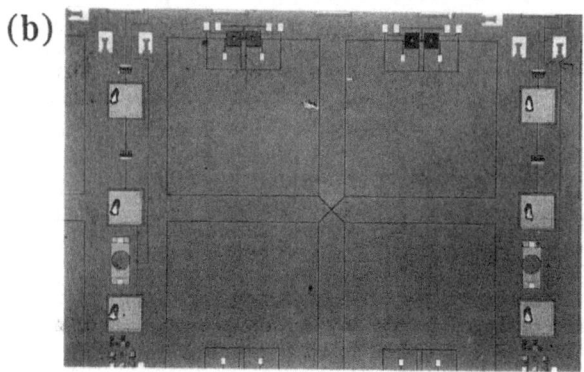

Figure 22. (a) Photograph of wire-wound second-derivative gradiometer for biomedical applications (courtesy BTi, inc.). (b) Second-order ($\partial^2 B_z/\partial x \partial y$) planar gradiometer fabricated in $Nb/Nb_2O_5/PbAuIn$ edge-junction technology. The device is 6.35mm square. (Courtesy M. B. Ketchen, W. J. Gallagher, R. L. Sandstrom and A. W. Kleinsasser, IBM [72])

There are two basic kinds of measurements made on the human brain. In the first, one detects spontaneous activity: a classic example is the generation of magnetic pulses by subjects suffering from focal epilepsy [77]. The second kind involves evoked response: for example, Romani et al. [78] detected the magnetic signal from the auditory cortex generated by tones of different frequencies. Extensive reviews of this work appear elsewhere in these proceedings.

There are several other applications of gradiometers. One kind of magnetic monopole detector [79] consists of a gradiometer: the passage of a monopole would link a flux h/e in the pick-up loop and produce a step-function response from the SQUID. Gradiometers have recently been of interest in studies of corrosion and in the location of fractures in pipelines and other structures. (See the chapters by Donaldson and Wikswo.)

5.2 SUSCEPTOMETERS

In principle, one easily can use the first-derivative gradiometer of Fig. 21(b) to measure magnetic susceptibility χ. One establishes a static field along the z-axis and lowers the sample into one of the pick-up loops. Provided χ is nonzero, the sample introduces an additional flux into the pick-up loop and generates an output from the flux-locked SQUID. Very sophisticated susceptometers are available commercially [80]. Room temperature access enables one to cycle samples rapidly, and one can measure χ as a function of temperature between 1.8 K and 400 K in fields up to 5.5 tesla. These systems are capable of resolving a change in magnetic moment as small as 10^{-8} emu.

Novel miniature susceptometers have been developed by Ketchen and co-workers [37, 81, 82]. One version is shown schematically in Fig. 23. The SQUID loop incorporates two pick-up loops wound in the opposite sense and connected in series. The two square pick-up loops, 17.5 μm on a side and with an inductance of about 30 pH, are deposited over a hole in the ground plane that minimizes the inductance of the rest of the device. The SQUID is flux biased at the maximum of V_Φ by means of a control current I_C in one of the pick-up loops. One can apply a magnetic field to the two loops by means of the current I_F; by passing a fraction of this current into the center tap I_C, one can achieve a high degree of electronic balance between the two loops. The sample to be studied is placed over one of the loops, and the output from the SQUID when the field is applied is directly proportional to the magnetization. At 4.2 K, the susceptometer is capable of detecting the magnetization due to as few as 3000 electron spins.

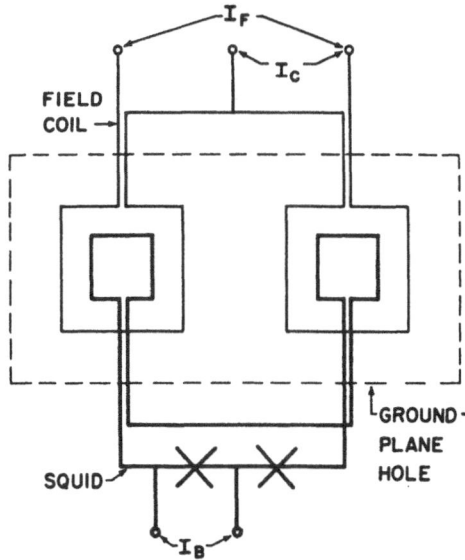

Figure 23. Thin-film miniature susceptometer (from ref. [81]).

Awschalom and co-workers [37, 82], have used a miniature susceptometer to perform magnetic spectroscopy of semiconductors with picosecond time-resolution. Linearly polarized pulses 4 ps in length are generated with a dye laser and split into a pump train and a weaker probe train. The time delay between the two trains can be varied, and each train is converted to circular polarization by a quarter-wave plate. The beams are chopped at 197Hz and passed down an optical fiber to the sample in the cryostat. The pump pulses induce a magneto-optical susceptibility χ_{op} which is subsequently measured by means of the much weaker probe pulses of intensity δI that induce a magnetization $\chi_{op}\delta I$. The magnetization is detected by the SQUID at the chopping frequency, and its output is lock-in detected. By varying the time delay between the pump and probe pulses, one can investigate the dynamics of the induced magnetization. One also can vary the dye laser frequency through the red region of the visible spectrum to study the energy dependence of the magnetization. This technique recently has been extended to temperatures down to 0.3K [82].

5.3 VOLTMETERS

Probably the first practical application of a SQUID was to measure tiny, quasistatic voltages [83]. One simply connects the signal source -- for example a low resistance through which a current can be passed -- in series with a known resistance and the input coil of the SQUID. The output from the flux-locked loop is connected across the known resistance to obtain a null-balancing measurement of the voltage. The resolution is generally limited by Nyquist noise in the input circuit, which at 4.2 K varies from about 10^{-15} V $Hz^{-1/2}$ for a resistance of 10^{-8} Ω to about 10^{-10} V $Hz^{-1/2}$ for a resistance of 100 Ω.

Applications of these voltmeters range from the measurement of thermoelectric voltages and of quasiparticle charge imbalance in nonequilibrium superconductors to noise thermometry and the high-precision comparison of the Josephson voltage-frequency relation in different superconductors.

5.4 THE DC SQUID AS A RADIOFREQUENCY AMPLIFIER

In recent years, the dc SQUID has been developed as a low-noise amplifier for frequencies up to 100 MHz or more [84]. To understand the theory for the performance of this amplifier, we need to extend the theory of Sec. 3.2 by taking into account the noise associated with the current J(t) in the SQUID loop. For a bare SQUID with $\beta = 1$, $\Gamma = 0.05$ and $\Phi = (2n+1)\Phi_0/4$, one finds the spectral density of the current to be [85]

$$S_J(f) \approx 11 \, k_B T/R. \tag{5.7}$$

Furthermore, the current noise is partially correlated with the voltage noise across the SQUID, the cross-spectral density being [85]

$$S_{VJ}(f) \approx 12 \, k_B T . \tag{5.8}$$

The correlation arises, roughly speaking, because the current noise generates a flux noise which, in turn, contributes to the total voltage noise across the junction, provided $V_\Phi \neq 0$.

If one imagines coupling a coil to the SQUID, the coil will "see" an impedance Z in the SQUID loop that can be written in the form [86]

$$\frac{1}{Z} = \frac{1}{j\omega L} + \frac{1}{R} . \tag{5.9}$$

The dynamic inductance L and dynamic resistance R are not simply related to L and R, but vary with bias current and flux; for example, $1/L$ is zero for certain values of Φ.

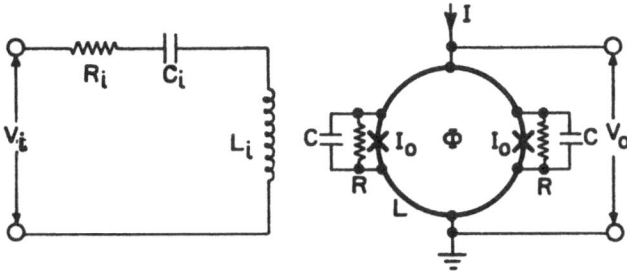

Figure 24. Tuned radiofrequency amplifier based on dc SQUID (from ref. [84]).

One can make a tuned amplifier, for example, by connecting an input circuit to the SQUID, as shown in Fig. 24. In general, the presence of this circuit modifies all of the SQUID parameters and the magnitude of the noise spectral densities [87]. Furthermore, the SQUID reflects an impedance $\omega^2 M_i^2/Z$ into the input circuit. Fortunately, however, one can neglect the mutual influence of the SQUID and input circuit, provided the coupling coefficient α^2 is sufficiently small, as it is under certain circumstances. For the purpose of illustration, we derive the noise temperature of the amplifier in Fig. 24. We assume given values of $S_V(f)$, $S_J(f)$, $S_{VJ}(f)$ and L_i, and find the values of C_i and R_i that optimize the noise temperature.

In the weak coupling limit, the noise current $J_N(t)$ induces a voltage $-M_i J_N$ into the input circuit, and hence a current $-M_i J_N / Z_i$, where

$$Z_i \approx R_i + j\omega L_i + 1/j\omega C_i \ . \tag{5.10}$$

Here, Z_i is the impedance of the input circuit and L_i and C_i are the series inductance and capacitance. The noise current in the input circuit, in turn, induces a flux in the SQUID loop and finally a voltage $-M_i^2 J_N V_\Phi/Z_i$ across the SQUID. Thus, the noise voltage across the SQUID in the presence of the input circuit is [84]

$$V_N'(t) = V_N(t) - M_i^2 J_N V_\Phi / Z_i \ , \tag{5.11}$$

where $V_N(t)$ is the noise voltage of the bare SQUID, which we assume to be unchanged by the input circuit in the limit $\alpha^2 \to 0$. The spectral density of $V_N'(t)$ is easily found to be

$$S_V'(f) = S_V(f) + \frac{\omega^2 M_i^4 V_\Phi^2 S_J(f)}{|Z_i|^2} - \frac{2\omega M_i^2 V_\Phi(\omega L_i - 1/\omega C_i)S_{VJ}(f)}{|Z_i|^2} \ . \tag{5.12}$$

We now suppose that we apply a sinusoidal input signal frequency $\omega/2\pi$, with a mean-square amplitude $\langle V_i^2 \rangle$. The mean-square signal at the output of the SQUID is

$$\langle V_0^2 \rangle = M_i^2 V_\Phi^2 \langle V_i^2 \rangle / |(Z_i)|^2. \tag{5.13}$$

The signal-to-noise ratio is

$$S/N = \langle V_0^2 \rangle / S_V'(f)B \tag{5.14}$$

in a bandwidth B. It is convenient to introduce a noise temperature T_N for the amplifier by setting S/N = 1 with $<V_i^2> = 4k_B T_N R_i B$. This procedure implies that the output noise power generated by the SQUID is equal to the output noise power generated by the resistor R_i when it is at a temperature T_N. We then can optimize T_N with respect to R_i and C_i for a given value of L_i, and find

$$R_i^{(opt)} = \frac{\alpha^2 \omega L_i L V_\Phi}{S_V} (S_V S_J - S_{VJ}^2)^{1/2},$$ (5.15)

$$\frac{1}{\omega C_i^{(opt)}} = \omega L_i \left(1 + \frac{\alpha^2 S_{VJ} L V_\Phi}{S_V}\right),$$ (5.16)

and

$$T_N^{(opt)} = \frac{\pi f}{k_B V_\Phi} (S_V S_J - S_{VJ}^2)^{1/2}.$$ (5.17)

We note from Eq.(5.16) that the optimum noise temperature occurs off-resonance. It often is more convenient in practice to use the amplifier at the resonant frequency of the tank circuit, given by $\omega^2 L_i C_i = 1$ (neglecting reflected components from the SQUID). In that case, one finds optimum values [84]

$$R_i^{(res)} = \alpha^2 \omega L_i L V_\Phi (S_J / S_V)^{1/2}$$ (5.18)

and

$$T_N^{(res)} = \frac{\pi f}{k_B V_\Phi} (S_V S_J)^{1/2}.$$ (5.19)

Using the results of Eqs.(3.7), (3.8), (5.7) and (5.8), we can write Eq.(5.18) in the form

$$\alpha^2 \omega L_i / R_i^{(res)} = \alpha^2 Q \approx 1.$$ (5.20)

This result shows that high-Q input circuits imply that α^2 is small, thereby justifying the assumption made at the beginning of this section. One also finds

$$T_N(f) \approx 18 f T / V_\Phi \approx 2 f \varepsilon(f)/k_B.$$ (5.21)

Thus, although $\varepsilon(f)$ does not fully characterize an amplifier, as noted earlier, within the framework of the model, it does enable one to predict T_N.

One can easily calculate the gain on resonance. For $\alpha^2 \ll 1$, an input signal V_i produces an output voltage $V_0 \approx (V_i / R_i^{(res)}) M_i V_\Phi$. The power gain is thus $G = (V_0^2/R_D)/(V_i^2/R_i)$, where R_d is the dynamic output resistance $(\partial V/\partial I)_\Phi$ of the SQUID. If we take $R_d \approx R$, we find

$$G \approx V_\Phi/\omega.$$ (5.22)

Hilbert and Clarke [84] made several radiofrequency amplifiers with both tuned and untuned inputs, flux biasing the SQUID near $\Phi = (2n + 1)\Phi_0/4$. There was no flux-locked loop. The measured parameters were in good agreement with predictions. For example, for an amplifier with $R \approx 8 \ \Omega$, $L \approx 0.4$ nH, $L_i \approx 5.6$nH, $M_i \approx 1$nH and $V_\Phi \approx 3 \times 10^{10}$ sec^{-1} at 4.2K, they found $G = 18.6 \pm 0.5$dB and $T_N = 1.7 \pm 0.5$K at 93MHz. The predicted values were 17dB and 1.1K, respectively.

We emphasize that in this theory and these measurements one is concerned only with the noise temperature of the amplifier itself. Nyquist noise from the resistor adds a contribution which, in the example just given, exceeds the amplifier noise. Thus, the optimization procedure just outlined does not necessarily give the lowest system noise, and one would use a different procedure when the value of T_N in Eq.(5.17) or Eq.(5.19) is well below T.

In concluding this section, we comment briefly on the quantum limit for the dc SQUID amplifier. At T = 0, Nyquist noise in the shunt resistors should be replaced with zero point fluctuations [Eq.(2.11)]. Koch *et al.* [88] performed a simulation in this limit and concluded that, within the limits of error, the noise temperature of a tuned amplifier in the quantum limit should be given by

$$T_N \approx hf/k_B\ln2. \qquad (5.23)$$

This is the result for any quantum-limited amplifier. The corresponding value for ε was approximately $\hbar$, but it should be emphasized that quantum mechanics does not impose any precise lower limit on ε [89]. A number of SQUIDs have obtained noise energies of $3\hbar$ or less, but there is no evidence as yet that a SQUID has attained quantum-limited performance as an amplifier.

5.5 MAGNETIC RESONANCE

SQUIDs have been used for two decades to detect magnetic resonance [90]. Most of the experiments involved the detection of magnetic resonance at low frequencies or the change in the static susceptibility of a sample induced by a resonance at high frequency. However, the development of the radiofrequency amplifier described in the previous section enables one to detect pulsed magnetic resonance directly at frequencies up to ~300 MHz.

Clarke, Hahn and co-workers have used the radiofrequency amplifier to perform nuclear quadrupole resonance [90] (NQR) and nuclear magnetic resonance [91] (NMR) experiments. They observed NQR in ^{35}Cl, which, in zero magnetic field, has two doubly degenerate nuclear levels with a splitting of 30.6856 MHz. The experimental configuration is shown in Fig. 25. The sample is placed in a superconducting pick-up coil, in series with which is an identical, counterwound coil. These coils are in series with an adjustable tuning capacitor C_i, the 4-turn input coil of a planar dc SQUID and 20 unshunted Josephson junctions. The resistor R_i represents contact resistance and losses in the capacitor. Radiofrequency pulses applied to the transmitter coil cause the nuclear spins to precess; after each pulse is turned off, the amplifier detects the precessing magnetization. The amplified signal is mixed down with a reference provided by the rf generator, and the mixed-down signal is passed through a low-pass filter, observed on an oscilloscope, and recorded digitally for further analysis.

The major difficulty with this technique, and indeed with other pulsed methods, is the saturation of the amplifier by the very large rf pulse. In the present experiments, the effects of this pulse are reduced in two ways. First, the gradiometer-like configuration gives a common-mode rejection that can be as high as 3×10^4. Second, the series of junctions in the input circuit acts as a Q-spoiler [90]. As the current begins to build in the tuned circuit, the junctions switch to the resistive state with a total resistance of about 1 kΩ, thereby reducing the Q to ~1. When the pulse is turned off, the transients die out very quickly and the junctions revert to their zero voltage state, rapidly restoring Q to its full value, usually several thousand. In this way, one can combine the benefits of a high-Q tuned circuit and a sensitive amplifier while retaining a relatively short dead-time after each pulse. In their initial experiments, Hilbert *et al.*

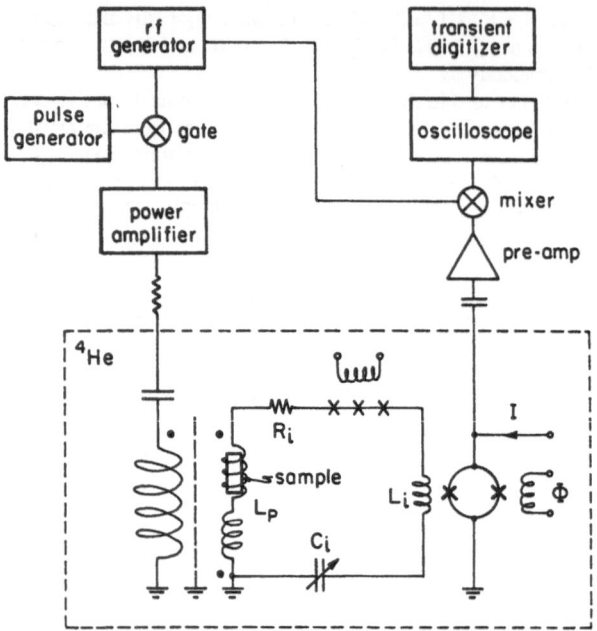

Figure 25. Circuit for NQR with dc SQUID amplifier (from ref. [90]).

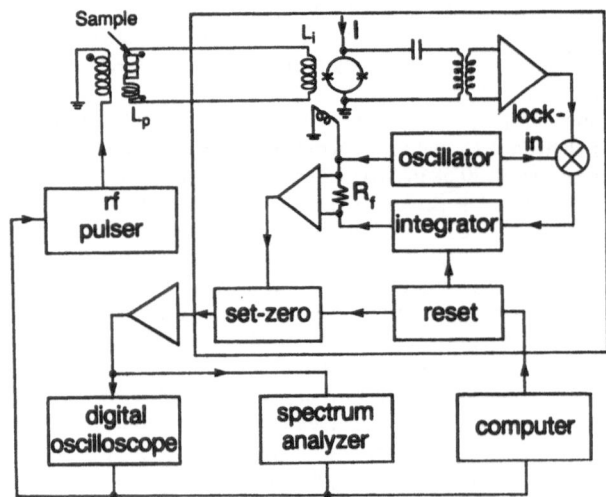

Figure 26. Configuration of NQR Fourier-transform spectrometer based on a dc SQUID amplifier (from ref. [95]).

[90] achieved a resolution for a single pulse of ~2×10^{16} spins (~2×10^{16} nuclear Bohr magnetons) in a bandwidth of 10kHz.

Subsequently, the Q-spoiler and SQUID amplifier were used to detect atomic polarization induced by precessing nuclear electric quadrupoles [92]. In this experiment, the $NaClO_3$ sample was placed in a capacitor that formed part of the tuned input circuit, and NQR induced in the usual way by radiofrequency pulses. The precessing electric quadrupole moments induce a net electric dipole moment in the neighboring atoms, provided the crystal is noncentro-symmetric. These dipole moments, in turn, produce an oscillating electric polarization in the crystal and hence a voltage on the capacitor that is amplified in the usual way. This technique yields information on the location and polarization of atoms near nuclear quadrupole moments.

The Q-spoiler and amplifier also have been used to detect nuclear magnetic resonance [91]. In these experiments one applies a magnetic field with an amplitude of several tesla to the crystal, and places the superconducting circuitry some distance away in a relatively low field. In yet another experiment, Sleator et al. [93] observed "spin noise" in ^{35}Cl. An rf signal at the NQR frequency equalized the populations of the two nuclear spin levels, and then was turned off to leave a zero-spin state. A SQUID amplifier (without a Q-spoiler) was able to detect the photons emitted spontaneously as the upper state decayed, even though the lifetime per nucleus for this process was ~10^6 centuries. The detected power was about 5×10^{-21} W in a bandwidth of about 1.3 kHz.

More recently, a dc SQUID with an untuned input circuit [94, 95] has been used to detect NQR in the frequency range 10 to 200kHz. The circuit configuration is shown in Fig. 26. The sample is placed inside one half of the turns of the superconducting pickup loop, which is wound in a gradiometer configuration and is connected to the input coil of the SQUID to form a superconducting flux transformer. The transmitter solenoid is wound coaxially around the sample and pick-up coil. The SQUID is flux modulated at 500kHz and operated in a flux-locked loop with a bandwidth of about 200kHz. During the application of the pulse that induces precession, the feedback loop is opened; after the pulse is turned off, the integrator is reset to zero before data acquisition begins. The use of an untuned input circuit not only results in a broad bandwidth but eliminates the Nyquist noise associated with a tuned circuit at the signal frequency. The effective noise temperature of the spectrometer was approximately 1mK for a bath temperature of 1.5K.

The spectrometer was used to study the zero-field NQR resonance of ^{14}N in powdered ammonium chlorate (NH_4ClO_4). Figure 27(a) shows the signal produced by two pulses 5ms apart, each pulse consisting of a single rf cycle at 45kHz. The averaged signal, consisting of 16,000 pulse sequences with a cycle rate of 3Hz, shows a free induction decay after the second pulse and the formation of a spin echo after 5ms. For the purpose of display, the signal has been demodulated with a frequency of 35kHz. Figure 27(b) shows the Fourier transform of a similar echo: we observe three sharp peaks at 17.4, 38.8 and 56.2kHz. Note that the frequencies of the lower two peaks sum to the highest frequency, indicating that we are exciting and observing simultaneously the three transitions of a three-level system. The longitudinal and transverse relaxation times at 1.5K were measured to be $T_1 = 63 \pm 6$ ms and $T_2 = 22 \pm 2$ ms, respectively. The sensitivity, resolution and broad bandwidth of the spectrometer, as well as the possibility of pulsed spin echo, relaxation and multidimensional experiments, make the technique attractive for low-frequency NQR and NMR studies of a wide range of solids, particularly in the polycrystalline or amorphous state.

5.6 GRAVITY WAVE ANTENNAS

A quite different application of SQUIDs is the detection of minute displacements, such as those of the bar in a gravity wave antenna [96, 97]. Several groups worldwide are using these antennas to search for the pulse of gravitational radiation that is expected to be emitted when a

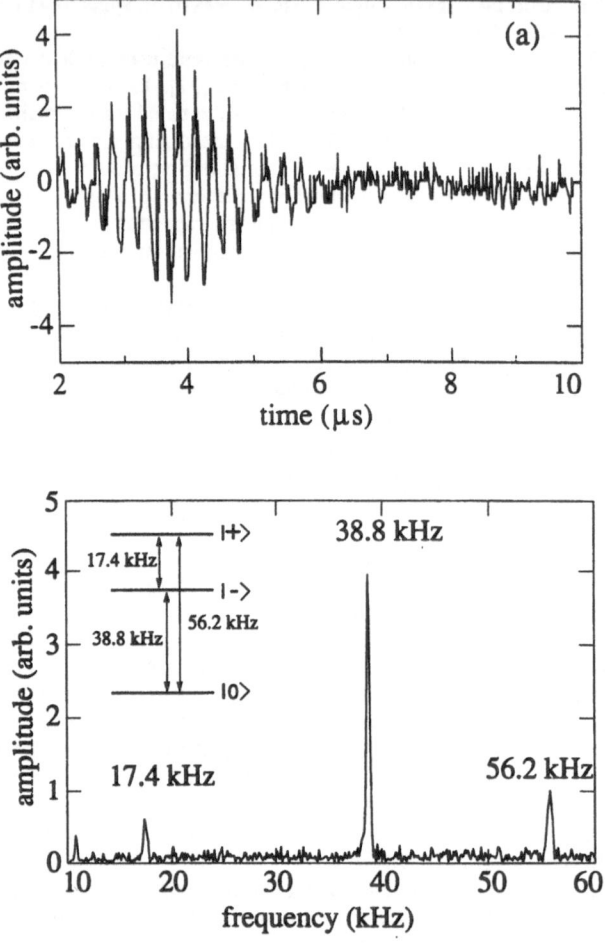

Figure 27. (a) Free induction decay after second pulse and spin echo of
NH_4ClO_4 at T = 1.5 K. For the purpose of display, the real-time signal has
been demodulated with 35kHz. (b) Fourier transform of spin echo. The
three resonant peaks are due to transitions between energy levels (shown
inset) of ^{14}N nuclei in the presence of electric field gradient in NH_4ClO_4
(see text). In inset, $|0> = |10>$ and $|\pm> = (|11> \pm |1 - 1>)/\sqrt{2}$ where $|I,m>$
is the eigenstate of ^{14}N nucleus (I = 1) with $I_z = m$.

star collapses. The radiation induces longitudinal oscillations in the large, freely suspended bar, but because the amplitude is very tiny, one requires the sensitivity of a dc SQUID to detect it. As an example, we briefly describe the antenna at Stanford University, which consists of an aluminum bar 3 meters long (and weighing 4800 kg) suspended in a vacuum chamber at 4.2 K. The fundamental longitudinal mode is at $\omega_a/2\pi \approx 842$ Hz, and the Q is 5×10^6. The transducer is shown schematically in Fig. 28. A circular niobium diaphragm is clamped at its perimeter to one end of the bar, with a flat spiral coil made of niobium wire mounted on each side. The two coils are connected in parallel with each other and with the input coil of a SQUID; this entire circuit is superconducting. A persistent supercurrent circulates in the closed loop formed by the two spiral coils. The associated magnetic fields exert a restoring force on the diaphragm so that by adjusting the current, one can set the resonant frequency of the diaphragm equal to that of the bar. A longitudinal oscillation of the bar induces an oscillation in the position of the diaphragm relative to the two coils, thereby modulating their inductances. As a result of flux quantization, a fraction of the stored supercurrent is diverted into the input coil of the SQUID, which detects it in the usual way.

The present Stanford antenna has a root-mean-square strain sensitivity $<(\delta\ell)^2>^{1/2}/\ell$ of 10^{-18}, where ℓ is the length of the bar, and $\delta\ell$ is its longitudinal displacement. This very impressive sensitivity, which is limited by thermal noise in the bar, is nonetheless adequate to detect events only in our own galaxy. Because such events are rare, there is very strong motivation to make major improvements in the sensitivity.

If the bar could be cooled sufficiently, the strain resolution would be limited only by the bar's zero-point motion and would have a value of about 3×10^{-21}. At first sight one might expect that the bar would have to be cooled to an absurdly low temperature to achieve this quantum limit, because a frequency of 842 Hz corresponds to a temperature $\hbar\omega_a/k_B$ of about 40nK. However, it turns out that one can make the effective noise temperature T_{eff} of the antenna much lower than the temperature T of the bar. If a gravitational signal in the form

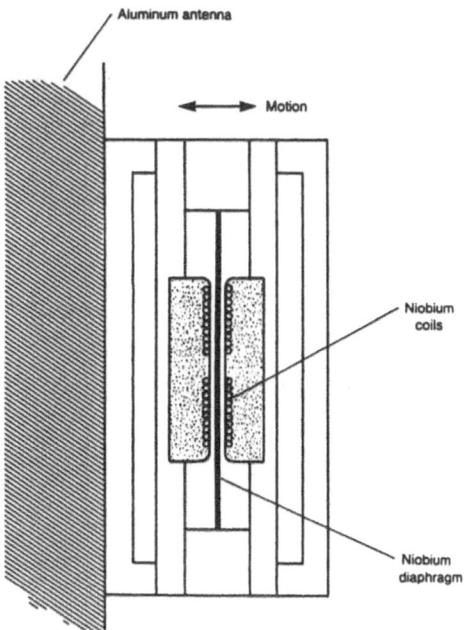

Figure 28. Transducer for gravity wave antenna. (Courtesy P.F. Michelson)

of a pulse of length τ_S interacts with an antenna that has a decay time Q/ω_a, then the effective noise temperature is given approximately by the product of the bar temperature and the pulse length divided by the decay time: $T_{eff} \approx \tau_S \omega_a T/Q$. Thus, one can make the effective noise temperature much less than the temperature of the bar by increasing the bar's resonant quality factor sufficiently. To achieve the quantum limit, in which the bar energy $\hbar\omega_a$ is greater than the effective thermal energy $k_B T_{eff}$, one would have to lower the temperature T below $Q\hbar/k_B\tau_S$, which is about 40 mK for a quality factor Q of 5×10^6 and a pulse length τ_S of 1 msec. One can cool the antenna to this temperature with the aid of a large dilution refrigerator.

Needless to say, to detect the motion of a quantum-limited antenna, one needs a quantum-limited transducer, a requirement that has been the major driving force in the development of ultra-low-noise dc SQUIDs. As we have seen, however, existing dc SQUIDs at low temperatures are now within striking distance of the quantum limit, and there is every reason to believe that one will be able to operate an antenna quite close to the quantum limit within a few years.

5.7 GEOPHYSICAL APPLICATIONS

Low-T_c SQUIDs have been used in a wide range of geophysical applications, including magnetotellurics, controlled source electromagnetic sounding, gravity gradiometers, rock magnetism and paleomagnetism, tectomagnetism and internal ocean waves [68]. We shall briefly describe two very different techniques, namely the gravity gradiometer and magnetotellurics.

The gravity gradiometer, which also makes use of a transducer to detect minute displacements, has been pioneered by Paik [98] and Mapoles [99]. The gradiometer consists of two niobium proof masses, each constrained by springs to move along a common axis (Fig. 29). A single-layer spiral coil of niobium wire is attached to the surface of one of the masses so that the surface of the wire is very close to the opposing surface of the other mass. Thus, the inductance of the coil depends on the separation of the two proof masses, which, in turn, depends on the gravity gradient. The coil is connected to a second superconducting coil which is coupled to a SQUID via a superconducting transformer. A persistent supercurrent, I, maintains a constant flux in the detector circuit. Thus, a change in the inductance of the pick-up coil produces a change in I, and hence, a flux in the SQUID that is related to the gravity gradient. More sophisticated versions of this design enable one to balance the restoring forces of the two springs electronically [99], thereby eliminating the response to an acceleration (as opposed to an acceleration gradient). Sensitivities of a few Eötvös $Hz^{-1/2}$ have been achieved at frequencies above 2 Hz. Instruments of this kind could

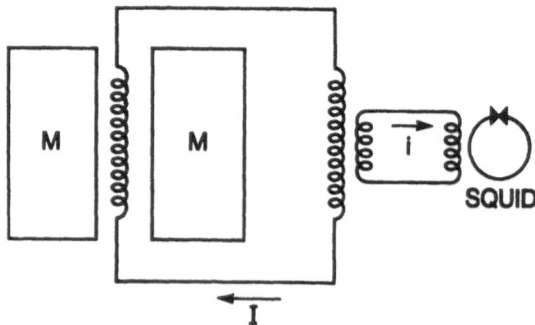

Figure 29. Gravity gradiometer showing two proof masses (M) on either side of a planar spiral coil (from ref. [99]).

be used to map the earth's gravity gradient, and may be used to test the inverse gravitational square law and in inertial navigation.

We turn now to magnetotellurics [100], in which one makes use of electromagnetic energy propagating to the earth from the ionosphere (<1Hz) and thunderstorm activity (>1Hz). The incident field is reflected by the earth, but components of the electric and magnetic fields, $\vec{E}(\omega)$ and $\vec{H}(\omega)$, decay into the ground with a characteristic length $\delta \approx 0.5 \, (\rho T)^{1/2}$ km, where ρ is the resistivity in Ωm and T is the period in seconds. The frequency range of interest is typically 10^{-4} to 10^{2}Hz; at 1Hz δ is usually between 1 and 5km. One measures simultaneously the horizontal components of the magnetic field, $H_x(t)$ and $H_y(t)$, and of the electric field, $E_x(t)$ and Ey(t), the Fourier components of which are related via the impedance tensor $\underset{=}{Z}(\omega)$:

$$E_x(\omega) = Z_{xx}(\omega)H_x(\omega) + Z_{xy}(\omega)H_y(\omega) \qquad (5.24)$$

and

$$E_y(\omega) = Z_{yx}(\omega)H_x(\omega) + Z_{yy}(\omega)H_y(\omega) \ . \qquad (5.25)$$

The magnetic fields are measured by magnetometers or induction coils and the electric fields by buried electrodes connected to sensitive amplifiers. In the conventional analysis scheme, one multiplies each equation in turn by the complex conjugate of one of the fields, and averages the equations over a number of data records to reduce noise. One then solves a subset of the equations for the elements of the impedance tensor. For a homogeneous earth, the elements Z_{xx} and Z_{yy} are zero. In general, however, all four elements are non-zero, and in the usual procedure one rotates the axes of $\underset{=}{Z}$ to minimize the quantity $|Z_{xx}(\omega)|^2 + |Z_{yy}(\omega)|^2$. One of the axes is then aligned along the direction of maximum translational invariance, and one assumes that the ground can be represented adequately by a 2-D model so that Z_{xx} and Z_{yy} are negligible. The results are presented as the apparent resistivities in the x- and y-directions of the rotated tensor, $\rho_{xy}(\omega) = 0.2|Z_{xy}(\omega)|^2 T$ and $\rho_{yx}(\omega) = 0.2|Z_{yx}(\omega)|^2 T$, where ρ_{xy} and ρ_{yx} are in Ωm, and Z_{xy} and Z_{yx} are in (mV/km)nT.

Unfortunately, estimates of $\underset{=}{Z}(\omega)$ obtained in this way are often unreliable because of noise that can produce large biases in the results. Gamble *et al.* [101] attempted to overcome this problem by using SQUID magnetometers with much lower noise than the induction coils used previously. They found, however, that the use of a quieter magnetometer did not lead to an improvement in the data. They then introduced the remote reference technique [101] in which in addition to the measurement of H_x, H_y, E_x and E_y at the magnetotelluric site one also measures simultaneously the magnetic fields H_{xr} and H_{yr} at a remote site several kilometers away using a second magnetometer. By multiplying Eqs. (5.24) and (5.25) in turn by $H_{xr}(\omega)$ and $H_{yr}(\omega)$ and averaging over many data records one obtains four equations that can be solved for the elements of the impedance tensor. For example, one finds

$$Z_{xy}(\omega) = \frac{<E_x \, H_{yr}^*> <H_x \, H_{xr}^*> - <E_x \, H_{xr}^*> <H_x \, H_{yr}^*>}{<H_x \, H_{xr}^*> <H_y \, H_{yr}^*> - <H_x \, H_{yr}^*> <H_y \, H_{xr}^*>} \ . \qquad (5.26)$$

Provided any noise sources at the magnetotelluric site and at the remote site are uncorrelated, the estimates for the impedance tensor will be unbiased by noise. Figure 30 shows the apparent resistivities obtained at a site in Bear Valley, California, and illustrates the very high quality of the data that can be obtained with this technique.

An important additional benefit of the remote reference scheme is that it enables one to place reliable confidence limits [102] on the apparent resistivities, an essential requirement if one is to carry out meaningful modeling. The probable errors can be as low as a few tenths

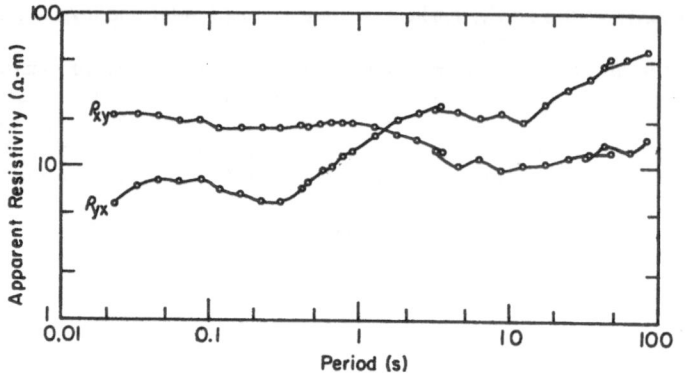

Figure 30. Apparent resistivities obtained at site in Bear Valley, California using remote reference magnetotellurics. (From ref. [101].)

of one percent, particularly at higher frequencies where a good deal of data is usually available. Once the apparent resistivities and other data have been collected at a series of sites, one performs an inversion to obtain the resistivity of the ground as a function of position and depth.

SQUIDs were used for magnetotellurics in the early 1980's but were abandoned when the price of oil dropped, and oil prospecting was curtailed. The need to use liquid helium in remote areas outside the continental United States also proved to be a serious impediment. However, the introduction of liquid nitrogen-cooled SQUIDs is likely to generate a renewed interest in their applications to geophysics.

6. High-T_c SQUIDs

The advent of the high-T_c superconductors has resulted in a world-wide effort to develop SQUIDs and flux transformers operating in liquid nitrogen at 77K. I cannot hope to do justice to the vast literature on the subject that has appeared in the last seven years, but I will give some idea of the current state of the art. I begin with a brief survey of the many types of Josephson junction that have been invented, and then discuss the design criteria for SQUIDs operating at 77K. The rest of the section is devoted to a description of practical dc and rf SQUIDs and flux transformers.

6.1 JOSEPHSON JUNCTIONS

The development of a reproducible junction technology has been a major preoccupation. Many structures have been made, all with nonhysteretic I-V characteristics, and I have divided them into three broad classes: *grain boundary* junctions, which may be natural or engineered, *barrier* junctions, which involve a barrier of a nonsuperconducting material, and *weakened* junctions, in which a region of the film is weakened in a controlled way. These various junctions are illustrated in Fig. 31.

To my knowledge, the first thin-film Josephson junction was made by Koch *et al.* [103], who made polycrystalline films of $YBa_2Cu_3O_{7-x}$ (YBCO) containing randomly oriented grains [Fig. 31(a)]. Using a photomask they ion implanted a region of the film to make it nonsuperconducting, leaving a microbridge of superconductor crossed by one or perhaps

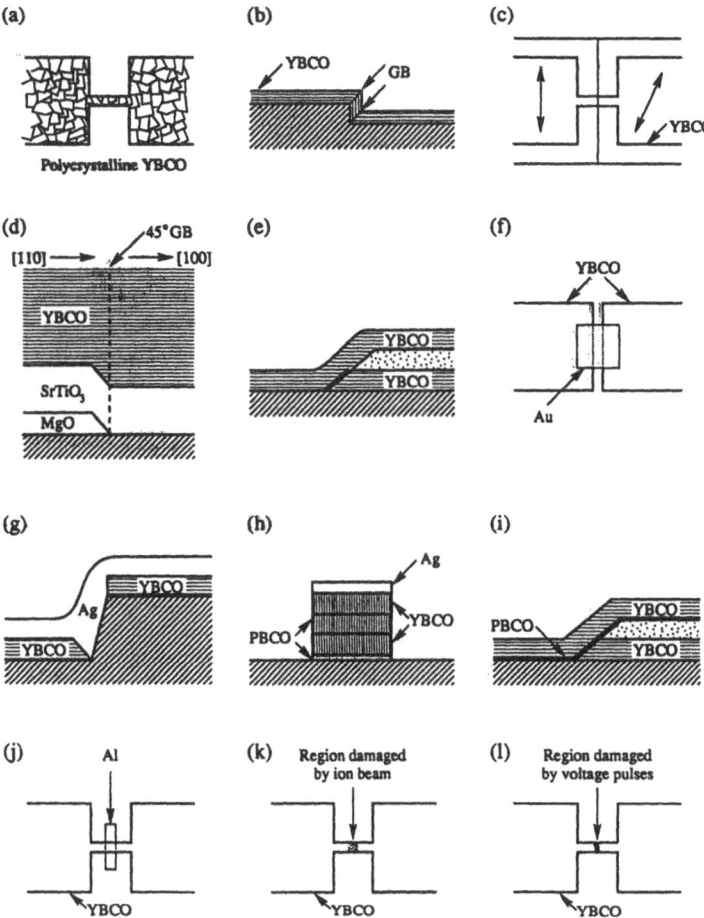

Figure 31. Schematic representation of twelve types of high-T$_c$ thin-film Josephson junction: (a) naturally occurring grain boundary junction [103], (b) step-edge grain boundary (GB) junction [106], (c) bicrystal grain boundary junction (arrows indicate [100] axes) [108], (d) bi-epitaxial grain boundary junction [109], (e) edge junction [110], (f) YBCO-Au-YBCO c-axis junction [111], (g) YBCO-Ag-YBCO a-axis junction [112], (h) trilayer YBCO-PBCO-YBCO junction [115], (i) trilayer YBCO-PBCO-YBCO edge junction [116], (j) poison stripe junction [106], (k) ion beam damage junction [117, 118], (l) electroshocked junction [119].

several grain boundaries that behaved as weak links or Josephson junctions. Subsequently, many groups have made similar junctions, usually by removing regions of the film with acid etching or ion-beam milling, and extended the technique to other materials such as BiSrCaCuO [104] and TlBaCaCuO [105]. Grain boundary junctions with not too high critical current densities often exhibit almost ideal resistively shunted junction behavior. However, they suffer from the drawback that one has little control of their properties and none of their location.

Several techniques have been devised to induce the growth of a grain boundary at a specific location. Simon et al. [106] deposited a film across a step etched in a substrate [Fig. 31(b)]. The film was then patterned to form a bridge across the step; provided the step edge is sharp enough, a grain boundary junction is formed at top and bottom. Subsequent transmission electron microscopy at Jülich [107] showed that if the step angle is steeper than 45°, the film has an a-axis orientation on the step, thus forming a grain boundary with each of the c-axis films on either side.

Another widely adopted technique involves bicrystal substrates [108], made by cutting a wedge from a crystal of (100) SrTiO$_3$ and fusing the two pieces together to form a grain boundary between two regions with a known in-plane misorientation [Fig. 31(c)]. When YBCO is deposited with an in situ process, the ab-axes mimic the orientation of the substrate, producing a grain boundary. One can pattern microbridges across any part of the grain boundary to produce junctions with relatively predictable characteristics. Related to the bicrystal junction is the bi-epitaxial process [109] [Fig. 31(d)], in which an appropriate seed layer is deposited on the substrate and then selectively removed to leave edges in predetermined locations. One deposits a SrTiO$_3$ buffer layer followed by a YBCO film which undergoes a 45° in-plane re-orientation wherever it crosses the edge of the seed layer. Again, one patterns microbridges across the 45° grain boundaries to form junctions.

In the second group of junctions, we begin with the edge junction [Fig. 31(e)] of Laibowitz et al. [110]. These authors first deposited a YBCO film and a (non-epitaxial) insulating layer, and then ion-milled an edge. After exposing this edge to an oxygen-fluorine plasma, they deposited a YBCO film to form an edge junction in which the supercurrent flowed along the ab-planes of the films. Both junctions and SQUIDs have been made with this technique, but the yield and reproducibility are apparently poor.

Making use of the proximity effect, Schwartz et al. [111] fabricated a junction by patterning a narrow slit in a YBCO film using electron-beam lithography and depositing a gold film across the slit [Fig. 31(f)]. This junction exhibited supercurrents at temperatures up to 16K. Using a different configuration [Fig. 31(g)], DiIorio et al. [112] fabricated proximity effect junctions that exhibited Josephson effects to temperatures well over 77K. A step is milled in a LaAlO$_3$ substrate and a YBCO film is sputtered at an angle to the substrate so as to leave the step edge uncovered. A silver film, deposited immediately afterwards, connects the two films so that currents flow in the ab-plane. Another class of proximity effect junctions, involving PrBaCuO (PBCO) as the barrier was pioneered by Rogers et al. [113]. In their early work, they grew a c-axis trilayer of YBCO-PBCO-YBCO, with a PBCO thickness of typically 50nm, and patterned it to form junctions [Fig. 31(h)]. Some of these junctions exhibited a supercurrent at temperatures up to 65K. However, given the weak proximity effect of c-axis films [114], it seems likely that the supercurrent was due to shorts through the PBCO. Subsequently, Barner et al. [115] fabricated a -axis trilayers which, after being patterned into junctions, exhibited supercurrents at temperatures as high as 80K. An alternative configuration for a-axis junctions [Fig. 31(i)] was adopted by Gao et al. [116] who fabricated edge junctions with PBCO barriers.

The third category contains "weakened" structures. Simon et al. [106] deposited a YBCO film across a thin aluminum strip which "poisons" the superconductor locally, thereby reducing its transition temperature [Fig. 31(j)]. In another method [Fig. 31(k)], one reduces the transition temperature of a narrow region by focusing high energy ions [117] onto a microbridge or using a mask patterned with electron-beam lithography [118] to define the area exposed to an ion beam. In a third technique [119], controlled electrical pulses were applied to microbridges at 77K, again producing a weakened region [Fig. 31(ℓ)]. Although these weakening techniques are controllable at least to some degree, they generally lead to flux-flow characteristics (probably because the weakened region is long compared with the coherence length) that are undesirable for most applications. However, if the region weakened by poisoning or ion-bombardment could be reduced to a few tens of nanometers, these techniques might prove to be very useful.

Which of the various junctions look most promising for SQUIDs? To date, the best SQUIDs have been made from step edge grain boundary or bicrystal grain boundary junctions. These structures generally exhibit RSJ-like current-voltage characteristics with I_0R products up to about 200μV at 77K. Resistances vary from the order of 1Ω to as high as 10-20Ω [120]. Although careful fabrication procedures have improved the yield of these junctions, particularly in the case of the bicrystal devices, the scatter in the critical current and resistance remains higher than one would like. There is ongoing work to improve the understanding and reproducibility of trilayer edge junctions with a variety of barrier materials [121] and one might hope to see steady progress in the yield of this class of junction. For the moment, the simplicity offered by the grain boundary junctions, which involve only a single YBCO layer, makes them very appealing for practical applications.

6.2 PREDICTIONS FOR WHITE NOISE

In designing dc SQUIDS for operation at 77K, we should bear in mind not only the constraints on the critical current and inductance, $I_0\Phi_02\pi >> k_BT$, but also the requirement that the thermally induced flux noise in the SQUID loop, $<\Phi_N^2>^{1/2} = (k_BTL)^{1/2}$, be much less than one flux quantum. For low-T_c SQUIDs the corresponding upper bound on the inductance, $L << \Phi_0^2/k_BT$, is sufficiently large that it is almost never an important constraint; however, this is not the case at 77K. Although computer simulations of course take into account the effects of thermal noise, a result by Enpuku et al. [122] is useful in predicting the reduction in V_Φ that results from the thermal flux noise:

$$V_\Phi = \frac{4I_0R}{\Phi_0(1+\beta)} [1 - (L/L_T)^{1/2}] \quad . \tag{6.1}$$

Here, $L_T \equiv \Phi_0^2/4\pi k_BT = 321$pH at 77K is a temperature-dependent inductance; Eq. (6.1) is valid for inductances rather less than 321pH. We see that the effects of flux noise can be very significant: for example, for $L = 100$pH, $(L/L_T)^{1/2} = 0.56$. Thus, there is a good reason to keep L well below 100pH for most applications at 77K. On the other hand, at 4.2K $L_T \approx$ 6nH and since dc SQUIDs typically have inductances of 100-200pH, the flux noise term in Eq. (6.1) is relatively unimportant. For a high-T_c SQUID with an inductance of 50pH and with $\beta = 1$ ($I_0 = 20$ pA), Eq. (6.1) predicts $V_\Phi \approx 100\mu$V/Φ_0 for $R = 2\Omega$. Assuming that Eq. (3.8) is approximately valid for the higher value of Γ (0.16), we find $S_\Phi^{1/2} \approx 2\times10^{-6}\Phi_0Hz^{-1/2}$ and $\epsilon \approx 2\times10^{-31}$ JHz^{-1}. This noise energy is an order-of-magnitude higher than that expected for low-T_c SQUIDs with comparable parameters operating at 4.2K, reflecting the increase in temperature.

In the case of the rf SQUID, if we use $I_0 = 50\mu$A, $LI_0 = 2.5\Phi_0$, $\omega_{rf}/2\pi = 30$ MHz and T = 77K in Eq. (4.10), we find $\epsilon \approx 3\times10^{-29}$ JHz^{-1}. This noise energy is not too different from that obtained for typical 20MHz devices operating at 4.2K, where the effective noise temperature T_a^{eff} [Eq. (4.12)] of the room temperature preamplifier and tank circuit is much higher than the bath temperature. When the SQUID is operated at 77K, there is no reason for T_a^{eff} to increase, and the system noise energy should be comparable with that for a 4.2K SQUID. Thus, although at 4.2K dc SQUIDs are clearly superior to rf SQUIDs operating at around 20 MHz, we expect the margin to be narrower for 77K devices.

With regard to 1/f noise, one expects both critical current fluctuations and flux noise to contribute. However, it is not possible to make any *a priori* predictions of the magnitude of these contributions.

6.3 DC SQUIDS

I shall give a brief summary of the fabrication and performance of bicrystal SQUIDs at Berkeley; other groups have made similar devices. We deposit the YBCO film on a $10 \times 10 mm^2$ bicrystal $SrTiO_3$ (STO) substrate with a 24° grain boundary using an excimer laser operating at 248nm and a repetition rate of 5Hz. The substrates are maintained at approximately 810°C in 210mTorr of O_2, and the films are grown at 0.05-0.07nm per pulse. Typical film thicknesses are 150-180nm. After completing the deposition we admit 0.8 atm of O_2 to the chamber and cool the substrate to 450°C in 20 min. We then allow it to cool to ambient temperatures in about 30 min. To enable us to make low-resistance electrical contacts to the YBCO film, we transfer the chip from the deposition chamber to an evaporator and deposit roughly 50nm of Ag through a shadow mask on the region of the film where the contact pads will be patterned. We pattern up to 12 SQUIDs on a single chip using conventional photolithography and a 500eV Ar ion mill. The junction width is 1-3µm. Subsequently, 32µm-diameter Al leads are attached to the Ag pads with a wedge bonder.

We measure the noise of the SQUIDs using the flux modulation scheme and flux-locked loop described in Sec. 3.4. The voltage across the SQUID is amplified by a cooled transformer which has a turns ratio of 1:15. The power spectrum of the noise for a relatively low inductance device, with an estimated inductance L of 10pH and measured with a static bias current, is shown in Fig. 32. The SQUID was enclosed in a 40 mm-diameter conetic shield and immersed in liquid nitrogen in a dewar surrounded by three concentric mu-metal shields. The magnetic shielding is not quite sufficient to eliminate the pickup of 60Hz and its odd harmonics, as can be seen in the figure; spikes at lower frequencies are due to microphonics. The flux noise at frequencies above about 5kHz is very low, about $1.5 \mu \Phi_0$ $Hz^{-1/2}$, corresponding to a noise energy of about 5×10^{-31} $JHz^{-1/2}$. However, at lower frequencies the spectral density increases as 1/f, and the rms noise at 1 Hz is an order of magnitude higher than at 5kHz. Also shown in Fig. 32 is the result of operating the SQUID with the bias reversal scheme described in Sec. 3.6. The 1/f noise power at 1Hz is reduced by two orders of magnitude, indicating that the 1/f noise was produced by critical current fluctuations rather than by the hopping of vortices. Similar reductions in the level of 1/f noise have been reported by other authors [123].

Although the magnetic flux noise is very low in this device, its small area implies that the magnetic field noise is high. To obtain low magnetic field noise, we must enhance the effective area. The following two sections are concerned with single and multilayer flux transformers that enhance the sensitivity of the SQUID to magnetic field.

6.4 SINGLE-LAYER MAGNETOMETERS

The directly-coupled magnetometer [124] consists of a single film of YBCO deposited on a bicrystal and patterned in the configuration shown in Fig. 33. A magnetic field B applied to the pickup loop of area A_p and inductance L_p induces a supercurrent $J = BA_p / L_p$ which, in turn, links a flux $\alpha_d LJ$ to the SQUID. Here, α_d is the fraction of the SQUID inductance L to which the current couples. The effective area of the magnetometer is thus

$$A_{eff}^{(d)} = A_s + \alpha_d A_p L / L_p, \qquad (6.2)$$

where A_s is the effective area of the bare SQUID. Neglecting A_s and using the estimated values for a particular device, $\alpha_d = 0.8$, $A_p = 47mm^2$, $L = 20pH$ and $L_p = 11nH$, we find $A_{eff}^{(d)}$

$\approx 0.068mm^2$. The measured value was somewhat higher, $0.086mm^2$. Figure 34(a) shows the magnetic field noise of a directly coupled magnetometer with $I_0 \approx 45\mu A$, $R \approx 3.4\Omega$ and the parameters listed above, operated at 77K in a flux-locked loop with bias reversal. The noise is white at frequencies down to below 1 Hz with the value $93fTHz^{-1/2}$. This value is listed in Table I.

Subsequently, we increased the effective area of the device by coupling it to a single-layer flux transformer [125] in the configuration of Fig. 35. The smaller loop is inductively coupled to the pickup loop of a directly coupled magnetometer in a flip-chip arrangement in which the two chips are pressed together with a mylar sheet between the YBCO films. We define $\mathcal{A}_i$ and $\mathcal{L}_i$ as the area and inductance of the small input loop, $\mathcal{A}_p$ and $\mathcal{L}_p$ as the area and inductance of the large pickup loop and α' as the coupling coefficient between $\mathcal{L}_i$ and L_p. The effective area is easily shown to be

$$A_{eff}^{(m)} \approx \alpha_d \frac{L}{L_p} \left[A_p + \mathcal{A}_p \frac{\alpha' (L_p \mathcal{L}_i)^{1/2}}{\mathcal{L}_i + \mathcal{L}_p} \right]. \tag{6.3}$$

For the directly coupled magnetometer described above and a transformer with the estimated parameters $\alpha = 0.9$, $\mathcal{L}_i = 10$ nH, $\mathcal{L}_p = 85nH$, and $\mathcal{A}_p = 1.33 \times 10^{-3}m^2$, we calculate an estimated effective area $A_{eff}^{(m)} \approx 0.26mm^2$. The measured value was $0.29mm^2$, yielding a transformer gain of 3.4. The measured magnetic field noise, shown in Fig. 34(b), is $31fTHz^{-1/2}$ at 1kHz and $39fTHz^{-1/2}$ at 1Hz.

Further improvements in the performance of the directly-coupled magnetometer were achieved by Lee et al. [120], who made pickup loops with larger linewidths, thereby lowering their inductance. A typical performance is $40fTHz^{-1/2}$ at 1kHz and $65fTHz^{-1/2}$ at 1Hz. An even lower noise was achieved by Cantor et al. [126] with the aid of a $20 \times 20mm^2$ bicrystal and a larger pickup loop: $14fTHz^{-1/2}$ at 1kHz and $26fTHz^{-1/2}$ at 1Hz.

6.5 MULTILAYER FABRICATION PROCEDURES

We turn now to a discussion of magnetometers involving multilayers, namely two YBCO films separated by an insulating layer of STO. At specified points it is necessary to open a via in the STO to enable one to make a superconducting connection between the YBCO films. Clearly, each layer has to be patterned separately. The fabrication of multilayer devices is much more demanding than single-layer devices, and for a long period of time patterned multilayers produced relatively high levels of l/f noise due to the thermally activated hopping of vortices. It is believed that the excess noise arose predominantly from the upper YBCO film which had a poorer crystalline quality than YBCO films deposited directly on a substrate. However, we have recently developed processing techniques that substantially reduce the l/f noise, and we begin with a brief summary of our current technique [127].

We first deposit a 10nm-thick STO buffer layer on a STO (100) substrate and follow it with a 120nm-thick YBCO film and a 15nm-thick "cap" of STO. The cap plays a crucial role in that it is subject to less damage than a YBCO film during photolithographic patterning. As a result, the subsequent layers of STO and YBCO grow with a higher degree of crystallinity and fewer defects than in the uncapped case. We pattern the first, capped layer with Ar ion milling at a 45˚ degree of incidence to the rotating substrate, remove the photoresist, and return the chip to the deposition chamber where we deposit 230nm of STO as the insulating layer. We discovered that the quality of this film was sufficiently high to reduce the diffusion of O_2 into the YBCO substantially; as a result, our standard O_2 annealing procedure yielded transition temperatures as low as 40K. To remedy this oxygen deficiency, after depositing the STO film we cool the bilayer over a period of 30 min to

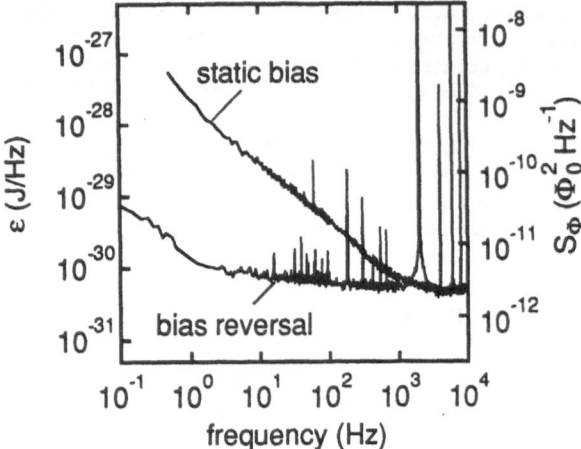

Figure 32. Flux noise spectral density $S_\Phi(f)$ of 10 pH SQUID at 77K with static bias and with bias reversal. Left-hand ordinate shows noise energy $\varepsilon(f) = S_\Phi(f) / 2L$.

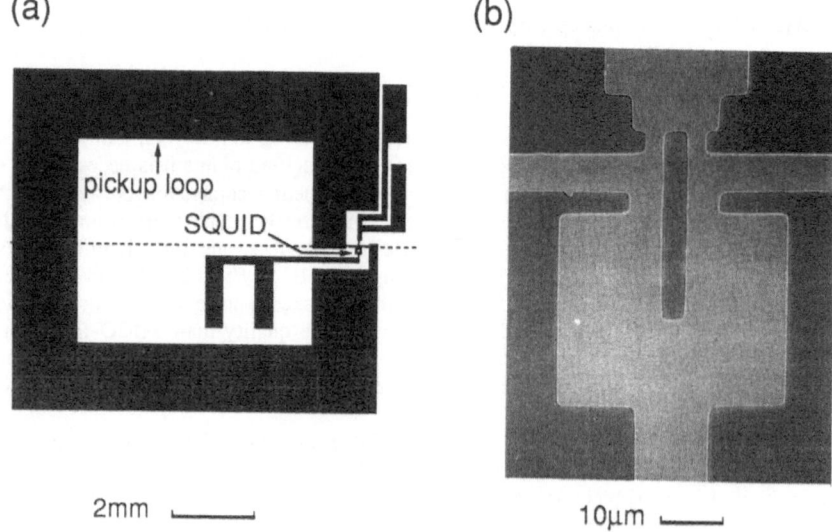

Figure 33. Single-layer magnetometer deposited on $10 \times 10mm^2$ bicrystal; outer dimensions of pickup loop are 7mm x 8mm. Pickup loop in (a) is connected to the SQUID in (b); square washer is 32μm x 32μm. Dashed line in (b) indicates grain boundary.

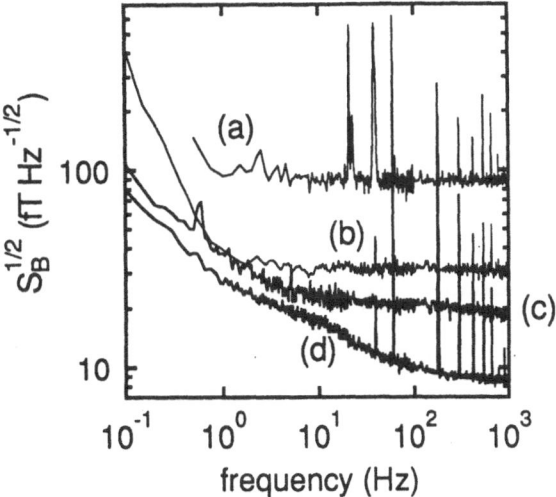

Figure 34. Magnetic field noise of four magnetometers, all operated at 77K with bias reversal: (a) directly coupled magnetometer, (b) directly coupled magnetometer with flip-chip 50 mm flux transformer, (c) fractional turn SQUID, (d) SQUID with flip-chip 16-turn flux transformer.

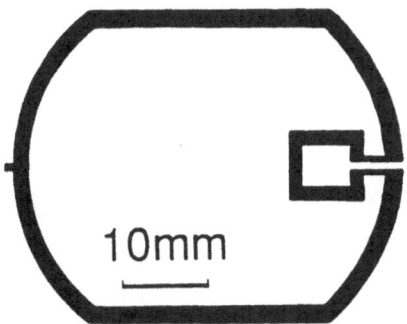

Figure 35. Configuration of single-layer flux transformer. The smaller loop is inductively coupled to a directly coupled magnetometer in a flip-chip arrangement.

500 °C in 0.8 atm of O_2, and maintain this temperature for 3 hours before turning off the heater. Even this protracted annealing step results in a somewhat suppressed transition temperature, typically 85K. We open vias in the STO, again using an Ar ion mill at a 45° angle of incidence and a rotating substrate. In the patterning of both the capped YBCO and STO films the angled ion mill and rotating substrate are important in producing smoothly beveled edges over which subsequent films can grow with relatively high quality. The final, 250nm-thick YBCO film is patterned at a normal angle of incidence on a stationary substrate.

6.6 FRACTIONAL TURN SQUIDs

The configuration of the fractional turn SQUID [128, 129] (multiloop magnetometer) is shown in Fig. 36(a). In the center is a patterned YBCO-STO-YBCO multilayer. Each pickup loop, patterned largely in the upper YBCO film, makes contact with the lower film in the cross-shaded region. The two bicrystal junctions, in series, connect the upper and lower YBCO films. The films are patterned in such a way that no narrow lines other than those forming the two junctions cross the grain boundary.

The effective inductance and area are given by [130]

$$L_{eff} = L_p / N^2 + L_s / N + L_j \qquad (6.4)$$

and

$$A_{eff} = A_p / N - A_s . \qquad (6.5)$$

Here, L_p and A_p are the inductance and area of the large, outer loop, L_s and A_s are the average inductance and area of one spoke of the "cartwheel," L_j is the inductance of the connections from the pickup loops to the junctions, and N is the number of loops. Drung *et al.* [131] have discussed the optimization of the design for operation at 77K, and concluded that the optimum value of N is between 15 and 20 for a device diameter of 7mm. We fabricated two 16-loop devices [shown in Fig. 36(b)], with two bicrystal junctions nominally

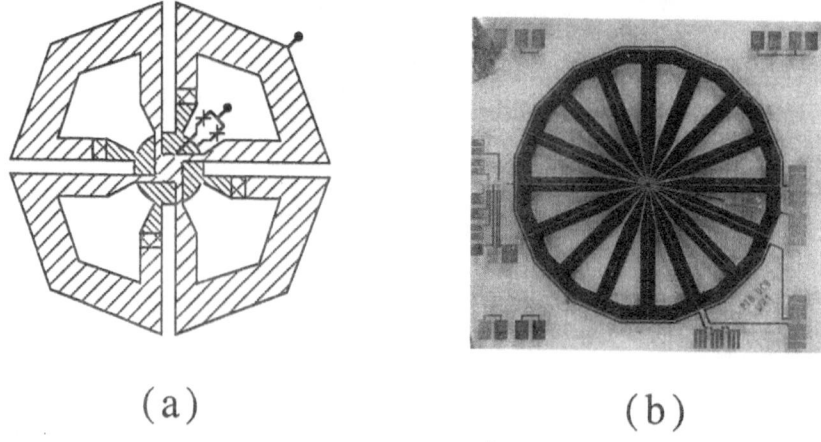

(a)   (b)

Figure 36. (a) Schematic layout and (b) photograph of multiloop magnetometer. The 16-loop device in (b) is 7 mm across.

2.5 μm wide in the first YBCO film [132]. For these devices we estimate $L_p = 12.2nH$, $A_p = 34.5 mm^2$, $L_s = 1.17nH$, $A_s = 0.39 mm^2$ and $L_j = 24 pH$, yielding an effective inductance of 145pH and an effective area of $1.77mm^2$. The measured effective areas were $1.84mm^2$ and $1.89mm^2$, in good agreement with our estimate. For the better of the two magnetometers the junction resistance R was 10Ω and the critical current I_0 (corrected for thermal noise rounding) was about 13μA. These values give an I_0R product of about 130μV (typical for our process) and $\beta_L = 2LI_0 / \Phi_0 \approx 1.8$.

After measuring the noise of the magnetometers we concluded that there was a non-negligible contribution from the Nyquist noise of the conetic shield. Consequently, we replaced it with a superconducting shield consisting of an yttria stabilized zirconia tube with a length of 125mm and inner and outer diameters of 25mm and 32.5mm, coated on both sides with a thick film of YBCO [133]. The resulting noise of the better magnetometer, plotted in Fig. 34(c), was 18fT $Hz^{-1/2}$ at 1kHz and 37 fT $Hz^{-1/2}$ at 1Hz.

6.7 MULTITURN FLUX TRANSFORMERS

A number of groups have successfully fabricated multilayer flux transformers in which a multiturn coil is inductivley coupled to the SQUID. These magnetometers fall into two classes: in one the transformer and SQUID are fabricated on separate substrates and coupled in a flip-chip arrangement, and in the other the magnetometer is fabricated on a single substrate.

We begin with the flip-chip devices; the configuration of the multiturn flux transformer is shown in Fig. 37(a). The multiturn input coil is inductively coupled to the SQUID, for which two designs are also shown in Fig. 37. The effective area of the magnetometer is given by

$$A_{eff} = A_s + A_p \frac{M_i}{L_i + L_p} , \qquad (6.6)$$

where $M_i = \alpha(LL_i)^{1/2}$ is the mutual inductance between the SQUID and the multiturn input coil of inductance L_i. Provided that α does not depend on L_i and that $A_s \ll A_{eff}$, for a given pickup loop A_{eff} is a maximum when $L_i = L_p$:

$$A_{eff}^{max} = \frac{\alpha}{2} A_p \left(\frac{L}{L_p} \right)^{1/2} . \qquad (6.7)$$

For the representative values $\alpha = 0.5$, $A_p = 80mm^2$, $L = 40pH$ and $L_p = 20nH$, we find $A_{eff}^{max} \approx 1 mm^2$.

In our most recent series of such devices [127, 134, 135] we used 16-turn input coils, generally patterning the pickup loop and input coil in the lower YBCO film and the crossover connecting the innermost turn to the pickup loop in the upper YBCO film. The pickup loop is 10mm on a side with a width of 1 mm, giving an area A_p of about $80mm^2$. The input coil lines were 10μm wide with a pitch of 14μm, and the crossovers were 50μm wide. Photographs of the input coil appear in Fig. 38.

We patterned 12 SQUIDs in the two configurations A/A and A/C shown in Figs. 37(b) and (c) [135]. The YBCO film, deposited at Conductus on a 24° STO bicrystal, was 250nm thick. Each SQUID had an outer dimension of 500μm, a slit width of 4μm and junction widths of 1-3μm. The resistance R of each junction ranged from 2.4-8.6Ω. To assemble the magnetometers, we carefully aligned the input coil of the transformer over the SQUID and clamped the two chips together with a 3μm-thick mylar sheet between them. The magnetic field noise of our best magnetometer, with a type A/A SQUID, is shown in Fig. 34(d). At 1kHz the noise is white with a value of 8.5fT $Hz^{-1/2}$, and at 1Hz it is 27fT $Hz^{-1/2}$; at frequencies just above 1Hz the rms noise falls off more slowly than $1/f^{1/2}$, and appears to have a contribution from a random telegraph signal.

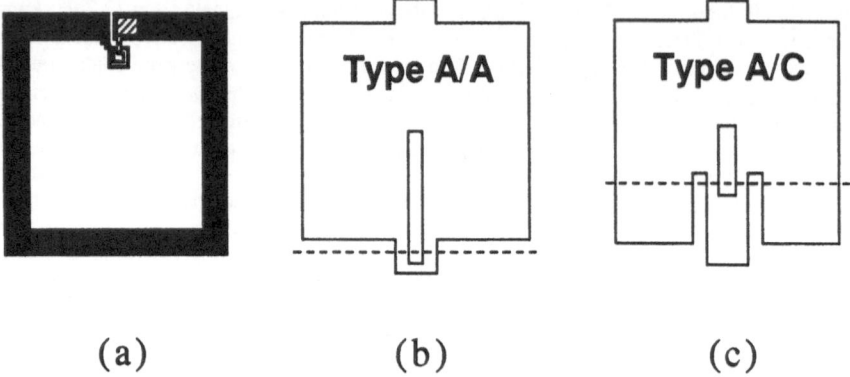

Type A/A

Type A/C

(a)   (b)   (c)

Figure 37. (a) Configuration of multiturn flux transformer (not to scale). (b) Type A/A SQUID, and (c) type A/C SQUID; each washer is 500μm x 500μm.

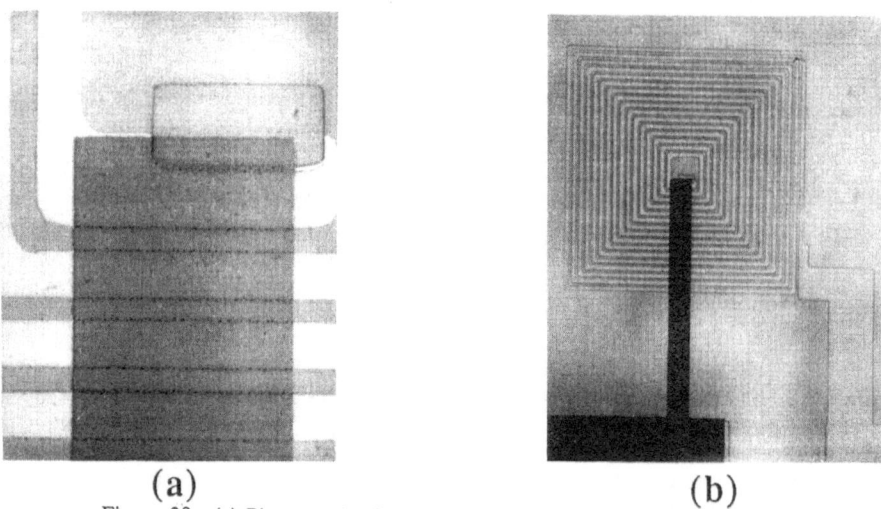

(a)   (b)

Figure 38. (a) Photograph of 16-turn input coil of flux transformer, with enlargement of central region shown in (b). Outer dimension of coil is 500μm.

We [134] and several other groups [136-139] have successfully integrated magnetometers in which the SQUID and flux transformer are deposited on the same substrate. The SQUID washer is often used as either a crossunder or a crossover for the flux transformer [140], so that only two superconducting layers are required. The process of manufacturing an integrated device of this kind is more challenging than for flip-chip mangetometers because one has to achieve not only a functioning flux transformer with low levels of flux noise but also junctions with near-optimum parameters. At the time of writing, the most sensitive integrated magetometer of this type was fabricated on a $10 \times 10 \text{mm}^2$ bicrystal by Dössel et al. [139], and achieved a magnetic field noise of 200fT $\text{Hz}^{-1/2}$ at frequencies down to 1Hz. To increase the yield, they used two SQUIDs with two input coils connected to the same pickup loop. One might expect substantial improvements in the performance of integrated devices in the near future.

6.8 RF SQUIDS

Several groups have successfully made rf SQUIDs that operate at 77K; of these, the most sensitive are those of the Jülich group [65, 141, 142], which I briefly describe.

For operation at 150MHz, the SQUIDs were made from YBCO films deposited on SrTiO_3 substrates in which shallow pits had been ion milled to produce step-edge junctions. The films were patterned as square washers with inner dimensions (d_1) of $20 \times 20 \ \mu\text{m}^2$ to $400 \times 400 \ \mu\text{m}^2$ and outer dimensions (d_2) of $6 \times 6 \ \text{mm}^2$ to $8 \times 8 \ \text{mm}^2$. The tank circuit had a Q of 40-60, and the measured values of V_Φ were greater than $40 \mu\text{V} / \Phi_0$ for 120 pH SQUIDs. The best of these devices, with an inner area less than $100 \times 100 \ \mu\text{m}^2$, exhibited a white noise of $40 \mu \Phi_0 \ \text{Hz}^{-1/2}$ at frequencies down to 1Hz, corresponding to a noise energy of 5×10^{-29} JHz^{-1}. It is particularly noteworthy that the large outer dimensions of these SQUIDs produces a substantial effective area, $A_{\text{eff}} \approx d_1 d_2$, due to the focusing of the externally applied magnetic flux [143], and thus yields a high sensitivity to magnetic fields, $\Phi_N(f) / d_1 d_2$. The best magnetic field sensitivity reported was 170fT $\text{Hz}^{-1/2}$ at 1 Hz. Subsequently, Zhang et al. [144] coupled an rf SQUID washer with inner and outer dimensions of 0.2mm and 8mm to a flux transformer with a $40 \times 40 \text{mm}^2$ pickup loop and achieved a magnetic field noise of 24ftHz-1/2 at frequencies down to 0.5Hz. The rf bias frequency was 150MHz.

The Jülich group have also achieved impressive results with rf SQUIDs operated at microwave frequencies. Zhang et al. [142] patterned a YBCO SQUID in the configuration of Fig. 39 and biased it at 3GHz. In an early version, an S-shaped microstrip resonator was

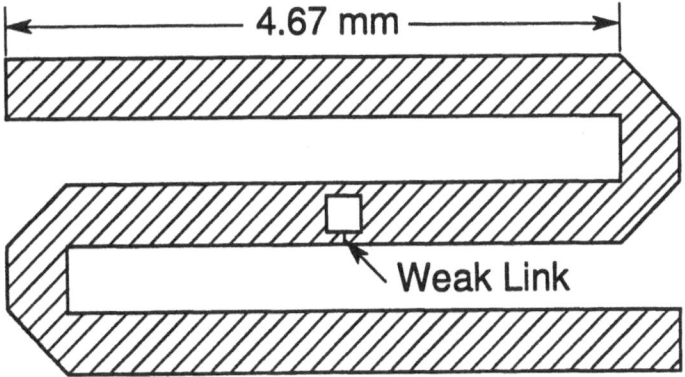

Figure 39. Configuration of microwave SQUID resonator with $100 \times 100 \mu\text{m}^2$ hole (re-drawn from ref. [142]).

patterned to contain a 100x100 μm^2 hole with two step-edge junctions along one edge; subsequently, the SQUID area was reduced to 10x100 μm^2, with the longer side parallel to the edge of the resonator. The microstrip configuration, formed by a copper ground plane on the underside of the substrate, was capacitively coupled to a 50Ω coaxial cable leading to the room temperature electronics. For the smaller SQUID, with an estimated inductance of 80 pH, the transfer function was about 100 μV / Φ_0 and the flux noise about 1.6×10^{-5} $\Phi_0 Hz^{-1/2}$ at frequencies down to about 0.1 Hz, corresponding to a noise energy of 6.4×10^{-30} JHz^{-1}. The noise energy is about a factor of 8 better than that of the 150 MHz SQUID described earlier. Mück [65] suggests that a factor of two improvement in the noise energy of the microwave SQUID could be achieved with a cooled preamplifier.

The flux-focusing factor of the microwave SQUID was about 8, leading to a magnetic field sensitivity of about 0.5 pT $Hz^{-1/2}$. The geometry of the stripline makes it difficult to achieve substantially higher flux focusing, so that a flux transformer will be necessary to improve the magnetic field sensitivity further.

6.9 MAGNETIC FLUX NOISE

A persistent problem with the high-T_c magnetometers has been the presence of 1/f magnetic flux noise at low frequencies. This phenomenon was investigated by Ferrari et al. [145, 146] who measured the flux noise generated by high-T_c films in a nearby low-T_c square washer SQUID. The assembly was enclosed in a vacuum can immersed in liquid helium. The SQUID was maintained at 4.2K, while the temperature of the chip could be increased by means of a heater. All films studied exhibited 1/f flux noise at temperatures below T_c, with a peak in the noise at T_c. In early polycrystalline YBCO films [145], the noise level was very high -- of the order of 10^{-5} Φ_0^2 Hz^{-1} at 4.2K and 1 Hz -- but the noise has diminished substantially as the quality of the films has improved. For example, in high quality epitaxial YBCO films [146] the noise at 77K is well below 10^{-10} Φ_0^2 Hz^{-1} at 1 Hz. At least in single layer in situ films, it is possible to achieve this noise level quite consistently: for example, in the bicrystal SQUID illustrated in Fig. 32, the noise is below 10^{-11} Φ_0^2 Hz^{-1} at 1 Hz.

Using a modified version of the theory presented in Sec. 3.6, Ferrari et al. [147] have described the origin of the noise in terms of the motion of flux quanta in the film. Each vortex hops independently between two pinning sites under thermal activation, a process that produces a random telegraph signal with a Lorentzian power spectrum of the form $\tau / (1 + \omega^2\tau^2)$, where $\tau = \tau_0 \exp [-U(T) / k_B T]$ is a characteristic time, τ_0 is an attempt frequency and U(T) is the height of the energy barrier separating the wells. One computes the spectral density of an ensemble of such independent processes by adding the individual Lorentzians; if the distribution of activation energies is broad on the scale of $k_B T$, the result is a 1/f power spectrum [147]. The spectral density of the noise increases approximately linearly with the magnetic field in which the sample is cooled.

As mentioned in Sec. 6.5, the levels of 1/f noise in multilayer structures have generally been higher than in single-layer films, because of the difficulty in maintaining sufficiently high qualtiy microstructure through several growth processes. Considerable progress has been made, however, and levels of 1/f flux noise in (say) three-layer structures will hopefully be reduced to levels approaching the noise of the SQUID before too long. The increase in 1/f noise in an ambient magnetic field -- in both single-layer and multilayer devices -- remains an extremely important issue. All of the noise spectra shown in Fig. 34 were obtained in zero field, and the 1/f noise is appreciably higher when the devices are operated in magnetic fields comparable with that of the earth. It may be hoped that the ever-improving quality of YBCO films will lead to a reduction in 1/f noise not only in zero magnetic field but also in fields of (say) 100μT. For the moment, one solution for biomagnetic measurements is to operate the magnetometers in a magnetically shielded room, as is currently the practice for low-T_c instruments. Of course, for geophysical measurements, this solution is not applicable.

An important alternative suggested by Koch *et al.* [148] is to cool the magnetometer in zero field and then to move it into the earth's magnetic field with a "flux dam" (weak link) in the pickup loop to limit the supercurrents induced and thus inhibit the entry of magnetic flux into the material. More work on the crucial issue of low-frequency noise in static magnetic fields is very much in order.

7. Concluding Remarks

Progress with high-T_c SQUIDs in the three years since the 1992 NATO ASI has been very impressive. We now have high-T_c multilayer magnetometers with $10 \times 10 mm^2$ pickup loops operating in liquid nitrogen achieving white noise levels of less than $10 fTHz^{-1/2}$ and below $30 fTHz^{-1/2}$ at 1Hz. A single-layer device on a $20 \times 20 mm^2$ bicrystal has achieved $14 fTHz^{-1/2}$ at 1kHz and $26 fTHz^{-1/2}$ at 1Hz -- and is readily available commercially. These performances compare favorably with those for commercially available low-T_c SQUIDs not so many years ago. It is important to appreciate that a substantial number of groups worldwide have successfully fabricated high-T_c magnetometers with good performance, and there is every reason to believe they will become widely used in the next two or three years. For applications requiring high sensitivity in a small area -- notably arrays of magnetometers for biomagnetic measurements -- it is likely that multilayer devices will be preferred. On the other hand, in applications involving a relatively small number of magnetometers, for example, in geophysics or nondestructive evaluation -- single-layer devices are likely to be perfectly adequate.

High-T_c SQUIDs are already beginning to make their way into a variety of applications. Miklich *et al.* [149] made a picovoltmeter by inductively coupling a 2-turn YBCO coil (patterned in a single-layer) to the pickup loop of a directly-coupled magnetometer. The coil is connected in series with a resistor and the voltage source to be measured. Dantsker *et al.* [150] constructed a 3-axis geophysical magnetometer involving directly-coupled magnetometers, and demonstrated that it functioned in the field. In a quite different application, Black *et al.* [151] used a high-T_c SQUID in a scanning mode to image a variety of objects, including the simulation of a crack induced in the underside of Al sheets by a rivet. A number of groups have obtained high quality magnetocardiograms. The most ambitious system to date is the 16-channel system operated at the Superconducting Sensor Laboratory in Japan [152]. Operating in a tiny shielded room just sufficient to accommodate a person, this system successfully recorded 16 channels of data.

Of course, challenges remain. One of the most important issues is the increase in the l/f flux noise in YBCO films operated in the earth's magnetic field. Hopefully, the continuing improvements in the quality of films will ameliorate this problem. Also, the reproducibility of Josephson junctions is still not as good as any of us would like. Now that high-T_c SQUIDs have white noise levels close to the predicted values, a factor-of-two departure from the design values of critical current or resistance can cause a significant deterioration in performance. Such variation in parameters means that the yield in fully integrated magnetometers with high performance is less than that with flip-chip magnetometers, where one can select the best SQUID from a set fabricated on a single chip. However, I have no doubt that continuing refinements of the fabrication process will improve the margins of junction critical current and resistance quite substantially.

Thus, I believe that we are about to enter a new era of SQUID applications brought about by the use of liquid nitrogen. Liquid nitrogen is of course appreciably cheaper than liquid helium, but more importantly it boils away much more slowly. As a result, one can readily achieve operating times of many months with a dewar that is still readily portable, or operate a system the size of a coffee can for at least a day. Not to be overlooked is the fact that SQUIDs at 77K can be operated significantly closer to objects at room temperature and pressure than helium temperature devices. This may have interesting consequences for both

54

biology and nondestructive evaluation. I believe that the next few years will see the most explosive growth ever in the use of SQUIDs -- not only high-T_c devices but also low-T_c devices as multi-channel systems have an increasing impact on biomedical measurements.

Acknowledgments

I am indebted to A. Barfknecht for supplying Fig. 7, D. Crum Fig. 22(a), W. J. Gallagher Fig. 22(b), P.F. Michelson Fig. 28, D. Paulsen Fig. 19, and F. C. Wellstood Fig. 10. For that part of the work carried out at Berkeley, I wish to thank the following for their hard work and dedication: Mark Colclough, Gene Dantsker, Gordon Donaldson, Non Fan, Mark Ferrari, Tom Gamble, Wolf Goubau, Erwin Hahn, Mike Heaney, Claude Hilbert, Christoph Heiden, Martin Hürlimann, Mark Johnson, Mark Ketchen, Jack Kingston, Reinhold Kleiner, Roger Koch, Dieter Koelle, Rasmus Kromann, Philippe Lerch, Frank Ludwig, John Martinis, Andy Miklich, David Nemeth, Charles Pennington, Alex Pines, Du Quan, Didier Robbes, Bonaventura Savo, Tim Shaw, Tycho Sleator, Claudia Tesche, Cristian Urbina, Dale Van Harlingen and Fred Wellstood.

This work was supported by the Director, Office of Energy Research, Office of Basic Energy Sciences, Materials Sciences Division of the U.S. Department of Energy under contract number DE-AC03-76SF00098.

General References

I draw the readers' attention to the comprehensive reviews:
  Ryhänsen, T., Seppä, H., Ilmoniemi, R., and Knuutila, J. (1989) SQUID magnetometers for low-frequency applications', *J. Low Temp. Phys.* 76, 287-386,
  Hämäläinen, M., Hari, R., Ilmoniemi, J., Knuutila, J., Lounasmaa, O.V. (1993) Magnetoencephalography-theory, instrumentation, and applications to noninvasive studies of the working human brain, *Rev. Mod. Phys.* 65, 413-497,
and to the recent book containing 7 chapters by different authors:
  Barone, A. (ed.) (1992) *Principles and Applications of Superconducting Quantum Interference Devices*, World Scientific Publishing Company, Singapore.

References

1. London, F. (1950) *Superfluids*, Wiley, New York .
2. Josephson, B. D. (1962) Possible new effects in superconductive tunneling, *Phys. Lett.* 1, 251-253; Supercurrents through barriers, (1965) *Adv. Phys.* 14, 419-451.
3. Jaklevic, R. C., Lambe, J., Silver, A. H., and Mercereau, J. E. (1964) Quantum interference effects in Josephson tunneling, *Phys. Rev Lett.* 12, 159-160.
4. Zimmerman, J. E., Thiene, P., and Harding. J. T. (1970) Design and operation of stable rf-biased superconducting point-contact quantum devices, and a note on the properties of perfectly clean metal contacts, *J. Appl. Phys.* 41, 1572-1580.
5. Mercereau, J. E. (1970) Superconducting magnetometers, *Rev. Phys. Appl.* 5. 13-20; Nisenoff. M. (1970) Superconducting magnetometers with sensitivities approaching 10^{-10} gauss, *Rev. Phys. Appl.* 5, 21-24.
6. Clarke, J. (1993) SQUIDs: theory and practice, in H. Weinstock and R. W. Ralston (eds.), *The New Superconducting Electronics*, Kluwer Academic Publishers, Dordrecht, pp. 123-180.
7. Stewart, W. C. (1968) Current-voltage characteristics of Josephson junctions, *Appl. Phys. Lett.* 12, 277-280.
8. McCumber, D. E. (1968) Effect of ac impedance on dc voltage-current characteristics

of Josephson junctions, *J. Appl. Phys.* **39**, 3113-3118 .

9. Ambegaokar, V. and Halperin, B. I. (1969) Voltage due to thermal noise in the dc Josephson effect, *Phys. Rev. Lett.* **22**, 1364-1366.

10. Clarke, J. and Koch, R. H. (1988) The impact of high-temperature superconductivity on SQUIDs, *Science* **242**, 217-223.

11. Likharev, K. K. and Semenov, V. K. (1972) Fluctuation spectrum in superconducting point junctions, *Pis'ma Zh. Eksp. Teor. Fiz.* **15**, 625-629. [(1972) *JETP Lett.* **15**, 442-445].

12. Vystavkin, A. N., Gubankov, V. N., Kuzmin, L. S., Likharev, K. K., Migulin, V. V., and Semenov, V. K. (1974) S-c-s junctions as nonlinear elements of microwave receiving devices, *Phys. Rev. Appl.* **9**, 79-109.

13. Koch, R. H., Van Harlingen, D. J., and Clarke, J. (1980) Quantum noise theory for the resistively shunted Josephson junction, *Phys. Rev Lett.* **45**, 2132-2135.

14. Tesche, C. D. and Clarke. J. (1977) dc SQUID: noise and optimization *J. Low. Temp. Phys.* **27**, 301-331.

15. Bruines, J. J.,P., de Waal, V. J., and Mooij, J. E. (1982) Comment on DC SQUID noise and optimization, by Tesche and Clarke *J. Low. Temp. Phys.* **46**, 383-386.

16. De Waal, V. J., Schrijner, P., and Llurba, R. (1984) Simulation and optimization of a dc SQUID with finite capacitance, *J. Low. Temp. Phys.* **54**, 215-232.

17. Ketchen, M. B., and Jaycox, J. M. (1982) Ultra-low noise tunnel junction dc SQUID with a tightly coupled planar input coil, *Appl. Phys. Lett.* **40**, 736-738.

18. Jaycox, J. M. and Ketchen, M. B. (1981) Planar coupling scheme for ultra low noise dc SQUIDs, *IEEE Trans. Magn.*, MAG-17, 400-403.

19. Barfknecht, A., Colclough, M. S., and de la Cruz, A. Conductus, Inc., Sunnyvale, California (unpublished).

20. Gurvitch, M., Washington, M. A., and Huggins, H. A. (1983) High quality refactory Josephson tunnel junction utilizing thin aluminum layers, *Appl. Phys. Lett.* **42**, 472-474.

21. De Waal, V. J., Klapwijk, T. M., and Van den Hamer, P. (1983) High performance dc SQUIDs with submicrometer niobium Josephson junctions, *J. Low. Temp. Phys.* **53**, 287-312 .

22. Tesche, C. D., Brown, K. H., Callegari, A. C., Chen, M. M., Greiner, J. H., Jones, H. C., Ketchen, M. B., Kim, K. K., Kleinsasser, A. W., Notarys, H. A., Proto, G., Wang, R. H., and Yogi, T. (1985) Practical dc SQUIDs with extremely low l/f noise, *IEEE Trans. Magn.* MAG-21, 1032-1035.

23. Pegrum, C. M., Hutson, D., Donaldson, G. B., and Tugwell, A. (1985) DC SQUIDs with planar input coils, *ibid*, 1036-1039.

24. Noguchi, T., Ohkawa, N., and Hamanaka, K. (1985) Tunnel junction dc SQUID with a planar input coil, in H. D. Hahlbohm. and H. Lubbig (eds.), *SQUID 85 Superconducting Quantum Interference Devices and their Applications*, Walter de Gruyter, Berlin, pp. 761-766.

25. Foglietti, V., Gallagher, W. J., Ketchen, M. B., Kleinsasser, A. W., Koch, R. H., and Sandstrom, R. L., (1989) Performance of dc SQUIDs with resistively shunted inductance, *Appl. Phys. Lett.* **55**, 1451-1453.

26. Muhlfelder. B., Beall. J. A., Cromar, M. W., and Ono, R. H. (1986) Very low noise tightly coupled dc SQUID amplifiers, *Appl. Phys. Lett.* **49**, 1118-1120.

27. Knuutila, J., Kajola, N., Seppä, H., Mutikainen, R., and Salmi. J. (1988) Design, optimization and construction of a dc SQUID with complete flux transformer circuits, *J. Low. Temp. Phys.* **71**, 369-392.

28. Wellstood, F. C., Heiden. C., and Clarke, J. (1984) Integrated dc SQUID magnetometer with high slew rate, *Rev. Sci. Inst.* **55**, 952-957.

29. Carelli, P. and Foglietti, V. (1982) Behavior of a multiloop dc superconducting

quantum interference device, *J. Appl. Phys.* **53**, 7592-7598.

30. Ketchen, M. B., Stawiasz, K. G., Pearson, D. J., Brunner, T. A. Hu, C-K, Jaso, M. A., Manny, M. P., Parsons, A. A., and Stein, K. J. (1992) Submicron linewidth input coils for low T_C integrated thin-film dc superconducting quantum interference devices, *Appl. Phys. Lett.* **61**, 336-338.

31. Clarke. J., Goubau, W. M., and Ketchen. M. B. (1976) Tunnel junction dc SQUID: fabrication, operation, and performance, *J. Low. Temp. Phys.* **25**, 99-144.

32. Ketchen, M. B., Goubau, W. M., Clarke, J., and Donaldson, G.B. (1978) Superconducting thin-film gradiometer, *J. Appl. Phys.* **44**, 4111-4116.

33. Wellstood, F.C., and Clarke, J. unpublished.

34. Wellstood, F. C., Urbina, C., and Clarke, J. (1987) Low-frequency noise in dc superconducting quantum interference devices below 1K, *Appl. Phys. Lett.* **50**, 772-774.

35. Roukes, M. L., Freeman, M. R., Germain, R. S., Richardson, R. C., and Ketchen, M. B. (1985) Hot electrons and energy transport in metals at millikelvin temperatures, *Phys. Rev. Lett.* **55**, 422-425.

36. Wellstood, F. C., Urbina, C., and Clarke, J. (1989) Hot-electron limitation to the sensitivity of the dc superconducting quantum interference device, *Appl. Phys. Lett.* **54**, 2599-2601.

37. Ketchen, M. B., Awschalom, D. D., Gallagher, W. J., Kleinsasser, A. W., Sandstrom, R. L., Rozen, J. R., and Bumble, B. (1989) Design, fabrication and performance of integrated miniature SQUID susceptometers, *IEEE Trans. Magn.* MAG-25, 1212-1215.

38. Koch, R. H., Clarke, J., Goubau, W. M., Martinis, J. M., Pegrum, C. M. and Van Harlingen, D. J. (1983) Flicker (l/f) noise in tunnel junction dc SQUIDs, *J. Low. Temp. Phys.* **51**, 207-224.

39. Rogers, C.T. and Buhrman, R.A. (1984) Composition of l/f noise in metal-insulator-metal tunnel junctions, *Phys. Rev. Lett.* **53**, 1272-1275.

40. Dutta, P. and Horn, P.M. (1981) Low-frequency fluctuations in solids: l/f noise, *Rev. Mod. Phys.* **53**, 497-516.

41. Savo, B., Wellstood, F. C. and Clarke. J. (1987) Low-frequency excess noise in Nb-Al_2O_3-Nb Josephson tunnel junction, *Appl. Phys. Lett.* **50**, 1757-1759.

42. Tesche, C. D., Brown, R. H., Callegari, A. C., Chen, M. M., Greiner, J. H., Jones, H. C., Ketchen, M. B., Kim, K. K., Kleinsasser, A. W., Notarys, H. A., Proto, G., Wang, R. H. and Yogi. T. (1984) Well-coupled dc SQUID with extremely low l/f noise, in U. Eckern, A. Schmid, W. Weber, H. Wühl (eds.), *Proc. 17th International Conference on low temperature physics LT-17*, North Holland, Amsterdam, pp. 263-264.

43. Foglietti, V, Gallagher, W. J., Ketchen, M. B., Kleinsasser, A. W., Koch, R. H., Raider, S. I. and Sandstrom, R. L. (1986) Low-frequency noise in low l/f noise dc SQUIDs, *Appl. Phys. Lett.* **49**, 1393-1395.

44. Biomagnetic Technologies, Inc. 9727 Pacific Heights Blvd., San Diego, CA 92121-3719.

45. Fujimaki, N., Tamura, H., Imamura, T. and Hasuo, S. A single-chip SQUID magnetometer, Digest of Tech. papers of 1988 International Solid-State conference, (ISSCC), San Francisco, pp. 40-41.

46. Drung, D. (1986) Digital feedback loops for dc SQUIDs, *Cryogenics* **26**, 623-627; Drung, D., Crocoll, E., Herwig, R., Neuhaus, M. and Jutzi, W. (1989) Measured performance parameters of gradiometers with digital output, *IEEE Trans. Magn.* MAG-25, 1034-1037.

47. Mück, M. and Heiden, C. (1989) Simple dc SQUID system based on a frequency modulated relaxation oscillator, *IEEE Trans. Magn.* MAG-25, 1151-1153.

48. Drung, D., (1991) Investigation of a double-loop dc SQUID magnetometer with

additional positive feedback, in H. Koch and H. Lübbig (eds.), *Superconducting Devices and their Applications*, Springer-Verlag, Berlin, pp. 351-356.

49. Seppä, H. (1991) DC SQUID electronics based on adaptive noise cancellation and a high open-loop gain controller, in H. Koch and H. Lübbig (eds.) *Superconducting Devices and their Applications*, Springer-Verlag, Berlin, pp. 346-350.

50. Clarke, J. (1977) Superconducting QUantum Interference Devices for low frequency measurements' in B. B. Schwartz. and S. Foner. (eds.), *Superconductor Applications : SQUIDs and Machines*, Plenum, New York, pp 67-124.

51. Giffard, R. P., Webb, R. A. and Wheatley, J. C. (1972) Principles and methods of low-frequency electric and magnetic measurements using rf-biased point-contact superconducting device, *J. Low. Temp. Phys.* 6, 533-610.

52. Kurkijärvi, J. (1972) Intrinsic fluctuations in a superconducting ring closed with a Josephson junction, *Phys. Rev.* B 6, 832-835.

53. Kurkijärvi, J. and Webb, W. W. (1972) Thermal fluctuation noise in a superconducting flux detector, in H. M. Long and W. F. Gauster (eds.), *Proc. Applied Superconductivity Conference*, Annapolis, MD, pp. 581-587.

54. Jackel, L.D. and Buhrman, R.A. (1975) Noise in the rf SQUID, *J. Low. Temp. Phys.* 19, 201-246.

55. Ehnholm, G. J. (1977) Complete linear equivalent circuit for the SQUID, in H. D. Hahlbohm and H. Lubbig (eds.), *SQUID Superconducting Quantum Interference Devices and their Applications*, Walter de Gruyter, Berlin, pp. 485-499; Theory of the signal transfer and noise properties of the rf SQUID, *J. Low. Temp. Phys.* 29, 1-27 (1977).

56. Hollenhorst, H. N. and Giffard, R. P. (1980) Input noise in the hysteretic rf SQUID: theory and experiment, *J. Appl. Phys.* 51, 1719-1725.

57. Kurkijärvi, J. (1973) Noise in the superconducting flux detector, *J. Appl. Phys.* 44, 3729-3733.

58. Giffard, R. P., Gallop, J. C. and Petley, B. N. (1976) Applications of the Josephson effects, *Prog. Quant. Electron* 4, 301-402.

59. Ehnholm, G. J., Islander, S. T., Ostman, P. and Rantala, B. (1978) Measurements of SQUID equivalent circuit parameters, *J. de Physique* 39, colloque C6. 1206-1207.

60. Giffard, R. P. and Hollenhorst, J. N. (1978) Measurement of forward and reverse signal transfer coefficients for an rf-biased SQUID, *Appl. Phys. Lett.* 32, 767-769.

61. Jackel, L. D., Webb, W. W., Lukens, J. E. and Pei, S. S. (1974) Measurement of the probability distribution of thermally excited fluxoid quantum transitions in a superconducting ring closed by a Josephson junction, *Phys. Rev.* B9, 115-118.

62. Long, A., Clark, T. D., Prance.,R. J. and Richards, M. G. (1979) High performance UHF SQUID magnetometer, *Rev. Sci. Instrum.* 50, 1376-1381.

63. Hollenhorst, J. N. and Giffard, R. P. (1979) High sensitivity microwave SQUID, *IEEE Trans. Magn.* MAG-15, 474-477.

64. Ahola, H., Ehnholm, G. H., Rantala, B. and Ostman, P. (1978) Cryogenic GaAs-FET amplifiers for SQUIDs, *J. de Physique* 39, colloque C6, 1184-1185; Cryogenic GaAs-FET amplifiers for SQUIDs, (1979) *J. Low Temp. Phys.* 35, 313–328.

65. Mück, Michael (1993) Progress in rf-SQUIDs, *IEEE Trans. Appl. Supercond.* 3, 2003-2010.

66. Zimmerman, J. E. (1971) Sensitivity enhancement of Superconducting Quantum Interference Devices through the use of fractional-turn loops, *J. Appl. Phys.* 42, 4483-4487.

67. Shoenberg, D. (1962) Superconductivity, Cambridge University Press, Cambridge.

68. For a review, see Clarke, J. (1983) Geophysical Applications of SQUIDs, *IEEE Trans. Magn.* MAG-19, 288-294.

69. De Waal, V. J. and Klapwijk, T. M. (1982) Compact Integrated dc SQUID gradiometer,

58

Appl. Phys. Lett. **41**, 669-671.

70. Van Nieuwenhuyzen, G. J. and de Waal, V. J. (1985) Second order gradiometer and dc SQUID integrated on a planar substrate, *Appl. Phy. Lett.* **46**, 439-441.

71. Carelli, P. and Foglietti, V. (1983) A second derivative gradiometer integrated with a dc superconducting interferometer, *J. Appl. Phys.* **54**, 6065-6067.

72. Koch, R. H., Ketchen, M. B., Gallagher, W. J., Sandstrom, R. L., Kleinsasser, A. W., Gambrel, D. R., Field, T. H., and Matz, H. (1991) Magnetic hysteresis in integrated low-T$_C$ SQUID gradiometers, *Appl. Phys. Lett.* **58**, 1786-1789.

73. Knuutila, J., Kajola, M., Mutikainen, R., Salmi, J. (1987) Integrated planar dc SQUID magnetometers for multichannel neuromagnetic measurements, *Proc. ISEC '87* pp. 261.

74. Hoenig, H. E., Daalmans, G. M., Bär, L. Bömmel, F., Paulus, A., Uhl, D., Weisse, H. J., Schneider, S., Seifert, H., Reichenberger, H., Abraham-Fuchs, K., (1991) Multichannel dc SQUID sensor array for biomagnetic applications, *IEEE Trans. Magn.* **MAG-27**, 2777-2785.

75. For reviews, see Romani, G. L., Williamson, S. J. and Kaufman, L. (1982) Biomagnetic instrumentation, *Rev. Sci. Instrum.* **53**, 1815-1845; Buchanan, D. S., Paulson, D. and Williamson, S. J. Instrumentation for clinical applications of neuromagnetism, *Adv. Cryo. Eng.* **33**, 97-106.

76. Neuromag, Helsinki, Finland.

77. Barth, D. S., Sutherling. W., Engel. J. Jr. and Beatty J. (1984) Neuromagnetic evidence of spatially distributed sources underlying epileptiform spikes in the human brain, *Science* **223**, 293-296.

78. Romani, G. L., Williamson, S. J. and Kaufman, L. (1982) Tonotopic organization of the human auditory cortex, *Science* **216**, 1339-1340.

79. Cabrera, B. (1982) First results from a superconductive detector for moving magnetic monopoles, *Phys. Rev. Lett.* **48**, 1378-1381.

80. Quantum Design, San Diego, CA and Conductus, Inc., Sunnyvale, CA.

81. Ketchen, M. B., Kopley, T. and Ling, H. (1984) Minature SQUID susceptometer, *Appl Phys. Lett.* **44**, 1008-1010.

82. Awschalom, D. D. and Warnock, J. (1989) Picosecond magnetic spectroscopy with integrated dc SQUIDs, *IEEE Trans. Magn.* **MAG-25**, 1186-1192.

83. Clarke, J. (1966) A superconducting galvanometer employing Josephson tunneling, *Phil. Mag.* **13**, 115-127.

84. Hilbert, C. and Clarke, J. (1985) DC SQUIDs as radiofrequency amplifiers, *J. Low Temp. Phys.* **61**, 263-280.

85. Tesche, C. D. and Clarke, J. (1979) DC SQUID: current noise, *J. Low Temp. Phys.* **37**, 397-403.

86. Hilbert, C. and Clarke, J. (1985) Measurements of the dynamic input impedance of a dc SQUID, *J. Low Temp. Phys.* **61**, 237-262.

87. Martinis, J. M. and Clarke, J. (1985) Signal and noise theory for the dc SQUID, *J. Low Temp. Phys.* **61**, 227-236, and references therein.

88. Koch, R.H., Van Harlingen, D. J. and Clarke, J. (1981) Quantum noise theory for the dc SQUID, *Appl. Phys. Lett.* **38**, 380-382.

89. Danilov, V. V., Likharev, K. K. and Zorin, A. B.(1983) Quantum noise in SQUIDs, *IEEE Trans. Magn.* **MAG-19**, 572-575.

90. Hilbert, C., Clarke, J., Sleator, T. and Hahn, E. L. (1985) Nuclear quadruple resonance detected at 30MHz with a dc superconducting quantum interference device, *Appl. Phys. Lett.* **47**, 637-639. (See references therein for earlier work on NMR with SQUIDS).

91. Fan, N. Q., Heaney, M.B., Clarke, J., Newitt, D., Wald, L. L., Hahn, E. L., Bielecke, A. and Pines, A.(1989) Nuclear magnetic resonance with dc SQUID preamplifiers, *IEEE*

Trans. Magn. **MAG-25**, 1193-1199.

92. Sleator, T., Hahn, E. L., Heaney, M.B., Hilbert, C. and Clarke, J. (1986) Nuclear electric quadrupole induction of atomic polarization, *Phys. Rev. Lett.* **57**, 2756-2759.

93. Sleator, T., Hahn, E. L., Hilbert, C. and Clarke, J. (1987) Nuclear-spin noise and spontaneous emission, *Phys. Rev. B.* **36**, 1969-1980.

94. Fan, N. Q. and Clarke, J. (1991) Low-frequency nuclear magnetic resonance and nuclear quadrupole resonance based on a dc superconducting quantum interference device, *Rev. Sci. Instrum.* **62**, 1453-1459.

95. Hürlimann, M.D., Pennington, C.H., Fan, N.Q., Clarke, J., Pines, A., and Hahn, E. L. (1992) Pulsed fourier-transform NQR of ^{14}N with a dc SQUID, *Phys. Rev. Lett.* **69**, 684-687.

96. For an elementary review on gravity waves, see Shapiro, S. L., Stark, R.F. and Teukolsky, S. J. (1985) The search for gravitational waves, *Am. Sci.* **73**, 248-257.

97. For a review on gravity-wave antennae, see Michelson, P. F., Price, J. C. and Taber, R. C. (1987) Resonant-mass detectors of gravitational radiation, *Science* **237**, 150-157.

98. Paik, H. J. (1981) Superconducting tensor gravity gradiometer with SQUID readout, in H. Weinstock and W. C. Overton, Jr. (eds.), *SQUID Applications to Geophysics*, Soc. of Exploration Geophysicists, Tulsa, Oklahoma, pp. 3-12.

99. Mapoles, E. (1972) A superconducting gravity gradiometer, in H. Weinstock and W. C. Overton, Jr. (eds.), *SQUID Applications to Geophysics*, Soc. of Exploration Geophysicists, Tulsa, Oklahoma, pp. 153-157.

100. For a review, see K. Vozoff, (1972) The Magnetotelluric Method in the Exploration of Sedimentary Basins, *Geophysics* **37**, 98-114.

101. Gamble, T. D., Goubau, W. M., and Clarke, J. (1979) Magnetotellurics with a Remote Magnetic Reference, *Geophysics* **44**, 53-68.

102. Gamble, T. D., Goubau, W. M., and Clarke, J. (1979) Error Analysis for Remote Reference Magnetotellurics, *Geophysics* **44**, 959-968.

103. Koch, R. H., Umbach, C. P., Clark, G. J., Chaudhari, P., and Laibowitz, R. B. (1987) Quantum interference devices made from superconducting oxide thin films, *Appl. Phys. Lett.* **51**, 200-202.

104. Face, D. W., Graybeal, J. M., Orlando, T. P., and Rudman, D. A. (1990) Noise and dc characteristics of thin-film Bi-Sr-Ca-Cu-oxide dc SQUIDs, *Appl. Phys. Lett.* **56**, 1493-1495.

105. Koch, R. H., Gallagher, W. J., Bumble, B., and Lee, W. Y. (1989) Low-noise thin-film TlBaCaCuO dc SQUIDs operated at 77K, *Appl. Phys. Lett.* **54**, 951-953.

106. Simon, R. W., Bulman, J. B., Burch, J. F., Coons, S. B., Daly, K. P., Dozier, W. D., Hu, R., Lee, A. E., Luine, J. A., Platt, C. E., Schwarzbek, S. M., Wire, M. S., and Zani, M. J. (1991) Engineered HTS microbridges, *IEEE Trans. Magn.* **MAG-27**, 3209-3214.

107. Jia, C. L., Kabius, B., Urban, K., Herrman, K., Cui, G. J., Schubert, J., Zander, W., Braginski, A. I., and Heiden, C. (1991) Microstructure of epitaxial YBa$_2$Cu$_3$O$_7$ films on step-edge SrTiO$_3$ substrates, *Physica C* **175**, 545-554.

108. Dimos, D., Chaudhari, P., Mannhart, J., and LeGoues, F. K. (1988) Orientation dependence of grain-boundary critical currents in YBa$_2$Cu$_3$O$_{7-\delta}$ bicrystals, *Phys. Rev. Lett.* **61**, 219-222.

109. Char, K., Colclough, M. S., Garrison, S. M., Newman, N., and Zaharchuk, G. (1991) Bi-epitaxial grain boundary junctions in YBa$_2$Cu$_3$O$_7$, *Appl. Phys. Lett.* **59**, 733-735.

110. Laibowitz, R. B., Koch, R. H., Gupta, A., Koren, G., Gallagher, W. J., Foglietti, V., Oh, B., and Viggiano, J. M. (1990) All high-T$_c$ edge junctions and SQUIDs, *Appl. Phys. Lett.* **56**, 686-688.

111. Schwarz, D. B., Mankiewich, P. M., Howard, R. E., Jackel, L. D., Straughn, B. L., Burhat, E. G., and Dayem, A. H. (1989) The observation of the ac Josephson effect in a Yba$_2$Cu$_3$O$_7$/Au/Yba$_2$Cu$_3$O$_7$ junction, *IEEE Trans. Magn.* **MAG-25**, 1298-1300.

112. DiIorio, M. S., Yoshizumi, S. Yang, K-Y, Yang, J., and Maung, M. (1991) Practical high-T_C Josephson junctions and dc SQUIDs operating above 85K, *Appl. Phys. Lett.* **58**, 2552-2554.

113. Rogers, C. T., Inam, A., Hedge, M. S., Dutta, B., Wu, X. D., and Venkatesan, T. (1989) Fabrication of heteroepitaxial $YBa_2Cu_3O_{7-x}$-$PrBa_2CuO_{7-x}$-$YBa_2Cu_3O_{7-x}$ Josephson devices grown by laser deposition, *Appl. Phys. Lett.* **55**, 2032-2034.

114. Kupriyanov, Yu M., and Likharev, K. K. (1991) Towards the quantitative theory of the high-T_C Josephson junctions, *IEEE Trans. Magn.* **MAG-27**, 2460-2463; Beasley, M. R. (1991) Tunneling and proximity effect studies of the high temperature superconductors, *Physica C* **185-189**, 227-233.

115. Barner, J. B., Rogers, C. T., Inam, A., Ramesh, R., and Bersey, S. (1991) All a-axis oriented $YBa_2Cu_3O_{7-y}$-$PrBa_2Cu_3O_{7-x}$-$YBa_2Cu_3O_{7-y}$ Josephson devices operating at 80K, *Appl. Phys. Lett.* **59**, 742-744.

116. Gao, J., Aarnink, W. A. M., Gerritsma, G. J., Veldhuis, D., and Rogalla, H. (1991) Preparation and properties of all high-T_C SNS-type edge dc SQUIDs, IEEE Trans. Magn. MAG-27, 3062-3065.

117. Zani, M. J., Luine, J. A., Simon, R. W., and Davidheiser, R. A. (1991) Focused ion beam high-T_C superconductor dc SQUIDs, *Appl. Phys. Lett.* **59**, 234-236.

118. Tinchev, S. S., Cui, G., Zhang, Y., Buchal, Ch., Schubert, J., Zander, W., Hermann, K., Sodtke, E.,Braginski, A. I., and Heiden, C. (1990) Properties of rf-SQUIDs fabricated from epitaxial YBCO films, LT-19 Satellite Conference on High Temperature Superconductors, Cambridge, England, August 1990 (unpublished).

119. Robbes, D., Miklich, A. H., Kingston, J. J., Lerch, Ph., Wellstood, F. C., and Clarke, J. (1990) Josephson weak links in thin films of $YBa_2Cu_3O_{7-x}$ induced by electrical pulses, *Appl. Phys. Lett.* **56**, 2240-2242; *erratum* **57**, 1169.

120. Lee, L.P., Longo, J., Vinetskiy, V., and Cantor J. (1995) Low noise $Yba_2Cu_3O_{7-\delta}$ direct-current superconducting quantum interference device magnetometer with direct signal injection, *Appl. Phys. Lett.* **66**, 1539-1541.

121. Char, K, Antognazza, L, and Geballe, T.H. (1993) Study of interface resistances in epitaxial $YBa_2Cu_3O_{7-x}$/barrier/$YBa_2Cu_3O_{7-x}$ junctions, *Appl. Phys. Lett.* **63**, 2420-2422; Verhoeven, Martin A. J., Gerritsma, Gerrit J., Rogalla, Horst, and Golubov, Alexander A. (1995) Ramp type HTS Josephson junctions with PrBaCuGaO barriers, *IEEE Trans. Appl. Supercond.* **5**, 2095-2098; Faley, M.I., Poppe, U., Jia, C.L., Dähne, U., Goncharov, Yu, Klein, N., Urban, K., Glyantsev, V.N., Kunkel, G., Siegel, M. (1995) Application of Josephson edge type junctions with a $PrBa_2Cu_3O_7$ barrier prepared with Br-etthanol etching or cleaning, *IEEE Trans. Appl. Supercond.* **5**, 2608-2611; Satoh, T., Kukpriyanov, M. Yu., Tsai, J.S., Hidaka, M., and Tsuge, H. Resonant tunneling transport in YBaCuO/PrBaCuO/YBaCuO edge-type Josephson junctions, *IEEE Trans. Appl. Supercond.* **5**, 2612-2615.

122. Enpuku, K., Shimomura, Y., and Kisu, T. (1993) Effect of thermal noise on the characteristics of a high-T_C superconducting quantum interference device, *J. Appl. Phys.* **73**, 7929-7934.

123. For example: Koch, R. H., Eidelloth, W., Oh, B., Robertazzi, R.P., Andrek, S.A., and Gallagher, W. J. (1992) Identifying the source of 1/f noise in SQUIDs made from high-temperature superconductors, *Appl. Phys. Lett.* **60**, 507-509; Keene, M. N., Satchell, J. S., Goodyear, S. W., Humphreys, R. G., Edwards, J. A., Chew, N. G., and Lander, K. (1995) Low noise HTS gradiometers and magnetometers constructed from $YBa_2Cu_3O_{7-x}$/$PrBa_2Cu_3O_{7-y}$ thin films, *IEEE Trans. Appl. Supercond.* **5**, 2923-2926.

124. Koelle, D., Miklich, A.H., Ludwig, F., Dantsker, E., Nemeth, D.T., and Clarke, J. (1993) DC SQUID magnetometers from single layers of $YBa_2Cu_3O_{7-x}$, *Appl. Phys. Lett.* **63**, 2271-2273.

125. Koelle, D., Miklich, A. H., Dantsker, E., Ludwig, F., Nemeth, D. T., Clarke, J., Ruby, W.

and Char, K. (1993) High performance dc SQUID magnetometers with single layer $YBa_2Cu_3O_{7-x}$ flux transformers, *Appl. Phys. Lett.* **63**, 3630-3632.

126. Cantor, R., Lee, L.P., Teepe, M., Vinetskiy, V., and Longo, J. (1995) Low-noise, single-layer $YBa_2Cu_3O_{7-x}$ dc SQUID magnetometers at 77K, *IEEE Trans. Appl. Supercond.* **5**, 2927-2930.

127. Ludwig, F., Koelle, D., Dantsker, E., Nemeth, D. T., Miklich, A. H., Clarke, John and Thomson, R. E. (1995) Low noise $YBa_2Cu_3O_{7-x}/SrTiO_3/YBa_2Cu_3O_{7-x}$ multilayers for improved superconducting magnetometers, *Appl. Phys. Lett.* **66**, 373-375.

128. Zimmerman, J. E. (1971) Sensitivity enhancement of superconducting quantum interference devices through the use of fractional-turn loops, *J. Appl. Phys.* **42**, 4483-4487.

129. Drung, D., Cantor, R., Peters, M., Scheer, H. J., and Koch H. (1990) Low-noise high-speed dc superconducting quantum interference device magnetometer with simplified feedback electronics, *Appl. Phys. Lett.* **57**, 406-408.

130. Drung, D., Cantor, R., Peters, M., Rhyänen, T., and Koch H. (199') Integrated dc SQUID magnetometer with high dV/dB, *IEEE Trans. Magn.* MAG-27, 3001-3004.

131. Drung, D., Knappe, S., and Koch, H. (1995) Theory for the multiloop dc superconducting quantum interference device magnetometer and experimental verification *J. Appl. Phys.* **77**, 4088-4098.

132. Ludwig, F., Dantsker, E., Kleiner, R., Koelle, D., Clarke, John, Knappe, S., Drung, D., Koch, H., Alford, Neil McN., and Button, T.W. (1995) Integrated high-T_C multiloop magnetometer, *Appl. Phys. Lett.* **66**, 1418-1420.

133. Button, T. W., Alford, N. McN., Wellhofer, F., Shields, T. C., Abell, F. S. and Day, M. (1991) The processing and properties of high-T_c thick films, *IEEE Trans. Magn.* **MAG-27**, 1434-1437.

134. Ludwig, F., Dantsker, E., Koelle, D., Kleiner, R., Miklich, A. H., Nemeth, D. T., Clarke, John, Drung, D., Knappe, S., and Koch, H. (1995) High-T_C multilayer magnetometers with improved l/f noise, *IEEE Trans. Appl. Supercond.* **5**, 2919-2922.

135. Dantsker, E., Ludwig, F., Kleiner, R., Clarke, John, Teepe, M., Lee, L. P., Alford, Neil McN., and Button, T. (1995) Addendum: Low noise $YBa_2Cu_3O_{7-x}$ /$SrTiO_3/YBa_2Cu_3O_{7-x}$ multilayers for improved superconducting magnetometers, *Appl. Phys. Lett.* **67**, 725-726.

136. Lee, L. P., Char, K., Colclough, M. S., Zaharchuk, G. (1991) Monolithic 77K dc SQUID magnetometer, *Appl. Phys. Lett.* **59**, 3051-3053.

137. David, B., Grundler, D., Krumme, J.-P., and Dössel, O. (1995) Integrated high-T_C SQUID magnetometer, *IEEE Trans. Appl. Supercond.* **5**, 2935-2938.

138. Hilgenkamp, J. W. M., Brons, G. C. S., Soldevilla, J. G., Isselsteijn, R. P., Flokstra, J., and Rogalla, H. (1994) Four layer monolithic integrated high-T_C SQUID magnetometer *Appl. Phys. Lett.* **64**, 3497-3499.

139. Dössel, O., David, B., Eckart, R., Grundler, D., and Krey, S. (1995) High-T_C SQUID magnetometers for magnetocardiography, in Dave H. A. Blank (ed.), *Proceedings of the HTS Workshop on Applications and New Materials*, University of Twente, Enschede, pp. 124-130.

140. Kromann, R., Kingston, J. J., Miklich, A. H., Sagdahl, L. T., and Clarke, John (1993) Integrated high-transition temperature magnetometer with only two superconducting layers, *Appl. Phys. Lett* **63**, 559-561.

141. Zhang, Y., Mück, M., Herrmann, K., Schubert, J., Zander, W., Braginski, A., and Heiden, C., (1992) Low noise $YBa_2Cu_3O_{7-x}$ rf SQUID magnetometer, *Appl. Phys. Lett.* **60**, 645-647.

142. Zhang, Y., Mück, M., Herrmann, K., Schubert, J., Zander, W., Braginski, A., and Heiden, C., (1992) Microwave rf SQUID integrated into a planar $YBa_2Cu_3O_{7-x}$, *Appl. Phys. Lett.* **60**, 2303-2305.

143. Ketchen, M. B., Gallagher, W. J., Kleinsasser, A. W., Murphy, S., and Clem, J. R. (1985) DC SQUID flux focuser, in H.D. Hahlbohm and H. Lübbig (eds), *Proceedings of SQUID '85*, Berlin, pp. 865-871.

144. Zhang, Y., Krüger, U., Kutzner, R., Wördenweber, R., Schubert, J., Zander, W., Strupp, M., Sodtke, E., and Braginski, A.I. (1994) Single layer $YBa_2Cu_3O_7$ radio frequency SQUID magnetometers with direct-coupled pickup coils and flip-chip flux transformers, *Appl. Phys. Lett.* **65**, 3380-3382.

145. Ferrari, M. J., Johnson, M., Wellstood, F. C., Clarke, J., Rosenthal, P. A., Hammond, R. H., and Beasley, M. R. (1988) Magnetic flux noise in thin-film rings of $YBa_2Cu_3O_{7-\delta}$, *Appl. Phys. Lett.* **53**, 695-697.

146. Ferrari, M. J., Johnson, M., Wellstood, F. C., Clarke, J., Inam, A., Wu, X. D., Nazar, L., and Venkatesan, T. (1989) Low magnetic flux noise observed in laser-deposited *in situ* films of $Yba_2Cu_3O_y$ and implications for high-T_C SQUIDs, *Nature* **341**, 723-725.

147. Ferrari, M. J., Johnson, M., Wellstood, F. C., Clarke, J., Mitzi, D., Rosenthal, P. A., Eom, C. B., Geballe, T. H., Kapitulnik, A., and Beasley, M. R. (1990) Distribution of flux-pinning in $YBa_2Cu_3O_{7-\delta}$ and $Bi_2Sr_2CaCu_2O_{8+\delta}$ from flux noise, *Phys. Rev. Lett.* **64**, 72-75.

148. Koch, R. H., Sun, J. Z., Foglietti, V., and Gallagher, W. J. (1995) Flux dam, a method to reduce extra low frequency noise when a superconducting magnetometer is exposed to a magnetic field, *Appl. Phys. Lett.* **67**, 709-711.

149. Miklich, A. H., Koelle, D., Ludwig, F., Nemeth, D. T., Dantsker, E., and Clarke, John (1995) Picovoltmeter based on a high transition temperature SQUID, *Appl. Phys. Lett.* **66**, 230-232.

150. Dantsker, E., Koelle, D., Miklich, A. H., Nemeth, D. T., Ludwig, F., Clarke, John, Longo, J. T., and Vinetskiy, V. (1994) High-T_C three-axis dc SQUID magnetometers for geophysical applications, *Rev. Sci. Instrum.* **65**, 3809-3813.

151. Black, R. C., Wellstood, F. C., Dantsker, E., Miklich, A. H., Koelle, D., Ludwig, F., and Clarke, J. (1995) High frequency magnetic microscopy using a high-T_C SQUID, *IEEE Trans. Appl. Sup.* **5**, 2137-2141.

152. Kado, H. (1994) Applied Superconductivity Conference, Boston, MA, October 1994 (unpublished).

ADVANCED SQUID READ-OUT ELECTRONICS

DIETMAR DRUNG
Physikalisch-Technische Bundesanstalt
Abbestrasse 2-12, D-10587 Berlin, Germany

Abstract. Superconducting quantum interference devices (SQUIDs) are the most sensitive devices for measuring weak magnetic fields. Beside the SQUID itself, the design of the read-out electronics decisively determines the performance of the whole sensor. In standard read-out electronics, a cooled impedance-matching transformer between SQUID and preamplifier, and a flux modulation technique are employed. Recently, several novel SQUID read-out concepts without flux modulation have been developed, mainly for biomagnetic multichannel systems. This chapter gives a description and comparison of the most important ones: concepts using multiple SQUIDs, SQUIDs with additional positive feedback, relaxation oscillation SQUIDs, and digital SQUIDs. Both the noise and the dynamic behavior are discussed, and a simple model for the achievable speed of a directly-coupled feedback loop without flux modulation is presented, which allows one to estimate the dynamic merits of the various SQUID concepts.

1. Introduction

A dc superconducting quantum interference device (SQUID) can be considered as a flux-to-voltage converter with a strongly nonlinear, periodic voltage vs. flux (V-Φ) characteristic. In order to provide a linear relation between the applied signal flux Φ_s and the corresponding output signal of the sensor, the SQUID has to be operated in a flux-locked loop. In the commonly used read-out concept [1] an ac flux is applied to the SQUID having typically a frequency between 100 and 500 kHz and a peak-to-peak amplitude of $\Phi_0/2$, where $\Phi_0 = 2.068 \times 10^{-15}$ Vs is the flux quantum. The resulting ac voltage across the SQUID is increased by a cooled impedance matching transformer, further amplified at room temperature, lock-in detected, integrated, and fed back as a current into a feedback coil in order to keep the flux in the SQUID at a constant level. Consequently, the voltage across the feedback resistor (i.e., the output signal of the SQUID sensor) is proportional to the applied flux even for very large signals $\Phi_s \gg \Phi_0$. Motivated by the need to simplify the read-out electronics of biomagnetic multichannel systems, several novel read-out concepts have recently been developed which do not require a flux modulation technique [2,3].

63

H. Weinstock (ed.), SQUID Sensors: Fundamentals, Fabrication and Applications, 63–116.
© *1996 Kluwer Academic Publishers.*

The most important performance parameters of a SQUID system are noise and dynamic behavior (bandwidth, maximum slew rate). Other relevant parameters are linearity (harmonic distortion), feedback range, number of wires between cryogenic and room temperature parts, and crosstalk in multichannel systems. Usually, the overall noise of a SQUID system is dominated by the intrinsic SQUID noise, while the dynamic behavior is limited by the flux-locked-loop electronics. The demands on the dynamic performance of biomagnetic multichannel systems are relatively low because biomagnetic signals are commonly very weak, and large interference signals are suppressed by operating these systems inside a magnetically-shielded room and/or by using gradiometer pickup coils [3]. Therefore, the dynamic behavior is neither an optimization criterion, nor is it tested or specified for most of the alternate SQUID read-out concepts. On the other hand, there is growing interest in fast SQUID systems, e.g., for low-noise amplifiers in magnetic resonance experiments [4], for picovolt-meters [5], for geophysical magnetometers [6], or for unshielded gradiometers based on magnetometers [7]. Therefore, to decide the overall quality of an alternate SQUID read-out concept, both the noise and the dynamic behavior should be considered.

In this chapter, I give an overview of the fundamentals and the performance of various SQUID read-out concepts. First, in Sec. 2, a simple model for noise, dynamic behavior and harmonic distortion of a directly-coupled feedback loop without flux modulation is presented, from which general design criteria for feedback loops can be derived. Other relevant topics such as integrator type, SQUID bias mode and feedback configuration also are discussed. In the following sections the most important alternate SQUID read-out concepts are described and compared on the basis of the above-mentioned model: concepts using multiple SQUIDs in Sec. 3, SQUIDs with additional positive feedback in Sec. 4, relaxation oscillation SQUIDs in Sec. 5, and digital SQUIDs in Sec. 6. Section 7 contains some concluding remarks.

2. Fundamentals

2.1. THE DC SQUID

The basic principles of SQUID sensors are described in the preceding chapter "SQUID fundamentals and selected applications" by John Clarke and in the literature (e.g., [8-10]). To calculate the behavior of the feedback loop, it is adequate to assume that the SQUID is simply a flux-to-voltage converter with a periodic V-Φ characteristic (see Fig. 1). The dotted line in Fig. 1 represents an example of a real V-Φ characteristic; the solid line is a straight-line approximation. The working point W is maintained if the voltage V across the SQUID is equal to the bias voltage V_b, and the total flux Φ threading the SQUID loop is equal to the bias flux Φ_b. The usable voltage swing $2\delta V$ of the V-Φ characteristic and the transfer coefficient $V_\Phi = (\partial V/\partial \Phi)_w$ at the working point should be as large as possible in order to simplify the SQUID read-out electronics. Note that commonly the peak-to-peak voltage swing ΔV is quoted, which is equal to the usable voltage swing $2\delta V$ if the working point lies symmetrically between the minimum and maximum voltage of the V-Φ characteristic. However, optimum

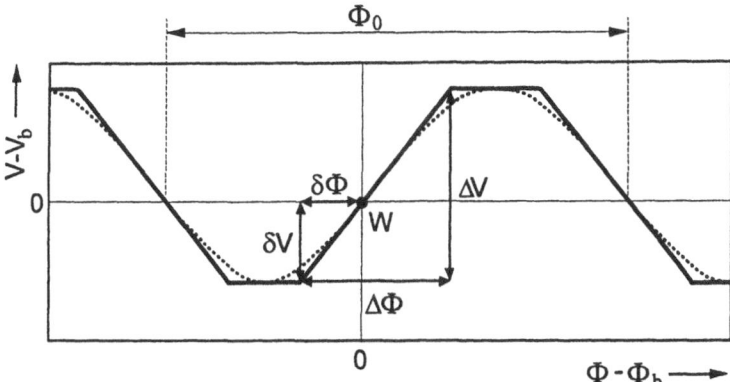

Figure 1. Idealized (solid line) and real (dotted line) V–Φ characteristic.

noise performance of low-T_c (transition temperature) SQUIDs is often found for an asymmetrically-positioned working point. Then $\delta V < \Delta V/2$ is the smaller value of the positive and negative voltage ranges, and $2\delta V$ is a more appropriate figure of merit than ΔV. As illustrated in Fig. 1, the linear flux ranges $2\delta\Phi$ and $\Delta\Phi$ are defined similarly, using the relation $|V_\Phi| = \delta V/\delta\Phi = \Delta V/\Delta\Phi$. We shall see later that δV and $\delta\Phi$ decisively determine the maximum system bandwidth and slew rate.

SQUIDs are usually optimized for minimum noise. Two widely-used figures of merit for SQUID noise are the energy sensitivity $\varepsilon(f)$ and the coupled energy sensitivity $\varepsilon_c(f)$, which correspond to the minimum magnetic energy per unit bandwidth that can be detected in the SQUID loop or in the input coil, respectively [8]. They are given by

$$\varepsilon(f) = S_\Phi(f)/2L, \quad \varepsilon_c(f) = \varepsilon(f)/k^2, \tag{1}$$

where $S_\Phi(f)$ is the flux-noise spectral density, L is the SQUID inductance, and k is the coupling coefficient between the input coil and the SQUID. The flux-noise spectral density is equal to the spectral density of the voltage noise across the SQUID divided by the square of the transfer coefficient: $S_\Phi(f) = S_V(f)/V_\Phi^2$. The voltage noise measured across the SQUID is only in part a "true" voltage noise, but it arises also from fluctuations of the circulating current in the SQUID loop ("true" flux noise), which are converted into voltage fluctuations via the flux-to-voltage transfer function. If the SQUID is operated in a flux-locked loop, any fluctuation of the SQUID voltage, independent of its origin, is compensated by a corresponding fluctuation of the feedback flux. The resulting feedback flux noise is interpreted as the SQUID flux noise, although it arises only in part from fluctuations of the flux in the SQUID loop.

Noise and transfer coefficient of a SQUID are determined by three basic parameters: the noise parameter Γ, the reduced inductance β, and the hysteresis parameter β_c defined as

$$\Gamma = 2\pi k_B T/I_0\Phi_0, \quad \beta = 2LI_0/\Phi_0, \quad \beta_c = 2\pi I_0 R^2 C/\Phi_0. \tag{2}$$

Here, k_B is the Boltzmann constant, T is the absolute temperature, and I_0, R and C are the critical current, the shunt resistance and the capacitance of one Josephson junction, respectively. The noise parameter Γ describes the rounding of the current vs. voltage (I–V) characteristic of a shunted Josephson junction due to Nyquist noise of the shunt resistor. $\Gamma = 0$ means no noise rounding, while for $\Gamma \gtrsim 1$, the I–V characteristic is so strongly disturbed by Nyquist noise that it approaches the resistor line R (or R/2 in the case of a symmetric dc SQUID) [11,12]. The hysteresis parameter β_c determines the degree of hysteresis in the I–V characteristic (nonhysteretic case $\beta_c \lesssim 1$) as well as the damping of the Josephson junction: for small values of the current $I \ll I_0$, the junction behaves like an inductance $L_j = \Phi_0/2\pi I_0$, forming an R-L-C parallel-resonant circuit with a quality factor $Q_j = \sqrt{\beta_c}$ [13]. For optimum SQUID performance, the reduced inductance β and the hysteresis parameter β_c should be close to unity [14–16].

The commonly used approximations for transfer coefficient and energy sensitivity are deduced from simulations of a symmetric low-T_c SQUID with $I_0 = 3.5\,\mu A$ ($\Gamma = 0.05$), L = 300 pH, and negligible junction capacitance: $|V_\Phi| \simeq R/L$ and $\varepsilon \simeq 9k_BTL/R$ [10,14]. The approximation for the energy sensitivity is valid in the white noise region. Note that throughout this chapter the omission of the extension "(f)" in a spectral density, e.g., in $\varepsilon(f)$ or $S_\Phi(f)$, denotes the white noise level, unless a frequency is explicitly specified. In simulations as in practice, a bias flux of $\simeq \Phi_0/4$ was chosen, as well as a bias current I_b through the SQUID that maximizes $|V_\Phi|$. The usable voltage swing is $2\delta V \simeq I_0R/2$ for $\beta = 1$ and $\Gamma = 0.05$. In the case of low-T_c SQUIDs it is often convenient to write the approximations for voltage swing and energy sensitivity in the following form:

$$2\delta V \simeq 0.14\,\Phi_0/\sqrt{LC}, \quad \varepsilon \simeq 16\,k_BT\sqrt{LC}. \tag{3}$$

Equation (3) implies that the junction critical current and the shunt resistance are chosen to fulfill the conditions $\beta \simeq 1$ and $\beta_c \simeq 1$. For a typical low-T_c SQUID at 4.2 K with L = 100 pH and C = 0.3 pF, Eq. (3) predicts $2\delta V \simeq 53\,\mu V$ and $\varepsilon \simeq 8\,h$ (h is Planck's constant) corresponding to a very low white flux-noise level $\sqrt{S_\Phi} \simeq 5\times10^{-7}$ $\Phi_0/\sqrt{Hz}$. In general, the performance of low-T_c SQUIDs is reasonably well predicted by theory. Deviations are mainly caused by parasitic resonances in the SQUID structure that tend to increase the noise due to mixing-down effects and to lower the voltage swing due to the need for stronger damping (i.e., smaller $\beta_c \lesssim 0.5$) [9].

The transfer coefficient of high-T_c SQUIDs degrades for large values of the SQUID inductance $L \gtrsim 40$ pH [17]. There are two reasons for this. First, due to the higher operating temperature, the root-mean-square (rms) noise flux in the SQUID loop becomes large and smears the V–Φ characteristic [18]. To a rough approximation, the SQUID corresponds to an inductance in parallel with the series connection of the two shunt resistors. The rms noise current in the inductance integrated over the frequency is $\sqrt{k_BT/L}$, corresponding to an rms noise flux of $\sqrt{k_BTL}$ (noise bandwidth = $\pi/2 \times$ 3-dB bandwidth [19]). The rms noise flux does not change if the junction capacitances are taken into account. To avoid large V_Φ degradation, the rms noise flux should be smaller than the linear flux range $\delta\Phi$ leading (for a sinusoidal V–Φ characteristic) to the condition $\sqrt{k_BTL} \lesssim \Phi_0/2\pi$ or $L \lesssim 100$ pH at 77 K.

The second reason for the V_Φ degradation in high-T_c SQUIDs is the strong noise rounding of the I–V characteristics for large values of Γ. As already mentioned, for small bias currents a Josephson junction behaves like an inductance $L_j = \Phi_0/2\pi I_0$ in parallel with R and C. Obviously, the rms noise current in the junction $\sqrt{k_BT/L_j} = I_0\sqrt{\Gamma}$ should be well below the critical current I_0 to avoid significant noise rounding: $\Gamma \ll 1$ or $I_0 \gg 3.2$ µA at 77 K. For optimized high-T_c SQUIDs ($\beta \simeq 1$) this condition requires $L \ll 320$ pH at 77 K.

To describe the degradation of the transfer coefficient quantitatively we extend the standard theory by introducing a factor α:

$$|V_\Phi| \simeq \alpha R/L. \tag{4}$$

Formulas for the factor α were recently published by Enpuku et al. [18,20]. However, these formulas were derived from computer simulations of a 77-K SQUID with $I_0 = 20$ µA ($\Gamma = 0.16$) and do not describe the dependence of α on Γ very precisely. To obtain a formula for a wider range of Γ, I fitted Eq. (4) to simulation results reported in the literature [14,15,18,21-24] and found for optimized high-T_c SQUIDs ($\beta \simeq 1$) with negligible junction capacitance

$$\alpha \simeq 1/(1+1.8\Gamma)^2. \tag{5}$$

Equation (5) agrees well with available simulation results over a wide range $0.05 \lesssim \Gamma \lesssim 1$. The usable voltage swing and the energy sensitivity are approximately given by

$$2\delta V \simeq 0.3\,\Phi_0\alpha R/L, \quad \varepsilon \simeq (7.5+1/\alpha^2)\,k_BTL/R. \tag{6}$$

The expression for $2\delta V$ is derived from Eq. (4) assuming that the V–Φ characteristic is slightly steeper than sinusoidal. The term $1/\alpha^2$ in the expression for ε takes into account that, for strong V_Φ degradation, the voltage noise across the SQUID approaches the Nyquist voltage noise of the two shunt resistors in parallel. A 50-pH SQUID at 77 K with high quality junctions ($I_0 = 20$ µA and $R = 10\ \Omega$) has, according to Eq. (6), a performance similar to that achieved experimentally with typical low-T_c SQUIDs: $2\delta V \simeq 74$ µV, $\varepsilon \simeq 82$ h, and $\sqrt{S_\Phi} \simeq 1.1\times10^{-6}\ \Phi_0/\sqrt{Hz}$.

2.2. BASIC DIRECTLY-COUPLED FEEDBACK LOOP

In the following sections, expressions for the most important performance parameters of a SQUID system are derived based on the simplified flux-locked-loop circuit depicted in Fig. 2. The SQUID is assumed to be an infinitely fast flux-to-voltage converter with a V–Φ characteristic approximated by straight lines (solid curve in Fig. 1). The SQUID is directly coupled to an ideal, noiseless preamplifier and integrator circuit (dotted box in Fig. 2) which has a voltage gain in the frequency domain of $G_I(f) = V_f(f)/V_e(f)$, where $V_e(f)$ is the input (error) voltage and $V_f(f)$ is the output voltage. The latter is converted into a current $I_f(f) = V_f(f)/R_f$ via a feedback resistor R_f and fed back into the SQUID as a flux via a mutual inductance M_f. For an infinite

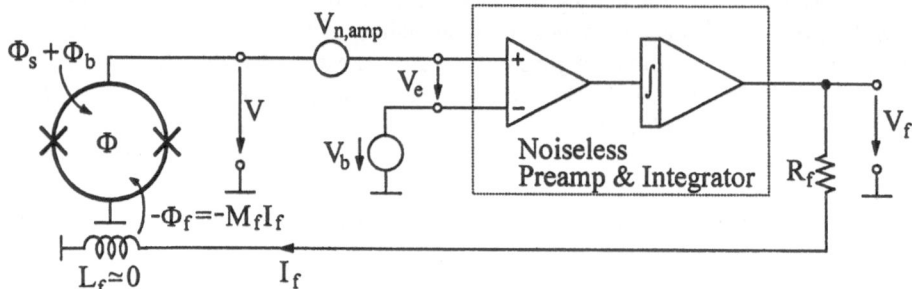

Figure 2. Idealized flux-locked-loop circuit.

integrator gain $G_I(f)$, the total flux Φ in the SQUID is kept constant, and the feedback loop provides a linear relationship between the applied signal flux $\Phi_s(f)$ and the system output voltage $V_f(f)$.

The voltage noise of the preamplifier is taken into account by a noise source $V_{n,amp}$ of spectral density $S_{V,amp}(f)$ placed between the SQUID and the preamplifier. This noise source includes contributions from Nyquist noise of the resistors in the pre-amplifier circuit referred to the input. If necessary, the preamplifier current noise $S_{I,amp}(f)$ can be considered as an equivalent voltage noise contribution $\Delta S_{V,amp}(f) = S_{I,amp}(f) R_{dyn}^2$, where R_{dyn} is the dynamic resistance of the SQUID at the working point. The effective flux-noise spectral density $S_{\Phi,t}(f)$ of the SQUID system is given by

$$S_{\Phi,t}(f) = S_\Phi(f) + S_{V,amp}(f)/V_\Phi^2 , \qquad (7)$$

where $S_\Phi(f)$ is the intrinsic flux-noise spectral density of the SQUID and $S_{\Phi,amp}(f) = S_{V,amp}(f)/V_\Phi^2$ is the preamplifier flux-noise contribution. According to Eq. (7), for a typical low-T_c SQUID with an intrinsic white flux noise of 10^{-6} $\Phi_0/\sqrt{Hz}$ and a transfer coefficient of 100 $\mu V/\Phi_0$, the white preamplifier voltage noise should be well below 0.1 nV/$\sqrt{Hz}$ to be negligible. Room-temperature amplifiers, however, have a typical noise level of 1 nV/$\sqrt{Hz}$, and consequently the transfer coefficient $|V_\Phi|$ has to be enhanced by more than one order of magnitude to make the preamplifier noise negligible. In the standard flux-modulation read-out scheme, this is done by using a cooled step-up transformer between SQUID and preamplifier, whereas in alternate read-out concepts often the voltage signal δV of the SQUID itself is increased and/or the linear flux range $\delta \Phi$ is decreased, since $|V_\Phi| = \delta V/\delta \Phi$.

A small-signal analysis around the working point W for the noise-free flux-locked loop circuit shown in Fig. 2 gives a closed-loop frequency response

$$A(f) = \frac{\Phi_f(f)}{\Phi_s(f)} = \frac{G(f)}{G(f)+1} , \qquad (8)$$

where

$$G(f) = \frac{\Phi_f(f)}{\Phi_e(f)} = V_\Phi G_I(f) M_f/R_f \qquad (9)$$

is the open-loop gain of the feedback loop. In Eqs. (8) and (9) the feedback flux $\Phi_f(f)$ = $V_f(f)M_f/R_f$ stands for the output signal of the SQUID system. It is subtracted from the signal flux $\Phi_s(f)$ to provide a non-inverting transfer function – indicated in Fig. 2 by negative signs. An ideal one-pole integrator with $G_I(f) \propto 1/jf$ has an open-loop gain

$$G(f) = \frac{1}{jf/f_1},\qquad(10)$$

where $j = \sqrt{-1}$ is the imaginary unit and f_1 is the unity-gain frequency of the feedback loop. The closed-loop frequency response $A(f)$ with one-pole integrator is identical to that of a first-order low-pass filter with a 3-dB cutoff frequency

$$f_c = f_1.\qquad(11)$$

Equation (11) is valid for ideal circuit components with negligible parasitic phase shifts and gain errors at high frequencies. Upper limits for the system bandwidth will be discussed in Sec. 2.4.

The error voltage across the SQUID becomes large at high frequencies and/or large signal amplitudes due to the finite open-loop gain of the feedback loop. The maximum output signal is obtained if the error voltage amplitude $\hat{V}_e(f)$ becomes equal to the usable voltage swing δV. From this condition and Eq. (9) the maximum amplitude $\hat{\Phi}_{f,max}(f)$ and maximum slew rate $\dot{\Phi}_{f,max}(f)$ of a sinusoidal feedback flux can be derived [6]:

$$\hat{\Phi}_{f,max}(f) = \delta\Phi\,|G(f)|,\quad \dot{\Phi}_{f,max}(f) = 2\pi f\,\hat{\Phi}_{f,max}(f) = 2\pi f\,\delta\Phi\,|G(f)|.\qquad(12)$$

For a one-pole integrator we obtain from Eqs. (10) and (12)

$$\hat{\Phi}_{f,max}(f) = \delta\Phi\,f_1/f,\quad \dot{\Phi}_{f,max}(f) = 2\pi f_1\delta\Phi.\qquad(13)$$

Thus, the maximum slew rate of a flux-locked loop with a one-pole integrator is frequency independent. According to Eq. (13) it increases linearly with the unity-gain frequency f_1 (i.e., with the system bandwidth f_c) and with the linear flux range $\delta\Phi$. Therefore, in terms of the slew rate, it is not desirable to increase V_{Φ} by lowering $\delta\Phi$ as is often the case in alternate SQUID read-out concepts. The maximum feedback flux amplitude increases with decreasing frequency. At low frequencies, however, it becomes frequency independent due to the finite feedback range which is given by the maximum current that can be passed through the feedback coil or the maximum voltage the output amplifier can provide without producing inadmissibly high non-linear distortions.

The dynamic range of a SQUID system is commonly defined as the feedback range (i.e., a *peak* value) divided by the white *rms* flux noise and is often interpreted as the signal-to-noise ratio in a 1-Hz bandwidth. However, the following should be kept in mind. First, the peak value of the flux noise is equal to the rms value multiplied by the crest factor [19], which lies typically between 2 and 4 dependent on the duration of the observation interval. Second, the dynamic range should be quoted as a function of frequency using the frequency dependent values $\hat{\Phi}_{f,max}(f)$ and $S_{\Phi,t}(f)$ rather than using the low-frequency feedback range $\hat{\Phi}_{f,max}(f \rightarrow 0)$ and the high-frequency white noise $S_{\Phi,t}$. Third, the dynamic range as defined above does not include the effect of nonlinear distortions which may considerably deteriorate the "true" signal-to-noise ratio.

2.3. HARMONIC DISTORTION

A figure of merit for system linearity is the total harmonic distortion (THD) of the output signal for an applied sinusoidal input signal. It is defined as the square root of the sum of the squared amplitudes of the output signal harmonics divided by the amplitude of the fundamental frequency of the output signal. In the following, simple analytical equations for the THD of the SQUID system depicted in Fig. 2 will be derived. For this, we approximate the V-Φ relation by the leading terms of a Taylor expansion at the working point W:

$$V_e = V_\Phi \Phi_e - \frac{c}{2} \frac{V_\Phi}{\delta \Phi} \Phi_e^2 - \frac{c'}{6} \frac{V_\Phi}{\delta \Phi^2} \Phi_e^3 . \tag{14}$$

Here, $V_e = V - V_b$ is the error voltage across the SQUID, $\Phi_e = \Phi - \Phi_b = \Phi_s - \Phi_f$ is the error flux in the SQUID, $V_\Phi = (\partial V/\partial \Phi)_w = \pm \delta V/\delta \Phi$ is the first derivative of the V-Φ characteristic at the working point, and

$$c = -\frac{\delta \Phi}{V_\Phi} (\partial^2 V/\partial \Phi^2)_w , \quad c' = -\frac{\delta \Phi^2}{V_\Phi} (\partial^3 V/\partial \Phi^3)_w > 0 \tag{15}$$

are the reduced second and third derivatives, respectively. Maximum linearity of a SQUID system is found if the working point is equal to the inflection point of the V-Φ characteristic (i.e., c = 0). For a small voltage difference $V_0 \ll \delta V$ between working point and inflection point, the absolute value of the second derivative is given by

$$|c| \simeq c' V_0/\delta V . \tag{16}$$

The reduced third derivative c' depends on the SQUID parameters. Unfortunately, to my knowledge c and c' have up to now been determined neither experimentally nor theoretically. Strongly noise-rounded V-Φ characteristics of high-T_c SQUIDs are approximately sinusoidal leading to $c' \simeq 1$. Low-T_c SQUIDs with small noise rounding are expected to exhibit larger values of c'. Recently, for a 400-pH low-T_c SQUID with additional positive feedback, an experimental dependence $\partial V/\partial \Phi \simeq V_\Phi \cos(1.8 V_e/\delta V)$ was found [25], from which $c' \simeq 3.2$ can be derived.

To estimate the THD we assume that a sinusoidal signal flux of frequency f and amplitude $\hat{\Phi}_s$ is applied to the SQUID. In the linear case c = c' = 0, this signal flux will cause a sinusoidal feedback flux of amplitude $\hat{\Phi}_f = |A(f)| \hat{\Phi}_s$ and a sinusoidal error flux in the SQUID of amplitude $\hat{\Phi}_e = \hat{\Phi}_s/|G(f)+1|$. In the following, we assume that the error flux remains exactly sinusoidal with unchanged amplitude $\hat{\Phi}_e$ even if the SQUID nonlinearity is taken into account – this can be achieved by applying appropriate additional input signals. Using the relations $2\cos^2\varphi = 1+\cos 2\varphi$ and $4\cos^3\varphi = 3\cos\varphi + \cos 3\varphi$, we find from Eq. (14) that the error voltage across the SQUID will exhibit additional components of frequency f, 2f and 3f, plus a dc component. These additional components are amplified by the feedback loop and fed back into the SQUID with corresponding amplitudes $\hat{\Phi}_{f,n} \ll \hat{\Phi}_f$, where n = 0-3 denotes the frequency in multiples of f. To maintain the original sinusoidal error flux of the linear case, we apply additional small-signal flux components with exactly the same phases and amplitudes $\hat{\Phi}_{s,n} = \hat{\Phi}_{f,n}$. Next, we split each additional feedback flux component into two signals. The larger one, which has an amplitude $|A(nf)| \, \hat{\Phi}_{s,n}$, is the linear response of the SQUID system to the additional signal flux of amplitude $\hat{\Phi}_{s,n}$; the smaller one of amplitude $\hat{\Phi}_{f,n}/|G(nf)+1|$ is a distortion flux caused by the system nonlinearity. Finally, we divide these distortion flux amplitudes by the feedback flux amplitude $\hat{\Phi}_f$ to obtain the reduced distortion flux amplitudes

$$a_0 = \frac{|c|}{4}\frac{1}{|G(f)|}X, \quad a_1 = \frac{c'}{8}\frac{1}{|G(f)+1|}X^2, \quad a_2 = \frac{|c|}{4}\frac{|A(2f)|}{|G(f)|}X, \quad a_3 = \frac{c'}{24}\frac{|A(3f)|}{|G(f)|}X^2. \quad (17)$$

Here, the subscripts denote the frequency in multiples of f.

$$X = \hat{\Phi}_s/\hat{\Phi}_{s,max}(f) = \frac{1}{|G(f)+1|}\hat{\Phi}_s/\delta\Phi \quad (18)$$

is the reduced signal flux amplitude, and $\hat{\Phi}_{s,max}(f) = \delta\Phi\,|G(f)+1|$ is the maximum signal flux amplitude corresponding to $\hat{\Phi}_{f,max}(f)$ in Eq. (12). Harmonic or inter-modulation distortions arising from the additional small input signals (which we have introduced to maintain the sinusoidal error flux) are negligibly small for small values of the reduced signal flux amplitude $X \ll 1$. Therefore, we neglect these additional input signals, as well as their linear responses at the output, i.e., we consider only the original sinusoidal input flux of amplitude $\hat{\Phi}_s$ and the corresponding feedback flux of amplitude $\hat{\Phi}_f$, plus the distortion flux components $a_n\hat{\Phi}_f$ according to Eq. (17). Further neglecting that the distortion flux component $a_1\hat{\Phi}_f$ slightly affects the amplitude of the fundamental frequency of the output signal, the THD is given by

$$THD = \sqrt{a_2^2 + a_3^2} \simeq \frac{c'}{24}\frac{X}{|G(f)|}\sqrt{X^2 + (6 V_0/\delta V)^2}. \quad (19)$$

The approximation on the right hand side of Eq. (19) is obtained from Eqs. (16) and (17) for $|A(2f)| \simeq |A(3f)| \simeq 1$. Equation (19) shows that for $V_0 = 0$ (i.e., working point = inflection point) the THD is minimum and scales with the square of the reduced signal flux amplitude X. For $V_0 \neq 0$ the THD scales only linearly with X for $X < 6V_0/\delta V$. In either case, the THD decreases strongly with increasing open-loop gain $|G(f)|$, and hence with decreasing frequency f due to the factor $1/|G(f)|$ in Eq. (19) and the $1/|G(f)+1|$ dependence of X according to Eq. (18). The theoretical dependence of the THD on $\hat{\Phi}_s$ and f will be compared with recently published experimental data in Sec. 4.3.

The described model also is useful to estimate the mixing-down effect of a high-frequency ($f \gg f_c$) sinusoidal signal flux applied to the SQUID [26]: $\Phi_{f,dc} = a_0\hat{\Phi}_{f,rf} = a_0|A(f)|\hat{\Phi}_{s,rf}$, where $\hat{\Phi}_{s,rf}$ is the amplitude of the high-frequency interference signal and $\Phi_{f,dc}$ is the resulting dc feedback flux change. From Eqs. (16) - (18) we obtain for $f \gg f_c$

$$\Phi_{f,dc} = \frac{|c|}{4}\,\hat{\Phi}_{s,rf}^2/\delta\Phi \simeq \frac{c'}{4}\frac{V_0}{\delta V}\,\hat{\Phi}_{s,rf}^2/\delta\Phi. \tag{20}$$

Equation (20) shows that the dc flux change scales with the square of the rf signal amplitude. To minimize the effect of rf signals on the SQUID system, the linear flux range should be maximized, and the working point should be as close as possible to the inflection point ($c \to 0$ or $V_0 \to 0$). For example, for a SQUID system with a sinusoidal V-Φ characteristic and $V_0/\delta V = 0.1$, Eq. (20) predicts that an rf signal of amplitude $\hat{\Phi}_{s,rf} = \delta\Phi/2 = 0.08\,\Phi_0$ applied to the SQUID will produce a dc flux change $\Phi_{f,dc} = 10^{-3}\,\Phi_0$.

In closing this section, we note that in the standard flux modulation scheme, the dc and 2f (for 2f $\ll$ modulation frequency) distortion components are eliminated if the SQUID is perfectly symmetric with identical Josephson junctions and shunt resistors. Bias current reversal schemes, which were developed to suppress the influence of critical-current fluctuations of the Josephson junctions [27], remove these distortions even in the case of asymmetric SQUIDs or in read-out schemes without flux modulation [28,29].

2.4. BANDWIDTH LIMITS

In the preceding sections it was shown that a high maximum slew rate and low harmonic distortions require a large linear flux range and a high open-loop gain (i.e., a high system bandwidth). In this section, three important limits for the achievable bandwidth will be discussed. The first one arises from the fact that the preamplifier broadband noise voltage is amplified, integrated and fed back into the SQUID as a noise flux. Approximating the V-Φ characteristic by straight lines (solid curve in Fig. 1) we obtain for the preamplifier noise flux

$$\Phi_{n,amp}^{peak}(f_c) \simeq 4\,\Phi_{n,amp}^{rms}(f_c) = 4\sqrt{(\pi/2)\,f_c\,S_{V,amp}}\,/V_\Phi, \tag{21}$$

where $\Phi_{n,amp}^{peak}(f_c)$ and $\Phi_{n,amp}^{rms}(f_c)$ are the peak and the rms value, respectively [3]. The peak value is equal to the rms value multiplied by the crest factor [19]; the crest factor of 4 used in Eq. (21) means that for Gaussian-distributed noise, the probability for exceeding the quoted peak value is 6.3×10^{-5}. The preamplifier voltage noise is low-pass filtered by the one-pole integrator feedback loop, resulting in an equivalent flux-noise bandwidth of $(\pi/2)f_c$. With the idealized straight-line approximation of the V-Φ characteristic, the output signal does not exhibit nonlinear distortions as long as the sum of dynamic error flux and preamplifier noise flux remains within the linear flux range. The corresponding maximum slew rate of the undistorted output signal

$$\dot{\Phi}_{f,max}^{lin}(f_c) = 2\pi f_c\big[\delta\Phi - \Phi_{n,amp}^{peak}(f_c)\big] \tag{22}$$

depends on the system bandwidth f_c. It has an absolute maximum for an optimum bandwidth

$$f_{c,max} \simeq 0.0044\,(2\,\delta V)^2/S_{V,amp}, \tag{23}$$

for which the preamplifier peak noise flux [Eq. (21)] amounts to $\frac{2}{3}$ of the linear flux range $\delta\Phi$. Equation (23) shows that for maximum system bandwidth, the voltage signal of the SQUID should be maximized and the white preamplifier voltage noise should be minimized.

The optimum bandwidth $f_{c,max}$ in Eq. (23) was derived for an idealized V-Φ characteristic. In practice, with a real V-Φ characteristic, the preamplifier noise flux will start to round the V-Φ characteristic, and hence to reduce $|V_\Phi|$ and to correspondingly increase the total flux-noise density $S_{\Phi,t}(f)$ for a system bandwidth $f_c < f_{c,max}$. Furthermore, the intrinsic SQUID noise rises if the working point fluctuates too much around the optimum working point due to broadband noise fed back into the SQUID. For $f_c = f_{c,max}$ the reduction of $|V_\Phi|$ and the increase of $S_{\Phi,t}(f)$ are noticeable, but still tolerable so that Eq. (23) is a reasonable estimate of the achievable bandwidth of the SQUID system. A bandwidth much above $f_{c,max}$ will result in a strongly noise-rounded V-Φ characteristic and a significantly higher noise level.

The second basic bandwidth limit results from the dead-time t_d in the feedback loop which causes phase shifts at high frequencies. The dead-time is essentially the total propagation delay on the lines between the SQUID and the room-temperature feedback electronics. For a 1-m separation, a typical value of t_d is 10 ns. The open-loop gain of a one-pole integrator feedback loop with dead-time is given by

$$G(f) = \frac{1}{jf/f_1} e^{-j2\pi ft_d}. \tag{24}$$

From Eqs. (8) and (24) with the help of the relation $e^{-j2\pi ft_d} = \cos(2\pi ft_d) - j\sin(2\pi ft_d)$, the absolute value of the closed-loop frequency response is found to be

$$|A(f)| = 1/\sqrt{1+\frac{f}{f_1}\left[\frac{f}{f_1} - 2\sin(2\pi ft_d)\right]}. \tag{25}$$

For low frequencies $f \to 0$, the closed-loop gain to leading order becomes:

$$|A(f)|_{f\to 0} = 1 - \frac{1}{2}(1 - 4\pi f_1 t_d)(f/f_1)^2. \tag{26}$$

Equation 26 shows that, dependent on the product $f_1 t_d$, the amplitude error at low frequencies may be negative or positive. It is minimum for an optimum value $f_1 t_d = 1/4\pi$, for which Eq. (26) gives $|A(f)|_{f\to 0} = 1$, implying that the low-frequency amplitude error is only determined by frequency terms of higher order than quadratic.

The frequency response [Eq. (25)] is plotted in Fig. 3 for three values of the product $f_1 t_d$. For negligible dead-time ($f_1 t_d = 0$) the frequency response is equal to that of a first-order low-pass filter with a 3-dB bandwidth $f_c = f_1$. The phase margin φ_m is 90° (φ_m is the deviation of the phase angle from the critical value $-180°$ at the unity-gain frequency f_1). The reduced bandwidth f_c/f_1 increases with increasing $f_1 t_d$ due to a reduction of the phase margin. For $f_1 t_d = 1/4\pi = 0.08$ the phase margin is 61° and the frequency response is optimally flat, i.e., has just no resonance peak. The corresponding system bandwidth and unity-gain bandwidth are

$$f_{c,max} = 0.18/t_d, \quad f_{1,max} = f_{c,max}/2.25, \tag{27}$$

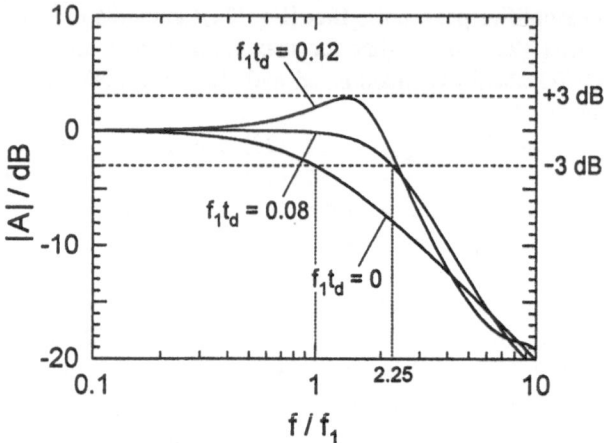

Figure 3. Frequency response of an ideal feedback loop with dead-time.

respectively. The optimum unity-gain bandwidth $f_{1,max}$, which determines slew rate and harmonic distortions at low frequencies, is approximately a factor of 2.25 smaller than $f_{c,max}$ in good agreement with the experimental factor of $\simeq 2.5$ recently obtained for wideband SQUID systems [30,31]. A further increase of the system bandwidth above the optimum value $f_{c,max}$ leads to a peak in the frequency response, but the reduced bandwidth f_c/f_1 is only slightly affected. For $f_1t_d = 0.12$ the phase margin is reduced to 47°, resulting in a 2.8-dB peak at $f/f_1 \simeq 1.4$. For $f_1t_d = 0.25$ the phase margin falls to zero, and the circuit will start to oscillate at the unity-gain frequency f_1. It should be noted here that the increase of the reduced bandwidth f_c/f_1 due to the dead-time does not lead to a corresponding increase of the maximum slew rate at low frequencies $f \ll f_c$, as the maximum slew rate is proportional to f_1 and not to f_c according to Eq. (13).

The third bandwidth limit is due to the finite 3-dB bandwidth $f_{c,amp}$ of the preamplifier. Usually, operational amplifiers with resistive feedback and a high closed-loop gain have a first-order low-pass behavior. If the closed-loop gain is reduced in order to increase the bandwidth, the second dominant pole in the open-loop frequency response of the preamplifier will become noticeable, leading to a second-order low-pass behavior with a resonance peak in the closed-loop frequency response. To simplify the following calculation we assume that the preamplifier gain is selected to provide an optimally flat frequency response just without a resonance peak. The corresponding quality factor is $1/\sqrt{2}$. It results in a preamplifier step response with a 4.3 % overshoot, a 10-90 % rise time of $0.34/f_{c,amp}$, and a time delay between the 50 % points of input and output signal of $0.228/f_{c,amp}$.

The reciprocal of the frequency response of a second-order low-pass filter is a polynomial of jf. For a quality factor of $1/\sqrt{2}$, this polynomial is equal to the leading terms of a Taylor expansion of a dead-time function $e^{j2\pi f t_d}$ at $f = 0$. Consequently, at low frequencies the preamplifier behaves like a series connection of a frequency-independent amplifier and a delay circuit with an equivalent dead-time

$$t_{d,amp} = 0.225/f_{c,amp}. \qquad (28)$$

For $f < f_{c,amp}/2$ the maximum gain and phase deviations between the two circuits are 3 % and 2.8°, respectively. Thus, to a first approximation, we replace the frequency response of the preamplifier by that of its equivalent delay circuit and use the results obtained for the flux-locked loop with dead-time. From Eqs. (27) and (28) we find

$$f_{c,max} \simeq 0.8\, f_{c,amp}. \qquad (29)$$

According to Eq. (29) the maximum bandwidth of a SQUID system is slightly lower than that of its preamplifier. The bandwidth may be somewhat enlarged above the quoted limit at the expense of a resonance peak in the closed-loop frequency response. A compensation circuit after the preamplifier with a frequency response $G'(f) = 1 + jf/f_1'$ and $f_1' \simeq f_{c,amp}$ may help to reduce the phase lag caused by the preamplifier and to improve the closed-loop frequency response correspondingly [30,31]. A simple realization is a capacitor in parallel to the resistor at the integrator input or a resistor in series to the integration capacitor [cf. one-pole integrator circuit shown inset in Fig. 5(a)].

The bandwidth limit [Eq. (29)] was derived for a preamplifier with a somewhat arbitrarily chosen but still realistic frequency response. With the selected preamplifier parameters the step response of the preamplifier exhibits a time delay between the 50 % points of input and output signal, which is equal to the equivalent dead-time [Eq. (28)]. This suggests that the effective dead-time $t_{d,eff}$ of the complete feedback loop may be determined by simply measuring the 50 % delay time of the open loop with the integration capacitors being short-circuited or replaced by small resistors to have a frequency-independent integrator gain. The effective dead-time includes the total propagation delay in the loop as well as the phase shifts of all electronic components. Although this is a strongly simplified model for the system dynamics, I expect that Eq. (27) with $t_{d,eff}$ instead of t_d will reasonably well predict the achievable bandwidth even if the preamplifier has a small resonance peak and/or a low-pass behavior of higher order than two. Therefore, to achieve a closed-loop bandwidth of 18 MHz, the effective dead-time should not exceed $\simeq 10$ ns. Usually, at least 5 ns are due to the total propagation delay on the lines between SQUID and preamplifier (2×2.5 ns for 0.75-m-long lines with speed of light) so that a minimum preamplifier bandwidth of 45 MHz is required according to Eq. (28). In practice, an even higher preamplifier bandwidth may be necessary because parasitic L-C resonance effects in the lines between SQUID and preamplifier lead to additional phase shifts which increase the effective dead-time.

The bandwidth limits Eqs. (23) and (27) are plotted in Fig. 4 vs. the usable voltage swing $2\delta V$ for a typical preamplifier voltage noise of 1 nV/$\sqrt{Hz}$ and a dead-time $t_d = 10$ ns. The preamplifier noise flux limits the bandwidth for $2\delta V \lesssim 64$ μV, while the dead-time is the limiting factor for $2\delta V \gtrsim 64$ μV. Recently, a direct-coupled SQUID system without flux modulation was optimized for maximum bandwidth [31]. An experimental bandwidth of 11 MHz was achieved with $2\delta V = 60$ μV and $t_d \simeq 7$ ns ($t_{d,eff} \simeq 18$ ns), close to that obtained from Eqs. (23) and (27) (solid circle in Fig. 4, see also Sec. 4.3). Finally we note that in the standard flux modulation read-out the

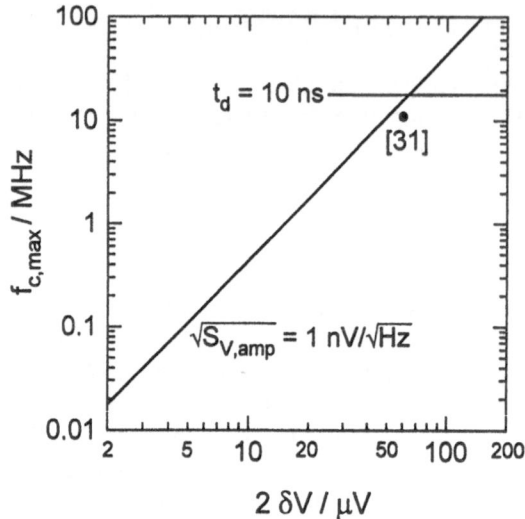

Figure 44. Maximum bandwidth vs. SQUID voltage swing. Lines are calculated from Eqs. (23) and (27), solid circle is measured [31].

preamplifier voltage noise referred to the SQUID is lowered by the gain of the impedance matching transformer placed between SQUID and preamplifier so that a higher bandwidth can be achieved for small values of the usable voltage swing. However, with flux modulation the bandwidth is severely limited by the modulation frequency of typically 100-500 kHz. Only very recently, a 5-MHz bandwidth SQUID system with a 16-MHz flux modulation has become commercially available [32].

To summarize, the transfer coefficient V_Φ of a SQUID has to be enhanced by about an order of magnitude to enable a direct coupling of the SQUID to the room-temperature preamplifier. For maximum slew rate and minimum harmonic distortions this should be done by increasing the voltage signal of the SQUID rather than by decreasing the linear flux range $\delta\Phi$. Although alternate SQUID read-out concepts usually suffer from a reduced $\delta\Phi$, their high system bandwidth (partially) compensates this, and the overall performance is often improved compared to the standard flux modulation read-out.

2.5. INTEGRATOR TYPES

To obtain the highest possible slew rate one selects the highest open-loop gain (band-width) for which the SQUID system remains stable and no intolerably high excess noise is observed [6]. At low frequencies the loop gain and hence the slew rate can be further increased by introducing a second pole in the frequency response of the open loop [33]. The two-pole integrator frequency response with dead-time t_d may be written in the form

$$G(f) = \frac{1}{jf/f_1} \frac{jf+f_2}{jf+f_2'} e^{-j2\pi f t_d}. \tag{30}$$

Open-loop gain $|G(f)|$ and phase of a one-pole and a two-pole integrator are plotted in Fig. 5 for $f_2 = f_1/4$ and $f_2' = f_2/400$. The corresponding electrical circuits are shown inset in Fig. 5(a). For high frequencies $f \gg f_2$, the behavior of the two-pole integrator is identical to that of the one-pole integrator. At low frequencies $f < f_2$, the gain vs. frequency curve of the two-pole integrator has a slope of -40 dB/decade instead of -20 dB/decade obtained with a one-pole integrator. Therefore, the two-pole integrator gain and hence the slew rate is higher by a factor of $\simeq f_2/f$ at low frequencies compared to that of the one-pole integrator, but the phase margin is reduced. At very low frequencies $f < f_2'$, the phase is shifted back to $-90°$, and the gain/slew rate enhancement factor approaches its maximum value f_2/f_2'. The unity-gain frequency of the two-pole integrator according to Eq. (30) deviates somewhat from f_1. However, it will be shown later that optimum loop stability requires $f_2 \lesssim f_1/4$, resulting in a unity-gain bandwidth

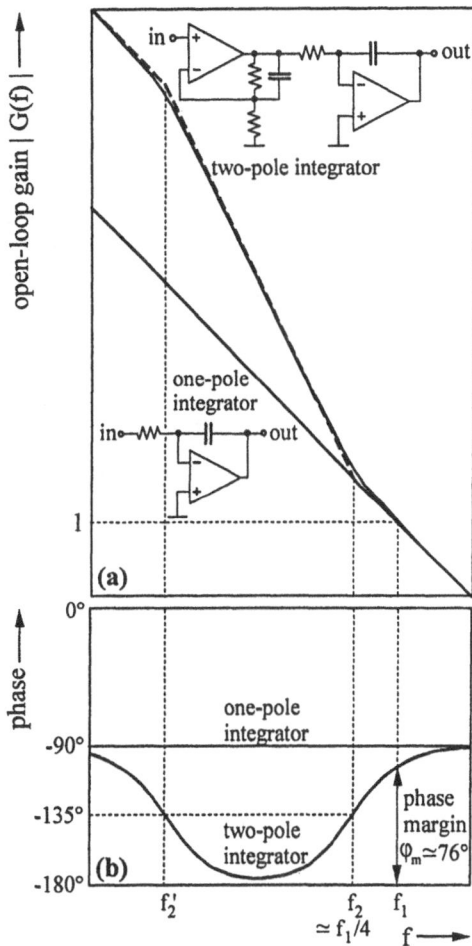

Figure 5. (a) Gain and (b) phase vs. frequency for a one-pole and a two-pole integrator. Insets in (a) show examples of electrical circuits providing the respective frequency response.

between f_1 and $1.03 f_1$. We neglect this small deviation and consider f_1 to be the unity-gain bandwidth.

The corner frequency $f_2' \ll f_2$ has negligible effect on the system dynamics so that $f_2' = 0$ can be assumed in the following calculations for simplicity. Using Eqs. (8) and (30) with $f_2' = 0$, we find for the closed-loop gain

$$|A(f)| = 1/\sqrt{1 + (f/f_1)^2 \frac{f^2 - 2ff_1\sin(2\pi ft_d) - 2f_1f_2\cos(2\pi ft_d)}{f^2 + f_2^2}} . \qquad (31)$$

For low frequencies $f \to 0$, the closed-loop gain to leading order becomes:

$$|A(f)|_{f \to 0} = 1 + \frac{f_1}{f_2}(f/f_1)^2. \qquad (32)$$

Equation (32) shows that the amplitude error of the two-pole integrator feedback loop is always positive at low frequencies, independent of the dead-time. This means that the two-pole integrator closed-loop gain always has a resonance peak. Further, in spite of the higher open-loop gain, the absolute value of the low-frequency amplitude error with a two-pole integrator is considerably higher than that for a one-pole integrator, at least by a factor of $2f_1/f_2$ for $t_d = 0$ according to Eqs. (26) and (32).

To offer some insight into the two-pole integrator dynamics, we briefly discuss the case with negligible dead-time. For $t_d = 0$, the error flux $\Phi_e(f)$ of the two-pole integrator feedback loop is equal to the high-pass filtered signal flux $\Phi_s(f)$. The corresponding filter is of second order, having a resonance frequency $\sqrt{f_1f_2}$ and a quality factor $\sqrt{f_2/f_1}$. A maximum quality factor of 0.5 was recommended in [33] giving $f_2 = f_1/4$. In this case, the phase margin is reduced by 14° to 76°, and the closed-loop frequency response exhibits a peak of 1.25 dB at a frequency $f_r = f_1/\sqrt{8}$. In the time domain, the step response has a 13.5 % overshoot after two time constants $(2/2\pi\sqrt{f_1f_2} = 0.64/f_1)$ and settles to the final value (± 10 %) after three time constants $0.95/f_1$. In contrast, the ideal one-pole integrator loop has a first-order low-pass behavior without overshoot and settles to the final value (± 10 %) in a much shorter time of $2.3/2\pi f_1 = 0.37/f_1$.

We turn now to a discussion of the behavior of the two-pole integrator loop including dead-time. As already mentioned, the closed-loop gain with two-pole integrator always exhibits a resonance peak. The resonance peak has a minimum for an optimum value of the unity-gain frequency f_1 if the corner frequency f_2 and the dead-time t_d are kept constant. The exact optimum can be determined only numerically; a sufficiently accurate, analytical solution [34] will be described in the following. First, the following approximations for the sine and cosine terms in Eq. (31) are used: $\sin\varphi \simeq \varphi$ and $\cos\varphi \simeq 1 - \varphi^2/2$. Then, the resonance frequency f_r and the peak in the closed-loop gain $|A(f_r)|$ are determined from the zero of the derivative $\partial(1/|A(f)|^2)/\partial f$. The resonance peak $|A(f_r)|$ is a function of the dead-time and of the loop parameters f_1 and f_2. Next, the optimum unity-gain frequency f_1, yielding a minimum resonance peak $|A(f_r)|_{min}$ for fixed f_2 and t_d, is determined from the zero of the derivative $\partial(1/|A(f_r)|^2)/\partial f_1$. For the minimum resonance peak, the simple expression $|A(f_r)|_{min} = 1/(1 - 2\pi f_2 t_d)^2$ is obtained. The optimum unity-gain frequency is a relatively complicated function of f_2 and t_d; however, for realistic values of the product $f_2 t_d$, it is approximately equal to the optimum value $f_1 = 1/4\pi t_d$ found for the one-pole integrator loop with dead-time. Finally, we approximate the dead-time t_d in the expression for $|A(f_r)|_{min}$ by $1/4\pi f_1$, and obtain

$$|A(f_r)|_{min} \simeq 1/(1 - \tfrac{1}{2}f_2/f_1)^2. \tag{33}$$

For the recommended corner frequency $f_2 = f_1/4$, Eq. (33) predicts a minimum resonance peak of 2.3 dB. The resonance peak with dead-time is higher than that without dead-time by about 1 dB.

Equation (33) was derived for the case for which the corresponding frequency response with a one-pole integrator $(f_2 \rightarrow 0)$ has just no resonance peak. In practice, however, if the bandwidth of the SQUID system is maximized, the one-pole closed-loop frequency response might already exhibit a resonance peak. In this case, Eq. (33) is expected to give a reasonable prediction for the ratio of the resonance peaks $|A(f_r)|$ obtained with a two-pole and a one-pole integrator. A recently developed 11-MHz bandwidth SQUID magnetometer [31] had small resonance peaks of 0.5 dB and 1.65 dB at $f \simeq 3$ MHz $\simeq \tfrac{2}{3}f_1$ in the one-pole and two-pole mode, respectively ($f_2 = 480$ kHz). The difference of 1.15 dB between the two peaks is in good agreement with the prediction of 0.95 dB calculated with Eq. (33). An earlier 5-MHz bandwidth magnetometer [30] had relatively strong resonance peaks of 1.9 dB and 4.2 dB at $f \simeq 2$ MHz $\simeq f_1$ for two corner frequencies f_2 of 120 kHz and 480 kHz, respectively. In spite of the stronger peaking, the experimental difference of 2.3 dB between the peaks is still in reasonable agreement with the theoretical prediction of 1.7 dB obtained from Eq. (33).

In summary, the two-pole integrator improves the slew rate and reduces the harmonic distortion at low frequencies, but it deteriorates the high-frequency performance of the SQUID system somewhat. In most cases the deterioration of the high-frequency performance will be tolerable so that the two-pole integrator is a good choice in applications where large low-frequency signals have to be tracked with low distortions, e.g., the power-line interference in unshielded magnetometer systems. Recently, a so-called $PI^{3/2}$ controller (i.e., a 1½-pole integrator) was developed by Seppä and Sipola [35], which provides a phase of $-135°$ and a gain slope of -30 dB/decade over five frequency decades. Compared to the two-pole integrator, the $PI^{3/2}$ controller has an improved phase behavior, but a reduced low-frequency gain. It is rarely used, perhaps because the realization of -30 dB/decade requires a relatively complicated R-C network, and the dimensioning is not as simple as for a one-pole or a two-pole integrator.

2.6. SQUID BIAS MODES

In the preceding sections we assumed the SQUID to be a nonlinear flux-to-voltage converter without considering its biasing. In this section, the two basic bias modes of a SQUID, current bias and voltage bias, will be discussed. In the commonly-used current bias mode [Fig. 6(a)], a constant current I_b is passed through the SQUID, and the voltage V across it is sensed, i.e., the SQUID acts as a flux-to-voltage converter. In contrast, in the voltage bias mode [Fig. 6(b)], the voltage across the SQUID is kept

80

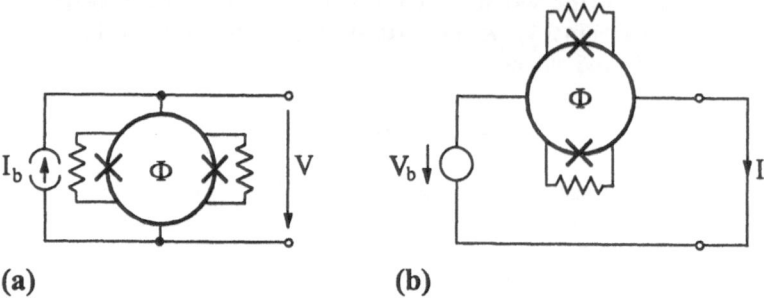

Figure 6. SQUID bias modes: (a) current bias and (b) voltage bias.

constant and the current I through the SQUID is detected. Here, the SQUID acts as a flux-to-current converter. A dc bias voltage source V_b is required to maintain the working point W. In practice, the bias voltage may be generated by passing a current V_b/R_V through a bias resistor with a very low resistance $R_V \ll R_{dyn}$, where $R_{dyn} = (\partial V/\partial I)_w$ is the dynamic resistance of the SQUID at the working point. Bias resistances R_V between 0.025 and 0.05 Ω have been used for SQUIDs with shunt resistances R of around 1 Ω [36,37].

A current-biased SQUID should be connected to a circuit with a high input impedance, e.g., a voltage amplifier at room temperature. In contrast, the stage after a voltage-biased SQUID should have a very low input impedance. It can be a virtual ground of an operational amplifier [38], the primary coil of an impedance-matching transformer [6], or a superconducting input coil of another SQUID [36,37]. Therefore, the voltage bias is particularly attractive for two-stage configurations where a sensor SQUID is inductively coupled to a preamplifier SQUID (Sec. 3.1). An advantage of voltage bias is that the corresponding transfer function, the I-Φ characteristic, appears to have a somewhat larger linear flux range than that for current bias, the V-Φ characteristic [39].

For ideal current (voltage) bias the effective load impedance "seen" by the SQUID including the impedance of the bias source should be much larger (smaller) than the dynamic resistance of the SQUID, respectively. In practice, this condition may not be well satisfied (particularly at high frequencies), so that one has a combination of both bias modes. In this case the small-signal behavior of the SQUID may be described by the complete derivative

$$dV = (\partial V/\partial \Phi)_w d\Phi + (\partial V/\partial I)_w dI = V_\Phi d\Phi + R_{dyn} dI, \qquad (34)$$

where dV, dΦ, and dI are infinitesimal deviations of the SQUID voltage, flux and current from the working point values, respectively. Using Eq. (34), the open-loop gain G(f) of the feedback loop can be calculated for any bias condition. Equation (34) also includes ideal current and voltage bias: it gives $dV/d\Phi = V_\Phi$ for dI = 0 (current bias), and $dI/d\Phi = -V_\Phi/R_{dyn}$ for dV = 0 (voltage bias). We see that the flux-to-voltage and flux-to-current transfer coefficients have different signs since R_{dyn} is always positive.

According to Eq. (34), the small-signal behavior of a SQUID around its working point may be characterized by the transfer coefficient V_Φ and the dynamic resistance R_{dyn}. As discussed in Sec. 2.2, V_Φ determines also the influence of preamplifier voltage noise on the total flux noise, Eq. (7). To describe the effects of preamplifier current noise it is often convenient to define a current sensitivity [40]

$$M_{dyn} = -(\partial\Phi/\partial I)_w = R_{dyn}/V_\Phi . \tag{35}$$

The current sensitivity M_{dyn}, which is the *inverted* current-to-flux transfer coefficient at constant SQUID voltage, has the same sign as V_Φ. Note that I defined it in [29,30,40,41] as the *noninverted* (bias) current-to-flux transfer coefficient, $+\partial\Phi/\partial I_{(b)} = R_{dyn}/V_\Phi$, inconsistently with the signs in the definitions of R_{dyn} and V_Φ. Therefore, in those Refs. $+\partial\Phi/\partial I_{(b)}$ should be read as $-\partial\Phi/\partial I_{(b)}$.

Equation (34) can be rewritten using Eq. (35):

$$dV = V_\Phi (d\Phi + M_{dyn} dI) . \tag{36}$$

Equation (36) describes an equivalent SQUID with zero dynamic resistance, where any current change dI is converted into a flux change $M_{dyn}dI$ via the current sensitivity and produces a corresponding voltage change via the transfer coefficient V_Φ. In this equivalent SQUID, M_{dyn} is the effective mutual inductance between the bias current paths and the SQUID loop. Consequently, small-signal current effects in a SQUID may be described using either the dynamic resistance R_{dyn} (which includes the effects of stray flux generated by the bias current) or the current sensitivity (apparent mutual inductance) M_{dyn}.

The current sensitivity can be measured directly in a flux-locked loop using current bias [40]. For this, a small current dI is superimposed on the bias current I_b. The corresponding voltage change $dV = R_{dyn}dI$ is compensated by the feedback loop via a flux change $d\Phi = -dV/V_\Phi$ from which the current sensitivity $M_{dyn} = -d\Phi/dI$ can be determined. To a rough approximation M_{dyn} scales with the SQUID inductance L. It is relatively weakly dependent on the working point [42]. For low-T_c SQUIDs, the reduced current sensitivity M_{dyn}/L was found to be typically between 1 and 2 for values of the SQUID inductance between 400 pH and 7 pH [14,40,42,43]. For high-T_c SQUIDs with L between $\simeq 30$ pH and 145 pH, a typical current sensitivity $M_{dyn} \simeq 2L$ was measured recently [44,45]. High-T_c SQUIDs with strong V_Φ degradation are expected to have a higher value of M_{dyn}/L because, in the limit of strong noise-rounding, R_{dyn} approaches R/2, but V_Φ falls to zero, and thus $M_{dyn} = R_{dyn}/V_\Phi$ rises.

Figure 7 depicts the small-signal equivalent circuits of a current-biased and voltage-biased SQUID directly coupled to a room-temperature preamplifier. The SQUID is represented by a voltage source $V_\Phi(d\Phi + \Phi_n)$ in series with a resistance R_{dyn}, where Φ_n is the intrinsic flux noise of the SQUID. Noise and impedance of the bias current (voltage) source are taken into account by a noise source $I_{n,b}$ ($V_{n,b}$) and a resistance R_I (R_V), respectively. The room-temperature preamplifiers involve ideal operational amplifiers with infinite voltage gain and input impedance. The amplifier noise is approximated by uncorrelated voltage and current noise sources $V_{n,amp}$ and $I_{n,amp}$ at

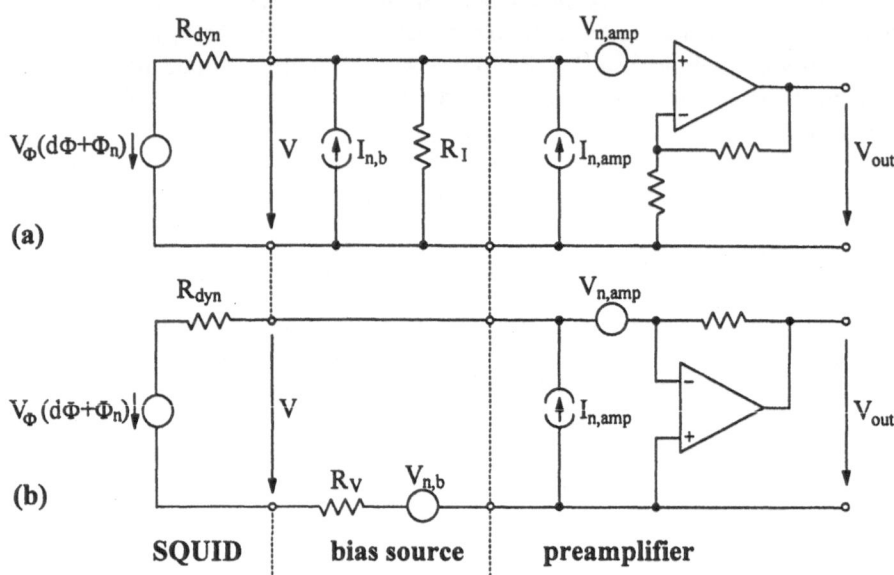

Figure 7. Small-signal equivalent circuits for (a) current bias and (b) voltage bias.

the amplifier input. These noise sources may include noise contributions from other electronic components in the feedback loop referred to the amplifier input.

For ideal current or voltage bias ($R_I \to \infty$ or $R_V \to 0$), the equivalent circuits in Fig. 7 give a total flux-noise spectral density

$$S_{\Phi,t}(f) = S_{\Phi}(f) + S_{V,amp}(f)/V_{\Phi}^2 + S_{I,amp}(f)M_{dyn}^2 . \qquad (37)$$

Here, $S_{V,amp}(f)$ and $S_{I,amp}(f)$ are the effective preamplifier voltage and current noise spectral densities, respectively. We note that for ideal bias sources, the total noise levels with current and voltage bias are identical, and that $S_{I,amp}(f)$ includes the current noise of the bias source for current bias, and $S_{V,amp}(f)$ includes the voltage noise of the bias source for voltage bias. For nonideal bias sources with finite source impedance, the total flux noise increases due to a reduction of the effective SQUID transfer coefficient. Equation (37) holds for finite values of the source resistance if V_{Φ} is replaced by $V_{\Phi}/(1+R_{dyn}/R_I)$ in the current bias case, or if M_{dyn} is replaced by $M_{dyn}(1+R_V/R_{dyn})$ in the voltage bias case.

2.7. FEEDBACK CONFIGURATIONS

Commonly, a SQUID is used as a sensitive null detector for magnetic flux. To make a magnetometer or gradiometer one usually connects one or more pickup coils to an input coil which is magnetically coupled to the SQUID [Fig. 8(a)]. The feedback electronics detects any change of the magnetic flux in the SQUID loop by sensing the voltage across or the current through the SQUID, and compensates it by passing a

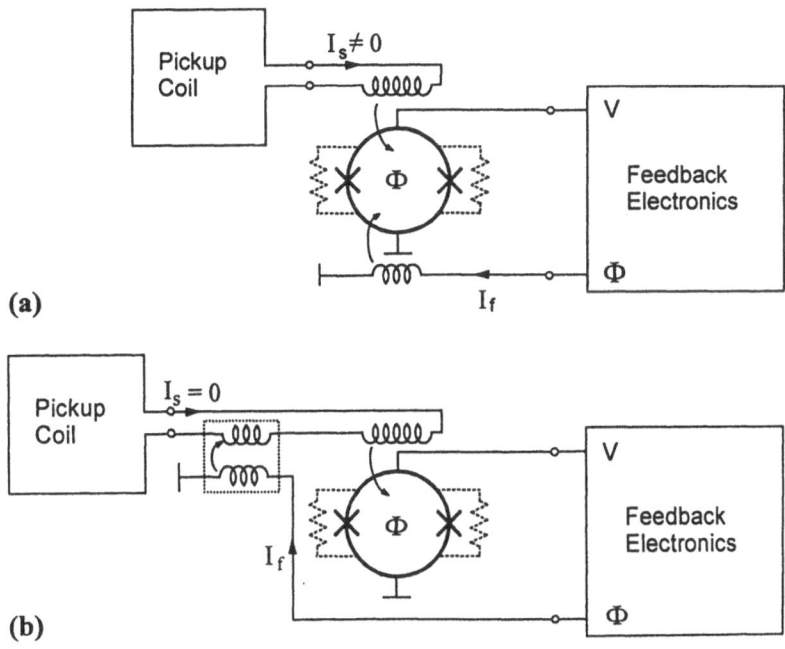

Figure 8. Feedback configurations: flux nulling (a) in the SQUID loop and (b) in the input circuit.

feedback current I_f through the feedback coil, thereby producing the feedback flux. The Josephson junctions in Fig. 8 are drawn with dashed shunt resistors to illustrate that any type of SQUID, hysteretic or nonhysteretic, may be used.

Figure 8(b) depicts an alternate feedback configuration involving flux nulling in the input circuit. It was developed to minimize crosstalk in biomagnetic multichannel systems [46]. In this configuration the secondary coil of a feedback transformer is connected in series to the input coil. The input circuit, consisting of pickup coil, input coil and feedback-transformer secondary coil, forms a superconducting loop. The read-out electronics keeps the total flux in this loop constant by passing a feedback current through the feedback-transformer primary coil. In this way, the signal current I_s in the input circuit is nulled, and the pickup coil "sees" an infinitely large load impedance. Without a circulating current in the input circuit, the signal field is not distorted, and hence, magnetic crosstalk in multichannel systems is eliminated. In the case of digital single-chip SQUIDs (Sec. 6.2), this feedback configuration allows one to increase the dynamic range substantially, at low frequencies to an almost infinitely large value limited by the maximum magnetic field the superconducting components can with-stand without malfunctioning [47]. The feedback transformer may be integrated into the SQUID chip — most preferably using a gradiometric ("double-washer") design both for the feedback transformer and the SQUID itself [48] to simplify magnetic shielding and to minimize stray coupling from the feedback-transformer primary coil directly into the SQUID loop.

84

A SQUID can be used to build a sensitive detector for any physical quantity that can be converted into magnetic flux, e.g., electrical voltage or mechanical displacement. As in the case of a magnetometer, either the flux in the SQUID loop or the flux (current) in the input circuit may be nulled. For SQUID-based instruments the latter method is preferable, because it minimizes feedback from the SQUID circuit into the signal source. For example, a SQUID-based picovoltmeter [5] has a high input impedance if the feedback electronics nulls the current in the input circuit by generating a compensation voltage using a small resistance in series to the input coil. However, the input impedance decreases with the open-loop gain, and consequently, a high system bandwidth is crucial to the achievement of a sufficiently high input impedance over a reasonable frequency range (e.g., dc to 1 kHz).

3. Multiple-SQUID Concepts

3.1. TWO-STAGE CONFIGURATION

A simple method to increase the transfer coefficient of a SQUID is to use another SQUID as a low-noise preamplifier. An example of such a two-stage configuration is shown in Fig. 9 – configurations with more than two stages are also possible, but will not be discussed here. The first SQUID, the sensor SQUID, is inductively coupled to the second one, the preamplifier SQUID. The sensor SQUID determines the lower limit of the overall system noise, so it is inevitably a standard dc SQUID with resistively-shunted junctions operated in the current- or voltage-bias mode. For the preamplifier, any SQUID arrangement may be used. A variety of amplifier SQUIDs is described in the literature: single SQUIDs with hysteretic junctions being operated either digitally [28] or in an analog mode [49]; single SQUIDs with resistively-shunted junctions read out either with flux modulation [50] or directly [51,52]; and series

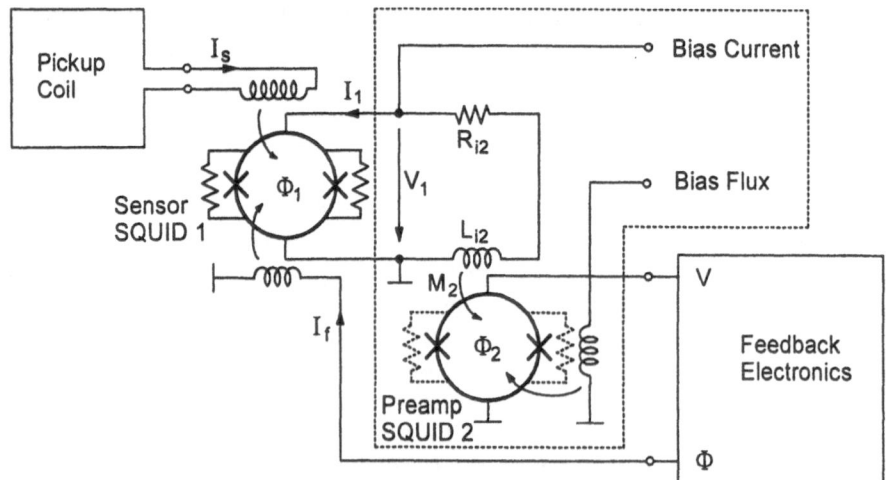

Figure 9. Two-stage configuration. The second stage is surrounded by a dashed frame.

SQUID arrays [36]. To set the working point of the amplifier SQUID, additional wires between the cryogenic region and room temperature are necessary.

An important parameter of a two-stage configuration is the low-frequency small-signal flux gain at the working point:

$$G_\Phi = (\partial \Phi_2 / \partial \Phi_1)_w = \frac{M_2}{R_{i2} + R_{dyn1}} V_{\Phi 1} . \tag{38}$$

Here, Φ_1 and Φ_2 are the flux in the sensor and amplifier SQUID, $V_{\Phi 1}$ and R_{dyn1} are the transfer coefficient and dynamic resistance of the unloaded sensor SQUID, and R_{i2} and M_2 are the input resistance and input coil mutual inductance of the amplifier SQUID, respectively. Equation (38) gives $G_\Phi = V_{\Phi 1} M_2 / R_{i2}$ for a sensor SQUID with current bias ($R_{i2} \gg R_{dyn1}$), and $G_\Phi = M_2 / M_{dyn1}$ with voltage bias ($R_{i2} \ll R_{dyn1}$). The flux gain decreases at high frequencies $f > f_{i2} = (R_{i2}+R_{dyn1})/2\pi L_{i2}$ because the preamplifier input coil L_{i2} and the resistances R_{i2} and R_{dyn1} form a first-order low-pass filter with a 3-dB cutoff frequency f_{i2}. Obviously, the effective flux-to-voltage or flux-to-current transfer coefficient of a two-stage configuration is enhanced by the flux gain compared to that of the preamplifier SQUID alone. The rms flux-noise contributions of preamplifier SQUID and room temperature preamplifier to the total system flux noise are proportional to the reciprocal of the flux gain. For a sufficiently high flux gain they can be made negligibly small compared to the intrinsic flux noise of the sensor SQUID.

The total flux noise at low frequencies $f \ll f_{i2}$ of an analog two-stage SQUID coupled directly to a room temperature preamplifier is given by

$$S_{\Phi,t}(f) = S_{\Phi 1}(f) + 4k_B T R_{i2}/V_{\Phi 1}^2 + \frac{1}{G_\Phi^2}[S_{\Phi 2}(f) + S_{V,amp}(f)/V_{\Phi 2}^2 + S_{I,amp}(f)M_{dyn2}^2] , \tag{39}$$

where the subscripts 1, 2, and "amp" denote the sensor SQUID, the preamplifier SQUID, and the room-temperature preamplifier, respectively. Equation (39) holds for a digital amplifier SQUID (comparator) if the preamplifier voltage-noise spectral density $S_{V,amp}(f)$ is set to zero. The term $4k_B T R_{i2}/V_{\Phi 1}^2$ in Eq. (39) accounts for Nyquist noise of the input resistance R_{i2}. It becomes negligible for $R_{i2} \lesssim R_{dyn1}$, i.e., for a voltage-biased sensor SQUID. The expression in brackets on the right hand side of Eq. (39) is the flux noise that one would obtain if the feedback flux would be coupled directly to the preamplifier SQUID without the sensor SQUID [cf. Eq. (37)]. The resulting rms flux-noise contribution diminishes with increasing flux gain. However, the effective linear flux range also decreases with increasing flux gain, $\delta\Phi \simeq \delta\Phi_2/G_\Phi$, and consequently, one must compromise between low noise and high slew rate.

Another problem arising from a high flux gain is that the feedback loop might lock to multiple working points. Assume that both the sensor and amplifier SQUID are biased at their nominal working points while the feedback loop is opened to reset the system. After closing the flux-locked loop, the system will automatically lock to the nominal working points and remain in this state provided that the sum of applied signal flux and interference signals does not exceed the maximum system slew rate. However, the flux in both SQUIDs during a reset is often undefined, e.g., if the magnetic shielding is imperfect, so that power-line interference and low-frequency field fluctuations produce too much flux. If it happens that the error flux in both

SQUIDs is about $\pm\Phi_0/2$ (i.e., that the transfer coefficients of both SQUIDs are inverted) when the system is reset, the feedback loop will lock to an undesirable state. Both SQUIDs will settle to working points with inverted transfer coefficient – the amplifier SQUID having zero output signal (required for loop stability) and an error flux $\Phi_{e2} \simeq \pm\Phi_0/2$, the sensor SQUID having an error voltage $V_{e1} = \Phi_{e2}R_{i2}/M_2$ (required to maintain the error flux in the amplifier SQUID). The sensor noise level deteriorates for $V_{e1} \neq 0$. It is possible to achieve this undesirable state only if the corresponding error voltage V_{e1} does not exceed the extrema of the V_1-Φ_1 characteristic of the sensor SQUID loaded with the series connection of R_{i2} and L_{i2}. For ideal current or voltage bias ($R_{i2} \gg R_{dyn1}$ or $R_{i2} \ll R_{dyn1}$), this condition may be written as

$$G_\Phi \lesssim \frac{\Phi_0/2}{\Delta\Phi_1 - \delta\Phi_1}, \tag{40}$$

where $\Delta\Phi_1 - \delta\Phi_1$ is the larger linear flux range of the sensor V_1-Φ_1 characteristic for current bias according to Fig. 1 or the similarly-defined linear flux range of the sensor I_1-Φ_1 characteristic for voltage bias. With a sinusoidal sensor transfer characteristic, multiple working points are avoided by lowering the flux gain to $G_\Phi \lesssim \pi$.

For high values of the flux gain, additional undesirable working points for the sensor SQUID with larger error voltages and correspondingly higher noise levels are possible. At these working points the error voltage of the sensor SQUID differs from that of the already-discussed cases by integer multiples of the value $\Phi_0 R_{i2}/M_2$ that produces an amplifier error flux of exactly one flux quantum. All these working points are not obtainable at low flux gains when the condition Eq. (40) is fulfilled. Recently, Foglietti [51] proposed a two-stage configuration where the inductance of the non-hysteretic amplifier SQUID was about one order of magnitude smaller than that of the sensor SQUID. In this way, $S_{\Phi2}$, $V_{\Phi2}$ and M_{dyn2} are improved considerably, so that a flux gain of about two may already be sufficient to make the noise contributions of the amplifier SQUID and room-temperature preamplifier negligible compared to the intrinsic sensor noise level. However, the reduction of the SQUID inductance (geometry) may be limited by the available lithography. The use of a series SQUID array (Sec. 3.2) instead of a single amplifier SQUID is another approach to lower the required flux gain. If a high flux gain cannot be avoided, the feedback electronics should automatically bias the SQUIDs close to their nominal working points during reset to avoid potential degradation of the system noise and dynamic performance.

In the two-stage configuration depicted in Fig. 9, the preamplifier SQUID is operated in a small-signal mode and affects, therefore, the effective linear flux range of the system. To avoid a reduction of the linear flux range at low frequencies, the transfer characteristic of the amplifier SQUID may be linearized by an additional feedback loop that nulls the error flux in the amplifier SQUID and provides an output voltage proportional to the sensor signal. Commercial rf SQUIDs [53], as well as dc SQUIDs with flux modulation read-out [54], were used in the past as amplifiers for low-noise dc SQUID sensors operated in a small-signal mode. Foglietti et al. [55] demonstrated an unmodulated flux-locked loop circuit where an "inner" feedback loop linearized the transfer characteristic of a nonhysteretic amplifier SQUID and an "outer" feedback loop was used to provide a linear relation between the flux in the sensor SQUID and the output signal of the system. This circuit tolerates a two-times higher flux gain than that given by Eq. (40) without locking to multiple working points because a system state with inverted sensor and amplifier transfer coefficients is not possible. However, the maximum system bandwidth is lower than that of the configuration in Fig. 9 because the system bandwidth (i.e., the bandwidth of the "outer" feedback loop) must be kept well below the bandwidth of the "inner" amplifier feedback loop, which should not exceed the limits discussed in Sec. 2.4.

3.2. SERIES SQUID ARRAYS

The relatively small voltage signal of a single SQUID can be enlarged considerably by connecting many identical SQUIDs in series [56,57]. Provided that all SQUIDs of such a series array are biased at the same working point and that the same flux Φ is coupled into each of the SQUIDs, the array behaves like a single SQUID with an improved dynamic range. For a series of N SQUIDs we find

$$V_\Phi = N V_{\Phi,i}, \quad \delta V = N \delta V_i, \quad \delta \Phi = \delta \Phi_i, \quad M_{dyn} = M_{dyn,i},$$

$$S_V(f) = N S_{V,i}(f), \quad S_\Phi(f) = S_{\Phi,i}(f)/N, \quad \epsilon(f) = N S_\Phi(f)/2L = \epsilon_i(f), \tag{41}$$

where the subscript i denotes a single SQUID of the array. Both the transfer coefficient V_Φ and the usable voltage swing δV scale with N, and consequently, the linear flux range $\delta \Phi$ is not decreased. The flux-noise spectral density $S_\Phi(f)$ scales with 1/N, but the effective energy sensitivity $\epsilon(f)$ is not improved [57]. The voltage-noise spectral density $S_V(f)$ across the array increases with N. Typically, the voltage-noise spectral density of a single SQUID is two orders of magnitude lower than that of the room-temperature preamplifier. Therefore, a series array of at least 100 SQUIDs is required to enable a direct read-out without significant deterioration of the total noise level.

Recently, Welty and Martinis [56] developed a 100-SQUID series array. A very large voltage swing $2 \delta V \simeq 2.5$ mV and a high array bandwidth >175 MHz were measured, making this device particularly attractive for wideband SQUID systems. Stawiasz and Ketchen [57] performed noise measurements on 100-SQUID series arrays with 0.5 μm^2 junctions, $2 \delta V \simeq 2.5$ mV, and $V_\Phi \simeq 10$ mV/Φ_0. They connected the arrays directly to a 0.8 nV/$\sqrt{Hz}$ room-temperature preamplifier without flux-locked loop and achieved a coupled white energy sensitivity $\epsilon_c = \epsilon/k^2 \simeq 56$ h. The inductance L of a single SQUID was 100 pH, and the coupling coefficient between the input coils and the SQUIDs was k $\simeq 0.9$. The total input-coil inductance $L_i \simeq 10$ nH would be optimum for a small integrated pick-up coil, e.g., with a total area of 5×5 mm^2 and a linewidth of 400 μm. An undesirable feature of all these large series SQUID arrays is that they exhibit amplitude modulation of their V-Φ characteristics for large values of the flux bias. This is attributed to small variations of the SQUID geometries as a result of fabrication and lithography inaccuracies. Furthermore, larger amounts of trapped flux may cause severe distortions of the V-Φ characteristics [56].

Series SQUID arrays are highly suited as the amplifier stage in a two-stage configuration. Using Eqs. (39) and (41) we see that the rms flux-noise contribution originating from voltage noise of the room-temperature preamplifier scales with 1/N.

Therefore, $N \simeq 10$ and a low flux gain $G_\Phi \lesssim 2$ to avoid multiple working points may be sufficient for a direct read-out. The smaller number of SQUIDs in a two-stage configuration lowers the demands on the fabrication process. Recently, Welty and Martinis [36] developed a two-stage device using a 10-pH voltage-biased sensor SQUID coupled to a preamplifier involving a 100-SQUID series array – a similar device is available commercially [58]. They achieved an output voltage swing $2\delta V_2 \simeq 3.5$ mV, an effective transfer coefficient $(\partial V_2 / \partial \Phi_1)_w \simeq 21$ mV/Φ_0, and a total white energy sensitivity $\varepsilon_{c,t} \simeq 310$ h measured without a flux-locked loop. The bandwidth of this two-stage device, however, was limited to 390 kHz due to the high input inductance of the amplifier SQUID array. The frequency-dependent characteristics (gain, phase and harmonic distortion) of a two-stage device involving a series SQUID array as an amplifier stage were recently studied by numerical simulation showing that a bandwidth in excess of 100 MHz should be attainable [59].

4. SQUIDs with Additional Positive Feedback

4.1. THEORY OF ADDITIONAL POSITIVE FEEDBACK

Additional positive feedback (APF) has been developed to simplify the read-out electronics of biomagnetic multichannel systems [60]. The APF circuit consists of a resistor R_a and a coil L_a in series, connected in parallel with the SQUID (Fig. 10). A small increase of the magnetic flux Φ in the SQUID produces a small change in the voltage V across it. The resulting current change in the APF circuit induces an additional flux in the SQUID via the mutual inductance M_a between the APF coil and the SQUID, thereby increasing the flux-to-voltage transfer coefficient V_Φ and reducing the flux noise contribution $S_{V,amp}(f)/V_\Phi^2$ arising from voltage noise of the room-temperature preamplifier. APF may be used with current or voltage bias; in both cases the increased V_Φ leads to the same improvement in the total flux noise as discussed in Sec. 2.6.

The V–Φ characteristic becomes strongly asymmetric with APF. It becomes steeper at the side with the selected working point and (undesirably) the linear flux range is reduced, while it becomes flatter at the opposite side where the additional feedback becomes negative. The I–Φ characteristic is not affected by APF, i.e., the transfer coefficient and linear flux range remain unchanged in the voltage bias mode. There-

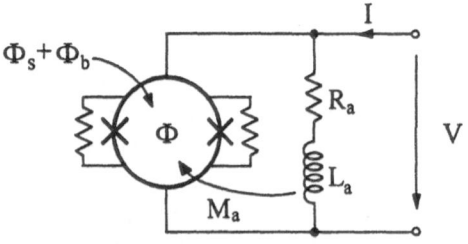

Figure 10. Basic APF circuit.

fore, the maximum slew rate with voltage bias is higher than with current bias (for a given system bandwidth approximately by the factor that APF enhances the steepness of the V–Φ characteristic). However, with voltage bias the inverting input of the preamplifier is connected directly to the SQUID in order to control the SQUID voltage [38], and a broadband noise current flows through the SQUID arising from pre-amplifier voltage and current noise (cf. Fig. 7). The rms noise current increases with the square root of the preamplifier bandwidth, and hence, it may be necessary to reduce the system bandwidth below that achieved with current bias in order to avoid a degradation of the transfer characteristic of the SQUID. Therefore, current bias seems to be preferable to voltage bias in applications where the highest possible bandwidth is required. On the other hand, voltage bias appears advantageous to achieve the highest possible slew rate at low frequencies.

Due to the current V_b/R_a flowing through the APF circuit, the bias current and the bias flux must be readjusted to maintain the original working point W of the SQUID:

$$I_b = I_{b,i} + V_b/R_a , \quad V_b = V_{b,i} , \quad \Phi_b = \Phi_{b,i} - V_b M_a/R_a . \qquad (42)$$

Throughout this section the subscript i denotes parameters of the SQUID without APF (i.e., with $R_a \to \infty$). At the working point, one obtains for the frequency-dependent transfer coefficient $V_\Phi(f)$ and dynamic resistance $R_{dyn}(f)$ with APF the relation

$$\frac{V_\Phi(f)}{V_{\Phi,i}} = \frac{R_{dyn}(f)}{R_{dyn,i}} = \frac{1 + jf/f_a}{1 - G_a + jf/f_a} , \qquad (43)$$

where

$$G_a = \frac{M_a - M_{dyn,i}}{R_a} V_{\Phi,i} \lesssim 1 , \quad f_a = R_a/2\pi L_a \qquad (44)$$

are the effective low-frequency APF gain and APF bandwidth, respectively. The low-frequency transfer coefficient V_Φ and dynamic resistance R_{dyn} are enhanced by a factor of $1/(1-G_a)$. We note that the bandwidth of the enlarged transfer coefficient is not equal to the APF bandwidth f_a, but equal to $(1-G_a)f_a$. Therefore, the system bandwidth with current bias will degrade if the APF gain is chosen very close to the critical value 1 in order to maximize V_Φ. This means that G_a should not be made unnecessarily high, but just high enough to make the effect of preamplifier voltage noise negligible compared to the intrinsic SQUID noise. In most cases $G_a \simeq 0.95$, i.e., a 20-fold enhancement of V_Φ, will be sufficient. At high frequencies $f > (1-G_a)f_a$, both $V_\Phi(f)$ and $R_{dyn}(f)$ decrease; they become equal to the values without APF, $V_{\Phi,i}$ and $R_{dyn,i}$, for frequencies above the APF bandwidth f_a. The current sensitivity $M_{dyn} = R_{dyn}/V_\Phi = M_{dyn,i}$ is not affected by APF over the entire frequency range because the frequency responses of $V_\Phi(f)$ and $R_{dyn}(f)$ are identical.

A SQUID with APF may be operated in a direct-coupled flux-locked loop without flux modulation using current or voltage bias. With current bias the APF gain G_a should not exceed the critical value 1 in order to avoid hysteresis in the V–Φ characteristic and, thereby, additional switching noise or system instability [38]. If G_a is slightly above unity at the optimum working point, it may be adjusted to $G_a < 1$ by selecting a slightly non-optimum working point with a smaller intrinsic transfer coefficient $V_{\Phi,i}$ [61]. Thus, the APF gain may be fine-tuned via the bias current I_b. With voltage bias, $G_a > 1$ is tolerable if the SQUID voltage is kept constant up to the highest frequency at which the frequency dependent APF gain is larger than unity. In practice, however, it is difficult to maintain voltage bias at high frequencies. Consequently, either the APF gain should not exceed unity in the entire frequency range as with current bias, or the APF bandwidth should be reduced to lower the APF gain at high frequencies [62].

The total flux-noise density with APF, both for current and voltage bias, is given by

$$S_{\Phi,t}(f) = S_\Phi(f) + S_{V,amp}(f)/|V_\Phi(f)|^2 + S_{I,amp}(f)M_{dyn}^2 . \tag{45}$$

In Eq. (45), the effective preamplifier voltage- and current-noise spectral densities $S_{V,amp}(f)$ and $S_{I,amp}(f)$ may include noise contributions from the bias voltage and current sources. The flux noise $S_\Phi(f)$ of a SQUID with APF is higher than the intrinsic flux noise $S_{\Phi,i}(f)$ without APF due to Nyquist noise in the APF resistor:

$$S_\Phi(f) = S_{\Phi,i}(f) + S_{V,a}(f)/V_{\Phi,i}^2, \quad S_{V,a}(f) = 4k_BTR_a\frac{G_a^2}{1+(f/f_a)^2} , \tag{46}$$

where $S_{V,a}(f)$ is the low-pass filtered Nyquist voltage noise of the APF resistor. The intrinsic transfer coefficient $V_{\Phi,i}$ should be maximized and R_a should be chosen as low as possible to minimize the flux-noise contribution $S_{V,a}(f)/V_{\Phi,i}^2$ arising from Nyquist noise in R_a. However, the APF resistor loads the SQUID, and hence, it reduces the usable voltage swing. To a rough approximation, the dynamic resistance of the SQUID without APF is $R/2$ at the extrema of the V-Φ characteristic, leading with APF to a voltage swing $\delta V \simeq \delta V_i/(1+R/2R_a)$. Therefore, one has to compromise between low APF resistor noise and large voltage swing δV. From Eqs. (45) and (46), the contribution of APF resistor noise to the energy sensitivity is found to be $4.5k_BTL/R$, assuming a low-T_c SQUID with $V_{\Phi,i} = R/L$ and realistic parameters $G_a = 0.95$ and $R_a = 2.5 R$. This is just one half of the intrinsic energy sensitivity of the SQUID, $\varepsilon_i \simeq 9k_BTL/R$, independent of the SQUID parameters or the temperature. For high-T_c SQUIDs with strong degradation of $V_{\Phi,i}$, the effect of APF resistor noise increases, and a proper choice of R_a is crucial to achieve a low overall noise level.

A small-signal equivalent circuit of a SQUID with APF is depicted in Fig. 11 illustrating Eqs. (43) - (46). According to Fig. 11, APF represents basically a low-noise small-signal amplification of the SQUID voltage. The gain of the equivalent amplifier is $1/(1-G_a)$, and its voltage noise is just the low-pass-filtered Nyquist noise of the APF resistor $S_{V,a}(f)$ given by Eq. (46). For example, using a 10-Ω APF resistor and an APF gain $G_a = 0.95$, the corresponding amplifier voltage noise is 0.046 nV/$\sqrt{Hz}$ at T = 4.2 K and 0.2 nV/$\sqrt{Hz}$ at 77 K. These noise figures are superior to room-temperature amplifiers, particularly if one considers that there is no $1/f$ noise component. However, it should be noted that recently a liquid-nitrogen-cooled amplifier involving bipolar transistors MAT-02 (Analog Devices) was reported that has an ultra-low white noise level of 0.2 nV/$\sqrt{Hz}$ and a $1/f$ corner frequency of about 8 Hz [63].

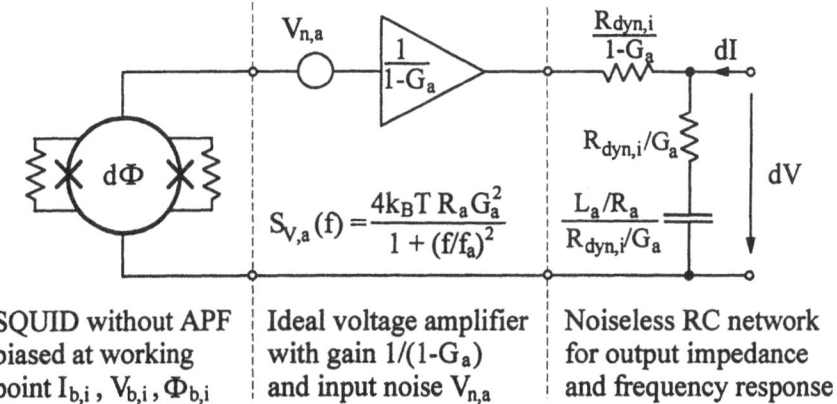

| SQUID without APF biased at working point $I_{b,i}$, $V_{b,i}$, $\Phi_{b,i}$ | Ideal voltage amplifier with gain $1/(1-G_a)$ and input noise $V_{n,a}$ | Noiseless RC network for output impedance and frequency response |

Figure 11. Small-signal equivalent circuit of a SQUID with APF.

The APF circuit may be realized using a wire-wound coil and a metal-film resistor, but for multichannel applications it is more desirable to integrate it into the SQUID chip. With fixed values of R_a and M_a the tolerances for the critical current of a current-biased APF SQUID were experimentally found to be about $\pm 10\,\%$ [61]. To enlarge the tolerances, the APF resistance may be made adjustable by integrating a resistor network and selecting the optimum value of R_a with a few Al bond wires [4,61] or by laser trimming [64]. The use of a cooled field-effect transistor (FET) as a tunable APF resistor [38] is an elegant method to adjust the APF gain manually or even automatically [39], but requires an additional wire between cryogenic and room temperature. The combination of a voltage-biased APF SQUID with a cooled FET as a tunable APF resistor is called adaptive noise cancellation [39]. The possibility to adjust the APF gain is particularly important in applications where the SQUID is optionally connected to different pickup coils via a tightly-coupled input coil. In this case the circulating current in the input circuit, which depends on the pickup coil inductance, screens partially the SQUID loop and hence the APF mutual inductance M_a, and the APF gain G_a will vary if the pickup coil inductance is not kept constant.

APF has been successfully used by several groups to read out low-noise low-T_c SQUIDs with integrated input coil [4,39,64,65] and integrated SQUID magnetometers [60,61,66,67]. Large transfer coefficients of 5 mV/Φ_0 together with total white energy sensitivities ε_t or $\varepsilon_{c,t}$ of 36 h were measured [29,66]. Flux-density noise levels $\sqrt{S_{B,t}}$ down to 1 fT/$\sqrt{Hz}$ at 1 kHz and 2 fT/$\sqrt{Hz}$ at 1 Hz have been achieved using 7.2×7.2 mm^2 multiloop magnetometers [68]. Wideband APF SQUID systems with a closed-loop bandwidth of up to 11 MHz [30,31] have been developed – see Sec. 4.3. Several biomagnetic multichannel systems with up to 256 APF SQUIDs were realized [41,67]; the first one, the PTB 37-channel magnetometer [69], has been operated with high reliability since June 1990. Recently, the APF scheme was successfully applied to high-T_c SQUID magnetometers [44,45], and noise levels $\sqrt{S_{B,t}}$ as low as 9.7 fT/$\sqrt{Hz}$ at 1 kHz and 44 fT/$\sqrt{Hz}$ at 1 Hz were demonstrated – see Sec. 4.4.

4.2. BIAS CURRENT FEEDBACK

To read out a SQUID directly without flux modulation, a bipolar preamplifier is recommended due to its lower $1/f$ voltage noise compared to FET amplifiers. An operational amplifier often used for direct-coupled read-out electronics, the LT1028 (Linear Technology), has a very low white voltage noise level of 1 $nV/\sqrt{Hz}$ with a $1/f$ corner frequency around 5 Hz, but a relatively high white bias-current noise level of about 3 $pA/\sqrt{Hz}$ with a $1/f$ corner frequeny of $\gtrsim 100$ Hz. The white preamplifier current noise has usually negligible effect on the total system noise, particularly in the case of high-T_c SQUIDs where a low SQUID inductance results in a low current sensitivity, and the intrinsic flux-noise level is relatively high. In contrast, the $1/f$ current noise of the preamplifier may severely degrade the low-frequency system noise performance as recently found experimentally for 400-pH low-T_c SQUIDs with APF [40]. Generally speaking, the preamplifier low-frequency current noise must be taken into consideration if SQUIDs with high inductance and/or low intrinsic noise are read out directly without flux modulation.

In the APF scheme, the SQUID *voltage* is converted into a current via the APF circuit and directly fed back as a flux in order to enhance V_Φ and to correspondingly reduce the effect of preamplifier *voltage* noise. The influence of preamplifier *current* noise, however, is not affected because the current sensitivity M_{dyn} remains constant. To reduce it, the *current* through the SQUID must be directly fed back as a flux. This technique, called bias current feedback (BCF) [40], is shown in Fig. 12 for a current- and voltage-biased SQUID with APF [3]. For BCF an additional coil, magnetically coupled to the SQUID via a mutual inductance M_b, is inserted between the SQUID and

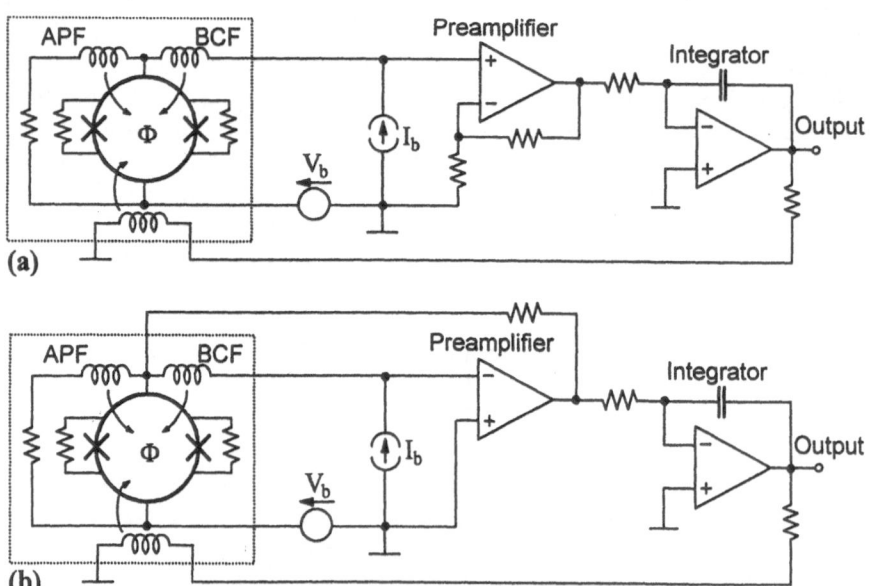

Figure 12. Flux-locked-loop circuit of a SQUID with APF and BCF using (a) current bias and (b) voltage bias.

the room-temperature preamplifier. A small positive fluctuation dI of the preamplifier input current or of the SQUID bias current (e.g., due to noise or temperature drifts) will generate a negative flux change $-M_b dI$ in the SQUID via M_b, as well as a positive voltage change $V_\Phi M_{dyn,i} dI$ via the dynamic resistance of the SQUID, where $M_{dyn,i}$ is the intrinsic current sensitivity without BCF, i.e., for $M_b = 0$. The effects of positive voltage and negative flux cancel (partially), and thus, the current sensitivity of the SQUID with BCF is reduced to

$$M_{dyn} = M_{dyn,i} - M_b . \qquad (47)$$

The current sensitivity and hence the dynamic resistance of the SQUID become zero for $M_b = M_{dyn,i}$. The SQUID resistance even becomes negative for $M_b > M_{dyn,i}$, but without impairing the stability of the flux-locked-loop circuits depicted in Fig. 12. BCF has no influence on the V-Φ characteristic and hence on V_Φ, since APF has no influence on the I-Φ characteristic. A filter resistor in parallel with the BCF coil is recommended for large values of M_b to by-pass high-frequency noise currents flowing in the SQUID-to-preamplifier interconnection line which would otherwise noise-round the transfer characteristic of the SQUID. "Inherent" BCF with $M_b = \pm L/2$ occurs if the bias current is asymmetrically fed into one junction instead of symmetrically into the SQUID inductance [67].

BCF was originally developed to reduce the low-frequency noise in a direct-coupled flux-locked loop involving a current-biased low-T_c SQUID with APF. Although not yet proven experimentally, BCF should be suited, as well, for any other type of SQUID using either current bias or voltage bias. With voltage bias, however, an additional wire between the SQUID and the room-temperature electronics is needed, as the circuit in Fig. 12(b) demonstrates. Note that BCF would make the I-Φ characteristic steeper at the working point and reduce its linear flux range if the feedback resistor of the preamplifier would be connected directly to the inverting input in order to save the additional wire between cryogenic and room temperature. Besides improving the low-frequency noise, BCF also might help to suppress the effect of rf interference coupled to the SQUID as a bias current via the lines between the SQUID and the room-temperature electronics [26]. In wideband systems it allows one to lower the SQUID resistance R_{dyn} in order to terminate approximately the SQUID-to-preamplifier interconnection line if R_{dyn} is too high without BCF (e.g., due to strong APF).

This discussion of BCF is closed with a comment on noise impedance matching. As stressed several times, a basic problem in reading out a SQUID is to increase its weak signal (without adding extra noise) to a level where the SQUID noise dominates the preamplifier noise. In the common flux-modulation read-out scheme, this signal enhancement is done by means of a cooled L-C resonant circuit or a cooled transformer. The effect of preamplifier voltage noise decreases with increasing SQUID signal, but that of preamplifier current noise increases because an x-fold signal enhancement is accompanied with an x^2-fold impedance enhancement [10]. The minimum overall noise is obtained when the effective source impedance $x^2 R_{dyn}$ is equal to the optimum source impedance of the preamplifier R_{opt} that minimizes its noise temperature [9]. However, it is important to note that the noise impedance-matching rule is *not* applicable to read-out schemes with a direct coupling between the SQUID and the preamplifier. For example, with APF the SQUID impedance R_{dyn} scales linearly with the transfer coefficient V_Φ, and the flux-noise contribution arising from preamplifier current noise remains constant. The minimum overall noise is obtained for $V_\Phi \to \infty$, i.e., for *infinite* source impedance $R_{dyn} \to \infty$. On the other hand, if the SQUID impedance is adjusted additionally by BCF, minimum noise is found for $V_\Phi \to \infty$ but with *zero* source impedance $R_{dyn} = 0$. This demonstrates clearly that the noise impedance-matching rule generally fails in the case of direct-coupled feedback loops.

4.3. WIDEBAND SYSTEMS

With APF, as with any other direct-coupled read-out concept, it is relatively easy to achieve a high system bandwidth due to the omission of flux modulation. Using the LT1028 for the preamplifier, fast SQUID systems with a bandwidth between several 100 kHz and $\simeq 1$ MHz have been realized [3,39,60,65]. The LT1028 is particularly well suited for low-frequency applications because it has a low white and 1/f noise, an offset voltage drift of a few 100 nV/°C, and an internal input-current cancellation circuitry that reduces the input bias current to below 100 nA and the temperature drift to about 0.1 nA/°C. However, above $\simeq 100$ kHz the voltage-noise spectral density increases with frequency, exhibiting peaks at about 400 kHz and 2.5 MHz [62]. Together with an inadequate phase at high frequencies (-135° at 2.5 MHz), this restricts the use of the LT1028 to systems with a bandwidth $\lesssim 1$ MHz.

Recently, the AD797 (Analog Devices) has become available. This operational amplifier has about the same current- and voltage-noise level and offset voltage drift as the LT1028, but roughly a 10-fold input bias current and a 100-fold input bias current drift due to the lack of input current cancellation circuitry. For use in wideband SQUID systems, the AD797 is preferable to the LT1028 because its phase behavior is superior at high frequencies and its voltage-noise spectral density is flat to beyond 1 MHz. The AD797 has made it possible to extend easily the bandwidth of APF SQUID systems into the MHz range while still maintaining very compact and simple feedback electronics [4,30,31].

A 5-MHz bandwidth low-T_c SQUID system was recently developed and extensively characterized [30]. This system consisted of a current-biased 400-pH multiloop magnetometer with APF and BCF integrated on a 7.2×7.2 mm^2 chip, and feedback electronics involving a AD797 preamplifier and a two-pole integrator. A low white flux noise level $\sqrt{S_{\Phi,t}} = 3.4 \times 10^{-6}$ $\Phi_0/\sqrt{Hz}$ was measured corresponding to a white flux-density noise level $\sqrt{S_{B,t}} = 1.6$ fT/$\sqrt{Hz}$. Below the 1/f corner frequency of $\simeq 7$ Hz, the system noise was dominated by intrinsic 1/f noise of the SQUID. Above 1 MHz the noise level increased slightly, approximately with $S_{\Phi,t}(f) \propto f^{1/3}$. The system slew rate $\dot{\Phi}_{f,max}(f)$ was determined by applying a sinusoidal signal flux. At high frequencies $f \gtrsim f_2 = 480$ kHz, it was roughly frequency independent, $\dot{\Phi}_{f,max}(f) \simeq 8 \times 10^5$ Φ_0/s. For $f < f_2$, it increased with decreasing frequency, having a maximum of 3×10^7 Φ_0/s at 7.6 kHz. Below 7.6 kHz, the maximum feedback flux amplitude was limited by the static feedback range of ± 620 Φ_0, and thus the system slew rate decreased with the frequency.

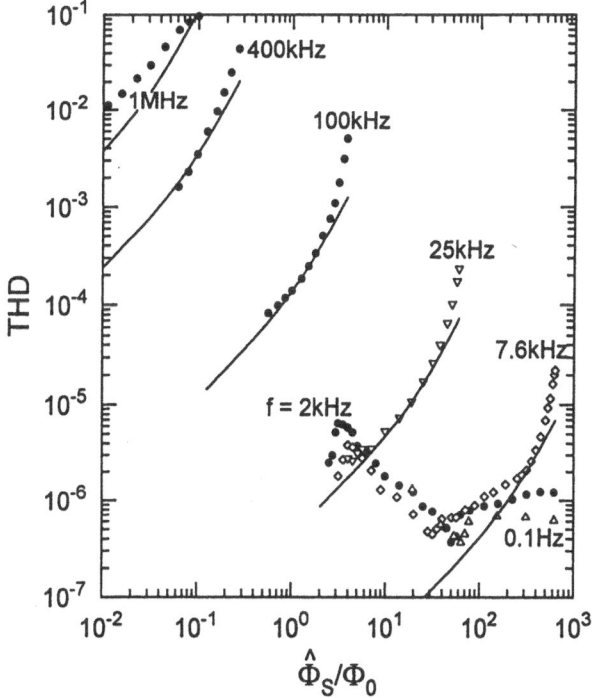

Figure 13. Total harmonic distortion vs. signal flux amplitude of a 5-MHz-bandwidth low-T_c SQUID magnetometer with APF [30]. Symbols are measured, solid lines are calculated from Eqs. (16) - (19) for c' = 2.5 and $V_0 = 1.8$ μV.

The linearity of the 5-MHz system was also investigated in detail. Figure 13 shows the experimental THD as a function of the signal flux amplitude $\hat{\Phi}_s$ for various frequencies between 0.1 Hz and 1 MHz. For each frequency, $\hat{\Phi}_s$ was varied from the maximum value given by the slew rate or by the static feedback range, down to a level where either the THD of the signal source or the system noise limited the measurement. At low frequencies f ≲ 5 kHz, very low THD's between 4×10^{-7} and 10^{-5} were obtained for $\hat{\Phi}_s \gtrsim 2 \, \Phi_0$. Above ≃5 kHz, the THD increased strongly with the frequency and, for each frequency, with the reduced signal flux amplitude X due to the nonlinearity of the V–Φ characteristic and the frequency dependence of the open-loop gain. However, two anomalous dependencies of the THD on $\hat{\Phi}_s$ also are found in Fig. 13: a roughly constant THD of ≲10^{-6} for f ≲ 2 kHz and $\hat{\Phi}_s \gtrsim 60 \, \Phi_0$, which is believed to be caused by nonlinearities of the integrator and self-heating effects in the feedback resistors; and a scaling of the THD with $1/\hat{\Phi}_s$ for $\hat{\Phi}_s \lesssim 60 \, \Phi_0$, which is not yet understood.

The theoretical dependence of the THD calculated from Eqs. (16) - (19) is plotted in Fig. 13 as solid lines for realistic parameters c' = 2.5 and $V_0 = 1.8$ μV = 0.06 δV. Good agreement between calculation and experiment is obtained for reduced signal amplitudes X ≲ 0.5, whereas for X ≳ 0.5 the calculation systematically underestimates the experimental data. Disagreements are expected for large values of X because higher-order terms of the V–Φ characteristic and the influence of the additional input signals (which were introduced to obtain a sinusoidal error flux) have been neglected in the model. Notable disagreements between measurement and calculation occur only in the regions of anomalous THD behavior and at the highest frequency f = 1 MHz for unknown reasons.

The dynamic performance of the described 5-MHz system was subsequently improved by using a SQUID with a higher APF bandwidth, further optimizing the feedback electronics, and shortening the wires between the SQUID and the room-temperature electronics from 1 m to 0.7 m in order to reduce the dead-time [31]. The upgraded system involved a 200-pH low-T_c multiloop magnetometer integrated on a 3.6×3.6 mm^2 chip. Its white noise level, $\sqrt{S_{\Phi,t}} = 3.3 \times 10^{-6} \, \Phi_0/\sqrt{Hz}$ or $\sqrt{S_{B,t}} = 6 \, fT/\sqrt{Hz}$, was relatively high mainly due to nonideal SQUID parameters and a relatively low transfer coefficient of 520 $\mu V/\Phi_0$. The feedback range was $\pm 930 \, \Phi_0$. The dynamic performance both with a one-pole and a two-pole integrator is depicted in Fig. 14: the small-signal frequency response in (a), the large-signal step response to a 500-kHz square-wave signal in (b), the system slew rate vs. frequency in (c), and the

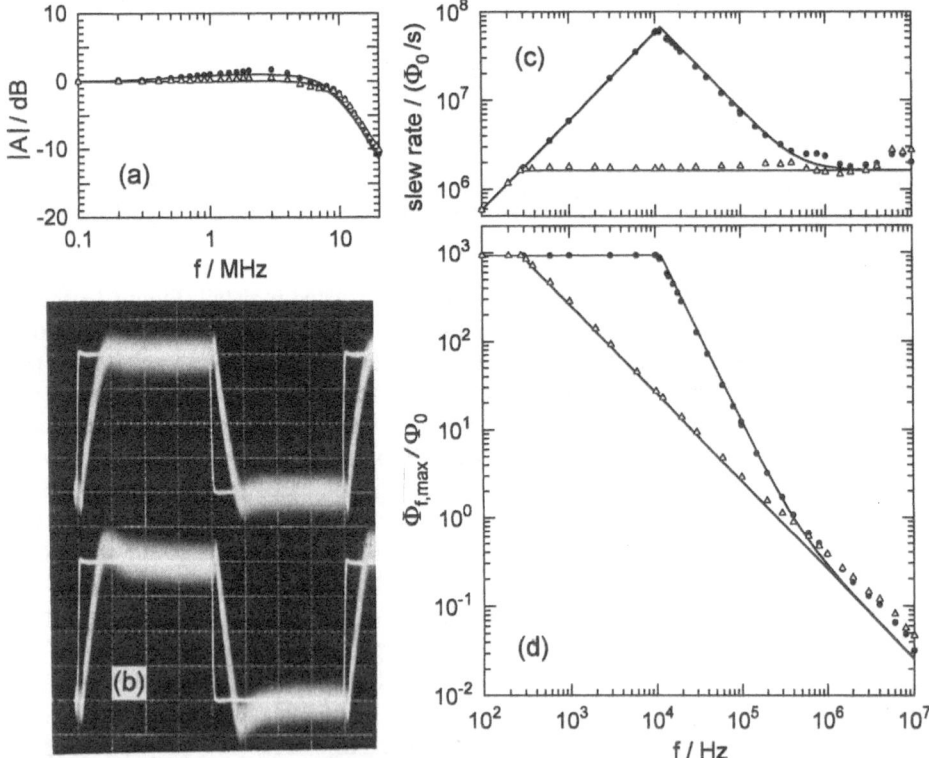

Figure 14. Dynamic behavior of a wideband SQUID system: (a) frequency response, (b) step response, (c) system slew rate vs. frequency, and (d) maximum feedback amplitude vs. frequency. Symbols are measured with one-pole (△) or two-pole (•) integrator; solid lines are calculated from Eqs. (12),(30),(31). In (b), noiseless traces are signal flux, noisy traces are feedback flux with one-pole (top) and two-pole (bottom) integrator; horizontal axes: 250 ns/div; vertical axes: 0.0625 Φ_0/div (courtesy H. Matz [70]).

maximum feedback amplitude vs. frequency in (d). The frequency response exhibits a very high 3-dB bandwidth of 11 MHz, but only small peaks of 0.5 dB and 1.65 dB with a one-pole and a two-pole integrator, respectively. The system slew rate of $\simeq 1.7 \times 10^6$ Φ_0/s for a one-pole integrator is independent of frequency for f > 280 Hz. With a two-pole integrator it is $\simeq 2 \times 10^6$ Φ_0/s at high frequencies f $\gtrsim$ f_2 = 480 kHz, and increases to 6×10^7 Φ_0/s at 10 kHz. The theoretical behavior of the system, calculated with Eqs. (12), (30) and (31) for $\delta\Phi$ = 0.058 Φ_0, f_1 = 4.5 MHz, f_2' = $f_2/340$ and $t_{d,eff}$ = 18 ns, also is shown in Fig. 14 (solid lines). Good agreement between theory and experiment is found except at high frequencies where gain and phase errors in the feedback electronics are expected to become significant.

4.4. BIAS CURRENT REVERSAL

The standard flux modulation read-out involving a static bias current I_b through the SQUID eliminates the influence of in-phase fluctuations of critical current and/or shunt resistance of the two Josephson junctions [27]. Out-of-phase fluctuations appear as corresponding fluctuations of the magnetic flux in the SQUID and may deteriorate the low-frequency noise performance [27,71]. However, flux changes caused by these out-of-phase fluctuations are inverted if the bias current is reversed, and the resulting V-Φ characteristic is shifted along the flux axis (in case of a symmetric SQUID by $\simeq \Phi_0/2$). Therefore, the influence of out-of-phase fluctuations can be eliminated by special bias-reversal schemes where the bias current is periodically switched between $+I_b$ and $-I_b$ [28,71-73].

The standard APF scheme with static bias current is sensitive to both in-phase and out-of-phase fluctuations. Fortunately, state-of-the-art low-T_c junctions exhibit low critical-current fluctuations of some 10^{-6} $I_0/\sqrt{Hz}$ at 1 Hz [72,74]. Furthermore, the influence of critical current fluctuations is often negligible compared to other sources of low-frequency noise such as intrinsic 1/f flux noise of the SQUID or environmental noise. Therefore, in most cases (especially for biomagnetic multichannel applications) low-T_c SQUIDs with APF do not require a bias-reversal technique. Present high-T_c junctions, however, have much higher critical-current fluctuations of the order of 10^{-4} $I_0/\sqrt{Hz}$ at 1 Hz [20]. Consequently, critical-current fluctuations are a dominant source of low-frequency noise in high-T_c dc SQUIDs, and a bias-reversal scheme must inevitably be used. Unfortunately, most of the bias-reversal schemes require symmetric V-Φ characteristics and are therefore not useful for APF SQUIDs. One exception is the second-harmonic detection (SHAD) scheme of Foglietti et al. [72], but this technique increases the white flux noise density S_Φ by a factor of two.

Recently, a simple bias-reversal scheme for dc SQUIDs with APF was developed that does not increase the white noise level [29]. The scheme is based on a similar one developed previously for a digital feedback loop [28]. Its function is illustrated in Fig. 15 with a set of V-Φ characteristics of an integrated low-T_c magnetometer. The V-Φ characteristics are strongly asymmetric due to APF with the optimum working points (open circles) at the steepest part. Figure 15(a) shows the case in which only the bias current is reversed, but neither a bias voltage nor a bias flux is used. In Fig. 15(b) a square-wave bias voltage is applied synchronously with the bias current, in order to

98

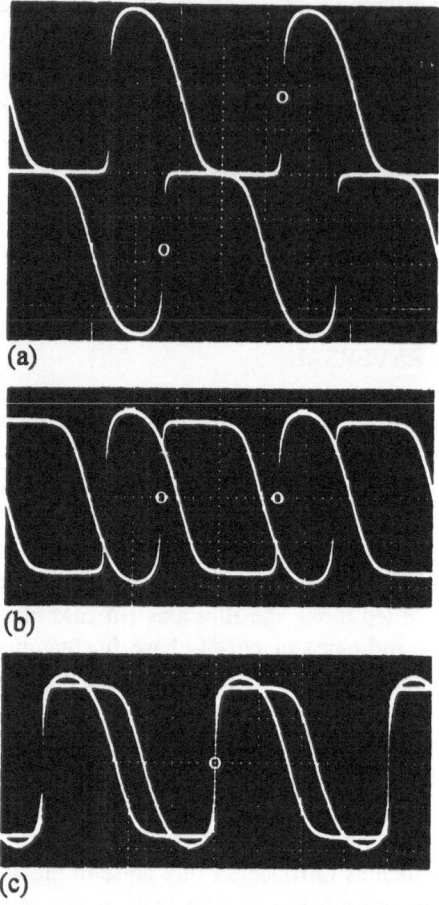

Figure 15. V–Φ characteristics of an integrated low-T_c SQUID magnetometer with APF for $I_b = \pm 5.3\ \mu A$. Working points are marked by open circles. (a) $V_b = \Phi_b = 0$, (b) $V_b = \pm 26\ \mu V$, $\Phi_b = 0$, (c) $V_b = \pm 26\ \mu V$, $\Phi_b = \pm 0.34\ \Phi_0$. Horizontal axes: 0.25 Φ_0/div; vertical axes: 15 μV/div (from [29]).

shift the V–Φ characteristics vertically and to place the working points onto the zero-voltage axis. Finally, in Fig. 15(c) both a square-wave bias voltage and a square-wave bias flux are applied, and now the working points for positive and negative bias current coincide. A small applied flux will give equal voltage changes at both working points, i.e., the behavior of the flux-locked loop with respect to signal flux (noise, dynamic behavior) will be identical to that with static bias. In contrast to that, a small change of the critical currents or the (shunt or APF) resistances will produce equal but opposite voltage changes at the two working points. Therefore, with bias reversal only a small symmetric square-wave signal will appear at the output but with no dc component, and thus the influence of critical current and resistance fluctuations is eliminated at low frequencies. The effect of preamplifier low-frequency noise, however, is not suppressed.

The frequency of the bias reversal should be larger than the 1/f corner frequency that one would obtain with static bias. A typical bias frequency for high-T_c SQUIDs with bicrystal or step-edge junctions is 10 kHz. The minimum bias frequency with APF is higher than that for bias reversal schemes involving flux modulation because the flux modulation removes the (often dominant) noise contribution arising from in-phase critical-current fluctuations. At high bias frequencies, the finite time required to reach the working points must be considered, and it is limited by the time constant L_a/R_a of the APF circuit.

The feedback electronics with APF and bias reversal is identical to that with static bias (Fig. 12), except that the dc bias sources I_b and V_b must be replaced by square-wave sources, and an additional square-wave bias flux is required. Figure 16 depicts an implementation with current bias and a minimum number of wires (two twisted pairs) between cryogenic and room temperature [29]. The SQUID voltage is read out in a two-terminal configuration. The bias current I_b is fed into both SQUID voltage leads in order to eliminate the influence of voltage drops caused by the finite wire resistance. These voltage drops would otherwise lead to a dependence of the SQUID working point on the liquid-helium or liquid-nitrogen level in the dewar via the temperature dependence of the wire resistance. The bias voltage is generated by passing the bias current through a cooled resistor R_V. Alternately, R_V may be placed at room temperature close to the inverting preamplifier input. A special design was developed for the differential preamplifier in which the total voltage noise is determined by only one operational amplifier [75].

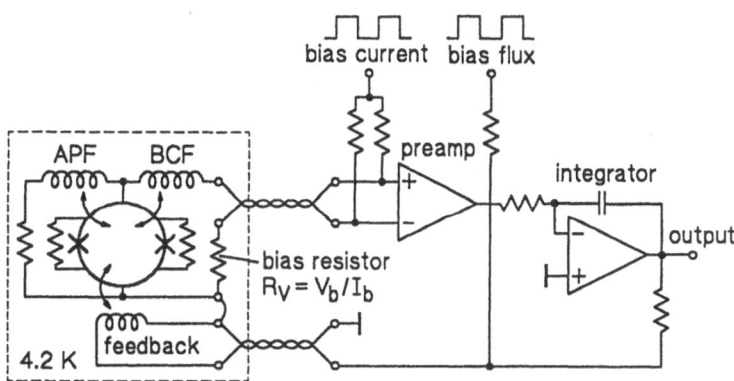

Figure 16. Flux-locked loop circuit of a current-biased SQUID with APF, BCF and bias reversal [29].

The described bias-reversal scheme should be useful to read out directly, without flux modulation, any type of analog dc SQUID both with symmetric and asymmetric transfer characteristic. The scheme was recently verified experimentally using a 200-pH low-T_c multiloop magnetometer with APF and BCF [29]. With static bias current this particular device exhibited a strong Lorentzian excess noise at low frequencies due to poor junction quality. With bias reversal the rms excess noise was reduced by two orders of magnitude, but the white noise level $\sqrt{S_{\Phi,t}} = 1.5 \times 10^{-6} \ \Phi_0/\sqrt{Hz}$ did not deteriorate.

More recently, APF with bias reversal was used to operate low-noise high-T_c magnetometers in a magnetically-shielded room [44,45]. All basic types of multilayer magnetometers were investigated: a flip-chip magnetometer involving a 10×10 mm^2 pickup coil and a small washer-type SQUID on separate chips which were pressed together, two 7-mm-diameter multiloop magnetometers, and one integrated magnetometer with a transformer-coupled 8.3×8.6 mm^2 pickup coil. The devices were fabricated at three institutions using $YBa_2Cu_3O_{7-x}$-$SrTiO_3$-$YBa_2Cu_3O_{7-x}$ multilayer processes, three of them with bicrystal junctions and one of the multiloop magnetometers with step-edge junctions. A wide range of junction parameters was covered in this study: critical currents between $\simeq 5.7$ and 100 µA and normal junction resistances between 1.8 and 13 Ω. Very low flux-density noise levels $\sqrt{S_{B,t}}$ between 9.7 and 31 fT/$\sqrt{Hz}$ were achieved at 1 kHz corresponding to energy sensitivities ε_t between 1600 and 14000 h. At 1 Hz, noise levels between 44 and 125 fT/$\sqrt{Hz}$ were found after subtracting the ambient noise inside the shielded room. Due to the low noise levels, high-quality magnetocardiograms were obtained without signal averaging. The device with the lowest noise, the integrated magnetometer, also was used to record very weak signals from the human brain and even from the peripheral nerve system with a signal-to-noise ratio comparable to that of biomagnetic low-T_c systems currently used [45].

5. Relaxation Oscillation SQUIDs

5.1. SINGLE RELAXATION OSCILLATION SQUIDS

The I–V characteristic of a hysteretic Josephson junction without a shunt resistor in parallel exhibits stable and unstable branches, indicated in Fig. 17 by solid and dotted lines, respectively. If the current I through the junction is initially zero and then increased, the junction will first stay in the zero-voltage state. If the critical current I_0

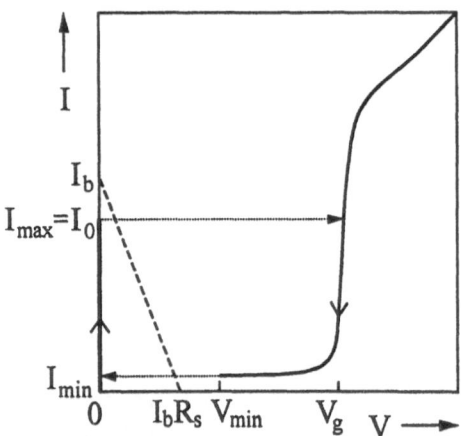

Figure 17. I–V characteristic of a hysteretic Josephson junction (solid line) with load line of a shunt resistor R_s in parallel to the junction (dashed line).

is exceeded, the junction will switch within a few ps into the voltage state along the dotted line in Fig. 17, and the voltage V across the junction will rise from zero to the gap voltage V_g ($\simeq 2.8$ mV for Nb-Al$_2$O$_3$-Nb junctions at 4.2 K) [13]. The junction will remain in the voltage state even for $I < I_0$ and not switch back into the zero-voltage state until the current is lowered below the minimum current I_{min}. Both the critical and the minimum current depend on the gradient of the current ramp [76]. They also are influenced by noise. Typically, the spread ΔI_0 in the critical current, given by the full width at half-maximum value of the switching probability distribution [77], is of the order of 1 μA at 4.2 K. The minimum current I_{min} and voltage V_{min} depend strongly on the damping of the junction, i.e., on the subgap leakage current, and thus, on the junction quality [76].

The I–V characteristic of a hysteretic dc SQUID resembles that of a single junction because a dc SQUID is basically a parallel connection of two junctions via the SQUID inductance L. The major difference is that the critical current of a SQUID, $I_{max}(\Phi)$, depends sensitively on the applied magnetic field, but that of a small (point contact) junction does not. In the voltage state, only the subgap branches differ somewhat, e.g., due to parasitic L-C resonance effects. The dependence of the minimum current I_{min} on the flux Φ is of secondary importance and will not be considered here. Therefore, to understand the function of a circuit based on a hysteretic dc SQUID, it is convenient to consider the SQUID as a single junction with a critical current controlled by the current in the input coil or directly by the applied magnetic field.

A relaxation oscillation SQUID (ROS) consists of a hysteretic dc SQUID that is shunted by a resistor R_s and an inductor L_s in series [77-80]. The shunt circuit is similar to the APF circuit shown in Fig. 10, except that the inductor should not be magnetically coupled to the SQUID. Relaxation oscillations can occur if the ROS is biased with a constant current $I_b > I_{max}(\Phi)$. This is illustrated in Fig. 17 by the dashed load line of the shunt resistor. If the load line does not intersect the stable subgap voltage branch, a stable working point in the voltage state is not possible, and the circuit will oscillate permanently. Thus, relaxation oscillations always will appear for $(I_b - I_{min})R_s < V_{min}$. If this condition is not fulfilled, either the relaxation oscillations stop and the SQUID settles to a stable working point with finite voltage or, if the shunt inductance L_s is within a certain range, the relaxation oscillations persist. The limits have been determined by numerical simulation of a single junction with realistic values of I_0, C and R_s [77].

During the relaxation oscillations, the voltage across the ROS oscillates between $V = 0$ and $V \simeq V_g$. Both the frequency and the duty cycle (and hence the average voltage $\overline{V}$ across the ROS) depend on the bias current I_b and on the critical current $I_{max}(\Phi)$. The latter is a function of the flux Φ in the SQUID and, consequently, a ROS can be used as a flux-to-frequency or flux-to-voltage converter.

Typically, a ROS remains most of the time in the zero-voltage state and produces a sequence of short voltage pulses. After each voltage pulse the current through the SQUID increases from the minimum current I_{min} towards the bias current I_b, with a time constant L_s/R_s, until it reaches the critical current $I_{max}(\Phi)$. For $I_{min} \ll I_b$ the time required is [78]

$$t_0 = -\frac{L_s}{R_s} \ln\left[1 - I_{max}(\Phi)/I_b\right] \simeq 1/f_{ro}. \tag{48}$$

The SQUID switches then into the voltage state for a short time t_V, during which the current through the SQUID decays from $I_{max}(\Phi)$ to I_{min}. The pulse repetition frequency f_{ro} is approximately equal to the reciprocal of t_0 because the pulse duration t_V is typically much smaller than t_0. In the extreme high-frequency case $L_s \rightarrow 0$, a ROS is roughly equivalent to a standard dc SQUID with resistively shunted junctions that produces single-flux-quantum voltage pulses with $\int V(t) dt = \Phi_0$ and a pulse repetition frequency equal to the Josephson frequency $f_j = \nabla/\Phi_0$ [14].

A strongly simplified description of a ROS is that the relaxation oscillations generate an ac bias current through the SQUID that samples the critical current $I_{max}(\Phi)$ periodically with frequency f_{ro}. During each sampling event, $I_{max}(\Phi)$ is sensed with an uncertainty given by thermal noise [77]. If the sampling events are un-correlated, the average uncertainty and thus the resulting flux noise $\sqrt{S_\Phi}$ will scale with $1/\sqrt{f_{ro}}$. A rough estimate for the typical energy sensitivity of a SQUID which is operated based on sampling of the critical current is

$$\varepsilon \simeq 0.23 \, k_B T \left(\frac{1}{f_b} + \frac{20}{f_p}\right), \tag{49}$$

where $f_p = 1/2\pi\sqrt{L_j C}$ is the plasma frequency, i.e., the resonance frequency of the resonant circuit formed by the equivalent junction inductance $L_j = \Phi_0/2\pi I_0$ and the junction capacitance C [13]. Equation (49) is a fit to experimental and theoretical results of ROSs with frequency read-out [81], double relaxation oscillation SQUIDs (Sec. 5.2) [82], and digital SQUIDs (Sec. 6) [83,84]. The bias frequency f_b is the relaxation frequency f_{ro} at the working point or, in digital SQUIDs, one or two times the clock frequency f_c. Equation (49) shows that the relaxation frequency should be chosen as high as possible in order to achieve low noise. However, at very high relaxation frequencies $\gtrsim 1$ GHz, the noise is limited by the plasma frequency [85,86]. This is taken into account in Eq. (49) by the second term $0.23 \, k_B T \times 20/f_p = 16 \, k_B T \sqrt{LC/\beta}$ which, for $\beta = 1$, is equal to the energy sensitivity of a standard dc SQUID with resistively-shunted junctions according to Eq. (3). Furthermore, at high relaxation frequencies, the noise level may degrade due to parasitic resonance effects in the SQUID structure, in particular the resonant circuit formed by the shunt inductance L_s and the SQUID capacitance $2C$. Although the effect of such resonances can be strongly reduced by proper damping [82], it can usually not be completely eliminated and must be considered in Eq. (49) by introducing additional noise terms. For low relaxation frequencies $f_{ro} = f_b \ll f_p/20$ and $T = 4.2$ K, Eq. (49) becomes

$$\varepsilon/h \simeq 20 \, \text{GHz}/f_b. \tag{50}$$

Usually, Eq. (50) is a realistic estimate of ε for $f_b \lesssim 1$ GHz.

We turn now to a brief discussion of practical ROSs. A simple frequency-modulated read-out scheme for ROSs (Fig. 18) was developed by Mück and coworkers [81,87]. The relaxation frequency $f_{ro}(\Phi)$ is first converted into a voltage via a f/V converter and then, after subtracting a voltage corresponding to the bias frequency f_b, integrated and

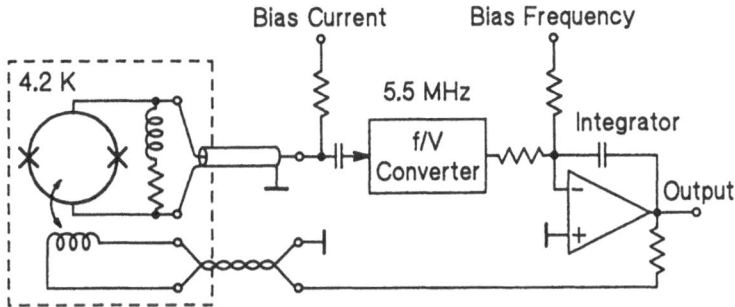

Figure 18. Flux-locked-loop circuit of a ROS using frequency modulation [87].

fed back into the SQUID to make $f_{ro}(\Phi) = f_b$. Consequently, the flux in the SQUID is kept constant and the feedback current is proportional to the applied signal flux. This scheme was successfully tested using an uncoupled SQUID with Nb nanobridges and $L \simeq 80$ pH. A low white flux noise level $\sqrt{S_{\Phi,t}} = 8 \times 10^{-6} \, \Phi_0/\sqrt{Hz}$ was achieved for $f_b = 5.5$ MHz and T = 4.2 K, corresponding to $\varepsilon_t \simeq 2600$ h [81]. Below about 1.5 Hz excess low-frequency noise was observed which scaled as $S_{\Phi,t}(f) \propto 1/f^2$. Due to the high critical current $I_0 \simeq 10$ mA and the resulting extremely high $\beta \simeq 800$, the frequency-modulation depth was only about 1% of the bias frequency. Uehara et al. [88] fabricated ROSs with Nb-Al_2O_3-Nb tunnel junctions and investigated their frequency stability at high relaxation frequencies $f_{ro} \simeq 1$ GHz. A very small relative linewidth of the order of 10^{-5} was measured, from which an intrinsic energy sensitivity $\varepsilon \simeq 4.5$ h $\simeq 2 \times 16 \, k_B T \sqrt{LC}$ was estimated. However, direct flux-noise measurements were not performed to confirm this very low noise level. Due to a near-optimum $\beta \simeq 1.4$, a high frequency-modulation depth of about ± 25 % was found at $f_{ro} \simeq 1$ GHz.

ROSs also have been operated with voltage read-out using conventional electronics with flux modulation [80] or direct-coupled electronics without flux modulation [77]. Large usable voltage swings $2\delta V$ of up to $\simeq 300$ μV and transfer coefficients $V_\Phi = \partial V/\partial \Phi$ as high as 4 mV/Φ_0 were obtained with uncoupled 20-pH ROSs for $f_{ro} \lesssim 100$ MHz [77]. Typically, however, the transfer coefficient was below 1 mV/Φ_0 – less than four times V_Φ of a resistively-shunted dc SQUID with the same inductance. The noise level of these 100-MHz ROSs was considerably higher than that for resistively-shunted dc SQUIDs. It may be lowered by increasing the relaxation frequency, but the transfer coefficient also tends to decrease with increasing f_{ro}, particularly if additional damping is required at high values of f_{ro}. Thus, we conclude that ROSs with voltage read-out are not superior to standard dc SQUIDs with resistively-shunted junctions.

5.2. DOUBLE RELAXATION OSCILLATION SQUIDS

A double relaxation oscillation SQUID (DROS) consists of two hysteretic dc SQUIDs in series shunted by an L_s-R_s circuit. In the original version, named balanced ROS [80,85], the signal flux Φ_s is applied to both SQUIDs. In the simplified version discussed here [77], Φ_s is applied only to one of the SQUIDs, the sensor SQUID – see

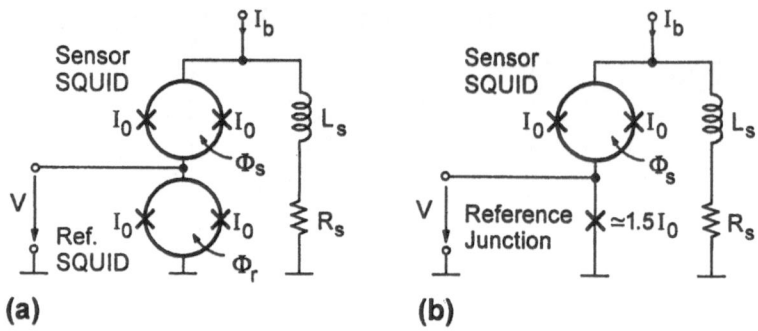

Figure 19. Basic DROS circuits (a) with reference SQUID and (b) with reference junction.

Fig. 19(a). The other one, the reference SQUID, is biased with a constant flux Φ_r to adjust its critical current to a value that lies typically in the middle between the maximum critical current $2I_0$ and the minimum critical current (e.g., $1.05I_0$ for $\beta = 1$). A constant bias current I_b is passed through the DROS to start relaxation oscillations. The output voltage V is tapped across the reference SQUID. During relaxation oscillations, only the SQUID with the smaller critical current will oscillate between the zero-voltage and the voltage state; the other SQUID always will stay in the zero-voltage state. Thus, a DROS behaves like a critical-current comparator. The output voltage of the DROS will be zero if the critical current of the reference SQUID is larger than that of the sensor SQUID; otherwise, the voltage will be nonzero with an average value depending on I_b. The transition between these two states is very sharp, leading to extremely high transfer coefficients V_Φ.

The reference SQUID is only required to provide a reference for the critical current. It should be well shielded against the signal flux to avoid a shift of its working point for large values of Φ_s. The use of a single Josephson junction with an appropriate critical current instead of a reference SQUID [Fig. 19(b)] removes this effect, but the critical current at the working point can then no longer be trimmed to minimize the noise [89]. The simplified DROS with reference junction needs fewer wires between cryogenic and room temperature than the DROS with reference SQUID. This is a useful feature for multichannel applications. However, the minimum of five wires per independent DROS unit is still larger than the minimum of four wires needed to operate SQUIDs with APF or series SQUID arrays (cf. Fig. 16).

Due to their high transfer coefficients, DROSs can be operated in a simple direct-coupled flux-locked loop without flux modulation. Adelerhof et al. [77] fabricated and characterized uncoupled DROSs with L = 20 pH and $f_{ro} \simeq 100$ MHz. Typical V_Φ values of 10-30 mV/Φ_0, maximum values of up to 80 mV/Φ_0, and minimum noise levels $\sqrt{S_{\Phi,t}} = 1{\times}10^{-6}$ $\Phi_0/\sqrt{\text{Hz}}$ and $\varepsilon_t = 160$ h were measured. Subsequently, the relaxation frequency was increased to 0.4-10 GHz requiring an additional damping resistor in parallel to the DROS [82,86]. The intrinsic noise level of the investigated DROSs decreased with increasing f_{ro}, leveling off to $\varepsilon = 13$ h at $f_{ro} \gtrsim 3$ GHz due to the detrimental effect of plasma oscillations [86]. The transfer coefficient tended also to decrease with increasing f_{ro}, from 7 mV/Φ_0 at $f_{ro} \simeq 1$ GHz to 2 mV/Φ_0 at $f_{ro} \simeq$

10 GHz. A minimum noise level $\varepsilon_t = 34$ h at $f_{ro} \simeq 1$-3 GHz was measured in a flux-locked loop without flux modulation.

Recently, two 3-channel DROS systems with wire-wound pickup coils were realized: a second-order gradiometer with a white noise level $\sqrt{S_{B,t}} = 36$ fT/$\sqrt{Hz}$ [89] and a first-order gradiometer with $\sqrt{S_{B,t}} = 4$ fT/$\sqrt{Hz}$ at 1 kHz and 10 fT/$\sqrt{Hz}$ at 1 Hz [90]. In the latter system, there was no interaction between the channels other than a low-frequency crosstalk of 0.5 %, implying that crosstalk at the relaxation frequency does not play a significant role. More recently, integrated planar first-order DROS gradiometers [91] and multiloop DROS magnetometers [92] were fabricated and characterized. The best energy sensitivity ε_t achieved so far with a coupled DROS is 275 h [90]; typical values are around 1000 h. The usable voltage swing $2\delta V$ of coupled DROSs is of the order of 100 μV, and the transfer coefficient is $\gtrsim 1$ mV/Φ_0.

6. Digital SQUIDs

6.1. DIGITAL ROOM-TEMPERATURE FEEDBACK LOOPS

In almost all applications, the output signal of a SQUID system is processed digitally, thus requiring a high-resolution analog-to-digital (A/D) conversion. The demands on the A/D converter are much lower if the amplified error signal of the SQUID (instead of the feedback signal) is digitized, then integrated using a digital signal processor (DSP), and fed back into the SQUID via a high-resolution D/A converter – see following chapter "SQUID gradiometers in real environments" by Jiri Vrba [93,94]. In this way, the demanding A/D conversion is replaced by a (perhaps less demanding) D/A conversion. Furthermore, the dynamic range easily can be increased considerably by utilizing the periodicity of the transfer characteristic of the SQUID. For this, the integrator is reset, and a flux quanta-counter is incremented or decremented each time the feedback flux exceeds $\pm\Phi_0$. This flux-quanta counting principle has been successfully implemented in digital feedback loops, but we note that it does not necessarily require a digital integrator [95].

In the above-mentioned digital feedback loops, the SQUID sensor is still conventionally operated in an analog mode. In the following, digital SQUID concepts will be discussed where the SQUID itself or, in multi-stage configurations, at least one of the following stages is operated in a digital mode.

Digital SQUID concepts are usually based on high-frequency critical-current detection. A single hysteretic Josephson junction can be operated as a latching current comparator. For this, a high-frequency clock current and the signal current are passed through the junction, which switches into the voltage state if the critical current I_0 is exceeded. The junction is reset when the sum of clock and signal current becomes smaller than the minimum current I_{min}. Thus, the junction can be operated so that it produces a sequence of voltage pulses if the signal current is above a certain threshold or otherwise remains in the zero-voltage state. To enhance the current sensitivity and to avoid direct feedback of the voltage pulses into the signal source, the single junction may be replaced by a hysteretic dc SQUID and with the signal current magnetically coupled to it via a multiturn input coil. The comparator SQUID converts the magnetic flux Φ produced by the signal current or by an applied magnetic field into a pulse probability $p(\Phi)$. The mathematical description of such a hysteretic comparator SQUID is equivalent to that of a nonhysteretic current-biased SQUID if the pulse probability $p(\Phi)$ is substituted for the SQUID voltage $V(\Phi)$. One obtains the following analogy for the basic small-signal parameters (cf. Sec. 2.6):

$$V_\Phi \triangleq (\partial p/\partial \Phi)_w, \quad R_{dyn} \triangleq (\partial p/\partial I)_w, \quad M_{dyn} = R_{dyn}/V_\Phi \triangleq -1/(\partial I_{max}/\partial \Phi)_w. \quad (51)$$

Here, $M_{dyn} = -(\partial \Phi/\partial I)_w$ is the inverted current-to-flux transfer coefficient at constant SQUID voltage or pulse probability, respectively, and $(\partial I_{max}/\partial \Phi)_w$ is the slope of the threshold characteristic $I_{max}(\Phi)$ vs. Φ at the working point.

Figure 20 shows a digital flux-locked loop where the SQUID is operated as a latching comparator [96]. A high-frequency unipolar clock current is passed through the comparator. The resulting voltage pulses of a few mV amplitude control an up/down counter (digital integrator) at room temperature. The counter increases if the SQUID switches into the voltage state during a clock cycle; otherwise, the counter decreases. The content of the counter (i.e., the digital output signal) is converted into the feedback current (i.e., the analog output signal) via a D/A converter. The feedback loop locks automatically at a switching probability $p_w = 50 \%$ if the amplitude of the clock current pulses lies between the minimum and the maximum critical current of the SQUID. The influence of preamplifier voltage noise is eliminated because only the occurence of the voltage pulses is detected, but neither their amplitude nor their phase carry information. The preamplifier current noise, however, still contributes to the total noise via M_{dyn}. The comparator flux noise decreases with increasing clock frequency f_c. Simulations as well as experiments are in reasonable agreement with Eq. (50) for clock frequencies $f_c = f_b$ between $\simeq 1$ and 700 MHz [83]. At high clock frequencies the noise deteriorates due to plasma oscillations, resonances in the comparator input coil structure, or reflections on the transmission lines between cryogenic and room temperature [34]. The reason is that all these parasitic effects lead to a correlation between subsequent sampling events (clock cycles), and hence, to a reduction of the effective sampling rate (clock frequency).

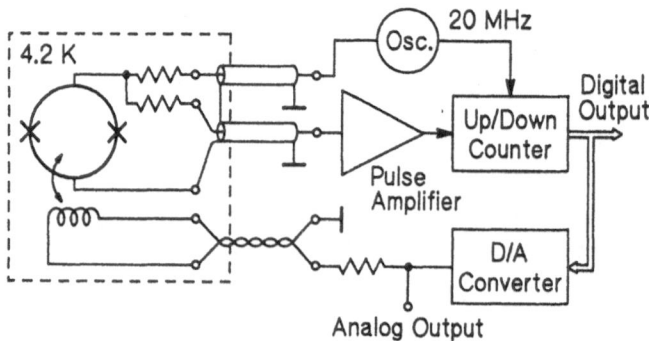

Figure 20. Flux-locked-loop circuit of a digital SQUID using pulse probability modulation [83].

Prototype digital SQUIDs with room-temperature feedback electronics were fabricated using a Nb-Nb$_2$O$_5$-PbInAu junction process and extensively characterized at the University of Karlsruhe [28,34]. The lowest comparator flux-noise level of 4.5×10^{-6} $\Phi_0/\sqrt{Hz}$ and coupled energy sensitivity of 4300 h were measured for f_c = 20 MHz. A two-stage configuration with a resistively-shunted sensor SQUID in front of the comparator was used in order to achieve a low system noise level. The sensor SQUID with L = 50-60 pH was realized with three different integrated pickup circuits: two 4-mm^2 [83] or two 31-mm^2 [97] loops arranged as a planar first-order gradiometer or a single 51-mm^2 magnetometer loop [98]. The lowest noise levels, $\sqrt{S_{\Phi,t}} \lesssim 10^{-6}$ $\Phi_0/\sqrt{Hz}$ and $\varepsilon_{c,t} \lesssim 70$ h, were measured for sensors with gradiometer pickup coils; for the larger gradiometer this corresponds to a minimum gradient noise level of $\simeq 4$ fT/cm$\sqrt{Hz}$. The magnetometers exhibited higher noise levels $\sqrt{S_{\Phi,t}} = 3.5 \times 10^{-6}$ $\Phi_0/\sqrt{Hz}$ or $\sqrt{S_{B,t}}$ = 8.5 fT/$\sqrt{Hz}$. A simple bias-current reversal scheme was developed for this two-stage device that effectively suppresses excess low-frequency noise arising from critical-current fluctuations [28]. A high system slew rate of 2×10^6 Φ_0/s at f = 7.5 kHz was achieved using a digital two-pole integrator, i.e., a counter with variable step size [97]. Similar digital two-stage devices with NbN-MgO-NbN junctions were fabricated and coupled to different integrated pickup coils in a flip-chip arrangement [99].

The linearity of the digital two-stage system was investigated using a 12-bit D/A converter with a full-scale feedback range of $\pm\Phi_0$ and flux-quanta counting [98,100]. A low-frequency "staircase" signal flux was applied to the sensor SQUID covering a range of $\pm\Phi_0$, and the error flux was determined at each step (0.4 s averaging time per step). The standard deviation of the error flux was 4.3×10^{-5} Φ_0, including contributions from the system noise. The resulting system linearity of at least 14.5 bit/Φ_0 was higher than that of the D/A converter alone because the D/A converter errors were averaged by the broadband noise flux over a range of about 10^{-2} Φ_0.

The described digital feedback loop also can be realized using a bipolar (sinusoidal) clock current if the comparator SQUID has an asymmetric threshold characteristic $I_{max}(\Phi)$ vs. Φ [101]. An example is given in Fig. 21. The shape of the threshold characteristic depends on the total SQUID inductance L, the junction critical currents, and the point where the bias current is inserted [102-104]. For unequal critical currents, the smaller one is defined here to be I_0 and is used to calculate the reduced inductance $\beta = 2LI_0/\Phi_0$. The larger critical current is I_{0a}. The critical current asymmetry factor [14]

$$a = \frac{I_{0a} - I_0}{I_{0a} + I_0} \tag{52}$$

is zero in the symmetric case ($I_{0a} = I_0$) and approaches unity for $I_{0a}/I_0 \to \infty$. The critical-current modulation depth ΔI_{max} (cf. Fig. 21) is mainly determined by I_0 and β. It depends relatively weakly on a and is independent of the asymmetry of the bias current feed [14]. A useful approximation is [103,105]

$$\Delta I_{max}/I_0 \simeq \frac{2}{1 + \beta - a/2} \text{ for } \beta \geq a/2, \quad \Delta I_{max}/I_0 \simeq 2 \text{ for } \beta \leq a/2. \tag{53}$$

Equation (53) also may be used to estimate the voltage swing of a resistively-shunted

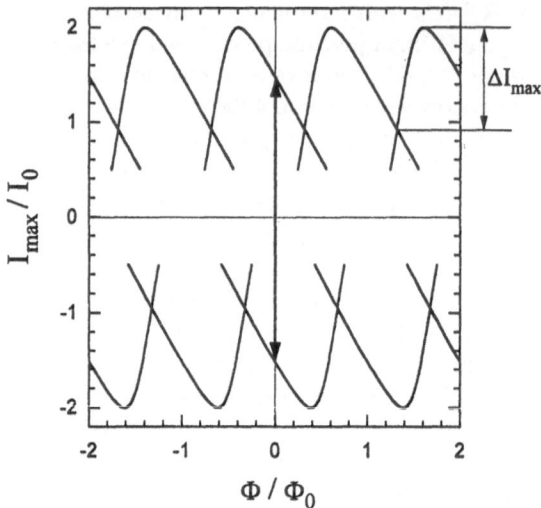

Figure 21. Threshold characteristic of a dc SQUID with $\beta = 0.8$, $a = 0$ and asymmetric bias current feed into one of the junctions.

dc SQUID. In the symmetric case we obtain $\Delta V \simeq \Delta I_{max} R/2 \simeq I_0 R/(1+\beta)$ in reasonable agreement with simulation results for low-T_c SQUIDs [14,16].

An adequate asymmetry for operation with a bipolar clock easily can be obtained for $\beta \simeq 0.8$ and identical junctions ($a = 0$) by applying the bias current to one of the junctions instead of to the center tap of the SQUID inductance (Fig. 21). The size and hence the critical current of the other junction may be increased in order to maintain the asymmetry of the threshold characteristic for smaller values of β [47]. Larger values of β can be used if the bias current is inserted less asymmetrically or, alternately, split and applied to both junctions via appropriate resistors [84].

The bipolar clock current is applied to the asymmetric comparator with an amplitude just equal to the critical current with zero flux – see arrow in Fig. 21. During each clock cycle, the comparator will produce a positive and a negative voltage pulse with a probability $p^+(\Phi)$ and $p^-(\Phi)$, respectively. For zero flux (i.e., at the nominal working point) the probabilities are equal: $p^+(0) = p^-(0) = p_w$. A small positive change in the flux Φ will increase the probability of positive pulses and decrease that of negative pulses. A negative change of Φ will have the opposite effect. The feedback loop will lock to $p^+(\Phi) = p^-(\Phi)$ and thus to a constant flux if the comparator voltage pulses are integrated (counted up or down dependent on their polarity) and fed back into the comparator as a flux. The pulse probability p_w at the working point can be adjusted via the clock-current amplitude, whereas the feedback loop locks inevitably to $p_w = 50\%$ with a unipolar clock. The comparator flux noise has a minimum at $p_w \simeq 50\text{-}70\%$ [84,106]. The switching threshold is rather sharp, leading to tight tolerances for the bipolar clock-current amplitude of the order of $\pm 1\%$ [84]. The use of a bipolar clock and the resulting bipolar comparator voltage pulses are the bases for the single-chip SQUIDs discussed in Sec. 6.2. An attractive feature is that the bipolar clock has the same effect as bias-current reversal for standard dc SQUIDs; thus, it suppresses the influence of low-frequency critical-current fluctuations. Furthermore, the bias frequency is twice the clock frequency, $f_b = 2f_c$, because the critical current is sampled twice during each clock cycle.

The effective dead-time in a digital feedback loop may be several clock cycles (about 5 for a 20-MHz system with room-temperature feedback circuits [28]), but certainly not less than one half of a clock cycle. The corresponding maximum system bandwidth according to Eq. (27) may perhaps be about one tenth of the clock frequency. This means that clock frequencies in excess of 100 MHz are required to improve the system bandwidth over that of analog wideband systems. The linear flux range $\delta\Phi$ of digital SQUIDs is very small due to their sharp switching threshold, typically of the order of $10^{-2}\,\Phi_0$ [34,84]. Consequently, one must increase the clock frequency even into the GHz range to obtain a system slew rate superior to that of high-speed analog systems. Only fast digital SQUIDs with on-chip feedback (i.e., single-chip SQUIDs) are potentially able to perform better than the fastest analog systems with room-temperature feedback electronics available today [30-32].

6.2. SINGLE-CHIP SQUIDS

The digital feedback loop shown in Fig. 20 may be integrated on the sensor chip together with some data preprocessing logic using Josephson-junction logic circuits. Although the required logic circuits are already available (except for the D/A converter), the complexity of such a single-chip SQUID would be rather high. A simplified digital feedback loop for operation at 4.2 K was developed at Fujitsu by Fujimaki et al. [84,107]. In this scheme, a superconducting storage loop with a hysteretic dc SQUID as a flux-quanta write gate performs the function of the up/down counter, and the D/A converter is realized by coupling the storage loop magnetically to the comparator SQUID [cf. Fig. 22(a)]. Driven by a high-frequency sinusoidal bias current, the comparator SQUID produces positive and negative voltage pulses which are used to incre-

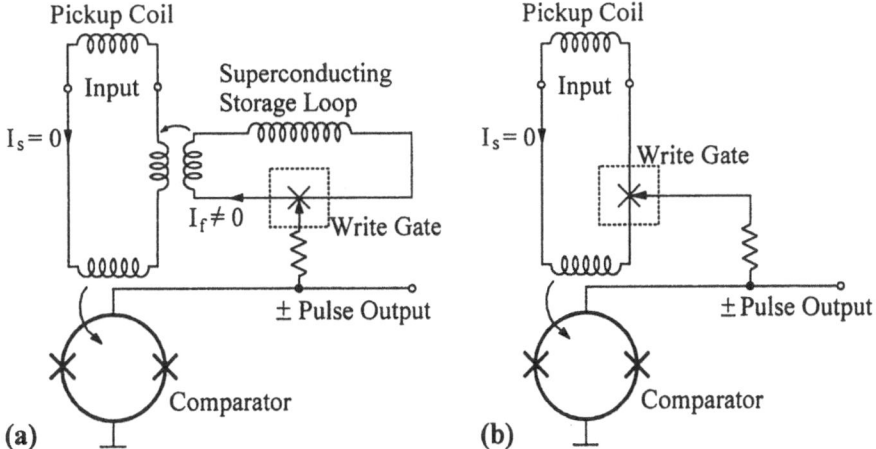

Figure 22. Basic single-chip SQUID circuits (a) with superconducting storage loop [84] and (b) with direct pulse feedback into the input circuit [47].

ment or decrement the flux in the storage loop in steps of Φ_0 via the write gate. Flux in the storage loop is equivalent to a circulating current I_f that flows through the feedback-transformer primary coil in order to close the loop. Some buffer gates and several damping resistors (not shown in Fig. 22) are required for optimum performance.

A sensor with a 0.5-μH input coil and the complete feedback electronics were integrated on a single 4×10 mm^2 chip [84]. The maximum clock frequency of 10 MHz resulted in a noise level $\sqrt{S_{\Phi,t}}$ = 6.2×10^{-6} $\Phi_0/\sqrt{Hz}$ ($\varepsilon_{c,t}$ = 4100 h) and a small-signal bandwidth of 1.1 kHz. The noise was white down to 0.2 Hz. The minimum step size of the feedback flux, Φ_{LSB} = 1.7×10^{-5} Φ_0, was calculated from the measured system slew rate $\Phi_{f,max}(f)$ = $\Phi_{LSB}f_c$ = 170 Φ_0/s at f_c = 10 MHz. For $f \lesssim 13$ Hz, the system slew rate was limited by the static feedback range of ± 2.2 Φ_0, corresponding to $\pm 1.3 \times 10^5$ Φ_0 in the storage loop. An array of 8 of these SQUIDs was operated with a Josephson multiplexer at a SQUID bias frequency of 5 MHz and a multiplexer clock frequency of 120 MHz [108]. Noise and dynamic performance of this 8-channel SQUID array would be just sufficient for biomagnetic measurements inside a magnetically-shielded room if large, wire-wound gradiometer pickup coils were connected to the input coils.

The dynamic range of the Fujitsu single-chip SQUID was rather low because Φ_{LSB} was chosen unnecessarily small. An improvement by two or three orders of magnitude should be possible by simply enlarging Φ_{LSB}. However, the dynamic range is limited in principle because the input flux is nulled via a nonzero feedback current flowing in the superconducting storage loop. A virtually infinite dynamic range can be obtained if the bipolar output pulses of the comparator are directly used to null the input flux without an intermediate storage loop [Fig. 22(b)]. This feedback concept was first proposed for rapid single-flux-quantum (RSFQ) logic [109,110] and realized at HYPRES using latching Josephson logic [47]. The input circuit is a superconducting loop consisting of the pickup coil, an input coil and a write gate. The latter involves two hysteretic dc SQUIDs [47] or two RSFQ pulse generators with following Josephson transmission lines [110,111]. A bipolar clock current is applied to the comparator that produces a sequence of bipolar voltage pulses. For each positive/negative voltage pulse, the flux in the input circuit is increased/decreased (or vice versa) by one flux quantum so as to keep the average total flux in the input circuit constant. The comparator voltage pulses represent the digitized over-sampled time derivative of the feedback flux. The pulses may be digitally integrated and low-pass filtered to obtain output data words with an appropriate rate and a large number of effective bits.

The feedback flux is quantized with a minimum step size of Φ_0 if referred to the input circuit. The corresponding flux Φ_{LSB} in the comparator SQUID is typically much smaller than Φ_0 due to a relatively low flux transfer from the input circuit into the comparator SQUID. For example, the flux transfer coefficient is $1/2n$ for an ideally-coupled n-turn input coil with inductance matching to the pickup coil and negligible write-gate inductance. The quantization noise is negligible if Φ_{LSB} does not exceed about twice the broadband rms noise flux in the comparator [112]. The effective comparator rms noise flux is lowered by the flux gain G_{Φ} if a two-stage configuration with a resistively-shunted sensor SQUID in front of the comparator is used. In this case, Φ_{LSB} in the sensor SQUID should not exceed twice the rms sum of the sensor noise flux and the effective comparator noise flux. (In a two-stage device, a low-pass filter between sensor and comparator should be used to avoid aliasing effects, and the sensor rms noise flux is integrated only over the filter bandwidth.) Typically, the maximum Φ_{LSB} is of the order of $10^{-2}\,\Phi_0$. The corresponding flux transfer coefficient of $\simeq 10^{-2}$ can be obtained with a 50-turn input coil. Fewer turns are required if an intermediate matching transformer between the input circuit and the front-end SQUID is used [37].

A prototype single-chip SQUID was fabricated at HYPRES using standard niobium technology, and proper operation was experimentally verified [47]. Subsequently, a three-stage version was developed consisting of a nonhysteretic voltage-biased sensor SQUID as a front-end device, a 100-SQUID series array as a preamplifier, and a sensitive comparator SQUID with eight parallel washers [37]. High-speed operation of a single-chip SQUID with a clock frequency of 500 MHz was experimentally demonstrated [111]. It is expected that the clock frequency can be further increased (perhaps by two orders of magnitude) in single-chip SQUIDs implemented with RSFQ logic circuits [110]. Using RSFQ logic one might even directly process the ultra-short voltage pulses which are produced by a dc SQUID with resistively-shunted junctions, and the rate of which is proportional to the voltage across the SQUID. The digital processing of the SQUID output signal potentially allows one to achieve a system noise level close to the intrinsic SQUID noise level, and the extremely high speed of RSFQ logic would make such a single-chip SQUID very fast.

7. Concluding Remarks

Several novel SQUID concepts have been developed in the past ten years. SQUIDs with APF have proven to be highly suited for multichannel and wideband applications. They achieve a low noise level and a high bandwidth using simple read-out electronics without flux modulation. All important performance parameters (noise, dynamic behavior, harmonic distortion) were investigated experimentally, and good agreement with theoretical predictions is usually found. SQUIDs with APF are employed in several biomagnetic multichannel systems with up to 256 channels. Recently, APF with bias reversal was used to read out high-T_c SQUID magnetometers, and high-quality biomagnetic signals from the human heart, brain and peripheral nerve system were recorded in a magnetically-shielded room.

The steepest V–Φ characteristics were demonstrated with DROSs. At low relaxation frequencies $f_{ro} \simeq 100$ MHz, the transfer coefficient V_Φ and the noise level ϵ_t are about 10 times larger than for SQUIDs with APF. The noise level can be reduced by increasing f_{ro}, but the transfer coefficient also tends to decrease with f_{ro}. The noise has a minimum at high relaxation frequencies $f_{ro} \gtrsim 1$ GHz, at which the performance of DROSs (V_Φ and ϵ_t) becomes comparable to that of SQUIDs with APF. The first practical 3-channel DROS gradiometer systems have been realized.

Series SQUID arrays may attain the highest slew rates because they provide a large V_Φ without reducing the linear flux range $\delta\Phi$. A low noise level can be obtained with simple direct-coupled read-out electronics if a sufficiently large number of SQUIDs is used. A total of about 100 SQUIDs is required in a single-stage configuration. Large arrays of 100 SQUIDs have been fabricated and characterized. All these arrays exhibited amplitude modulation of their V~Φ characteristics for large values of the flux bias, which might be an obstacle for unshielded operation. The demands on the fabrication process may be lowered by using a two-stage configuration consisting of a single sensor SQUID coupled to a series of about 10 preamp SQUIDs with a flux gain $G_\Phi \simeq 1$.

Digital read-out electronics can be completely integrated into the SQUID chip. The number of cables between helium bath and room temperature can be minimized using digital preprocessing and multiplexing at 4.2 K. However, high-frequency transmission lines are required which usually have a higher thermal conductance (i.e., a stronger effect on the helium evaporation rate of the dewar) than the low-frequency twisted pairs used to operate SQUIDs with APF. Furthermore, the high complexity of single-chip SQUIDs puts greater demands on the fabrication process. The first prototypes of single-chip SQUIDs with a wide dynamic range have been fabricated and tested, but reliable low-noise operation in practical applications has yet to be proven.

Acknowledgments

The author thanks Frank Ludwig and Hartmut Matz for stimulating discussions and helpful comments on the manuscript. This work was partially supported by the German BMBF under Contract Nos. 13N6398 and 13N6435.

References

1. Clarke, J., Goubau, W.M., and Ketchen, M.B. (1976) Tunnel junction dc SQUID: fabrication, operation, and performance, *J. Low Temp. Phys.* **25**, 99-144.
2. Drung, D. (1991) dc SQUID systems overview, *Supercond. Sci. Technol.* **4**, 377-385.
3. Drung, D. (1994) Recent low temperature SQUID developments, *IEEE Trans. Appl. Supercond.* **4**, 121-127.
4. Thomasson, S.L. and Gould, C.M. (1995) 1 MHz bandwidth true NMR SQUID amplifier, *J. Low Temp. Phys.* **101**, 243-248.
5. Polushkin, V., Drung, D., and Koch, H. (1994) A broadband picovoltmeter based on the direct current superconducting quantum interference device, *Rev. Sci. Instrum.* **65**, 3005-3011.
6. Wellstood, F., Heiden, C., and Clarke, J. (1984) Integrated dc SQUID magnetometer with a high slew rate, *Rev. Sci. Instrum.* **55**, 952-957.
7. Koch, R.H., Rozen, J.R., Sun, J.Z., and Gallagher, W.J. (1993) Three SQUID gradiometer, *Appl. Phys. Lett.* **63**, 403-405.
8. Ketchen, M.B. (1987) Integrated thin-film dc SQUID sensors, *IEEE Trans. Magn.* **23**, 1650-1657.
9. Ryhänen, T., Seppä, H., Ilmoniemi, R., and Knuutila, J. (1989) SQUID magnetometers for low-frequency applications, *J. Low Temp. Phys.* **76**, 287-386.
10. Clarke, J. (1993) SQUIDs: theory and practice, in H. Weinstock and R.W. Ralston (eds.), *The New Superconducting Electronics*, Kluwer Acadamic Publishers, Dordrecht, pp. 123-180.
11. Falco, C.M., Parker, W.H., Trullinger, S.E., and Hansma, P.K. (1974) Effect of thermal noise on current-voltage characteristics of Josephson junctions, *Phys. Rev. B* **10**, 1865-1873.
12. Voss, R.F. (1981) Noise characteristics of an ideal shunted Josephson junction, *J. Low Temp. Phys.* **42**, 151-163.
13. Van Duzer, T and Turner, C.W. (1981) *Principles of Superconductive Devices and Circuits*, Elsevier, New York.

14. Tesche, C.D. and Clarke, J. (1977) dc SQUID: noise and optimization, *J. Low Temp. Phys.* **27**, 301-331.
15. Bruines, J.J.P., de Waal, V.J., and Mooij, J.E. (1982) Comment on: "dc SQUID: noise and optimization" by Tesche and Clarke, *J. Low Temp. Phys.* **46**, 383-386.
16. De Waal, V.J., Schrijner, P., and Llurba, R. (1982) Simulation and optimization of a dc SQUID with finite capacitance, *J. Low Temp. Phys.* **54**, 215-232.
17. Koelle, D., Miklich, A.H., Ludwig, F., Dantsker, E., Nemeth, D.T., and Clarke, J. (1993) dc SQUID magnetometers from single layers of $YBa_2Cu_3O_{7-x}$, *Appl. Phys. Lett.* **63**, 2271-2273.
18. Enpuku, K., Shimomura, Y., and Kisu, T. (1993) Effect of thermal noise on the characteristics of a high T_c superconducting quantum interference device, *J. Appl. Phys.* **73**, 7929-7934.
19. Motchenbacher C.D. and Fitchen, F.C. (1973) *Low-Noise Electronic Design*, Wiley, New York.
20. Enpuku, K., Tokita, G., Maruo, T., and Minotani, T. (1995) Parameter dependencies of characteristics of a high-T_c dc superconducting quantum interference device, *J. Appl. Phys.* **78**, 3498-3503.
21. Foglietti, V., Koch, R.H., Sun, J.Z., Laibowitz, R.B., and Gallagher, W.J. (1995) Characterizing, modeling, and optimizing high-T_c superconducting quantum interference devices, *J. Appl. Phys.* **77**, 378-381.
22. Sun, J.Z., Yu-Jahnes, L.S., Foglietti, V., Koch, R.H., and Gallagher, W.J. (1995) Properties of YBaCuO thin film single-level dc SQUIDs fabricated using step-edge junctions, *IEEE Trans. Appl. Supercond.* **5**, 2107-2111.
23. Keene, M.N., Satchell, J.S., Goodyear, S.W., Humphreys, R.G., Edwards, J.A., Chew, N.G., and Lander, K. (1995) Low noise HTS gradiometers and magnetometers constructed from $YBa_2Cu_3O_{7-x}$ /$PrBa_2Cu_3O_{7-y}$ thin films, *IEEE Trans. Appl. Supercond.* **5**, 2923-2926.
24. Kleiner, R. (1995) unpublished.
25. Drung, D. (1992) Investigation of a double-loop dc SQUID magnetometer with additional positive feedback, in H. Koch and H. Lübbig (eds.), *Superconducting Devices and Their Applications*, Springer-Verlag, Berlin, pp. 351-356.
26. Koch, R.H., Foglietti, V., Rozen, J.R., Stawiasz, K.G., Ketchen, M.B., Lathrop, D.K., Sun, J.Z., and Gallagher, W.J. (1994) Effects of radio frequency radiation on the dc SQUID, *Appl. Phys. Lett.* **65**, 100-102.
27. Koch, R.H., Clarke, J., Goubau, W.M., Martinis, J.M., Pegrum, C.M., and Van Harlingen, D.J. (1983) Flicker (1/f) noise in tunnel junction dc SQUIDs, *J. Low Temp. Phys.* **51**, 207-224.
28. Drung, D., Crocoll, E., Herwig, R., Neuhaus, M., and Jutzi, W. (1989) Measured performance parameters of gradiometers with digital output, *IEEE Trans. Magn.* **25**, 1034-1037.
29. Drung, D. (1995) Low-frequency noise in low-T_c multiloop magnetometers with additional positive feedback, *Appl. Phys. Lett.* **67**, 1474-1476.
30. Drung, D., Matz, H., and Koch, H. (1995) A 5-MHz bandwidth SQUID magnetometer with additional positive feedback, *Rev. Sci. Instrum.* **66**, 3008-3015.
31. Drung, D., Jurthe, S., Knappe, S., Matz, H., and Peters, M. (1995) unpublished.
32. Quantum Magnetics, 11558 Sorrento Valley Road, San Diego, CA 92121-1131.
33. Giffard, R.P. (1980) Fundamentals for SQUID applications, in H.D. Hahlbohm and H. Lübbig (eds.), *Superconducting Quantum Interference Devices and their Applications*, Walter de Gruyter, Berlin, pp. 445-471.
34. Drung, D. (1988) *Sensor und A/D-Wandlerstufe auf einem Chip zur Präzisionsmessung von Magnetfeldgradienten mit Josephson-Kontakten*, Ph.D. thesis, University of Karlsruhe.
35. Seppä, H. and Sipola, H. (1990) A high open-loop gain controller, *Rev. Sci. Instrum.* **61**, pp. 2449-2451.
36. Welty, R.P. and Martinis, J.M. (1993) Two-stage integrated SQUID amplifier with series array output, *IEEE Trans. Appl. Supercond.* **3**, 2605-2608.
37. Radparvar, M. and Rylov, S. (1995) An integrated digital SQUID magnetometer with high sensitivity input, *IEEE Trans. Appl. Supercond.* **5**, 2142-2145.
38. Seppä, H., Ahonen, A., Knuutila, J., Simola, J., and Vilkman, V. (1991) dc SQUID electronics based on adaptive positive feedback: experiments, *IEEE Trans. Magn.* **27**, 2488-2490.
39. Seppä, H. (1992) dc SQUID electronics based on adaptive noise cancellation and a high open-loop gain controller, in H. Koch and H. Lübbig (eds.), *Superconducting Devices and Their Applications*, Springer-Verlag, Berlin, pp. 346-350.

114

40. Drung, D. and Koch, H. (1993) An electronic second-order gradiometer for biomagnetic applications in clinical shielded rooms, *IEEE Trans. Appl. Supercond.* **3**, 2594-2597.
41. Drung, D. (1995) The PTB 83-SQUID system for biomagnetic applications in a clinic, *IEEE Trans. Appl. Supercond.* **5**, 2112-2117.
42. Ryhänen, T., Seppä, H., Cantor, R., Drung, D., Koch, H., and Veldhuis, D. (1992) Noise studies of uncoupled dc SQUIDs, in H. Koch and H. Lübbig (eds.), *Superconducting Devices and Their Applications*, Springer-Verlag, Berlin, pp. 321-325.
43. Seppä, H., Kiviranta, M., Satrapinski, A., Grönberg, L., Salmi, J., and Suni, I. (1993) A coupled dc SQUID with low 1/f noise, *IEEE Trans. Appl. Supercond.* **3**, 1816-1819.
44. Drung, D., Dantsker, E., Ludwig, F., Koch, H., Kleiner, R., Clarke, J., Krey, S., Reimer, D., David, B., and Doessel, O. (1996) Low noise $YBa_2Cu_3O_{7-x}$ SQUID magnetometers operated with additional positive feedback, *Appl. Phys. Lett.* **68**, 1856-1858.
45. Drung, D., Ludwig, F., Müller, W., Steinhoff, U., Trahms, L., Koch, H., Shen, Y.Q., Jensen, M.B., Vase, P., Holst, T., Freltoft, T., and Curio, G. (1996) Integrated $YBa_2Cu_3O_{7-x}$ magnetometers for biomagnetic measurements, *Appl. Phys. Lett.* **68**, 1421-1423.
46. ter Brake, H.J.M., Fleuren, F.H., Ulfman, J.A., and Flokstra, J. (1986) Elimination of flux-transformer crosstalk in multichannel SQUID magnetometers, *Cryogencs* **26**, 667-670.
47. Radparvar, M. (1994) A wide dynamic range single-chip SQUID magnetometer, *IEEE Trans. Appl. Supercond.* **4**, 87-91.
48. Shinada, K., Munaka, T., Ueda, M., Fujiyama, Y., Nagamachi, S., and Yamada, Y. (1995) Double washer dc SQUIDs with short weak link junctions, in *Extended Abstracts of 5th International Superconductive Electronics Conference*, Nagoya, Japan, Sept. 1995, pp. 364-365.
49. Gershenson, M. (1991) Design of a hysteretic SQUID as the readout for a dc SQUID, *IEEE Trans. Magn.* **27**, 2910-2912.
50. Koshelets, V.P., Matlashov, A.N., Serpuchenko, I.L., Filippenko, L.V., and Zhuravlev, Yu.E. (1989) dc SQUID preamplifier for dc SQUID magnetometer, *IEEE Trans. Magn.* **25**, 1182-1185.
51. Foglietti, V. (1991) Double dc SQUID for flux-locked-loop operation, *Appl. Phys. Lett.* **59**, 476-478.
52. Maslennikov, Yu.V., Beljaev, A.V., Snigirev, O.V., Kaplunenko, O.V., and Mezzena, R. (1995) A double dc SQUID based magnetometer, *IEEE Trans. Appl. Supercond.* **5**, 3241-3243.
53. Ketchen, M.B. and Tsuei, C.C. (1980) Low frequency noise in small-area tunnel junction dc SQUIDs, in H.D. Hahlbohm and H. Lübbig (eds.), *Superconducting Quantum Interference Devices and their Applications*, Walter de Gruyter, Berlin, pp. 227-235.
54. Wellstood, F.C., Urbina, C., and Clarke, J. (1987) Low-frequency noise in dc superconducting quantum interference devices below 1 K, *Appl. Phys. Lett.* **50**, 772-774.
55. Foglietti, V., Giannini, M.E., and Petrocco, G. (1991) A double dc SQUID device for flux locked loop operation, *IEEE Trans. Magn.* **27**, 2989-2992.
56. Welty, R.P. and Martinis, J.M. (1991) A series array of dc SQUIDs, *IEEE Trans. Magn.* **27**, 2924-2926.
57. Stawiasz, K.G. and Ketchen, M.B. (1993) Noise measurements of series SQUID arrays, *IEEE Trans. Appl. Supercond.* **3**, 1808-1811.
58. HYPRES, Inc., 175 Clearbrook Road, Elmsford, NY 10523.
59. Takeda, E. and Nishino, T. (1995) Design of SQUID sensor having SQUID amplifier, in *Extended Abstracts of 5th International Superconductive Electronics Conference*, Nagoya, Japan, Sept. 1995, pp. 361-363.
60. Drung, D., Cantor, R., Peters, M., Scheer, H.J., and Koch, H. (1990) Low-noise high-speed dc superconducting quantum interference device magnetometer with simplified feedback electronics, *Appl. Phys. Lett.* **57**, 406-408.
61. Drung, D. and Koch, H. (1994) An integrated dc SQUID magnetometer with variable additional positive feedback, *Supercond. Sci. Technol.* **7**, 242-245.
62. Kiviranta, M. and Seppä, H. (1995) dc SQUID electronics based on the noise cancellation scheme, *IEEE Trans. Appl. Supercond.* **5**, 2146-2148.
63. Ukhansky, N.N., Gudoshnikov, S.A., Vengrus, I.I., and Snigirev, O.V. (1995) Low noise liquid-nitrogen-cooled preamplifier for a high-T_c SQUID, in *Extended Abstracts of 5th International Superconductive Electronics Conference*, Nagoya, Japan, Sept. 1995, pp. 346-348.

64. Takada, Y., Tsukada, K., and Adachi, A. (1995) A high reliable SQUID gradiometer with controllable additional positive feedback, in *Extended Abstracts of 5th International Superconductive Electronics Conference*, Nagoya, Japan, Sept. 1995, pp. 366-367.
65. Clarke, J. (1994) Low frequency quadrupole resonance with SQUID amplifiers, *Z. Naturforsch.* **49a**, 3-13.
66. Ryhänen, T., Cantor, R., Drung, D., and Koch, H. (1991) Practical low-noise integrated dc superconducting quantum interference device magnetometer with additional positive feedback, *Appl. Phys. Lett.* **59**, 228-230.
67. Kazami, K., Takada, Y., Uehara, G., Matsuda, N., and Kado, H. (1994) Evaluation of Drung-type magnetometers for multi-channel systems, *Supercond. Sci. Technol.* **7**, 249-252.
68. Drung, D., Knappe, S., and Koch, H. (1995) Theory for the multiloop dc superconducting quantum interference device magnetometer and experimental verification, *J. Appl. Phys.* **77**, 4088-4098.
69. Koch, H., Cantor, R., Drung, D., Erné, S.N., Matthies, K.P., Peters, M., Ryhänen, T., Scheer, H.J., and Hahlbohm, H.D. (1991) A 37 channel dc SQUID magnetometer system, *IEEE Trans. Magn.* **27**, 2793-2796.
70. Matz, H. and Drung, D. (1995) unpublished.
71. Simmonds, M. B. and Giffard, R. P. (1983) Apparatus for reducing low frequency noise in dc biased SQUIDs, U. S. Patent No. 4 389 612.
72. Foglietti, V., Gallagher, W.J., Ketchen, M.B., Kleinsasser, A.W., Koch, R.H., Raider, S.I., and Sandstrom, R.L. (1986) Low-frequency noise in low 1/f noise dc SQUIDs, *Appl. Phys. Lett.* **49**, 1393-1395.
73. Dössel, O., David, B., Fuchs, M., Kullmann, W.H., and Lüdeke, K.-M. (1991) A modular low noise 7-channel SQUID-magnetometer, *IEEE Trans. Magn.* **27**, 2797-2800.
74. Savo, B., Wellstood, F.C., and Clarke, J. (1987) Low-frequency excess noise in Nb-Al$_2$O$_3$-Nb Josephson tunnel junctions, *Appl. Phys. Lett.* **50**, 1757-1759.
75. Drung, D., Cantor, R., Peters, M., Ryhänen, T., and Koch, H. (1991) Integrated dc SQUID magnetometer with high dV/dB, *IEEE Trans. Magn.* **27**, 3001-3004.
76. Zappe, H.H. (1973) Minimum current and related topics in Josephson tunnel junction devices, *J. Appl. Phys.* **44**, 1371-1377
77. Adelerhof, D.J., Nijstad, H., Flokstra, J., and Rogalla, H. (1994) (Double) relaxation oscillation SQUIDs with high flux-to-voltage transfer: Simulations and experiments, *J. Appl. Phys.* **76**, 3875-3886.
78. Vernon, F.L. and Pedersen, R.P. (1968) Relaxation oscillations in Josephson junctions, *J. Appl. Phys.* **39**, 2661-2664.
79. Gutmann, P. (1979) dc SQUID with high energy resolution, *Electr. Lett.* **15**, 372-373.
80. Gudoshnikov, S.A., Maslennikov, Yu.V., Semenov, V.K., Snigirev, O.V., and Vasiliev, A.V. (1989) Relaxation-oscillation-driven dc SQUIDs, *IEEE Trans. Magn.* **25**, 1178-1181.
81. Mück, M. and Heiden, C. (1989) Simple dc SQUID system based on a frequency modulated relaxation oscillator, *IEEE Trans. Magn.* **25**, 1151-1153.
82. Adelerhof, D.J., Kawai, J., Uehara, G., and Kado, H. (1995) High sensitivity double relaxation oscillation superconducting quantum interference devices with large transfer from flux to voltage, *Rev. Sci. Instrum.* **66**, 2631-2637.
83. Drung, D., Crocoll, E., Herwig, R., Knüttel, A., Neuhaus, M., and Jutzi, W. (1987) Experimental gradiometer with a digital feedback loop, in *Extended Abstracts of 1st International Superconductive Electronics Conference*, Tokyo, Japan, Aug. 1987, pp. 21-24.
84. Fujimaki, N., Gotoh, K., Imamura, T., and Hasuo, S. (1992) Thermal-noise-limited performance in single-chip superconducting quantum interference devices, *J. Appl. Phys.* **71**, 6182-6188.
85. Gudoshnikov, S.A., Kaplunenko, O.V., Maslennikov, Yu.V., and Snigirev, O.V. (1991) Noise in relaxation-oscillation-driven dc SQUIDs, *IEEE Trans. Magn.* **27**, 2439-2441.
86. Adelerhof, D.J., van Duuren, M.J., Flokstra, J., Rogalla, H., Kawai, J., and Kado, H. (1995) High sensitivity magnetic flux sensors with direct voltage readout: Double relaxation oscillation SQUIDs, *IEEE Trans. Appl. Supercond.* **5**, 2160-2163.
87. Mück, M., Rogalla, H., and Heiden, C. (1988) A frequency-modulated read-out system for dc SQUIDs, *Appl. Phys. A* **47**, 285-289.
88. Uehara, G., Morooka, T., Kawai, J., Mizutani, N., and Kado, H. (1993) Characteristics of the relaxation oscillation SQUID with tunnel junctions, *IEEE Trans. Appl. Supercond.* **3**, 1866-1869.

116

89. Lee, Y.H., Kim, J.M., Kwon, H.C., Park, Y.K., Park, J.C., van Duuren, M.J., Adelerhof, D.J., Flokstra, J., and Rogalla, H., (1995) 3-channel double relaxation oscillation SQUID magnetometer system with simple readout electronics, *IEEE Trans. Appl. Supercond.* 5, 2156-2159.
90. van Duuren, M.J., Adelerhof, D.J., Lee, Y.H., Flokstra, J., Rogalla, H., Kawai, J., and Kado, H. (1995) DROS magnetometers operated in a multi-channel system, in D. Dew-Hughes (ed.), *Applied Superconductivity 1995*, IOP Publishing, Bristol, pp. 1507-1510.
91. Lee, Y.H., Kwon, H.C., Kim, J.M., Park, Y.K., and Park, J.C. (1995) Integrated planar gradiometer based on double relaxation oscillation SQUID, in *Extended Abstracts of 5th International Superconductive Electronics Conference*, Nagoya, Japan, Sept. 1995, pp. 178-180.
92. Adelerhof, D.J., Wichers, H., Brons, C., van Duuren, M., Veldhuis, D., Flokstra, J., and Rogalla, H. (1995) Direct coupled multi-loop DROS magnetometer with direct voltage readout, in *Extended Abstracts of 5th International Superconductive Electronics Conference*, Nagoya, Japan, Sept. 1995, pp. 371-373.
93. Vrba, J., Betts, K., Burbank, M., Cheung, T., Fife, A.A., Haid, G., Kubik, P.R., Lee, S., McCubbin, J., McKay, J., McKenzie, D., Spear, P., Taylor, B., Tillotson, M., Cheyne, D., and Weinberg, H. (1993) Whole cortex, 64 channel SQUID biomagnetometer system, *IEEE Trans. Appl. Supercond.* 3, 1878-1882.
94. Zimmermann, E., Brandenburg, G., Clemens, U., and Halling, H. (1995) Kompensationsregelung für extrem nichtlinearen Sensor mit digitalem Signalprozessor, in *Proceedings of DSP Deutschland 95*, München, Germany, Sept. 1995, pp. 133-142.
95. Sarwinski, R.E. (1977) Superconducting instruments, *Cryogenics* 12, 671-679.
96. Drung, D. (1986) Digital feedback loops for dc SQUIDs, *Cryogenics* 26, 623-627.
97. Matz, H., Drung, D., Crocoll, E., Herwig, R., Krämer, G., Neuhaus, M., and Jutzi, W. (1990) High slew rate gradiometer prototype with digital feedback loop of variable step size, *Cryogenics* 30, 330-334.
98. Matz, H., Drung, D., Crocoll, E., Herwig, R., Krämer, G., Neuhaus, M., and Jutzi, W. (1991) Integrated magnetometer with a digital output, *IEEE Trans. Magn.* 27, 2979-2982.
99. Eschner, W., Fath, U., Höfer, G., Hundhausen, R., Kratz, H., Ludwig, W., Rothmund, W., and Wülker, M. (1993) Magnetic field sensors with digital feedback read-out, *IEEE Trans. Appl. Supercond.* 3, 1824-1827.
100. Matz, H. (1993) Impact of noise on linearity of SQUID feedback loops at high slew rate, *IEEE Trans. Appl. Supercond.* 3, 3054-3058.
101. Igarashi, Y., Goto, T., Hayashi, T., Fujimaki, N., Kawabe, K., Shimura, T., and Hayashi, H. (1989) Improved SQUID magnetometer with an external feedback circuit, in S.J. Williamson, M. Hoke, G. Stroink, and M. Kotani (eds.), *Advances in Biomagnetism*, Plenum Press, New York, pp. 645-648.
102. Tsang, W.-T. and Van Duzer, T. (1975) dc analysis of parallel arrays of two and three Josephson junctions, *J. Appl. Phys.* 46, 4573-4580.
103. Peterson, R.L. and Hamilton, C.A. (1979) Analysis of threshold curves for superconducting interferometers, *J. Appl. Phys.* 50, 8135-8142.
104. Wunsch, J., Jutzi, W., and Crocoll, E. (1982) Parameter evaluation of asymmetric interferometers with two Josephson junctions, *IEEE Trans. Magn.* 18, 735-737.
105. Drung, D. (1996) unpublished.
106. Fujimaki, N., Tamura, H., Imamura, T., and Hasuo, S. (1989) Thermal noise-limited sensitivity of the pulse-biased SQUID magnetometer, *J. Appl. Phys.* 65, 1626-1630.
107. Fujimaki, N., Tamura, H., Imamura, T., and Hasuo, S. (1988) A single-chip SQUID magnetometer, *IEEE Trans. Electron Devices* 35, 2412-2418.
108. Gotoh, K., Fujimaki, N., Imamura, T., and Hasuo, S. (1993) 8-channel array of single-chip SQUIDs connection to Josephson multiplexer, *IEEE Trans. Appl. Supercond.* 3, 2601-2604.
109. Rylov, S.V. (1991) Analysis of high-performance counter-type A/D converters using RSFQ logic/memory elements, *IEEE Trans. Magn.* 27, 2431-2434.
110. Likharev, K.K. and Semenov, V.K. (1991) RSFQ logic/memory family: A new Josephson-junction technology for sub-THz-clock-frequency digital systems, *IEEE Trans. Appl. Supercond.* 1, 3-28.
111. Yuh, P.-F. and Rylov, S.V. (1995) An experimental digital SQUID with large dynamic range and low noise, *IEEE Trans. Appl. Supercond.* 5, 2129-2132.
112. Sripad, A.B. and Synder, D.L. (1977) A necessary and sufficient condition for quantization errors to be uniform and white, *IEEE Trans. Acoust., Speech, Signal Processing* 25, 442-448.

SQUID GRADIOMETERS IN REAL ENVIRONMENTS

, JIRI VRBA
CTF Systems Inc.
15-1750 McLean Ave, Port Coquitlam, B.C., Canada

Abstract. Use of SQUIDs for stationary magnetic field detection (e.g., in Magneto-EncephaloGraphy, MEG) is discussed. Such devices operate in the presence of environmental noise and must be provided with shielding and/or noise cancellation techniques. First, shielded and unshielded environments are characterised and some specific noise sources are examined in detail. Then, gradiometers of various orders and their construction by hardware and software methods are described, and the gradiometer errors (represented by common mode and eddy current vectors) are analysed. Noise cancellation by high-order spatial gradiometers is examined in detail, and the gradiometer performance in shielded and unshielded environments is evaluated experimentally using a whole cortex MEG system. Successful operation of high-order gradiometers in unshielded environments is demonstrated on examples of human MEG experiments. Adaptive noise cancellation also is examined, and frequency independent and frequency dependent methods are described. The adaptive and gradiometer noise cancellation performances are compared, and it is shown that the adaptive methods are effective only under special circumstances when the noise character is time independent, while the gradiometers are quite universal and work well even when the noise character is changing.

1. Introduction

SQUID (Superconducting QUantum Interference Device) sensors are used in a variety of applications - biomagnetic, nondestructive testing, stationary geophysical, mobile, etc. - and in a variety of environments (ranging from unshielded to extremely well shielded using superconducting shields). The largest number of commercial high sensitivity SQUID sensors are presently employed in MEG applications (MagnetoEncephaloGraphy) [1] and of these the majority are installed in systems which are used inside modestly shielded environments, e.g., [2 to 4]. However, biomagnetic SQUID sensors also are used in highly shielded environments [5 to 7], and in unshielded environments [8 to 10].

Biomagnetism represents a stationary application of SQUIDs, i.e., the SQUID sensor system is static relative to environmental fields and the only motions are undesirable vibrations which are usually minimized by the design. In the opposite extreme are the applications where the SQUID system is mobile during detection (military, geophysical) [11]. In these applications, the system is exposed to large variations of the environmental fields and gradients. There are other fundamental differences between the

H. Weinstock (ed.), SQUID Sensors: Fundamentals, Fabrication and Applications, 117–178.

biomagnetic and mobile applications: biomagnetometers operate in near field, i.e., the distance from the detector to the source is usually less than 10 cm. This makes it possible to utilize high-order gradiometers which for near sources act almost like magnetometers, but behave like gradiometers for distant noise sources. On the other hand, mobile systems usually operate in far field, i.e., the distance between the detector and the source is more than several meters. The mobile detectors utilize only magnetometers and 1st order gradiometers, since higher order gradiometers would excessively attenuate the signal.

Both stationary biomagnetic and mobile applications utilize the utmost sensitivity of the SQUID detector, and both applications require the low-frequency noise to be minimized. In this presentation, the stationary MEG applications will be used to demonstrate how the magnetic environment and the mode of operation affect the SQUID detector design.

In section 2, the magnetic environments will be discussed with illustrative examples of the environmental noise variability and power line effects. In section 3, gradiometers will be explained. First-order gradiometers will be treated in greater detail since their behaviour has many features common to gradiometers of any order. Following this, both hardware and software higher-order gradiometers will be discussed and the required balancing levels will be determined. In section 4, the operation of higher-order gradiometers in both shielded and unshielded environments will be demonstrated in the case of a whole-cortex MEG system. The levels of noise cancellation, the effects of shielded-room distortions, motion, and vibrations will be discussed. Examples of 3rd-order gradiometer operation in an unshielded environment will be shown. In section 5, noise elimination by adaptive methods will be examined. The conditions under which adaptive methods are expected to be effective will be determined, and frequency independent and frequency dependent adaptation will be demonstrated. One example will be shown where the combination of high-order gradiometer and adaptive-noise cancellation was used successfully.

The notation used in this presentation is as follows: Capital letters will denote the actual fields and gradients applied to the SQUID detector. Thus B, $G = G^{(1)}$, $G^{(2)}$, $G^{(3)}$, etc., will denote the environmental field, 1st, 2nd and 3rd gradients, respectively. These quantities have dimensions of T, T/m, T/m^2, T/m^3 etc., and are vectors or tensors of rank 1, 2, 3 and 4, respectively. Lower-case letters will denote the SQUID detector outputs. Thus, b, $g = g^{(1)}$, $g^{(2)}$, $g^{(3)}$, etc., denote the outputs of magnetometer, 1st-, 2nd- and 3rd-order gradiometers. These quantities have dimensions of Φ_o, where Φ_o is the flux quantum. The SQUID detector outputs are associated with the applied fields and gradients via the SQUID detector gain and, in the case of gradiometers, appropriate baselines. Thus, e.g., magnetometer and 3rd order gradiometer outputs are given as

$$b = \alpha_B \cdot B \qquad (1.1)$$

$$g^{(3)} = \alpha_{G3} \cdot G^{(3)} \cdot d_1 \cdot d_2 \cdot d_3 \qquad (1.2)$$

where α_B and α_{G3} are the magnetometer and 3rd-order gradiometer gains, and d_1, d_2, and d_3 are the baselines corresponding to 1st-, 2nd- and 3rd-order gradiometers. The gains α_B

and α_{G3} are in units Φ_o/fT. The 1st- and 2nd-order gradiometer gains are defined similarly.

2. Magnetic Environments

Biomagnetic measurements are carried out in a world which is polluted by magnetic noise from, e.g., moving cars and small magnetic objects. In this section, these magnetic environments will be reviewed. As examples, the change of the magnetic noise as a function of traffic patterns will be shown, and the effects of power lines and the Earth's field will be examined.

2.1. DESCRIPTION OF MAGNETIC ENVIRONMENTS

To appreciate the magnitude of environmental disturbances relative to biomagnetic fields, examples of both are displayed on the same diagram in Figure 2.1.

The severity of the magnetic noise depends on whether the measuring instrumentation is located in a busy urban area or a relatively quiet rural area. The magnetic disturbances can be eliminated either by shielding or by noise cancellation techniques. The shielding can be accomplished by μ-metal rooms, which are usually supplemented by thick Al or Cu layers to enhance their high-frequency attenuation. Roughly speaking, two types of μ-metal rooms are in use: rooms with a modest low-frequency shielding factor (less than 100) [12 to 15] and rooms with a large low-frequency shielding factor ($\approx 10^4$) [16, 17].

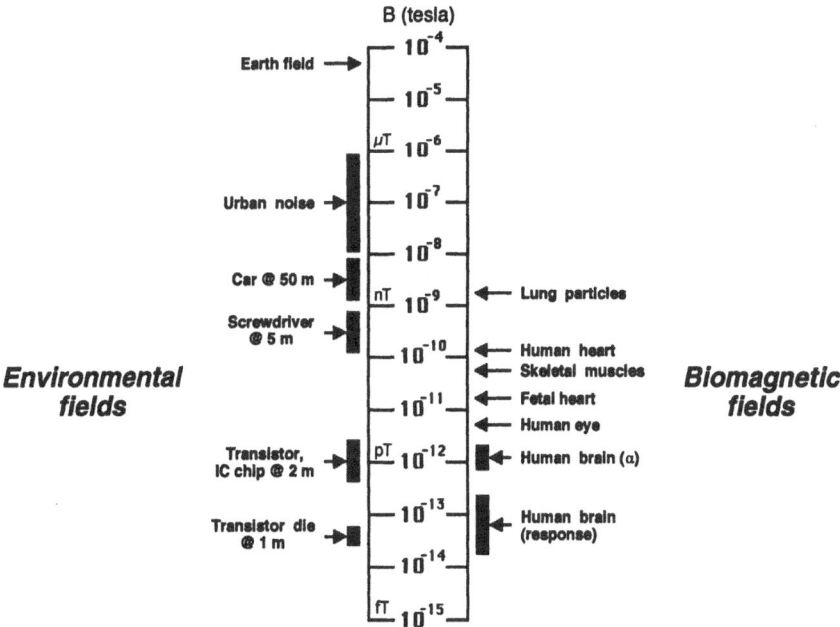

Figure 2.1. Comparison of selected biomagnetic fields and environmental disturbances.

120

Sometimes, the shielding is accomplished by eddy currents using a very thick layer of high-conductivity metal [18]. However, the eddy-current shielded rooms do not provide any significant shielding at low frequencies. The shielding also can be provided by superconducting shields. One example is the proposed superconducting helmet [19], which is projected to provide a shielding factor of about 3,500. Another example is a whole-body high-T_c superconducting shield, already in operation [7, 20], which provides a shielding factor of approximately 10^8.

The magnitudes of environmental magnetic fields at various locations are shown in Figure 2.2. The environmental noise has an approximately 1/f character and in an unshielded urban environment the noise levels at 1 Hz are in the range from about 100 pT to 10 nT rms/√Hz (Figure 2.2.a, b, f). At power line frequencies, the noise amplitudes can be several hundred nT. Also, the noise spectra contain various environmental and vibrational lines, as evidenced in Figure 2.2.a.

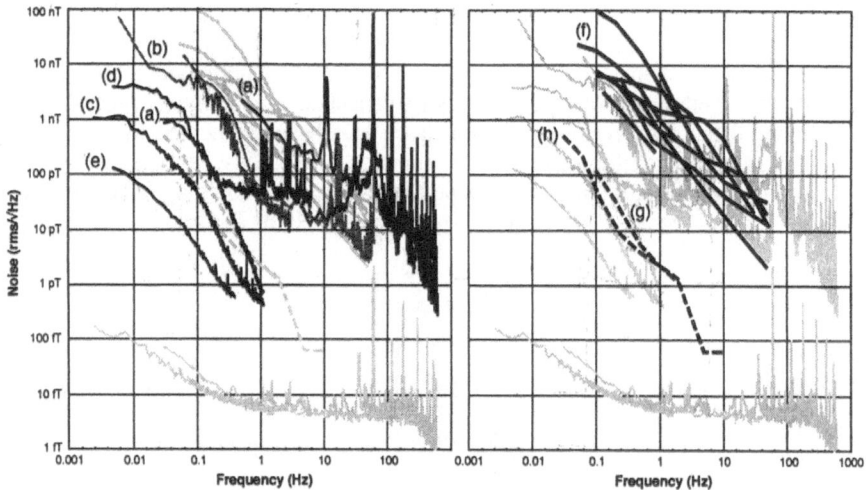

Figure 2.2. Magnitudes of the environmental magnetic field noise. (a) measured at CTF Systems Inc. (two curves correspond to collections at different times), (b) measured at the Simon Fraser University, Brain Behaviour Laboratory, (c) inside shielded room located in a "normal" urban environment, (d, e) inside a shielded room located about 50 m from a subway station (d - during the day, e - at night), (f) noise from [15, 16, 21 to 24], (g) geophysical noise [25], (h) noise measured approximately 6 km from CTF Systems Inc. in 1982. For ease of comparison, the noise plots which are not being discussed are shown in light gray.

The low-frequency magnetic field noise levels inside modestly shielded μ-metal rooms [13, 14] are shown by curves c, d and e in Figure 2.2. Curve c corresponds to a shielded room which is not close to any very large noise source. The low-frequency magnetic noise for that room is roughly 1 to 2 orders of magnitudes smaller than the noise in the unshielded environment. The curves d and e correspond to the magnetic noise inside a shielded room installed about 50 m from a subway line. Figure 2.2.d corresponds to the noise during the day when the trains are running, and Figure 2.2.e corresponds to the noise at night when there is no train traffic. The difference between these two noise levels is more than 1 order of magnitude and indicates that even inside shielded rooms, the noise environment can be quite variable.

By comparison, the magnetic field noise inside the whole body superconducting shield has been shown to be no more than 3 fT rms/√Hz for frequencies above 1 Hz [20] (limited by instrumental resolution).

2.2. CHANGES OF THE MAGNETIC NOISE DUE TO CAR TRAFFIC

The variability of environmental noise as a function of time is shown in Figure 2.3. In Figure 2.3.a the noise spectra as measured during the periods of light and peak traffic are shown by solid black lines (f_s = 32 samples/sec, points per trial = 4096, number of trials = 4). During the period of light traffic (at night), the noise exhibits roughly 1/f character, while during the peak period, the noise exhibits a roughly $1/f^4$ character in the vicinity of the low-frequency onset.

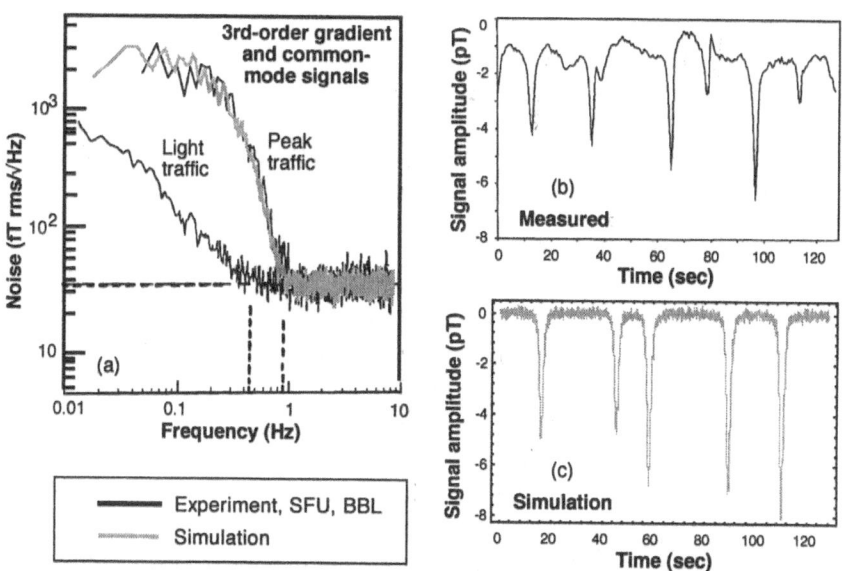

Figure 2.3. Dependence of environmental noise on the traffic pattern. Measurement was performed with a 3rd-order hardware gradiometer (RF SQUIDs, 1980, at the Simon Fraser University, Brain Behaviour Laboratory). The detected traffic noise is due to the field and 1st-order gradient common-mode signals. (a) frequency spectra of the measured and simulated noise, (b) measured time signal, (c) simulated time signal.

The time signal measured during the peak traffic period is shown in Figure 2.3.b. The spikes in the data correspond to individual cars passing on a road at a distance of approximately 20 m from the instrument. The traffic pattern also was simulated by assuming the following parameters: car velocity v = 14 m/sec, spacing between the cars 250 m with 80% variability, car dipole moments M = 10^{17} fT·cm³ with 50% variability, dipole-moment orientations distributed within a cone whose axis was oriented along the direction of the Earth's field, and cone angle 30 degrees. For this particular hardware gradiometer, the field and 1st-order gradiometer common-mode signals were shown to be responsible for the observed car noise. The magnitudes of the common-mode vectors - see Section 3.1.2 - were measured as $C_B = 10^{-4}$ and $C_{G1} = 2 \times 10^{-2}$. The car signals were simulated using these values, and the resulting time data are shown in Figure 2.3.c. The magnitudes of the simulated and observed car peaks are in rough agreement.

The frequency spectrum of the simulated signal is shown by the gray line in Figure 2.3.a. The agreement between the measured and simulated frequency spectra is excellent and the $1/f^4$ noise character at the onset of the low-frequency noise is reproduced well by the simulation.

The shape of this low-frequency noise is typical of the noise measured in the vicinity of a road or any passage where magnetic objects are moving. This shape also can be derived analytically. Assume a dipole source and neglect the vector and tensor character of the magnetic fields and gradients. Then the magnitude of the k-th gradient can be written as

$$G^{(k)}(t) = \frac{a_k \cdot M}{R^{3+k}} \qquad \text{where} \qquad a_k = \frac{(2+k)!}{2} \qquad (2.1)$$

where M is the dipole moment, and R is the distance from the dipole. Assume that the dipole is moving along a straight path, with the point of closest approach at a distance D_o from the gradiometer. If the dipole velocity is v and the time is measured from the point of closest approach, then the distance to the dipole can be expressed as

$$R = D_o \cdot \sqrt{1 + \left(\frac{v \cdot t}{D_o}\right)^2} \qquad (2.2)$$

Substituting Eq.2.2 into Eq.2.1, the square of the Fourier transform of the k-th gradient for a dipole moving along a straight path can be determined as [26]

$$F\left\{G^{(k)}\right\}^2 = \frac{\pi \cdot D_o^{-2 \cdot (1+k)} \cdot (M \cdot \omega)^2 \cdot \left(\frac{D_o \cdot \omega}{v}\right)^k \cdot \left[K_{1+\frac{k}{2}}\left(\frac{D_o \cdot \omega}{v}\right)\right]^2 \cdot \left[(2+k)!\right]^2}{2^{k+2} \cdot v^4 \cdot \Gamma\left(\frac{3}{2}+\frac{k}{2}\right)^2} \qquad (2.3)$$

where $\omega = 2\pi \cdot f$ is the frequency, $K_n(z)$ is the modified Bessel function of the second kind and $\Gamma(z)$ is the Gamma function. Since the car signals are not overlapping, it is possible to compute the spectral power for 1 car and then add the powers of 5 cars (Figure 2.3) to match the observed signals. When this is done, and the appropriate common mode vectors and normalizations are applied, the analytically computed and the observed frequency spectra are seen to agree well (as shown in Figure 2.4). The characteristic shape of the spectrum is determined by the Bessel function $K_n(z)$, and the onset of the low-frequency noise is related to the transit time of the car as observed by the gradiometer (the measure of the transit time, D_o/v, is in the argument of the function K_n, which dominates the behaviour of Eq.2.3).

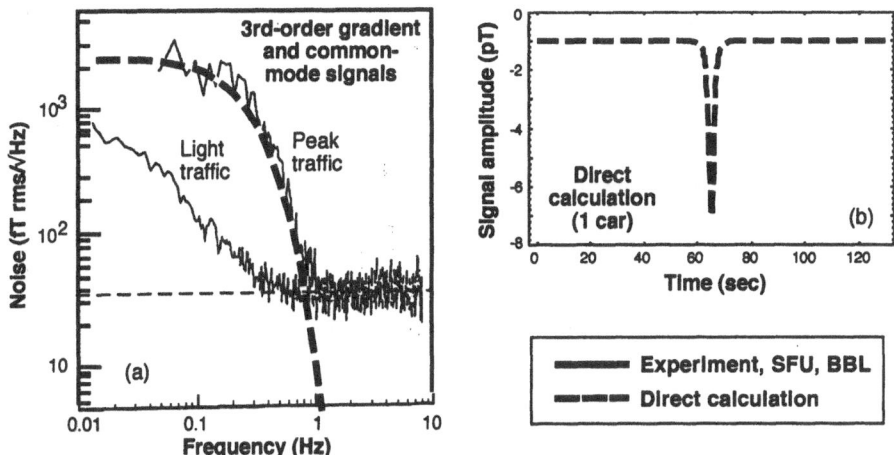

Figure 2.4. Comparison of analytically-determined and measured spectra of car traffic at a distance of 20 m from a 3rd-order hardware gradiometer. (a) superposition of the measured and analytically-computed spectra, (b) time trace computed for 1 car. $D_o = 20\,m$, $v = 14\,m/sec$, $M = 10^{17}\,fT\cdot cm^3$.

2.3. POWER LINE MAGNETIC FIELDS

Sometimes, the measured noise spectra do not exhibit $1/f^n$ character, but are nearly flat for frequencies less than the power-line frequency (Figure 2.2). This is caused by the steps in the amplitudes of the power-line magnetic fields - the steps are a consequence of switching electrical equipment on and off and abruptly, changing the current flow in the power lines. A typical example of such power-line steps is shown in Figure 2.5.

In addition to the amplitude steps, the power lines also produce numerous frequency lines [27]. For example, in British Columbia, Canada, the power lines produce frequency lines around 30 Hz which are a consequence of the natural resonance of 500-kV transmission lines compensated by a serial capacitor, lines around 2 Hz which are the result of locking smaller local power stations to the grid frequency, and a line around 0.5 Hz which is a resonant frequency of 500-kV lines connecting the British Columbia grid to the US grid.

If the amplitude of the power-line signal is ΔB_o and its frequency is f, then the maximum slew rate of the power-line signal is

$$S_\Phi = \alpha \cdot 2\pi \cdot f \cdot \Delta B_o \qquad [\Phi_o/sec] \qquad (2.4)$$

where α is the gain of the SQUID detector - see Eqs.1.1 and 1.2 - and the slew rate is expressed in Φ_o/sec. For magnetometers, the power-line signal amplitude is directly equal to the power-line magnetic field, $\Delta B_o = B_{\text{power line}}$. For gradiometers, the power-line signal amplitude is caused either by the field common mode, $\Delta B_o = C \cdot B_{\text{power line}}$, or directly by the gradient of the power-line field, $\Delta B_o = G \cdot d$, where d is the gradiometer baseline. The gradient of the power-line field can be estimated from the magnitude of its magnetic field by assuming that the power-line currents are carried by parallel wires.

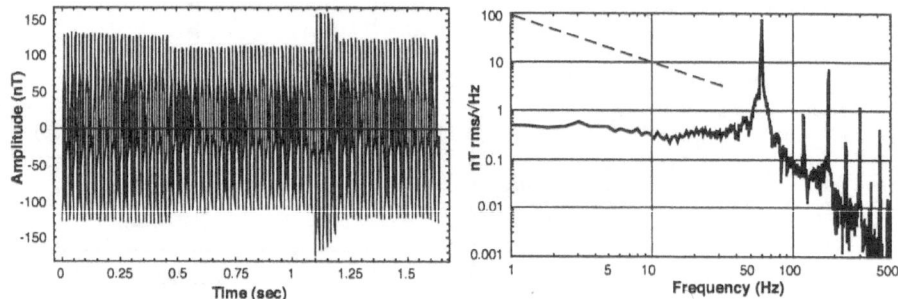

Figure 2.5. Example of power-line steps and their effect on the frequency spectrum. (a) measured steps in the amplitude of the power line magnetic field, (b) resulting frequency spectrum. Component B_3, $f_s = 1250$ samples/sec, points = 2048, low-pass filter = 300 Hz (4th order).

For parallel wires, the field decays as the inverse of distance squared, and the gradient and the field can be related by $G \cdot d/B \approx 2 \cdot d/R$, where R is the distance from the wires. The frequencies considered for the slew rate calculations are 60 and 180 Hz, because the experimentally-observed amplitudes of the first and third power-line harmonics often exhibit comparable amplitudes. The slew rates due to power-lines are summarized in Table 2.1, where C is the common mode vector magnitude, discussed in Section 3.

Table 2.1: Slew rates due to power-line fields. $B_{\text{power line}} = 200$ nT, $G \cdot d = 4$ nT (estimated), $C = 10^{-2}$, d = 5 cm, R = 5 m, $\alpha_B = 10^{-7} \, \Phi_o/\text{fT}$, $\alpha_{GI} = 3 \times 10^{-6} \, \Phi_o/\text{fT}$.

		Gradiometer		
Frequency	Magnetometer	due to C	due to grad.	Units
f = 60 Hz	3,700	2,300	4,500	Φ_o/sec
f = 180 Hz	22,600	6,800	13,600	Φ_o/sec

The parameters and slew rates in Table 2.1 are representative of MEG instruments operated in unshielded environments. Thus even in unshielded environments, the slew rates are relatively low and do not represent a significant challenge from the point of view of the SQUID detector design.

The dynamic range imposed by the power-line fields is examined next. The dynamic range can be defined as the ratio of the applied peak-to-peak signal to the system resolution in a 1-Hz bandwidth. Using the parameters of Table 2.1 and assuming that the magnetometer and gradiometer resolutions in the 1-Hz bandwidth are 100 fT rms/√Hz and 5 fT rms/√Hz, respectively, the dynamic ranges due to the power-line fields are

magnetometers:	$D_{B\ power\ line} = 22$ bits	(2.5.a)
gradiometers:	$D_{G1\ power\ line} = 21$ bits	(2.5.b)

This dynamic range is quite large. However, it is well within the capabilities of the digital SQUID electronics described in Section 4.1.

2.4. THE EARTH'S FIELD

The magnitudes of the Earth's field and its 1st gradients are listed in Table 2.2. The values are only approximate, and the geological and laboratory 1st gradients can vary significantly depending on local conditions - the laboratory gradient was measured at the CTF laboratories in 1982.

Table 2.2: Approximate magnitudes of the DC environmental fields and 1st gradients

	Origin	Magnitude	Units
Magnetic field, B_E	Earth dipole	50	μT
	Earth dipole *	0.02	nT/m
1st gradient, G_E	Geology	2	nT/m
	Laboratory	200	nT/m

* $G_E = 3 \cdot B_E/R$, where $R \approx 6{,}380$ km

Large DC fields affect the SQUID detector only when there is relative motion between the SQUID detector and the field, e.g., when the system vibrates. The vibrations can be either rotational or translational. Assume that the rotational vibrations are caused by small rotations θ and the translational vibrations are caused by small shifts Δd. For a magnetometer this results in rotational and translational field noise given by

$$\Delta B_R \approx B_E \cdot \theta \qquad (2.6.a)$$

$$\Delta B_T \approx G_E \cdot \Delta d \qquad (2.6.b)$$

Rotational noise is proportional to the magnitude of the environmental magnetic field, and the translational noise is proportional to the magnitude of the gradient of the local fields. For gradiometers, vibrational noise can be caused either by field vibrational noise via the common mode vector C, or directly, by vibrations in the local gradients. Gradiometer rotational and translational noise due to the common mode vector is given by

$$\Delta G_R^C \approx C \cdot \Delta B_R = C \cdot B_E \cdot \theta \qquad (2.7.a)$$

$$\Delta G_T^C \approx C \cdot \Delta B_T = C \cdot G_E \cdot \Delta d \qquad (2.7.b)$$

and gradiometer vibrational noise due to the environmental gradients is

$$\Delta G_R^{grad} \approx \theta \cdot G_E \cdot d \qquad (2.8.a)$$

$$\Delta G_T^{grad} \approx G^{(2)} \cdot \Delta d \cdot d \qquad (2.8.b)$$

where $G^{(2)}$ is the 2nd gradient of the local fields. If the SQUID system is operated in a shielded room with attenuation factor ψ ($\psi \le 1$), then vibrational noise caused by the field (i.e. ΔB_R or ΔG_R^C) is reduced proportionally to the DC attenuation factor, and vibrational noise caused by the gradients (i.e., ΔB_T, ΔG_T^C and ΔG_R^{grad}) depends on the residual DC gradients in the shielded room, which may be quite different from the outside gradients multiplied by the room attenuation factor, ψ.

Assume that the vibrational amplitudes are $\theta \approx 10^{-5}$ rad (= 10 μm/m) and $\Delta d = 10$ μm, and the system is operated without any shielding, $\psi = 1$. The magnitudes of the vibrational noise for these conditions are summarized in Table 2.3.a.

Table 2.3.a: Magnitudes of the vibrational noise in an unshielded environment. $\psi = 1$, $C = 10^{-2}$, $d = 5$ cm, $\theta = 10^{-5}$ rad, $\Delta d = 10$ μm. Magnitudes of B_E and G_E are as in Table 2.2.

Rotational vibrations		Translational vibrations	
Noise type	Noise ampl. (fT)	Noise type	Noise ampl. (fT)
ΔB_R	5×10^5	ΔB_T (Earth)	0.2
ΔG_R^C	5,000	ΔB_T (Geology)	20
ΔG_R^{grad} (Earth)	10^{-2}	ΔB_T (Laboratory)	2,000
ΔG_R^{grad} (Geology)	1	ΔG_T^C (Earth)	2×10^{-3}
ΔG_R^{grad} (Laboratory)	100	ΔG_T^C (Geology)	0.2
		ΔG_T^C (Laboratory)	20

ΔG_T^{grad} was not computed because $G^{(2)}$ is not well known.

Now assume that the vibrational amplitudes are the same, and the system is operated in a modestly shielded room with a residual DC attenuation factor of $\psi = 10^{-3}$ and a residual DC gradient of ≈ 10 nT/m. The magnitudes of the vibrational noise under these conditions can be computed using Eqs.2.6 to 2.8 with B_E and G_E replaced by their corresponding shielded-room values. These results are summarized in Table 2.3.b.

Table 2.3.b: Magnitudes of the vibrational noise in a shielded room. $\psi = 10^{-3}$, $C = 10^{-2}$, $d = 5$ cm, $\theta = 10^{-5}$ rad, $\Delta d = 10$ μm, $G_{Room} \approx 10$ nT/m and $B_{room} = \psi \cdot B_E$ and the magnitude of B_E is as in Table 2.2.

Rotational vibrations		Translational vibrations	
Noise type	Noise ampl. (fT)	Noise type	Noise ampl. (fT)
ΔB_R	500	ΔB_T	100
ΔG_R^C	5	ΔG_T^C	1
ΔG_R^{grad}	5		

ΔG_T^{grad} was not computed because $G^{(2)}$ is not precisely known.

If the required SQUID detector resolution is considered to be 5 fT, then the rotational vibrations in Tables 2.3 produce unacceptable noise for the magnetometers and for the gradiometers via the common-mode signal, regardless of whether the detection system is shielded or not. The rotational vibrations due to the gradients produce excessive noise only when the system is rotating in the laboratory gradients without any shielding. The translational vibrations in Tables 2.3 produce excessive noise for the magnetometers vibrating in the geological or laboratory gradients without any shielding and also inside shielded rooms. Gradiometer translational-vibration noise due to the common-mode vector is too large only if the gradiometer is subjected to translational vibrations in the laboratory gradients without any shielding.

Obviously, good balancing and formation of high-order gradiometers will reduce significantly vibrational noise. In addition, sensors and references must be mounted rigidly together, since the noise caused by relative vibrations between them could not be removed by balancing or by higher-order gradient formation.

3. Gradiometers

Gradiometers have been used in the great majority of SQUID devices produced to date [28]. Their popularity is based on their ability to reject environmental noise - section 4.2. The gradiometer is a flux transformer consisting of a system of subtractive coils which are either attached to a single SQUID sensor (hardware gradiometers) or combined in software or firmware (software gradiometers).

The simplest flux transformer is a magnetometer, which consists of a single coil connected inductively to a SQUID sensor (Figure 3.1.a). The magnetometer measures the projection of the magnetic field along the coil normal.

When two magnetometers of opposite polarities are connected together, they form a 1st-order gradiometer (Figure 3.1.b). Similarly, two 1st-order gradiometers of opposite polarities connected together will produce a 2nd-order gradiometer (Figure 3.1.c), and two 2nd-order gradiometers of opposite polarities connected together will form a 3rd-order gradiometer (Figure 3.1.d), where this form of the 3rd-order gradiometer has been obtained by altering the distances between the gradiometer coils.

Examples in Figure 3.1 correspond to the so-called "hardware" gradiometers, i.e. gradiometers whose coils are connected together in such a way that the gradient signal is delivered to a single SQUID sensor. It also is possible to form a k-th-order gradiometer

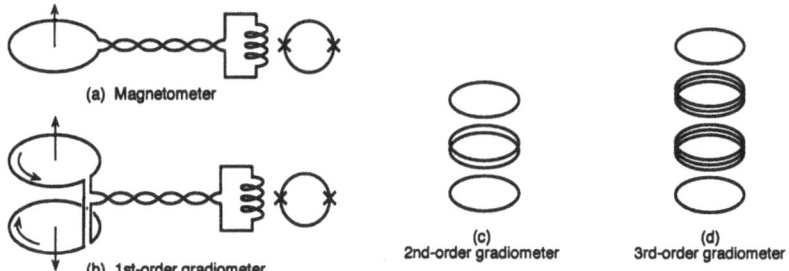

Figure 3.1. Examples of flux transformers. (a) magnetometer, (b) 1st-order gradiometer, (c) 2nd-order gradiometer, (d) 3rd-order gradiometer.

by measuring signals from magnetometers and/or hardware gradiometers with order less than k and combining these lower-order gradiometer outputs electronically or in software or firmware. Such a k-th-order gradiometer would be called a software gradiometer.

3.1. FIRST-ORDER GRADIOMETERS

First-order gradiometers will be described in greater detail, since they exhibit many basic features which are common to gradiometers of any order. The discussion will be independent of the way the gradiometers are realized, and thus, it will not be necessary to distinguish between hardware or software gradiometers (even though to aid comprehension, hardware gradiometers will be used in a number of examples).

3.1.1. Tensor Character of 1st-Order Gradiometer Signal

The 1st gradient is a spatial derivative of the magnetic field. As there are 3 components of the magnetic field and 3 spatial directions, the 1st gradient is a 9-component tensor. However, because $\text{div} B = 0$ the 1st-gradient tensor is traceless and because $\text{curl} B \approx 0$, the tensor is symmetrical. Consequently, there are only 5 linearly independent components of the gradient tensor [29]. One possible choice of the 5 components is shown on the right-hand side of Eq.3.1

$$G = \begin{pmatrix} G_{11} & G_{12} & G_{13} \\ G_{21} & G_{22} & G_{23} \\ G_{31} & G_{23} & G_{33} \end{pmatrix} = \begin{pmatrix} G_{11} & G_{12} & G_{13} \\ G_{12} & G_{22} & G_{23} \\ G_{13} & G_{23} & -G_{11}-G_{22} \end{pmatrix} \qquad (3.1)$$

Physically, the components of the 1st-gradient tensor may be realized as shown in Figure 3.2. The axial component G_{33} is realized by an axial gradiometer aligned along the x_3 axis (Figure 3.2.a). The off-axis components G_{32} and G_{23} are realized by planar gradiometers, oriented with the baseline along the x_2 axis and coil normal to the x_3 axis (Figure 3.2.b), and with the baseline along the x_3 axis and coil normal to the x_2 axis (Figure 3.2.c). Note that due to the 1st gradient tensor symmetry, the components G_{23} and G_{32} yield identical outputs.

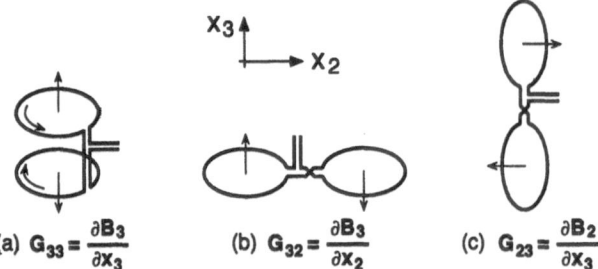

$$(a)\ G_{33} = \frac{\partial B_3}{\partial x_3} \qquad (b)\ G_{32} = \frac{\partial B_3}{\partial x_2} \qquad (c)\ G_{23} = \frac{\partial B_2}{\partial x_3}$$

Figure 3.2. Physical realization of 1st-gradient tensor components. (a) component G_{33}, axial gradient, (b) component G_{32}, off- axis component, (c) component G_{23}, off-axis component.

In practice, gradiometers are not perfect, either because of manufacturing errors or due to the presence of conducting or superconducting objects in the vicinity. The gradiometer deviation from perfection can be described by means of common-mode and eddy-current vectors.

3.1.2. Common-Mode Vector

The field common-mode vector describes the gradiometer residual sensitivity to magnetic field. The origins of the common-mode vector can be either mechanical (imperfect construction, as in Figure 3.3.a and b), or the common-mode vector can be induced by the presence of a superconducting object near the gradiometer (Figure 3.3.c).

Figure 3.3. Sources of common-mode vector: (a) inequality of the coil areas; (b) coil tilt; (c) presence of a superconducting object near the gradiometer.

The mechanical sources can be evaluated easily. Consider a gradiometer configuration as in Figure 3.4. This gradiometer consists of 2 coils at locations $\mathbf{u}_1$ and $\mathbf{u}_2$ with coil orientations $\mathbf{p}_1$ and $\mathbf{p}_2$ (unit vectors), coil areas P_1 and P_2, and number of turns N_1 and N_2. Define a coil area error parameter ξ as

$$N_j \cdot P_j = \xi_j \cdot P \qquad j = 1, 2 \tag{3.2}$$

where P is an effective coil area. The signal applied to the gradiometer is the sum of fluxes applied to both coils, Φ, divided by the effective coil area, P,

$$\Delta B = \frac{\Phi}{P} = \xi_1 \cdot (\mathbf{p_1} \cdot \mathbf{B_1}) + \xi_2 \cdot (\mathbf{p_2} \cdot \mathbf{B_2}) \tag{3.3}$$

where $\mathbf{B}_1$ and $\mathbf{B}_2$ are vectors of magnetic field at the locations $\mathbf{u}_1$ and $\mathbf{u}_2$. A Taylor expansion can be used to express the field at a point $\mathbf{u}$ in terms of the field at the origin, $\mathbf{B}°$, and the 1st gradient, G, as $\mathbf{B} = \mathbf{B}° + G \cdot \mathbf{u} + \dots$. Define the baseline vector $\mathbf{d} = \mathbf{u}_1 - \mathbf{u}_2$ and shift the origin of coordinates to the center of the baseline vector. Then Eq.3.3 may be rewritten as

$$\Delta B = C \cdot B + p \cdot G \cdot d \tag{3.4}$$

where

$$C = \xi_1 \cdot \mathbf{p}_1 + \xi_2 \cdot \mathbf{p}_2 \tag{3.5.a}$$

$$p = \frac{-\xi_1 \cdot \mathbf{p_1} + \xi_2 \cdot \mathbf{p_2}}{2} \tag{3.5.b}$$

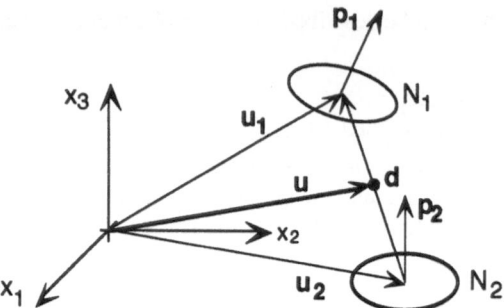

Figure 3.4. Configuration of a 1st-order gradiometer used for evaluation of the common-mode vector.

The common-mode vector is now expressed in terms of the coil area errors ξ_1 and ξ_2, and the individual coil orientations $\mathbf{p}_1$ and $\mathbf{p}_2$. As an example, examine an ideal gradiometer where $\mathbf{p}_2 = -\mathbf{p}_1$ and $\xi_1 = \xi_2$. For such a gradiometer $\mathbf{C} = 0$ and $\mathbf{p} = \mathbf{p}_2$. As a by-product of this common-mode vector analysis, the gradiometer response to the applied gradient is shown to be equal to the gradient tensor projection on the baseline and coil vectors $\mathbf{d}$ and $\mathbf{p}$ (last term, $\mathbf{p \cdot G \cdot d}$, in Eq.3.4).

3.1.3. Eddy-Current Vector

If there is a normal-metal object near the gradiometer, then the time-varying applied fields will excite currents in it, which in turn will generate magnetic fields and affect the gradiometer. This effect of the time-varying fields on the gradiometer performance can be described in terms of an eddy-current vector $\mathbf{E}$.

The normal-metal object can be thought of as an R-L circuit. Assume that a periodic spatially-uniform magnetic field is applied to the R-L circuit (with the gradient tensor equal to zero).

$$B = B_o \cdot \sin \omega \cdot t \tag{3.6}$$

This field will generate current i_1 in the circuit

$$i_1 = I_1 \cdot \sin(\omega \cdot t + \varphi) \tag{3.7.a}$$

where

$$\tan \varphi = \omega_1 / \omega \tag{3.7.b}$$

$$\omega_1 = R/L \tag{3.7.c}$$

$$I_1 = \frac{\omega}{\sqrt{\omega^2 + \omega_1^2}} \cdot B_o \cdot \frac{A_1}{L} \tag{3.7.d}$$

and R and L are the resistance and inductance of the R-L circuit, and A_1 is the inductor area. If a magnetometer coil is in the vicinity of such R-L circuit, the current i_1 will generate a magnetic field B_1 in the magnetometer coil (where B_1 is an average magnetic field applied to the magnetometer coil)

$$B_1 = B_0 \cdot \kappa \cdot \frac{\omega}{\sqrt{\omega^2 + \omega_1^2}} \cdot \sin(\omega \cdot t + \varphi) \qquad (3.8)$$

where the coupling constant κ is given by

$$\kappa = \frac{M \cdot A_1}{L \cdot A_2} \qquad (3.9)$$

with M being the mutual inductance between the R-L circuit and the magnetometer coil, and A_2 is the area of the magnetometer coil. The magnetometer is now exposed to the original field B and the field B_1, so the total field acting on the magnetometer is

$$B_T = B + B_1 = \left(1 + \kappa \cdot \frac{\omega^2}{\omega^2 + \omega_1^2}\right) \cdot B + \kappa \cdot \frac{\omega_1}{\omega^2 + \omega_1^2} \cdot \dot{B} \qquad (3.10)$$

This total field is frequency dependent and is composed of a term proportional to the applied field and a term proportional to the time derivative of the applied field ($\dot{B}$). Consider now a gradiometer positioned in the vicinity of a conducting object (Figure 3.5). This gradiometer can be thought of as two coils coupled to the conducting object by the coupling constants κ_1 and κ_2, and exposed to total fields B_{T1} and B_{T2}.

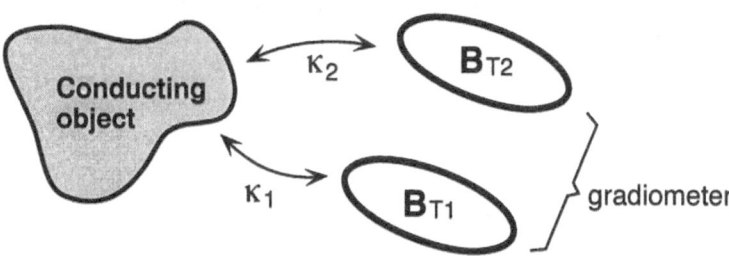

Figure 3.5. Gradiometer in the vicinity of a conducting object.

Further, assume that the gradiometer is imperfectly constructed and its common-mode vector is C_o. For simplicity, assume that the total signal applied to the gradiometer is

$$\Delta B \approx (1 + C_o) \cdot B_{T1} - B_{T2} \qquad (3.11)$$

Substituting for the B_T's from Eq.3.10, the signal seen by the gradiometer can be written as

$$\Delta B = C \cdot B + E \cdot \dot{B} \qquad (3.12)$$

where

$$C = C_o + \left(\Delta\kappa + C_o \cdot \kappa_1\right) \cdot \frac{\omega^2}{\omega^2 + \omega_1^2} \qquad (3.13)$$

$$E = \left(\Delta\kappa + C_o \cdot \kappa_1\right) \cdot \frac{\omega_1}{\omega^2 + \omega_1^2} \qquad (3.14)$$

$$\Delta\kappa = \kappa_1 - \kappa_2 \qquad (3.15)$$

Thus, the gradiometer output consists of the common-mode term (proportional to the applied field) and the eddy-current term (proportional to the time derivative of the applied field). Both vectors C and E are frequency dependent, and it is important to note that the normal-metal object introduces not only the eddy-current term, but also a frequency-dependent common-mode term.

In the high-frequency limit, when $\omega \gg \omega_1$ (or $\omega_1 \to 0$, or $R \to 0$), the material behaves like a superconductor, and the vectors C and E become

$$\omega \gg \omega_1: \qquad C = C_o + C_s \qquad (3.16.a)$$

$$E = 0 \qquad (3.16.b)$$

$$C_s = \Delta\kappa + C_o \cdot \kappa_1 \qquad (3.17)$$

where C_s is the magnitude of the common-mode vector which would be induced if the object were superconducting. In the high-frequency limit, the conducting object affects only the common-mode vector. The situation is opposite in the low-frequency limit, where the conducting object induces only the eddy-current vector

$$\omega \ll \omega_1: \qquad C = C_o \qquad (3.18.a)$$

$$E = \frac{C_s}{\omega_1} \qquad (3.18.b)$$

An acceptable magnitude of the eddy-current vector can be estimated by requiring that the eddy-current signal is smaller than the common-mode signal when the common-mode vector is balanced to a required level C_B:

$$E \cdot \dot{B} < C_B \cdot B \qquad (3.19)$$

Assuming that the applied field is periodic, the time derivative can be computed easily. Then, considering only the magnitudes, the eddy-current vector is acceptable if

$$E < \frac{C_B}{\omega} \qquad (3.20)$$

In practice, the condition in Eq.3.20 may be difficult to achieve, and the elimination of the eddy-current signal by methods similar to the common-mode balancing must be performed.

As an example, consider a brass cylinder with radius a = 2 cm, length h = 10 cm, wall thickness t = 0.5 cm, and resistivity $\rho = 6.4 \times 10^{-8}$ $\Omega \cdot$m, coupled to an axial gradiometer with baseline d = 8 cm, gradiometer common-mode vector $C_o = 2 \times 10^{-3}$, where the brass cylinder is positioned coaxially with the gradiometer such that the cylinder center is at a distance $\ell = 60$ cm from the closest gradiometer coil. Such a brass cylinder results in $C_s = 3.8 \times 10^{-5}$ and a characteristic frequency of $f_1 = 207$ Hz ($\omega_1 = 2\pi \cdot f_1$).

To understand the character of the common-mode and eddy-current vectors induced by the normal-metal object, the frequency dependence of various quantities is plotted as a function of normalized frequency $x = \omega/\omega_1$ in Figure 3.6 for the brass cylinder discussed above. The quantities plotted are E (Eq.3.14), $\omega \cdot$E, C_B, and ΔC, where C_B is the ultimate level of the common-mode balance and ΔC is the common-mode vector induced by the conducting object (Eq.3.13)

$$\Delta C = C - C_o = \left(\Delta \kappa + C_o \cdot \kappa_1\right) \cdot \frac{\omega^2}{\omega^2 + \omega_1^2} \tag{3.21}$$

The eddy-current vector E is frequency independent for x << 1, and its magnitude decreases as $1/\omega^2$ for x >> 1. As a result, $\omega \cdot$E, which is proportional to the magnitude of the eddy-current signal, increases as ω for x << 1, and decreases as $1/\omega$ for x >> 1. The common-mode vector ΔC increases as ω^2 for x << 1 and is independent of frequency for x >> 1. The required ultimate balance, C_B, is frequency independent and is crossed by the $\omega \cdot$E term at x = 0.026 (frequency f $\approx$ 5.6 Hz). This result implies that for frequencies less than 5.6 Hz, the eddy-current signal due to the brass cylinder is negligible, while for frequencies greater than 5.6 Hz, it is dominant and cannot be neglected.

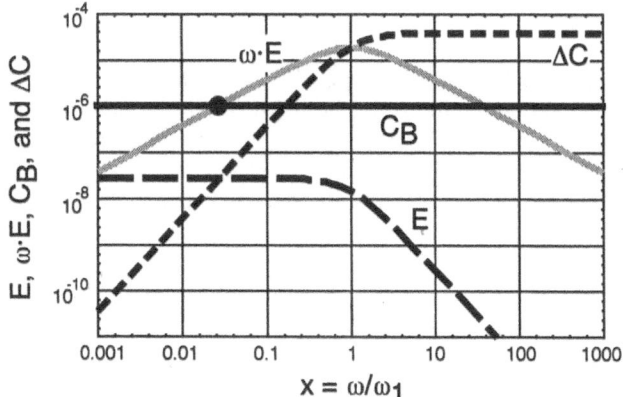

Figure 3.6. Frequency dependence of the common-mode and eddy-current vectors induced by a brass cylinder (discussed above). $C_s = 3.8 \times 10^{-5}$, $f_1 = 207$ Hz, $C_o = 2 \times 10^{-3}$, $C_B = 10^{-6}$.

3.1.4. Phase Shifts

Consider a system consisting of one gradiometer and a 3-component magnetometer. Such a system can measure the gradient and use the 3-component magnetometer to subtract the common-mode signal. Assume that there is a phase shift φ between the gradiometer and the magnetometers (Figure 3.7), the applied magnetic field is spatially uniform and periodic (as in Eq.3.6), eddy-current effects are absent and the gradiometer output is caused by the common-mode signals only.

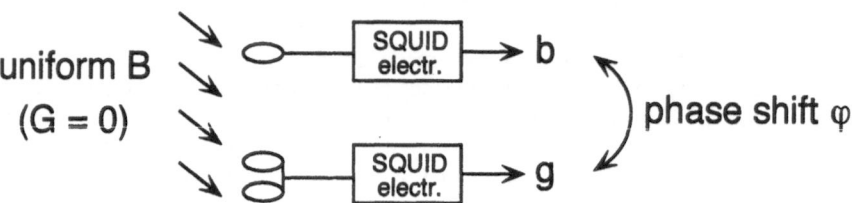

Figure 3.7. Phase shift between SQUID channels.

Then, the gradiometer output is $g = \alpha_G \cdot C_o \cdot B_o \cdot \sin \omega \cdot t$, and the magnetometer output is phase shifted as $b = \alpha_B \cdot B_o \cdot \sin(\omega \cdot t + \varphi)$. Assume that the magnetometer output is used to balance out completely the gradiometer common-mode signal and that the phase shift φ is small. Denote the balanced gradiometer output by s. Then

$$s = g - \frac{\alpha_G}{\alpha_B} \cdot C_o \cdot b \approx \alpha_G \cdot \tilde{E} \cdot \dot{B} \tag{3.22}$$

where α_B and α_G are the magnetometer and gradiometer gains, and $\tilde{E}$ is defined by

$$\tilde{E} = \frac{C_o \cdot \varphi}{\omega} = C_o \cdot \delta t \tag{3.23}$$

where the phase shift has been related to the time delay δt by $\varphi = \omega \cdot \delta t$. Eqs.3.22 and 3.23 indicate that if the balancing is performed between the phase-shifted channels, an effective eddy-current signal will be introduced. An acceptable magnitude of the phase shift or time delay between the channels may again be estimated from Eq.3.19. Substituting Eq.3.23 into Eq.319, the acceptable time delay between the channels is

$$\delta t \leq \frac{C_B}{C_o} \cdot \frac{1}{\omega} \tag{3.24}$$

Consider an example which might be relevant to biomagnetism: if $f = 100$ Hz, $C_B \approx 10^{-6}$, $C_o \approx 2 \times 10^{-3}$, then $\delta t \leq 0.8$ μsec. For comparison, the delay between two first-order feedback loops with approximately 30 kHz bandwidth and 10% bandwidth difference, is about 0.5 μsec. Thus, the effect of the phase shifts may not be entirely negligible if the system is not designed properly.

3.1.5. General Form of the Gradiometer Signal

The eddy-current and phase-shift analyses have been performed assuming that the applied fields are uniform, i.e., that the gradients are zero. If the effect of applied gradients also is considered, Eq.3.4, the general form of the gradiometer output is given by

$$g = \alpha_G \cdot \left(\mathbf{C} \cdot \mathbf{B} + \mathbf{E} \cdot \dot{\mathbf{B}} + \mathbf{p} \cdot \mathbf{G} \cdot \mathbf{d} \right) \qquad (3.25)$$

Note that the common-mode and the eddy-current terms in Eq.3.25 also could be unified using a complex notation. The first two terms in Eq.3.25 are undesirable. They usually are minimized by the system design, and are balanced-out by the common-mode and eddy-current vector balancing. The balancing procedures can be executed either in hardware, electronically or in software. The three methods are shown in Figure 3.8.

In the hardware method (Figure 3.8.a and b), only the common-mode vector usually is balanced. The balancing can be carried out either by bringing small pieces of superconductor (e.g., superconducting loops) to the gradiometer vicinity to balance out the imperfections [30], or the gradiometer flux-transformer leads can be made to contain small auxiliary axial gradiometers oriented in three orthogonal directions [30, 31]. These auxiliary gradiometers have sliding superconducting shields which expose the correct polarity coil to the environment, and thus inject the correct signal to balance the common-mode vector. In electronic balancing [32, 33] (Figure 3.8.c), the gradiometer signal is measured together with a 3-component magnetometer signal. The magnetometer signal and its derivatives are then fed into the gradiometer output and subtracted so as to yield a balanced gradiometer output. Software balancing (Figure 3.8.d) is similar to electronic balancing, except that the subtraction of the common-mode and eddy-current terms is done in software.

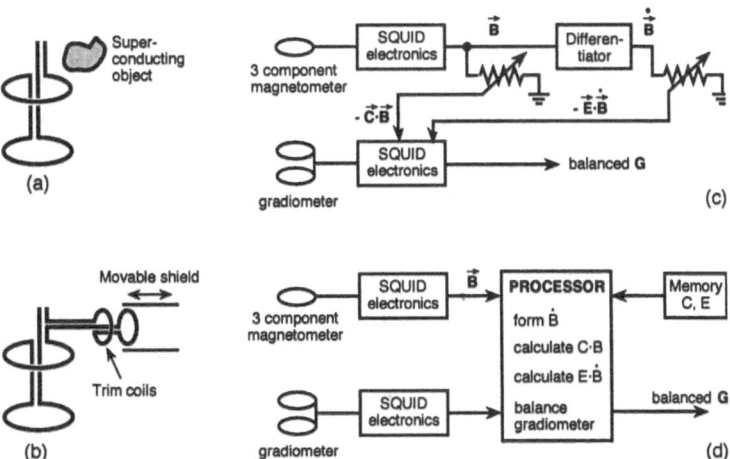

Figure 3.8. Methods of gradiometer balancing: hardware balancing by (a) superconducting vanes; (b) trim coils; (c) electronic method; (d) software method.

Figure 3.8 shows the basic balancing methods, but does not say anything about how the balancing coefficients are derived. The coefficients can be derived by applying uniform magnetic fields of different frequencies to the gradiometer system, measuring the system response and calculating or adjusting the vectors C and E. The application of uniform field changes to the system can be accomplished either by some configuration of uniform field coils [34, 37] or by moving the system in a sufficiently uniform field [38], or the coefficients can be determined adaptively by monitoring the environmental noise for sufficiently long periods of time [32, 39, 44]. Generally, the finesse of the balancing method is better than the uniformity of the applied field. However, the balancing precision is only as good as the degree of field uniformity. For example, the finesse of the hardware balancing vanes (Figure 3.8.a) is usually better than 10^{-6} and the finesse of software balancing is limited only by computer precision (and system stability). On the other hand, the practical achievable level of field uniformity in a typical laboratory is only about 10^{-4} to 10^{-5}.

3.2. GRADIOMETER REALIZATION

Gradiometers can be realized either in hardware or in software (or firmware). The two methods will be discussed below.

3.2.1. Hardware gradiometers

Hardware gradiometers consist of a number of coils connected to a common SQUID sensor. The coils are mutually additive or subtractive such that the resulting signal is equivalent to the required gradiometer order. Examples of possible coil arrangements for magnetometer, 1st-, 2nd- and 3rd-order axial gradiometers are shown in Figure 3.9 [40].

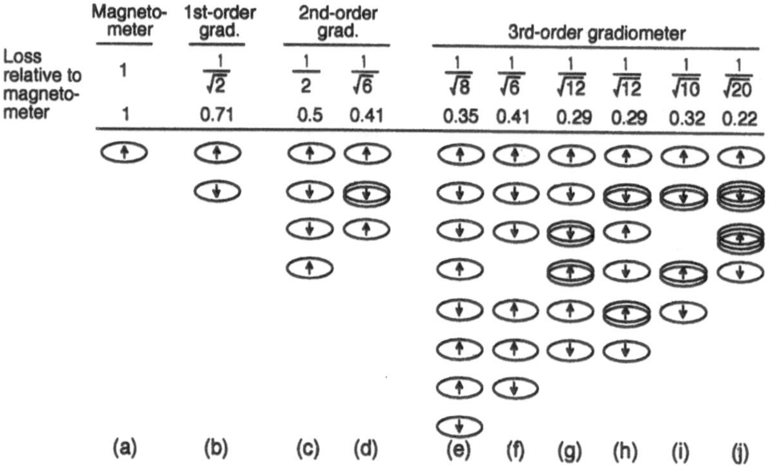

Figure 3.9. Coil arrangements for hardware magnetometer, 1st-, 2nd-, and 3rd-order gradiometers.

It was explained in connection with Figure 3.1 that a k-th-order hardware gradiometer can be obtained by connecting together two (k-1)th-order gradiometers of opposite polarity. The spacing between the gradiometer coils can be varied, resulting in different configurations for 2nd- or 3rd-order gradiometers, as shown in Figure 3.9. Consider the effects on 3rd-order hardware axial gradiometers obtained by changing the gradiometer length by an integer number of the baseline units, as shown in Figures 3.9.f to j. When coils of the same polarity are pushed to the same location, a multiturn coil will result, and when two coils of opposite polarity are moved into the same position, the coils become absent at that location. Each gradiometer in Figure 3.9 has different matching requirements and a different parasitic inductance (where the flux-transformer coil closest to the head or other magnetic region of interest is called the sensing coil, and the remaining coils are called the parasitic coils, since they provide parasitic inductance and degrade the gradiometer sensitivity). Signal losses due to parasitic inductance relative to a magnetometer, assuming that all coils have identical diameters, are shown for each gradiometer. For example, the 3rd-order hardware gradiometer in Figure 3.9.j is most practical because its length is short [40]. However, its loss due to parasitic inductance is $1/\sqrt{20} \approx 0.22$.

Flux-transformer gain can be optimized by matching SQUID input-coil inductance to flux-transformer inductance and by manipulating the inductance of the parasitic coils relative to the sensing-coil inductance. Consider a 3rd-order gradiometer flux-transformer, as in Figure 3.10. The diameter of the sensing coil has been determined on the basis of detection requirements, and the diameter of the parasitic coils has been enlarged [49].

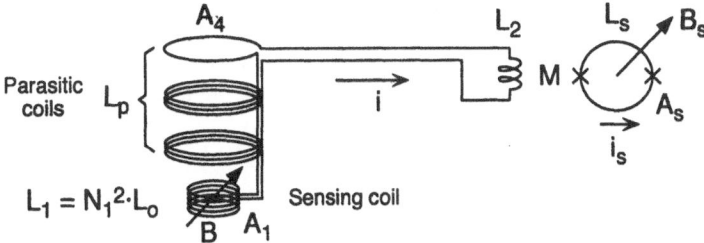

Figure 3.10. Flux-transformer connection to a SQUID sensor. The flux-transformer is asymmetric; the diameter of the parasitic inductance coils is larger than the diameter of the sensing coil. The effective area of the coils is equalized by increasing the number of turns of the sensing coil.

To compensate for the larger diameter of the parasitic coils, the number of turns of the sensing coil has been increased to N_1, where

$$A_4 = A_1 \cdot N_1 \tag{3.26}$$

Neglecting the inductance of the leads, the current in the flux-transformer circuit is

$$i = \frac{N_1 \cdot A_1 \cdot B}{L_1 + L_p + L_2} \tag{3.27}$$

The flux injected into the SQUID sensor is then

$$\Phi_s = B_s \cdot A_s = M \cdot i \tag{3.28}$$

where the geometrical mutual inductance can be expressed as $M = k\sqrt{(L_2 \cdot L_s)}$, where L_2 is the in-situ input-coil inductance, and L_s is the bare SQUID inductance - note that the flux-transformer treatment in a weak-coupling limit is similar. The flux-transformer field gain, r, can be defined as the average field injected into the SQUID sensor divided by the applied field

$$r = \frac{B_s}{B} = \frac{A_1}{A_s} \cdot \frac{N_1 \cdot k \cdot \sqrt{L_2 \cdot L_s}}{L_1 + L_p + L_2} \qquad (3.29)$$

The flux-transformer gain can be optimized by matching the coupling inductance L_2 to the total flux-transformer inductance, i.e., $L_2 = L_1 + L_p$. Then

$$r_{max} = \frac{k}{2} \cdot \frac{A_1}{A_s} \cdot \sqrt{\frac{L_s}{L_o}} \cdot \frac{1}{\sqrt{1 + \frac{L_p}{L_o \cdot N_1^2}}} \qquad (3.30)$$

where L_o is the inductance of a 1-turn coil with diameter equal to the sensing-coil diameter. If the diameter of the sensing coil is fixed (due to detection constraints), then according to Eq.3.30, the flux-transformer gain can be maximized by making the coupling constant k large, or by increasing the number of turns on the sensing coil, which is equivalent to increasing the diameter of the parasitic coils - see Eq.3.26. Note that the inductance L_p increases slower than N_1^2. These conclusions are valid for a gradiometer of any order. The flux-transformer optimization steps are summarized in Figure 3.11 for a 3rd-order gradiometer.

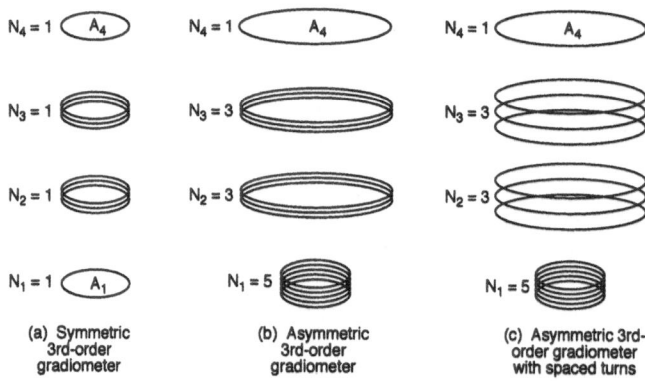

Figure 3.11. Optimization of flux-transformer gain for the example of 3rd-order gradiometer: (a) symmetric gradiometer; (b) asymmetric gradiometer; (c) asymmetric gradiometer with spaced turns.

The symmetric gradiometer in Figure 3.11.a has the largest parasitic losses (as shown in Figure 3.9). For the asymmetric gradiometer in Fig.3.11.b, the parasitic losses are reduced by increasing the parasitic-coil diameters and the number of turns on the sensing coil. The parasitic losses can be reduced further by increasing the turn spacing of the parasitic-coils, as in Figure 3.11.c (and also by slightly increasing the turn spacing of the sensing coil).

3.2.2. Software gradiometers

Software gradiometers can be implemented using hardware sensors and references. For a k-th-order software gradiometer, the sensor and the references need to be of order k-1 or lower. Implementing the gradiometers in software seems to be the most practical method for multi-channel systems, as one reference system can serve an arbitrary number of sensors [41].

Software formation of a 1st-order gradiometer from magnetometers. In this case the primary sensor is a magnetometer, and the reference system is a three-component magnetometer, as in Figure 3.12 [42 to 44]. For simplicity, it is assumed that all reference magnetometers have identical gains.

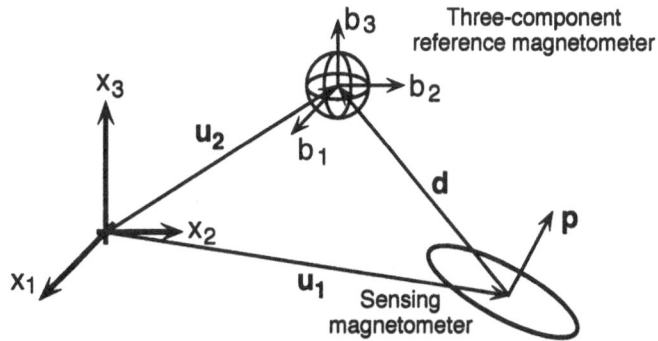

Figure 3.12. Software formation of a 1st-order gradiometer from magnetometers.

The reference magnetometer outputs are given by

$$b_k = \alpha_B \cdot B_k, \qquad \text{where } k = 1, 2, 3 \qquad (3.31)$$

and B_k is one of the three orthogonal components of the applied magnetic field, $\mathbf{B}$. The reference magnetometer outputs, b_k, form a vector of reference outputs, $\mathbf{b}$. The sensing magnetometer output is given by the projection of the applied magnetic field on the sensing-magnetometer-coil normal, $\mathbf{p}$

$$s = \alpha_s \cdot (\mathbf{p} \cdot \mathbf{B}) \qquad (3.32)$$

where α_s is the sensing magnetometer gain. To form a software gradiometer, the reference-magnetometer outputs are projected on the direction $\mathbf{p}$, the projection is scaled by the sensor-to-reference magnetometer gain ratio, and the result is subtracted from the sensor output. Then, similar to the derivation of Eq.3.4 (using the Taylor expansion for the magnetic field at a location $\mathbf{u}$, defining the gradiometer baseline as $\mathbf{d} = \mathbf{u}_2 - \mathbf{u}_1$, and shifting the origin to the center of the baseline), the software 1st-order gradiometer output, $g^{(1)}$, can be derived as

$$g^{(1)} = s - \frac{\alpha_S}{\alpha_B} \cdot (p \cdot b) = \alpha_S \cdot p \cdot G \cdot d \qquad (3.33)$$

This output is the same as Eq.3.4, except that the common-mode term is absent, because the software gradiometer considered here is assumed to be perfect.

Examples of ideal 1st-order gradiometers are shown in Figure 3.13. Figure 3.13.a corresponds to an axial gradiometer where $d = d \cdot (0, 0, 1)$, $p_2 = -p_1 = (0, 0, 1)$ and $g^{(1)} = \alpha_{G1} \cdot p \cdot G \cdot d = \alpha_{G1} \cdot d \cdot G_{33}$. Figure 3.13.b corresponds to an off-axis gradiometer where $d = d \cdot (0, 1, 0)$, $p_2 = -p_1 = (0, 0, 1)$ and $g^{(1)} = \alpha_{G1} \cdot d \cdot G_{23}$. In Figure 3.13.c the coils and the baseline are not aligned, $d = d \cdot (0, 0, 1)$, $p_2 = -p_1 = (0, 1/\sqrt{2}, 1/\sqrt{2})$ and $g^{(1)} = \alpha_{G1} \cdot (G_{23} + G_{33})/\sqrt{2}$. In this last case, the gradiometer output is a linear combination of the 1st-gradient tensor components.

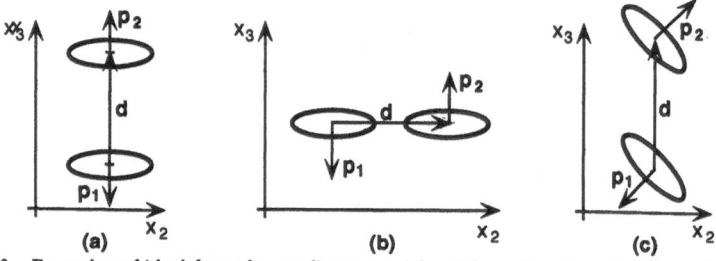

Figure 3.13. Examples of ideal 1st-order gradiometers; (a) axial gradiometer; (b) off-axis (planar) gradiometer; (c) gradiometer with tilted coils.

Software formation of a 2nd-order gradiometer [41]. Consider two ideal 1st-order gradiometers located at positions u and u', as in Figure 3.14. The gradiometers have coil vectors p and p', and baselines d and d'.

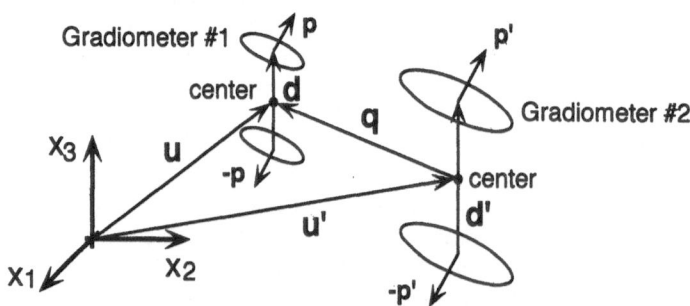

Figure 3.14. Software formation of 2nd-order gradiometer.

Use the Taylor expansion to express the gradient at a point $\mathbf{u}$ (in terms of the gradient at the origin) as $G(\mathbf{u}) = G + G^{(2)}{\cdot}\mathbf{u}$, where G is the 1st-gradient tensor measured at the origin, and $G^{(2)}$ is the 2nd-gradient tensor. The outputs of the ideal 1st-order gradiometers can then be written as (Eq.3.4)

$$g = \alpha_G{\cdot}\mathbf{p}{\cdot}(G + G^{(2)}{\cdot}\mathbf{u}){\cdot}\mathbf{d} \qquad (3.34.a)$$

$$g' = \alpha_{G'}{\cdot}\mathbf{p'}{\cdot}(G + G^{(2)}{\cdot}\mathbf{u'}){\cdot}\mathbf{d'} \qquad (3.34.b)$$

The software 2nd-order gradiometer output, $g^{(2)}$, with gain equivalent to the gain of gradiometer g, can be formed by scaling the output of g' by the ratio of the gains $\alpha_G/\alpha_{G'}$ and by the ratio of the baselines d/d', and by subtracting the scaled result from g. Assuming that the vectors $\mathbf{p}$ and $\mathbf{p'}$, and $\mathbf{d}$ and $\mathbf{d'}$ are parallel, defining the 2nd-order gradiometer baseline as $\mathbf{q} = \mathbf{u} - \mathbf{u'}$ and shifting the origin to the center of the 2nd-order gradiometer baseline, the 2nd-order gradiometer output can be expressed as

$$g^{(2)} = g - \frac{\alpha_G}{\alpha_{G'}} \cdot \frac{d}{d'} \cdot g' = \alpha_{G2} \cdot \mathbf{p} \cdot G^{(2)} \cdot \mathbf{q} \cdot \mathbf{d} \qquad (3.35)$$

where α_{G2} is the 2nd-order gradiometer gain. The output of the software 2nd-order gradiometer is then a projection of the 2nd-gradient tensor onto the gradiometer vectors $\mathbf{p}$, $\mathbf{q}$, and $\mathbf{d}$, where $\mathbf{p}$ is a unit vector, and $\mathbf{q}$ and $\mathbf{d}$ have lengths associated with them.

The two 1st-order gradiometers used for the construction of the software 2nd-order gradiometer can be constructed either in hardware, or either one or both 1st-order gradiometers can be formed in software. If the sensing 1st-order gradiometer is software-formed from magnetometers, then the above procedure describes the formation of a software 2nd-order gradiometer based on a magnetometer as a sensor.

Examples of ideal 2nd-order gradiometers are shown in Figure 3.15. Figure 3.15.a corresponds to a 2nd-order gradiometer where all three principal vectors are parallel, $\mathbf{p} \parallel \mathbf{d} \parallel \mathbf{q}$, and $\mathbf{p} = (0, 0, 1)$. The 2nd-gradiometer output is then $g^{(2)} = \alpha_{G2}{\cdot}\mathbf{p}{\cdot}G^{(2)}{\cdot}\mathbf{q}{\cdot}\mathbf{d} = \alpha_{G2}{\cdot}q{\cdot}d{\cdot}G_{333}$. The gradiometer in Figure 3.15.b corresponds to the case where $\mathbf{p} \parallel \mathbf{d} \perp \mathbf{q}$, and $\mathbf{p} = (0, 0, 1)$, $\mathbf{q} = q{\cdot}(0, 1, 0)$ and the 2nd-order gradiometer output $g^{(2)} = \alpha_{G2}{\cdot}q{\cdot}d{\cdot}G_{233}$. The gradiometer in Figure 3.15.c has a tilted baseline $\mathbf{q}$, and $\mathbf{p} \parallel \mathbf{d}$, $\mathbf{p} = (0, 0, 1)$, $\mathbf{q} = q{\cdot}(0, 1/\sqrt{2}, 1/\sqrt{2})$, and the software 2nd-order gradiometer output is $g^{(2)} = \alpha_{G2}{\cdot}q{\cdot}d{\cdot}(G_{233} + G_{333})$. In this case the gradiometer output is a linear combination of the 2nd-gradient tensor components.

In general, a software 2nd-order gradiometer is formed using a reference system consisting of a 5-component 1st-order tensor gradiometer and a 3-component vector magnetometer. The sensor can be either a magnetometer or a hardware 1st-order gradiometer (Figure 3.16). If the sensor is a magnetometer, then the software formation of the 2nd-order gradiometer proceeds in two steps: (1) a software 1st-order gradiometer with vectors $\mathbf{p}$ and $\mathbf{d}$ is generated, and (2) the software 2nd-order gradiometer is constructed by projecting the 1st gradient tensor onto the vectors $\mathbf{p}$ and $\mathbf{d}$, scaling it suitably and subtracting the result from the output of the software 1st-order gradiometer. If the sensor is a 1st-order gradiometer, then step 1 above is omitted, and the 2nd-order gradiometer is directly constructed as in step 2. Of course, if the sensor is a magnetometer, steps 1 and 2 can be combined, and performed together as one step.

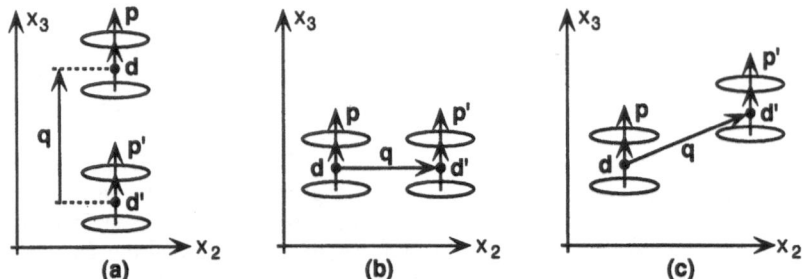

Figure 3.15. Examples of ideal 2nd-order gradiometers; (a) axial gradiometer G_{333}; (b) gradiometer G_{233}; (c) 2nd-order gradiometer with tilted baseline.

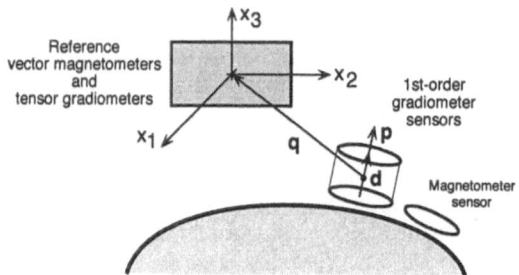

Figure 3.16. General method of software formation of 2nd-order gradiometer.

General form of gradiometer signals. An imperfect k-th-order gradiometer will have residual sensitivity to field and all gradients of order lower than k. Similarly, such a gradiometer also will have a residual eddy-current sensitivity to derivatives of field and of all gradients with order lower than k. However, to simplify the discussion, only the eddy-current contribution due to the field derivative will be considered. Then the outputs of 1st-, 2nd- and 3rd-order gradiometers will be

$$g^{(1)} = \alpha_{G1} \cdot \left(\mathbf{C_B} \cdot \mathbf{B} + \mathbf{E} \cdot \dot{\mathbf{B}} + \mathbf{p} \cdot G^{(1)} \cdot \mathbf{d_1} \right) \tag{3.36}$$

$$g^{(2)} = \alpha_{G2} \cdot \left(\mathbf{C_B} \cdot \mathbf{B} + \mathbf{E} \cdot \dot{\mathbf{B}} + \mathbf{C_{G1}} \cdot \mathbf{y_1} \cdot \mathbf{d_1} + \mathbf{p} \cdot G^{(2)} \cdot \mathbf{d_1} \cdot \mathbf{d_2} \right) \tag{3.37}$$

$$g^{(3)} = \alpha_{G3} \cdot \left(\mathbf{C_B} \cdot \mathbf{B} + \mathbf{E} \cdot \dot{\mathbf{B}} + \mathbf{C_{G1}} \cdot \mathbf{y_1} \cdot \mathbf{d_1} + \mathbf{C_{G2}} \cdot \mathbf{y_2} \cdot \mathbf{d_1} \cdot \mathbf{d_2} + \mathbf{p} \cdot G^{(3)} \cdot \mathbf{d_1} \cdot \mathbf{d_2} \cdot \mathbf{d_3} \right)$$

$$\tag{3.38}$$

where d_1, d_2, and d_3 have been used to denote the baselines of the 1st-, 2nd- and 3rd-order gradiometer; C_B, C_{G1} and C_{G2} are 3-, 5- and 7-component common-mode vectors; y_1 and y_2 are 5- and 7-component vectors with elements equal to the linearly independent components of the 1st- and 2nd-gradient tensors; and $G^{(1)}$, $G^{(2)}$ and $G^{(3)}$ are the 1st-, 2nd- and 3rd-gradient tensors. The outputs of gradiometers with order higher than 3 can be expressed similarly.

3.3. REQUIRED MAGNITUDE OF THE COMMON-MODE VECTORS

This section will give a brief description of how to determine the required magnitudes of the common-mode vectors C_B, C_{G1}, C_{G2}, etc. To simplify the notation, the tensor character of the gradiometers will be omitted in this section and the last terms in Eqs.3.36 to 3.38 will be expressed as scalar products with the unit length vector p also omitted. For example, the last term in Eq.3.38, $p \cdot G^{(3)} \cdot d_1 \cdot d_2 \cdot d_3$, will be written as $G^{(3)} \cdot d_1 \cdot d_2 \cdot d_3$, where $G^{(3)}$ now denotes the relevant 3rd-gradient terms. Also, the present analysis will assume that the noise source is a dipole with field and gradients given by Eq.2.1. Note that the magnetic field can be considered a 0-th gradient and is obtained from Eq.2.1 by setting $k = 0$.

Typically, the gradients due to external objects are important only if the objects are sufficiently close such that their gradients are larger than the SQUID detector noise. The distance at which the k-th gradient signal is equivalent to the detector noise will be denoted by R_I. A gradiometer will be considered adequately balanced if for distances smaller than R_I, all common-mode signals are smaller than the gradiometer response (Figure 3.17.a). In such cases it can be said that the gradiometer is in its intrinsic regime.

Specifically, consider an example of a 3rd-order gradiometer. The distance R_I is determined from

$$G^{(3)} \cdot d_1 \cdot d_2 \cdot d_3 = n \qquad (3.39)$$

where n is the detector noise. The conditions for the 3rd-order gradiometer to be in its intrinsic regime are then (for $R \leq R_I$)

$$C_B \cdot B \leq G^{(3)} \cdot d_1 \cdot d_2 \cdot d_3 \qquad (3.40.a)$$

$$C_{G1} \cdot G^{(1)} \cdot d_1 \leq G^{(3)} \cdot d_1 \cdot d_2 \cdot d_3 \qquad (3.40.b)$$

$$C_{G2} \cdot G^{(2)} \cdot d_1 \cdot d_2 \leq G^{(3)} \cdot d_1 \cdot d_2 \cdot d_3 \qquad (3.40.c)$$

A 3rd-order gradiometer in its intrinsic regime is illustrated graphically in Figure 3.17.a. For distances larger than R_I the amplitudes of the 3rd gradient and of all common-mode signals are smaller than the system noise.

The expressions for the maximum allowed common-mode vector magnitudes and for the distance R_I are shown in Table 3.1. These expressions were derived from Eqs.3.39 and 3.40 by substituting for fields and gradients from Eq.2.1 and putting $R = R_I$.

144

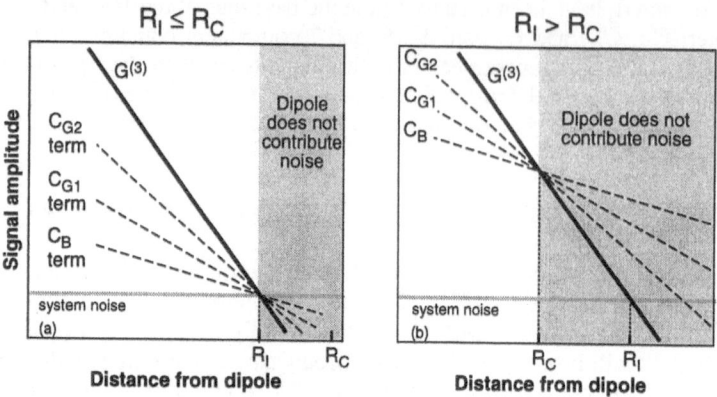

Figure 3.17. Schematic representation of the common-mode signals for a 3rd-order gradiometer in its intrinsic regime. (a) $R_I \le R_C$, (b) $R_I > R_C$. The distance R_C is defined below.

Table 3.1. Conditions for gradiometers of 1st-, 2nd- and 3rd-order to be in their intrinsic gradiometer regime.

Parameter	Gradiometer type		
	1st-order grad.	2nd-order grad.	3rd-order grad
C_B	$\sqrt[4]{\dfrac{n\cdot(3\cdot d_1)^3}{M}}$	$\sqrt[5]{\dfrac{n^2\cdot(12\cdot d_1\cdot d_2)^3}{M^2}}$	$\sqrt{\dfrac{60\cdot n\cdot d_1\cdot d_2\cdot d_3}{M}}$
C_{G1}	——	$\sqrt[5]{\dfrac{2^8\cdot n\cdot d_2^4}{3\cdot M\cdot d_1}}$	$\sqrt[3]{\dfrac{n\cdot(20\cdot d_2\cdot d_3)^2}{3\cdot M\cdot d_1}}$
C_{G2}	——	——	$\sqrt[6]{\dfrac{n\cdot(5\cdot d_3)^5}{12\cdot M\cdot d_1\cdot d_2}}$
R_I	$\sqrt[4]{\dfrac{3\cdot M\cdot d_1}{n}}$	$\sqrt[5]{\dfrac{12\cdot M\cdot d_1\cdot d_2}{n}}$	$\sqrt[6]{\dfrac{60\cdot M\cdot d_1\cdot d_2\cdot d_3}{n}}$

The intrinsic distance R_I and the magnitudes of the common-mode vectors depend on the noise level n in Eq.3.39, which in turn depends on the following factors: system peak-to-peak noise, n_{pp}, shielded-room attenuation, ψ, and the effect of averaging on the environmental noise. The system peak-to-peak noise, n_{pp}, depends on the observation bandwidth, BW, the system white noise, w_{rms} (e.g., in fT rms/√Hz), the number of averages performed during the data collection, N_{ave}, and may be written as

$$n_{pp} = \frac{5\cdot w_{rms}\cdot\sqrt{BW}}{\sqrt{N_{ave}}} \tag{3.41}$$

where the factor of 5 is used to effect the conversion between the rms and peak-to-peak noise.

If environmental disturbances are assumed to be random (e.g., random passage of various cars or objects), and if they are only occasional (i.e., not in all trials or if the majority of events are located at clearly separated times in different trials), then they will average as $1/N_{ave}$. If, on the other hand, environmental noise is caused by many overlapping random sources then the environmental noise will average as $1/\sqrt{N_{ave}}$. Consider the case where environmental averaging is like $1/N_{ave}$ (modification to the case of $1/\sqrt{N_{ave}}$ averaging is straightforward). The observed averaged environmental noise, n_{obs}, is then given as

$$n_{obs} = \frac{G^{(3)} \cdot d_1 \cdot d_2 \cdot d_3}{N_{ave}} \tag{3.42}$$

The environmental noise will be considered negligible if it is smaller by a factor of $1/\alpha$ (for $\alpha > 1$) than the peak-to-peak system noise, n_{pp}. Further, if the detection system is in a shielded room with an attenuation factor ψ (for $\psi < 1$), the condition for negligible environmental noise is

$$\psi \cdot \frac{G^{(3)} \cdot d_1 \cdot d_2 \cdot d_3}{N_{ave}} \leq \frac{n_{pp}}{\alpha} \tag{3.43}$$

Substituting for n_{pp} from Eq.3.41 and comparing Eq.3.43 and Eq.3.39, the noise level, n, used for the determination of R_I and acceptable common-mode vector magnitudes, can be expressed as

$$n = \frac{5 \cdot w_{rms} \cdot \sqrt{BW \cdot N_{ave}}}{\alpha \cdot \psi} \tag{3.44}$$

The noise level n from Eq.3.44 now may be used in the expressions in Table 3.1 to determine the intrinsic distance R_I and the maximum allowable magnitudes of the common-mode vectors. An example of such a calculation is shown in Table 3.2.

Table 3.2. Example of the maximum allowable common mode vector magnitudes and the intrinsic distances for a 3rd-order gradiometer. $w_{rms} = 5$ fT rms/$\sqrt{Hz}$, $d_1 = 5$ cm, $d_2 = 20$ cm, $d_3 = 15$ cm, $M = 10^{17}$ fT·cm^3, $\alpha = 3$. The dipole moment M is a typical value for a car.

parameter	lower limit	upper limit	units
bandwidth	1	100	Hz
N_{ave}	1	100	—
$1/\psi$	1	70	—
C_B	2.7×10^{-6}	2.3×10^{-4}	—
C_{G1}	5.9×10^{-4}	1.1×10^{-2}	—
C_{G2}	2.3×10^{-2}	1.0×10^{-1}	—
R_I	32.0	7.3	m

Considering moving sources, one more factor should be taken into account during the evaluation of common-mode vectors. If the disturbing source is located either very far from the detector, or if it moves very slowly, then the frequency range of the noise signal may be lower than the lowest frequency of interest, and the effect of such sources can be neglected. The maximum frequency of the k-th gradient of a moving dipole can be determined as the 1/2-power point of the frequency spectrum in Eq.2.3. However, this procedure is rather cumbersome because of the complex character of Eq.2.3. The frequency of the moving dipole can be approximated as $f = (2\pi\cdot\tau)^{-1}$, where τ is the dipole transit time between the two 1/2-power points of the gradiometer signal in the time domain. Using Eqs.2.1 and 2.2 to determine τ, the frequency of the k-th gradient of the moving dipole is

$$f = \frac{1}{2\pi\cdot\tau} = \frac{v}{4\pi\cdot D_o}\cdot\frac{1}{\sqrt{\frac{1}{2^{3+k}-1}}} \tag{3.45}$$

where v is the dipole velocity, and D_o is the distance of the dipole path from the gradiometer. If it is assumed that frequencies below some cut-off frequency f_C are of no interest, then Eq.3.45 can be inverted to calculate the cut-off distance, R_C. If the dipole with a given velocity is more distant than this cut-off distance, then the frequency content of its signal is lower than the cut-off frequency f_C, and the dipole effect can be neglected. R_C is given by

$$R_C = \frac{v}{4\pi\cdot f_C}\cdot\frac{1}{\sqrt{\frac{1}{2^{3+k}-1}}} \tag{3.46}$$

If $R_C > R_I$, then the cut-off distance is in the range where the dipole is not seen, and the frequency consideration does not affect the common-mode vector conditions. This case is illustrated in Figure 3.17.a. If $R_C < R_I$, then the cut-off distance is in the range where the dipole is visible, and the frequency considerations will relax the required common-mode vector requirements. In this case, all common-mode signals should be smaller than the gradiometer response for $R < R_C$ $(< R_I)$, as shown in Figure 3.17.b.

As an example, consider $f_c = 0.1$ Hz and a car moving at city speed, $v = 14$ m/sec. The cut-off distance for a 3rd-order gradiometer is $R_C \approx 32$ m, which is larger than or equal to the intrinsic distances in Table 3.2. Therefore, the cut-off frequency argument does not affect the determination of the maximum acceptable common-mode vector magnitudes. If, however, a small object carried by a walking person is considered, then $v \approx 1$ m/sec and the 3rd-order gradiometer cut-off distance is $R_C \approx 2.3$ m. If the small-object moment is assumed to be $M = 10^{14}$ fT·cm^3 (corresponding for example to a screwdriver or a file), then the intrinsic distance is in the range 2.3 m $< R_I <$ 10 m. In this case, the condition $R_I \geq R_C$ is reached, and the cut-off frequency argument will relax the common-mode vector requirements.

Note that the above common-mode vector analysis has been performed assuming that the system vibrations are negligible.

3.4. MEASUREMENT OF COMMON-MODE VECTOR MAGNITUDES

In order to assess the performance of software gradiometers, residual magnitudes of their common-mode vectors were measured. In principle, the common-mode vectors could be measured by applying uniform fields and gradients, and detecting the corresponding gradiometer responses. In practice, it is difficult to assemble a set of coils with sufficient precision and uniformity over the required volume [34 to 37]. Therefore, an alternate technique has been developed in which the magnitudes of the residual common-mode vectors were estimated by measuring the gradiometer response to a rotating dipole positioned at distances ranging from 3 to 50 m from the gradiometer. Examples of measured and simulated signals and of measured noise in an unshielded environment are shown in Figure 3.18. Noise is shown only for distances > 7 m because for shorter distances the measurement was affected by large broad shoulders of the rotating magnet signal line.

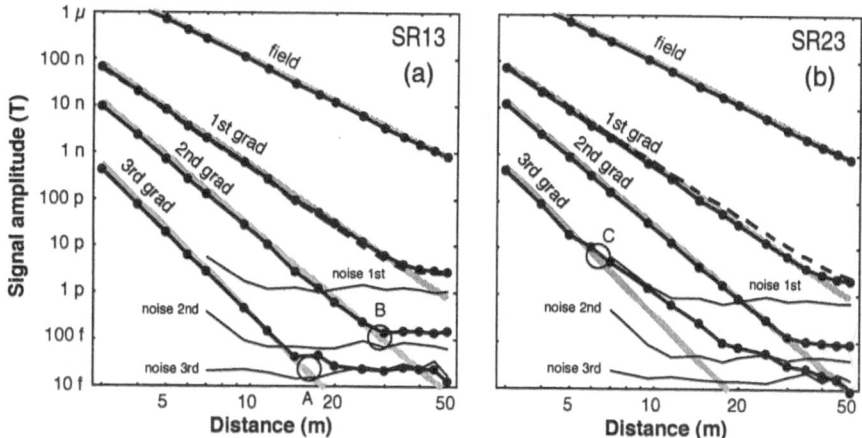

Figure 3.18. Magnetometer, 1st-, 2nd- and 3rd-order gradiometer signal amplitudes as a function of distance from a rotating magnet in an unshielded environment for channels SR13 (a) and SR23 (b). Dashed line - measured 1st-gradiometer signal amplitude (not balanced); heavy solid lines - measured magnetometer and balanced 1st-, 2nd- and 3rd-order gradiometer signal amplitudes; gray lines - simulations; thin black lines - peak noise amplitudes of 1st-, 2nd- and 3rd-order gradiometers in 1 Hz bandwidth. Magnet parameters: dipole moment $M = 4.36 \times 10^{16}$ fT·cm³, rotation frequency = 3 Hz, rotation axis - vertical. MEG system was tilted by 15 degrees in a vertical plane containing the magnet positions.

Simulations in Figure 3.18 (gray lines) correspond to ideal gradiometers (C_B, C_{G1} and $C_{G2} = 0$ in Eqs.3.36 to 3.38) and take into account the correct geometry between the sensing system and the rotating magnet. The measured magnetometer and software balanced and/or formed gradiometer outputs are shown by dots at the measured distances and are connected by heavy solid lines. The degree of agreement between the measured and simulated gradiometer responses can be used to estimate the magnitudes of the residual common-mode vectors.

Figure 3.18.a is an example where all gradiometers are in their intrinsic gradient regime (as in Figure 3.17.a). In this case, the measured and simulated responses exhibit a good match, and the errors associated with the common-mode vectors cannot be seen. The upper bounds on the residual common-mode vector magnitudes can be estimated from the intersection of the gradient lines with the appropriate noise levels (e.g., points A and B in Figure 3.18.a for 3rd- and 2nd-order gradiometers).

Figure 3.18.b is an example where the 1st- and 2nd-order gradiometers are in their intrinsic regimes, while the 3rd-order gradiometer is not. The measured 3rd-order gradiometer response starts deviating from the simulated response at a distance of about 6 m from the rotating magnet (point C). For distances greater than the point C and smaller than about 20 m, the 3rd- and 1st-order gradiometer responses are parallel, indicating that the 1st-gradient common-mode term is dominant in that region. The estimated magnitude of the 1st-gradient common-mode vector in this case is $C_{G1} \approx 2 \times 10^{-3}$, still within the range of the required values in Table 3.2.

In Figure 3.19 the measured common-mode vector magnitudes for the 3rd-order gradiometers, corresponding to all 64 channels, are compared with their required ranges (from Table 3.2), and with the estimated limitation of the measuring method. It is found that the measured common-mode vector magnitudes are close to their required values, and good noise cancellation is thus realized. Besides demonstrating good gradient formation, this result also confirms that the gradient coefficients are very stable. The gradient formation coefficients were originally determined with the dewar vertical, while this experiment was carried out with the dewar tilted at 15 degrees and at a completely different location.

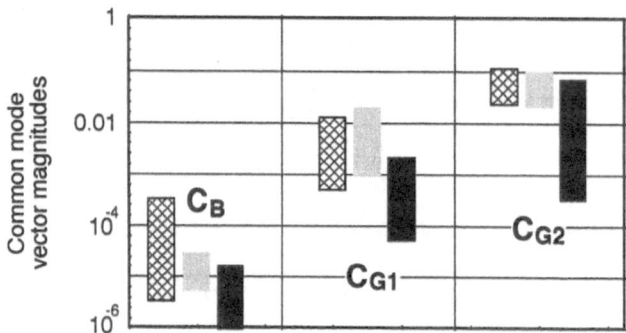

Figure 3.19. Common-mode vector magnitudes for 3rd-order gradiometers in a 64 channel system, estimated from a rotating magnet signal. Black bars - estimated limits of the measuring method; cross-hatched bars - ranges of the residual common-mode vectors required for 3rd-order gradiometers to be in their intrinsic regime (from Table 3.2); gray bars - measured range of common-mode vectors for all 64 channels (for which in many cases the upper-bound estimate was substituted for the measured value).

Similar analyses for 2nd- and 1st-order software gradiometers also show that their measured residual common-mode vector magnitudes are within the range of their required values.

4. SQUID Systems in Various Environments

In this section the operation of a stationary MEG system in unshielded and shielded environments will be discussed. In unshielded environments it is essential to use effective noise cancellation methods, while in modestly shielded environments additional noise cancellation significantly improves system performance. Noise cancellation will be divided into three parts: noise cancellation by high-order gradiometers; adaptive noise cancellation; and combined high-order gradiometer and adaptive noise cancellation.

Noise cancellation methods and experiments will be discussed against the backdrop of a 64-channel whole cortex MEG system [8 to 10], and the construction of such a system will be reviewed first. Note that similar system architecture also is used for MCG (Magneto-CardioGraphy) applications [45].

4.1. DESCRIPTION OF A WHOLE-CORTEX MEG SYSTEM

A whole-cortex MEG system [45] was constructed with the following goals in mind: (a) the system was required to have good noise cancellation, such that it could be operated in unshielded environments where the levels of environmental noise are not excessively large, and operated in shielded rooms where noise cancellation should work in combination with the shielded room to produce superior rejection of environmental noise; (b) the noise cancellation method should be general enough to allow software formation of high-order gradiometers (2nd- or 3rd-order), as well as flexible employment of adaptive methods; (c) the system should provide whole cortex coverage; and (d) the system architecture should accommodate a large number of channels. For example, systems under construction contain 143 MEG sensors, 26 reference channels, 64 EEG channels, 16 ancillary channels, 8 general purpose A/D and D/A channels, and 16 trigger-input and 16 trigger-output channels).

A block diagram of this system is shown in Figure 4.1. The brain magnetic signals are detected by sensors which are 1st-order hardware gradiometers (but they could also be magnetometers or 2nd-order hardware gradiometers). Using the reference system, the sensor outputs can be configured by software or firmware into balanced 1st-, 2nd-, or 3rd-order gradiometers. The sensors have a coil diameter of 2 cm and baseline of 5 cm. All flux-transformers and SQUID sensors are housed in a liquid helium dewar which has a tail configured into a helmet shape. The separation of the sensors from the room-temperature surface of the dewar in the helmet region is 1.6 cm. The layout of the 1st-gradient sensors for the 143 channel system is shown in Figure 4.2.a, and the finished helmet is shown in Figure 4.2.b.

The signals are transmitted to DC SQUID amplifiers at room temperature which can be positioned up to 6 m from the dewar. From the amplifiers, the signals are transmitted to the main electronics rack which contains digital SQUID electronics, digital EEG electronics and ancillary channel electronics. The signals from all channels are pre-processed by a DSP processing unit before they are transmitted to a data-collection computer. The function of the DSP processing unit is to provide filtering, resampling, real time firmware formation of high-order gradiometers, and adaptive noise cancellation when applicable. The system output corresponds to a selected gradiometer order. However, the gradiometer order can always be changed by software post-processing, since all reference channels also are saved with the data.

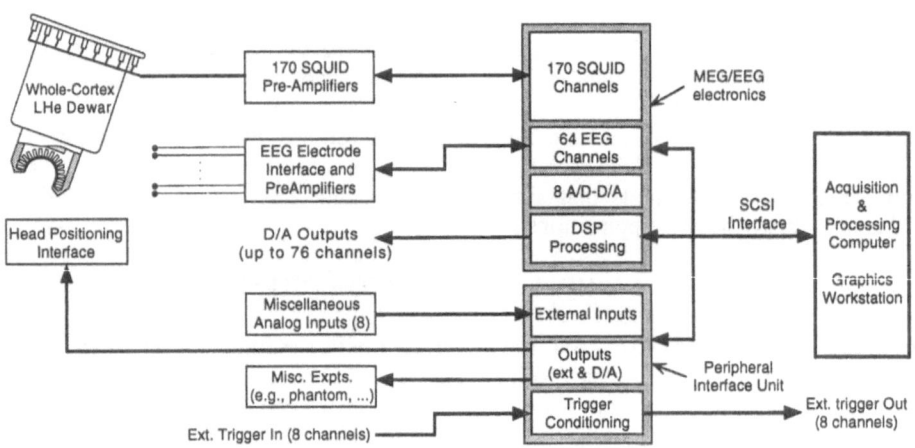

Figure 4.1. Block diagram of the whole-cortex MEG system.

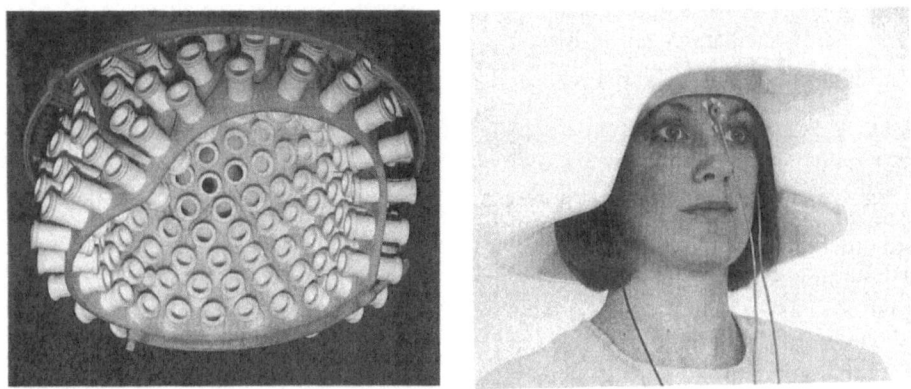

Figure 4.2. Photographs of the 1st-order gradiometer sensors in the helmet area, and a view of the dewar tail around a subject's head. The coils attached to the subject's head are used for accurate determination of the head position relative to the sensors.

The Peripheral Interface Unit can amplify either internally generated or external signals and pass them through current drivers for head positioning, phantom experiments, etc. It also provides trigger input/output and trigger conditioning.

After the DC amplifiers, the signals are digitized, and the feedback loop is closed digitally using a DSP processor. The system utilizes the inherent quantization of the SQUID output. The feedback loop is locked at a certain minimum of the periodic SQUID transfer function and maintains lock for SQUID signals in the range of $\pm 1\ \Phi_o$ around the lock point. When the applied signal exceeds this range, the loop lock is released, the loop re-locks in an adjacent minimum of the transfer function, and the electronics adds an up or down count as the lock point shifts along the transfer function. The signal from within the $\pm 1\ \Phi_o$ range is then combined with the counts total to yield a 32-bit data word. A resolution of 20 bits has been assigned to $1\ \Phi_o$. It is important for data analysis to have both MEG and EEG signals measured simultaneously and subjected to nearly identical transfer functions. Therefore, the EEG signals are amplified and digitized (with the sample and hold triggered by an MEG trigger), and both EEG and MEG signals are subjected to common DSP processing to assure as close as possible a similarity of their transfer functions. A schematic diagram of the SQUID electronic loop is shown in Figure 4.3.

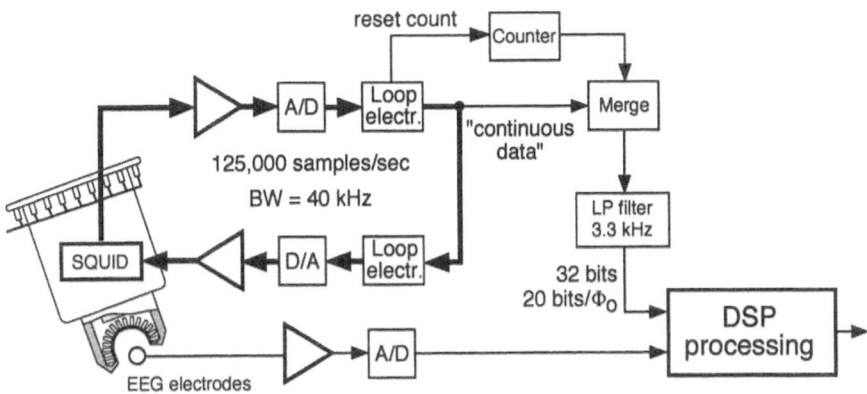

Figure 4.3. A schematic diagram of the digital SQUID electronics loop, and incorporation of EEG into the data collection system.

4.2. NOISE CANCELLATION BY HIGH-ORDER GRADIOMETERS

In this section noise cancellation by high-order gradiometers will be discussed, and their performance will be demonstrated. First, the concept of high-order gradiometer noise cancellation will be outlined, and then the noise cancellation performance will be discussed separately for unshielded and shielded environments. The achieved noise levels will be shown, and examples of MEG system operation in an unshielded environment will be given.

4.2.1. Principle of Noise Cancellation by High-Order Gradiometers

Noise cancellation by high-order gradiometers can be explained by assuming that the noise source has, e.g., a dipole character (since the noise cancellation principles for other types of sources are similar). The magnetic field generated by the dipole source

152

depends on the distance R from the source as R^{-3}. The spatial gradients are obtained from the magnetic field by spatial differentiation, which means that increasing the gradient order by 1 increases the distance decay exponent also by 1. The distance dependence of a k-th-order gradient is thus proportional to R^{-3-k}.

When the distance from the dipole source increases, the corresponding signal decays faster if the gradient order is higher. Thus, for a high-order gradiometer, the dipole disturbance is attenuated faster than for lower-order gradiometers. This is demonstrated in Figure 4.4 for a dipole source magnitude roughly equivalent to a passenger car. The vector and tensor characters of the magnetic field and gradients has been neglected in Figure 4.4.

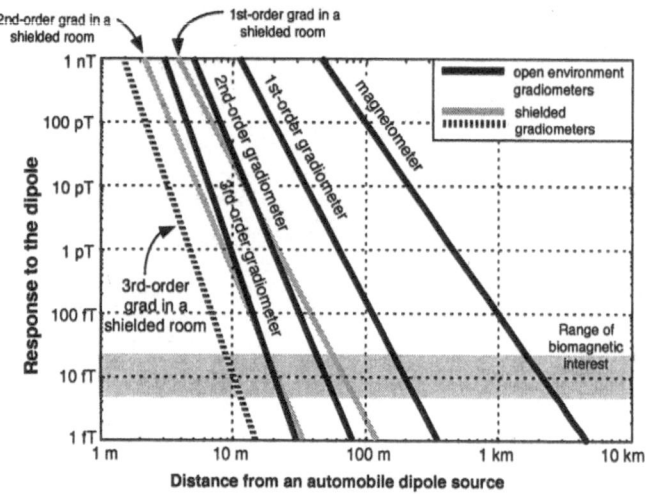

Figure 4.4. Response of spatial gradiometers to a dipole source. Dipole moment is equivalent to the moment of a passenger car, $M = 10^{17}$ fT·cm^3, gradiometer baseline unit = 5 cm, shielding ratio = 70.

The gray band around the 10-fT level indicates the approximate range of the required biomagnetic resolution (in a 1 Hz bandwidth). In the example in Figure 4.4, a magnetometer would be disturbed by a car up to distances of approximately 2 km. As the gradient order is increased, the car disturbance is decreased, and finally for a 2nd- or 3rd-order gradiometer, the car represents a disturbance only if it is closer than about 80 or 20 m from the detector, respectively.

Also shown in Figure 4.4 is the effect of a moderately shielded environment on the dipole signals. If it is assumed that the shielded room does not distort the spatial character of the magnetic fields, then the addition of a shielding factor of 70 to the gradiometer noise cancellation is roughly equivalent to increasing the gradiometer order by 1. A shielded 2nd-order gradiometer therefore behaves as an unshielded 3rd-order gradiometer, while a shielded 3rd-order gradiometer could tolerate cars as close as 10 m from the detection instrument.

Noise elimination by high-order gradiometers can be termed "spatial filtering" [46], and it works by assuming that the noise sources are distant while the signal sources are near.

To conclude this section, the shielding factor required to reduce the magnetic disturbance of large (e.g., cars, $M \approx 10^{17}$ fT·cm³) and small (e.g., various tools, $M \approx 10^{14}$ fT·cm³) sources to insignificant levels as a function of distance, is shown in Figure 4.5. The dashed horizontal lines indicate the low-frequency shielding factors of modestly shielded μ-metal rooms (≈ 70) and of high-attenuation μ-metal rooms ($\approx 10^4$) [12 to 17].

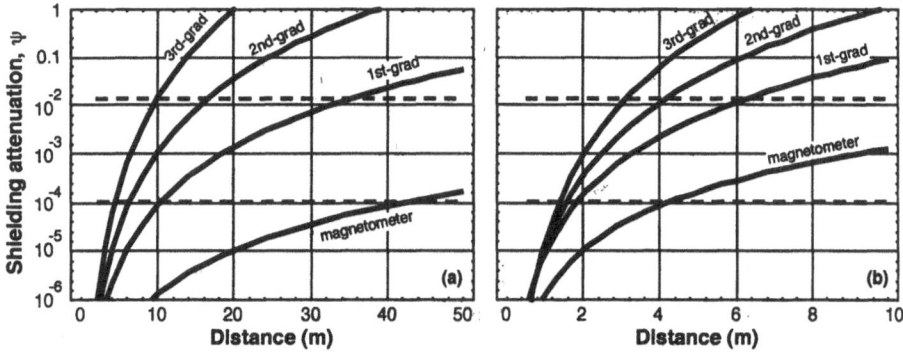

Figure 4.5. Shielding attenuation required to attenuate dipole signals to the level of the system noise as a function of distance and gradiometer order. Bandwidth = 100 Hz, $N_{ave} = 1$, $w_{rms} = 8$ fT rms/√Hz, $d_1 = 5$ cm, $d_2 = 20$ cm, $d_3 = 15$ cm, $\alpha = 3$ (Eq.3.4.3). (a) typical passenger car, $M \approx 10^{17}$ fT·cm³; (b) small tools, $M \approx 10^{14}$ fT·cm³. Dashed lines indicate shielding factors of low- and high-attenuation μ-metal rooms.

For example, an unshielded 3rd-order gradiometer ($\psi = 1$) could tolerate cars at a distance of ≈ 20 m. If the same performance would be required from a 1st-order gradiometer, then the shielding ratio would have to be $\psi \approx 10^{-3}$. Figure 4.5 also suggests that a whole-body superconducting shield with an attenuation of $\approx 10^{-8}$ [20] (off-scale in Figure 4.5) can support the use of magnetometers.

4.2.2. *Noise Cancellation in an Unshielded Environment*
Noise cancellation by high-order gradiometers was measured in an unshielded environment using a rotating magnet. An example of one such measurement is shown in Figure 4.6, where the peak-to-peak signal amplitude is shown as a function of gradiometer order for all channels of the MEG system. The scatter of the traces is caused by the difference in relative geometry between the rotating magnet and the variously oriented sensors on the helmet surface.

There are only small differences between the "not balanced" and "balanced" 1st-order gradiometers because the rotating magnet is quite close to the detection system, and the field common-mode signal does not play an important role in this case. The majority of traces in Figure 4.6 are parallel to the calculated signal, indicating that the formation of the high-order gradiometers is close to the expected behavior.

154

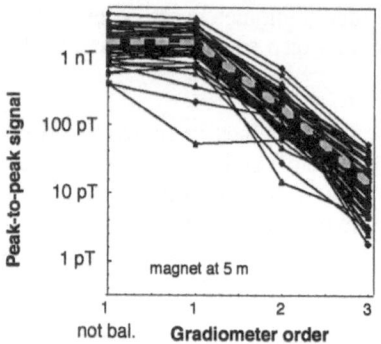

Figure 4.6. Attenuation of a rotating magnet field by spatial gradiometers. Thin solid lines - traces corresponding to all channels of the 64-channel system; heavy dashed line - average signal based on expected behaviour. Distance between the magnet and the center of the detection system = 5 m. Run Oct 4:4 (93).

The signal magnitude for different gradient orders and for all sensors is shown in a logarithmic histogram in Figure 4.7. Note that for the magnetometers there are only 3 readings (3 components of the magnetic field), while for the gradiometers there are 66 readings (in this case a 66-channel system was used).

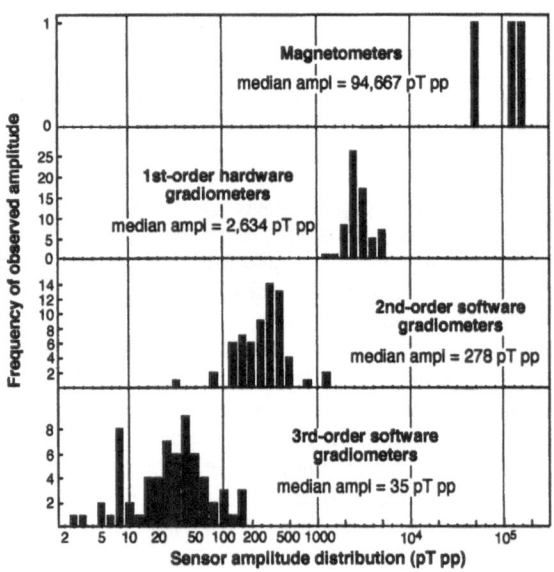

Figure 4.7. Cancellation of a rotating magnet signal by gradiometers in an unshielded environment. $f_s = 125$ samples/sec, bandwidth = DC to 20 Hz, magnet rotating in the horizontal plane, magnet distance from the center of the detection system = 5 m. The bin boundaries b_1 and b_2 satisfy $log_{10}b_1 - log_{10}b_2 = 0.1$.

For the magnet distance considered in Figure 4.7, the increase of the gradiometer order by one roughly decreases the magnet signal by a factor of 10. The median signal amplitudes and the cancellation ratios relative to 1st-order gradiometer and magnetometer performance are summarized in Table 4.1.

Table 4.1: Summary of a rotating magnet signal cancellation, unshielded system. Magnet distance from the detection system center = 5 m.

Sensor type	Median pp amplitude (pT)	Cancellation relative to:	
		1st-order grad	magnetometer
Magnetometer	94,667	---	1
1st grad.	2,634	1	36
2nd grad.	278	9.5	340
3rd grad.	35	75.3	2,705

For the magnet distance of 5 m, the cancellation ratio of the 3rd-order gradiometer relative to the 1st-order gradiometer is about 75 times, and relative to a magnetometer it is about 2,700 times.

4.2.3. *Noise Cancellation in Shielded Rooms*

The MEG system was positioned inside a moderately-shielded room [13], approximately 1.5 m from the door, with the center of the detection system approximately 1.1 m above the room floor.

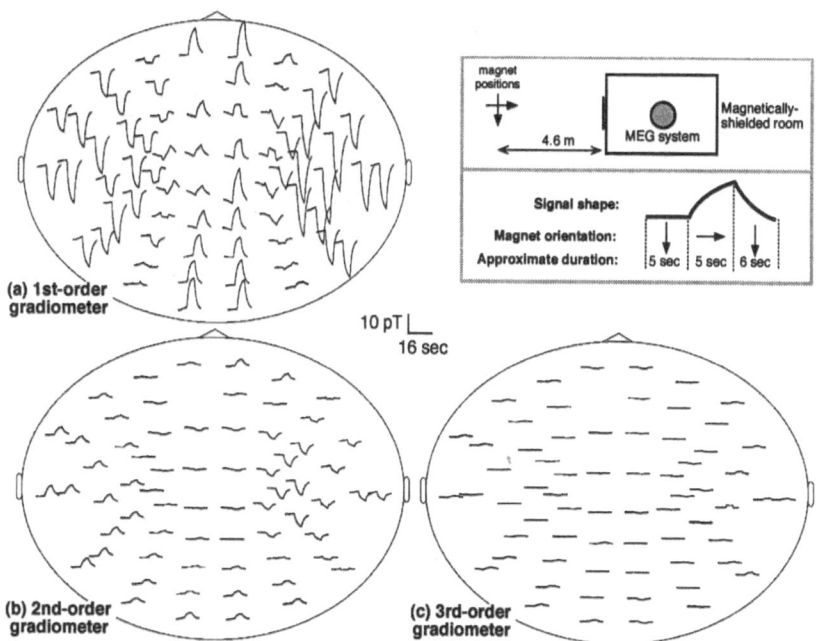

Figure 4.8. Detection of a magnet outside a shielded room by a MEG system inside the shielded room. f_s = *125 samples/sec, bandwidth = DC to 30 Hz, collection time = 16 sec. (a) Balanced 1st-order gradiometer; (b) 2nd-order gradiometer; (c) 3rd-order gradiometer, all on the same scale.*

156

A magnet was located in front of the shielded room at a distance of 4.6 m from the shielded-room door. Because the shielded room has a very long time constant, the magnet motion was performed in three discrete steps to allow the room relaxation to take place: for the first 5 sec, the magnet was left resting on the concrete floor pointing at the shielded room, then the magnet was abruptly rotated by 90 degrees and kept in the new position for another 5 sec, and finally, the magnet was abruptly rotated to the original position and kept there for 6 sec. The magnet signals were detected inside the shielded room and 1st-, 2nd- and 3rd-order gradiometers were constructed using software. The results are shown in Figure 4.8.

Similar to the unshielded environment case, as the gradiometer order increases, the magnitude of the detected signal decreases. The signal attenuation as a function of the gradiometer order is shown for all channels on a logarithmic histogram in Figure 4.9, and the median signal amplitudes and the cancellation ratios relative to 1st-order gradiometer and magnetometer performance are summarized in Table 4.2. Note that, for the 3rd-order gradiometers, the signal amplitude for a large number of sensors is equal to the peak-to-peak noise in the measurement bandwidth of 30 Hz (the lowest frequency bin used in Figure 4.9).

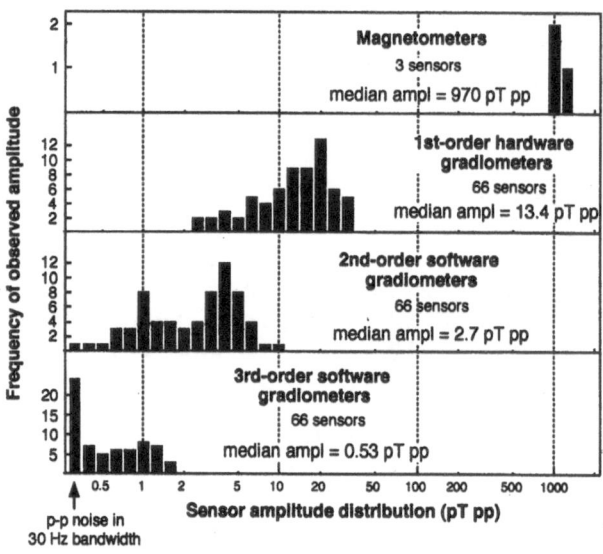

Figure 4.9. Cancellation of a rotating magnet signal by gradiometers with MEG system in a shielded room. $f_s = 125$ samples/sec, bandwidth = DC to 20 Hz, magnet rotating in the horizontal plane, magnet distance from the front of the shielded room = 4.6 m. The bin boundaries b_1 and b_2 satisfy $log_{10}b_1 - log_{10}b_2 = 0.1$.

For 3rd-order gradiometers, the cancellation of the magnet field including the low-frequency cancellation of the shielded room is on the order of 10^5, and the cancellation ratios relative to magnetometers and 1st-order gradiometers are comparable to that in an unshielded environment. Note that the ultimate cancellation ratios due to the 3rd-order gradiometer in Table 4.2 are slightly skewed, because the signals are very small and close to the system peak-to-peak noise. A graphical comparison of the magnet signal rejection by shielded and unshielded gradiometers is shown in Figure 4.10.

Table 4.2: Summary of the rotating magnet signal cancellation for a shielded MEG system. Magnet distance from the shielded room door = 4.6 m. Low-frequency cancellation of the shielded room ≈ 50.

Sensor type	Median pp amplitude (pT)	Cancellation relative to:		Cancel. including shielded room
		1st-order grad	magnetometer	
Mag.	970	---	1	50
1st grad	13.4	1	72.4	3.6×10^3
2nd grad	2.7	5.0	359	1.8×10^4
3rd grad	0.53	25.3	1,830	0.9×10^5

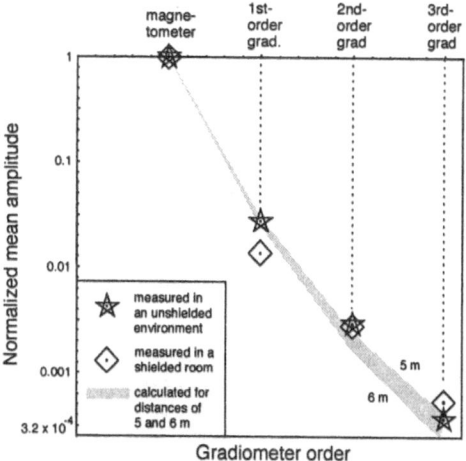

Figure 4.10. Comparison of signals detected by gradiometers of different orders in shielded and unshielded environments. Signals are normalized by the magnetometer signal. The distance between the MEG system and the magnet is ≈ 6 m for the shielded system and ≈ 5 m for the unshielded system. The data are averaged over all channels.

The cancellation ratios relative to the magnetometer for shielded and unshielded environments are approximately equal and agree quite well with the calculated ratios - the distances between the magnet and the sensor system are slightly different for the shielded and unshielded systems, but this difference produces only a small variation of the cancellation ratios, as shown by the calculated curve in Figure 4.10.

The similarity of the shielded and unshielded results suggests that when the MEG system is located near the center of the shielded room, the room does not significantly distort the spatial character of the environmental signals, even when the signal source is as close as 4.6 m from the shielded room.

When the detection system is located near the shielded room floor, then the room distorts the spatial character of the environmental fields. This is demonstrated in Figure

158

4.11, where the MEG system was located 15 cm above the room floor and the magnet signals were applied as in Figure 4.8.

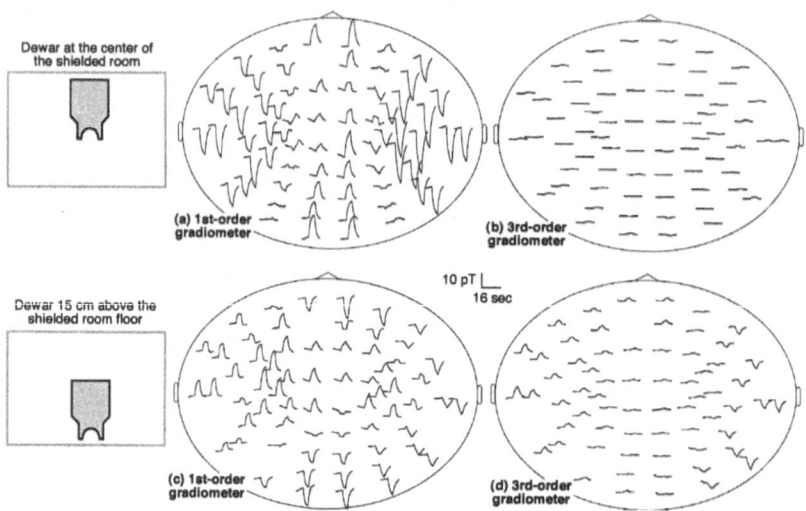

Figure 4.11. Comparison of the magnet signal cancellation when the dewar is close to the center of the shielded room and 15 cm above the room floor. Data collection parameters and magnet signal as in Figure 4.8: (a, b) the MEG system is close to the room center; (c, d) the MEG system is 15 cm above the room floor.

In Figure 4.11.a and c, the detected 1st gradients of the magnet signals are shown. For the MEG system close to the room floor, the signals are slightly smaller than when the MEG system is close to the room center, indicating a slightly larger attenuation at the room floor. The detected 3rd gradients are shown in Figure 4.11.b and d. In this case, the 3rd gradients are much larger at the room floor than in the room center, suggesting that close to the room floor, the spatial character of the external signals is disturbed by the room.

The effect of room motion and room vibrations on the signals detected inside the room also were measured. Room motion was introduced by a person leaning against the room wall. Changes of field and 1st gradients in the room measured by the references are shown in Figure 4.12.

Room motion introduced field changes of approximately 20 to 30 pT and 1st-gradient changes of approximately 250 fT/cm. The effect of the room motion on the sensors is shown in Figure 4.13. The motion signal is quite large when the sensors are operated in 1st-order gradiometer mode, and the effect of the motion is entirely eliminated when 3rd-order gradiometers are formed using software.

Similarly, the effect of room vibrations was investigated. The room was repeatedly struck with a fist and the resulting fields and 1st gradients measured inside the room are shown in Figure 4.14. The vibrational field signals were about 10 pT and the vibrational gradients were about 100 fT/cm. Note that striking the room also excited low-frequency signals similar to room motion, as in Figure 4.12. The effect of the room vibrations on the sensors is illustrated in Figure 4.15. Again, the vibrational signal is quite large when the sensors are operated in their 1st-order gradiometer mode and the vibrational signal is completely eliminated when the 3rd-order gradiometers are software formed.

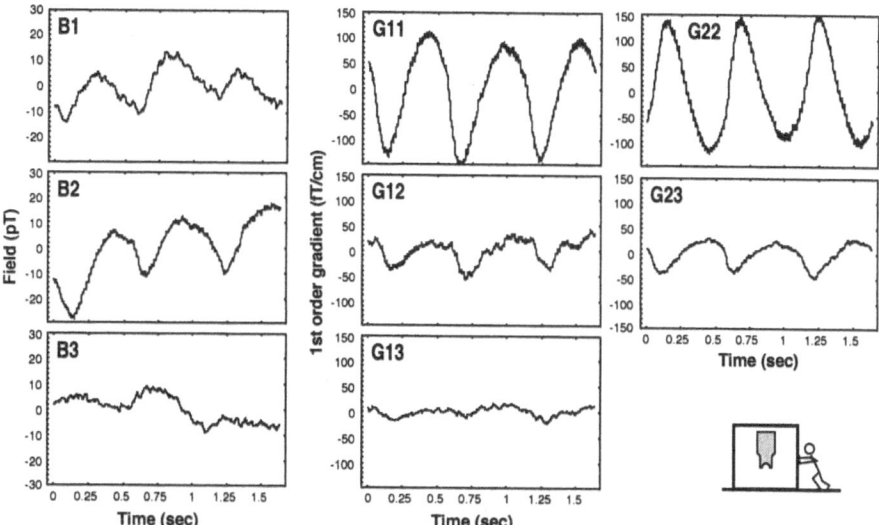

Figure 4.12. Field and 1st gradients measured when the shielded room was slightly moved by a person leaning periodically against the side wall. The MEG system was located near the center of the room. f_s = 1250 samples/sec, points = 2048, bandwidth = DC to 25 Hz, file: Dec 2:2 (92).

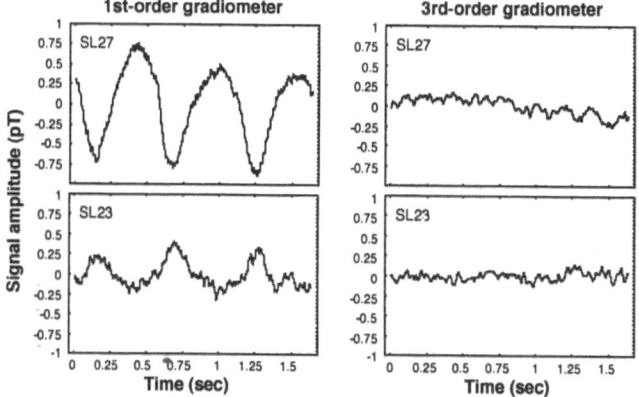

Figure 4.13. Sensor outputs in balanced 1st- and 3rd-order gradiometer mode measured when the shielded room was slightly moved by a person leaning against the side wall. The MEG system was located near the center of the room. f_s = 1250 samples/sec, points = 2048, bandwidth = DC to 25 Hz, file: Dec 2:2 (92).

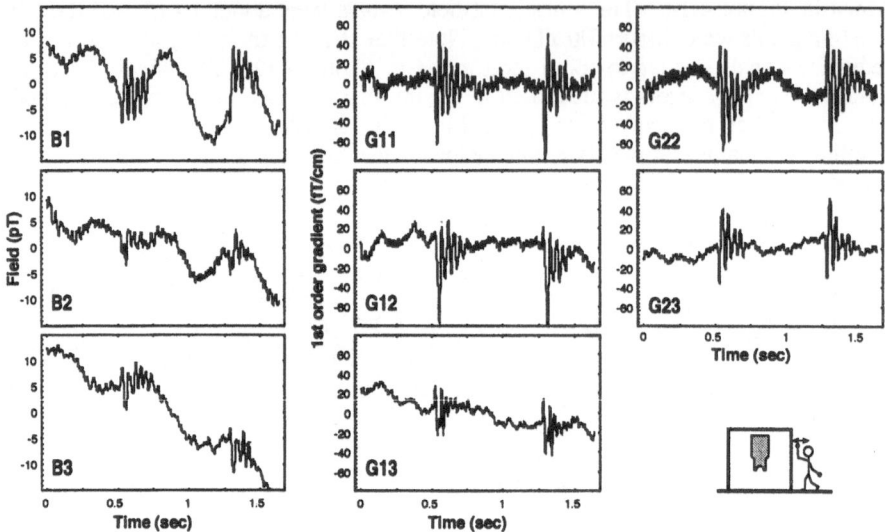

Figure 4.14. Field and 1st gradients measured during room vibrations. The MEG system was located near the center of the room. $f_s = 1250$ samples/sec, points = 2048, bandwidth = DC to 25 Hz, file: Dec 2:3 (92).

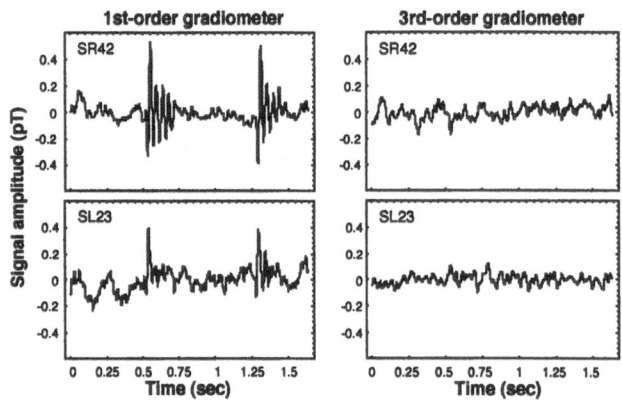

Figure 4.15. Sensor outputs in balanced 1st- and 3rd-order gradiometer mode measured during the room vibrations. The MEG system was located near the center of the room. $f_s = 1250$ samples/sec, points = 2048, bandwidth = DC to 25 Hz, file: Dec 2:3 (92).

4.2.4. Summary of the Noise Performance

The formation of high-order gradiometers was shown to reduce significantly environmental noise. An example of noise levels achieved in an unshielded environment is shown in Figure 4.16.a. For a 3rd-order gradiometer, the achieved noise level is in the vicinity of 10 fT rms/√Hz for frequencies above about 0.3 Hz. Generally, the white noise appears to be below the 10 fT rms/√Hz level. However, various vibrational and environmental lines below about 50 Hz exceed 10 fT rms/√Hz. The power line signals, even though strongly attenuated relative to the environmental magnetic fields, are very prominent.

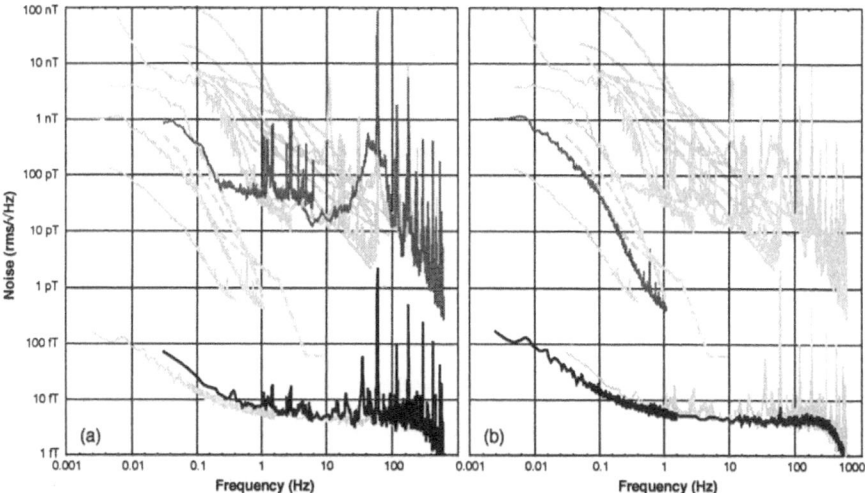

Figure 4.16. Noise levels in 3rd-order gradiometer mode (black lines). The darker gray lines show the magnitude of magnetic fields corresponding to the 3rd-order gradiometer plots. (a) Unshielded environment; (b) shielded environment. For ease of comparison, the plots from Fig.2.2 and plots which are not being discussed are shown by gray shading. The plots above approximately 200 fT rms/√Hz all correspond to the magnitude of magnetic field.

The noise levels achieved in the shielded room are shown in Figure 4.16.b. In this case the 3rd-order gradiometer noise is below 10 fT rms/√Hz for frequencies above about 0.2 Hz, and the white-noise levels are about 5 fT rms/√Hz. The effects of higher-order gradiometers on the noise inside a shielded room are shown in greater detail in Figure 4.17. The same data collection was post-processed to yield 1st-, 2nd- and 3rd-order gradiometer outputs [47].

The time plots show considerable differences between the gradiometer orders (Figure 4.17.b). The excursions of the 1st-order gradiometer are about 3 pT, less than about 1 pT for the 2nd-order gradiometer and less than 150 fT for the 3rd-order gradiometer. This reduction of the noise is reflected in the frequency spectra in Figure 4.17.a, where the low-frequency noise decreases as the gradiometer order increases. Note that if only a magnetometer were operated in a shielded room, its noise spectrum would exhibit an onset of low-frequency noise at frequencies of about 20 Hz or more.

Details of the noise improvement by gradiometer formation in a shielded room are shown in Figure 4.18. In Figure 4.18.a a 1st-order gradiometer output is shown for one channel. The spectrum exhibits a white-noise level of about 3.5 fT rms/√Hz. However, there are distinct effects related to the 60 Hz power-line signal and its harmonics, and there is an increased noise level in the range from about 6 to 30 Hz, probably due to

162

vibrations in the residual field of the shielded room. The spectrum also exhibits a hint of low-frequency noise. However, the length of the record is not sufficient to show the low-frequency noise reliably.

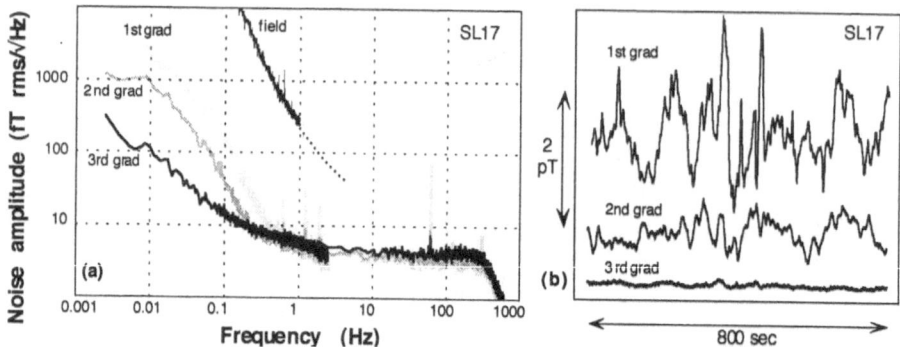

Figure 4.17. Effect of higher-order gradiometers on the noise measured inside a shielded room. (a) Frequency spectra corresponding to the time plots. Note that the magnetometer spectra were truncated at about 200 fT rms/√Hz. Below this level the magnetometer outputs are limited by the magnetometer resolution. (b) Time plot of the 1st-, 2nd- and 3rd-order gradiometers. Collection parameters: time data - fs = 5 samples/sec, duration = 819.2 sec, file: Feb 10:0 (93), frequency data (low frequency): fs = 5 samples/sec, points per trial = 4096, trials = 36, file: Feb 10:0 (93), frequency data (high frequency): fs = 1250 samples/sec, points per trial = 2048, trials = 30, file: Feb 2:28 (93).

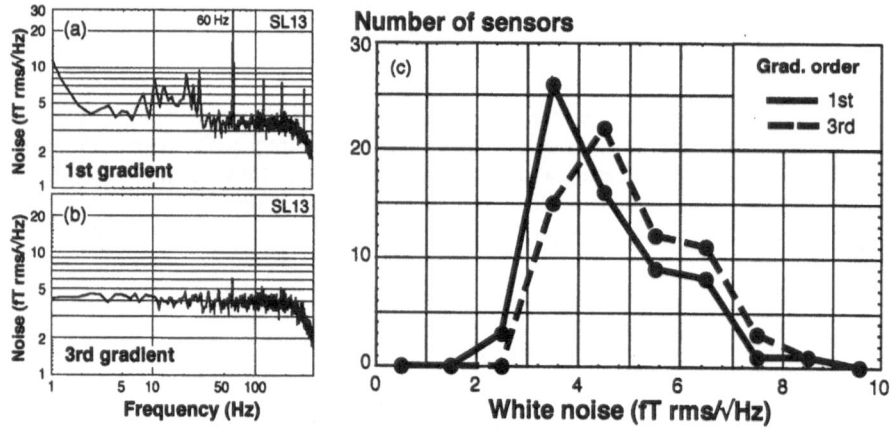

Figure 4.18. Effect of 3rd-order gradiometer on sensor noise in a shielded room: (a) balanced 1st-order gradiometer output; (b) after formation of the 3rd-order gradiometer; (c) comparison of the balanced 1st- and 3rd-order gradiometer white-noise levels for all 64 channels.

In Figure 4.18.b the same signal is shown after a software 3rd-order gradiometer was formed. The power-line signals are nearly eliminated, the excess noise in the 6 to 30

Hz region is removed, and there is no hint of low-frequency noise. The level of the white noise has increased only slightly, by less than 1 fT rms/√Hz.

The white noise increase due to the software 3rd-order gradiometer is shown for all 64 channels in Figure 4.18.c. The noise interval from 0 to 10 fT rms/√Hz is divided into 1 fT rms/√Hz bins and the numbers of channels with white noise corresponding to each bin are plotted for 1st- and 3rd-order gradiometers. The shift between the 1st- and 3rd-order gradiometer curves is only about 0.5 fT rms/√Hz, indicating that the process of the 3rd-order gradiometer formation does not increase the sensor noise significantly. This feature has been achieved by designing the reference system with sufficiently high resolution.

4.2.5. *Examples of the MEG System Operation in an Unshielded Environment*
Operation of an MEG system in shielded environments has been described in numerous papers [e.g., 2 to 7]. In this section the benefits of high-order gradiometers will be demonstrated using examples of MEG measurements in unshielded environments. Additional descriptions of the work performed in unshielded environments can be found in [8, 10].

In Figure 4.19, an Auditory Evoked Field (AEF) experiment performed in unshielded laboratories at CTF and the SFU Brain Behaviour Laboratory (BBL) are shown. The figure shows the effect of gradient formation.

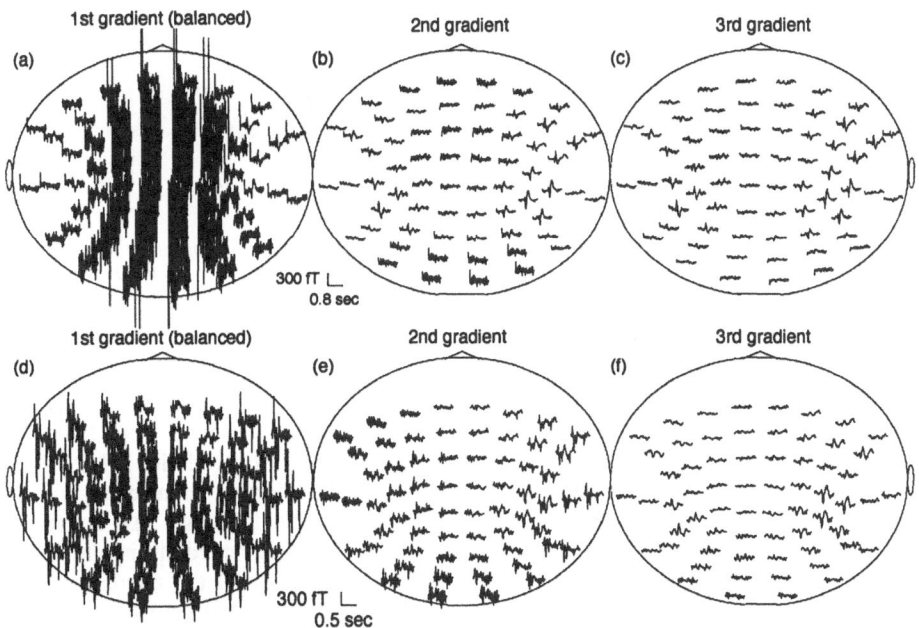

Figure 4.19. AEF experiments in an unshielded environment: (a, b, c) at CTF laboratories, f_s = 625 samples/sec, duration = 0.82 sec, bandwidth = DC to 40 Hz, 100 averages, subject A, file: Oct 6:15 (93); (d, e, f) BBL at SFU, f_s = 250 samples/sec, duration = 0.5 sec, bandwidth = DC - 40 Hz, 60 averages, subject B, file: May 27:4 (94). The stimulus in both cases was a 1 kHz tone of 50-msec duration into the right ear.

164

In both examples there is no signal discernible when the detection is performed with 1st-order gradiometers. The signal becomes visible when 2nd-order gradiometers are formed, even though environmental noise is still large. The signal becomes quite clean when viewed in the 3rd-order gradiometer mode. Besides illustrating the effectiveness of high-order gradient formation in software, these results also indicate the high degree of stability of the gradient-formation coefficients as both data sets were obtained using the same coefficients for gradiometer formation, even though the MEG system was transported (by truck) between the two laboratories.

An example of spontaneous activity measured without averaging in an unshielded CTF laboratory is shown in Figure 4.20. The activity with eyes closed and eyes open is compared with the background noise with no subject - all traces are in the 3rd-order gradiometer mode. The white-noise level with no subject is below 10 fT rms/√Hz, and there are several environmental lines which were not completely eliminated. With the subject's head in the helmet, the background noise level is increased to about 30 fT rms/√Hz (head noise). In addition, when the eyes are closed, a spontaneous activity peak at approximately 8.5 Hz is visible.

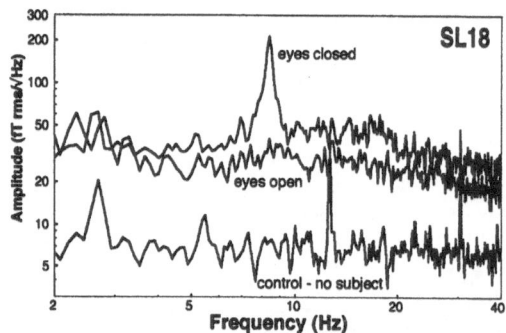

Figure 4.20. Spontaneous activity and system noise measured in an unshielded environment (urban laboratory). $f_s = 250$ samples/sec, points per trial $=2050$, trials $= 20$, collection bandwidth $= DC$ to 50 Hz, files: no subject - Oct 7:15 (93), eyes closed - Oct 7:17 (93), eyes open - Oct 7:18 (93). All traces are in the 3rd-order gradiometer mode.

4.3. ADAPTIVE NOISE CANCELLATION

Environmental noise also can be removed using adaptive noise cancellation. Like software (or firmware) gradiometer formation, a linear combination of reference signals is subtracted from each sensor signal so as to remove environmental noise. To determine the adaptive coefficients, a system of sensors and references is observed during the application of some unwanted signals (or noise), and the subtraction of the references from the sensors is adjusted to minimize the effect of such signals, e.g., [32, 39]. Adaptive noise cancellation is a useful tool only when the noise has a time independent character; adaptive methods do not perform well when the character of the noise is rapidly changing because the optimum adaptive coefficients also change; in contrast, the coefficients for gradiometer formation are independent of the noise character. For the noise sources with variable character, continuous re-adaptation methods could be used. However, our experiments suggest that such continuous methods also may affect the MEG signal.

In the following section, a general discussion of the adaptive method will be presented, followed by examples of frequency-independent and frequency-dependent adaptation.

4.3.1. General Discussion of the Method

Adaptive noise cancellation can be accomplished by: (1) monitoring an array of sensors and subtracting the common correlated noise from their outputs; or (2) by using an array of sensors, some incomplete reference system and subtracting the reference outputs from the sensor outputs; or (3) by using a complete reference system and subtracting it from the sensor outputs. In this discussion, a complete reference system means that its elements consist of complete vector or tensor devices (e.g., a 3-component magnetometer or 5-component 1st-order gradiometer, etc.). An incomplete reference system would consist of one or two magnetometer components and/or less than five independent 1st-gradient tensor components.

A general system for adaptive subtraction is shown schematically in Figure 4.21. The sensor channel, S, can be a detector of any type (magnetometer, 1st-order gradiometer, etc.). The reference system, R, consists of a combination of detectors which can be of different types (e.g., magnetometers, 1st-order gradiometers, etc.). The adaptive procedure finds a coefficient vector, ξ, such that the expression

$$S - \xi \cdot R \qquad (4.1)$$

is minimized. The coefficients ξ can be either frequency dependent or frequency independent. A simple example of a system with frequency-independent coefficients is a software 1st-order gradiometer based on a magnetometer sensor (as in Section 3.2.2).

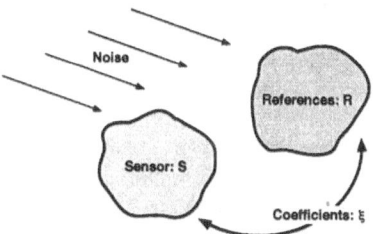

Figure 4.21. General configuration for adaptive noise cancellation. S - sensor, can be a magnetometer, 1st-order gradiometer, or higher-order gradiometer. R - references, a combination of a number of channels of (possibly) different character (e.g., magnetometers, 1st-order gradiometers, etc.).

In general, the adaptive method does not yield coefficients which are equivalent to the software-gradiometer coefficients even if the reference system is complete. This can be demonstrated simply by an example where there is one magnetometer sensor and one magnetometer reference aligned coaxially along the z axis as in a hardware axial 1st-order gradiometer - see Figure 4.22. If the sensor magnetometer output is denoted by b_s,

166

the reference magnetometer output by b_r, and the single adaptive coefficient by ξ, then the signal after adaptation, b_a, is

$$b_a = b_s - \xi \cdot b_r \tag{4.2}$$

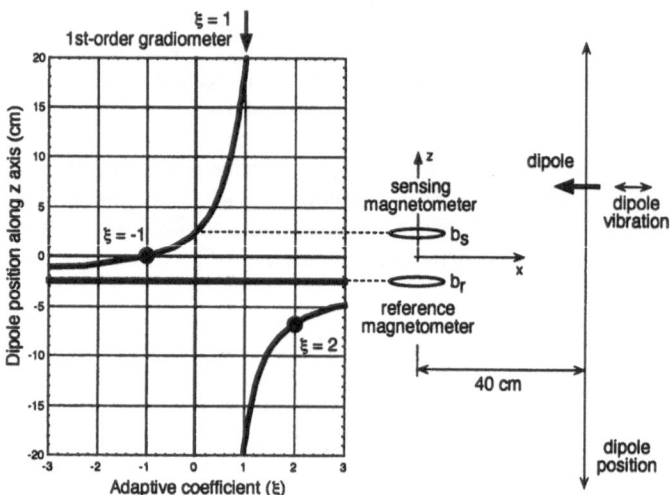

Figure 4.22. An example of adaptive subtraction between two magnetometers aligned along the z axis. The signal source is a dipole moving along a straight path in the x-z plane and parallel to the z axis. The dipole is oriented along the x axis and is randomly vibrating in the x direction.

The input for the determination of the adaptive coefficient in this example is a dipole moving along a straight path parallel to the z axis and in the x-z plane. The dipole is oriented along the x axis and is randomly vibrating in the x direction.

This system of 2 magnetometers would yield a 1st-order gradiometer only if the adaptation coefficient was $\xi = 1$. However, as the dipole is moving along its path, the adaptation coefficient is changing in the range from $-\infty$ to $+\infty$. The value $\xi = 1$ is reached only if the dipole is very far from the gradiometer. As the dipole moves past the reference coil, the adaptation coefficient changes from $-\infty$ to ∞ or vice versa (because the reference signal goes through zero and the sensor signal is non-zero). When the dipole is opposite the sensor coil, the adaptation coefficient is zero (because the sensor signal is zero and the reference signal is non-zero). This example demonstrates how the adaptive method is critically dependent on the character of the noise.

For noise with time-independent character, the adaptive procedure can produce better noise cancellation than a gradiometer can. However, for noise with a variable character, the adaptive procedure is inferior to gradiometer formation. This is demonstrated in Figure 4.23, where the noise of the adapted system is plotted as a function of the dipole position along its path for three values of the adaptation coefficient: $\xi = 1$, -1, and 2 (where $\xi = 1$ corresponds to a true 1st-order gradiometer).

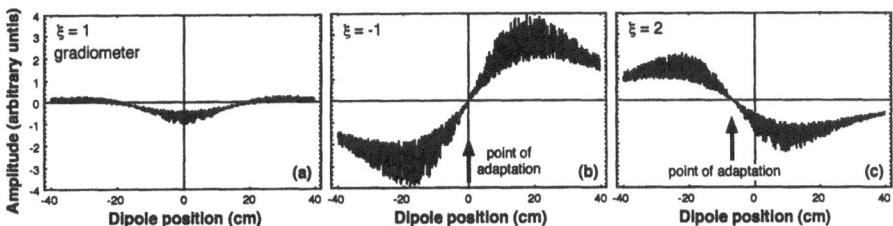

Figure 4.23. Noise of the system in Figure 4.22 for different values of the adaptation coefficients: (a) ξ = 1 (1st-order gradiometer); (b) ξ = -1; (c) ξ = 2. The dipole is moving along a straight path in the x-z plane and parallel to the z axis. The dipole is oriented along the x axis and is randomly vibrating in the x direction.

At the point of adaptation (i.e., the point on the dipole path where the adaptation coefficient was determined), the dipole noise cancellation is perfect, and the adaptive system performs better than the gradiometer system in Figure 4.23.a. However, when the dipole is positioned even slightly off the adaptation point, the system noise rapidly increases and is larger than the corresponding gradiometer noise. Thus, in this simple example, the gradiometer yields the best overall performance, and it will be shown by example in the next section that this conclusion is valid in general. The adaptive method works well when the noise character is constant (i.e., when the dipole would remain at the point of adaptation in the above example). If the noise character is changing (e.g., when the dipole moves away from the point of adaptation), adaptive methods are inferior to noise cancellation by gradiometers.

4.3.2. Frequency-Independent Adaptation
For frequency-independent adaptation, the adaptive coefficients are frequency-independent constants. To demonstrate the characteristics of such an adaptive procedure and the differences between the adaptive and high-order gradiometer noise cancellation methods, an experiment was performed in an unshielded environment where a magnet with moment $M \approx 4.8 \times 10^{13}$ fT·cm^3 was moved randomly at a distance of ≈ 7 m from an MEG system, in different positions and orientations. Specifically, in position 1, the magnet was located on the x_1 axis, oriented along the x_2 axis and moved parallel to the x_2 axis. In position 2, the magnet was located on the x_2 axis, oriented along the x_1 axis and moved parallel to the x_3 axis.

In Figure 4.24.a the "as collected" data (1st-order gradiometer, no balancing) for the magnet moved in position 1 are shown, while in Figure 4.24.b the same data are shown after the 3rd-order gradiometers were formed, and the scale was expanded ten times. In Figures 4.24.c to f, the data after adaptive noise cancellation are shown on the same scale as the 3rd-order gradiometer data. In Figures 4.24.c and d the adaptive references were identical to the references used for the formation of a 2nd-order gradiometer. In Figure 4.24.c, the determination of the adaptive coefficients was performed on the data itself (i.e., self-adaptation), and the results are only slightly worse than for the 3rd-order gradiometer results. In Figure 4.24.d, the adaptive coefficients were determined using data obtained when the magnet was moved in position 2 and then applied to the (shown) data corresponding to magnet motion in position 1. The results are rather poor, and the adaptation does not work well because the character of the noise was changed (since the magnet was moved to a different position).

168

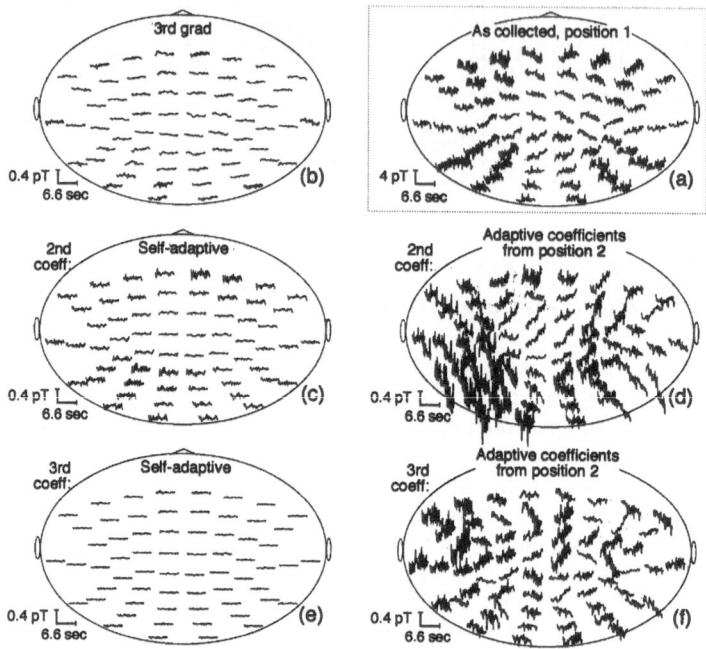

Figure 4.24. Comparison of noise cancellation by 3rd-order gradiometer formation and by adaptive subtraction. Unshielded environment, $f_s = 250$ samples/sec, duration = 6.6 sec, bandwidth = DC to 3 Hz, file: Feb 10:10 (95). Magnet in position 1. (a) as collected (1st-order gradiometer, not balanced); (b) 3rd-order gradiometers; (c) adaptive noise cancellation on the data itself, using references for formation of 2nd-order gradiometers; (d) adaptive noise cancellation using coefficients determined with magnet in position 2, using references for formation of 2nd-order gradiometers; (e) adaptive noise cancellation on the data itself, using references for formation of 3rd-order gradiometers; (f) adaptive noise cancellation using coefficients determined with magnet in position 2, using references for formation of 3rd-order gradiometers.

In Figure 4.24.e and f, the references used for adaptation were identical to the references used for the formation of the 3rd-order gradiometer. In Figure 4.24.e, the determination of the adaptive coefficients was performed on the data itself (i.e., self-adaptation), producing very good results, even better than for the 3rd-order gradiometer results. In Figure 4.24.f, the adaptive coefficients were determined from the data obtained when the magnet was moved in position 2 and then applied to the present data for magnet motion in position 1. Similar to the situation for Figure 4.24.d, the results are again poor, because the character of the noise has changed.

The results in Figure 4.24 indicate that it is necessary to have a sufficient number of references for successful noise removal by the adaptive method - the results in Figure 4.24.c are not as good as the results in Figure 4.24.e. Thus it is seen that the adaptive procedure works well only if the noise character remains constant; if the noise character changes, the adaptive procedure is ineffective.

In order to determine how fast the noise character changes, an experiment was conducted in an unshielded environment where the environmental noise (without additional moving magnets) was measured for a period of 1 sec, and the measurements were repeated every 1.65 sec. The adaptive and 3rd-order gradiometer noise cancellations were compared. The results are summarized in Figure 4.25.

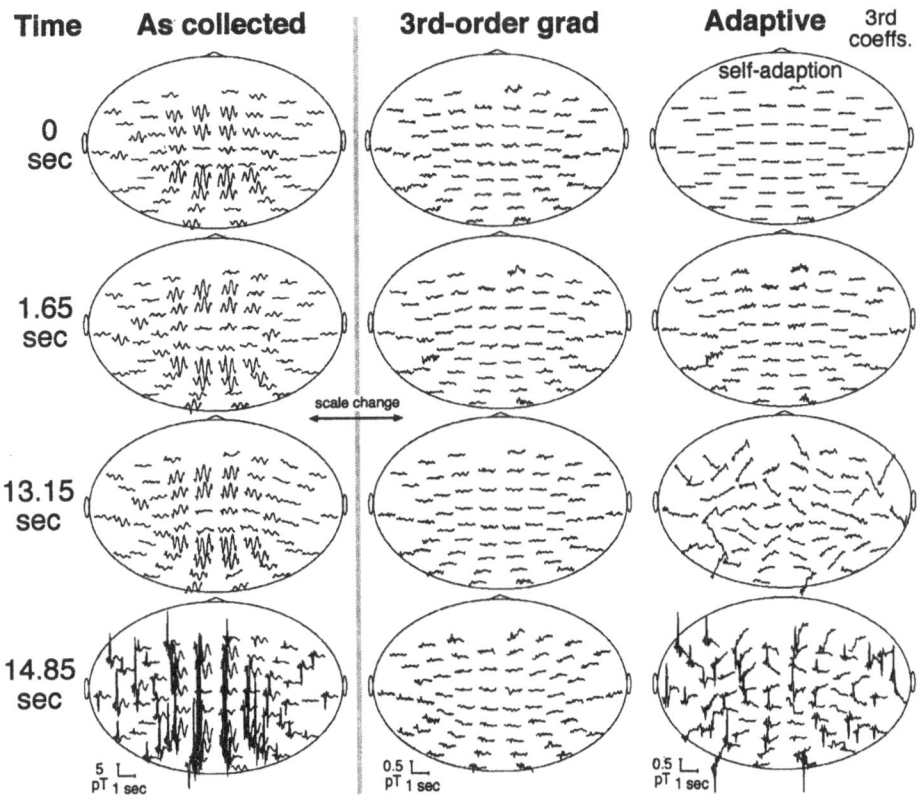

Figure 4.25. Comparison of noise cancellation by 3rd-order gradiometer and by adaptive subtraction as a function of time. Unshielded environment. $f_s = 1250$ samples/sec, duration = 1 sec, bandwidth = DC to 10 Hz, file: Oct 6:14 (93). Adaptive references are the same as the references used for 3rd-order gradiometer formation. Column 1 - data as collected (1st-order gradiometer, no balancing); column 2 - 3rd-order gradiometer; column 3 - adaptive noise cancellation.

In column 1 of Figure 4.25, the as-collected data are shown for times t = 0, 1.65, 13.15, and 14.85 sec. For the first 3 times shown, the as-collected data seem to be stationary; there are only minor differences between the records. At the 4th time (t = 14.85 sec), however, there was a large magnetic disturbance present. Column 2 in Figure 4.25 shows the same data as for column 1, but after formation of a 3rd-order gradiometer - note that the scale was expanded by a factor of 10. In this case, all four times yield similar noise, although the noise is slightly larger for the 4th time segment when the magnetic disturbance was present. Column 3 in Figure 4.25 shows the same data as for column 1, but after adaptive noise cancellation. The adaptive coefficients were determined using the data for t = 0 and then were applied to all four times. For the initial time, the adaptive noise cancellation yields a better result than for the 3rd-order

170

gradiometer formation. However, the noise at the second time (after only 1.65 sec) is the same for both adaptive and 3rd-order gradiometer noise cancellations. At the third time (t = 13.15 sec) the adaptive noise cancellation is much worse than that for the 3rd-order gradiometers; and at the fourth time (when the magnetic disturbance was present) the adaptive noise cancellation fails completely.

The above results show that the self-adaptation method (i.e., the determination of the adaptive coefficients from the data itself) yields better noise cancellation than the 3rd-order gradiometer (for t = 0 sec). However, the quality of the adaptive noise cancellation quickly deteriorates. After 1.65 sec, the adaptive and 3rd-order gradiometer cancellations yield comparable results, and after more than 10 sec the adaptive cancellation fails, while the 3rd-order gradiometer noise cancellation provides consistent, time-independent results.

4.3.3. Frequency-Dependent Adaptation

Assume that the transfer functions between references and sensors are frequency-dependent ($H_{iy}(f)$ in Figure 4.26), and also that the references can be mutually correlated. The problem is solved by using the ordered, conditioned input-output model [48]. In the ordered, conditioned model, the references are ordered by the magnitude of the ordinary coherence function between each reference, and the sensor output and the linear effects of the references $R_1, R_2, ...R_{k-1}$ are removed from R_k by optimum linear least squares prediction techniques.

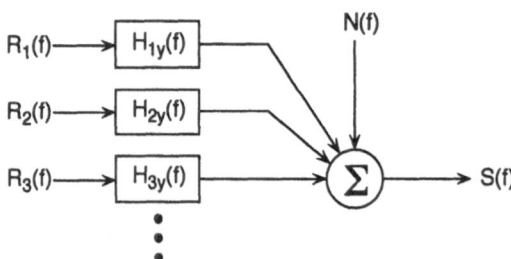

Figure 4.26. Frequency-dependent linear input-output model. The inputs (references R) can be mutually correlated. The sensor output, S, is modeled by adding the references multiplied by appropriate frequency-dependent transfer functions and some random noise, N.

A comparison between frequency-dependent and frequency-independent adaptive noise cancellation is shown in Figure 4.27. In this example the sensor is a magnetometer, and the references are 3 orthogonal magnetometers. The system was employed to measure AEF signals and was subjected to large vibrations. There is no evidence of any AEF signal before the application of the noise cancellation techniques.

When frequency-dependent adaptive noise cancellation is applied, as shown in Figure 4.27.b, practically all vibrational noise is removed, and the AEF signal is detected - the magnetometer was not positioned at the AEF signal maximum, and therefore the signal is rather small. If, however, only the frequency-independent adaptive method is used, as shown in Figure 4.27.c, only a small part of the vibrational noise is removed, and the

AEF signal is not resolved. In this example, the frequency-dependent adaptive technique is essential for good noise elimination.

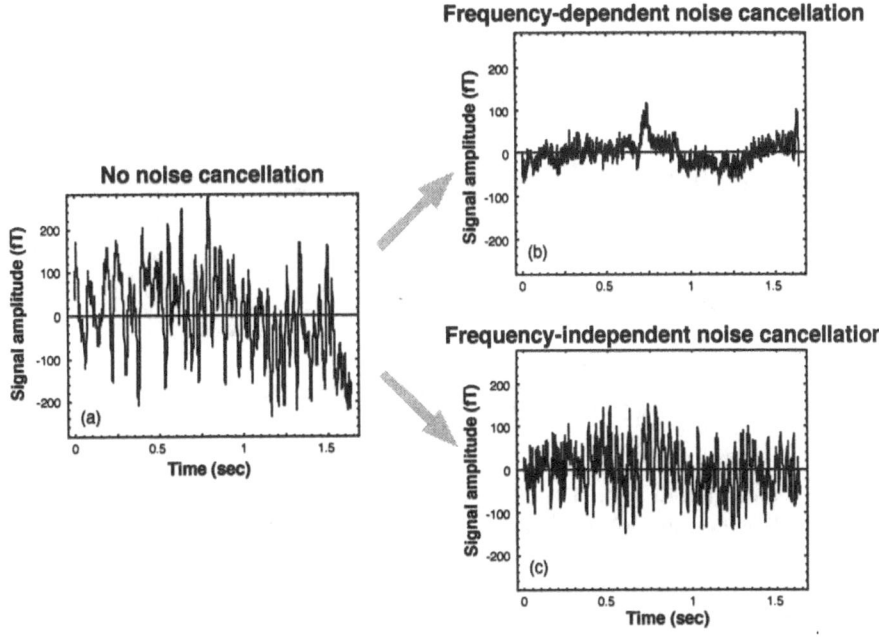

Figure 4.27. Comparison of frequency-dependent and frequency-independent adaptive noise cancellation techniques. AEF experiment. Sensor - magnetometer, references - 3 orthogonal magnetometers. f_s = 1250 samples/sec, point per trial = 2048, bandwidth = DC to 300 Hz, data is an average of 50 trials, transfer functions determined from 30 trials. (a) Data as collected, no noise cancellation; (b) after frequency-dependent noise cancellation; (c) after frequency-independent noise cancellation.

Note, however, that even though the frequency-dependent adaptive technique seems to be more successful in some cases than the frequency-independent technique, the time variation of the coefficients and the transfer functions is a problem for both techniques, and in both cases it requires frequent re-determination of the adaptive coefficients or transfer functions.

4.4. COMBINED HIGH-ORDER GRADIOMETER AND ADAPTIVE NOISE CANCELLATIONS

In this section, an example is shown in which a shielded room [14] is located near a subway line with a station located approximately 50 m from the room (nearly at the point of closest approach). The passage, stopping, or starting of the train generates magnetic fields and gradients within the shielded room: the magnitude of the train's magnetic field in the shielded room is ≈ 6 nT and the magnitude of the balanced 1st-order gradiometer signals can be as much as 20 pT, depending on the detection channel - see Figure 4.28.a.

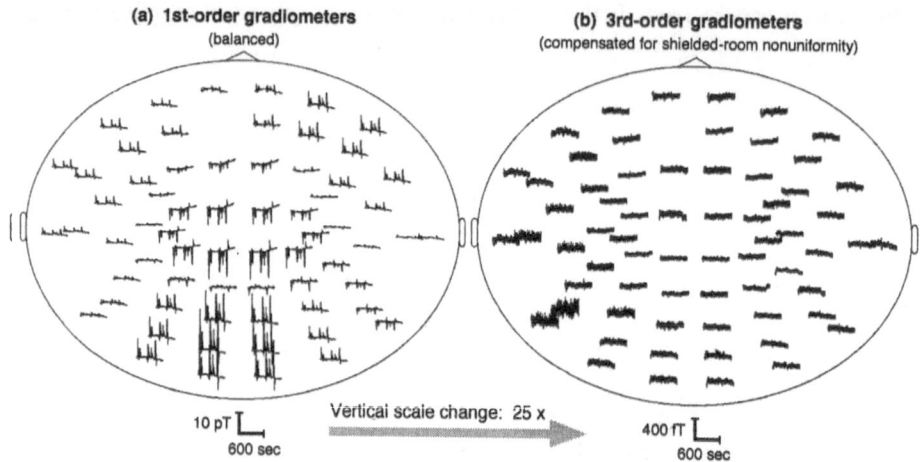

(a) 1st-order gradiometers
(balanced)

(b) 3rd-order gradiometers
(compensated for shielded-room nonuniformity)

10 pT ⌐
600 sec

Vertical scale change: 25 x

400 fT ⌐
600 sec

Figure 4.28. Signals detected inside a shielded room due to a subway train about 50 m away. Fs = 12.5 samples/sec, duration = 656 sec, bandwidth = DC to 4 Hz.

Train signal reduction by spatial filtering is shown in Figure 4.29.A for one detection channel (SL14). For this channel the 1st-order gradiometer train signal is ≈ 10 pT. Formation of the 2nd-order gradiometer reduces this signal to ≈ 4 pT (Figure 4.29.A.b), and the formation of the 3rd-order gradiometer reduces it to ≈ 500 fT (Figure 4.29.A.c and enlarged Figure 4.29.A.e), but it does not eliminate it completely. The incomplete removal of the train signal by the 3rd-order gradiometer is a result of small distortions of the field introduced by the shielded room (thus acting as a nearby noise source).

A rough estimate of the train fields and gradients also suggests that the residual 3rd gradient is not caused directly by the train. Assume that the train signal has a dipolar character, the room shielding factor is ≈ 70, the distance to the dipole is ≈ 50 m, the 1st-order gradiometer baseline is $d_1 = 5$ cm, and the measured field in the room is B = 6 nT. Then the expected magnitude of the 1st-order gradiometer output inside the shielded room is 18 pT, in rough agreement with the observation. A similar argument for higher-order gradiometers and accounting for the system dimensions, but ignoring distortions by the shielded room, yields the expected magnitude of the 3rd-order gradiometer output inside the shielded room to be about several fT, not 500 fT as observed in Figure 4.29.A.e. The above results are only weakly dependent on the character of the magnetic source. Therefore, the dipole approximation is adequate for this discussion.

The proposed model for the train detection inside the shielded room is as follows: the train generates a large, nearly uniform field at the location of the shielded room. This field is shielded by the room with attenuation factor ψ and also is slightly distorted by the room. These distortions then generate spatial gradients in the room with the same time dependence as the applied magnetic field and are responsible for the residual train signal in Figure 4.29.A.e - possible sources of distortions in the shielded room are various constructional nonuniformities, e.g., the door region.

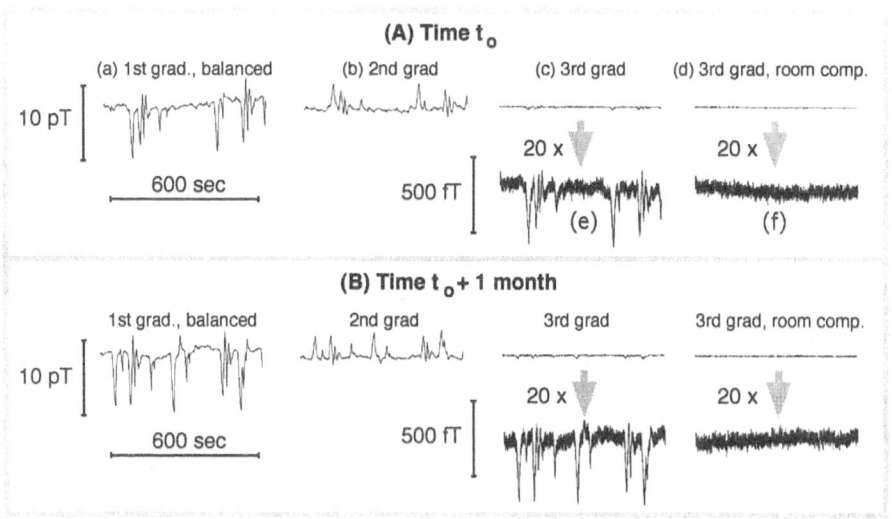

Figure 4.29. Signals detected inside a shielded room due to a subway train about 50 m away. Channel SL14, fs = 12.5 samples/sec, duration = 656 sec, bandwidth = DC to 4 Hz. Files: Jun 3:9 (93) and Jul 2:9 (93).

Since the gradients due to the room distortion have the same time dependence as the applied magnetic field, they can be eliminated by additional adaptive cancellation using magnetometers as references. Noise removal by this additional adaptive procedure is shown in Figure 4.29.A.d and enlarged in Figure 4.29.A.f for channel SL14, and in Figure 4.28.b for all channels. The subway signal is effectively removed by this procedure. However, it is crucial to perform the noise cancellation by both the 3rd-order gradiometers and adaptive subtraction with magnetometers as references. Adaptive subtraction alone or applied to only 1st- or 2nd-order gradiometer outputs, does not remove the train signals.

This noise removal procedure is quite stable in time, since the geometry of the shielded room and the subway is fixed. This is demonstrated in Figure 4.29.B, where the subway signals were measured after a 1 month interval and were eliminated using the identical gradient formation and adaptive coefficients. The sensors, reference system, and the gradient formation coefficients also were shown to be very stable. The gradient-formation coefficients were determined during the MEG system manufacturing process, after which the system was warmed up, disassembled, shipped to the final location, and reassembled. It was found, nevertheless, that the same gradient-formation coefficients were still applicable.

5. Conclusions and Summary

The use of SQUIDs for stationary magnetic field detection in the near field region has been discussed. Such devices operate in the presence of environmental noise and must be provided with shielding and/or noise cancellation techniques. Specifically, the MEG application has been used to illustrate performance under a variety of noise situations.

Various noise environments were described, and it was shown that both shielded and unshielded instruments are exposed to large variations of noise levels. An example of such a variation is the noise caused by car traffic. Traffic noise was analyzed in more detail, and it was shown to be in agreement with a simple model of dipoles moving past the instrument. Unshielded systems also are subject to large power-line fields. The character of these signals was discussed, and it was shown that in urban environments the slew rates and dynamic range imposed by the power lines can be handled by the present generation of SQUID instrumentation.

In addition to various noise sources, SQUID instruments are immersed in large dc fields and gradients from the Earth and local anomalies. In principle, the dc fields and gradients should not affect a stationary system. However, even stationary systems move slightly due to various unavoidable vibrations, and these vibrations relative to the dc fields introduce motion noise. The magnitude of the motion noise was analyzed, and it was shown that even in shielded rooms, good gradiometer balance and the use of higher-order gradiometers are required for elimination of vibrational noise. However, magnetometer detection is possible in shields with very large attenuation, e.g., in superconducting shields.

It has been assumed that the signal sources are very near to the detector - less than 10 cm - and that the noise sources are fairly distant. Therefore, noise cancellation by spatial filtering using high-order spatial gradiometers can be used successfully. The gradiometer construction, imperfections, and methods for characterizing and handling the imperfections were described in detail for 1st-order gradiometers, and the construction of higher-order gradiometers by hardware and software methods was outlined.

The evaluation of software high-order gradiometers was discussed in the context of a 64-channel whole-cortex MEG system. The cancellation of a dipole source signal by 1st-, 2nd- and 3rd-order gradiometers in shielded and unshielded environments was measured. The results from the two environments were compared, and were found to be similar and in good agreement with the expectations based on a simple dipole model. The results confirmed that the software gradiometers were constructed correctly and that their common mode vectors were sufficiently small.

The finding that noise cancellation as a function of gradiometer order is approximately the same for shielded and unshielded gradiometers was true when the MEG system was located roughly in the center of a shielded room. This implies that the distortions of the spatial character of the external fields by the shielded room are small. However, when the gradiometer was positioned close to a room wall (or the floor), the distortions by the room were evident. In this case, the shielded room introduced large additional spatial gradients which could be directly measured.

The effect of slight motions and vibrations of the shielded room also was measured. It was found that room motion can introduce very large signals in magnetometers and 1st-order gradiometers installed inside the room. However, these motion signals are partially canceled by 2nd-order gradiometers, and are completely canceled by 3rd-order gradiometers.

It was shown that even when operated without a shield, environmental noise can be canceled by 3rd-order gradiometers, such that the resulting system white-noise level is below 10 fT rms/√Hz for frequencies above about 0.5 to 1 Hz and below the power-line frequency, with the exception of several strong environmental or vibrational lines which can exceed the 10 fT rms/√Hz level. The power-line signals, even though attenuated by about 5 orders of magnitude (relative to the magnetic field) by 3rd-order gradiometers are always much larger than 10 fT rms/√Hz. In shielded rooms the white-noise level of the same system is about 5 fT rms/√Hz, and a noise level of 10 fT rms/√Hz is reached at a frequency of about 0.2 Hz. Low-frequency noise in shielded rooms is significantly reduced by high-order gradiometers.

The noise introduced by the formation of high-order software gradiometers was shown to be negligible for a properly-designed system. On average, the white-noise level increase corresponding to the change of the gradiometer order from 1st to 3rd was only about 0.5 fT rms/√Hz in the 64-channel whole-cortex MEG system.

In addition to noise cancellation by high-order gradiometers, adaptive noise removal techniques also were examined. It was shown that adaptive noise cancellation can perform well if the noise character is constant in time. However, when the noise character is changing, adaptive noise cancellation does not work. This has been demonstrated by measuring the signals generated by random motion of a magnet located at different positions relative to the MEG system. Adaptive noise cancellation worked well when the magnet motion was constrained to the vicinity of a given position. When the magnet was moved to a different position, a completely different set of adaptation coefficients was required. For the unshielded environment at Simon Fraser University, it was shown that the characteristic time during which the adaptation coefficients remain constant is only a few seconds.

Frequency-dependent and frequency-independent adaptive techniques also were examined. The frequency-dependent adaptive technique uses frequency-dependent transfer functions instead of frequency-independent coefficients. It was shown that under certain circumstances, the frequency-dependent techniques can provide much better noise cancellation than the frequency-independent techniques. However, both techniques require frequent re-determination of the cancellation parameters when the noise character is changing.

Finally, noise cancellation by combined adaptive and 3rd-order gradiometer techniques was applied to noise generated by a subway train passing at a distance of approximately 50 m from the shielded room. The train fields were slightly distorted by the shielded room, and these distortions resulted in additional gradients with the same temporal dependence as that of the applied train field (as measured by magnetometers). The additional distortions were removed adaptively using the magnetometers as references. However, the use of 3rd-order gradiometers was essential; the adaptive method alone did not remove the train signals. In this experiment the adaptive method worked well, at least over periods of several months, because the geometry of the train track and the shielded room was fixed, and the noise character was constant.

Acknowledgements

I would like to express my thanks to Dr. A. A. Fife, Dr. P. R. Kubik and Mr. B. Taylor of CTF Systems Inc. and to Dr. W. Ludwig of CTF Systems Europe GmbH for reviewing the manuscript and for their constructive comments. I would like to thank my colleagues at CTF Systems Inc., K. Betts, M. Burbank, T. Cheung, A. A. Fife, G. Haid, P. R. Kubik, S. Lee, J. McCubbin, J. McKay, D. McKenzie, V. Mirochnikov, C. Schroyen, P. Spear, B. Taylor, M. Tillotson and S. White for their enthusiastic collaboration and help with various experiments and analysis; Dr. H. Weinberg and Dr. D. Cheyne of Simon Fraser University, Brain Behaviour Laboratory for guidance and discussions during human experiments; Dr. H. Wilson and Mr. R. Erickson of EDRD, Victoria, B.C. for support and many discussions regarding the operation of 1st-order gradiometers under mobile conditions; Dr. H. Matsuba and Mr. A. Yahara of the Furukawa Electric Co., Ltd., Yokohama, Japan for support, helpful discussions and collaboration on various aspects of magnetometer operation; and Mr. I. Tamura and Mr. A. Matani of Osaka Gas Co., Osaka, Japan for help and support during the shielded room experiments.

The work was supported in part by the Canadian Departments of National Defence, Supply and Services, Transport Canada, and the BC Science Council.

References

1. Wikswo, J. P. Jr. (1995) SQUID Magnetometers for Biomagnetism and Nondestructive Testing: Important Questions and Initial Answers. Applied Superconductivity Conference, Boston, October 1994. *IEEE Trans. Appl. Sup.* **5**,
2. Nakasato, N., Fujita, S., Matani, A., Tamura, I., Fujiwara, S. and Yoshimoto, T. (1995) Clinical Application of the Whole Head MEG: Auditory Evoked Response in Patients with Intracranial Structural Lesions. To be published in: *Biomagnetism: Fundamental Research and Clinical Applications*, C. Baumgartner, L. Deecke, G. Stroink, S. Williamson (Eds), Amsterdam, Elsevier/IOS-Press, 1995.
3. Nakasato, N., Fujita, S., Seki, K., Kawamura, T., Matani, A., Tamura, I., Fujiwara, S. and Yoshimoto, T. (1995) Functional localization of bilateral auditory cortices using an MRI-linked whole head magneto-encephalography (MEG) system, *Electroencephalogr. Clin. Neurophysiol.* **94**, 183-190.
4. Ribary, U., Llinas, R., Lado, F., Mogilner, A., Jagow, R., Nomura, M. and Lopez, L. (1992) The spatial and temporal organization of the 40 Hz response in human brain, in Hoke, M., Erne, S. N., Okada, Y. C. and Romani, G. L. (eds.), *Biomagnetism: Clinical Aspects, Proceedings of the 8th International Conference on Biomagnetism*, Munster, 19-24 August 1991, pp. 159-163.
5. Drung, D., Absmann, Curio, G., Mackert, B.-M., Matthies, K.-P., Matz, H., Peters, M., Scheer, H.-J. and Koch, H. (1995) The PTB 83-SQUID System for Biomagnetic Applications in a Clinic. Applied Superconductivity Conference, Boston, October 1994. *IEEE Trans. Appl. Sup.* **5**,
6. Ahlfors, S. P., Ilmoniemi, R. J., Kajola, M. J., Knuutila, J. E. T. and Simola, J. T. (1995) Whole-Head Distribution of Visual Evoked Magnetic Fields, To be published in: *Biomagnetism: Fundamental Research and Clinical Applications*, C. Baumgartner, L. Deecke, G. Stroink, S. Williamson (Eds), Amsterdam, Elsevier/IOS-Press, 1995.
7. Matsuba, H., Shintomi, K., Yahara, A., Irisawa, D., Imai, K., Yoshida, H. and Seike, S. (1995) Superconducting Shield Enclosing a Human Body for Biomagnetism Measurement. To be published in: *Biomagnetism: Fundamental Research and Clinical Applications*, C. Baumgartner, L. Deecke, G. Stroink, S. Williamson (Eds), Amsterdam, Elsevier/IOS-Press, 1995.
8. Cheyne, D., Vrba, J., Crisp, D., Betts, K., Burbank, M., Cheung, T., Fife, A. A., Haid, G., Kubik, P. R., Lee, S., McCubbin, J., McKay, J., McKenzie, D., Spear, P., Taylor, B., Tillotson, M. and Weinberg, H.,

Basar, E. and Tsutada, T. (1992) Use of an unshielded 64 channel whole-cortex MEG system in the study of normal and pathological brain function, *Proceedings of the Satellite Symposium on Neuro-science and Technology* , pp 46-50, 14th Annual International Conference of the IEEE Engineering in Medicine and Biology Society, Lyon, France, November.

9. Vrba, J., Betts, K., Burbank, M., Cheung, T., Fife, A. A., Haid, G., Kubik, P. R., Lee, S., McCubbin, J., McKay, J., McKenzie, D., Spear, P., Taylor, B., Tillotson, M., Cheyne, D. and Weinberg, H. (1993) Whole cortex, 64 channel SQUID biomagnetometer system, *IEEE Trans. Appl. Sup.* **3**, 1878-1882.

10. Vrba, J., Taylor, B., Cheung, T., Fife, A. A., Haid, G., Kubik, P. R., Lee, S., McCubbin, J. and Burbank, M. B. (1994) Noise Cancellation by Whole-Cortex SQUID MEG System. Applied Superconductivity Conference, Boston, October 1994. *IEEE Trans. Appl. Sup.* **5**, 2118-2123.

11. Clem, T. R. (1994) Superconducting Magnetic Sensors Operating from a Moving Platform. Applied Superconductivity Conference, Boston, October 1994. *IEEE Trans. Appl. Sup.* **5**,

12. Amuneal Manufacturing Corp., 4737 Darrah Street, Philadelphia, PA 19124, USA.

13. Vacuumschmelze GmbH, Hanau, Germany; Shielded Room model AK-3.

14. Tokin Corporation, 6-7-1 Koriyama Tihakuku, Sendai-City, Miyagi-pref, 982, Japan.

15. Sullivan, G. W. and Flynn, E. R. (1987) Performance of the Los Alamos Shielded Room, in K. Atsumi, M. Kotani, S. Ueno, T. Katila, S. J. Williamson (eds.), *Biomagnetism 87, 6th International Conference on Biomagnetism*, Tokyo, Japan, August 27-30, Tokyo Denki University Press, Tokyo, pp.486-489.

16. Kelha, V. O. (1981) Construction and performance of the Otaniemi magnetically shielded room., in S. N. Erne, H. D. Hahlbohm, H. Lubbig (eds.), *Biomagnetism, Proceedings of the Third International Workshop on Biomagnetism*, Berlin, May 1980, Walter de Gruyter, Berlin, New York, 1981.

17. Erne, S. N., Hahlbohm, H.-D., Scheer, H. and Trontelj, Z. (1981) The Berlin Magnetically Shielded Room (BMSR) Section B - Performances, in S. N. Erne, H. D. Hahlbohm, H. Lubbig (eds.), *Biomagnetism, Proceedings of the Third International Workshop on Biomagnetism*, Berlin, May 1980, Walter de Gruyter, Berlin, New York, 1981.

18. Stroink, G., Blackford, B., Brown, B. and Horacek, M. (1981) Aluminum Shielded Room for Biomagnetic Measurements. *Rev. Sci. Instrum.* 52(3), 463-468.

19. Flynn, E. R., Los Alamos National Laboratory, private communication

20. Furukawa Electric Co., Ltd., 2-4-3 Okano, Nishi-ku, Yokohama 220, Japan

21. Carelli, P., Modena I. and Romani G. L. (1982) Detection Coils, in S. J. Williamson, G. L. Romani, L. Kaufman, and I. Modena (eds.), *NATO ASI Biomagnetism, An Interdisciplinary Approach*, Sep 1-12, Rome, Italy, Plenum Press, New York and London.

22. Katila, T. (1989) Principles and applications of SQUID sensors, in S. J. Williamson, M. Hoke, G. Stroink, M. Kotani (eds.), *Advances in Biomagnetism, Proceedings of the 7th international conference on biomagnetism* held in August 1989 in New York, New York, Plenum Press, New York and London, pp.19-32.

23. Cantor, R., Drung, D., Erne, S.N. and Koch H. (1991) Electronic Gradiometric Balancing Capabilities of dc SQUID Magnetometer System for Biomagnetism, *8th International Conference on Biomagnetism*, Munster, August 18-24, 1991, Germany

24. Katila, T. (1981) Instrumentation for biomedical applications, in S. N. Erne, H. D. Hahlbohm, H. Lubbig (eds.), *Biomagnetism, Proceedings of the Third International Workshop on Biomagnetism*, Berlin, May 1980, Walter de Gruyter, Berlin, New York, 1981.

25. Fraser-Smith, A. C. and Buxton, J. L. (1975) Superconducting Magnetometer Measurements of Geomagnetic Activity in the 0.1 to 14 Hz Frequency Range, *J. Geophys. Res.* **80**, 3141-3147.

26. Mathematica, Wolfram Research Inc., 100 Trade Center Drive, Champaign, Illinois 61820-7237, USA

27. Hughes, B., PowerTech Labs Inc., Surrey, B.C., Canada, private communication.

28. Williamson, S. J. and Kaufman, L. (1981) Biomagnetism, *Journal of Magnetism and Magnetic Materials* **22**, 129-201.

29. Wynn, W. M., Frahm, C. P., Carroll, P. J., Clark, R. H., Wellhoner, J. and Wynn, M. J. (1975) Advanced Superconducting Gradiometer/Magnetometer Arrays and a Novel Signal Processing Technique. *IEEE Trans. Mag.* **MAG-11**, 701-707.

30. Sarwinski, R. E. (1977) Superconducting Instruments, *Cryogenics* **17**, 671-679.

31. Brenner, D., Kaufman, L. and Williamson, S. J. (1977) Application of a SQUID for Monitoring Magnetic Response of the Human Brain, *IEEE Trans. Mag.* **MAG-13**, 365-368.

32. Williamson, S. J., Pelizzone, M., Okada, Y., Kaufman, L., Crum, D. B. and Marsden, J. R. (1984) Magnetoencephalography with and Array of SQUID Sensors, in Collan, H., Berglund, P. and Krusius, M. (eds.), *Proceedings of the Tenth International Cryogenic Engineering Conference*, Butterworth, Guildford, pp.339-348.

33. Matlashov, A., Zhuravlev, Yu., Lipovich, A., Alexandrov, A., Mazaev, E., Slobodchikov, V. and Washiewski, O. (1989) Electronic Noise suppression in multi-channel neuromagnetic system, in S. J. Williamson, M. Hoke, G. Stroink, M. Kotani (eds.), *Advances in Biomagnetism*, Proc. 7th Int. Conf. on Biomagnetism, New York, N.Y., pp. 725-728, August 1989.

34. Zijlstra, H. (1967) Experimental Methods in Magnetism, in E. P. Wohlfarth (ed.), *Series of Monographs on Selected Topics in Solid State Physics*, North-Holland Publishing Company, Amsterdam, John Wiley & Sons, Inc., New York, 1967.

35. Alldred, J. C. and Scollar, I. (1967) Square Cross Section Coils for the Production of Uniform Magnetic Fields, *J. Sci. Instrum.* **44**, 755-760.

178

36. Garrett, M. W. (1951) Axially Symmetric Systems for Generating and Measuring Magnetic Fields. Part I, *J. Appl. Phys.* **22**, 1091-1107.
37. Merritt, R., Purcel, C. and Stroink, G. (1983) Uniform magnetic field produced by three, four and five square coils, *Rev. Sci. Instr.*, **54**, 879-882.
38. Vrba, J. (1979) Gradiometer Balancing by Motion, CTF Internal Report, CI-182-0779.
39. Williamson, S. J., Robinson, S. E. and Kaufman, L. (1987) Methods and Instrumentation for Biomagnetism, *BioMagnetism '87, Proceedings of the Sixths International Conference on Biomagnetism*, Tokyo, Japan, August 27-30, 1987.
40. Vrba, J., Fife, A. A., Burbank, B. M., Weinberg, H. and Brickett, P. A. (1982) Spatial Discrimination in SQUID Gradiometers and 3rd Order Gradiometer Performance, *Can. J. Phys.* **60**, 1060-1073.
41. Vrba, J. Haid, G., Lee, S., Taylor, B., Fife, A. A. ,Kubik, P., McCubbin, J. and Burbank, M. B. (1991) Biomagnetometers for Unshielded and Well Shielded Environments, *Clin. Phys. Physiol. Meas.* **12**, Suppl. **B**, 81-86.
42. Drung, D. (1991) Performance of an Electronic Gradiometer in Noisy Environments, in H. Koch and H. Lubbig (eds.), *SQUID'91, Superconducting Devices and their Applications*, Berlin, June 18-21, 1991, Springer Proceedings in Physics.
43. W. Becker, V. Dickmann, R. Jurgens and C. Kornhuber. (1993) First experiences with a multichannel software gradiometer recording normal and tangential components of MEG, *Physiol. Meas.* **14**, A45-A50.
44. Dieckmann, V., Jurgens R., Becker, W., Elias, H., Ludwig, W. and Vodel,W. (1995) RF- to DC-SQUID Upgrade of a 28-Channel Magnetoencephalography (MEG) System, submitted to *Measurement Science and Technology*.
45. CTF Systems Inc., 15-1750 McLean Avenue, Port Coquitlam, B.C., Canada, V3C 1M9.
46. Bruno, A. C., Dolce, C. S., Soares, S. D. and Ribeiro, P. C. (1989) Spatial Fourier Technique for Calibrating Gradiometers, in S. J. Williamson, M. Hoke, G. Stroink, M. Kotani (eds.), *Advances in Biomagnetism, Proceedings of the 7th international conference on biomagnetism* held in August 1989 in New York, New York, Plenum Press, New York and London, pp.709-712.
47. Vrba, J., Betts, K., Burbank, B. M., Cheung, T., Cheyne, D., Fife, A. A., Haid, G., Kubik, P. R., Lee, S., McCubbin, J., McKay, J., McKenzie, D., Mori, K., Spear, P., Taylor, B., Tillotson, M. and Xu, G. (1995) Whole Cortex 64 Channel System for Shielded and Unshielded Environments, to be published in: *Biomagnetism: Fundamental Research and Clinical Applications*, C. Baumgartner, L. Deecke, G. Stroink, S. Williamson (Eds), Amsterdam, Elsevier/IOS-Press, 1995.
48. Bendat, J. S. and Piersol, A. G. (1986) *Random Data*, John Willey & Sons, New York, Chichester, Brisbane, Toronto, Singapore.
49. Zimmerman, J. E. (1977) SQUID instruments and shielding for low level magnetic measurements, *J. Appl. Phys.* **48**, 702-710.

DC SQUIDS: DESIGN, OPTIMIZATION AND PRACTICAL APPLICATIONS

ROBIN CANTOR

Conductus, Inc.
969 West Maude Avenue
Sunnyvale, California, USA 94086

ABSTRACT. The dc Superconducting Quantum Interference Device (SQUID) is the most sensitive sensor of magnetic flux available, with an enormous frequency response extending from dc to a few GHz. It has been shown that the energy resolution of low-inductance SQUIDs operating at 4.2 K can approach the quantum limit. Unfortunately, the small inductance required makes it difficult to couple external signals to the SQUID. Excellent coupling can be achieved using thin-film and lithographic techniques to integrate an input coil on top of a washer-shaped SQUID inductance. This approach has significantly advanced SQUID technology for numerous applications. Unless the SQUID is heavily damped, however, the parasitic elements (capacitance and inductance) that are introduced can lead to resonances in the SQUID dynamics. These resonances manifest themselves as strong irregularities in the current-voltage (I-V) and voltage-flux (V-Φ) characteristics, leading to excess noise and making operation using conventional flux modulation techniques extremely difficult. Overly damping the SQUID may reduce the excess noise, but doing so also diminishes the amplitude of the SQUID output signal, placing more stringent demands on the readout electronics.

Several years ago, an innovative design optimization procedure was described which takes these parasitic effects into account. Using this procedure, it has been shown that a properly designed SQUID, optimized to avoid the adverse effects of parasitic resonances, has very smooth I-V and V-Φ characteristics and can therefore be operated throughout a wide range of bias points with very low noise. In addition, since the SQUID is designed to operate near the hysteretic limit, the flux-to-voltage transfer function can be very high. This facilitates coupling the low-impedance SQUID output signal to the room-temperature readout electronics.

In this chapter, the basic SQUID design optimization procedure is reviewed. Included is a summary of measurements on expanded models of actual dc SQUIDs. This is followed by a review of low-T_c SQUID fabrication techniques and SQUID readout electronics. Several SQUID applications are discussed, ranging from laboratory instruments to applications in biomagnetism. The chapter concludes with a discussion of (HTS) SQUID design and applications.

H. Weinstock (ed.), SQUID Sensors: Fundamentals, Fabrication and Applications, 179–233.

1. Introduction

The dc Supercoducting Quantum Interference Device (SQUID) is the most sensitive detector of magnetic flux currently available, with an enormous signal bandwidth extending from dc to several GHz. For this reason, dc SQUIDs are used in a wide variety of measurement applications where the physical quantity to be measured can be converted to magnetic flux. Examples include current, voltage, magnetic field or field gradient, gravitational field and magnetic susceptibility.

A dc SQUID consists of two resistively-shunted Josephson junctions connected in parallel using superconducting wire. A schematic diagram of a dc SQUID is shown in Fig. 1. Each junction (represented by a cross) is shunted by an external resistor R in order to eliminate the hysteresis in the junction I-V characteristics, and C is the intrinsic capacitance of each junction. A more detailed discussion of Josephson junctions may be found in a companion chapter by John Clarke in this volume.

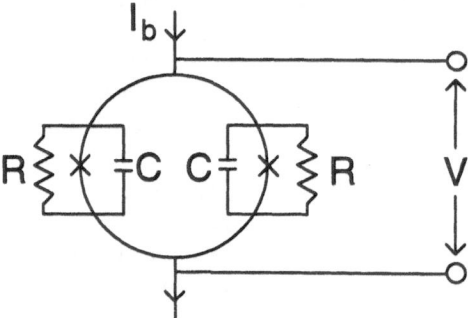

Figure 1. Schematic diagram of a dc SQUID.

The dc SQUID may be operated by applying a constant bias current I_b and measuring the average voltage V across the SQUID, as shown above, or by applying a constant bias voltage across the SQUID and measuring the current through the SQUID. When operated using constant current bias, the average voltage across the SQUID is a periodic function of the magnetic flux threading the SQUID loop; the periodicity of the voltage modulation is the magnetic flux quantum $\Phi_0 = h/2e$, where h is Planck's constant, and e is the charge of the electron. Thus, in this mode, the dc SQUID is a flux-to-voltage transducer. In Fig. 2, the experimentally measured voltage-current characteristics of a current biased dc SQUID with integral and half-integral flux through the SQUID are shown, along with a represenative voltage-flux characteristic for one value of the constant bias current.

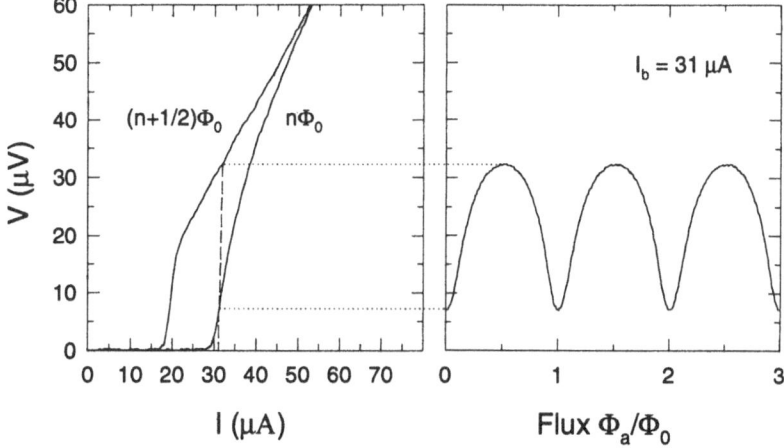

Figure 2. Measured dc SQUID voltage-current characteristics for integral and half-integral flux through the SQUID of an integrated SQUID magnetometer -see Sec. 4- and the corresponding voltage-flux characteristic for a bias current of 31 μA.

When operated using constant current bias, a dc flux bias of roughly $\Phi_0/4$ may be applied so the SQUID is at the steepest part of the voltage-flux characteristic where the flux-to-voltage transfer function is greatest and fairly linear for small signals. At this point, the SQUID is most sensitive to changes in the applied flux. In this small signal mode, however, the linear range is limited to a fraction of Φ_0. For this reason, most SQUID applications employ a feedback technique, and the SQUID is operated as a null detector. In this way, much higher resolution and dynamic range may be obtained. This is discussed in more detail in Sec. 3 and in a companion chapter by Dietmar Drung in this volume.

2. Practical dc SQUID Design

Several years ago, Knuutila *et al*[1] introduced an optimization procedure for the design of thin-film dc SQUIDs with integrated coupling circuits which considers the parasitic effects introduced by the input coil. This procedure was further developed by Cantor *et al*[2,3] to build integrated SQUID magnetometers and gradiometers. In this section, this optimization procedure for the design of integrated dc SQUIDs is reviewed in detail.

2.1. OPTIMIZATION MODEL

An often-used figure of merit for dc SQUIDs is the noise energy per unit bandwidth, usually referred to as the energy resolution of the SQUID. In this section, we derive expressions for the energy resolution that contain design-dependent parameters only. One may then use standard minimization algorithms to optimize the energy resolution. We discuss first a simple model of an uncoupled SQUID. This is then generalized to describe a coupled SQUID with an integrated input coil. The effects of parasitic resonances and parasitic capacitance are reviewed, and several approaches for the design of SQUIDs with high parasitic capacitance are discussed.

2.1.1. Uncoupled dc SQUIDs

An uncoupled or "bare" dc SQUID consists of two resistively-shunted Josephson junctions connected in parallel using superconducting wire. Each of the two Josephson junctions may be represented by the resistively-shunted-junction (RSJ) model, consisting of an ideal Josephson junction with junction capacitance C shunted by a resistance R (R is the resistance of the external shunt resistor in parallel with the junction quasiparticle resistance; usually, the junction quasiparticle resistance is much larger than the external shunt resistance). The critical current of each junction is I_c, and the inductance of the SQUID loop is L. The junctions are assumed to be identical, and the SQUID is assumed to be biased by a current I_b. A schematic diagram of the dc SQUID model is shown in Fig. 3.

In the absence of excess noise due to resonances, one may model the SQUID voltage noise as arising from two nearly uncorrelated noise currents in the SQUID, one circulating around the SQUID loop and the other through the SQUID[4]. Then, neglecting the noise of the room-temperature preamplifier, the spectral density of the voltage noise power across the SQUID may be written as[4]

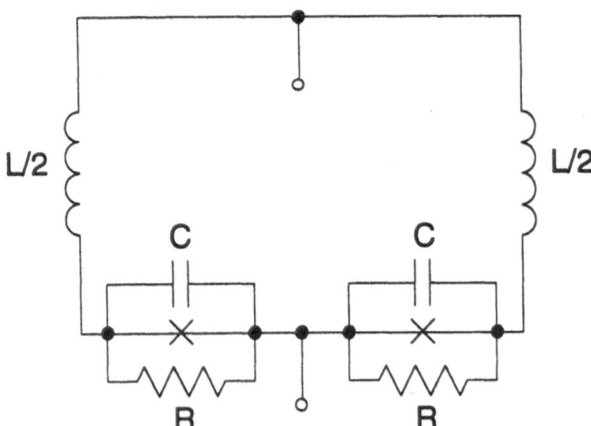

Figure 3. Schematic diagram of the uncoupled dc SQUID model.

$$S_V = \left(\frac{\partial V}{\partial \Phi}\right)^2 \frac{4k_B T L^2}{2R} + 4k_B T R_{\text{dyn}}, \tag{1}$$

where $R_{\text{dyn}} = \partial V / \partial I$ is the dynamic resistance of the SQUID, and $\partial V / \partial \Phi$ is the flux-to-voltage transfer function. Here, a general theory[5,6] of noise in a nonlinear resistance is used; i.e., the first term of the Taylor expansion of the power dissipation, proportional to $\partial V / \partial \Phi \equiv R_{\text{dyn}}$, is used.

An important figure of merit for the dc SQUID is the noise energy per unit bandwidth, commonly referred to as the energy resolution. The energy resolution is defined as $\varepsilon = S_\Phi(f)/(2L)$, where the flux power spectral density $S_\Phi(f)$ is given by $S_\Phi(f) = S_V(f)/(\partial V / \partial \Phi)^2$. Using Eq. (1), the flux power spectral density is

$$S_\Phi = 2k_B T \left\{ \frac{L^2}{R} + \frac{2R_{\text{dyn}}}{(\partial V / \partial \Phi)^2} \right\}, \tag{2}$$

and the energy resolution may then be written as

$$\varepsilon = \frac{k_B T}{L} \left\{ \frac{L^2}{R} + \frac{2R_{\text{dyn}}}{(\partial V / \partial \Phi)^2} \right\}. \tag{3}$$

In order to carry out an optimization of the SQUID energy resolution, one needs to relate the dynamic resistance R_{dyn} and the transfer function $\partial V/\partial \Phi$ to the SQUID design parameters.

If self-screening effects are neglected, that is, if the SQUID inductance is sufficiently small so that the total flux linking the SQUID is approximately equal to the applied flux Φ, then it can be shown[7] that the average voltage across the current-biased SQUID is approximately given by

$$V = \frac{IR}{2} \left[1 - \left(\frac{2I_c}{I} \cos \frac{\pi \Phi}{\Phi_0} \right)^2 \right]^{1/2}. \tag{4}$$

In the following, it is useful to define $\beta = 2LI_c/\Phi_0$. Then, Eq. (4) is valid in the limit $\beta \ll 1$. In this limit, the ideal SQUID behaves like a single Josephson junction with resistance $R/2$, an effective flux-dependent critical current $I_{c,\text{eff}} = 2I_c \cos(\pi\Phi/\Phi_0)$, and the total modulation depth of the critical current is $\Delta I_{c,\text{eff}} = 2I_c$. For flux-locked-loop operation, the SQUID is typically flux- and current-biased such that $\Phi_a \approx \Phi_0/4$

and $I_b \approx 2I_c$. Then, Eq. (4) may be differentiated at these values to determine the dynamic resistance and the transfer function. According to simulations[8], however, the energy resolution is optimal for $\beta \sim 1$. To relax the assumption that $\beta \ll 1$, RSJ model simulations[8] indicate that $\Delta I_{c,eff}(\beta \approx 1) \approx (1/2) \cdot \Delta I_{c,eff}(\beta \ll 1) \approx I_c$. Thus, $\partial V/\partial \Phi \approx \partial(R_{dyn}\Delta I_{c,eff})/\partial \Phi$ is roughly a factor of two smaller for $\beta \sim 1$. Then, differentiating Eq. (4),

$$R_{dyn} = \frac{\partial V}{\partial I} \approx \frac{R}{\sqrt{2}} \tag{5}$$

and

$$\frac{\partial V}{\partial \Phi}(\beta \approx 1) \approx \frac{R}{2L}. \tag{6}$$

From simulations, it is found that the transfer function at the optimal point of operation of the SQUID is more accurately given by[4,9]

$$\frac{\partial V}{\partial \Phi} = \frac{7}{\pi} \frac{I_c R}{1+\beta} \left\{ 1 - 3.57 \frac{\sqrt{k_B TL}}{\Phi_0} \right\}, \tag{7}$$

where T is the operating temperature of the SQUID. The term in brackets in Eq. (7) accounts for the effects of thermal noise which become significant for large inductances and high temperatures. For $T \leq 4.2$ K, this term is usually negligible. Thus, the simple approximation for $\partial V/\partial \Phi$ in Eq. (6) is rather good. Furthermore, it is found experimentally[10] that $\partial V/\partial \Phi \approx R_{dyn}/(\sqrt{2}L)$, which is consistent with Eqs. (5) and (6).

Using the approximation for R_{dyn} and the approximation $\partial V/\partial \Phi \approx 2I_c R/(1+\beta)$, which is valid for low temperatures, the energy resolution becomes

$$\varepsilon \approx k_B T \frac{L}{R} \left\{ 1 + \frac{\sqrt{2}(1+\beta)^2}{\beta^2} \right\}. \tag{8}$$

Setting $\beta = 1$ in Eq. (7), the optimal energy resolution is $\varepsilon \approx 6.7 k_B T L/R$. Defining the Stewart-McCumber parameter $\beta_c = (2\pi/\Phi_0) I_c R^2 C$, Eq. (8) may be re-written as

$$\varepsilon \approx k_B T \sqrt{LC} \sqrt{\frac{\pi\beta}{\beta_c}} \left\{ 1 + \frac{\sqrt{2}(1+\beta)^2}{\beta^2} \right\} \equiv \gamma k_B T \sqrt{LC} . \qquad (9)$$

On the basis of simulations[4], the expression for the energy resolution in Eq. (9) is a reasonable approximation for $\beta_c \leq 0.7$. Note that the energy resolution is now expressed in terms of design-dependent parameters only.

The dynamics of the SQUID are determined by two key parameters. The parameter β is a measure of the modulation depth of the maximum current at zero voltage as a function of applied flux. For large β ($\beta \gg 1$), the modulation depth becomes very small, while for small β ($\beta < 1$), the modulation depth approaches the limit $2I_c$. As noted above, the energy resolution is optimal for $\beta = 1$. The Stewart-McCumber or hysteresis parameter β_c is a measure of the damping of the Josephson junctions. For $\beta_c \leq 0.7$, the SQUID current-voltage (*I-V*) characteristics are non-hysteretic and the average voltage across the SQUID for a given external flux is single valued. For $\beta_c > 0.7$, the *I-V* characteristics are hysteretic; because of thermal noise, however, the experimentally measured curves may appear to be single-valued for much higher values of β_c[11]. The hysteretic behavior leads to excess noise and a degradation of the energy resolution not included in the simple RSJ model approximation in Eq. (9). Thus, the energy resolution is optimal roughly for $\beta_c \sim 1$, giving $\varepsilon \approx 12 k_B T \sqrt{LC}$. This is about 25% less than predicted by simulations carried out assuming that the junctions are strongly over-damped[8] (for $\beta_c \ll 1$, $\varepsilon \approx 16 k_B T \sqrt{LC}$), where the nearly unlimited bandwidth increases the mixing-down effect of noise[4].

Based on the above, the design of an uncoupled dc SQUID is straightforward: one needs to reduce the temperature T, the SQUID inductance L and the junction capacitance C. The operating temperature is generally fixed for a given technology (*e.g.*, 4.2 K or 77 K). It is difficult to make the SQUID inductance much smaller than 10 pH. Once L is determined, the condition on β fixes the critical current. The junction fabrication technology limits the minimum junction capacitance that can be achieved. For overlap junctions, the junction capacitance is typically 0.5 - 1 pF. One complication that may arise is that a strong *LC* resonance may be formed if the SQUID inductance and junction capacitance are not properly chosen[11]. Once the junction capacitance is determined by the fabrication technology, the condition that $\beta_c \leq 0.7$ determines the junction shunt resistance R. Thus, all of the SQUID parameters are defined.

A significant consequence of the small inductance of typical dc SQUIDs, however, is that the effective flux capture area is very small. Thus, an uncoupled, single-loop dc SQUID has rather poor magnetic field resolution, but on the other hand the spatial resolution can be very good. In this way, SQUID magnetic microscopes have been built which enable high resolution magnetic imaging[12]. Alternately, configured as a two-loop gradiometer, one may form a miniature susceptometer for studying the behavior of minute magnetic particles[13]. For other applications where it is not feasible to use the SQUID itself to detect the signal of interest, however, one needs a means of coupling the signal to the SQUID inductance.

2.1.2. COUPLED DC SQUIDS

Since a bare SQUID has a low effective flux capture area and therefore poor magnetic field resolution, these devices are insufficient for many practical applications. For this reason, it is advantageous to use a separate signal or pickup coil that can have a much larger flux capture area. The pickup coil is connected to a multiturn spiral input coil that can be transformer coupled to the SQUID inductance. For best coupling, this is ideally done by patterning the SQUID inductance in the shape of a washer with the input coil integrated on top[14]. A second (modulation) coil is usually integrated on top of the SQUID washer as well, in order to couple feedback and a flux modulation signal to the SQUID. This is essential for operation using conventional flux-locked-loop readout electronics.

Excellent coupling to the SQUID can be achieved using an integrated input coil, but in so doing a parasitic capacitance is introduced across the SQUID inductance which can lead to a sizable degradation of the SQUID energy resolution[4]. Furthermore, the input coil and the SQUID washer as a ground plane form a microwave transmission line which can have high-Q resonances near the intended operating frequency $f_{op} = 0.3f_j$ of the SQUID[4], where the Josephson frequency $f_j = I_cR/\Phi_0$. The frequency f_{op} corresponds to the voltage at which the noise performance of a current and dc flux-biased SQUID is optimal. The transmission-line resonance is either a half-wavelength or quarter-wavelength resonance, depending on whether the input coil is floating or grounded, respectively. The Q-value of the half-wavelength resonance of the washer also is enhanced by the groundplane effect of the multiturn input coil[4]. These high-Q resonances couple to the SQUID dynamics and produce strong irregularities in the I-V and V-Φ characteristics, significantly increasing the SQUID noise and making it difficult to find a suitable working point for proper operation of the SQUID[4,11]. Thus, in designing a SQUID with integrated coupling circuits, it is essential that the microwave resonances and adverse effects of the parasitic capacitance be properly taken into account.

Typically, the operating frequency f_{op} lies below the resonant frequency of the washer f_w and above the stripline resonant frequency f_s. For designs requiring an input coil with many turns, the long length of the input coil pushes the stripline resonance to (low) frequencies well below f_{op}, but the large size of the washer causes the washer resonance to move towards f_{op}. Thus, a design compromise is necessary. Based on previous work[2], it has been found that a reasonable compromise is to choose the SQUID layout such that

$$4f_s < f_{op} < f_w / 4. \tag{10}$$

This condition ensures that both the stripline resonance, as well as the washer resonance, are well away from the intended operating frequency of the SQUID.

The SQUID characteristics are found experimentally to depend markedly on the impedance connected across the input coil[15]. The resonant frequency of the input circuit, consisting of the input coil in parallel with the load inductance and the parasitic capacitance, is generally low, but simulations and experiments suggest that the resonance may be thermally activated[4]. It has been demonstrated that the input circuit resonance can be damped in a nearly noise-free manner by the insertion of a series $R_x C_x$ shunt across the input coil[16]. This damping element also may be used to suppress the transmission line resonance of the input coil[2]. The resonance of the microstripline is suppressed by preventing the reflection at the "open" end of the line, i.e., where the stripline leaves the washer. This is determined by the reflection coefficient $\Gamma = (Z_L - Z_0)/(Z_L + Z_0)$. Here, Z_L is the impedance of the termination at the end of the stripline, and Z_0 is the nominal impedance of the transmission line. Thus, the $R_x C_x$ shunt is designed to fulfill two purposes. The microstripline is properly terminated by matching R_x to the nominal impedance of the transmission line, thereby inhibiting the microstripline resonance, and the resonant frequency of the input circuit is fixed at a frequency of the order of 10 - 100 MHz by the shunt capacitor C_x. The capacitor C_x also blocks the low frequency components of the noise current due to R_x. The resonant frequency of the input circuit is clearly well above the signal frequencies to be measured, yet well below the intended operating frequency of the SQUID, which is of the order of 10 GHz. The $R_x C_x$ shunt also determines the Q-value of the input circuit resonance. A moderate Q (typically, $Q \sim 2$) ensures that the resonance is well damped, but not over-damped to prevent the Josephson oscillations from mixing noise due to R_x down to low frequencies.

If the coupled SQUID is designed so that the various resonances are well away from the intended operating frequency of the SQUID and damped, the equivalent flux noise of the coupled SQUID can still be obtained from the simple model approximation given in Eq. (2), provided the SQUID inductance and transfer function are modified to take the effects of screening into account. The flux noise of the SQUID is then given by[1]

$$S_\Phi = 2k_B T \left\{ \frac{L_{DC}^2}{R} + \frac{2R_{dyn}}{(\partial V/\partial \Phi)^2} \right\}. \tag{11}$$

The flux noise now depends on the screened inductance $L_{DC} = (1 - k_i^2 s_{in})L$, where k_i is the coupling constant between the input coil and the SQUID, and

$s_{in} = L_i / (L_i + L_\ell)$ is a factor which describes the screening of the SQUID inductance by the input coil. Here, L_i and L_ℓ are the inductances of the input coil and load, respectively. Conversely, the input coil is screened by the SQUID inductance, leading to an effective input inductance $L_{i,eff} = (1 - k_i^2 s)L_i$, where it can be shown[1] that the screening factor $s = \beta s_{in} k_i^2 / (6 + 2\beta + \beta s_{in} k_i^2)$. The screening effect of the input coil also reduces the transfer function. Neglecting the temperature-dependent noise flux factor in Eq. (7), it can be shown[1] that the transfer function for the coupled SQUID is given by

$$\frac{\partial V}{\partial \Phi} = \frac{7}{\pi} \frac{2(3 + \beta)}{2(3 + \beta) + \beta s_{in} k_i^2} \frac{I_c R}{1 + \beta}. \tag{12}$$

Then, using Eqs. (5) and (12), the energy resolution for the coupled SQUID becomes

$$\varepsilon = k_B T \sqrt{LC} \left(\frac{\pi \beta}{\beta_c}\right)^{1/2} \left\{ (1 - s_{in} k_i^2)^2 + \frac{4\sqrt{2}\pi^2}{49} \left(\frac{1 + \beta}{\beta}\right)^2 \left(\frac{2(3 + \beta) + s_{in} k_i^2 \beta}{2(3 + \beta)}\right)^2 \right\}. \tag{13}$$

Again, the energy resolution may be expressed in the form $\varepsilon = \gamma k_B T \sqrt{LC}$, where the dimensionless factor γ contains all of the design-dependent parameters.

The functional form of the energy resolution given in Eq. (13) is a complicated expression, but since it contains only design-dependent parameters, the energy resolution may be optimized using conventional multivariable minimization algorithms. Typically, for a fixed load inductance, we allow the washer hole size, the junction critical current density, and the width of an optional slit groundplane to vary for stepwise increasing values of the number of turns in the input coil. The remaining parameters, such as the junction area, film thicknesses and linewidths, and material constants are fixed. The dependence of the junction capacitance on the critical current-density is taken into account as the critical current is varied. The hysteresis parameter β_c also is fixed at the limiting value $\beta_c = 0.7$ by appropriately adjusting the junction shunt resistance. We note that it is possible to design the SQUID for proper operation in this limit if the microwave resonances are well away from the intended operating frequency of the SQUID and are damped as discussed above.

For designs such that the parasitic capacitance $C_p > C$, however, simulations and experiments suggest an additional degradation of the energy resolution not included in Eq. (13)[11,17]. For large SQUID inductances $L \sim 300$ pH, the dependence of ε on C_p is found from simulations to be roughly $\varepsilon \propto \sqrt{1 + 2C_p / C}$ for $C_p / C \leq 2$; for small inductances $L \sim 40$ pH, the degradation is found to be much worse. In both cases, the degradation saturates and the energy resolution improves for $C_p/C > 2$, becoming roughly twice the value for $C_p/C \approx 0$ in the limit $C_p/C \gg 1$. Thus, for the purpose of model calculations, it is useful to define the effective energy resolution, with the effect of parasitic capacitance taken into account, as

$$
\begin{aligned}
\varepsilon_p &= \varepsilon\sqrt{1+2C_p/C} \\
&= \gamma k_B T \sqrt{LC(1+2C_p/C)}
\end{aligned}
\qquad \left(C_p/C \le 2\right), \qquad (14)
$$

$$
\varepsilon_p = \sqrt{5}\varepsilon \qquad\qquad \left(C_p/C > 2\right). \qquad (15)
$$

Several design approaches are available for situations where the parasitic capacitance is unavoidably high. These are summarized below.

(a) *Resistive Damping.* In situations where the parasitic capacitance is large because of design requirements, such as the need for an input coil with many turns to match a high load inductance, it has been shown that the insertion of a damping resistor R_d across the SQUID inductance improves performance[18]. For $R_d = R$, the additional resistor effectively damps the LC_p resonance and smoothes the SQUID characteristics, leading to an enhancement of the transfer function and a reduction of the flux noise[11]. The improvement is partially offset, however, by the additional flux noise arising from the current noise owing to the damping resistor R_d. Nevertheless, this is a reasonable design approach for high C_p devices.

(b) *Intermediate Transformer Coupling.* A means of keeping the parasitic capacitance manageable is to design the input circuit using a small SQUID washer and a few turns in the input coil. The SQUID is then coupled to the load by an integrated, intermediate coupling transformer which matches the low SQUID inductance to the load inductance[1,19]. This elegant scheme is particularly attractive if the load inductance is very high. The matching transformer must be carefully designed, however, to avoid microwave resonances. The SQUID also must be designed carefully; if the SQUID input coil is made too short, as can easily occur as the number of turns is reduced, the input coil microstripline resonance moves to higher frequencies and may interfere with the operation of the SQUID. A disadvantage of this coupling scheme is that the energy transfer function from the load inductance to the SQUID is in practice a factor of two lower than without the transformer, if ideal inductance matching is achieved in both cases[1].

(c) *Double-Loop SQUIDs.* Another design approach exists for devices with high parasitic capacitance. It recently has been shown that, depending on the design parameters, the double-loop SQUID shown schematically in Fig. 4 may exhibit a

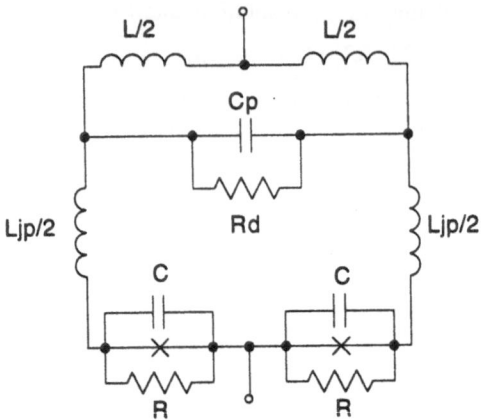

Figure 4. Schematic diagram of a double-loop dc SQUID.

high frequency resonance above the normal range of operating frequencies[11]. For this structure, the total SQUID inductance is given by $L_{tot} = L + L_{jp}$, where the parasitic inductance L_{jp} is much smaller than the coupling inductance L. The parasitic capacitance C_p and damping resistor R_d also are shown in the figure. At high frequencies, the large parasitic capacitance C_p is effectively a short, and a resonance occurs at the frequency $f_r = (1/2\pi)(L_{jp}C/2)^{-1/2}$. Inside this high-frequency resonance, the SQUID characteristics are very smooth, and the energy resolution is nearly the same as that for a SQUID having the same total inductance, but with $C_p = 0$. This is an important design alternative for devices with high parasitic capacitance, provided a double-loop geometry can be used. An example of such a device is the multi-loop SQUID magnetometer described by Drung *et al*[20].

2.2. EXPANDED MODELS OF DC SQUIDS

The resonances introduced by the presence of the input coil may be studied using an expanded model of the coupled SQUID[21,22]. Enpuku *et al*[22] used this technique to analyze the resonances in an integrated dc SQUID magnetometer described in [3]. Using an expansion factor $F = 500$, the geometrical parameters of the model were as follows: the input coil had $N = 12$ turns, the width w_i of each turn and the spacing s_i between turns was $w_i = s_i = 1.5$ mm, the total length of the input coil $\ell = 2.34$ m, the side length of the washer hole $d = 9$ mm, and the length of the slit $b = N(w_i + s_i) = 36$ mm. The model was fabricated from standard, nominally 0.5-mm thick, quartz epoxy printed circuit board with Cu electrodes having a thickness $t = 35$ μm on each side.

The skin depth at frequency f for a normal metal is given by $\delta = 1/\sqrt{\pi f \sigma \mu_0}$, where σ is the electrical conductivity of the metal and μ_0 is the permeability of free space. Using $\sigma = 5 \times 10^7$ S/m for the electrical conductivity of Cu, the skin depth δ_{cu} of the

Cu electrodes in the expanded model is less than 14 μm for $f > 20$ MHz. Thus, in this frequency range, $\delta_{Cu} < t$, and the separation of the Cu electrodes is much greater than $2\delta_{Cu}$, so the behavior of the electomagnetic fields in the expanded model is a reliable representation of the behavior in the SQUID.

A drawing of the model used is shown in Fig. 5. The electromagnetic properties of the model are determined experimentally by measuring the frequency dependence of the impedance Z_{AB} between terminals A and B in Fig. 5 with a network analyzer. Measurements over the range 2 to 500 MHz correspond to 1 to 250 GHz for the SQUID.

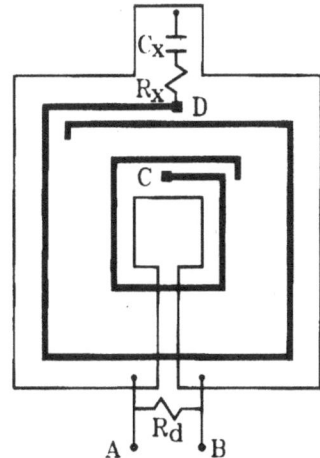

Figure 5. Schematic drawing of an expanded model of a dc SQUID.

Using various expanded models, several key points were investigated. First, all of the measurements described below were found to be in excellent agreement with calculations based on an earlier theoretical model developed by Enpuku *et al*[23]. This model is therefore useful for determining the electromagnetic properties of a given SQUID design.

Second, it should be noted that the inner end of the input coil (terminal C) is shorted to the washer. For an actual SQUID, this reduces the minimum number of photolithographic mask steps required to fabricate the device to four. More importantly, the stripline resonant frequency is two times lower for a grounded coil than for an open coil. This follows from the fact that, with the inner end of the coil grounded, the fundamental is a $\lambda/4$ resonance; with the input coil floating, the fundamental is a $\lambda/2$ resonance. This effect was confirmed by the expanded model measurements. Pushing the fundamental stripline resonance to a lower frequency makes it easier to design the SQUID such that the stripline resonance is well away from the operating frequency of the SQUID.

Third, it was shown that the resonances in the expanded model may be suppressed using external damping circuits, consisting of either a series R_xC_x shunt between the washer and terminal D or a damping resistor R_d. As shown in Fig. 6(a), the R_xC_x shunt effectively suppresses the stripline resonances corresponding to peaks in Z_{AB}; the large peak corresponding to the washer resonance is still present. The suppression of the stripline resonances is consistent with experimental observations carried out previously[2] using two identical SQUIDs, except that one SQUID had an integrated R_xC_x shunt installed, the other did not. As shown in Fig. 6(b), the resistor R_d is effective in suppressing the resonance of the washer, but the stripline resonances are still present. In principle, one may use both damping elements to suppress both the stripline and washer resonances, but the additional resistor R_d can contribute to the total noise. One may avoid any problems arising from the washer resonance simply by designing the SQUID such that this resonance is moved to frequencies much higher than the intended operating frequency of the SQUID.

Fourth, the properties of gradiometric washer configurations were determined, since such configurations are often used in practical applications. To form the gradiometer models, two washers were connected either in series or in parallel. Each washer was coupled to a multiturn input coil, with the two input coils connected in series as usual. For such configurations, the possible interaction of the rf fields between the two washers becomes an important issue. If the interaction is small, the rf properties of the gradiometer should be similar to those of the individual washers; if the interaction is strong, however, the rf properties of the gradiometer should be quite different from those of the magnetometer.

Expanded models of series and parallel gradiometers were constructed using nearly the same washer and input-coil parameters as those used to model the single-washer SQUID discussed above. For both model types, the inner terminal of each input coil was shorted to the washer. The impedances of both gradiometer models measured without damping are shown in Fig. 7.

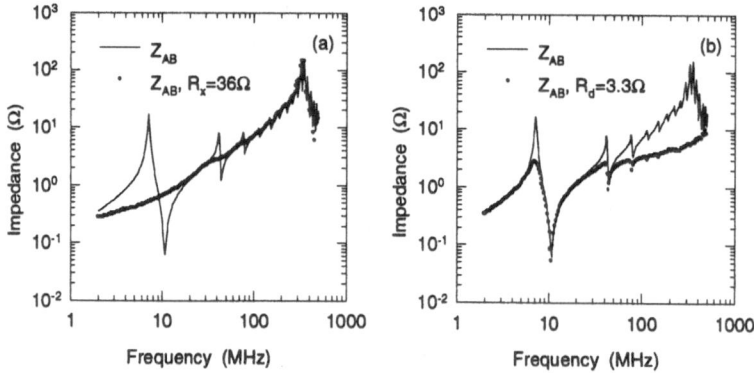

Figure 6. Absolute value of the measured impedance Z_{AB} vs. frequency for coupled SQUID models without (—) and with (•) (a) R_xC_x shunt and (b) damping resistor R_d.

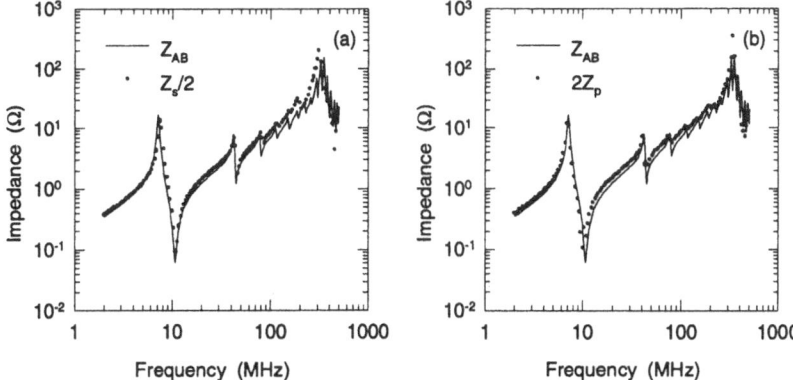

Figure 7. Absolute value of the measured impedance (a) $Z_s/2$ for the series gradiometer model and (b) $2Z_p$ for the parallel gradiometer model vs. frequency. The solid curves in (a) and (b) are the measured impedance of the single-washer model Z_{AB}.

In Fig. 7(a), the dotted curve is a factor of one-half times the measured impedance of the series gradiometer, $Z_s/2$; the solid line is the impedance of the single washer model Z_{AB}. The magnitude of $Z_s/2$ is almost the same as Z_{AB}. At low frequencies, the resonant peaks in $Z_s/2$ corresponding to the input coil resonances occur at nearly the same frequencies as those for the single washer. Only at very high frequencies near the large peak corresponding to the washer resonance does a small deviation between the two curves appear. For the gradiometer, the frequency of the washer resonance is slightly lower, and the fine structure around the washer resonant frequency disappears. Except for the disappearance of the fine structure, the small difference between Z_{AB} and $Z_s/2$ at high frequencies is most likely due to small parameter differences between the two cases, such as the increased parasitic inductance in the gradiometer model. The disappearance of the fine structure is more likely due to a weak rf interaction between the two washers. According to the above results, there is no significant rf interaction between the two washers for the series gradiometer. The impedance of the series gradiometer Z_s can therefore be expressed simply in terms of the impedance of the single washer as $Z_s = 2 Z_{AB}$.

The results for the parallel gradiometer Z_p are shown in Fig 7(b). The dotted curve is twice the measured impedance of the parallel gradiometer, $2Z_p$, and the solid curve is the impedance of the single washer model Z_{AB}. As before, the two curves are nearly the same. Thus, the rf interaction between the two washers in the parallel configuration also is negligible, and the impedance of the parallel gradiometer Z_p can be expressed as $Z_p = Z_{AB}/2$. These results for the series and parallel configured washer models are useful in designing gradiometric SQUIDs.

2.3 FABRICATION

Considerable effort world-wide has been directed over the years towards the development of process technology for LTS SQUIDs. As a consequence of this effort, it is now possible in many laboratories to mass produce extraordinarily high quality and robust Josephson junctions. Nearly all of the processes currently in use are based on the deposition of a Nb/I/Nb trilayer, where the insulating barrier I is usually thermally-oxidized Al[24]. In some cases, Si[25] and nitrided Si[26] also have been used. The main differences in the various process technologies concerns the way in which the Josephson junctions are defined. A complete summary of all the fabrication methods and results obtained is well beyond the scope of this section. Here, we summarize a simple, representative process technology that has been used to fabricate the devices described in Sec. 4. This process is very simple, requiring only four mask layers.

Deposit Trilayer. A Nb/I/Nb trilayer is deposited over an entire oxidized-Si wafer and patterned to define the SQUID washer, capacitors C_x, bond pads, modulation coil and pickup loops in the case of an integrated magnetometer or gradiometer. This layer can easily and quickly be patterned using a wet etch. Linewidths down to 4 μm can reliably be achieved in this way. An etch undercut of roughly 0.5 μm produces nicely tapered edges which improves the step coverage by the Nb wiring layer.

Anodize top Nb layer. The Josephson junctions (typically 2×2 μm² to 5×5μm²) along with several high critical-current vias to provide zero-voltage connections to the base electrode are defined by protecting the junctions and vias with photoresist and then anodizing the remaining areas of the upper Nb film (Selective Niobium Anodization Process, or SNAP[25]). The high critical-current vias are used, for example, to connect the junction shunt resistors and for routing the input-coil current return path through the SQUID washer. The upper Nb films of the capacitors C_x are anodized as well. The anodized Nb is Nb_2O_5, which has a dielectric constant $\varepsilon_r = 29$[27]. The high dielectric constant of Nb_2O_5 can enhance the parasitic capacitance across the SQUID inductunce. To reduce the parasitic capacitance, an additional insulation layer such as SiO_2 or Si_3N_4 may be deposited and patterned by a lift-off technique using the anodization mask as a stencil. In this case, the capacitor dielectric consists of a bilayer.

Deposit Resistor Layer. The Pd shunt resistors and damping resistors R_x are deposited and patterned using a lift-off stencil. The Pd layer also is used to passivate the wire bond pads. The use of Pd ensures that the devices will work properly at temperatures well below 1 K. Sheet resistances of 0.5 to 2 Ω/square are typical. Since it is not necessary to bury the resistors using a protective passivation layer, the resistors may be trimmed post fabriction by lightly etching in Ar plasma -the etch rate of Pd is much less than that of Nb[28]. This is sometimes helpful during device development.

Deposit Nb Wiring Layer. A Nb wiring layer is deposited and patterned to define the input coil (2 to 3 μm linewidth), capacitor top plates and interconnects. The Nb wiring layer may be patterned using a lift-off stencil or reactive ion etching.

Even with the limited number of process steps described above, it is possible to fabricate intricate designs including high performance magnetometers and gradiometers. The process may be extended to include an additional Nb wiring layer. In this case, two extra mask layers are required: one to pattern a deposited insulator layer, the other to pattern the Nb layer.

3. Readout Electronics

Conventional SQUID readout electronics use flux modulation and a cooled impedance-matching network, consisting of a transformer or a resonant circuit, to match the low SQUID output impedance to the typically high input impedance of a room-temperature preamplifier. The preamplifier is followed by a phase-sensitive detector and linearizing electronics for flux-locked-loop (FLL) operation. An example of this technique is discussed in more detail in Sec. 4. A drawback of such schemes based on phase detection is that the bandwidth cannot be arbitrarily extended, because the noise temperature of the preamplifier increases as the modulation frequency is increased. The modulation and demodulation circuits and impedance-matching network also add somewhat to the expense of the electronics, and the low-temperature impedance matching network adds volume to the SQUID package. Important advantages of ac measurement techniques, however, include the rejection of dc signal drift, the elimination of some types of low-frequency noise, and ease of operation. For example, it is straight-forward to implement a simple autotune algorithm which enables the user to tune the electonics simply and quickly. These are important considerations for commercial SQUID systems.

During the past few years, new types of direct-coupled readout electronics have been developed which simplify the room-temperature electronics required for reading out low-noise dc SQUIDs. The first such alternative readout scheme developed is the direct offset integration technique with additional positive feedback (APF)[29] -see Fig. 8 and Chapter 2 by D. Drung in this volume. The SQUID output is connected directly to the input of a low-noise preamplifier at room temperature. The limitation imposed by the comparatively large voltage noise of the room-temperature preamplifier is circumvented in the following way. The SQUID output also is connected to a positive feedback coil, the amount of feedback being determined by the resistor in the feedback circuit. The positive feedback substantially increases one slope of the V-Φ characteristic, thereby enhancing the flux-to-voltage transfer function $\partial V/\partial \Phi$ while decreasing the other slope. In this way, the voltage noise at the input of the preamplifier $S_V^{1/2}(f) = S_\Phi^{1/2}(f) \cdot |\partial V/\partial \Phi|$, where $S_\Phi^{1/2}(f)$ is the flux noise of the SQUID, can be made to exceed the preamplifier noise. Since an impedance-matching

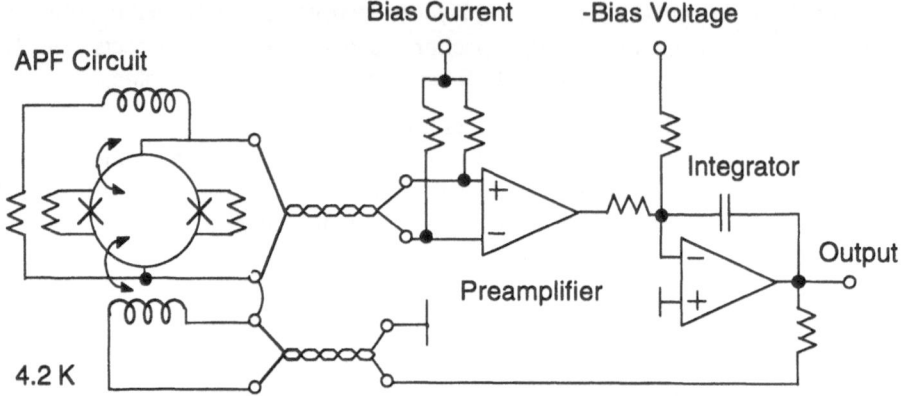

Figure 8. SQUID system based on the direct-offset integration technique with additional positive feedback (from [30]).

network is not used, flux modulation is no longer necessary, and a significantly higher bandwidth can be achieved. For FLL operation, the SQUID is dc flux-biased. A minimum of four (three of which must be very low resistance) wires to room temperature are required, and the electronics are very simple. A disadvantage of this approach, however, is the distortion of the V-Φ characteristics, which can reduce the dynamic range.

In a variation of this scheme[31], the SQUID is voltage-biased rather than current-biased, so that an applied flux causes a periodic modulation of the current through the SQUID. In the simplest version -see Fig. 9- the SQUID output is directly connected to the negative terminal of a low-noise preamplifier at room temperature. The SQUID is then biased by applying a voltage to the positive terminal of the preamplifier, which, via the current through resistor R_p, forces the voltage across the SQUID to the same value. Since the voltage across the positive feedback coil is now fixed, positive feedback, as described above, cannot take place. Consequently, the current-flux (I-Φ) characteristics are not distorted, and the dynamic range is unaffected. Flux modulation is not required, so that very high bandwidths can be obtained, as before. With this scheme, the large preamplifier voltage noise can be eliminated, because the voltage noise signal appears across the SQUID, as well as across the positive feedback circuit. By varying the resistance in the positive feedback circuit, the amount of the voltage noise coupled to the SQUID can be made equal to the preamplifier voltage noise. In this way, an exact noise cancellation can be achieved on one slope of the I-Φ characteristic, while the noise adds on the other slope. The noise cancellation can most easily be adjusted at room temperature using a cooled GaAs FET to control the amount of feedback. In a refinement of this technique[32], a proportional-integrator controller and square-wave flux and offset voltage modulation are used to provide automatic control of the noise cancellation. In this case the FET is in a nearly

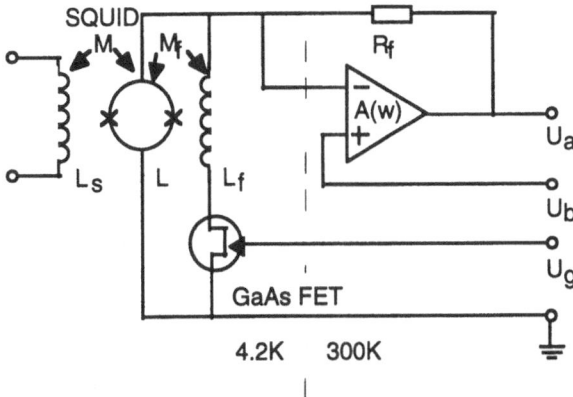

Figure 9. SQUID readout circuit based on voltage bias. The voltage across the SQUID is controlled by tuning the voltage U_b (from [31]; the negative feedback circuit is not shown).

voltageless state, so the power dissipated by the FET is negligible. The SQUID electronics are more complicated, however, and the FET adds to the volume of the SQUID package, both of which are disadvantages of this approach. Also, at least two additional wires to room temperature are required (one to the gate of the FET, the other because the negative feedback resistor R_f must be connected as close as possible to the negative terminal of the SQUID), but all six can be moderate resistance, low thermal conductivity leads.

In another scheme, Martinis and Welty at NIST describe a two-stage SQUID amplifier that can be operated with highly simplified readout electronics[33]. In this elegant design shown schematically in Fig. 10, the output signal of a low-noise SQUID in the input stage coherently modulates a series array of identical, non-hysteretic SQUIDs in the second stage via a a common modulation line. A substantial voltage swing can be realized in this way with a reasonable number of series-connected SQUIDs in the second stage. The second stage is designed to have sufficient gain so that the output noise is dominated by the amplified noise of the input SQUID and exceeds the noise of the room-temperature preamplifier. In this way, the output signal of the second stage may be directly coupled to the preamplifier at room temperature. Flux modulation and a cooled impedance matching network are not needed. The bandwidth can therefore be very high, and, since the voltage swing of the second stage is much larger than that of the input SQUID, the large intrinsic dynamic range of the input SQUID is preserved. Thus, with this approach, applications requiring high bandwidth and dynamic range can be satisfied with very simple room-temperature electronics.

198

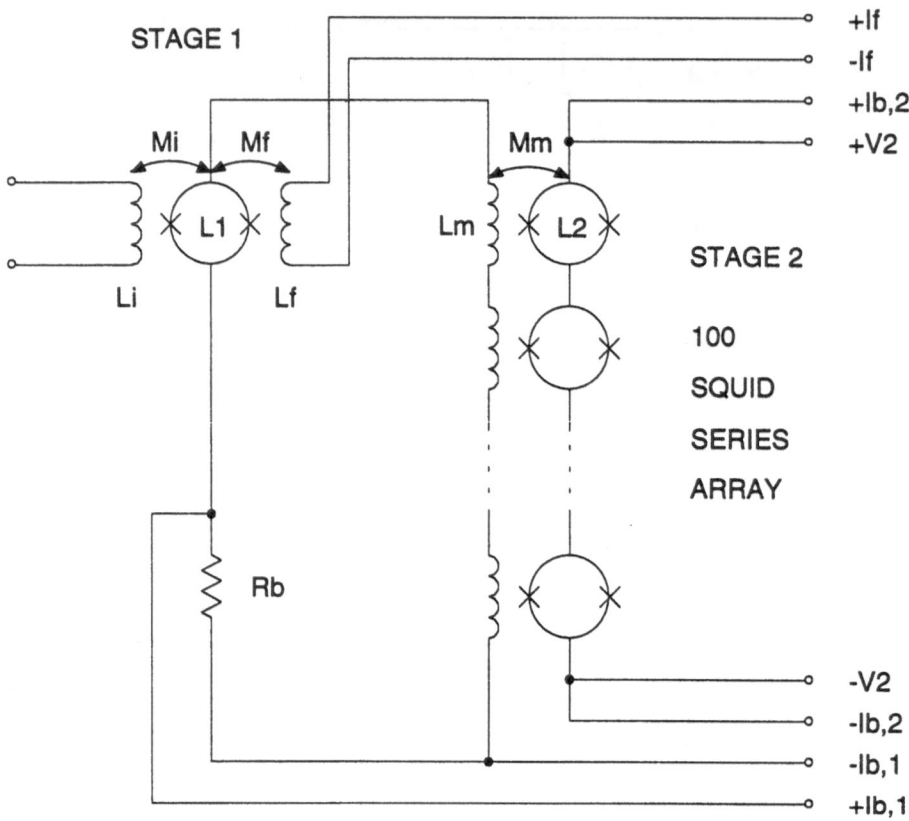

Figure 10. Schematic drawing of a two-stage SQUID amplifier.

4. Integrated dc SQUID Magnetometers

For designs where the parasitic capacitance can be kept small, the discussion in Sec. 2.1.2 describes a clear procedure for device optimization: the expression for the energy resolution given in Eq. (13) is minimized by varying the design-dependent parameters using standard algorithms, subject to the constraints on the microwave resonances given in Eq. (10). Depending on the application, however, the energy resolution may not be the appropriate functional form to use for carrying out the optimization procedure. For a magnetometer, for example, the relevant figure of merit for most applications is the rms magnetic field noise $S_B^{1/2}(f) = S_\Phi^{1/2}(f)/A_{\text{eff}}$, where A_{eff} is the effective flux capture area of the magnetometer. In this case, the magnetometer design may be optimized by minimizing the magnetic field noise using the expression

$$S_B^{1/2}(f) = \sqrt{2\varepsilon L} / A_{\text{eff}} \equiv \sqrt{2\varepsilon L} B_\Phi / \Phi_0 , \qquad (16)$$

where the field-to-flux conversion efficiency B_Φ is given by

$$B_\Phi = \Phi_0 \frac{L_p + L_{i,\mathrm{eff}}}{M_i A_{p,\mathrm{eff}}}.$$
(17)

Here, L_p is the inductance of the pickup loop, and $A_{p,\mathrm{eff}}$ is the effective area of the pickup loop including flux-focussing effects. Similarly, for a gradiometer the application-specific functional form to be minimized is the magnetic field gradient noise, $S_B^{1/2}(f)/b_g$, where b_g is the baseline of the gradiometer. In practice, the design parameters which minimize the application-specific functional form for the optimization procedure are often different from those which minimize the energy resolution, although the the two minima are quite broad and do not differ substantially (by less than a factor of 10).

The first integrated magnetometer designed using the optimization procedure described in Sec. 2.1.2. was fabricated on a nominally 4-mm × 4-mm chip by Cantor et al.[2] The parameters for this prototype design were obtained by minimizing the energy resolution. The design of the SQUID was chosen so that the microwave transmission-line resonances of the input coil are well damped, and the resonant frequency of the SQUID washer is well away from the intended operating frequency of the SQUID. Further, with an integrated $R_x C_x$ shunt in the SQUID input circuit, the SQUID characteristics were smooth over the entire operating range of the SQUID, and the noise minimum (<5 fT$/\sqrt{\mathrm{Hz}}$ at 1 kHz) was very broad. Both of these features simplify operation in the flux-locked-loop mode. The broad minimum of the flux noise is quite clear in the noise map for this magnetometer[34], which is shown in Fig. 11.

The magnetometer in [2] was subsequently improved by carrying out a minimization of the magnetic field noise[3]. For this device as well, very smooth characteristics were obtained, even though the SQUID was near the hysteretic limit with $\beta_c = 0.6$. A photograph of the single-washer SQUID is shown in Fig. 12. The inner turn of the input coil is connected to the washer, and the washer serves as the current return path for the input signal. In this way, the fundamental stripline resonance occurs at half the frequency than if the input coil were floating, and the two additional layers required to make a cross-over are not necessary. The connection to the input coil is made using a Josephson junction, which limits the maximum current that can flow in the input circuit. This prevents the formation of large screening currents if the magnetometer is exposed to large changes of the applied magnetic field; the size of this junction must therefore be chosen such that the screeing current set up by the maximum magnetic field signal to be measured does not cause the junction to go normal.

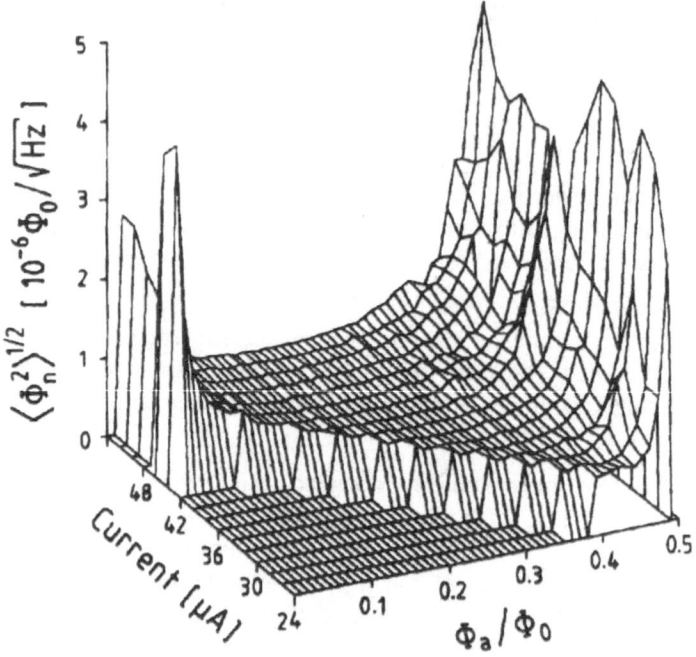

Figure 11. Noise map of the magnetometer described in [2]. Shown is the rms flux noise at 1 kHz measured open loop as a function of bias current and dc flux (from [34]). For this device, $B_\Phi = 6.1$ nT/Φ_0.

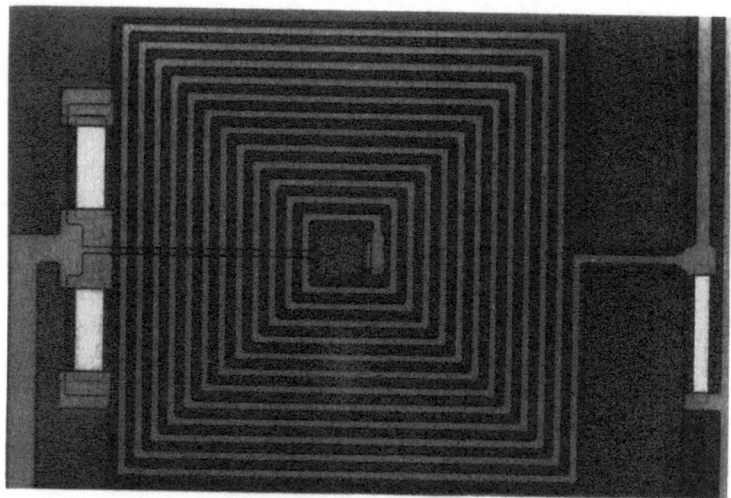

Figure 12. Photograph of the single-washer SQUID used to build the 4-mm × 4-mm integrated magnetometer described in [3].

For the device shown in Fig. 12, the SQUID inductance $L = 51$ pH, and the measured white flux noise at 1 kHz $S_{\Phi}^{1/2}(f) = 0.61$ $\mu\Phi_0 / \sqrt{Hz}$. A representative spectrum of the rms field noise is shown in Fig. 13(a). The noise minimum is again very broad, which is evident from the noise map data shown in Fig. 13(b). This figure shows the voltage noise referred to the SQUID as a function of the flux-to-voltage transfer function measured open loop using a SQUID amplifier at over 1000 working points. The working points correspond to a wide range of current and dc flux bias parameters. The straight lines in the figure mark several values of the flux noise. As can be seen from the data in the figure, the SQUID operates with very low noise at most working points, including many where the flux-to-voltage transfer function is very high ($\gg$100 $\mu V/\Phi_0$). This is a consequence of the design optimization which enables the SQUID to be operated near the hysteretic limit. The high transfer function simplifies operation using the direct-coupled readout electronics schemes discussed in Sec. 3, in that low-noise performance may be achieved without resorting to the use of APF.

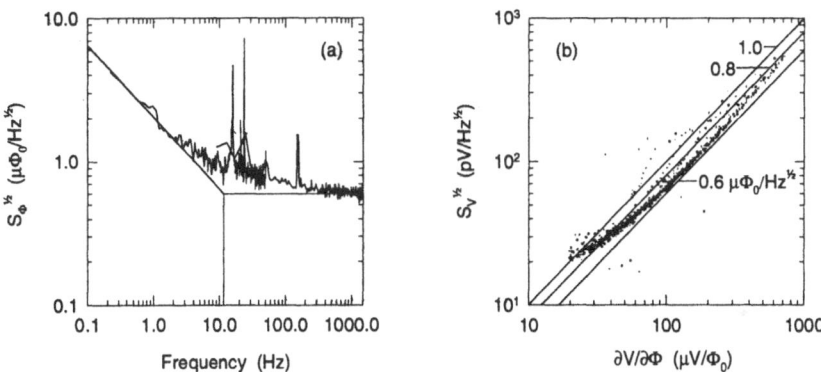

Figure 13. (a) The rms flux noise of the 4-mm × 4-mm magnetometer described in [3] measured open loop using a SQUID amplifier. Below about 10 Hz, the noise has a $1/f$ dependence. For this device, $B_{\Phi} = 5.6$ nT/Φ_0. (b) The rms voltage noise referred to the SQUID output versus the flux-to-voltage transfer function. Each point corresponds to a separate noise measurement at a current and dc flux biased working point.

The low SQUID inductance and optimal design described in [3] resulted in very low flux noise, but, owing to the small chip size (4-mm × 4-mm) and low mutual inductance M_i, of 0.5 nH, the field-to-flux conversion efficiency was only 5.6 nT/Φ_0. This results in a field noise at 1 kHz of 3.4 fT/$\sqrt{Hz}$, but the low intrinsic flux noise ($<$1 $\mu\Phi_0 / \sqrt{Hz}$) places stringent demands on the room-temperature electronics if conventional flux-modulation readout electronics are used. In addition, the flux noise increases at low frequencies, becoming about 2 $\mu\Phi_0 / \sqrt{Hz}$ at 1 Hz, thereby degrading the performance of the magnetometer at low frequencies. Interestingly, this value for the flux noise at 1 Hz is frequently observed for many different SQUIDs, independent of the junction technology used.

More recently, an improvement in magnetometer performance was achieved using a larger chip size, which increases the flux capture area of the pickup loop, along with a much higher SQUID inductance in order to increase the mutual inductance of the input coil[35]. The reason for using a higher mutual inductance is as follows. Since $L_{i,\text{eff}} \approx L_p$ to satisfy inductance matching, $B_\Phi \propto \left(L_p/A_{p,\text{eff}}\right)/M_i$. Note that a minimization of the magnetic field noise using Eq. 16 actually leads to a slight mismatch of $L_{i,\text{eff}}$ and L_p; that is, by making L_p somewhat larger than $L_{i,\text{eff}}$, a reasonable increase in A_{eff} (reduction in B_Φ) is achieved. Once the chip size is determined, the ratio $L_p/A_{p,\text{eff}}$ depends only on the linewidth of the pickup loop. For reasonable values of L_p, this ratio is a slowly varying function so that, once the chip size has been fixed, B_Φ depends mostly on the mutual inductance M_i, which in turn depends on the SQUID inductance L and the design of the input circuit. Thus, by increasing both the size of the chip and the mutual inductance, a significantly improved field-to-flux conversion efficiency may be obtained.

The larger magnetometer is fabricated on a nominally 8-mm × 8-mm chip with a 7.75-mm × 7.75-mm pickup loop. A relatively high SQUID inductance, L = 190 pH, is used, giving a high mutual inductance of M_i = 1.7 nH. The SQUID uses a series-configured gradiometric washer design. In this way, the parasitic capacitance is kept to a reasonable value even though only anodic Nb oxide is used as the insulator between the input coil and the SQUID washers. An integral $R_x C_x$ shunt is installed in the input circuit, ensuring that the SQUID characteristics are very smooth -the characteristics for one such magnetometer are shown in Fig. 2. A photograph of the SQUID is shown in Fig. 14.

Owing to the optimized designs of the SQUID and pickup loop, a high field-to-flux conversion efficiency of 0.83 nT/Φ_0 was achieved. Using a picovoltmeter, the intrinsic rms white flux noise of the magnetometer was measured to be 1.6 $\mu\Phi_0/\sqrt{\text{Hz}}$ at 1 kHz. The corresponding rms magnetic field noise is better than 1.4 fT/$\sqrt{\text{Hz}}$ -see Fig. 15(a). Because of the high SQUID inductance, the intrinsic magnetic flux noise is higher than what is generally observed for magnetometers using much lower SQUID inductances. The high flux noise is compensated by the large effective area of the magnetometer and facilitates coupling to the room-temperature readout electronics.

There are several important advantages of the magnetometer design described in [35]. The SQUID can be operated using conventional readout electronics with flux modulation, but without the cooled impedance-matching network typically used with such techniques; a single transformer at room temperature is used to couple the SQUID output signal to the room-temperature preamplifier. This greatly simplifies

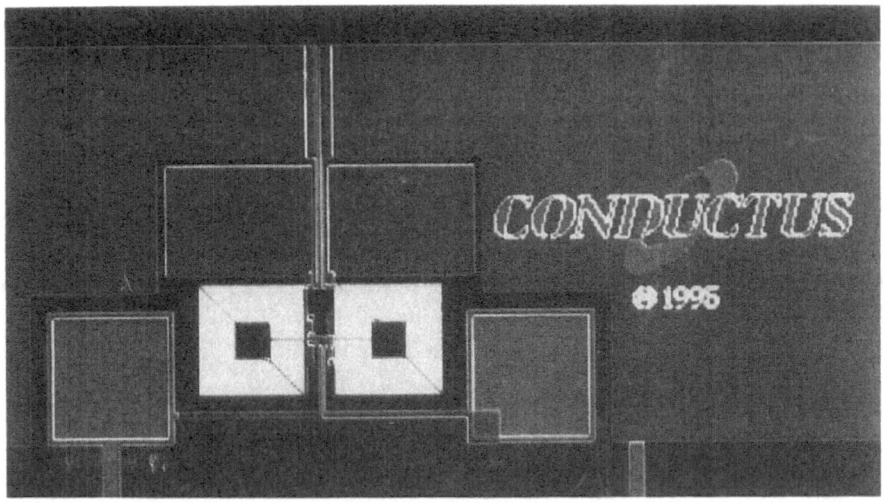

Figure 14. Photograph of the gradiometric SQUID used to design the 8-mm × 8-mm magnetometer described in [35].

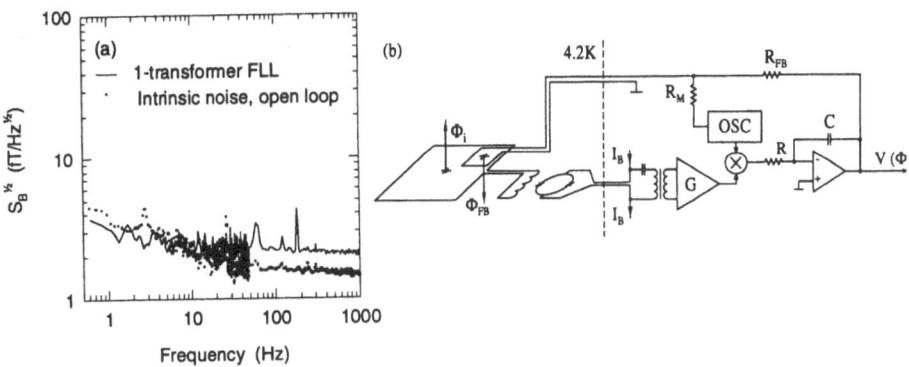

Figure 15. (a) The the intrinsic rms field noise measured open loop using a picovoltmeter (dotted line), and the rms magnetic field noise measured in flux-locked loop (FLL) using flux modulation and a two-wire connection to the room-temperature preamplifier (solid line). (b) Block diagram of the SQUID and FLL electronics.

and reduces the size of the SQUID magnetometer package. Further, common bias current and signal leads are used so that each magnetometer requires only four wires to room temperature: ±Signal and ±Modulation. The signal leads are very fine, low-resistance copper, while the modulation leads may be made from higher resistance, low-thermal-conductivity wire. The rms field noise measured using a warm transformer only and a two-wire connection to the SQUID is shown in Fig. 15(a). A block diagram of the electronics is shown is Fig. 15(b). The small number of electrical leads required per channel is an important consideration for multi-channel applications. Thus, the SQUID magnetometer and readout electronics employing flux modulation offer many of the same advantages realized by the direct-coupled readout electronics summarized in Sec. 3, in addition to the well-known advantages of ac measurement techniques: problems owing to dc signal drift are eliminated, and tuning the SQUIDs is greatly simplified and can be easily automated.

5. Coupled SQUIDs for Laboratory Applications

5.1. DESIGN REQUIREMENTS

To be useful for a wide variety of applications, a dc SQUID with transformer coupling of the input and modulation signals must meet several general requirements. Several of the key requirements are summarized below.

(a) *Inductance Matching.* The SQUID input circuit should have a reasonably high inductance in order to satisfy inductance matching to arbitrary external circuits. Values ranging from 0.6 to 1 µH are typical.

(b) *Insensitivity to ambient fields.* The SQUID itself should be insensitive to ambient fields which may be present in the measurement environment. This is generally accomplished by surrounding the SQUID as much as possible by a superconducting shield, and by designing the SQUID with a gradiometric configuration. This is easily accomplished using two washers which may be connected in parallel or in series with the two Josephson junctions. In either case, the SQUID will have zero response to a uniform external field. The response to external field gradients also will be negligible if the separation between the holes in the two washers is small, as is usually the case.

(c) *Feedback coupling.* The SQUID may be operated as a null detector using two different negative feedback modes[36], which may be referred to as flux-lock and current-lock modes. In the flux-lock mode, the feedback signal is coupled to the SQUID: a flux applied to the SQUID by an external signal in the input coil is cancelled by an opposing flux in the feedback coil. In this case, if the feedback and input coils are not sufficiently isolated, the feedback signal (and modulation signal if conventional readout electronics are used) may couple to the input circuit and interact with the load connected to the SQUID. Similarly, for experiments where the load inductance is not fixed, the feedback coupling changes as the load is varied because of screening effects. For these applications, it is desirable to design the SQUID such that there is negligible coupling of the feedback and modulation signals to the input circuit. For example, the symmetric design of Simmons[37] uses a four-hole, symmetric washer configuration rather than a two-hole washer configuration in order to provide separate secondaries for the input and feedback circuits. For a wide variety of applications where the load is fixed, the coupling of the feedback signal to the load does not pose a significant problem. In the current-lock mode, the feedback signal is coupled to the SQUID input. In this case, the feedback signal opposes any change of the current in the input circuit. If the load consists of a superconducting pickup loop, for example, this feedback mode maintains zero current in the pickup loop and input coil circuit. In this case, the current-lock mode offers the advantage that very little distortion of the field to be measured is introduced.

(d) *Negligible coupling of the bias and modulation signals to the SQUID.* The electrical connections to the SQUID for the current bias and modulation signals should be routed to prevent the magnetic fields arising from these signal currents from coupling to the SQUID. Substantial self-coupling may affect proper operation of the SQUID.

(e) *Low noise performance.* In order to succesfully realize the extreme sensitivity of the dc SQUID, the SQUID design must take into account the adverse effects of the parasitic capacitance and the microwave resonances invariably present in integrated thin-film transformer circuits. A commonly-used figure of merit for dc SQUID performance is the noise energy per unit bandwidth, usually referred to as the energy resolution. To achieve optimal energy resolution, the SQUID inductance is generally made as small as possible, subject to the constraint that the SQUID input circuit is inductively matched to the anticipated load. For this reason, gradiometric SQUID designs are configured with the SQUID washers in parallel with the Josephson junctions. The total inductance is then one-half the single-washer inductance. As discussed below, however, a series-configured design offers several advantages over a parallel-configured design for some applications.

As discussed in Sec. 4, the functional form to be used with the SQUID optimization procedure is an application-specific parameter. For many laboratory applications, dc SQUIDs are used as low-noise current-to-voltage transducers with low input impedance. For these applications, the relevant figure of merit is the equivalent rms current noise referred to the SQUID input. The rms current noise is given by

$$S_I^{1/2}(f) = \frac{S_\Phi^{1/2}(f)}{M_i} , \qquad (18)$$

where $S_\Phi^{1/2}(f)$ is the rms flux noise of the SQUID, and M_i is the mutual inductance of the input coil. To be useful for experiments requiring high current sensitivity, it is therefore desirable not only to achieve a reasonable energy resolution, but also to design the SQUID input circuit to obtain the highest possible mutual inductance with respect to the SQUID. To assure immunity against ambient noise, it is desirable to use a gradiometric washer configuration. Then, the question arises as to which gradiometric configuration for the SQUID washers is preferrable: parallel or series?

To answer this question a very simple model may be used. Let L_x and n_x, where $x = s$ or p, denote the inductance per washer and the number of turns in the input coil coupled to each washer for the series and parallel configurations, respectively. Then, assuming the coupling constant which describes the coupling of each input coil to its respective washer is unity and that the slit inductance of each washer is negligible, the total inductances for the two configurations may be approximated as given in Table 1 below.

TABLE 1. Approximate total inductances for series and parallel washer configurations (for $k_i = 1$ and neglecting the slit inductance L_{sl}.)

	Series	Parallel
Total SQUID inductance	$2L_s$	$L_s/2$
Total input inductance	$2n_s^2 L_s$	$2n_p^2 L_p$
Total mutual inductance	$2n_s L_s$	$n_p L_p$

If we assume that we wish to match to the same load inductance and desire the same mutual inductance for both configurations, we have

$$2n_s^2 L_s = 2n_s^2 L_s \tag{19}$$
$$2n_s L_s = n_p L_p \tag{20}$$

Then, it follows that $n_s = 2n_p$ and $L_p = 4L_s$. Thus, the total SQUID inductance turns out to be the same for both configurations, so if one ignores possible excess noise due to parasitic effects and microwave resonances, one would expect the flux noise of both configurations to be comparable. The parasitic capacitance in the parallel configuration, however, is likely to be about four times higher than in the series configuration. Excessive parasitic capacitance can degrade the performance of the SQUID. Also, the resonant frequency of the washers in the parallel configuration occurs at a much lower frequency because of the large size of the washers in that case. For a square washer with a square hole in the center, the inductance is given by $1.25\mu_0 d$, where d is the side length of the hole. This means that the opening in each of the washers in the parallel configuration is four times larger than in the series configuration. Even though twice as many input coil turns are required in the series configuration, the much larger hole size required in the parallel configuration causes each washer to have a larger overall dimension. This can push the washer resonance dangerously close to the intended SQUID operating frequency. Furthermore, the large washer openings in the parallel configuration significantly increase the flux capture area of the SQUID. This means that large screening currents may be generated in response to uniform ambient fields. These screening currents may lead to flux trapping in the Josephson junctions or otherwise compromise the operation of the SQUID. The flux capture area of the series configuration is much smaller, but more importantly, the screening currents completely cancel. Thus, the series configuration may be less susceptible to flux trapping. For several reasons, therefore, the series configuration appears to be more favorable than the parallel configuration.

This can be seen more clearly in the following example. We consider here the optimization of a SQUID for a 1-μH load inductance. The current noise defined in Eq. 18 is minimized for a given number of turns in the input coil using the model described in Sec. 2.1.2. Both series and parallel gradiometric (square) washer configurations are considered.

The optimized input inductance L_i and washer hole width d are shown in Fig. 16(a) as functions of the number of turns N in the input coil. Similarly, the parasitic capacitance C_p is shown in Fig. 16(b). Because of the model used for the energy resolution for large C_p -see Eq. 15- the input inductance of the two configurations is not significantly different, even though the parasitic capacitance differs substantially. For the series configuration, however, the width of the washer hole is significantly smaller. The large size of the parallel configured washers will cause this design to be much more sensitive to the formation of large screening currents and possibly to increased flux trapping. For the series configuration, as long as the SQUID is well balanced, these screening currents cancel.

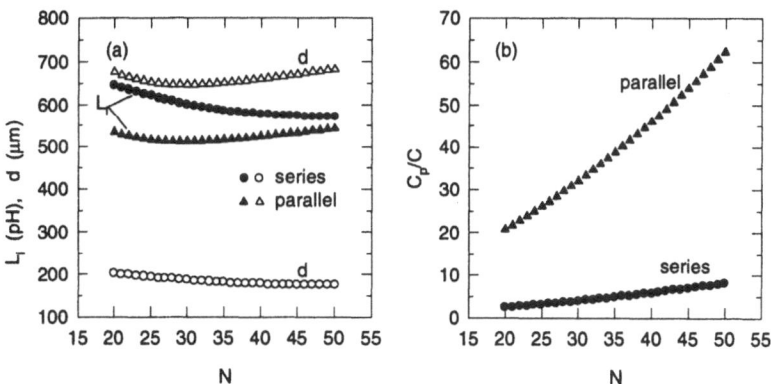

Figure 16. (a) The input inductance L_i (filled symbols) and washer hole width d (open symbols), and (b) the ratio of the parasitic capacitance to the junction capacitance C_p/C as functions of the number of turns in the input coil N for the series (circles) and parallel (triangles) configurations.

The optimized effective input inductance $L_{i,eff}$ and the mutual inductance M_i are shown over the same range of N in Figs. 17(a) and 17(b), respectively. Matching to a 1-μH load inductance requires input coil designs with 23 and 45 turns for the parallel and series configurations, respectively, with corresponding mutual inductances of 24 nH and 25 nH, respectively. Since the input inductances are not significantly different, the input coil of the parallel configuration requires far fewer turns. This may seem to be advantageous, but it is necessary to examine the behavior of the microwave resonances.

The fundamental of the stripline resonance at frequency f_s is calculated from the total length of the input coil. If the input and feedback coils are designed to be floating, the total length corresponds to $\lambda/2$ in both cases. The washer resonance at frequency f_w may be calculated using a previously-developed equivalent circuit model of the coupled SQUID[23]. The frequency f_{op} corresponding to the optimal point of operation is given by $f_{op} = 0.3f_J$, where the Josephson frequency $f_J = I_c R/\Phi_0$. These frequencies are plotted as functions of N in Figs. 18(a) and 18(b). Previous results suggest that it is necessary to maintain roughly a factor of 4 difference between the intended operating frequency and the stripline and washer resonant frequencies[2]; that is, $4f_s < f_{op} < f_w/4$. Because of the large number of turns required to match the load inductance, $f_{op} \gg f_s$ for both configurations, but in the parallel configuration the washer resonance is dangerously close to the intended operating frequency of the SQUID. This is another drawback of the parallel design.

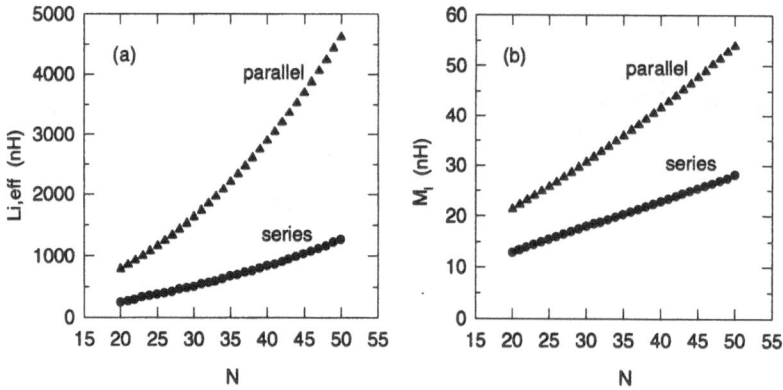

Figure 17. (a) The effective input inductance $L_{i,eff}$ and (b) mutual inductance M_i as functions of the number of turns N in the input coil for the series (circles) and parallel (triangles) configurations.

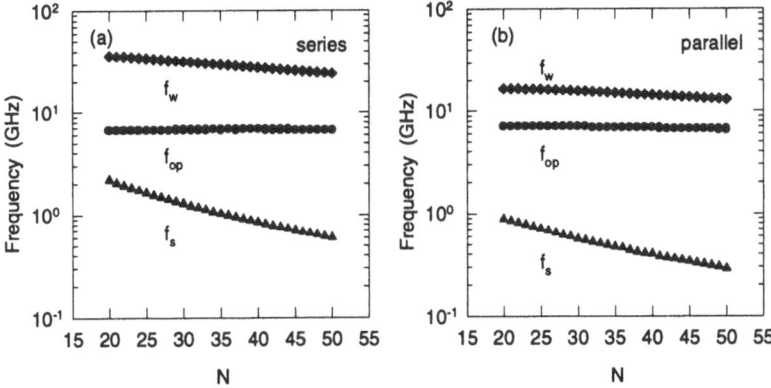

Figure 18. The resonant frequencies of the SQUID washer (diamonds) and input-coil microstripline (triangles) and the intended operating frequency (circles) as functions of the number of turns N in the input coil for the series (a) and parallel (b) configurations.

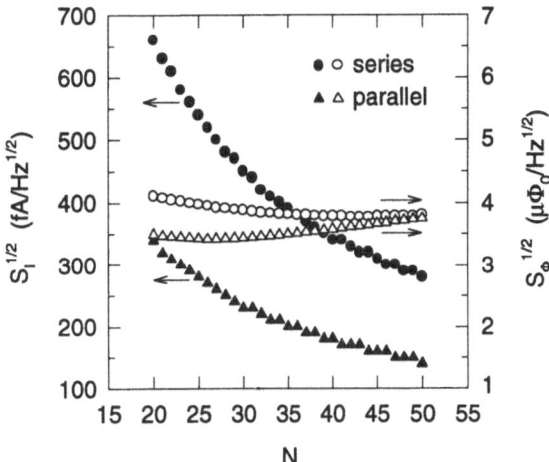

Figure 19. The equivalent current noise (filled symbols) and flux noise (open symbols) as functions of the number of turns N in the input coil for the series (circles) and parallel (triangles) configurations.

The minimized equivalent current noise $S_I^{1/2}(f)$ and magnetic flux noise $S_\Phi^{1/2}(f)$ are shown in Fig. 19. The current noise decreases steadily as N increases because of the increasing mutual inductance; for the values of N required to satisfy inductance matching to the load, however, the current noise of the parallel and series configurations are nearly the same, being 300 fA$/\sqrt{\text{Hz}}$ and 310 fA$/\sqrt{\text{Hz}}$, respectively. The flux noise also is comparable, with broad minima of 3.4 $\mu\Phi_0/\sqrt{\text{Hz}}$ and 3.8 $\mu\Phi_0/\sqrt{\text{Hz}}$ around $N = 26$ and $N = 43$ for the parallel and series configurations, respectively.

The energy resolutions ε_p and ε are shown in Fig. 20. The energy resolutions ε_p, calculated with the effect of parasitic capacitance taken into account, exhibit broad minima of 461 $\hbar$ and 501 $\hbar$ around $N = 25$ and $N = 38$ for the parallel and series configurations, respectively. The minima are broad and close to, but not coincident with, the flux noise minima. In terms of noise performance, the parallel and series configurations are therefore seen to be comparable. As mentioned above, however, the parallel configuration is more sensitive to the formation of large screening currents which can lead to increased flux trapping, and the resonant frequency of the parallel-configured washers may interfere with proper operation of the SQUID. For a well-balanced series configuration, the screening currents cancel.

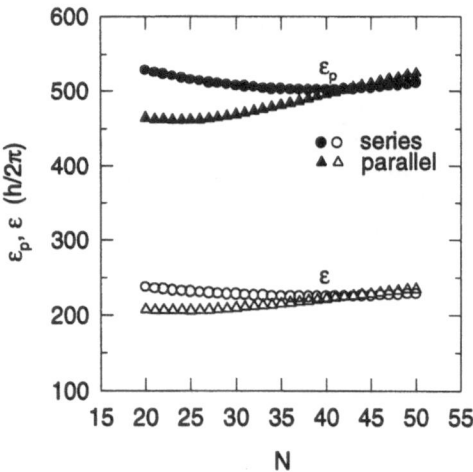

Figure 20. The energy resolutions ε_p ($C_p \neq 0$; filled symbols) and ε ($C_p = 0$; open symbols) as functions of the number of turns N in the input coil for the series (circles) and parallel (triangles) configurations.

5.2. SENSOR PACKAGING

An example of a standard low-T_c SQUID package for laboratory applications is shown in Fig. 21. The SQUID chip is attached to a printed circuit (PC) board using cryogenic epoxy. In order to ensure adequate cooling when operated in vacuum, the chip rests on copper striplines that are thermally anchored to the package mounting flange. The chip is encapsulated in a G-10 fiberglass assembly, along with a heater (surface mount resistor) mounted near the SQUID on the PC board to enable heating above T_c if necessary. The PC board also contains gold-plated copper pads for wire bond connections to the chip and a pair of screw terminals for attaching superconducting wire to the input circuit. The screw terminals are niobium, and niobium wire is used to make superconducting connections from the terminals to the chip. The niobium wire is attached using a gap welding technique. This ensures a robust, superconducting joint with a critical current well in excess of 100 mA. The mounting flange is fitted with a 10-pin connector to provide the necessary electrical connections to the package. A guide tube is provided through which external connections to the input circuit may be routed. Although the SQUID is designed to be insensitive to external signals, the package includes a niobium shield (not shown in Fig. 21) that is screwed onto the mounting flange to further protect the SQUID from electromagnetic interference.

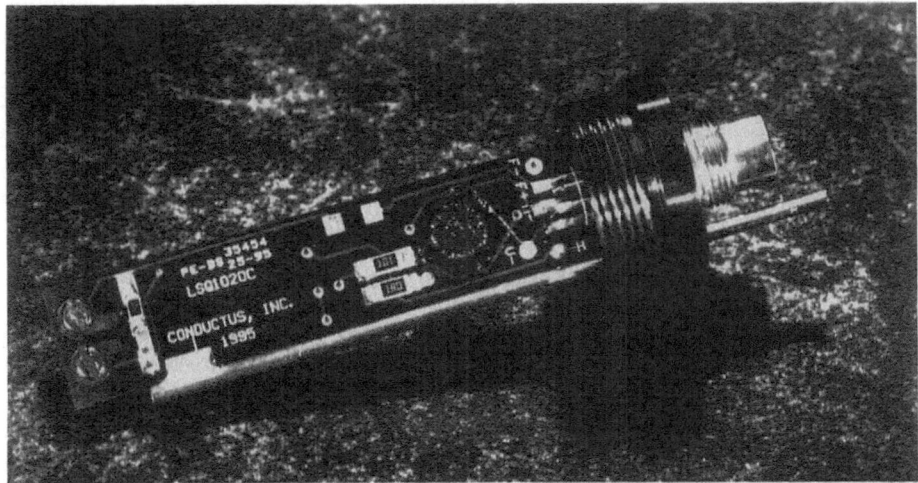

Figure 21. Photograph of a standard low-T_c SQUID package for laboratory applications. Superconducting connections to the SQUID input are made using the screw terminals. The SQUID chip is mounted on the back side of the pc board inside a hermetically sealed, glass epoxy (G-10) assembly. A separate niobium shield with threaded insert (not shown) screws onto the mounting flange and shields the SQUID against external electromagnetic interference.

6. High-T_c SQUIDs

The discovery of high-T_c superconductors has led to a considerable effort world-wide to develop SQUIDs that operate in liquid nitrogen at 77 K. These new sensors offer many advantages over their low-T_c counterparts. Liquid nitrogen is significantly cheaper, more readily available and easier to transport than liquid helium. Further, the latent heat of liquid nitrogen is much higher than that of liquid helium, which simplifies the construction of dewars with very long hold times (months) or that are very compact and portable yet sufficient to operate a sensor for a full day in the field. For these reasons, HTS sensors are very attractive for many applications, ranging from geophysics to nondestructive testing to biomagnetism.

During the past few years, considerable progress has been made in the development of robust and reliable Josephson junctions and thin-film high-T_c SQUIDs with low magnetic flux noise at 77 K. Much of the key work has been summarized by John Clarke and Alex Braginski in companion chapters in this volume. To ensure low noise operation, most of these SQUIDs are necessarily low-inductance devices; typical values of the SQUID inductance L are about 100 pH or less. As noted in Sec. 2, the low effective flux capture area A_s of these devices results in a magnetic field resolution that is insufficient for many applications. As before, it is advantageous to use a separate signal or pickup coil that can have a much larger flux capture area. The pickup coil is usually connected to a multiturn spiral input coil that is transformer coupled to the SQUID inductance. The flux transformer and SQUID may be fabricated on separate chips and then pressed together in a "flip-chip" configuration[38,39,40], or they may be integrated on a single chip as described in Sec. 4. Integrated magnetometers now have been built with impressive magnetic field noise performance. For example, Shen et al[41] describe a fully integrated magnetometer for operation at 77 K. Operated inside the Berlin Magnetically Shielded Room, the rms magnetic field noise measured by Drung et al[42] was 9.7 fT / $\sqrt{Hz}$ at 1 kHz and 53 fT / $\sqrt{Hz}$ at 1 Hz. This sensitivity was sufficient to record magnetocardiograms.

As discussed in Sec. 2.1.2, the design of flux transformers is straightforward, but the fabrication of multilayer structures using high-T_c superconductors is not. A general complication is that, even though high-T_c flux transformers with excellent electrical transport properties have been fabricated, substantial low-frequency $1/f$ noise is observed for frequencies f below 1 kHz when coupled to the SQUID[43,44]. For these reasons, transformer-coupled high-T_c SQUIDs still have limitations for low frequency applications.

Alternately, fractional-turn SQUIDs have been fabricated whereby the SQUID is made from several (16) loops connected in parallel[45],[46]. In this way, the SQUID inductance is kept to a tolerable level, and the effective area of the device is quite high. Another approach is to take advantage of the flux-focusing effect of a large superconducting washer[47] in order to enhance the effective area of the SQUID without increasing the SQUID inductance. Using this approach, Zhang et al[48] report a magnetic field noise of 170 fT $/ \sqrt{\text{Hz}}$ for frequencies down to 1 Hz for a large washer rf SQUID operating at 77 K.

A simpler approach is to couple the signal from the pickup loop to the SQUID by direct injection[49,50,51], a configuration in which the SQUID inductance is connected in parallel with the pickup loop. Although the inductance mismatch is substantial, reasonable coupling and a significant enhancement of the effective area can be achieved. Koelle *et al*[51] report a field noise of 93 fT $/ \sqrt{\text{Hz}}$ at 1 Hz for a direct-coupled dc SQUID magnetometer at 77 K. A significant advantage of this approach is that the complete device can be fabricated from a single layer of high-T_c superconductor.

6.1. DIRECT-COUPLED SQUID MAGNETOMETERS

In this section, we consider a simple model for the design of a direct-coupled SQUID magnetometer. The relevant figure of merit for a magnetometer is the rms magnetic field noise $S_B^{1/2}(f) = S_\Phi^{1/2}(f)/A_{\text{eff}}$. Thus, the magnetic field noise is optimized by minimizing the flux noise and maximizing the effective area.

The magnetic flux noise $S_\Phi^{1/2}(f) = S_V^{1/2}(f)/(\partial V/\partial \Phi)$, where $S_V^{1/2}(f)$ is the rms voltage noise across the SQUID and $\partial V/\partial \Phi$ is the flux-to-voltage transfer function. In the absence of excess noise due to poor film quality, junction parameter fluctuations or resonances, the voltage noise may be estimated using the expression

$$S_V^{1/2}(f) = 4\sqrt{k_B T R} , \qquad (20)$$

where R is the junction resistance -see the chapter by John Clarke in this volume. The flux-to-voltage transfer function is given in Eq. 7. For a SQUID operating at 77 K, the noise-flux term can become significant. Then, the rms flux noise is given by

$$S_\Phi^{1/2}(f) = \frac{4\pi \sqrt{k_B T R}(1+\beta)}{7 I_c R \left[1 - 3.57\sqrt{k_B T L}/\Phi_0\right]} . \qquad (21)$$

The effective flux capture area of a directly-coupled magnetometer is given by $A_{\text{eff}} = A_s + k(A_p/L_p)L \approx k(A_p/L_p)L$, where A_s is the typically small effective area of the SQUID loop, A_p is the effective area of the pickup loop with inductance L_p, and k is a constant which describes the coupling efficiency. For direct-coupled magnetometers, the screening currents induced in L and L_p may oppose or aid each other depending on the layout. In the former case, it is desirable to keep A_s as small as possible. Once the SQUID inductance L has been fixed, the effective area is optimized by maximizing the ratio A_s/L_p and the coupling constant k. The coupling constant is optimized by maximizing the fraction of the SQUID inductance seen by the injected signal current.

A consequence of designing the direct-coupled magnetometer such that there are no wire-bond pads inside the pickup loop is that the screening currents induced in L and L_p oppose each other. In this case, it is desirable to keep A_s as small as possible. A simple way of designing the SQUID with low intrinsic area is to use a pair of narrow coplanar striplines. An example of such a SQUID is shown in Fig. 22. For this device, the SQUID is formed from a 67-μm long, 20-μm wide strip of YBa$_2$Cu$_3$O$_{7-x}$ (YBCO) with a 57-μm long, 4-μm wide slit along its length. The 6-μm long, 2-μm wide grain boundary junctions are located at the outer end of the slit, along with two, 5-μm wide injection leads that are connected to the pickup loop. Four pads are located at the edge of the substrate for making wire-bond connections to the SQUID.

For the device shown in Fig. 22, the SQUID inductance may be written as the sum of two parts, $L = L_{sl} + L_j$, where L_{sl} is the slit inductance and L_j is the parasitic inductance associated with the junctions. Standard inductance formulas for coplanar lines may be used to calculate these inductances[52] -see Appendix. For the parameters given above, we calculate $L_{sl} = 41.8$ pH and $L_j = 8.4$ pH, which include kinetic inductance[53] contributions of 4 pH and 2.5 pH, respectively. For these calculations, a YBCO film thickness of 300 nm and London penetration depth of 250 nm have been used. Then, the total inductance $L = 50.2$ pH, and the coupling constant is given approximately by $k \approx L_{sl}/L = 0.83$. The quantity kL may be measured directly

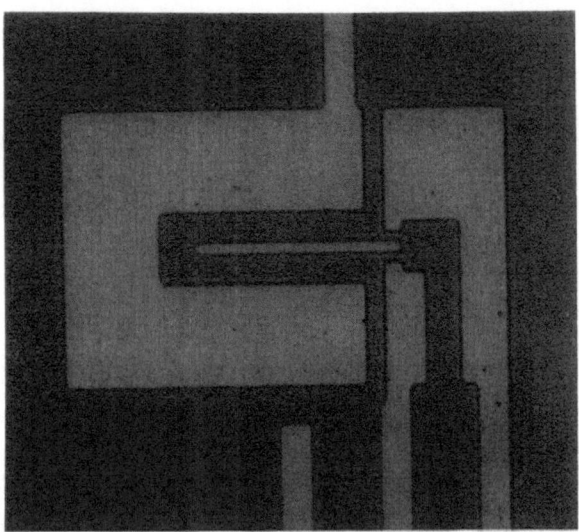

Figure 22. Photograph of a high-T_c dc SQUID. The two vertical strips are the bias leads, one of which is connected to the signal injection lead from the pickup loop.

by cutting the pickup loop and injecting an external current into the SQUID. These measurements typically agree with the inductance calculations by 5-10%.

For a square pickup loop with outer side length a and inner side length d, it can be shown that the ratio A_p/L_p gradually approaches a maximum in the limit $(a-d)/2 > d$. In this limit, $A_p = ad$ and $L_p = 1.25\mu_0 d$, so that $A_p/L_p = a/1.25\mu_0 = 0.637a[mm]$ mm^2/nH. Using the expression for the coupling constant k given above, the effective area of the magnetometer is given by

$$A_{eff} \approx 0.637a[mm]L_{sl}[nH] \ \ mm^2. \tag{22}$$

Given the pickup loop outer side length a, junction I_cR product, junction inductance L_j and β, Eqns. 21 and 22 may be used to calculate the rms field noise as a function of the slit inductance. For our bicrystal grain-boundary junctions, we find that the I_cR product is typically in the range of 100 to 150 μV. The junction inductance L_j may be calculated as described above for the given geometry. For optimal performance, we set $\beta = 1$. Using $a = 9.3$ mm, $L_j = 8.4$ pH and $I_cR = 100$ μV, the calculated field noise is shown in Fig. 23 as a function of the slit inductance L_{sl}. The field noise minimum, 32 fT $/\sqrt{Hz}$, is seen to be very broad. Also shown in Fig. 23 is the critical current for each value of the total inductance L determined by the condition that $\beta = 1$. The limitation here is that the Josephson coupling energy must be larger than the thermal energy. It has been shown[54] that a factor of 5 times larger should be sufficient,

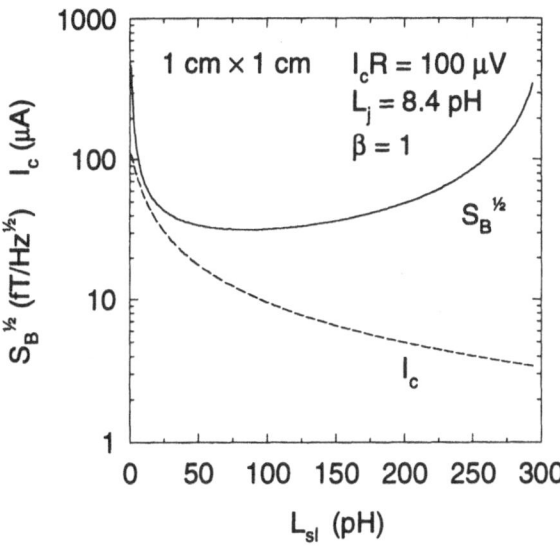

Figure 23. Calculated rms field noise and critical current for $\beta = 1$, constant I_cR product and $T = 77$ K as functions of the slit inductance.

which gives a minimum critical current of 16 µA at 77 K. Using the condition $\beta = 1$, this implies that the corresponding maximum SQUID inductance is about 65 pH. We have found it possible to work successfully with somewhat lower critical currents, so that slightly higher inductances are feasible.

6.2. EXPERIMENTAL RESULTS

The magnetometers are fabricated from c-axis oriented YBCO films deposited by laser ablation on 10×10 mm^2 SrTiO$_3$ bicrystal substrates with a 24° misorientation angle. The films are epitaxially grown using various heteroepitaxial and homoepitaxial buffer layers to a thickness of 150 to 350 nm. From SEM observations and AFM measurements, these films have a roughness of less than 4 nm. Distinct large grains or a-axis material are not detected. The films are patterned by ion milling through a photoresist mask. The contact pads are covered by a gold layer which is ion-beam deposited through a photoresist lift-off stencil.

We report here the results for a magnetometer fabricated from a YBCO film having a thickness of approximately 300 nm. The design parameters of the SQUID are the same as described in Sec. 6.1, and a photograph of the SQUID layout is shown in Fig. 22. For this magnetometer, $a = 9.3$ mm, with $d = 3$ mm in order to satisfy the condition $(a-d)/2 > d$. For these parameter values, $A_p \approx 28$ mm^2 and $L_p \approx 5.1$ nH. The pickup loop inductance includes a small contribution of about 0.4 nH owing to the slit in the pickup loop. Using $L_{sl} = 0.042$ nH, we calculate $A_{eff} \approx 0.23$ mm^2. It should be noted that the effective area is sensitive to the YBCO film thickness, because the kinetic inductance contribution to the slit inductance depends on the thickness of the film. We have determined A_{eff} directly by measuring the magnetometer response in a uniform magnetic field provided by a Helmholtz coil. We find that the field-to-flux conversion efficiency $B_\Phi = 11.1$ nT/Φ_0, which yields $A_{eff} \approx 0.19$ mm^2, in reasonable agreement with the calculation. The small difference is most likely due to a missing strip of pickup loop occupied by a row of four 1×1 mm^2 contact pads near the edge of the substrate and to the slit in the pickup loop, both of which reduce A_p, and therefore A_{eff}.

For the device described here, the critical current per junction $I_c = 11$ µA and normal resistance per junction $R = 9.7$ Ω, giving an I_cR product of 107 µV. Since the total SQUID inductance is estimated to be 50.2 pH, we calculate $\beta = 0.53$. The measured rms field noise of this magnetometer is shown in Fig. 24. For these measurements, the SQUID is operated using ac current bias in order to eliminate excess noise at low frequencies (below 1 kHz) caused by critical current fluctuations in the junctions[55],[56]. The field noise at 1 kHz is measured to be 41 fT$/\sqrt{\text{Hz}}$, increasing to about 58 fT$/\sqrt{\text{Hz}}$ at 1 Hz. The measured white field noise is in rather good agreement with the calculated noise of 36 fT$/\sqrt{\text{Hz}}$ for $L_{sl} = 42$ pH shown in Fig. 23.

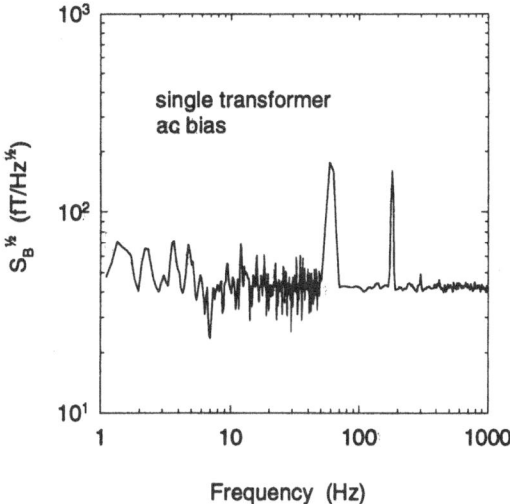

Figure 24. Measured rms field noise of a 10-mm × 10-mm direct-coupled magnetometer at 77 K. The noise was measured using a warm coupling transformer only -see Sec. 4- and ac current bias at 64 kHz.

The discrepancy may be accounted for by the reduction of A_{eff} owing to the presence of the large contact pads at the edge of the substrate and the slit in the pickup loop, the increase of L_p owing to the slit in the pickup loop, and the low value of β. The corrections to A_{eff} and L_p have not been explicitly taken into account in the simple model calculation discussed in Sec. 6.1.

In addition to the laser ablation technique, we have investigated other means of fabricating YBCO films for the production of dc SQUIDs and devices such as direct-coupled magnetometers. Films deposited by rf sputtering using a bulk YBCO target are found to be high quality and quite smooth, but SQUIDs fabricated from these films exhibit significant low frequency noise that cannot be eliminated using ac current bias methods. More recently, we have built a system for the deposition of uniform, large-area YBCO films by thermal evaporation of Y, Ba, and Cu. This technique enables the growth of homogeneous YBCO and other oxide films on large areas, as has already been demonstrated for wafers as large as 8 inches[57]. We have found the noise performance of SQUIDs fabricated from thermally evaporated YBCO films to be comparable to that observed for SQUIDs fabricated from laser ablated films.

A key component of the thermal evaporation system built at Conductus is the substrate heater assembly. The heater, which has been developed at Conductus following the work at the Technical University in Munich, allows the substrates to rotate between a deposition zone and an oxidation pocket[58]. An oxygen pressure of 20 mTorr is kept inside the heater pocket while a pressure three orders of magnitude lower

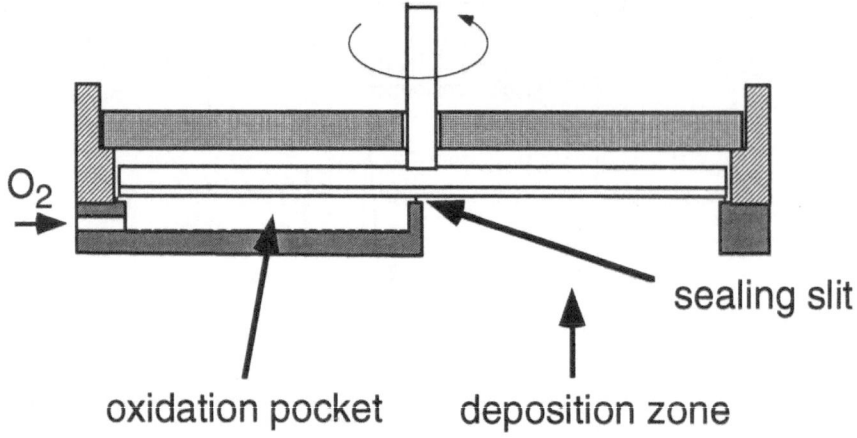

O$_2$ →

sealing slit

oxidation pocket   deposition zone

Figure 25. Schematic of the rotary substrate heater used for reactive evaporation of YBCO. The rotation of the substrates through the two spatially separated zones allows for sequential deposition and oxidation.

is maintained in the evaporation chamber. This pressure differential is achieved using a low conductance slit between the oxygen pocket and the heater aperture; -see Fig. 25. The low chamber pressure allows for large-area compositional uniformity.

Three resistive sources with independant rate control are used to co-evaporate Y, Ba and Cu. The substrates are mounted on a platen that can support up to nine 2"-diameter wafers. The evaporation rate of the metal species is controlled using collimated quartz crystal monitors which are fed back to the source power supplies. The metal atom stoichiometry can be tuned by adjusting the appropriate atomic fluxes. The total YBCO deposition rate is 0.2 nm/s, and the substrate temperature during YBCO deposition is 700°C. The superconducting critical temperature of the thermally evaporated YBCO films is typically 87 K and varies by less than 0.5 K across a 2" wafer. The thickness variation across a wafer is ±2%. Typical critical currents at 77 K are in the range of 2 to 4×10^6 A/cm^2.

We have fabricated several direct-coupled magnetometers, similar to the one described above except for a higher SQUID inductance, from thermally evaporated films of YBCO. For one such device, the critical current per junction $I_c = 80$ µA and normal resistance per junction $R = 2$ Ω, giving an I_cR product of 160 µV. Using the estimated 67 pH SQUID inductance, we calculate $\beta = 5.2$. The field calibration for this magnetometer is 10 nT/ Φ_0, and the rms field noise, measured using ac as well as dc bias, is shown in Fig. 26. With ac current bias, the field noise at 1 kHz is measured to be 73 fT / $\sqrt{\text{Hz}}$, increasing to about 83 fT / $\sqrt{\text{Hz}}$ at 1 Hz. The noise performance is expected to improve if the critical current is reduced such that the β parameter is closer to the optimal value $\beta = 1$.

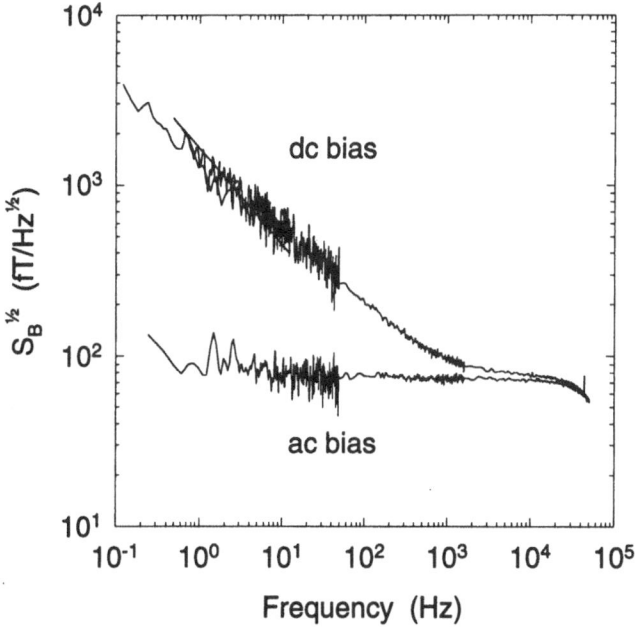

Figure 26. Measured rms field noise at 77 K of a 10-mm × 10-mm direct-coupled magnetometer fabricated from a single-layer, thermally-evaporated YBCO film. The noise was measured using a warm coupling transformer only -see Sec. 4- and ac current bias at 128 kHz.

From the above, the high magnetic field resolution that is currently achievable with single-layer, direct-coupled SQUID magnetometers operating at 77 K is attractive for many practical applications. One key issue that remains, however, is the feasibility of operation in unshielded environments. This is discussed in more detail in the following section.

6.3 OPERATION IN AMBIENT FIELDS

For operation in unshielded environments, it is important to bear in mind that SQUID magnetometers are vector sensors and therefore have fundamental limitations on how they may be operated in ambient fields. A SQUID magnetometer is sensitive to the total magnetic flux Φ passing through the magnetometer pickup loop: $\Phi = \int \mathbf{B} \cdot \mathbf{n} da$, where $\mathbf{B}$ is the applied magnetic field vector, $\mathbf{n}$ is the unit vector perpendicular to the plane of the magnetometer, and the integral is carried out over the area of the pickup loop. For a uniform field B_0 perpendicular to the magnetometer, for example, $\Phi = B_0 A$, where A is the area of the pickup loop. Thus, a SQUID magnetometer is sensitive to the magnetic field vector that is perpendicular to the plane of the pickup loop; therefore, such devices are very sensitive to motion in applied fields.

For example, suppose the applied magnetic field vector is uniform and makes an angle θ with respect to the unit vector perpendicular to the plane of the pickup loop. The normal component of the magnetic field is $B_\perp = B_0\cos\theta$, and the change in $B_\perp$ due to a small angular displacement is given by $dB_\perp = -B_0\sin\theta d\theta$. For a magnetometer with B_Φ = 10 nT/Φ_0 sensitivity and feedback electronics with 200 Φ_0 maximum full scale range (least sensitive range, Conductus Programmable Feedback Loop Model PFL-100), the corresponding magnetic field full scale range is 2 μT. Assuming the Earth's field is uniform with B_0 = 50 μT (0.5 G) and θ = 45°, this full scale range corresponds to a maximum angular displacement $d\theta_{max}$ = 2/[50(0.7)] = 0.057 rad or 3.3°. If the SQUID magnetometer is operated on the most sensitive range with 2 Φ_0 full scale range, the maximum angular displacement $d\theta_{max}$ = 0.033°. This constraint becomes tighter as the SQUID sensitivity is improved (B_Φ << 10 nT/Φ_0).

Further, even if the SQUID magnetometer is operated in a stationary position, small time-dependent angular fluctuations $d\theta(\omega)$ can lead to noise which exceeds the noise floor of the sensor. Let B_n be the rms magnetic field noise of the magnetometer. Then, for a sensor with B_n = 100 fT/$\sqrt{\text{Hz}}$, the maximum angular fluctuations that can be tolerated is given by $d\theta(\omega) = B_n/B_0\sin\theta \approx 3 \times 10^{-9}$ rad/$\sqrt{\text{Hz}}$. Thus, particularly at low frequencies (f < 100 Hz), the mechanical stability of the sensor in a stationary position is extremely important.

In view of the above, the operation of high-T_c magnetometers in ambient fields poses many technical challenges, even if the intrinsic noise of the magnetometer can be made arbitrarily low. One elegant solution to minimize the effects of uniform background fields is the 3-SQUID gradiometer configuration described by Koch et al[59]. In this configuration, the feedback loop output of a reference magnetometer is fed back not only to the reference magnetometer but also to two primary or sensor magnetometers through identical feedback coils coupled to each primary magnetometer. The voltage difference of the two feedback loop outputs for the primary magnetometers divided by the gradiometer baseline (the distance between the two primary magnetometers) gives a measure of the magnetic field gradient. Alternately, one may use intrinsic, integrated gradiometers where the pickup loops and SQUID are fabricated from thin films of YBCO on the same or different substrates in a flip-chip configuration[60,61]. To avoid problems owing to screening currents set up by changes of the background magnetic field, the pickup loops and the SQUID should be designed using series configurations as discussed in Sec. 5.1. Such devices require multi-layer fabrication technology and tend to exhibit excess low frequency noise, particularly in applied magnetic fields, as discussed above.

When a magnetometer is cooled in nominally "zero" field, there will inevitably be some flux vortices trapped in the superconducting films comprising the sensor (i.e., pickup and SQUID loops). Thermal fluctuations can cause these vortices to move, leading to excess noise at low frequencies (<100 Hz or so). For high-quality, single-layer YBCO films, this effect is usually not a significant noise source. If a static dc magnetic field is applied, however, a screening current arises in the pickup loop (and SQUID loop) in order to oppose the change of the magnetic flux through the pickup loop. The screening current exerts a Lorentz force on the flux vortices trapped in the superconducting films, and may cause the vortices to move leading to excess low frequency noise. The magnetic field itself can create additional vortices which further enhances the excess noise. This effect may be more acute in multi-layer YBCO structures; the flux vortices may not be as strongly trapped at cross-overs where the superconductivity of the YBCO is weakened.

We have found it possible to successfully operate the single-layer magnetometers described above in static magnetic fields[62]. A small noise increase (<2×) is observed for dc fields up to 100 µT (1 G), *provided the sensor is defluxed in the static magnetic field*. In order to heat the magnetometers quickly when they are exposed to a change of the background magnetic field, the magnetometers are hermetically sealed in a button-shaped, glass-epoxy (G-10) package that contains an integral heater. The heater may be activated remotely using the SQUID control electronics. Sample noise data for two different single-layer magnetometer styles are shown in Fig. 27. The rms flux noise at 1 Hz increases rapidly as the magnetic field is increased, but if the sensors are defluxed once the field has been applied, the noise is the same as measured in zero field for applied fields up to around 20 to 50 µT depending on the sensor.

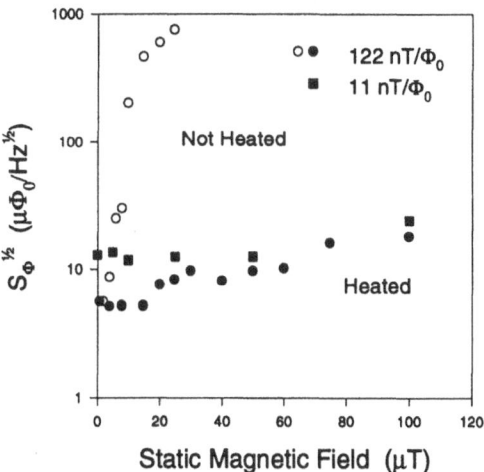

Figure 27. Measured rms flux noise at 1 Hz of two different magnetometer styles as a function of a static applied magnetic field.

It should be noted that, in practice, the magnetometers are not operated in a dc mode as was used to obtain the data shown in Fig. 27. The SQUID is usually operated as a null detector in a closed-loop feedback mode. Two different feedback modes are possible, as described in Sec. 5.1: flux-lock and current-lock. In the flux-lock mode, the flux coupled to the SQUID by an external flux through the pickup loop is cancelled by an opposing flux from the feedback coil that is coupled to the SQUID. In the current-lock mode, the feedback signal is coupled to the SQUID input circuit (e.g., the pickup loop for a direct-coupled magnetometer). For a magnetometer, the feedback signal opposes any change of the current in the pickup loop. In this way, the excess noise caused by vortex motion may be somewhat reduced. The packaged magnetometers described above are designed to operate with current-lock feedback.

7. Concluding Remarks

Over the past few years, considerable progress has been made in advancing both low-T_c and high-T_c dc SQUID technology. Guidelines for the design of coupled dc SQUIDs have been derived and tested which simplify the development of SQUID sensors for many applications. The technology for the fabrication of high-T_c SQUIDs, both single-layer and multi-layer designs, continues to be improved and refined. High-T_c SQUID magnetometers are now available commercially from several suppliers with sufficient sensitivity for practical applications. It is indeed notworthy that the magnetic field noise performance of the best high-T_c magnetometers is now within a factor of ten higher than observed for currently available low-T_c integrated magnetometers. This remarkable achievement represents an improvement of three orders of magnitude since the first high-T_c magnetometer was built in 1987.

Applications in biomagnetism such as magnetoencephalography (MEG) require the utmost sensitivity. For this reason, this application area most likely will continue to be the domain of low-T_c sensors for the foreseeable future. On the other hand, multi-layer high-T_c SQUID magnetometers have already been shown to have sufficient sensitivity for applications in magnetocardiography (MCG) when operated in a magnetically shielded environment. For biomagnetic (MCG) measurements in unshielded environments, high performance gradiometers with good balance will be needed.

For applications in geophysics and nondestructive evaluation, single-layer high-T_c devices already have sufficient sensitivity and may be adequate for many of these applications. The convience and lower cost of liquid nitrogen, and the fact that liquid nitrogen boils away much more slowly than liquid helium, makes high-T_c SQUID sensors very attractive for these applications. Interesting scientific and potentially significant commercial opportunities range from borehole and portable surface instrumentation for geophysical prospecting, to magnetic microscopy using miature SQUIDs closely spaced to samples at room temperature for nondestructive evaluation.

For high-T_c SQUID applications, many challenges remain, especially the problem with the increased noise typically observed when these SQUIDs are operated in unshielded environments. Recent results for single-layer SQUIDs, however, are very encouraging. This should help accelerate the use of high-T_c SQUIDs in those application areas where adequate performance has already been demonstrated.

Acknowledgements

We gratefully thank our colleagues at Conductus for their contributions to the work described here: A. Barfknecht, K. Char, E. Corpuz, A. Dela Cruz, A. Garachtchenko, L. Lee, M. Lopez, A. Mathai, V. Matijasevic, A. Matlashov, N. Peters, W. Ruby, J. Schmidt, R. Simon, M. Teepe, L. Valdez and V. Vinetskiy. The authors also would like to thank the following colleagues for their contributions, stimulating discussions and encouragement: J. Clarke, D. Drung, K. Enpuku, S. Erné, V. Foglietti, H. Koch, D. Koelle, F. Ludwig, J. Rowell, T. Ryhänen, and H. Seppä.

Appendix

In this section, we review the calculation of the key SQUID parameters used in the optimization procedure discussed in Sec. 2.

Junction Parameters. The total capacitance of each Josephson junction may be written as the sum of three parts,

$$C = C_j + C_{jp} + C_R, \qquad (A1)$$

where C_j is the capacitance of the junction, C_{jp} is the parasitic capacitance owing to the overlap of the counter-electrode outside the junction area, and C_R is the capacitance of the thin-film shunt resistor in the case the shunt resistor is deposited on top of the SQUID washer -see, for example, Fig. 12. To calculate the junction capacitance one needs to know the dependence of the specific capacitance of the junction C_j' on the critical-current density J_c. Since $J_c \propto \exp(-t)$ and $C_j' \propto 1/t$, where t is the thickness of the tunnel barrier, it follows that $1/C_j' = a - b\ln J_c$, where a and b are constants that depend on the barrier material. Then, $C_j = C_j' \cdot A_j$, where A_j is the area of the junction. The parasitic contribution is given by

$$C_{jp} = \frac{\varepsilon_0 \varepsilon_r A_{jp}}{h}, \qquad (A2)$$

where A_{jp} is the area of the overlap of the counter-electrode outside the junction, ε_0 is the permittivity of free space, and ε_r is the dielectric constant of the insulator with thickness h. Similarly, the capacitance of the shunt resistor is given by

$$C_R = \frac{\varepsilon_0 \varepsilon_r A_R}{2h},$$
(A3)

where A_R is the area of the shunt resistor. The factor of one-half arises because one end of the shunt resistor is grounded to the washer. Typically, the total junction capacitance ranges roughly from 0.5 to 1 pF.

The optimal critical current I_c depends on the SQUID inductance. In the absence of an input coil, the critical current is chosen such that $\beta = 2LI_c/\Phi_0 = 1$, where L is the total SQUID inductance -see below. With an input coil and superconducting load coupled to the SQUID, the SQUID inductance is reduced owing to the screening effect of the input circuit -see below. The screened SQUID inductance should be used to calculate I_c using the condition $\beta = 1$. The β parameter calculated using the optimal value for the critical current and the unscreened inductance is typically about 2 to 2.5.

Once the junction capacitance has been determined for a given value of the critical current I_c, the junction shunt resistance R may be calculated to achieve a particular value of the Stewart-McCumber parameter $\beta_c = (2\pi/\Phi_0)I_c R^2 C$. This may be done by varying the length of the resistor for a fixed resistor width and sheet resistance, or by varying the sheet resistance for a fixed resistor geometry. In the former case, an iterative process must be used since the changing resistor length also changes the capacitance C_R. Usually, β_c is in the range 0.3 to 0.7.

SQUID Inductance. For most applications, the SQUID inductance ranges from roughly 100 to 200 pH. The SQUID washer is usually square, though octagonal and circular washers also have been used. In the following expressions, d is the inner side length or diameter of the hole in the washer, a is the outer side length or diameter of the washer, and $b = (a-d)/2$ is the length of the washer slit. The SQUID inductance may be expressed as the sum of three parts,

$$L = L_h + L_{sl} + L_j,$$
(A4)

where L_h is the geometric inductance of the hole in the washer, L_{sl} is the inductance of the slit, and L_j is the parasitic inductance of the junction area. For $d/b < 1$, the hole inductance is given by

$L_h = 1.25\mu_0 d$	(square washer with square hole)[14],	(A5)
$L_h = 1.05\mu_0 d$	(octagonal washer with octagonal hole)[63],	(A6)
$L_h = \mu_0 d$	(circular washer with circular hole)[64].	(A7)

Since the hole inductance for the octagonal and circular geometries are nearly the same, in the following discussion only the circular geometry is included.

The slit inductance depends on whether or not a groundplane is used. In the absence of a groundplane, the input coil partially screens the slit. According to simulations and measurements on coupled SQUIDs[65] the slit inductance per unit length $L_{sl} = 0.3$ pH/μm. In the absence of an input coil, the slit inductance is slightly higher, about 0.4 pH/μm. If a slit groundplane is used, one may model the slit as two striplines[11] -see below- of length b; then, the slit inductance is twice the stripline inductance.

The parasitic inductance of the junction area is more difficult to estimate. Typically, this contribution should not be more than a few pH. This is usually negligible in comparison to the total SQUID inductance, unless one is trying to design a SQUID with extremely low inductance (<10 pH or so). In the following, this quantity is neglected.

The SQUID inductance is reduced because of screening effects if a superconducting load is attached to the input circuit. The screened inductance is given by

$$L_{dc} = (1 - k_i s_{in})L ,$$ (A8)

where k_i is the coupling constant -see below- and the screening parameter $s_{in} = L_i/L_i + L_p$. Here, L_i is the inductance of the input coil coupled to the SQUID and L_p is the inductance of the pickup loop or load inductor.

For high-T_c SQUIDs, the kinetic inductance contribution to the total inductance may not be negligible. For a thin-film stripline having thickness h and width b carrying a total current I, the kinetic inductance per unit length is given by $L_k' = \mu_0 (\lambda^2/I^2)\int J^2 dA$, where λ is the London penetration depth and the integral of the current density J is carried out over the cross-section of the film. For YBCO films at 77 K, $\lambda \approx 250$ nm[66]. Asuming J is uniform across the film, which is a rough approximation for $h < \lambda$ and $b >> h$, the kinetic inductance contribution may be estimated using67

$$L_k' \approx \mu_0 \frac{\lambda^2}{hb} \ell$$ (A9)

where ℓ is the mean path along the length of the film.

Input Circuit Inductances. The N-turn input coil has linewidth w_i and turns spacing s_i. The total length of the coil ℓ is given by

$$\ell = 4N[d + (N+1)s_i + Nw_i] \qquad \text{(square coil)}, \qquad (A10)$$

$$\ell = 2\pi N(d + w_i)/2 + \frac{w_i + s_i}{2}\sqrt{\frac{N}{2\pi}} \quad \text{(circular spiral coil)}. \qquad (A11)$$

The input coil and the washer as groundplane form a microstripline. Chang[68] has derived a set of expressions to calculate the stripline inductance L'_{strip} and capacitance C'_{strip} per unit length. The stripline capacitance is given by

$$C'_{strip} = \frac{\varepsilon_0 \varepsilon_r w_i}{h} K(w_i, h, t_i) \qquad (A12)$$

where t_i is the thickness of the input coil, and $K(w_i, h, t_i)$ is the field fringe factor -see [68]. For usual linewidths (2 μm to 5 μm) and film thicknesses (200 to 300 nm), the fringe factor is roughly in the range 1.1 to 1.6.

The expression for the stripline inductance is rather complicated; a good approximation, however, is given by[69]

$$L'_{strip} = \mu_0 \frac{h + \lambda_{L,w} + \lambda_{L,i}}{w_i + 2(h + \lambda_{L,w} + \lambda_{L,i})} \qquad (A13)$$

where $\lambda_{L,w}$ and $\lambda_{L,i}$ are the London penetration depths of the washer and input coil, respectively. The second term in the denominator is an aproximation which takes the fringe effects into account.

Then, the input coil inductance L_i is given by[65]

$$L_i = N^2(L_h + L_{sl}/3) + \ell L'_{strip} \qquad \text{(single washer)}, \qquad (A14)$$

$$L_i = 2[N^2(L_h + L_{sl}/3) + \ell L'_{strip}] \quad \text{(grad. washers, series or parallel)}. \qquad (A15)$$

The mutual inductance M_i is given by[65]

$$M_i = N(L_h + L_{sl}/2) \qquad \text{(single washer or grad. washers, parallel)}, \qquad (A16)$$

$$M_i = 2N(L_h + L_{sl}/2) \qquad \text{(gradiometric washers, series)}, \qquad (A17)$$

and the coupling constant k_i is

$$k_i = \frac{M_i}{\sqrt{LL_i}} \qquad \text{(A18)}$$

Typically, M_i ranges from roughly 1 to 20 nH and k_i ranges from 0.85 to 0.95.

The screening effect of the SQUID washer reduces the input coil inductance slightly. The screened inductance is given by[1]

$$L_{i,\text{eff}} = \left(1 - k_i^2 s\right)L_i, \qquad \text{(A19)}$$

where

$$s = \frac{\beta s_{\text{in}} k_i^2}{6 + 2\beta + \beta s_{\text{in}} k_i^2}. \qquad \text{(A20)}$$

The screened input coil inductance is usually chosen to match the inductance of the load or pickup loop inductance -see below.

The parasitic capacitance of the input coil[23] is given by $C_{strip}/8$. In the event a groundplane is used, it is necessary to add the contribution from the groundplane, C_{gp}. The capacitance introduced by the groundplane may be calculated assuming the groundplane consists of two striplines of length b. Then, the total parasitic capacitance is given by

$$C_p = \ell C'_{strip} / 8 + 0.5 C_{gp} \qquad \text{(single washer)}, \qquad \text{(A21)}$$
$$C_p = \ell C'_{strip} / 16 + 0.25 C_{gp} \quad \text{(gradiometric washers, series)}, \qquad \text{(A22)}$$
$$C_p = \ell C'_{strip} / 4 + C_{gp} \qquad \text{(gradiometric washers, parallel)}. \qquad \text{(A23)}$$

Ideally, the SQUID design should be chosen such that $C_p < C$[11]. This can be difficult, especially if the SQUID is being designed to match a high load inductance (much greater than a few nanohenry). If a high C_p is unavoidable, it may be desirable to use a double-loop design[11] -see Sec. 2.1.2.

The input coil stripline impedance Z is

$$Z = \sqrt{\frac{L_{strip}}{C_{strip}}}. \qquad \text{(A24)}$$

The value of the shunt resistor R_s is determined by setting $R_s = Z$. Typical values for Z are roughly in the range 5 to 15Ω. The shunt capacitance C_s is chosen such that the Q of the input circuit is about 2. Typically, $C_s \sim 25$ to 100 pF.

The resonant frequency of the stripline depends on whether or not the innermost turn is grounded to the washer. The phase velocity is $v = \left(L'_{strip}C'_{strip}\right)^{-1/2}$. Then, the stripline resonant frequency f_s is given by

$$f_s = \frac{v}{4\ell} \qquad \text{(input coil grounded to washer)} \qquad \text{(A25)}$$

$$f_s = \frac{v}{2\ell} \qquad \text{(input coil floating)} \qquad \text{(A26)}$$

If a groundplane is not used, the resonant frequency of the washer f_w may be calculated using a simple model[23] in the case $C_p \gg C$. According to this model,

$$f_w = \frac{c}{2\ell_{eff}\sqrt{\varepsilon_r(1+(\lambda_w+\lambda_i)/h)}} \qquad \text{(A27)}$$

where the effective length $\ell_{eff} = 4(d+4b/3)$. More generally, one may use the equivalent circuit models described by Enpuku *et al* in [22] and [23].

According to simulations, the optimal operating frequency of the SQUID $f_{op} = 0.3f_J$, where $f_J = I_cR/\Phi_0$. In addition to damping the SQUID resonances with the R_sC_s shunt, the SQUID design should be chosen such that $4f_s < f_{op} < f_w/4$. This keeps the input coil and washer resonances well away from the intended operating point[2].

Thin-Film Pickup Loop Inductance. The pickup loop has linewidth w_p and inner side-length d. The inductance and effective area of a square loop are given by

$$L_p = \frac{2}{\pi}\mu_0(d+w_p)\left[ln(1+d/w_p)+0.5\right] \quad (d/w_p > 10)[70], \qquad \text{(A28)}$$

$$L_p = \frac{2}{\pi}\mu_0 d\left[ln(5+d/w_p)+0.25\right] \qquad (1 < d/w_p < 10)[71], \qquad \text{(A29)}$$

$$L_p = 1.25\mu_0 d \qquad (d/w_p < 1)[14], \qquad \text{(A30)}$$

and[47]

$$A_{p,\square} = ad. \qquad \text{(A31)}$$

From the above, $(A_p/L_p)_\square$ is maximum for $d/w_p < 1$, giving

$$(A_e/L_p)_\square = a/(1.25\mu_0) = 0.637a \text{ mm}^2/\text{nH} \quad (a \text{ in mm}).$$ (A32)

For a circular loop the inductance and effective area are

$$L_p = \mu_0 \frac{d}{2}\left[\ln(8d/w_p) - 2\right] \qquad (d/w_p > 10)[70],$$ (A33)

$$L_p = \mu_0 d \qquad (d/w_p < 1)[64],$$ (A34)

and[47]

$$A_{p,O} = (2/\pi)ad.$$ (A35)

As before, $(A_e/L_p)_O$ is maximum for $d/w_p < 1$, giving

$$(A_e/L_p)_O = 2a/\pi\mu_0 = 0.507a \text{ mm}^2/\text{nH} \quad (a \text{ in mm}).$$ (A36)

Depending on the size of the chip and linewidth of the pickup loop, the pickup loop inductance typically ranges from 10 to 100 nH.

The inductance of the connecting leads to the pickup loop may be calculated using standard relations for coplanar striplines. Let w_c and s_c be the linewidth and separation of the coplanar lines. Then, the coplanar stripline inductance L_{cop} is given by[71]

$$L_{cop} = \frac{\mu_0}{\pi}\ln\left[\frac{4(w_c + s_c)}{w_c} - \frac{w_c}{w_c + s_c}\right]$$ (A37)

For the special case that $w_c = s_c$, the coplanar stripline inductance $L_{cop} = 0.8$ pH/μm. For most applications, the inductance of the connecting leads is usually small in comparison with the pickup loop inductance and may be neglected.

Wire-Wound Pickup Loop Inductance.

The inductance of a coil made from wire of radius r wound on a form of radius R may be calculated using Eq. (A33),

$$L_p = \mu_0 R\left[\ln(8R/r) - 2\right] \qquad (R/r > 10).$$ (A38)

In this limit, the loop inductance is approximately given by

$$L_p \approx 7.93 R[\text{mm}] \text{ nH} \qquad (R/r > 10).$$ (A39)

230

The inductance of the twisted-pair leads L_{leads} also contributes to the total pickup loop inductance. Typically, the lead inductance $L_{leads} \approx 0.5\ell$ nH/mm, where ℓ is the length of the twisted pair in mm.

References

1. Knuutila, J., Kajola, M., Seppä, H., Mutikainen, R. and Salmi, J. (1988) Design, optimization, and construction of a dc SQUID with complete flux transformer circuits, *J. Low Temp. Phys.* **71**, 369-392.
2. Cantor, R., Ryhänen, T., Drung, D., Koch, H. and Seppä, H. (1991) Design and optimization of DC SQUIDs fabricated using a simplified four-level process, *IEEE Trans. Magn.* **MAG-27**, 2927-2931.
3. Cantor, R., Enpuku, K., Ryhänen, T. and H. Seppä (1993) A high performance integrated DC SQUID magnetometer, *IEEE Trans. Appl. Superconductivity* **3**, 1800-1803.
4. Ryhänen, T., Seppä, H., Ilmoniemi, R., and Knuutila, J. (1989) SQUID magnetometers for low-frequency applications, *J. Low Temp. Phys.* **76**, 287-386.
5. Gupta, M. S. (1978) Thermal fluctuations in Driven Nonlinear Resistive Systems, *Phys. Rev.* **A18**, 2725-2737.
6. Gupta, M. S. (1982) Thermal noise in nonlinear resistive devices and its circuit representation, *Proc. IEEE* **70**, 788-792.
7. Tinkham, M. (1975) *Introduction to Superconductivity*, McGraw-Hill, New York.
8. Tesche, C.D. and Clarke, J. (1977) dc SQUID: noise and optimization, *J. Low Temp. Phys.* **29**, 301-331.
9. Enpuku, K., Shimomura, Y. and Kisu, T. (1993) Effect of thermal noise on the characteristics of a high-T_c superconducting quantum inteference device, *J. Appl. Phys.* **73**, 7929-7934.
10. Ryhänen, T., Seppä, H., Cantor, R., Drung, D., Koch, H., and Veldhuis, D. (1992) Noise studies of uncoupled dc SQUIDs, in *Superconducting Devices and their Applications,* edited by H. Koch and H. Lübbig, (Springer Verlag, Berlin,) pp. 321-325.
11. Ryhänen, T., Seppä, H. and Cantor, R. (1992) Effect of parasitic inductance and capacitance on the dynamics and noise of direct current superconducting quantum interference devices, *J. Appl. Phys.* **71**, 6150-6166.
12. Ketchen, M.B. and Kirtley, J.R. (1995) Design and performance aspects of pickup loop structures for miniature SQUID magnetometry, *IEEE Trans. on Applied Superconductivity* **5**, 2133-2136.
13. Ketchen, M., Pearson, D.J., Stawiasz, K., Hu, C.-K., Kleinsasser, A.W., Brunner, T., Cabral, C., Chandrashekhar, V., Jaso, M., Manny, M., Stein, K., Bhushan, M. (1993) Octagonal washer dc SQUIDs and integrated susceptometers fabricated in planarized sub-μm Nb-AlO$_x$-Nb technology, *IEEE Trans. on Applied Superconductivity* **3**, 1795-1799.
14. Jaycox, J.M. and Ketchen, M.B. (1981) *IEEE Trans. Magn.* **MAG-17**, 400-403.
15. Knuutila, J., Ahonen, A. and Tesche, C. (1987) Effects on dc SQUID characteristics of damping of input coil resonances, *J. Low Temp. Phys.* **68**, 269-284.
16. Seppä H. and Ryhänen, T. (1987) Influence of the signal coil on the dc SQUID dynamics, *IEEE Trans. Magn.* **MAG-23**, 1083-1086.
17. Ryhänen, T., Cantor, R., Drung, D., Koch, H. and Seppä, H. (1990) Effect of parasitic capacitance on dc SQUID performance, *IEEE Trans. Magn.* **27**, 3013-3016.
18. Enpuku, K., Yoshida, K. and Kohijo, S. (1986) Noise characteristics of a dc SQUID with resistively shunted inductance. II. Optimal damping, *J. Appl. Phys.* **60**, 4218-4223.
19. Muhlfelder, B., Johnson, W. and Cromar, M.W. (1983) Double transformer coupling to a very low noise SQUID, *IEEE Trans. Magn.* **19**, 303-307; Muhlfelder, B., Beall, J.A., Cromar, M.W., Ono, R. and Johnson, W. (1983) Well-coupled, low noise, dc SQUIDs, *IEEE Trans. Magn.* **21**, 427-429; Muhlfelder, B., Beall, J.A., Cromar, M.W., and Ono, R. (1986) Very low noise, tightly coupled, dc SQUID amplifiers, *Appl. Phys. Lett.* **49**, 1118-1120.
20. Drung, D., Cantor, R., Peters, M., Ryhänen, T. and Koch, H. (1991) Integrated dc SQUID magnetometer with high dV/dB, *IEEE Trans. Magn.* **27**, 3001-3004.

21. Ryhänen, T. and Seppä, H. (1989) unpublished.

22. Enpuku, K., Cantor, R., and Koch, H. (1992) Modelling the dc superconducting quantum interference device coupled to the multiturn input coil. II, *J. Appl. Phys.* **71**, 2338-2346.

23. Enpuku, K., Cantor, R., and Koch, H. (1992) Modelling the dc superconducting quantum interference device coupled to the multiturn input coil. III, *J. Appl. Phys.* **72**, 1000-1006.

24. Gurvitch, M., Washington, M.A. and Huggins, H.A. (1983) High quality refractory Josephson tunnel junctions utilizing thin aluminum layers, *Appl. Phys. Lett.* **42**, 472-474.

25. Kroger, H., Smith, L.N. and Jillie, D.W. (1981) Selective niobium anodization process for fabricating Josephson tunnel junctions, *Appl. Phys. Lett.* **39**, 280-282.

26. Cantor, R., Drung, D., Peters, M. and Koch, H. (1990) Plasma nitridation process for the fabrication of all-refractory Josephson junctions, *J. Appl. Phys.* **67**, 3038-3042.

27. Henkels, W.H. and Kircher, C.J. (1977) Penetration depth measurements on Type II superconducting films, *IEEE Trans. Magn.* **MAG-13**, 63-66.

28. Maissel, L. (1970) Application of sputtering to the deposition of films, in L.I. Maissel and R. Glang (eds.), *Handbook of Thin Film Technology*, McGraw-Hill, New York, pp. 4-1 - 4-43.

29. Drung, D., Cantor,, R. Peters, M., Scheer, H.J. and Koch, H. (1990) Low-noise high-speed dc superconducting quantum interference device magnetometer with simplified feedback electreonics, *Appl. Phys. Lett.* **57**, 406-408.

30. Drung, D. (1991) DC SQUID systems overview, *Supercond. Sci. Technol.* **4**, 377-385.

31. Seppä, H., Ahonen, A., Knuutila, J., Simola, J. and Vilkman, V. (1991) DC-SQUID electronics based on adaptive positive feedback: experiments, *IEEE Trans. Magn.* **27**, 2488-2490.

32. H. Seppä (1992) DC-SQUID electronics based on adaptive noise cancellation and a high open-loop gain controller, in *Superconducting Devices and Their Applications, Springer Proceedings in Physics*, edited by H. Koch and H. Lübbig (Springer, Berlin) p. 346-350.

33. Welty, R.P. and Martinis, J.M. (1993) Two-stage integrated SQUID amplifier with series array output, *IEEE Trans. Appl. Superconductivity* **3**, 2605-2608.

34. Ryhänen, T. (1992) *Theoretical aspects, design, and characterization of dc superconducting quantum interference devices*, Ph.D. Thesis, Helsinki University of Technology, Finland.

35. Cantor, R., Vinetskiy, V. and Matlashov, A. (1996) A low-noise, integrated dc SQUID magnetometer for applications in biomagnetism, presented at Biomag'96, Santa Fe, NM, 16-21 February.

36. Clarke, J., Goubau, W.M. and Ketchen, M. (1976) Tunnel junction dc SQUID: fabrication, operation, and performance, *J. Low Temp. Phys.* **25**, 99-144.

37. Simmonds, M.B. (1990) High symmetry dc SQUID system, US Patent No. 5053834.

38. Miklich, A.H., Kingston, J.J., Wellstood, F.C. and Clarke (1991) Sensitive $YBa_2Cu_3O_{7-x}$ thin film magnetometer, *J. Appl. Phys. Lett.* **59**, 988-990.

39. Koelle, D., Miklich, A.H., Ludwig, F., Dansker, E., Nemeth, D.T. Clarke, J., Ruby, W. and Char, K. (1993) High performance dc SQUID magnetometers with single layer $YBa_2Cu_3O_{7-x}$ flux transformers, *Appl. Phys. Lett.* **63**, 3630-3632.

40. Miklich, A.H., Koelle, D., Dantsker, E., Nemeth, D.T., Kingston, J.J., Kromann, R.F. and Clarke, J. (1993) Bicrystal YBCO dc SQUIDs with low noise, *IEEE Trans. Appl. Sup.* **3**, 2434-2437.

41. Shen, Y.Q., Sun, Z.J., Kromann, R., Holst, T., Vase, P. and Freloft, T. (1995) Integrated high-T_c superconducting magnetometer with multiturn input coil and grain boundary junctions, *Appl. Phys. Lett.* **67**, 2081-2083.

42. Drung, D., Ludwig, F., Müller, W., Steinhoff, U., Trahms, L. Shen, Y.Q., Jensen, M.B., Vase, P., Holst, T., Freloft, T., and Curio, G. (1996) Integrated $YBa_2Cu_3O_{7-x}$ magnetometer for biomagnetic measurements, *Appl. Phys. Lett.* **68**, 1421-1423.

43. Ferrari, M., Kingston, J.J., Wellstood, F.C., and Clarke, J. (1991) Flux noise from superconducting $YBa_2Cu_3O_{7-x}$ flux transformers, *Appl. Phys. Lett.* **58**, 1106-1108.

44. Ludwig, F., Dantsker, E., Nemeth, D.T., Koelle, D., Miklich, A.H. Clarke, J., Knappe, S., Koch, H. and Thomson, R.E. (1994) Fabrication issues in optimizing $YBa_2Cu_3O_{7-x}$ flux transformers for low $1/f$ noise, *Supercond. Sci. Tech.* **7**, 273-276.

45. Drung, D., Knappe, S. and Koch, H. (1995) Theory for the multiloop dc superconducting quantum interference device magnetometer and experimental verification, *J. Appl. Phys.* **77**, 4088-4098.

46. Ludwig, F., Dantsker, E., Kleiner, R., Koelle, D., Clarke, J., Knappe, S., Drung, D., Koch, H., McNeal, A. and Button, T.W. Integrated high-T_c multiloop magnetometer, *Appl. Ohys. Lett.* **66**, 1418-1420.

47. Ketchen, M.B., Gallagher, W.J., Kleinsasser, A.H., Murphy, S. and Clem, J.R. (1985) DC SQUID flux focuser, in H. Hahlbohm and H. Lübbig (eds.), *SQUID '85 Superconducting Quantum Interference Devices and their Applications*, Springer-Verlag, Berlin, pp. 865-871.

48. Zhang, Y., Mück, M., Hermann, K., Schubert, J., Zander, W., Braginski, A.I. and Heiden, C. (1993) Sensitive rf SQUIDs and magnetometers operating at 77 K, *IEEE Trans. Appl. Superconductivity* **3**, 2465- 2468.

49. Matsuda, M., Murayama, Y., Kiryu, S., Kasai, N., Kashiwaya, S., Koyanagi, M. and Endo, T. (1991) Directly-coupled dc SQUID magnetometers made of Bi-Sr-Ca-Cu oxide films, *IEEE Trans. Magn.* **27**, 3043-3046.

50. Knappe, S., Drung, D., Schurig, T., Koch, H., Klinger, M. and Hinken, J. (1992) A planar $YBa_2Cu_3O_7$ gradiometer at 77 K, *Cryogenics* **132**, 881-884.

51. Koelle, D., Miklich, A.H., Ludwig, F., Dantsker, E., Nemeth, D.T. and Clarke, J. (1993) DC SQUID magnetometers from single layers of $YBa_2Cu_3O_{7-x}$, *Appl. Phys. Lett.* **63**, 2271-2273.

52. K. C. Gupta, R. Garg, and I. J. Bahl, (1979) *Microstrip Lines and Slotlines*, Artech House, Dedham, MA, pp. 263-265.

53. Van Duzer, T. and Turner, C.W. (1981) *Principles of Superconductive Devices and Circuits*, Elsevier, New York.

54. Clarke J. and Koch, R. (1988) *Science* **242**, 217-223.

55. Simmonds, M.B. and Gifford, R.P. (1983) Apparatus for reducing low frequency noise in dc biased SQUIDs, US Patent No. 4389612.

56. Koch, R.H., Clarke, J. Goubau, W.M., Martinis, J.M., Pegrum, C.M. and Van Harlingen, D.J. (1983) Flicker ($1/f$) noise in tunnel junction dc SQUIDs, *J. Low Temp. Phys.* **51**, 207-224.

57. Kinder, H., Berberich, P., Prusseit, W., Semerad, R. and Utz, B (1995) Very large area double sided evaporation of high quality YBCO films, *Proceedings of ISTEC-MRS*, 182-184.

58. Berberich, P., Utz, B., Prussiet, W. and Kinder, H. (1994) Homogeneous high quality $YBa_2Cu_3O_7$ films on 3" and 4" substrates, *Physica C* **219**, 497-504.

59. Koch, R.H., Rozen, J.R., Sun, J.Z. and Gallagher, W.J. (1993) Three SQUID Gradiometer, *Appl. Phys. Lett.* **63**, 403-405.

60. Keene, M.N., Satchell, J.S., Goodyear, S.W., Humphreys, R.G., Edwards, J.A., Chew, N.G. and Laner, K. (1995) Low noise HTS gradiometers and magnetometers constructed from $YBa_2Cu_3O_{7-x}$/ $PrBa_2Cu_3O_{7-y}$ thin films, *IEEE Trans. Appl. Superconductivity* **5**, 2923- 2926.

61. Bär, L.R., Daalmans, G.M., Barthel, K.H., Ferchland, L., Selent, M., Kühnl, M. and Uhl, D. (1995) Single layer and integrated YBCO gradiometer coupled SQUIDs, in H. Hayakawa (ed.), *Fifth International Superconductive Electronics Conference Extended Abstracts*, unpublished.

62. Schmidt, J.M., Lee, L.P., Matlashov, A., Teepe, M., Vinetskiy, V. and Cantor, R. (1996) Low-noise, single layer YBCO dc SQUID magnetometers for shielded and unshielded operation, presented at Biomag'96, Santa Fe, NM, 16-21 February.

63. Ketchen, M.B., Stawiasz, K.G., Pearson, D.J., Brunner, T.A., Hu, C.-K., Jaso, M.A., Manny, M.P., Parsons, A.A. and Stein, K.J. (1992) Sub-µm linewidth input coils for low T_c integrated thin-film dc superconducting quantum interference devices, *Appl. Phys. Lett.* **61**, 336-338.

64. Ketchen, M.B. (1987) Integrated thin-film dc SQUID sensors, *IEEE Trans. Magn.* **MAG-23**, 1650-1657.

65. Ketchen, M.B. (1991) Design considerations for dc SQUIDs fabricated in deep sub-micron technology, *IEEE Trans. Magn.* **27**, 2916-2919.

66. Hardy, W.N., Bonn, D.A., Morgan, D.C., Liang, R. and Zhang, K. (1993) Precision measurements of the temperature dependence of λ in $YBa_2Cu_3O_{6.95}$: strong evidence for nodes in the gap function, *Phys. Rev. Lett.* **70**, 3999-4002.

67. Lee, J.Y. and Lemberger, T. (1993) Penetration depth $\lambda(T)$ of YBa$_2$Cu$_3$O$_{7-x}$ films determined from the kinetic inductance, *Appl. Phys. Lett.* **62**, 2419-2421.

68. Chang, W.H. (1979) The inductance of a superconducting strip transmission line, *J. Appl. Phys.* **50**, 8129-8134.

69. Jutzi, W. (1981) Applications of the Josephson technology *Advances in Solid State Physics* **21**, 403-432.

70. Grover, F.W. (1962) *Inductance calculations, working formulas and tables*, Dover, New York.

71. Drung, D. (1988) *Sensor und A/D-Wandlerstufe auf einem Chip zur Präzisionsmessung von Magnetfeldgradienten mit Josephson-Kontakten*, Ph.D. Thesis, Universität Fridericiana Karlsruhe, Germany.

FABRICATION OF HIGH-TEMPERATURE SQUID MAGNETOMETERS
Correlations with Design and Performance

A.I. BRAGINSKI
Institut für Schicht- und Ionentechnik (ISI), Forschungszentrum Jülich (KFA)
D-52425 Jülich, Germany

Abstract
Fabrication of SQUID magnetometers from epitaxial thin films of high-temperature superconductors and multilayered film structures including perovskite insulator films is reviewed in this chapter. Emphasized are patterning and processing methods, fabrication of Josephson junctions having resistively-shunted current-voltage characteristics, attained reproducibility and spreads of junction critical current and normal resistance. Subsequently, guidelines are given for designing and fabricating single and multilayered SQUID sensor structures, especially flux transformers and multiloop SQUIDs. Included are correlations with attained performance characteristics of magnetometers: their effective area, white and low-frequency flux noise, and the resulting magnetic field resolution at high and low signal frequencies. Additionally described are alternative planar rf SQUID tank-circuit resonators and the resulting properties of rf SQUIDs.

1. Introduction

In this chapter, we address the fabrication technology of SQUIDs, and especially SQUID magnetometer sensors, which are fabricated from high-temperature superconductors (HTS) and capable of operation at temperatures up to that of liquid nitrogen (77K). The scope is limited to superconducting sensor structures. Fabrication of other components of a SQUID system is not considered, with the exception of tank circuits for rf SQUIDs. Processing methods, and effects of material properties and processing on SQUID sensor performance characteristics and designs, are presented and discussed.

We begin with a succinct overview of material properties and general processing methods, followed by a more detailed description of Josephson junctions used in SQUIDs. Subsequently, we review some formulae useful in designing layouts, and also the resulting characteristics of single-layer SQUID and magnetic flux pickup or concentrator structures. Approaches to fabrication of multilayered pickup structures used in the flip-chip configuration are described, and total integration with SQUIDs addressed, also in the context of trilayer junction technology. We conclude with an

H. Weinstock (ed.), SQUID Sensors: Fundamentals, Fabrication and Applications, 235–288.
© 1996 Kluwer Academic Publishers.

outlook discussing current and anticipated development trends. The fundamentals of SQUID operation and theory, and definitions of terms we use here, can be found in Chapter 1.

2. Materials and Processing

2.1. MATERIALS

2.1.1. Superconducting Films and Epitaxial Substrates

Initially, HTS SQUIDs were fabricated from polycrystalline bulk or thin-film YBa$_2$Cu$_3$O$_{7-\delta}$ (YBCO), and utilized grain-boundary weak links naturally occurring in this material. These early devices exhibited extremely high 1/f noise, *i.e.*, low sensitivity, caused mainly by critical current fluctuations and vortex hopping betweeen potential minima at defects, *e.g.*, near grain boundaries. Today, YBCO is still the standard material for SQUID fabrication. It remained the natural material of choice, because the enormous effort invested worldwide in developing high-quality epitaxial thin films has concentrated on c-axis YBCO (c-axis normal to film plane). Only in films of YBCO (and other 123 compounds) are adequate levels of flux pinning readily obtainable up to 77K. Currently SQUID sensors are made almost exclusively from YBCO thin films, which have low defect concentrations (not counting 90° twins and very low-angle grain boundaries), and are virtually free of naturally, *i.e.*, uncontrollably occuring weak links. As a consequence of the electronic anisotropy of the material, the c-axis orientation is necessary to secure high critical-current densities, J$_c$, in the film plane, and usable Cooper-pair transport across weak links in the film. Higher-angle grain boundaries are such weak links, and are still predominantly used, but in a controllable way - see Sections 3.2.1, 3.2.2. Since 1987, mostly as a consequence of YBCO thin-film development, the 1/f SQUID noise resulting from J$_c$ fluctuations and vortex hopping was reduced, such that the low-frequency energy resolution of SQUIDs was improved by 7 to 8 orders of magnitude. In this chapter, we thus concentrate on fabrication of thin-film SQUIDs, although bulk YBCO ceramic structures are still occasionally used, especially for flux-focusing structures.

Typical film and multilayer deposition methods used for SQUID fabrication are pulsed laser deposition (PLD) and high-pressure (usually magnetron) sputtering. Both of these methods produce smooth YBCO and dielectric films having good crystal perfection and, in the case of YBCO, requisite superconducting properties. Co-evaporation also is capable of producing films of at least comparable quality [1], but is much less used, due to difficult process control and the relatively large investment required. The description of all these methods is, unfortunately, outside the scope of this chapter. Relatively recent overviews are available [2,3,4]. The only significant new development gaining broad acceptance, but not covered in those overviews in detail, is the off-axis PLD for fabrication of large-area films free of micrometer-size surface particulates, the so-called "boulders" [5]. Although commercial high-quality YBCO films are available, their use for SQUID fabrication is not very practical, at least when the substrate requires patterning prior to film deposition, or more than one layer is needed.

The general qualitative requirements that YBCO films should meet, and general methods for their processing into electronic devices, are described in [6]. Typical properties of YBCO films on commercially available, epitaxially-polished single-crystal substrates ($SrTiO_3$, $LaAlO_3$, $NdGaO_3$ of (001) plane orientation) usable for SQUID fabrication, are summarized in Table 1. These can be considered as guidelines for specifications. Further improvements are desirable and probably possible.

The ranges of typical parameters indicated in Table 1 encompass the spread in fabrication, the substrate off-axis misorientation (usually less than 0.5°) and the effect of the lattice parameter mismatch between the film and substrate. For example, on $NdGaO_3$, which has the lowest mismatch, $\Delta \omega$ is typically about 0.1° or even narrower, while it is up to twice that value on $SrTiO_3$ and up to 0.3° on $LaAlO_3$.

The quality of substrate epitaxial polishing is critical for attaining the requisite properties, and a result of suppliers' proprietary polishing and cleaning methods. In

TABLE 1. Physical and electric properties of epitaxial c-axis YBCO films 200 nm thick

Parameter	Symbol	Unit	Range
Surface roughness		nm	10 - 20
Half-width of X-ray rocking curve (005 line)	$\Delta \omega$	degrees (°)	0.1 - 0.3
Minimum channeling yield	χ_{min}	%	2 - 3
Resistivity at 300K	ρ	$\mu \Omega$cm	150 - 300
Critical temperature (offset, inductive measurement)	T_c	K	87 - 90
Transition width (inductive measurement)	ΔT_c	K	0.5 - 0.8
Crit. current density at 77K	J_c	MA/cm^2	2 - 5

SQUID fabrication, commercial substrates of adequate quality usually require only standard cleaning methods, which involve ultrasonic rinsing in, e.g., acetone, methanol and distilled water, and blow-off with dry nitrogen. MgO substrates can and are being used for experimental SQUID fabrication. However, we don't consider them standard because of their large lattice parameter mismatch with YBCO and limited surface stability. Standard commercial substrate size is currently 10 x 10 mm^2, with 0.5 or 1.0 mm thickness. The same standard is used for $SrTiO_3$ bicrystalline substrates - see Section 3. When ordering substrates, it is advisable to specify tight thickness tolerances, and, especially for SQUIDs with step-edge junctions (discussed in Section 3.2.2), precise edge orientation along the main crystallographic axes, i.e., [100]. Custom-made $SrTiO_3$

substrates and bicrystals with area up to 20 x 20 mm^2 or even 30 x 30 mm^2, are manufacturable. Large wafers of $LaAlO_3$ are commercially available with diameters up to 10 cm.

Other single crystal substrates can be matched better to YBCO by first depositing a suitable buffer layer. One example is a CeO_2 buffer on sapphire. On r-cut (1$\bar{1}$02) sapphire, a CeO_2 buffer grows with mixed (00$\bar{1}$) and (111) orientations, but fortunately well-oriented, very high-quality (001) YBCO grows on both orientations of CeO_2 [7]. The only problem is that large thermal expansion coefficient mismatch might limit the critical thickness of YBCO, above which cracks in YBCO may occur upon thermal cycling. While this issue is not completely settled, it appears that with a suitable thickness of CeO_2, crack-free YBCO can be deposited up to thicknesses of 200 to 300 nm, thus approaching the London penetration depth along the c-axis, $\lambda_L(77K) \cong 240$ nm, and sufficient for SQUID use. We dwell on this example, since with an increasing number of fabricated SQUID sensors, and with the implementation of large-area planar-gradiometer structures, large-wafer-scale fabrication will become necessary. The $LaAlO_3$ is not suitable for large-area patterning of fine structures, such as junctions. Upon thermal cycling in processing, micron-scale movements of twins occur, caused by a second-order crystallographic transformation (from rhombohedral to cubic) near 500°C [8]. Large wafers of a material free of crystallographic transformation below room temperature will be needed. The CeO_2-buffered sapphire might be a much more economical solution than, *e.g.*, the very expensive and still twin-prone $NdGaO_3$, which exhibits a crystallographic transformation near 950°C. Also, homoepitaxial layers, *e.g.*, $SrTiO_3$ on $SrTiO_3$, can be gainfully employed to improve surface quality and affect the properties of the subsequent YBCO layer.

2.1.2. Multilayers
The typical multilayer used for SQUID fabrication is a YBCO/insulator/YBCO trilayer, usually fabricated *ex-situ*, *i.e.*, not in one uninterrupted sequence of depositions in the same chamber. The insulator is typically one of the substrate materials listed in Section 2.1.1., and its thickness is of the same order as that of YBCO when used for multilayered flux transformers, planar gradiometers or multiloop SQUIDs. In trilayered heteroepitaxial junctions used in SQUIDs, the insulator is replaced by one of several possible barrier materials - see Section 3.3.4 - in the form of a very thin film, typically 10 to 50 nm. In any event, absence of pinholes and microshorts, as well as thickness uniformity are essential, but are requirements difficult to meet. The specific fabrication processes will be described in conjunction with devices. Here, we address only the two generic issues of (a) insulator integrity and (b) oxygen concentration in YBCO, which determines its superconducting properties.

The insulator integrity is compromised mainly by particulates, *e.g.*, the PLD-generated "boulders" on the surface of the lower YBCO layer, and generally, by surface roughness significantly exceeding the values of Table 1. Defects in the lower layer, which are caused by an *ex-situ* patterning process, propagate to the insulator and also may cause microshorts. Insulators growing in a two-dimensional (2D) mode, and well matched to the lattice parameter of YBCO, are more suitable than those that are mismatched, and growing in 3D "island" mode, such as the case for MgO.

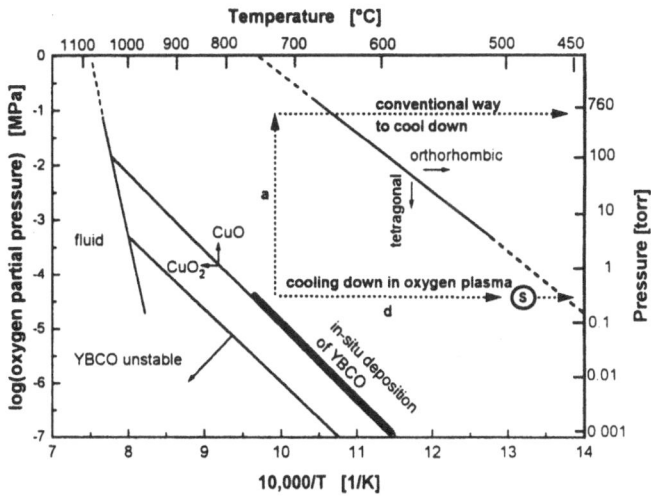

Figure 1. Oxygen pressure versus temperature phase diagram for YBCO phases. Cooling-down paths are indicated for both the conventional way of cooling down in O_2 after deposition and for low-pressure cooling in oxygen plasma [9].

The issue of oxygen concentration (*i.e*, the value of δ) in the lower YBCO layer is a direct consequence of insulator integrity. The diffusivity of oxygen in monocrystalline nonsuperconducting cuprates having an orthorhombic or tetragonal structure [such as $PrBa_2Cu_3O_7$ (PBCO)] is high, and sufficient to insure proper oxygenation of the lower YBCO layer, if PBCO or doped PBCO is used as an "insulator". Undoped PBCO is sometimes acceptable, in spite of its limited resistivity. In contrast, cubic perovskites, insulating and metallic, are more stable and usually exhibit very low oxygen diffusivities when free of crystalline defects. This is, for example, the case of monocrystalline $LaAlO_3$ and $NdGaO_3$, and even of $SrTiO_3$. Historically, once trilayers with these insulators attained reasonable crystal perfection, it became progressively more difficult to preserve adequate oxygen concentration in the lower YBCO layer of the trilayer. Consequently, additional oxygen and oxygen plasma-annealing processes became necessary. An effective method of *in situ* plasma annealing, after sputter deposition of the upper YBCO layer, was demonstrated by Ockenfuss *et al.* for trilayers with $NdGaO_3$ [9]. After deposition, the oxygen plasma is ignited by low-level rf power and maintained during cooling at the partial pressure of deposition, *i.e.*, at a few hundreds mtorr, as illustrated in Figure 1. During plasma processing, the surface of the trilayer is protected from any additional deposition by sufficiently increasing the distance from the target, and/or by inserting a shutter. To illustrate the effectiveness of this method, Figure 2 compares inductively-measured transitions of a lower YBCO layer at several processing stages: (a) as-deposited single layer cooled in 1 atmosphere of O_2, (b) after deposition of $NdGaO_3$ and cooling in 1 atmosphere of O_2, (c) after post-annealing at 600°C at 1 atmosphere of O_2, and (d) after the additional 500°C post-annealing in oxygen plasma ignited inside the sputtering system. One can see that the plasma treatment largely, but not fully, restored the original transition. The slightly higher transition (e) is that of another YBCO underlayer processed *in situ* as shown in Figure 1. Several days of annealing in equilibrium-pressure O_2 at the temperature of the tetragonal-to-orthorhombic transformation also can be effective, but may lead to excessive interdiffusion of cations.

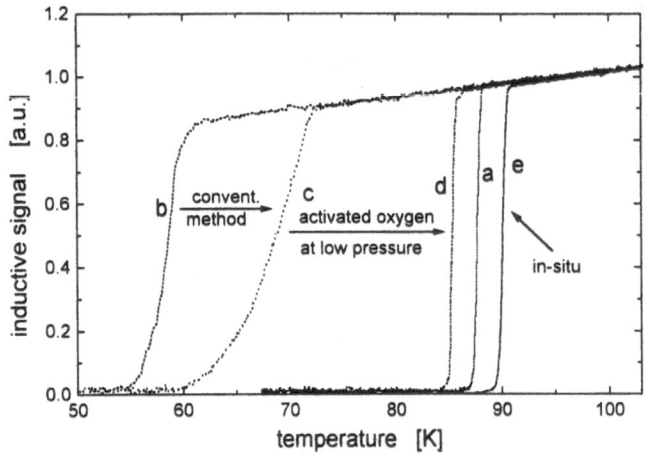

Figure 2. Comparison of inductively-measured transitions in YBCO for various oxygenation treatments: (a) as deposited single layer cooled down in 1 atmosphere of O_2, (b) after deposition of $NdGaO_3$ overlayer and cooling down in 1 atmosphere of O_2, (c) after post-annealing at 600°C at 1 atmosphere of O_2, and (d) after additional 500°C post-annealing in oxygen plasma ignited inside the sputtering system, (e) another YBCO underlayer processed *in situ* in oxygen plasma, following the lower path of Figure 1 [9].

After oxygenation, it is often observed that the upper YBCO layer exhibits superconducting properties inferior to those of the lower layer. This can be explained by less favorable conditions for the upper layer growth on the relatively rough insulator layer.

2.2. GENERAL PROCESSING METHODS

Processing methods generally used in fabricating thin-film HTS circuits, and largely applicable to SQUID fabrication, were reviewed in [6]. Methods of pattern transfer and etching were described, and their use illustrated by various approaches to HTS Josephson-junction fabrication. Three years later, most of that review is still up-to-date, so that there is no need to repeat it here in full. Specific fine points, which require an update, and pertinent novel developments will be addressed later, when discussing fabrication of SQUID components.

At this point, we should state only that the most standard method of pattern transfer for SQUID fabrication is that of optical photolithography with linewidth resolution down to approximately 1 micrometer. Generally available commercial photoresists and equipment can insure an adequate pattern quality. Photoresist masks for etching of YBCO and insulators are the most common, although metallic, ceramic and other masks also can be used for special purposes - see Section 3.2.2. Lift-off patterning methods require masks which can withstand the high temperatures of film deposition, but until now were not very popular despite their usefulness. Electron beam patterning by direct writing on PMMA masks is rarely used for SQUID processing, although significant progress in producing undamaged deep-submicron linewidth YBCO bridge structures was demonstrated in recent years [10,11].

The YBCO and insulator film etching method of choice in SQUID fabrication is that of ion beam etching (IBE) at low beam energies of 250 to 500 eV. It requires adequate cooling to prevent or at least minimize oxygen loss during etching. Chemical etching methods are used widely for patterning of coarse metallization layers, e.g., gold is commonly etched by a water-diluted (KI + I) etchant [12]. The modified "JPL etch" consisting of bromine in ethanol [13] is sometime used for anisotropic patterning of YBCO in conjunction with PMMA masks (discussed in Section 3.3.4), since that resist is not soluble in alcohol [14].

Other methods of YBCO circuit fabrication, such as the inhibit method [15,16], are sometimes used for submicron patterning, and show also some potential for patterning of multilayer structures.

3. Josephson Junctions

3.1. COMMON PROPERTIES

A resistively-shunted, nonhysteretic Josephson junction is the most essential component of any SQUID. Key to high-performance SQUID technology is a suitable junction technology. We thus devote a significant part of this chapter to HTS junction fabrication and properties. Of the many alternative HTS (mostly YBCO) thin film junction types known ([6] and references therein), most commonly used in SQUIDs are currently the bicrystal and step-edge junctions. Both are controlled grain-boundary devices. Also, some research groups and one manufacturer have been using SNS junctions, where N is a normal-conducting interlayer. The interlayer can be either a heteroepitaxially-grown perovskite or a noble metal. Alternately, a local region of the superconductor is normalized by ion implantation to create a weak link.

All such junctions, when properly fabricated, exhibit, at least at higher reduced temperatures, current-voltage (I-V) characteristics and dynamic behavior approximating the resistively-shunted-junction (RSJ) model. The shunt resistance is internal, since quasiparticle current flows readily across the weak link. Deviations from RSJ behavior are caused, *e.g.*, by inhomogemeity of critical parameters across the crossection. With widths (w) typically between 1 and 5 micrometers, critical currents (I_c) and junction resistances (R_n), are obtainable in a range suitable for SQUID operation at 77K, with the $I_c R_n$ product (critical voltage V_c) in the range of 100 to 500 μV and resistances between much less than one and several tens of ohms. Currently, Rn $\cong$ 1 Ω is most typical. This insures critical currents many times higher than the thermal fluctuation current, which at 77K is $I_f = (2e/h)k_B T \cong 0.042T \cong 3.2$ μA when T is in kelvins [17]. In addition to controlling I_c by geometry (junction cross section, interlayer thickness), one also can trim it and the $I_c R_n$ product by changing the oxygen concentration (δ) in the weak link. For a given junction, scaling of the critical voltage with J_c and ρ_n occurs, thus providing a useful guideline for the I_c and R_n correlation [18]:

$$V_c \propto (J_c)^{q/q+1} \text{ and } V_c \propto (1/\rho_n)^q . \qquad (3.2.1\text{-}1)$$

In grain-boundary junctions, q values between 1 and 1.5 are typical. The most effective trimming method is to perform plasma annealing at an elevated temperature of 500 to 600°C, and pressure in the 10 to 100 Pa range. Oxygen plasma increases δ up to near 7, while argon plasma reduces it.

The I_c (and R_n) fluctuations, which are one of two sources of major low-frequency 1/f-type noise originating in all HTS junctions, depend upon the junction type. A well-documented, objective comparison of 1/f noise magnitude (spectral energy density of voltage noise S_v) in the various types of junctions is not yet in hand. Rather, large variations from junction to junction can be observed for each type. To a large extent, these variations result from microstructural and current density inhomogeneities, and depend on fine, not entirely controllable, details of the fabrication.

An objective comparison of noise performance also has been difficult due to the measurement methodology. The 1/f noise usually has been measured in SQUIDs rather than in individual junctions, and a careful separation of fluctuation and flux hopping noise components by measuring dc SQUIDs in the flux-locked loop mode, both with and without current bias reversal, was, until very recently, rarely performed. Historically, SNS junctions with perovskite N interlayers exhibited much higher fluctuation noise than did the grain-boundary junctions. The early claims that step-edge junctions have by far the lowest 1/f noise were the consequence of their use mostly in rf SQUIDs, in which most of the fluctuation noise may be automatically eliminated when measuring in the flux-locked loop mode. More on the subject of fluctuation noise sources and their separation, including references, can be found in Chapter 1.

All of the known junction types exhibit an excessive variation of critical parameters (I_c and R_n) on chip, between chips and from one fabrication batch to another, in spite of occasional literature claims to the contrary. However, there is sufficient evidence that the best reproducibility and narrowest spreads, of typically $\pm$ 15 to $\pm$ 30 % on chip and, say, $\pm$ 50 to > $\pm$ 100 % between chips and batches, are readily obtainable in junctions on commercial $SrTiO_3$ bicrystal substrates. This is the main reason that bicrystal junctions also are favored by SQUID manufacturers. In step-edge junctions, I_c on-chip spreads of $\pm$ 50 to > $\pm$ 100 % are typical, although "best results" on chip can be as narrow as, *e.g.*, $\pm$ 7 % or 1σ of 5 %. An optimized process yielding $\pm$ 30% was reported recently [19]. Variations between wafers can be much larger. In most other junction types, variations of up to an order of magnitude, even on the same chip, have been common, although narrow spreads have been claimed. With these broad variations and limited control, it is fair to state that a reliable manufacturing technology of HTS junctions is not yet in hand. Bicrystal junctions are closer to that goal than any other type. Fortunately, I_c and R_n can be trimmed practically by adjusting δ via annealing, at least when the numbers of fabricated SQUIDs are not very high. Hence, the less than satisfactory state-of-the-art in junction technology is not yet a major barrier for SQUID fabrication. Below, we describe only those junction types which actually have been used in high-performance SQUIDs, beyond just a proof of functionality.

3.2. GRAIN-BOUNDARY JUNCTIONS

3.2.1 Bicrystal Junctions

Bicrystal junctions are the best characterized, due to the pioneering work of the IBM group [20,21,22] and many follow-up studies. An overview can be found in [18]. The substrate-defined junction predictability led to the sale of a commercial product by a bicrystal-substrate manufacturer [23]. Currently, most standard commercial bicrystal substrates are those of (001) $SrTiO_3$, 10x10 mm^2. They are manufactured with one tilt grain boundary, and preferably with one of two standard low-free-energy tilt angles of φ = 24° and 36.8°, either symmetrical or asymmetrical. The grain boundary is replicated in the epitaxial layer (YBCO or a buffer) deposited on top. In the YBCO film, the critical current density across the boundary, J_c, decreases roughly exponentially with increasing φ. In YBCO films 150 to 200 nm thick deposited on 24° substrates, typical J_c values at 77K are between 10^4 and 10^5 A/cm^2, and temperature-independent resistivities between $\rho_n = 10^{-8}$ and 10^{-9} ohmcm^2, depending upon preparation conditions. Typical 4-μm wide junctions have V_c = 100 to 200 μV, and are well suited for dc SQUIDs, with resistances of the order of 1 ohm, and critical currents $I_c \gg I_f$. The boundary separates two equal substrate areas and is positioned parallel to substrate edges, as shown in Figure 3. Of course, SQUIDs also have been made with other grain boundary angles and on noncommercial substrates using other materials, *e.g.*, the yttria-stabilized zirconia (YSZ) [24].

The nature of the weak link at the bicrystal grain boundary has been the subject of a long debate. The currently prevailing view, with strong supporting evidence, is that supercurrent transport occurs by resonant tunneling via localized states in very thin oxide-like barrier regions at the boundary [18]. Quasiparticle transport via localized states provides the internal shunt which is responsible for the RSJ-type behavior. The minute pair-current-carrying regions are separated by highly resistive and capacitive regions on a scale of nanometers [25,26,27].

244

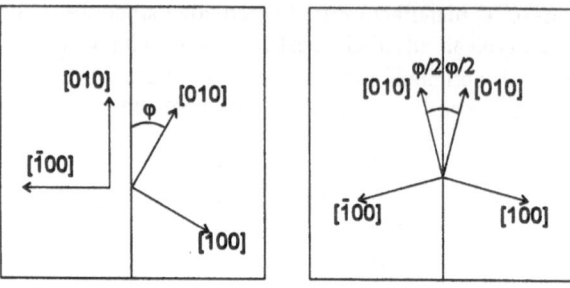

Figure 3. Schematic representation of symmetric and assymetric bicrystals.

Clearly, bicrystal junctions are the easiest to fabricate for SQUID purposes. The burden of weak-link definition is shifted to the substrate, while the topological limitation by the presence of a straight single boundary across the entire substrate is inconvenient (e.g., for the layout of a flux pickup structure a in single-layer SQUID), but not very severe. At a given φ, characteristics of a bicrystal junction depend on the quality of the polished and sintered grain-boundary interface, and on the surface finish. Also, it has been observed that in junctions with symmetrical boundaries I_c values can be up to an order of magnitude higher, and I_c spreads are a factor of 2 to 3 lower than for assymetrical ones. When depositing YBCO directly on standard quality commercial substrates of given φ and symmetry, one needs only to control I_c and R_n (principally) by chosing the film thickness (d) setting the junction width by patterning, and trimming the oxygen content δ by an appropriate annealing process. High-temperature trimming also may affect the grain boundary interface in the substrate. The method of deposition, which usually influences the kinetics of film growth, has no major effect on the characteristics. Yields of junctions operating at 77K are very high, approaching 100%. Some effect on I_c and R_n may be obtained by depositing an additional homo- or heteroepitaxial buffer layer on the substrate prior to YBCO deposition [19]. With a buffer on φ = 24° $SrTiO_3$ substrates, single-layer dc SQUIDs were reported with R_n up to 20 ohms at the typical V_c = 100 to 200 μV [28]. This very effectively reduced the SQUIDs white noise spectral density $S_\Phi \propto 1/R_n$, although I_c exceeded I_f by only a small factor of 2 to 3. Similar results were obtained simply by trimming δ without any buffer [29]. It is remarkable that with I_c values approaching I_f, record-low S_Φ's can be obtained. To conclude: bicrystal junctions are certainly recommended to any novice in the HTS SQUID fabrication trade, and especially to those who have limited technological experience and means, but are striving for quick, reliable and high-performance results.

3.2.2. Step-Edge Junctions

The step-edge junction (SEJ) represents the next step up in technological complexity (compared to the bicrystal junction), but brings topological freedom to the SQUID layout, and is inherently much less expensive in terms of substrate cost, but more so in labor investment. The idea and first demonstration are due to Simon *et al.* [30], and the understanding of the device grain-boundary structure and properties is based on the work by Jia *et al.* [31,32], Herrmann *et al.* [33,34], and many follow-up studies, dealing mostly with junctions fabricated from PLD films. Detailed microstructural investigations of such YBCO films grown on steep steps (angle with plane $\Theta > 45°$) in lattice-parameter-matched cubic single-crystal substrates ($SrTiO_3$, $LaAlO_3$, $NdGaO_3$) show that, at the two step edges, two nearly 90° tilt grain boundaries of (103)(103) type occur in YBCO by virtue of epitaxy. These boundaries, marked by arrows in Figure 4, act as two weak links in series. When the step is steep enough, their structure becomes dissimilar, as shown in the left part of Figure 4, and the lower boundary, now largely of (001)(010) symmetry, usually has a much lower J_c than the upper. Hence, very steep steps are desired for SQUID use, so that only the weaker link will define the SQUID characteristics, while the upper one will merely contribute some additional kinetic inductance, L_k.

The step height (h) should exceed the film thickness. With PLD films, typically $d/h = 2/3$ to 1. Sputtered films require a lower $d/h \cong 1/2$. Much lower d/h ratios can lead to film discontinuity over the step. Also, $d/h > 1$ is not suitable, since with increasing film thickness both weak links can be shorted by grain-boundary-free overgrowths.

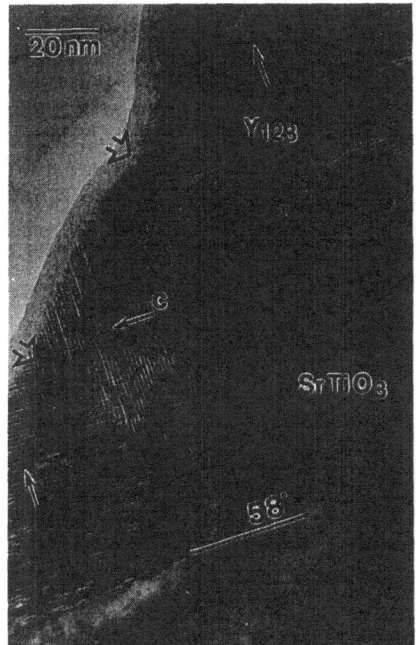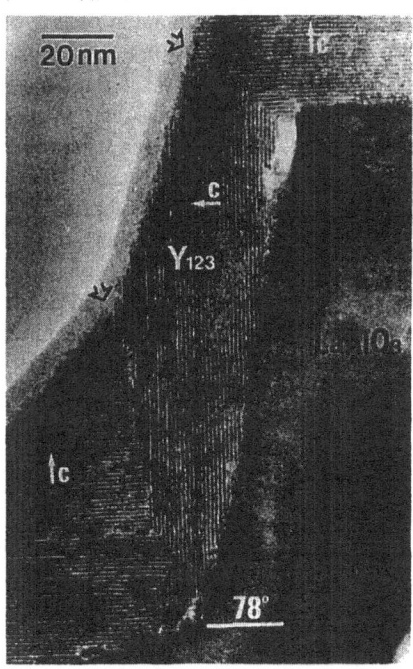

Figure 4. Microstructure of YBCO film grown by PLD on steep step in single crystal substrates. Left: $\Theta \cong 58°$ with similar grain boundaries ($SrTiO_3$ substrate). Right: $\Theta \cong 78°$ with dissimilar boundaries (LaAlO3 substrate) [31,32].

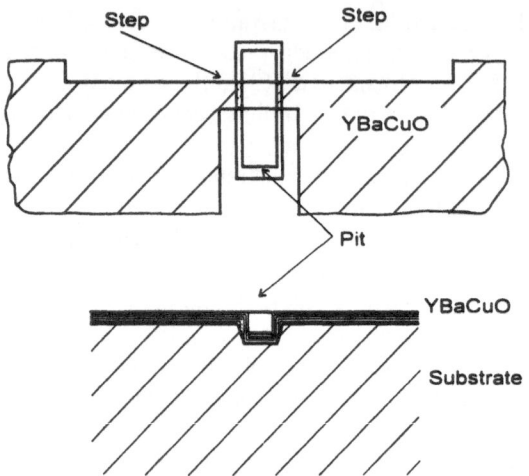

Figure 5. Schematic top view and section of a YBCO
microbridge crossing a local pit in the substrate.

The SEJ can be placed at any position on the substrate, so that there is little topological limitation, except that steps should be aligned along major cubic axes of the substrate, and, to obtain local steps only in the immediate vicinity of the junction bridge, a pocket or pit must be IB etched in the substrate. Figure 5 shows schematically the top view and section of a YBCO microbridge crossing such a pit. At first glance, this geometrical configuration is good for a dc SQUID. Unfortunately, even with wafer-holder rotation during step fabrication by IBE, the I_c and R_n of two junctions on opposite steps facing each other are usually quite different, due to the effect of plasma-plume orientation in PLD [35], while for a dc SQUID they should be identical. Hence, the two SQUID junctions must be positioned on the same edge, close to each other. In an rf SQUID, one only needs one junction. With a bridge across the pit, one cannot avoid fabricating two face-to-face SEJs with 4 weak links in series, but the difference in their characteristics comes at premium. The stronger SEJs only contribute additional kinetic inductance.

Critical parameters, J_c and ρ_n, of both weak links (averaged over the entire cross section) are much less-well defined than in bicrystal junctions, and can be only very roughly placed in the range of J_c(77K) between 10^4 and 10^5 A/cm^2, with ρ_n between 10^{-8} and 10^{-9} ohmcm^2, thus similar to the values for bicrystals. There is considerable evidence that analogous 90° tilt boundaries in planar films do not act as weak links, and that the weak-link behavior in SEJs is the consequence of high defect densities occurring at boundaries formed at step edges [34]. At such defect densities, transport across the boundary may occur by a mechanism similar to that for bicrystal junctions. Alternately, there also is evidence of extremely inhomogeneous distribution of J_c and ρ_n in the cross section, and of junction behavior similar to that for a point contact. Scaling of I_c and R_n with junctions' cross section is erratic at best, although perfectly obeyed for any single junction. For SQUID use, trimming of I_c and R_n by suitable annealing procedures is generally required to attain the desired value of the SQUID parameter, which is usually defined as $\beta_L = 2I_cL_s/\Phi_0$ for dc SQUID and $\beta_L = 2\pi I_cL_s/\Phi_0$ for rf SQUID, where, in both cases, L_s is the SQUID loop inductance and I_c is the critical current of a single junction. Similarly, as in bicrystal junctions, trimming can result in high junction resistances, and, when inductances of 50 to 100 pH are used, very low dc SQUID flux noise [29].

Since microstructural defects define junction characteristics, the latter must be extremely dependent upon processing parameters. The preparation of well-defined, microstructurally reproducible steps in the substrate is key to any process reproducibility. This requires extreme process discipline, with attention to minute details. It is characteristic that when the step fabrication process works well, high fabrication yields of 70% to 90%, and junctions with narrow spreads of I_c and R_n on chip (T = 77K), can be obtained. However, a slight change in uncontrolled variables can result in a zero yield for long periods of time.

The stability of unprotected SEJs over time, and especially over many temperature cycles, shows great variability from junction to junction, and also depends upon processing details. In the author's group, some rf SQUIDs have been reliably operating for four years, and over very many temperature cycles, as long as they were protected from moisture condensation after removal from liquid N_2. Other junctions deteriorated in times varying between many months to only a few weeks, or after only a few cycles. The faster deteriorating SQUIDs could in many cases be rejuvenated by oxygen plasma annealing, but these again experienced rapid deterioration. Exposure to moisture usually has been irreversibly fatal. The best stability has been obtained with epitaxial insulator encapsulation. Crack-free epoxy encapsulation also can be effective.

The standard method of step fabrication is to ion-beam etch through a lithographically patterned mask at normal beam incidence, and usually with wafer-holder rotation. Initially, metallic masks, especially of Nb had been used. Niobium could be reactively ion etched (RIE) using SF_6 or CF_4, and removed either by RIE or a chemical etch which does not attack the substrate. For Nb on $LaAlO_3$, an etch consisting of 1 part HF, 9 parts HNO_3 in 20 parts of H_2O is suitable [35]. To harden the mask and thus reduce the usually high metal milling rate, oxygen can be introduced during step IBE [36]. Simple optical resist masks also have been in use, but usually resulted in uneven erosion, with less steep and rounded steps, $\Theta = 50°$ to $60°$. Their use was motivated by simplicity. Also, a slight surface reaction of Nb with, *e.g.*, $SrTiO_3$ could be avoided. To illustrate recent progress, and because of importance of substrate-step fabrication also for another junction type, we describe here two improved step-fabrication procedures, both intended to improve the straight edge definition, *i.e.* to reduce waviness and prevent any chemical attack of the substrate surface. Reducing waviness is important to prevent a-axis grain nucleation at edges and to maximize uniformity.

(1) Diamond-like thin-film carbon masks (DLC) grown by plasma-enhanced chemical vapor deposition were introduced by Sun *et al.* [37]. The DLC ion-beam etch rate is very slow (1.5 nm/min at 500 eV, 0.3 mA/cm^2 of Ar), *i.e.*, the edge definition can be excellent, while it easily can be patterned and removed by oxygen RIE, which does not

attack the substrate. Mask patterning is done through a 100-nm-thick-Au or 50-nm-Al mask deposited on top of DLC. This mask is removed during IBE of the step. A sharp mask profile and a relatively smooth, well defined step are obtained, as shown in Figure 6a,b. Unfortunately, DLC films are not generally available.

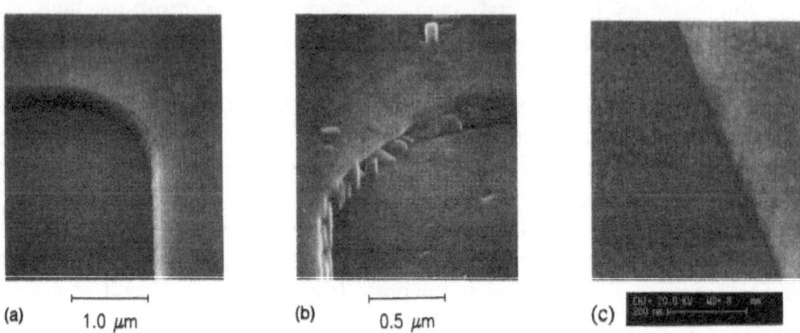

(a) 1.0 μm   (b) 0.5 μm   (c)

Figure 6. Scanning electron microscope (SEM) photographs of edges in substrates: (a) step flank in DLC, (b) step flank in YBCO, after IBE using a DLC mask [37], (© 1995 IEEE), (c) step fabricated using an Au + resist mask (note the 0.2 μm scale bar) [38].

(2) To be able to use a simple resist mask without uneven erosion, a very thin 25 to 30 nm layer of gold is deposited on YBCO prior to spinning on and patterning the resist [38]. The Au protects the substrate from any resist attack and eliminates optical interference during exposure, which occurs when the mask and substrate are transparent to exposing light. Gold covered by resist erodes little and more uniformly than the resist alone. The ion-beam etching is done without rotation, with the beam carefully aligned with the edge to be milled, and positioned at 45° to the normal of the film plane. After the completed IBE and resist removal, Au is etched away by the KI + I etch, which does not attack $SrTiO_3$ and $LaAlO_3$. This process produces smooth edges - see Figure 6c - and is currently widely used the in author's laboratory.

Steps can be fabricated not only on substrates, but also on insulator layers, as might be required in some integrated multilayer SQUID designs. Usually, the IB etch rates in these layers are much higher than in bulk single crystals [39]. This results in better edge definition.

Once a step is properly fabricated, the subsequent film deposition, processing and I_c trimming steps are nearly the same as for a bicrystal junction. At first glance, the definition of junction characteristics is also via the same parameters as for bicrystals. In reality, in addition to these parameters, the kinetics of film growth across the step has a strong influence on the grain-boundary microstructure and homogeneity. As already mentioned above, orientation of the plasma plume in PLD has an effect on I_c, R_n values and spreads [35]. By far the highest $I_c R_n$ and narrowest spreads were obtained when the plume axis was normal to the step face, rather than oriented along the step or normal to the film plane.

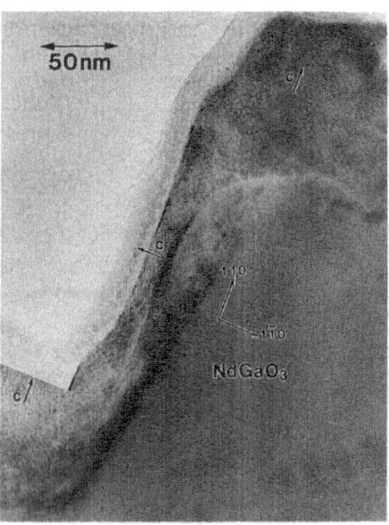

Figure 7. Microstructure of YBCO deposited on the step by low-rate sputtering [40].

As already stated, most experience in leading laboratories is with fast-growing PLD films far from equilibrium. However, low-rate, thermalized sputtering of films, leads to a quite different close-to-equilibrium microstructure of YBCO on steps.   An example of this is shown in Figure 7 for a YBCO film on a NdGaO₃ substrate [40]. The film across the step contains two (001)(010) grain boundaries and a thin flank. This results in higher junction resistance. Films 200-nm thick make possible an R_n of the order of 10 to 100 ohm, depending upon the junction's width. Such junctions are obviously of interest, but have not yet been tested in SQUIDs. Depending on the substrate type, growth conditions, film and step quality, grain boundaries also may form with such perfection that weak links cannot be obtained.

In poorly lattice-parameter-matched YBCO films on MgO steps, the film grows on the flank with c-axis normal to the flank plane, so that the angle of both grain boundaries is defined by the step angle [41]. Although interesting exploratory dc SQUID work was done using SEJs on MgO [41,42], we don't consider it representative enough to warrant a detailed discussion of such junctions.

In this Section, we have been dwelling at length on step fabrication details, but these details are crucial. With the "right" steps and use of PLD, the rest of the process is simple, and usable junctions are readily obtainable. The use of sputtered films also is possible, but more tricky. Indeed, we can draw the following conclusion. Laboratories and manufacturers with a specialized, rigorous step-fabrication process in hand have a good chance to fabricate state-of-the art SQUIDs with SEJs. All others better beware. In longer-term perspective, SEJ's probably will be replaced by an easier-to-control junction alternative, even if the corresponding fabrication process is more complicated.

3.2.3. Biepitaxial Junctions

These junctions, inspired by bicrystal junctions, contain nearly planar 45° tilt grain boundaries and are free of topological limitations, since the boundary is nucleated at a patterned edge of an extremely thin epitaxial template film with major crystal axes rotated by 45° with respect to those of the substrate [43,44]. In high-quality template films, other angles do not appear possible. The biepitaxial junctions were conceived and used for SQUIDs, even in the early manufacturing phase. However, at such a high tilt angle, the I_c(77K) is too low to attain a high yield and an optimum SQUID performance. Hence, these are of no current interest for SQUIDs operating at 77K.

3.3. SNS JUNCTIONS

3.3.1. Common Features

Physically, SNS devices are those where two superconducting electrodes or banks (S) are weakly coupled via a normal conducting interlayer (N), which is artificially fabricated. The concept and mechanism of an SNS device are quite well defined in low temperature superconductor (LTS) junctions, where the Cooper-pair transport across the N interlayer occurs via the proximity effect in a normal metal or semiconductor. Such internally-shunted devices exhibit nearly ideal RSJ characteristics, but, especially with normal-metal interlayers, have R_n's much too low to be of use in LTS SQUIDs. Hence, planar SIS tunnel junctions, where I is an insulator, are generally used with controllable external metallic shunts. This makes it possible to optimize white noise by chosing an appropriately high R_n value.

In the current state of HTS technology, various types of SNS junctions may or may not be proximity-effect coupled, and the N interlayer itself is not always the weakest link which defines the device properties. The only thing that these various types have in common is that a sufficiently thin nonsuperconducting interlayer is intentionally fabricated to link two superconducting electrodes. The characteristics are qualitatively always of the RSJ type, but in many cases do not show characteristic signatures of the proximity effect [45]. Within this general SNS category, a great variability of current-voltage and 1/f noise characteristics has been reported, depending strongly upon the junction type. Overall, the pool of reported data, especially on noise and I_c spreads, is less comprehensive and reliable than for the case of grain-boundary junctions.

3.3.2. Junctions with Implanted Weak Links

These are the only SNS devices currently suitable for SQUIDs, which are fabricated using single-layer technology. Another single-layer approach, where the normal weak link exhibiting RSJ behavior is fabricated by local irradiation of a microbridge with a focused electron beam (direct writing), does not yet show the necessary long-term and cycling stability, although quick progress in this direction is being made [46].

A practical implanted device pioneered by Tinchev [47] is currently used by one rf SQUID manufacturer [48,49]. A YBCO-film microbridge, 1 to 10 μm wide, is implanted locally with 100 keV oxygen ions (O^+). A fluence between 10^{13} and 10^{14} O^+ cm^{-2} reduces the T_c of YBCO by 15-20K, so that at 77K an SNS link is obtained if the cross section of the microbridge is uniformly implanted and the length of the normalized area, L_n, is not much greater than the normal coherence length, ξ_n. This appears possible, since the short ξ_n of normal YBCO diverges at temperatures barely exceeding T_c. To obtain a sufficiently short L_n, implantation occurs through a submicron (nominally 0.1 to 0.8-µm wide) slit positioned across the microbridge. The slit is e-beam patterned in a PMMA mask 0.8 to 1.0 µm thick, i.e., about 3 times thicker than the mean projected range of 100 keV oxygen ions. The mask slit width defines L_n. These junctions were developed and tested only in rf washer SQUIDs, hence no I-V characteristics, no spreads, and no junction 1/f noise data were reported in [47,49]. The manufacturer has been stressing the robust stability in time and upon temperature cycling, but without giving any specific data. Recently, a more systematic investigation of similar junctions implanted either with O+ or Ar+, and also focused Ga+ (implanted without a mask), has been conducted by Seidel et al. [50]. From preliminary published data, it appears that the 1/f noise in such junctions is higher than in grain boundary junctions. It is not yet clear whether this is due only to current fluctuations or also to vortex hopping in the still superconducting banks bordering the region normalized by implantation. Furthermore, there are suspicions that the industrially-fabricated junctions are not uniformly implanted, contain residual superconducting filaments, and exhibit flux-flow microbridge behavior.

The implanted devices appear attractive from the point of view of relatively easy manufacturability and the absence of any topological limitations. With a uniform large-area implantation beam, one can control precisely the ion fluence over the whole wafer, and thus the local T_c reduction in many SQUID junctions on that wafer. The only delicate processing step is the controllable e-beam lithography of deep-submicron slits in PMMA masks. Furthermore, measurement of resistance, or even I_c in situ, i.e., inside the implantation chamber, makes it possible during fabrication to easily and automatically trim I_c, and thus the β_L of a SQUID, to the desired value. This may prove to be an advantage for a more mature SQUID technology, at least for single-layer sensors. Developments in this direction can be anticipated.

3.3.3. Step-Edge Junctions with Noble Metal Interlayers
These junctions represent an LTS bilayer device concept directly transplanted to HTS by DiIorio et al. for use in dc SQUIDs [51,52]. Systematic studies and reliable data were reported by Ono and collaborators [53,54,55,56], and their results are representative of the state of the art. The principle of fabrication is easy to understand with reference to Figure 8, which shows schematically the essential steps of the process. Steep steps in the substrate or insulator overlayer [39] are fabricated as described in Section 3.2.2 (Figure 8a,b), with typical heights in the range of 100±50 nm. The YBCO film is deposited directionally across the step at an angle such that the step shadow results in a YBCO discontinuity on the step and exposure of free a-b plane film edges (Figure 8c). In the original work, off-axis magnetron sputtering was used [51], while Ono et al. used PLD, which makes directionality of deposition more reliable. In the next process step, a noble metal interlayer of Ag [51], Au [53-56], or Au-Ag alloy [53] is in situ (i.e., without

breaking the vacuum) deposited from the opposite direction to fill the gap and contact the exposed a-b planes of both banks (Figure 8d). The gap, L_n, is of the order of the relatively long normal coherence length in a noble metal, ξ_n, with $L_n/\xi_n > 1$, so that a Josephson current between electrodes is obtained. The metal also covers the film c-axis surface thus providing a quasiparticle shunt over the current transfer length. There is little supercurrent transport across that YBCO/Au c-axis interface. In the last fabrication step, the metal overlayer is patterned to control that shunt. Since I_c is almost unaffected, a complete removal of the shunt from the surface by a directional IBE (Figure 8e) leaves the metal only on the step, and results in highest attainable junction resistance and $I_c R_n$ product [55]. The use of steps, directional deposition and IBE introduce relatively mild topological limitations for a SQUID layout.

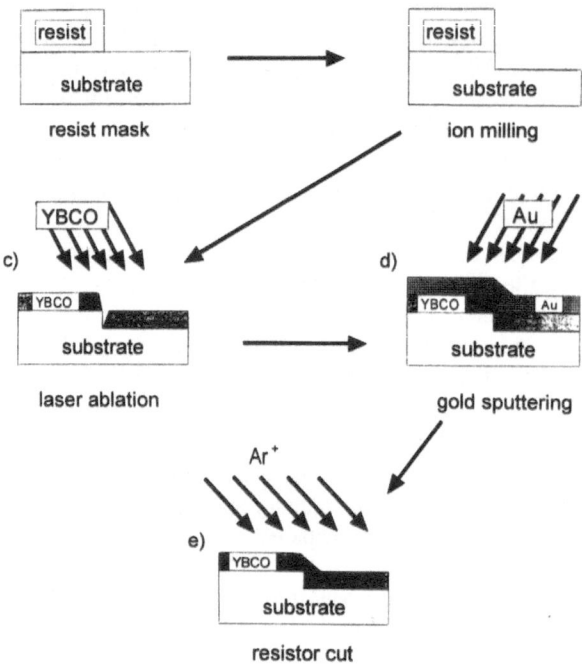

Figure 8. Schematic steps in fabrication on SNS step-edge junctions: (a) patterned resist mask on substrate, (b) IBE of step in the substrate, (c) YBCO film deposited across the step at an angle to exploit the shadow effect; (d) Au overlayer deposited at an opposite angle to fill the gap in YBCO and cover the film surface, (e) the overlayer removed by directed IBE at an angle to exploit the shadow effect.

The SNS SEJ's behavior qualitatatively conforms to the RSJ model, but often with considerable deviations, such as a large excess current in I-V characteristics. Also, its shunt-free, temperature-independent resistance exceeds, by 2 to 3 orders of magnitude, the resistance of the metal link, and the temperature dependence of I_c is nearly linear, as in grain-boundary junctions. Microscopic interpretations of junction nature vary, and include, *e.g.*, the SINIS [55] and SNcS models [38], where I is the S/N interface "insulator" causing the high R_n; and c is a point-contact-like constriction at one (lower) interface.

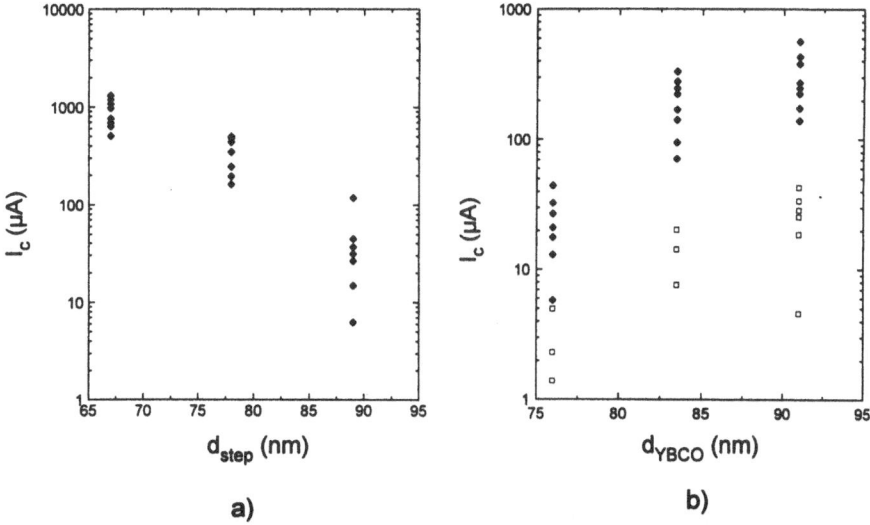

Figure 9. Dependence of I_c(4K) upon (a) step height (d_{step}), and (b) film thickness (d_{YBCO}) in simultanously fabricated SNS SEJs on one chip. In (b) open squares give I_c(70K) [56].

For SQUID use, the debate over the junction mechanism is academic, but the high and apparently controllable junction resistance is of definite interest, since $S_\Phi \propto 1/R_n$. For shunt-free junctions a couple micrometers wide, resistances between 10 and 100 Ω have been attained, with the 10-Ω level claimed to be reproducible and resulting in V_c(77K) approaching 1 mV [55,56]. The possibility of controlling I_c by varying L_n through step height (d_{step}) and film thickness (d_{YBCO}), illustrated in Figure 9 [56], also is attractive, especially if h > d, although the bilayer fabrication process steps, which rely upon directed deposition and ion milling, represents a significant step in increased technological complexity when compared to grain-boundary junctions. Such controls are needed, since, in contrast to grain-boundary junctions, it is not possible to trim δ in the weak link to control I_c over a wide range by special annealing procedures. Nevertheless, except for the original work [51,52] and its follow-up, SNS SEJs have not been used in serious SQUID development work or manufacturing. The main reason for this is the wide variability of properties on the chip, between chips and between fabrication batches.

Figure 9 gives an example of I_c spreads on chip, which approach an order of magnitude even at 4K. Average J_c values at 77K may span values between 0 and over 10^4 Acm^{-2}. Similarly, average resistivities are nominally of the order of 10^{-8} to 10^{-7} Ωcm^2, comparable to or higher than in grain-boundary junctions, but also with considerable variability. Relatively narrow spreads originally reported [51,52] have not been confirmed by others.

We refrain from a detailed analysis of this variability problem, and only point out that a major technological difficulty resides in reproducibly obtaining the YBCO discontinuity and a-b plane exposure at the lower edge. Existence of "parasitic" grain-boundary SEJs in parallel to the SNS channel cannot be excluded, especially in off-axis magnetron-sputtered films, in spite of successful "proofs of principle" showing no conduction in the absence of a metal interlayer in a number of such samples [51,38]. The low-rate, high-pressure sputtering process is not very directional, due to thermalization, and use of a shadow technique is risky. With PLD films, the process appears somewhat more reliable, but not to the point of acceptable spreads in I_c and R_n. Although many process improvement ideas have been tested, there is none thus far which really narrows the spreads or permits precise I_c tuning, as is necessary to insure the value $\beta_L \geq 1$ required for SQUIDs. For high-performance SQUID fabrication, SNS step-edge junctions will be an alternative of choice only if an effective I_c control method is found.

3.3.4. Edge Junctions with Epitaxial Oxide Interlayers

The edge or ramp junction represents an LTS trilayer concept which is quite useful for HTS cuprate film technology. It permits one to grow SNS and SIS epitaxial trilayers using c-axis films while retaining a Josephson current component linking the a-b planes of both YBCO electrodes. Planar trilayers of c-axis YBCO films can rely only on "atomic" steps in films to secure a Josephson current between a-b planes. These are in reality single- and multicell steps (with a height of $1.16n$ nm, $n = 1,2,3,...$), and the J_c is irreproducible, even if the thin interlayer is continuous, which is quite difficult to achieve. Planar trilayers of a,b-plane YBCO films have an orientation impractical for SQUIDs, are unstable, and extremely difficult to fabricate such that continuity of the thin interlayer is maintained. An epitaxial edge junction, with the edge at the bottom YBCO film exposing a-b plane terminations at both YBCO electrodes (as shown in Figure 10), is a convenient intermediate solution between these two extremes. Edge junctions with YBCO electrodes and a normal PrBa$_2$Cu$_3$O$_{7-\delta}$ (PBCO) interlayer, were pioneeered by Gao et al. [57,58]. Shortly thereafter, Hunt et al. fabricated edge junctions with a normal YBCO interlayer [59], to insure the best possible lattice parameter and thermal expansion matching in the trilayer, while Chin and Van Duzer used Nb-doped SrTiO$_3$ [60]. In succeeding years, edge junctions with several other normal interlayer types were investigated and also occasionally tested in dc SQUIDs. As investigated interlayer materials, we can list, for example, ruthenates CaRuO$_3$ [61], SrRuO$_3$ [62], and substituted (doped) YBCO: YBa$_2$Co$_x$Cu$_{3-x}$O$_{7-\delta}$ [63], Y$_{1-x}$Ca$_x$Ba$_2$Cu$_3$O$_{7-\delta}$ [64] and also Y$_{1-x}$Pr$_x$Ba$_2$Cu$_3$O$_{7-\delta}$ [65,66]. Recently, the PrBa$_2$Cu$_{3-x}$Ga$_x$O$_{7-\delta}$ interlayer has been investigated [67]. Many other edge-junction studies can be found in the literature.

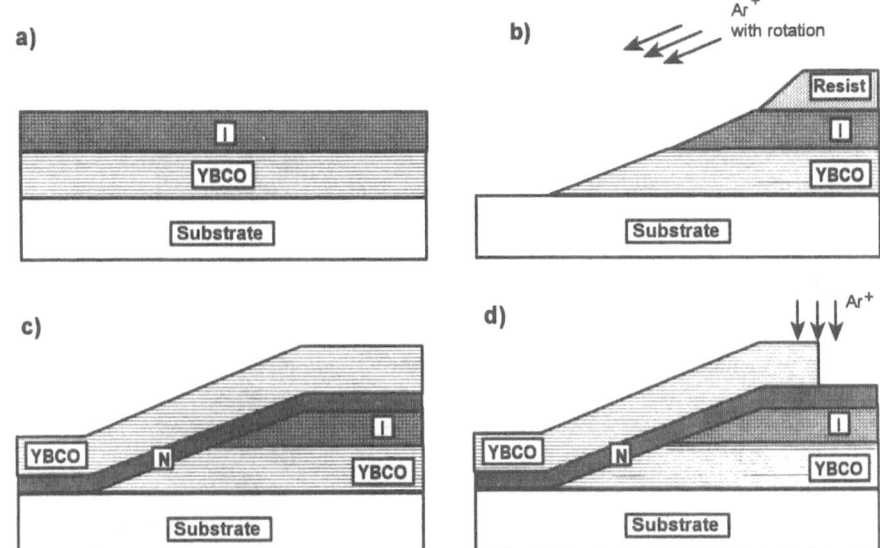

Figure 10. Edge junction fabrication steps: (a) YBCO/insulator bilayer is deposited, (b) a step is ion-milled *ex-situ* at a shallow angle (or etched), (c) barrier and counterelectrode layers are deposited in-situ, (d) counterelectrode is patterned.

Resistance of junctions with ruthenate interlayers is clearly dominated by interfacial resistance, and resonant tunneling via localized states appears to be the transport mechanism for both quasiparticles and Cooper pairs [68]. In contrast, junctions with substituted YBCO interlayers exhibit no significant interfacial resistance and behave as proximity-effect devices [45]. Their resistances are too low to be of interest for low-noise SQUID sensors. The behavior of junctions with PBCO interlayers depends upon the fabrication process. Those with damaged interface(s) show characteristics suggesting resonant-tunneling transport [69]. Generally, resonant tunneling through localized states introduces a high level of $1/f$ current-fluctuation noise. However, in dc SQUIDs this can be suppressed by bias reversal techniques. Current fluctuations should be of little consequence for rf SQUIDs.

Edge-junction fabrication steps are illustrated in Figure 10. To start (step a), an SI bilayer is deposited, where S is the YBCO base electrode layer, and I is a relatively thick and pinhole-free insulator film, e.g., $NdGaO_3$. Sometimes a relatively thick PBCO layer is used as a high-resistance "insulator" to insure easy oxygen diffusion to the bottom electrode at the cost of some external shunt resistance which reduces V_c. The bilayer edge and the bottom electrode layout are usually fabricated by IBE at an angle to the film normal, or by anisotropic wet etching in a bromine solution in ethanol [14] (step b). The angle of the edge should not exceed about 30° to prevent any grain-boundary formation in the counter electrode.

Ion milling causes edge (flank) surface damage, while bromine-ethanol leaves the surface relatively undamaged. Annealing in oxygen or oxygen plasma is routinely performed prior to the barrier deposition in order to heal the IBE damage. However, recent work showed that milling at a standard energy of 500 eV causes cation disorder in a few-nanometer-thick surface layer, which cannot be fully healed by high-temperature (800°C) oxygen annealing [70]. Much lower energies may cause less residual damage. Milling also should be optimized to obtain a smooth and uniform step profile - see Section 3.2.2. The IBE at an angle to the normal and with holder rotation introduces no topological limitation for a SQUID layout. Anisotropic wet etching also is free of any such limitation.

As an alternative to etching, when depositing the bilayer through a shadow mask, a very shallow damage-free edge with exposed a-b plane terminations and bottom electrode layout can be obtained in a completely *in situ* process [71, 66]. This once required mechanical shadow masks, unsuitable for fine linewidth resolution, and ingenious mechanical implements rather inconvenient for quantity fabrication. However, a new photolithographic microshadow mask technique can remove these limitations [72].

Once the base electrode is patterned by one of the alternate methods, the epitaxial thin interlayer and top electrode are sequentially deposited *in situ* (step c) and patterned (step d). Finally, vias to the base electrode are opened by IBE of the insulator - see Section 5.1.2 - and covered with YBCO interconnect or (for test purposes) metallization. The whole process of junction fabrication is much more complicated than those described in previous Sections. If the interlayer and insulator are relatively impervious to oxygen, as, *e.g.*, in the case of ruthenates and $NdGaO_3$, one of the main challenges is to insure the oxygenation of the base electrode (discussed in Section 2.1.2). The "sanity test" for edge quality is to fabricate junctions without an interlayer. Their J_c should approach that of electrodes. However, this requirement, usually met when using any of the base electrode processing alternatives, does not guarantee the intrinsic (electronic) quality of the S/N interface, *e.g.*, due to disorder and ion interdiffusion between different compound materials.

At a given edge junction cross section (wd) and edge angle, the I_c can be controlled by the interlayer thickness, d_n. The L_n is proportional to, but not necessarily identical with, d_n, since for shallow edge angles, L_n connecting a-b planes might be $>> d_n$. The edge angle affects I_c, but is not practical as a control parameter. Resistance can be controlled by d_n only if the interface resistance is not dominant. The range of d_n depends upon the interlayer material., and is typically 10 to 50 nm for PBCO. In the literature, the PBCO's ξ_n value determined from exponential $J_c(d_n)$ dependences varies greatly [6]. We quote the lowest values of $\xi_n = 1.5$ and 4 nm, giving $L_n/\xi_n >> 1$ [67,73].

Thus far, of dc SQUIDs fabricated with heteroepitaxial edge junctions, the highest performance devices probably have been those using quasiplanar junctions with PBCO barriers, where edges were etched in Br-ethanol [73]. In these devices, R_n scales with d_n and has a metallic temperature dependence. Since the interface resistance is negligible, good reproducibility of I_c and R_n can be expected. Indeed, I_c spreads of $\pm$

5% on chip have been claimed, but without sufficient statistical data. The I-V characteristics are qualitatively of the RSJ type. At a barrier thickness of 20 nm, an I_cR_n(77K) of 200 μV is typical, with R_n of the order of 1 ohm for a 5-μm junction width. Somewhat higher resistances can be obtained readily by reducing the junction cross section. A further R_n increase without I_c reduction appears possible through Ga doping [67]. The current fluctuation noise is high, but in dc SQUIDs with bias reversal excellent noise performance was obtained [74]. High stability of junction properties over time and upon temperature cycling has been claimed. It should certainly be superior to that of weak links exploiting defect structures, such as in grain-boundary junctions.

Until now, we have concentrated only on junctions with normal-conducting interlayers. However, YBCO edge junctions with nominally insulating barriers also were investigated. The RSJ-like properties of many of these could be explained by the presence of pinhole or microshort arrays in the barrier. In some cases, claims of intrinsic conduction through the barrier also were made. Recent representative examples of junctions with nominally insulating barriers which were tested in SQUIDs with acceptable noise results (when using bias reversal), include those with $SrTiO_3$ and $NdGaO_3$ barriers [75,76]. It remains to be seen whether such junctions have a good chances to be really reproducible, and whether they represent a base for a realistic SQUID technology. For now, one must remain rather skeptical.

As already observed above, the technology of edge junctions is complicated, and the possibility for I_c and R_n trimming is next to none, cuprate barriers excepted. Consequently, only very precise interlayer or barrier control might lead to reproducible fabrication of high-performance SQUIDs. Statistical evidence that this is possible is sorely needed. An argument in favor of edge junctions for use in SQUIDs without shielding is that the electrodes screen well the weak links from external perpendicular fields, B_z. This results in a weaker dependence of I_c upon the applied magnetic field. In multilayered integrated SQUIDs, which are destined to be the technology of the future, the additional required technological complexity of edge-junction fabrication may not be very significant.

4. Single-Layer SQUID and Flux Pickup Structures for Magnetometers

4.1. WASHER AND DIRECT-COUPLED SQUIDS

4.1.1. Effective Area and Magnetic field Resolution

All viable HTS thin-film SQUID designs borrow from concepts already proven in LTS technology. The planar-washer SQUID structure with a slit, first introduced by Ketchen and Jaycox [77,78] concentrates the flux into the SQUID loop and is the simplest, single-layer magnetometer structure directly suitable for HTS devices. Two variants of a square-washer structure currently used are shown in Figure 11, where (a) is a variant where junctions are positioned on the periphery of the washer, and (b) where junctions are in the center. In case (a), the slit parasitic inductance, L_{sl} adds to that of the hole.

This is acceptable for small dc SQUID washers. To minimize L_{sl}, an SIS overlap slit has been used in LTS dc SQUIDs [77]. This is not yet done easily in HTS technology.

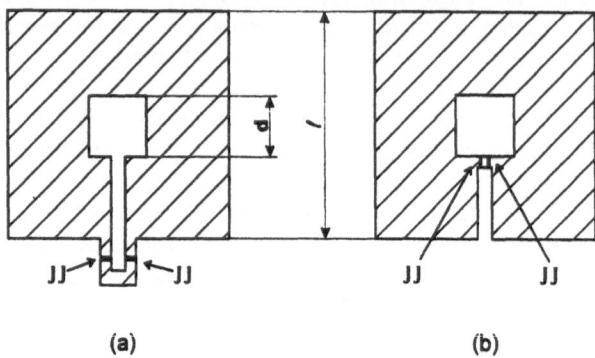

Figure 11. A square washer SQUID layout (not to scale): (a) with junctions near the outer edge, (b) with junctions near the center.

Ketchen *et al.* calculated the effective pickup area of a circular and square washer: $A_{eff} = \Phi/B_a$, where B_a is the applied magnetic field intensity, and showed that, at a fixed SQUID hole, it scales linearly with the washer's outer dimension [79]. Obviously, the price of simplicity is reduced efficiency, since A_{eff} is not proportional to the geometrical area of the structure. Josephson junctions bridging the slit were neglected in the calculation. For a circular washer

$$A_{eff} = \frac{8}{\pi^2} A_h(r_0/r_i) \qquad (4.1.1-1)$$

where r_0 is the outer radius, r_i is the radius of the SQUID loop (hole), and $A_h = \pi r_i^2$ is the SQUID hole area. For a square washer and hole shown in Figure 11

$$A_{eff} = aA_h(l/d) = adl \qquad (4.1.1-2)$$

where l is the outer edge of the washer, d is the inner edge of the loop (hole), $A_h = d^2$, and a is of the order of unity. For thin-film niobium square washers, experiments confirmed the linear dependence with good accuracy, and $a \cong 1.1$ [79]. For rectangular washers and holes, with an aspect ratio not deviating too much from 1, the square root of the washer or hole area can be substituted in Eq. (2). The original Ketchen washer was not meant to serve as a magnetometer, but as a groundplane implement to aid in coupling flux from the planar stripline input coil of a multilayered LTS thin-film flux transformer to the SQUID hole.

An alternate single-layer SQUID layout of equivalent function is the direct-coupled SQUID, also called an inductively-shunted SQUID. It was first introduced for HTS SQUID by Matsuda *et al.* [80]. The large superconducting washer is replaced by a large flux pickup loop of inductance L_p and geometric area A_p, analogous to that used in thin-film LTS flux transformers. The coil is connected to a very small washer SQUID, such that the pickup and SQUID loop inductances are connected in parallel, and the signal current induced in the pickup is directly injected in the vicinity of the SQUID junction(s). Figure 12 shows equivalent circuits of dc and rf SQUID. An example of a reasonably optimized direct-coupled dc SQUID layout is shown later in Figure 17a,b. The effective area of such a structure is [81]

$$A_{eff} \cong A_s + \alpha_d A_p (L_s/L_p) \qquad (4.1.1-3)$$

where A_s is the effective area of the small washer and $\alpha_d < 1$ is the coupling coefficient given by the fraction of L_s to which the signal supercurrent is coupled. Since $L_s \ll L_p$, with L_p of the order of 10 to 20 nH on a 1-cm^2 chip, this layout cannot be more effective than that for the large washer, even if α_d is made to approach 1. The only way to reduce the tremendous mismatch (of two to three orders of magnitude) between the large pickup and small SQUID inductance, is to reduce L_p. To a small extent (factor $\leq$ 2), this can be done by reducing the inner perimeter of the pickup loop [28]. A wide pickup loop also is preferable to prevent confinement of the screening supercurrent (induced in the loop) to a narrow path between edges possibly damaged by processing and contributing extra flux-hopping noise. Obviously, screening currents should not exceed the I_c defined by the loop's cross section (I_c limitation by a weak link may be purposely introduced to limit the self-field of the loop - see Section 4.1.4). Effectively, the layout of a very wide loop approaches that of a large washer. In both cases (of the pickup loop and the washer) the effective area is determined by the screening supercurrent induced near the outer perimeter of the loop or washer. Nevertheless, we shall show below that in higher ambient fields the direct coupling has some advantages over the washer.

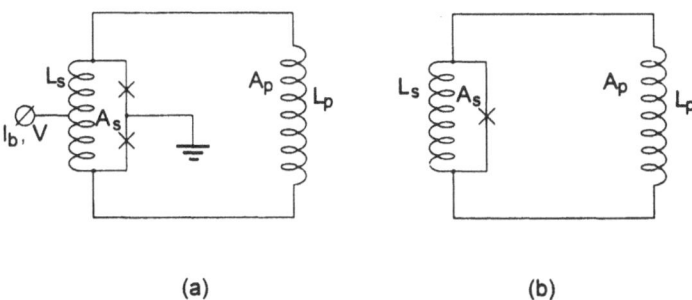

(a)   (b)

Figure 12. Equivalent circuits of directly-coupled dc (a) and rf (b) SQUID. A layout of a dc SQUID, and a detail showing connection of the pickup loop to a small washer with slit hole and bicrystal junctions is shown in Figure 17a,b [81].

4.1.2. Field-to-Flux Transformation, Magnetic Field Resolution, SQUID Inductance

For future use, we note here that the field-to-flux transformation coefficient $\partial B / \partial \Phi$ in units of T/Φ_0, where $\Phi_0 = 2.0679 \times 10^{-15}$ Wb is the flux quantum, represents the customary figure of merit for a magnetometer structure, with $\partial B / \partial \Phi \equiv 1/A_{eff}$. For consistency, when comparing the performance of various structures, we shall refer to $\partial B / \partial \Phi$ values obtainable for 1 cm^2, i.e., the area of a standard HTS substrate chip. With an effective layout covering most of that area, a single-layer washer, or directly coupled structure discussed below, typically has a $\partial B / \partial \Phi$ between 2 and 30 nT/Φ_0, depending on the SQUID hole size, i.e., on its inductance L_s. With the spectral energy density of flux noise, S_Φ in units of Φ_0/Hz, the magnetic field resolution of a magnetometer is

$$B_N \ [T/Hz^{1/2}] = (\partial B / \partial \Phi) S_\Phi^{1/2} \qquad (4.1.2\text{-}1)$$

We note also that the geometrical inductance of a square SQUID hole is conveniently estimated from [77]

$$L_s \ [pH] \cong 1.25 \mu_0 A_h^{1/2} \cong 1.57 A_h^{1/2} \ [\mu m] \qquad (4.1.2\text{-}2)$$

This expression is an asymptotic result of a three-dimensional calculation, and it is valid to a good accuracy when $(l - d)/d > 2$, i.e., for a loop (washer) wide in comparison to the hole size. The expression does not include the parasitic inductance of the junction(s).

For more arbitrary hole shapes, acccurate inductance values, accounting for both geometrical and kinetic components of L_s, require a sophisticated three-dimensional numerical computation [82]. Inductance of narrow slits and slit holes also can be estimated from formulas for coplanar lines [83]. In this case, kinetic inductance must be estimated separately. For narrow slits, the kinetic inductance may be relatively high and approaching the geometrical inductance.

We should mention that for sensitive HTS dc SQUID magnetometers, typical values of L_s range from 40 to 100 pH, and in rf SQUIDs up to $\geq$ 300 pH. It is easy to justify that difference when considering the dependence of SQUID's peak-to-peak voltage modulation, V_{pp}, and transfer function, $\partial V / \partial \Phi$ upon L_s in the absence of thermal (white) noise. For a dc SQUID $V_{pp} \propto 1/L_s$ while for rf SQUID $V_{pp} \propto 1/L_s^{1/2}$. However, neglecting noise is not realistic at 77K. As shown by Enpuku *et al.* [84] small inductances in HTS dc SQUID are even more essential for obtaining sufficiently high values $\partial V / \partial \Phi$, which lead to low flux noise: $S_\Phi = S_V/(\partial V / \partial \Phi)^2$, where S_V is the rms voltage noise. More on this subject can be found in Chapter 1. For rf SQUIDs, no calculations or simulations accounting for a significant thermal noise are available thus far. Experiments show less steep white noise dependence on L_s up to $\geq$ 300 pH [85] probably due to still significant lumped noise contribution of the tank circuit, transmission cable to room-temperature read-out electronics and the preamplifier. This contribution is less L_s-dependent than the intrinsic SQUID noise [86]. Hence, larger L_s values are acceptable. Generally, for the same energy resolution at an optimum β_L and

I_c/I_f ratio, HTS SQUID inductance, whether dc or rf, should be 20 times smaller than for LTS SQUIDs. Such small inductances would make coupling of the flux extremely inefficient, so that one accepts compromise ratios of I_c/I_f. Experience and simulations [87] show that ratios of 2 to 3 are fully acceptable.

4.1.3. Experimental Results

All results quoted in this Section were obtained at 77K. The first sensitive HTS square-washer magnetometers were introduced by Zhang *et al.* [88, 85] as rf SQUIDs based on step-edge grain boundary junctions, with very large washer and large hole inductances. The layout of Figure 11b was used since the consequence of large washer size is a prohibitively large slit inductance. Junctions - one of them active - were fabricated at the center hole, by etching a small, about 5-μm wide pit in the $SrTiO_3$ or $LaAlO_3$ substrate, and patterning a 10-μm long microbridge across it, as shown in Figure 5. This made it possible to avoid the contribution of slit inductance to L_s. For such a large washer structure, L_{sl} would be dominant, and much too high for SQUID operation. An additional consideration was that pulsed-laser-deposited YBCO film properties were found to be better near the center than at the washer edges. A local pit, rather than a step across the whole substrate, was used to prevent additional 1/f flux noise which was found experimentally when a long step crossed the whole washer body and resulted in an extended weak link. This noise was attributed mainly to current fluctuations in that link.

A 8x8 mm^2 outer-edge washer with a 10-μm-wide slit and a 150x150 μm^2 hole corresponding to $L_s \cong 230$ nH, had a measured $\partial B / \partial \Phi \cong 2.4$ nT/Φ_0 ($A_{eff} \cong 0.86$ mm^2) vs a calculated value of 1.4 nT/Φ_0. The discrepancy could, at least in part, be attributed to flux leakage through the long slit. The relatively large hole facilitated efficient flux coupling to the SQUID, at a cost of higher flux noise. White noise $S_\Phi^{1/2}$ $\cong 70$ $\mu\Phi_0/Hz^{1/2}$ down to 1-Hz signal frequency was measured with multilayered magnetic shielding and closed flux-locking loop giving $B_N \leq 170$ $fT/Hz^{1/2}$. The rf tank circuit frequency was 150 MHz, $\beta_L \cong 1$, but the inductive coupling to the tank circuit was subcritical, with $k^2Q \cong 0.5$, where k is the coupling coefficient, and Q is the quality factor of the tank coil. With additional improvements in the tank circuit, a similar washer with A_h = 200x200 mm^2 ($L_s \cong 315$ pH) had an experimental $\partial B / \partial \Phi \cong 1.5$ nT/Φ_0. White noise of $S_\Phi^{1/2} \cong 80$ $\mu\Phi_0/Hz^{1/2}$ at 77K resulted in an improved B_N = 120 $fT/Hz^{1/2}$ above 0.5 Hz [89]. Thus far, this has been the lowest recorded value for a 8x8 mm^2 large-washer rf SQUID with a conventional wire-wound tank-circuit coil, which invariably resulted in $k^2Q < 1$. Many such first-generation rf SQUIDs, sometime with additional flux concentrators (discussed in Section 4.2), have been installed in multiple small SQUID systems used by author's and collaborating groups to develop various HTS SQUID applications [90]. The B_N value given above can be considered as typical rather than best data, since individual β_L trimming was used.

The washer area of 8x8 mm^2 represents the near-maximum on a 1 cm^2 substrate chip when film deposition and photolithographic patterning are made on a single-chip rather than a large wafer diced after the complete fabrication. Film quality at the substrate edges is usually degraded, and spinning of photoresist on a single-chip results in its

262

accumulation near edges. The latter requires removing the edge-accumulated resist by an additional exposure through a 8x8 or 9x9 mm^2 mask, thus reducing somewhat the usable substrate area.

Tanaka *et al.* developed a large 11x11 mm^2 washer dc SQUID [91] with step-edge junctions. This SQUID has been used in the first 16 channel HTS SQUID demonstration system [92]. The step extended over the entire substrate, with slits and bridging junctions on opposite sides of the 25x25 mm^2 hole, as shown in Figure 13a. The flux leakage through both slits was reduced by sandwiching the washer with a flip-chip YBCO flux shielding plate, 11x11 mm^2, having a 50x50 mm^2 hole and a slit positioned at 90° to the thus shielded slit in the washer, as shown in Figure 13b. Without the plate, $\partial B / \partial \Phi \cong 59$ nT/Φ_0 (Aeff $\cong 0.035$ mm^2, 13% of calculated value), while with the plate $\partial B / \partial \Phi \cong 32$ nT/Φ_0. The small hole caused this very weak flux coupling. However, the correspondingly low L_s resulted in a white noise of only 10 $\mu\Phi_0$/Hz$^{1/2}$, so that open-loop $B_N \cong 300$ fT/Hz$^{1/2}$. At 1 Hz, a $B_N \cong 1.4$ fT/Hz$^{1/2}$ was obtained without bias reversal.

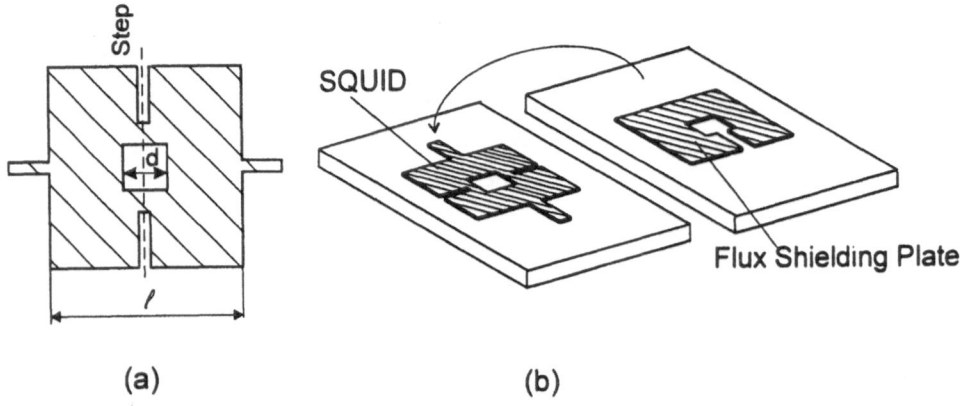

(a)   (b)

Figure 13. Example of a large-washer dc SQUID layout: (a) SQUID with two step-edge junctions; the step is extending across the whole substrate, but is placed inside of slits, (b) stereographic projection of the washer of (a) and a flux shielding plate with slit at right angle to washer's slits (© 1996 Jpn. Soc. Appl. Phys) [91].

Similarly, direct-coupled designs tested with dc SQUIDs had usually a very low L_s, hence a high $\partial B / \partial \Phi$ [81,93]. The especially successful design of [93], which is shown below in Figure 17a,b, had 9.3x9.3 mm^2 outer dimensions of the wide pickup loop, $L_s = 55$ pH, $L_p = 4.7$ nH, and a measured $\partial B / \partial \Phi \cong 10$ nT/Φ_0 (vs calculated 8.3 nT/Φ_0 with $\alpha_d \cong 0.76$ from experiment). Very low white noise of 3.5 $\mu\Phi_0$/Hz$^{1/2}$, and a noise of 6.5 $\mu\Phi_0$/Hz$^{1/2}$ at 1Hz were obtained. These noise figures resulted in a white $B_N = 35$ fT/Hz$^{1/2}$, and a resolution of 65 fT/Hz$^{1/2}$ at 1 Hz. Of course, also in these measurements the SQUID was characterized with multilayer magnetic shielding and bias reversal was used. The impressively low noise was obtained using bicrystal junctions with a very high resistance: $R_n \cong 14$ ohm, and the resulting $\beta_L \cong 0.5$. Most recently, confirming data were obtained for a similar design with step-edge junctions having R_n up to 10 ohm and SQUID $\beta_L \cong 0.5$, where white noise of $2.5~\mu\Phi_0/Hz^{1/2}$ and a resolution of $B_N \cong 20~fT/Hz^{1/2}$ were measured [29]. The same design as in [93], but with a larger pickup area of about 20x20 mm^2, and 20 ohm junctions, attained the present world record for single-layer HTS SQUID magnetic field resolution with $B_N \cong 10$ $fT/Hz^{1/2}$ in white noise frequency range and 26 $fT/Hz^{1/2}$ at 1 Hz, thus approaching the performance of average LTS magnetometers [28]. The authors of [93, 28] presented a statistical histogram of field noise (B_N) for first 50 standard size (10x10mm^2) chips that were fabricated [94]: neglecting two "risers", all 48 SQUIDs had a B_N between 50 and 300 $fT/Hz^{1/2}$, with about 30% below 100 $fT/Hz^{1/2}$. This spread included the learning curve, and represented a truly remarkable performance.

4.1.4. Effects of Stronger Ambient Magnetic Fields

In this Section, we consider the effect of static ambient and rf fields, and concentrate on two issues which are important for shielding-free operation: (1) field-exposure-induced 1/f noise and hysteresis, (2) field dependence of I_c. Additional 1/f noise caused by thermally-excited flux hopping is due to vortices trapped in the superconductor cooled in the Earth and ambient fields, especially when the SQUID's position is changed afterwards. The 1/f noise increase is also observed in a cold SQUID cooled in near-zero field and then exposed to these fields, e.g., by removing the shielding. Systematic measurements of 1/f noise vs static field B_0 applied inside the magnetic shielding to YBCO films and bicrystal junction SQUIDs have consistently shown a quasilinear increase of $S_\Phi(1~Hz)$ with B_0 [95]. In the Earth's field alone, an increase of up to an order of magnitude can be expected. Also in rf SQUIDs with step-edge junctions, an increase of $S_\Phi(1~Hz)$ by up to an order of magnitude is typical when cooling in the earth's field without magnetic shielding, and not altering the device's position afterwards. For example, the large washer rf SQUID with flip-chip flux focuser has a B_N of ≥ 40 $fT/Hz^{1/2}$ at 1 Hz, when measured in multilayered shielding [88, 89]. However, analogous flip-chip SQUIDs used in a first-order gradiometer cooled and characterized without shielding had a B_N at the same frequency typically of about 400 to 500 $fT/Hz^{1/2}$ [96,97]. The excess 1/f noise could be attributed to hopping of vortices trapped in both the SQUID and the focuser. Application of Earth's field to a zero-field-cooled SQUID can cause a still higher increase in 1-Hz noise of between one and two orders of magnitude.

The natural way to fight static field effects is to use near-perfect quality YBCO films thicker than λ_L, with the highest possible concentration of flux pinning centers. This is especially difficult when the films are patterned by etching, which almost invariably produces edge defects facilitating flux penetration. Also layout design can make a difference, since narrow lines and sharp corners, e.g., of the square hole, concentrate the field and facilitate flux entry, as shown by examining a square SQUID washer by SQUID microscopy [98]. Round holes and pickup structures are preferable, while narrow windings of, *e.g.*, a pickup coil, should be avoided. Sun *et al.* investigated the issue of vortex entry in detail and calculated the penetration field as a function of a pickup coil width and radius, with the coil exposed to the self-field created by screening supercurrent of intensity J_C [99]. They showed that vortex entry is the cause of magnetic hysteresis in dc SQUIDs.

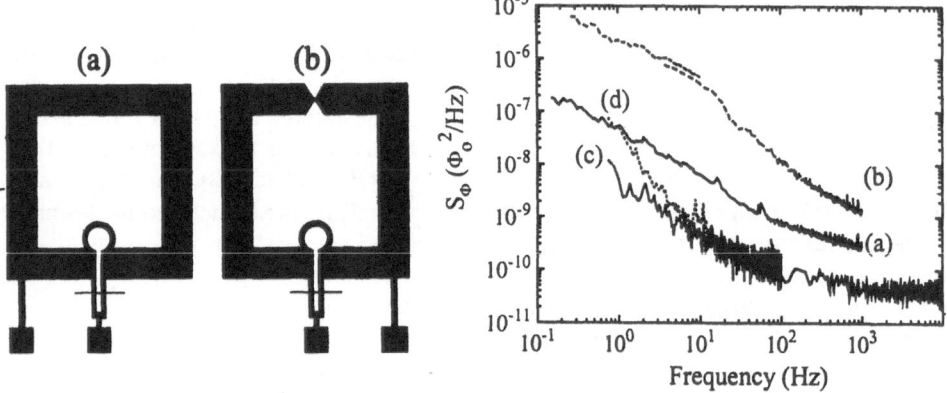

Figure 14. Schematic layout of direct-coupled dc SQUID magnetometers without and with a flux dam, and spectra of excess 1/f noise induced in a direct-coupled magnetometer by magnetic field exposure. Here, curve (a) and (b) are for the magnetometer without a flux-dam, curve (c) and (d) are from a device with a flux dam. Solid lines represent zero-field-cooled noise spectra, while dashed lines are the spectra after field exposure [100] (© 1995 IEEE).

Sun *et al.* proposed also a (so called) flux dam which reduces field-exposure-induced noise in direct-coupled HTS magnetometers [100,101] similar to those discussed above [81], with either step-edge or bicrystal grain-boundary junctions. For example, they exposed zero-field-cooled magnetometers to 0.05 mT for 30 min, waited for 2 hours and re-measured the noise spectra. The noise power below 100 Hz increased by up to two orders of magnitude, as shown in Figure 14b, curves (a) and (b) [100]. Introduction of a narrow constriction or a weak link into the pickup loop, shown in Figure 14a, limited the buildup of screening supercurrent in the loop, and significantly reduced the excess noise. Noise data obtained with the dam (in an otherwise identical layout) when following the same exposure sequence as above, are compared in Figure 14b, curves (c) and (d). One can assume that the flux enters and exits at the weak spot in the loop, rather than entering and being trapped in film regions coupling the hopping vortex noise to the SQUID. The I_C of the weak link in the loop, and its field dependence determine the upper limit of the dam effectiveness.

Recently, Schöne *et al.* demonstrated the possibility of additional 1/f noise reduction by "demagnetization" using a high-frequency (between 50 MHz and 5 GHz) magnetic field [102]. After applying a magnetic field transient to a zero-field-cooled niobium washer SQUID, they observed the typical increase in 1/f noise which could be removed by slowly increasing the rf-field power to a low-level maximum, and then slowly decreasing it to zero. The mechanism of this "demagnetization" is not yet clear, but a practical significance for HTS SQUIDs could be considerable.

Another detrimental effect of ambient static fields, which are much stronger than the measured signal, is the reduction in junction I_c, which is the natural consequence of the periodic I_c (B_0) dependence analogous to the Fraunhofer diffraction pattern. This reduction occurs particularly when the junction is positioned in the focused field at the center hole of a large washer, as in [85]. The dominant field component, B_z, is normal to to the washer plane, but, $e.g.$, in the case of grain-boundary junctions, is at an angle approaching 0° to the junction (barrier) plane. Consequently, the junction is not shielded, even if the YBCO film is thicker than λ_L. Experiments using step-edge junctions have confirmed that in direct-coupled SQUIDs, with pickup loops not too wide, the field-concentration effect is much weaker than in large washers [103]. With a pickup loop 8 mm in outer diameter and a 4-μm wide junction, the field independence of up to 0.3 mT, much above the Earth's field, was obtained. The junction width had to be reduced to about 0.3 μm in order to obtain a similar result with a washer SQUID of comparable size (8x8 mm^2). Although technologically very inconvenient, reducing the junction's width to submicron dimensions is an effective remedy against I_c suppression.

In contrast to grain-boundary junctions and also step-edge SNS junctions, trilayered edge junctions with electrodes thicker than λ_L are largely shielded from B_z. In dc SQUIDs with 3-μm wide quasiplanar PBCO barrier-edge junctions, Faley et $al.$ were able to demonstrate an impressive independence of I_c and 1/f noise at 1 Hz from orthogonal dc fields exceeding the Earth's field by an order of magnitude [74]. This result provides a strong argument for using such junctions in SQUIDs operating without shielding.

The ambient rf fields, $e.g.$, due to nearby radio, TV or radar transmitters, also may have a detrimental effect on SQUID operation by significantly increasing the low-frequency flux noise, or even making operation impossible [104,105], especially without magnetic and electromagnetic shielding. The effects of rf fields were studied sytematically only for the case of dc SQUID. In the case of rf SQUID, these effects can be even more detrimental, if the rf frequency is within or near the bandwidth of the tank circuit. The principal approach to rf attenuation is adequate rf shielding of the entire SQUID holder and the cable connecting it to the preamplifier. A high-resistivity shield material should be used to minimize the additional flux noise generated by screening currents in the shield. For example, for an rf SQUID gradiometer operating without magnetic shielding, the rf shield was fabricated from perforated lead foil [97].

4.1.5 Single-Layer planar Gradiometer Structures

The first YBCO dc SQUID gradiometer device was tested by Knappe et $al.$ [106], while more recent investigations of basically the same structure on bicrystal substrates were reported by Daalmans et $al.$ [107]. A simplified layout of their single-layer, planar first-order dc SQUID gradiometer fabricated from YBCO on a bicrystal substrate are shown in Figure 15. In the common central strip, signals and screening currents induced by the Earth's field in the two loops are subtracted. The central strip is direct-coupled to a dc

SQUID with a slit-shaped loop. The loop is not shielded so that a small magnetometric signal is superposed on the gradiometric one. As one might have expected, the authors observed a partial suppression of I_c, a related reduction in SQUID voltage modulation, and an increase in $1/f$ noise, all due to the coupling of ambient dc and rf fields to the SQUID hole. The advantage of this layout is that the contact pads can be symmetricaly positioned inside both pickup loops, which minimizes unbalance and contribution of the bias-current field. The gradiometer base is the distance between centers of the two large pickup loops. The use of a bicrystal substrate introduces very wide weak links in one pickup loop, which contribute additional current-fluctuation $1/f$ noise. Use of, *e.g.*, step-edge junctions across a local pit could remove that disadvantage.

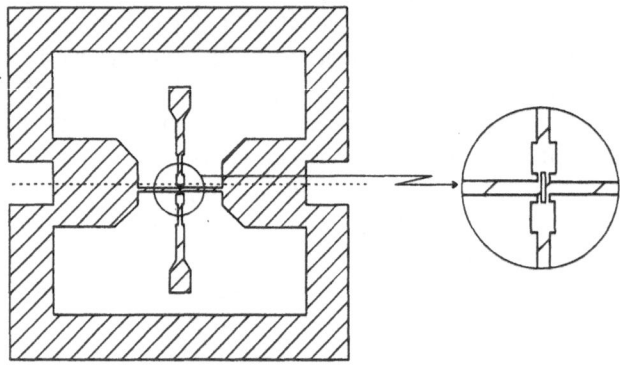

Figure 15. Single-layer gradiometer layout with direct coupling to a dc SQUID on bicrystal substrate. Dashed line indicates the grain boundary. The magnified segment including the small slit-hole washer is shown in the right-side circle. The layout is a simplified version of one used in [107].

A similar design can also be used for an rf SQUID gradiometer. However, the necessary inductive coupling to the tank-circuit coil requires a much larger area of the SQUID. This must result in a correspondingly stronger magnetometric signal.

4.2. SINGLE-LAYER FLIP-CHIP FLUX FOCUSERS AND FLUX TRANSFORMERS

A single-layer slitted flux focuser (concentrator) of larger dimensions than the washer SQUID, and in close proximity to it, can improve flux coupling to the SQUID and enhance the field-to-flux transformation. This offers the simplest way to improve the B_N of a SQUID. Using the method of Ketchen *et al.* [79], Matlashov *et al.* calculated the effect of spacing, z, between the washer and focuser, and of misalignment of their axes, x [108]. Figure 16 shows the dependences of the normalized focusing coefficient K/K_0 on x/D_0 and z/D_0, where $K = A_{eff}/A_h$, and $D_0 > d$ is the diameter of the flip-chip focuser's hole. For small normalized spacing and misalignment, K/K_0 can readily attain 0.8 to 0.9. The authors used that simple approach to fabricate LTS (niobium) dc SQUIDs without multilayered flux transformers, and attained a quite competitive performance.

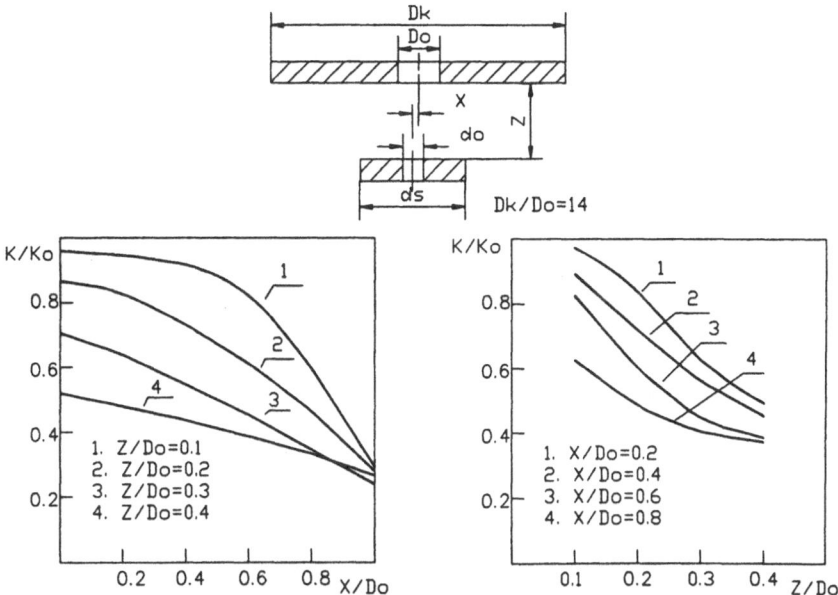

Figure 16. Normalized flux concentration coefficient of a flip-chip single-layer flux focuser versus the normalized misalignment of washer and focuser axes, x/D_0, and the normalized focuser-to-washer spacing, z/D_0, where D_0 is the diameter of the focuser's hole [108] (© 1996 IEEE).

An analogous approach was adapted by Tavrin *et al.* for HTS rf SQUID use [96]. In this case, circular bulk focusers fabricated of polished, 85% dense ceramics were used. The slit was obtained by cutting the disk in two equal parts, polishing the cut surfaces and glueing them together. The thin-film washer and bulk focuser then were sandwiched and clamped together to secure a very small spacing. With an outer diameter of 22 mm, $D_0 = 5$ mm, and thickness of 3.5 mm, an enhancement of $\partial B / \partial \Phi$ by a factor of about three, i.e. to 0.5 nT/Φ_0, was obtained when using the above described large washer with $L_s \cong 315$ pH, and $\partial B / \partial \Phi \cong 1.5$. This was in good quantitative agreement with [108]. The large D_0 made the small normalized z/D_0 easy to attain and simultanously permitted the placement of the rf tank circuit coil inside the hole, *i.e.*, directly against the washer's surface, to attain the maximum possible coupling coefficient k. With improved k^2Q, the author's group obtained a $B_N = 35$ to 40 fT/Hz$^{1/2}$ above 2 - 3 Hz when measuring in a very high-quality magnetic shielded room for biomagnetic measurements (BMSR) at the PTB in Berlin [109]. With thin-film concentrators of similar dimensions, an enhancement factor of 2.1 to 2.8 was obtained.

Similar to focusers are the single-layer flip-chip planar flux transformers. An example is shown in Figure 17 together with a direct-coupled dc SQUID magnetometer. The inductance mismatch in such a simple flux transformer is somewhat reduced by making the input coil relatively large (corresponding to a large D_0 in a focuser) and coupling it closely to the pickup coil (b) of a direct-coupled dc SQUID. In this case, the effective area is given by [110]

$$A_{eff} \cong (\alpha_d L_s/L_c)[A_c + k_m(L_cL_i)^{1/2}A_p/(L_p + L_i)] \qquad (4.2\text{-}1)$$

where L_c is now the direct-coupled pickup inductance of the SQUID with pickup area A_c, L_i is the single-turn input coil inductance and k_m is the mutual coupling coefficient between L_c and L_i. The input coil should have the same layout and inductance as the direct-coupled coil, $L_i = L_c$, with minimum spacing obtained by clamping them together. When coupling to a washer, the input-coil outer dimensions should not exceed those of the washer. Relatively high k_m values of 0.6 to 0.9 can be obtained in clamped structures.

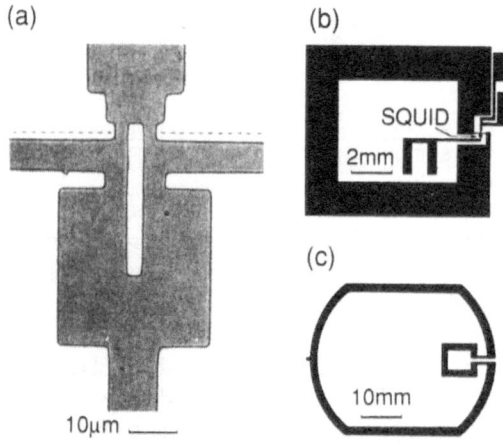

Figure 17. A single-layer flux transformer (c) inductively coupled to the pickup coil (b) of a direct-coupled, optimized dc SQUID magnetometer (a). The input coil layout in (c) closely matches the layout of the pickup coil of (b) [110] (© 1996 AIP).

In a first demonstration, Koelle *et al.* coupled a flux transformer with a pickup area of $A_p = 1330$ mm^2 (*i.e.*, the slanted-circular 45x35 mm^2 pickup loop of Figure 17c) to one of two alternate direct-coupled magnetometers (Figure 17a,b). The pickup areas were $A_c = 47$ and 52 mm^2, SQUID inductances $L_s = 20$ and 40 pH, and effective areas $A_{eff} = 0.075$ and 0.144 mm^2 [110]. The transformer increased the effective areas to 0.29 and 0.46 mm^2, with enhancement factors of about 3.9 and 3.2, respectively. The coupling to the SQUID and approximately linear scaling of effective area with outer dimensions are, indeed, similar to the situation for focusers. At white noise-frequencies Koelle *et al.* obtained the best $B_N = 31$ fT/Hz$^{1/2}$, while at 1 Hz, with bias reversal, resolution deteriorated to 39 fT/Hz$^{1/2}$. These best data were measured with a 20 pH SQUID having R_N between 2 and 3 ohm and $S_\Phi = 3.9$ $\mu\Phi_0$/Hz$^{1/2}$. Similarly, Zhang *et al.*

coupled a transformer with a 40x40 mm^2 pickup loop to their typical large washer rf SQUID and obtained $B_N \cong 25$ fT/Hz$^{1/2}$ at frequencies above 0.5 Hz [89]. All these impressive results recently have been surpassed by a 20x20 mm^2 direct-coupled dc SQUID with a very high R_n [28]. Nevertheless, flip-chip single-layer flux transformers and focusers are currently of interest due to their technological simplicity, and the opportunity to match with the relatively best small SQUIDs in a batch, which increases the overall yield. Also the 1/f flux noise in these structures, seems to be less effectively coupled to the SQUID [110] than in the case of an integrated, multilayered SQUID with flux transformer. Consequently, even relatively noisy single-layer flux transformers or focusers can be used. In the immediate future, such flip-chip structures are likely to be used in some practical applications.

Flip-chip, single-layer flux transformers also can be inductively coupled to single-layer planar gradiometer pickup loops, in order to increase the pickup area and the base length. However this does not eliminate operational problems due to the absence of SQUID shielding in the structure [107].

5. Multilayered Flux Concentrating Structures and SQUID Magnetometers

5.1. FLUX TRANSFORMERS WITH A MULTITURN INPUT COIL

5.1.1. Design Rules and Experimentally Attained HTS Performance
The principles of operation of a SQUID with a flux transformer having a microstrip spiral input coil with n turns [77] are discussed in Chapter 1. The large inductance of a small multiturn coil permits one to match it to the pickup loop inductance, and the stripline with a washer serves as the groundplane to the SQUID loop. For the sake of completeness, we give the expression for the effective area:

$$A_{eff} = A_{seff} + M_i A_p/(L_i + L_p) \cong M_i A_p/(L_i + L_p) \qquad (5.1.1-1)$$

where A_{seff} is the (negligible compared to A_{eff}) effective area of a bare washer SQUID. In the expression, the mutual inductance M_i, and the input coil inductance L_i are [77]

$$M_i = n(L_s - L_{sl}) \qquad (5.1.1-2)$$

$$L_i = n^2(L_s - L_{sl}) + L_{str} \qquad (5.1.1.-3)$$

Here, L_{sl} includes the parasitic inductance of the junction(s) and L_{str} is the stripline inductance, which is very small for n > 20. The coupling coefficient between the input coil and the SQUID is

$$k^2 = (1 - L_{sl}/L_s)/[1 + 1/n^2(1 - L_{sl}/L_s)] \cong 1 - L_{sl}/L_s \qquad (5.1.1-4)$$

with a sufficiently small separation between the groundplane washer and the stripline spiral. With small parasitic inductances (of the slit and junctions), and a large n, coupling is very tight and k → 1, nearly independent of the L_s value. Even in a flip-chip configuration, respectable k values up to 0.7 are attainable. By chosing a large n one also can minimize the impedance mismatch between the input and pickup coils by attaining $L_i \cong L_p$. An optimum signal energy transfer to the SQUID with high A_{eff}/A_p is thus possible even at very small L_s values, which, in turn, insure a high transfer function and low white noise. This is then a nearly ideal magnetometer concept. The only problems are the spurious parasitic resonances of L_i with stray capacitances, which are difficult to control.

For a comparison with single-layer concepts discussed earlier, one can show easily that the ratio of the effective areas of a direct-coupled single-layer SQUID and a SQUID with a multiturn input coil and an identical A_p can, at best, approach $(L_s/L_p)^{1/2}$. With typical values of $L_s = 40$ pH and a single-turn $L_p = 20$ nH on a 1x1 cm^2 substrate, the best possible ratio is approximately 0.09.

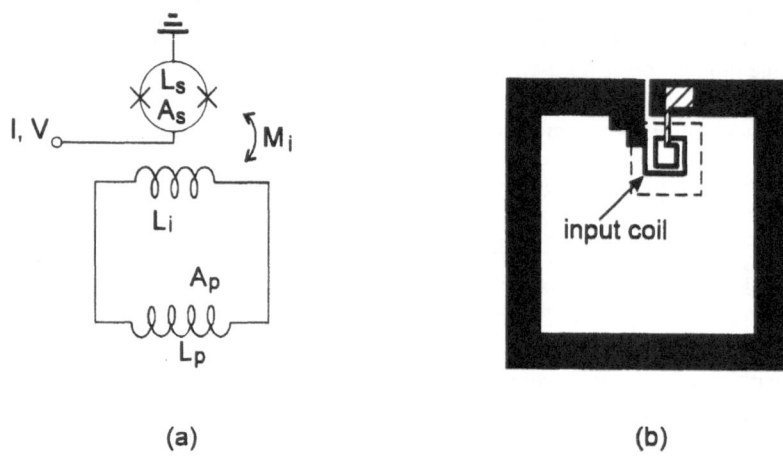

(a)   (b)

Figure 18. Equivalent circuit of a flux transformer coupled to a dc SQUID (a), and a quasi-schematic layout of a trilayer transformer with spiral input coil having only two turns to simplify the drawing (b). Dotted line in (b) shows the contour of the SQUID washer. Cross-hatched detail represents the top YBCO layer.

A multiturn spiral coil requires a return path from the center, which is insulated from the winding turns. Hence, a multilayer (at least trilayer) fabrication technology is required. Figure 18 shows the equivalent circuit, and a quasi-schematic representation of a layout with only two turns. The cross-hatched crossover layer (return path) is on top. Modern Nb film technology permits one to fabricate such structures easily and nearly defect-free, so that they are used in standard LTS SQUID magnetometer designs and contribute only little additional 1/f flux-hopping noise. However, the HTS (YBCO) multilayer technology is much less mature, and flux hopping at defects in the input coil can readily dominate the low-frequency noise. An acceptably low 1/f noise and the resulting magnetic field resolution clearly surpassing that of best single-layer SQUID were shown only recently by Dantsker *et al.* in a flip-chip magnetometer [111], and by

Drung *et al.* [112] with Shen *et al.* [113] in a fully-integrated dc SQUID magnetometer using additional positive feedback (APF) electronics.

In both cases, high-resistance bicrystal junctions were used, and a single-turn pickup coil with outer dimensions of about 9x9 mm^2 (on a 10x10 mm^2 SrTiO$_3$ substrate). The input coil in the flip-chip magnetometer [110] had 16 turns. The SQUID slit hole was partly shielded by the crossover strip, which reduced the bare SQUID inductance, *e.g.*, from 70 to 30 pH, while increasing V_{pp} and thus lowering the white flux noise. In [112], a 10.5-turn input coil was used with a much higher SQUID inductance of 130 pH, which resulted in a higher effective area. Table 3 shows the parameters and performance (with excellent shielding) attained in these best flip-chip and integrated devices. For comparison, also included is the best multilayered multiloop SQUID which will be described in Section 5.1.3. It has taken about five years since the first functional multilayered HTS SQUIDs were demonstrated, both as flip-chip [114,115] and integrated multiturn flux transformers [116], to demonstrate an acceptably low 1/f noise in such structures. Many published incremental advances, not cited here, contributed to the progress. The goal still to be reached is to attain sufficient reproducibility and yield.

TABLE 3. Performance of the best multilayered YBCO dc SQUIDs

Parameter/Type	R_n Ω	L_s (bare) pH	A_{seff} (bare) mm^2	A_{eff} mm^2	Gain A_{eff}/A_{seff}	V_{pp} μV	$S_\phi^{1/2}$ (1kHz) μ ϕ_0Hz$^{-1/2}$	B_N (1 kHz) fT Hz$^{-1/2}$	B_N (1 Hz)
Flip-chip flux transformer [111]	5.8 8.6	40 70	0.011 0.024	0.68 1.2	62 50	90 108	7.3 4.9	9.5 8.5	27 27
Integrated flux transformer [112]	9	130	?	1.7	?	11	8.0	9.7	53
Multiloop SQUID [128]	10	145	-	1.77	-	20	15	18	37

We note that the record performance data depend upon the shielding and electronics used. For example, a polycrystalline YBCO double-tube shield was found to contribute less 1/f noise than a metallic (Conetic) foil [111] did. The low-frequency data of Table 3 were all obtained with either a YBCO shield or inside the BMSR, using electronics with bias reversal, and in two cases also with APF.

5.1.2. Trilayer Fabrication Methods

Fabrication methods described here are applicable to any trilayer magnetometer structure, either with a flux transformer or a multiloop SQUID reviewed in the following Section. We follow multiturn-coil flux-transformer fabrication steps as an example. The transformers can be fabricated as either crossover or crossunder structures, *i.e.*, with the return path over or under the spiral. We first describe the relatively standard HTS fabrication process which resulted in the best performances to date, and subsequently alternate promising approaches which have not been optimized.

The first requirement for success, *i.e.*, for the absence of shorts between the spiral and the return path, is to fabricate a sufficiently smooth lower-layer film, free of particulates. The film should be *in-situ* capped by a thin, 10 to 20 nm, layer of an insulator, $SrTiO_3$ or $NdGaO_3$, to prevent damage during the subsequent patterning step, which may propagate to the insulator [117,118]. Although $SrTiO_3$ has been predominantly used until now, its high dielectric permittivity makes it undesirable for flux transformers, as it enhances spurious resonances. Most high-quality structures have been fabricated by PLD, but functional devices also were made from sputtered films. To minimize roughness of the first layer to a few nanometers, it is often made thinner than λ_L, between 100 and 150 nm.

The *ex-situ* patterning of the capped layer, to delineate either the input and pickup coils or the return path with SQUID washer and junctions, must be done such that all edges be beveled. The bevel angle should be low enough to prevent nucleation of grain boundaries in the upper YBCO layer. After standard resist (e.g., AZ 5214) patterning and bakeout at 90°C, the usual way is to: (1) add a short, 10 - 15 min, bakeout at about 120°C, which causes the resist to flow at the edges, (2) ion-beam etch at a shallow angle to the plane, 45° or less, with holder rotation [117,118].

After patterning, the bilayer is coated with a much thicker layer of the same insulator, say, 150 to 250 nm. Subsequently, two patterned vias are opened in the insulator, *e.g.*, to reach the free ends of the input and pickup coils (Figure 18). The preferred method for fabricating high-J_c vias is to again make beveled via edges and IBE holes through both layers, insulator and the lower YBCO layer, to reach the substrate. This should fully expose the a-b planes of the lower YBCO layer. Just before the next YBCO deposition step, the exposed edges must be cleaned by bromine-ethanol or very low energy IBE, as in the case for edge junctions. To obtain a sufficiently high I_c, equal or higher than that of the input coil, the perimeter of such vias must be sufficiently large. This is usually obtained by meandering their contour [117]. The earlier suggested method of fabricating vias by contacting c-planes [6], results in weak-link formation and should not be used.

The top YBCO layer is usually deposited to a thickness greater than λ_L, say 250 nm. To secure sufficient oxygenation of the lower layer after the deposition, the in-situ cooling in oxygen plasma described in Section 2.1.2 [9] is recommended after this and every earlier layer deposition step. As this may not be easy to implement after PLD, conventional annealing in oxygen must be used instead. In the case of best-performance flip-chip transformers fabricated by PLD [111, 118], the following cool-down process program was used: cool down from a deposition temperature of 780°C to 500°C in 0.8 atm of O_2 for 30 minutes, hold at that temperature for a time t_h and continue cooling to room temperature in the same environment. The hold times after deposition were: $t_h = $ 30 min tor the cap layer, 3 hours for the thick insulator, and 1 hour for the top YBCO layer. The heating before deposition also was in 0.8 atm O_2. However, the initial properties of the lower YBCO bare film were never fully recovered.

The patterning of the top layer can be made by any standard method, with pattern transfer using normal-incidence IBE or chemical etching. After etching, additional annealing in oxygen plasma helps removing the etching damage. With a sufficiently low bevel angle, the degradation of J_c at crossovers is tolerable. In the case of an entirely integrated SQUID and transformer structure, the top YBCO layer was fully *in-situ* coated with a Au-Ag alloy protective cap [119]. In this top crossover YBCO layer, patterning delineated the SQUID washer, the SNS step-edge junctions (with the Au-Ag cap serving as the weak-link interlayer) and the bias leads with contact pads.

The microshadow mask technique is a promising alternative approach to fabricating the patterned and fully in-situ capped lower bilayer with beveled edges [72]. A working flip-chip flux transformer was demonstrated using this method. The micromask is fabricated on the substrate as shown in Figure 19. A resist layout is patterned on the substrate (a), and a CaO/ZrO_2 double layer is deposited on top (b). After the resist lift-off (c), a short, 10-20 sec dip in water undercuts the lower CaO layer, thus forming a overhang-edge micromask on the substrate (d). After the dip, the chip is immediately immersed in an acetone stop bath, loaded to the deposition chamber, and outgassed at 200°C for 30 min to remove water incorporated into CaO. The YBCO and insulator cap are deposited through the micromask (e), with the bevel angle controlled primarily by the CaO thickness. Finally, the micromask is removed by a water rinse (f).

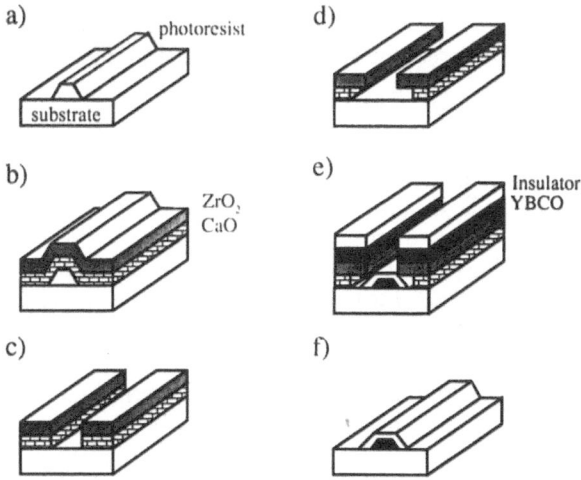

Figure 19. Patterning using a micromask: (a) resist layout patterned on substrate, (b) $CaO/ZrO2$ mask layer deposited, (c) resist lifted off, (d) CaO layer undercut by water dip, (e) YBCO with $SrTiO_3$ cap deposited through the micromask, (f) micromask removed by a water dip [72].

Another alternate technique, which was suggested earlier [6], is the planarization of the lower layer by using a CaO/ZrO_2 mask lift-off technique to fabricate trenches in the substrate filled by the (capped) YBCO. The substrate with the imbedded structure is polished prior to the deposition of a thicker insulator layer. This initially promising concept did not work due to defects originating at the interface between edges of the trench and the trench-filling layer. The defects propagated through the thick insulator layer, so that a degraded contour of the lower layer was replicated in the upper one [120].

A novel technique permitting one to fabricate completely planar SIS structures was proposed and demonstrated by Ma [121]. Implantation of impurities such as Si^+ or B^+ into a YBCO film covered with patterned photoresist permits one to inhibit the exposed areas. After annealing at 300° to 500°C, these areas remained insulating, but regained the perovskite crystalline structure. YBCO films grown on top of inhibit-patterned layers exhibited $T_c = 85$ to 88K and $J_c > 1 \times 10^{-6}$ A/cm^2, similar to properties of the lower YBCO layer. Crossover structures with an epitaxial insulator also were demonstrated. It is difficult to predict whether or not this technique can produce planar patterned trilayers of sufficient quality to make possible their use in flux transformers.

In closing we observe that, to date, fabrication has been primarily of trilayered magnetometer structures, with much less effort devoted to planar multilayered gradiometers. Recently, functional short-baseline first-order washer gradiometers with flip-chip flux transformers were successfully tested [122]. Also, functional first-order gradiometer flip-chip structures with a long baseline of about 3 cm were sputter-deposited on $LaAlO_3$ wafers 50 mm in diameter [123]. Standard processing techniques described above, with $NdGaO_3$ as the insulator, were used in this case in conjunction with a large-washer rf SQUID. Shielding of the input coil and washer by a bulk YBCO cup was necessary to suppress the strong magnetometric signal.

5.2 MULTILOOP SQUID

A viable alternative to the multiturn flux transformer is the multiloop or fractional-turn SQUID, an approach which entirely eliminates the mismatch between the pickup coil and SQUID inductance. It was originally proposed by Zimmerman and demonstrated as a bulk niobium device [124]. Much more recently, Drung *et al.* developed a planar, Nb thin film equivalent of it [125,126], and used such LTS dc SQUID sensors to construct large biomagnetic multichannel systems. Figure 20 shows an equivalent circuit (a), and a simplified schematic layout (b) of such a planar structure, currently called a "Drung's wheel". Its theory and optimization with respect to thermal noise can be found in [127]. Several large pickup loops connected in parallel have a sufficiently low inductance to serve directly as the SQUID inductance, so that the mismatch problem disappears. The effective pickup inductance $L_p = L_s$, and the effective area of a magnetometer with n loops are

$$L_s = (L_l/n^2) + (L_{sp}/n) + L_c \qquad (5.2\text{-}1)$$

$$A_{eff} = A_l/n - A_{sp} \qquad (5.2\text{-}2)$$

where L_l and A_l are the inductance and effective geometric area of the large perimeter loop (including all parallel loops), and L_{sp} and A_{sp} are the values for each "spoke". The additional inductance L_c is that of connections to the junctions. To minimize the parasitic capacitance of spokes, and thus the transmission line resonances, a coplanar rather than stripline design is used. This is at the expense of the effective area, and adds an additional "parasitic" inductance L_{sp}/n. Since an LTS dc SQUID can operate with relatively large inductances, say 400 pH, only eight parallel loops were used in the original design, with a wheel diameter of about 8 mm.

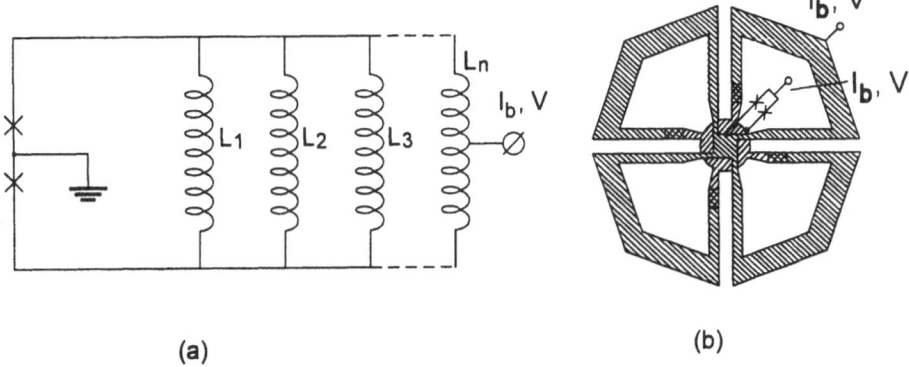

Figure 20. "Drung's wheel": (a) equivalent circuit, (b) a simplified layout of a multiloop dc SQUID. Oppositely shaded regions represent upper and lower YBCO films while cross-shaded regions indicate contact between these layers [128] (© 1996 AIP).

In the HTS version, the need to reduce the SQUID inductance dictates a larger n, but, unfortunately, at a greater expense to A_{eff} (Eq. 5.2-2). A compromise n = 16, and a wheel diameter of 7 mm, were chosen in [127]. This design requires as many crossovers and vias plus one extra via. In the center area is the patterned trilayer with junctions in the lower layer. The larger part of each pickup loop is in the upper layer, connecting to the lower layer in the cross-shaded areas shown in Figure 20b. This design was used by Ludwig *et al.* [128], who fabricated and tested it with 24° high-resistance bicrystal junctions on SrTiO$_3$ substrate. The fabrication process was the same as that described in Section 5.1.2 [118]. The wheel was positioned on the substrate such that the grain boundary was in the coplanar gap of two opposite spoke pairs. No narrow lines, other than those forming the two junctions, crossed the boundary. The design parameters were: L_l = 12.2 nH, A_l = 34.5 mm2, L_{sp} = 1.17 nH, A_{sp} = 0.39 mm^2, and L_c = 24 pH. The resulting SQUID inductance was 145 pH and the effective area was 1.77 mm^2. The best performance measured in a double YBCO shield with bias reversal is included in Table 3. It is quite comparable to that of the best integrated and flip-chip SQUIDs with flux transformer, except for a somewhat higher white noise.

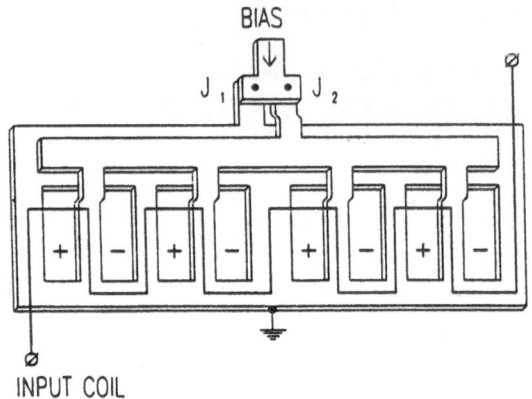

BIAS

INPUT COIL

Figure 21. Simplified schematic layout of self-shielding SQUID with
a meander single-layer input coil [133] (© 1996 IEEE).

The same design also was used by others with other types of junctions. For example, with step-edge junctions in the YBCO upper layer, *i.e.*, steps patterned in the thick $SrTiO_3$ insulator, a $B_N = 30$ fT/Hz$^{1/2}$ in white-noise region was obtained when measured in the BMSR using reverse bias and APF [129,130].

The fractional-loop concept also was applied to thin-film niobium gradiometers [131,132], and extended by Masalov *et al.* to a self-shielded or "astatic" dc SQUID magnetometer [133], which by itself is a first-order gradiometer not sensitive to the homogeneous (Earth's) field. A simplified layout is shown in Figure 21. The gradiometer is inductively tightly coupled to a meander input coil of a single-layer flux transformer such that summation of input signals (induced in gradiometer loops of opposite polarity) occurs. With two separate single-layer flux transformers the gradiometric mode can be preserved. The concept combines the advantage of a very simple flux transformer with cancellation of the magnetometric signal. The layouts proposed by Masalov *et al.* appear to offer an especially attractive solution for HTS magnetometers and gradiometers since HTS junctions will be protected from the ambient magnetic field. However, HTS experimental results are not yet available.

6. Auxiliary Structures

The only type of auxiliary planar SQUID structures we mention here is the resonant tank circuit for an rf SQUID. In contrast to dc SQUID [77], planar rf SQUIDs have not been optimized in low-temperature conventional superconductor technology. Commercial rf SQUIDs had a toroidal niobium SQUID and a toroidally-wound tank circuit insuring a tight coupling to the SQUID loop. In laboratory experiments, tank circuits used with planar washer SQUIDs have consisted of a copper-wire-wound 3-dimensional coil, several millimeters in diameter, or a planar spiral coil, and a capacitor. Inductive coupling between the washer and such coils is relatively weak. At typical tank frequencies, between, say, f_T = 20 and 300 MHz, it has been difficult to attain a sufficiently high Q of the coils to secure the critical coupling parameter, $k^2Q \geq 1$. Consequently, the transfer signal V_{pp} was lower, and the white noise level was higher than the optimum for a given f_T. Additionally, the bulky tank coil blocked access to the washer, thus making the use of efficient multilayered flip-chip transformers practically impossible (unless a deep pit holding the tank coil was drilled in the substrate).

It is shown in Chapter 1 that the intrinsic (white) noise energy of an rf SQUID is inversely proportional to f_T. Hence, high tank frequencies are preferred if the noise of the electronics remains comparable. In attempting to optimize a microwave rf SQUID, Mück et al. found a high-frequency solution to the problems described above, such that both $k^2Q > 1$ and integration with a flux transformer could be attained. They demonstrated a planar niobium rf SQUID with an integrated tank circuit operating at a frequency close to 3 GHz [134]. This tank circuit consisted of a high-Q $\lambda/2$ microstrip resonator with a SQUID hole and a Nb/AlO$_x$/Nb shunted tunnel junction in it. The resonator was capacitively coupled to a planar antenna patterned on the same substrate. Via a directional coupler, the antenna linked the SQUID to the microwave oscillator and the readout electronics. Readout was thus performed in the reflection mode.

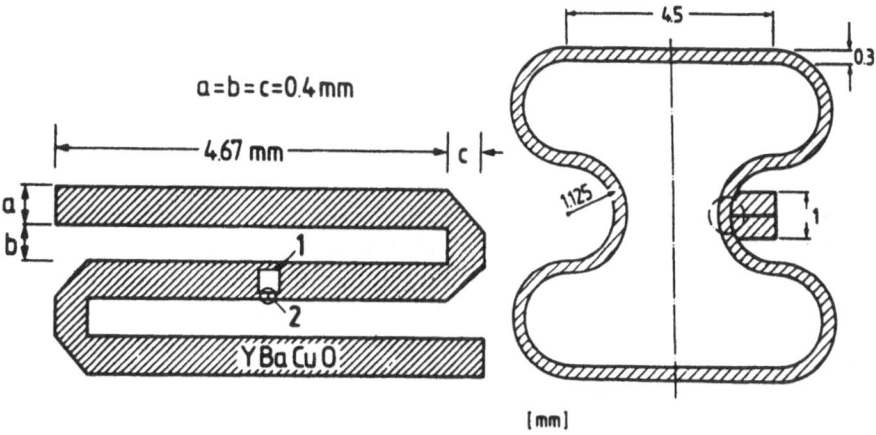

Figure 22. Simplified layouts of single-layer microwave SQUID's. Left: a $\lambda/2$ s-shaped resonator. Right: a λ microstrip resonator. In the left drawing "1" indicates the SQUID loop, and "2" the (step-edge) junction (marked by a small circle) [136].

The concept described above also was adopted for a YBCO single-layer planar microwave SQUID with step-edge junction(s) and a 16-mm-long microstrip resonator, which was s-shaped to fit a 10x10 mm^2 LaAlO$_3$ substrate [135]. A copper groundplane was used for simplicity. The layout is shown in Figure 22 (left). Unfortunately, such an integrated microstrip structure has a low A_{eff}. Introduction of additional concentrator wings adjacent to the SQUID loop proved rather ineffective [136]. The concept was thus extended to a λ ring resonator (right side of Figure 22) which served simultaneously as the direct-coupled pickup coil for the SQUID washer [136]. This layout was not optimized, yet it showed a significant improvement in A_{eff}.

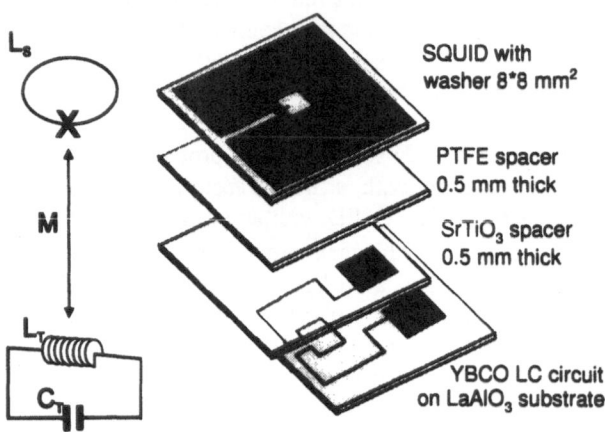

Figure 23. The equivalent circuit and a drawing showing a hairpin LC resonator with patch capacitors (bottom of the stack), spacers and a large washer (top). Normal groundplane under the resonator is not shown [137].

Very recently, flip-chip planar LC resonators consisting of YBCO microstrip hairpin resonators with two patch capacitors were demonstrated [137]. An example of one of the tested layouts is shown in Figure 23. In this case, the small loop of the LC resonator insures coupling to the large washer. With a copper groundplane, similar resonator layouts on a 10x10 mm^2 LaAlO$_3$ substrate, patterned in 200-nm-thick YBCO film had resonant frequencies between 0.6 and 1.3 GHz, and unloaded Q factors between 2700 and 700 respectively. For example, near 0.61 GHz a loaded Q (with the SQUID attached) of about 800 and $k^2Q > 1$ were obtained. Due to the high Q, the required coupling to the 8x8 mm^2 washer could be easily attained and adjusted using additional teflon and SrTiO$_3$ spacers between the tank circuit and the washer (Figure 23). With an rf SQUID inductance of only 35 pH, a transfer function of 150 μV/Φ_0, and white noise of 8×10^{-6} Φ_0/Hz$^{1/2}$ were obtained, resulting in a $B_N = 130$ fT/Hz$^{1/2}$. The demonstrated planar tank circuits promise the optimization of planar rf SQUIDs with multilayered pickup structures to attain magnetometer sensitivity equivalent to that of best dc SQUIDs. Access to the SQUID permitting one the use of a flip-chip or integrated multilayer flux transformer is entirely possible with a proper resonator layout.

Another tank-circuit concept, which permits one to use rf SQUIDs with multilayered flux transformers is that of a dielectric resonator first proposed by Kornev *et al.* [138]. An improved HTS version of an rf SQUID with a SrTiO$_3$ resonator was reported by

Zhang et al [139]. It showed a promise of very high sensitivity, but was bulky and rather expensive to implement.

7. Outlook

Available methods of thin-film SQUID fabrication are in principle sufficiently good to make further, incremental improvements in HTS magnetometer performance if the quality of films is high enough. Higher-critical temperature films would be an important asset if their crystalline quality, and the flux-pinning strength up to 77K, attain the same levels as in the case of YBCO. Further development of novel methods for patterning of multilayered and integrated flux pickup structures, such as micromasking, inhibit patterning and planarization is quite desirable as it might minimize processing damage. Emulation of conventional LTS solutions, e.g., by developing low-noise HTS superconducting wire for pickup coils, might be very advantageous for axial gradiometers, but is not absolutely necessary. The Josephson-junction types in use and methods of fabricating these are adequate for small device batches. However, further improvements in controllability, reproducibility and stability of critical parameters with time and upon temperature cycling are essential. They will be especially crucial for a reduction in fabrication and service maintenance cost, if commercial fabrication of larger device quantities becomes a reality. A trend away from grain-boundary junctions and toward broader implementation of multilayered SNS junctions is predicted.

Until 1995, the best HTS SQUID magnetometer performance in terms of sensitivity, i.e., of 1/f noise and low-frequency magnetic field resolution, was that of devices with single-layer flux pickup structures, either galvanically connected or inductively coupled in a flip-chip configuration. These represent the first generation of useful SQUIDs and are available commercially. Reasonably high performance is only possible with adequate, i.e., high-quality, magnetic shielding. The most recent progress in fabricating low-noise multilayered flux transformers permits one to expect that in the coming years 1/f-noise levels and field resolutions approaching those of commercial LTS SQUID systems (also operating with magnetic shielding) will become commonplace as a second generation of HTS SQUIDs.

The biggest challenge for the future remains the minimization of 1/f noise in structures exposed to ambient magnetic fields in the absence of external shielding. Painstaking progress in the quality of films and patterned structures, that is in the reduction of defect concentration, is required, but is not necessarily sufficient to attain that goal. Ingenious design improvements, use of shielded junctions and structures, planar astatic and gradiometric arrangements also will be necessary. One can hope that magnetometer systems capable of operating with very high sensitivity in the absence of external shielding will represent the third, ultimate generation of HTS SQUIDs. The open question is: in what signal bandwidth will their operation be satisfactory? One can be quite hopeful that the narrow-band performance will be adequate. A relatively wide-band performance, which is required, e.g., in medical diagnostics, might prove to be extremely difficult to attain.

Acknowledgements

I am thanking all authors who kindly granted their permissions to reproduce figures from their publications. I am most indebted to all my collaborators whose results constitute part of this chapter. Special thanks are due to Drs. E. Sodtke and Y. Zhang for their critical reading of the whole chapter and most useful comments. Research performed at KFA Jülich (and quoted here) was supported in part by German Ministry of Education, Science, Research and Technology (BMBF). Last, but not least, I like to gratefully acknowledge the corrections and helpful remarks made by the Editor.

References

1. Berberich, P., Utz, B., Prusseit, W. and Kinder, H. (1994) Homogeneous high quality YBCO films on 3" and 4" substrates, Physica C **219**, 497-504.
2. Phillips, J. (1993) Materials issues affecting the new superconducting electronics, in H. Weinstock and R.W. Ralston (eds), *The New Superconducting Electronics*, Kluwer Academic Publishers, Dordrecht, pp. 59-88.
3. Somekh, R.E. and Barber, Z.H. (1992) The deposition of ceramic superconductors, in R. Kossowsky, B. Raveau, D. Wohlleben and S.K. Patapis (eds)*Physics and Materials Science of High Temperature Superconductors, II*, Kluwer Academic Publishers, Dordrecht, pp. 443-469.
4. Chrissey, D.B. and Hubler, G.K., editors (1994) *Pulsed Laser Deposition of Thin Films*, John Wiley and Sons, New York.
5. Holzapfel , B., Roas, B., Schultz, L., Bauer, P. and Saemann-Ischenko, G. (1992) Off-axis deposition of YBCO thin films, Appl. Phys. Lett. **61**, 3178-3180.
6. Braginski, A.I. (1993) Thin film structures, in H. Weinstock and R.W. Ralston (eds), *The New Superconducting Electronics*, Kluwer Academic Publishers, Dordrecht, pp. 89-122.
7. Zaitsev, A.G., Kutzner, R. and Wördenweber, R. (1995) Growth of high-quality YBCO on CeO_2 buffer of mixed (001)/(111) orientation on sapphire, Appl. Phys. Lett. **67**, 2723-2725.
8. O'Bryan, H.M., Gallagher, P.K., Berkstresser, G.W. and Brandle, C.D. (1990) Thermal analysis of rare earth gallates and aluminates, J. Mater. Res. **5**, 183-189.
9. Ockenfuss, G., Wördenweber, R., Scherer, T.A., Unger, R. and Jutzi, W. (1995) In-situ low pressure oxygen annealing of $YBa_2Cu_3O_{7-\delta}$ single and multilayer systems, Physica C, **243**, 24-28.
10. Schneider, J., Kohlstedt, H. and Wördenweber, R. (1993) Nanobridges of optimized YBCO thin films for superconducting flux-flow type devices, Appl. Phys. Lett. **63**, 2426-2428.
11. van der Harg, A.J.M., van der Drift, E. and Hadley, P. (1995) Deep-submicron structures in YBCO: fabrication and measurement, IEEE Trans. Appl. Supercond. **5**, 1448-1451.
12. Eidelloth, W. and Sandstrom, R.L. (1991) Wet etching of gold films compatible

with high T_C superconducting thin films, Appl. Phys. Lett. **59**, 1632-1634.

13. Vasquez, R.P., Hunt, B,D. and Foote, M.C. (1988) Non-aqueous chemical etch for YBCO (1988), Appl. Phys. Lett. **53**, 2692-94.

14. Faley, M.I., Poppe, U., Soltner, H., Jia, C.L., Siegel, M. and Urban, K. (1993) Josephson junctions, interconnects, and crossovers on chemically etched edges of YBCO, Appl. Phys. Lett. **63**, 2138-2140.

15. Ma, Q.Y., Yang, E.S., Treyz, G.V. and Chang, C.-A. (1989) Novel method of patterning YBCO superconducting thin films, Appl. Phys. Lett. **55**, 896-898.

16. Copetti, C., Gassig, U., Zander, W. Schubert, J. and Buchal, Ch. (1992) Improved inhibit-patterning of YBCO thin films, Appl. Phys. Lett., **61**, 3041-304.

17. Likharev, K.K. (1986) *Dynamics of Josephson Junctions and Circuits*, Gordon and Breach, Pbl.., New York, p. 17.

18. Gross, R. (1994) Grain boundary Josephson junctions in the HTS, in S.L. Shinde and D. Rudman (eds.) *Interfaces in High-T_C Superconducting Systems*, Springer Verlag, New York, pp. 176-209.

19. Pettiette-Hall, C.L., Luine, J.A., Murduck, J.M., Burch, J.F., Hu, R., Sergant, M. and St. John, D. (1995) YBCO step-edge junctions on various substrates, IEEE Trans. Appl. Supercond. **5**, 2087-2090; Also: Murduck, J.M., Pettiette-Hall, C.L. and Luine, J.A. (1994) Improved step-edge junction uniformity and circuit yield, Extended Abstracts of Int. Workshop on HTS Electron Devices in Whistler, BC, **FED-136**, Tokyo, pp. 126-127.

20. Dimos, D., Chaudhari, P., Mannhart, J. and LeGoues, F.K. (1988) Orientation dependence of grain-boundary critical currents in YBCO bicrystals, Phys. Rev. Lett. **61**, 219-222.

21. Dimos, D., Chaudhari, P. and Mannhart, J. (1990) Superconducting transport properties of grain boundaries in YBCO bicrystals, Phys. Rev. B **41**, 4038-4049.

22. Gross, R., Chaudhari, P., Kawasaki, M. and Gupta, A. (1991) Superconducting transport characteristics of YBCO grain boundary junctions, IEEE Trans. on Magnetics **27**, 3227-3230.

23. Wako Bussan Co., Ltd., Tokyo, Japan

24. Ivanov, Z.G., Nilsson, P-A., Winkler, D., Alarco, J.A., Claeson, T., Stepantsov, E.A. and Tzalenchuk, A.Ya. (1991) Weak links and dc SQUIDs on artificial nonsymmetric grain boundaries in YBCO, Appl. Phys. Lett. **59**, 3030-3032.

25. Fischer, G.M., Mayer, B., Schulze H. and Gross, R. (1995) Critical current density distribution and magnetic flux states in YBCO bicrystal grain boundary junctions, IEEE Trans. Appl. Supercond. **5**, 2184-2187.

26. Froehlich, O.M., Schulze, H., Beck, A., Gerdemann, R., Mayer, B., Gross, R. and Huebener, R.P. (1995) Supercurrent density correlation function of YBCO grain boundary Josephson junctions, IEEE Trans. Appl. Supercond. **5**, 2188-2191.

27. Moeckly, B., Ph.D. Dissertation, Cornell Univ. 1994 (unpublished).

28. Cantor, R., Lee, L.P., Teepe, M., Vinetskiy, V. and Longo, J. (1995) Low-noise, single-layer YBCO dc SQUID magnetometers at 77K, IEEE Trans. Appl. Supercond. **5**, 2927-2930.

29. Dillmann, F. and Siegel, M., (KFA, 1995) unpublished (subm. to Appl. Phys. Lett.).

30. Simon, R.W., Burch, J.F., Daly, K.P., Dozier, W.D., Hu, R., Lee, A.E., Luine, J.A.,

Manasevit, H.M., Platt, C.E., Schwarzbeck, S.M., St. John, D., Wire, M.S. and Zani, M.J. (1990) Progress towards a YBCO circuit process, in R.D. McConnell and R. Noufi (eds), *Science and Technology of Thin Film Superconductors 2*, Plenum Press, New York, 549-558.

31. Jia, C.L., Kabius, B., Urban, K. Herrmann, K., Cui, G.J., Schubert, J., Zander, W., Braginski, A.I. and Heiden, C. (1991) Microstructure of epitaxial YBCO films on step-edge SrTiO$_3$ substrates, Physica C **175**, 545-554.

32. Jia, C.L., Kabius, B., Urban, K. Herrmann, K., Schubert, J., Zander,W.and Braginski, A.I. (1992) The microstructure of epitaxial YBCO films on steep steps in LaAlO$_3$ substrates, Physica C **196**, 211-226.

33. Herrmann, K., Zhang, Y., Mück, H. M., Schubert, J., Zander, W., and Braginski, A.I. (1991) Characterization of YBa$_2$Cu$_3$O$_7$ Step-Edge Junctions, Supercond. Sci. & Techn. **4**, 583-586.

34. Herrmann, K., Kunkel, G., Siegel, M., Schubert, J., Zander, W., Braginski, A.I., Jia, C.L., Kabius, B. and Urban, K. (1995) Correlation of YBCO step-edge junction characteristics with microstructure, J. Appl. Phys. **78**, 1131-1139.

35. Luine, J., Bulman, J., Burch, J., Daly, K., Pettiette-Hall, C. and Schwarzbek, S. (1992) Characteristics of high performance YBCO step-edge junctions, Appl. Phys. Lett. **61**, 1128-1130.

36. Berkowitz, S.J., De Obaldia, E., Galloway, M.L., Morales, G., Ono, R.H., Beall, J.H., Vale, L.R. and Rudman, D.A. (1993) Etching and annealing of substrates for superconducting multilayers and devices, IEEE Trans. Appl. Supercond. **3**, 2950-2952.

37. Sun, J.Z., Gallagher, W.J., Callegari, A.C., Foglietti, V. and Koch, R.H. (1993) Improved process for high-T$_c$ superconducting step-edge junctions, Appl. Phys. Lett. **63**, 1561-1563.

38. Bode, M. (1995) Ph.D. Dissertation, Univ. of Giessen, Germany (unpublished).

39. Missert, N., Harvey, T.E., Ono, R.H. and Reintsema, C.D. (1993) High-T$_c$ multilayer step-edge Josephson junctions and SQUIDs, Appl. Phys. Lett. **63**, 1690-1692.

40. Glyantsev, V., Divin, Yu, Jia, C.L., Poppe, U. and Siegel, M. (KFA, 1995) High-resistance YBCO step-edge junctions, unpublished.

41. Edwards, J.A., Satchell, J.S., Chew, N.G., Humphreys, R.G., Keene, M.N. and Dosser, O.D. (1992) YBCO thin-film step junctions on MgO substrates, Appl. Phys Lett. **60**, 2433-2435.

42. Ramos, J., Seitz, M., Daalmans, G.M., Uhl, D., Ivanov, Z. and Claeson, T. (1993) Noise properties of single-layer YBCO step-edge dc SQUIDs on MgO substrates, Physica C **220**, 51-54 and references therein.

43. Char, K., Colclough, M.S., Garrison, S.M., Newman, N. and Zaharchuk, G. (1991) Biepitaxial grain-boundary junctions in YBCO, Appl. Phys. Lett. **59**, 733-735.

44. Wu, X.D., Luo, L., Muenchausen, R.E., Springer, K.N. and Foltyn, S. (1992) Creation of 45° grain-boundary junctions by lattice engineering, Appl. Phys. Lett. **60**, 1381-1383.

45. Delin, K.A. and Kleinsasser, A.W. (1995) Comparative study of the stationary properties of high-T$_c$ proximity-coupled Josephson junctions, IEEE Trans. Appl.

Supercond. **5**, 2976-2979.

46. Pauza, A.J., Moore, D.F., Campbell, A.M., Broers, A.N. and Char, K. (1995) Electron beam damaged high-T_C junctions - stability, reproducibility and scaling laws, IEEE Trans. Appl. Supercond. **5**, 3410-3413.

47. Tinchev, S.S. (1990) Investigation of rf SQUIDs made from epitaxial YBCO films, Supercond. Sci. & Tech., **3**, 500-503.

48. FIT (Forschungsgesellschaft für Informationstechnik mbH), Bad Salzdetfurth, Germany.

49. Tinchev, S.S. and Hinken, J.H. (1992) Two-loop YBCO rf SQUID magnetometer, in H. Koch and H. Lübbig (eds.) *Superconducting Devices and their Applications*, Springer-Verlag, Berlin Heidelberg, pp. 102-105.

50. Schmidl, F. Dörre, L., Linzen, S., Wunderlich S., Machalett, F., Hübener, U., Schneidewind, H. Seidel, P. (1995) Realization of YBCO thin film dc SQUIDs using step-edge junctions on silicon substrates as well as ion-beam modified microbridges, H.A. Blank (ed.) *Proc. 2nd Workshop on HTS Applications*, Twente, pp.131-136.

51. DiIorio, M.S., Yoshizumi, S., Yang, K.-Y., Zhang, J. and Maung, M. (1991) Practical high-T_C junctions and SQUIDs operating above 85K, Appl. Phys. Lett. **58**, 2552-2554.

52. DiIorio, M.S., Yoshizumi, S., Yang, K.-Y., Zhang, J. and Power, B. (1993) Low-noise dc SQUIDs at 77K, IEEE Trans. Appl. Supercond. **3**, 2011-2017.

53. Ono, R.H., Beall, J.A., Cromar, R.W., Harvey, T.E., Johansson, M.E., Reintsema, C.D. and Rudman, D.A. (1991) High-T_C SNS Josephson microbridges with high-resistance normal metal links, Appl. Phys. Lett. **59**, 1126-1128.

54. Ono, R.H., Vale, L.R., Kimminau, K.R., Beall, J.A., Cromar, M.W., Reintsema, C.D., Harvey, T.E., Rosenthal, P.A. and Rudman, D.A. (1993) High-T_C SNS junctions for multilevel integrated circuits, IEEE Trans. Appl. Supercond. **3**, 2389-2392.

55. Rosenthal, P.A., Grossman, E.N., Ono, R.H. and Vale, L.R. (1993) Superconductor-normal metal-superconductor junctions with high characteristic voltage, Appl. Phys. Lett. **63**, 1984-1986.

56. Reintsema, C.D., Ono, R.H., Barnes, G., Borcherdt, L., Harvey, T.E., Kunkel, G., Rudman, D.A. and Vale, L.R. (1995) The critical current and normal resistance of high-T_C step-edge SNS junctions, IEEE Trans. Appl. Supercond. **5**, 3405-3409.

57. Gao, J., Aarnink, W.A.M., Gerritsma, G.J. and Rogalla, H. (1990) Controlled preparation of all high-T_C SNS-type edge junctions and dc SQUIDs, Physica C **171**, 126-130.

58. Gao, J., Boguslavskii, Y., Klopman, B.B., Terpstra, D., Wijbrans, R., Gerritsma, G.J. and Rogalla, H. (1992) YBCO/PrBCO/YBCO Josephson ramp junctions, J. Appl. Phys. **72**, 575-583.

59. Hunt, B.D., Foote, M.C. and Bajuk, L.J. (1991) All-high-T_C edge-geometry weak links utilizing YBCO barrier layers, Appl. Phys. Lett. **59**, 982-984.

60. Chin, D.K. and Van Duzer, T. (1991) Novel all-high-T_C epitaxial Josephson junctions, Appl. Phys. Lett. **58**, 753-755.

61. Char, K., Colclough, M.S., Geballe, T.H. and Myers, K.E. (1993) High T_C superconductor-normal-superconductor Josephson junctions using $CaRuO_3$ as the

metallic barrier, Appl. Phys. Lett. **62**, 196-198.

62. Antognazza, L., Char, K., Geballe, T.H., King, L.L.H. and Sleight, A.W. (1993) Josephson coupling of YBCO through a ferromagnetic barrier $SrRuO_3$, Appl. Phys. Lett. **63**, 1005-1007.

63. Char, K., Antognazza, L. and Geballe, T.H. (1994) Properties of $YBCO/YBa_2Cu_{2.79}Co_{0.21}O_{7-x}/YBCO$ edge junctions, Appl. Phys. Lett. **65**, 904-906.

64. Antognazza,L., Moeckly, B.H., Geballe, T.H. and Char, K. (1995) Properties of high T_c Josephson junctions with $Y_{0.7}Ca_{0.3}Ba_2Cu_3O_{7-\delta}$ barrier layers, Phys. Rev. B **52**, 4559-4567.

65. Polturak, E., Koren, G., Cohen, D., Aharoni, E. and Deutscher, G. (1991) The proximity effect in $YBCO/Y_{0.6}Pr_{0.4}Ba_2Cu_3O_7/YBCO$ SNS junctions, Phys. Rev. Lett. **67**, 3038-3041.

66. Stölzel, C., Siegel, M., Adrian, G., Krimmer, C., Söllner, J., Wilkens, W., Schulz, G. and Adrian, H. (1993) Transport properties of $YBCO/Y_{0.3}Pr_{0.7}Ba2Cu_3O_{7-\delta}/$-YBCO Josephson junctions, Appl. Phys. Lett. **63**, 2970-2972.

67. Verhoeven, M.A.J., Gerritsma, G.J. and Rogalla, H., Ramp type HTS junctions with PrBaCuGaO barriers, EEE Trans. Appl. Supercond. **5**, 2095-2098.

68. Dömel, R., Horstmann, C., Siegel, M. and Braginski, A.I. (1995) Resonant tunneling transport across $YBCO-SrRuO_3$ interfaces, Appl. Phys. Lett. **67**, 1775-1777.

69. Satoh, T., Kupriyanov, M.Yu., Tsai, J.S., Hidaka, M. and Tsuge, H. (1995) Resonant tunneling transport in YBCO/PBCO/YBCO edge-type Josephson junctions, IEEE Trans. Appl. Supercond. **5**, 2612-2615.

70. Jia, C.L., Faley, M.I., Poppe, U. and Urban, K. (1995) The effect of chemical and ion-beam etching on the atomic structure of interfaces in YBCO/PBCO Josephson junctions, Appl. Phys. Lett. **67**, 3635-3637.

71. Koren, G., Aharoni, E., Polturak, E. and Cohen, D. (1991) Properties of all YBCO Josephson edge junctions prepared by *in situ* laser ablation deposition, Appl. Phys. Lett. **58**, 634-636.

72. Strikovsky, M.D., Kahlmann, F., Schubert, J., Zander, W., Glyantsev, V., Ockenfuss, G. and Jia, C.L. (1995) Fabrication of YBCO thin-film flux transformers using a novel microshadow mask technique for *in situ* patterning, Appl. Phys. Lett. **66**, 3521-3523.

73. Faley, M.I., Poppe, U., Jia, C.L. and Urban, K (1995) Proximity-effect in edge-type junctions with PBCO barriers prepared by Br-ethanol etching, IEEE Trans. Appl. Supercond. **5**, 2091-2094.

74. Faley, M.I., Popppe, U., Urban, K., Hilgenkamp, H., Hemmes, H., Aarnink, W., Flokstra, J. and Rogalla, H. (1995) Noise properties of dc-SQUIDs with quasiplanar YBCO Josephson junctions, Appl. Phys. Lett. **67**, 2087-2089.

75. Laibowitz, R.B., Sun, J.Z., Foglietti, V., Koch, R.H., Altman, R.A. and Gallagher, W.J. (1995) Properties of multilevel ramp edge junctions and SQUIDs with laser-ablated $SrTiO_3$ barriers, IEEE Trans. Appl. Supercond. **5**, 2620-2623.

76. Grundler, D., Krumme, J.-P., David, B. and Doessel, O. (1995) Multilevel devices

of YBCO with NdGaO₃ barrier, IEEE Trans. Appl. Supercond. **5**, 2751-2754.

77. Jaycox, J.M. and Ketchen, M.B. (1981) Planar coupling scheme for ultra low noise dc SQUIDs, IEEE Trans. on Magnetics, **MAG-17**, 400-403.

78. Ketchen, M.B. and Jaycox, J.M. (1982) Ultra-low noise tunnel junction dc SQUID with a tightly coupled planar input coil, Appl. Phys. Lett. **40**, 736-738.

79. Ketchen, M.B., Gallagher, W.J., Kleinsasser, A.W., Murphy, S. and Clem, J.R. (1985) dc SQUID flux focuser, in H.D. Hahlbohn and H. Lübbig (eds.) *SQUID '85: Superconducting QUantum Interference Devices and their Applications*, Walter de Gruyter, Berlin, pp. 865-871.

80. Matsuda, M., Murayama, Y., Kiryu, S., Kasai, N., Kashiwaya, S., Koyanagi, M., Endo, T. and Kuriki, S. (1991) Directly-coupled dc SQUID magnetometers made of BSCCO oxide films, IEEE Trans. on Magnetics **27**, 3043-3046.

81. Koelle, D., Miklich, A.H., Ludwig, F., Dantsker, E., Nemeth, D.T. and Clarke, J. (1993) dc SQUID magnetometers from single layers of YBCO, Appl. Phys. Lett. **63**, 2271-2273.

82. Toepfer, H. (1991) Inductance determination in superconducting structures, in W. Krech, P. Seidel and H.-G. Meyer (eds) *Superconductivity and Cryoelectronics*, World Scientific, Singapore, pp. 170-177.

83. Gupta, K.C., Garg, R. and Bahl, I.J. (1979) *Microstrip Lines and Slotlines*, Artech House, Dedham, MA, pp. 263-265.

84. Enpuku, K., Shimomura, Y. and Kisu, T. (1993) Effect of thermal noise on the characteristics of a high T_c SQUID, J. Appl. Phys. **73**, 7929-7934.

85. Zhang, Y., Mück, M., Herrmann, K., Zander, W., Schubert, J., Braginski, A.I. and Heiden, Ch. (1993) Sensitive rf SQUIDs and magnetometers operating at 77K, IEEE Trans. Appl. Supercond. **3**, 2465-2468. In this work the L_s estimates were too low by a factor of 1.25.

86. Jackel, L.D. and Buhrman, R.A. (1975) Noise in the rf SQUID, J. Low Temp. Phys. **19**, 201-245.

87. Koch, R.H. (IBM) Unpublished simulation results, private information.

88. Zhang, Y., Mück, M., Herrmann, K., Zander, W., Schubert, J., Braginski, A.I. and Heiden, Ch. (1992) Low-noise YBCO rf SQUID magnetometers, Appl. Phys. Lett. **60**, 645-647.

89. Zhang, Y., Krüger, U., Kutzner, R., Wördenweber, R., Schubert, J., Zander, W., Sodtke, E., Braginski, A.I. and Strupp, M. (1994) Single layer YBCO rf SQUID magnetometers with direct-coupled pickup coils and flip-chip flux transformers. Appl. Phys. Lett. **65**, 3380-3382.

90. Zhang, Y., Tavrin, Y., Krause, H.-J., Bousack, H., Braginski, A.I., Kalberkamp, U., Matzander, U., Burghoff, M. and Trahms, L. (1995) Applications of high-temperature SQUIDs, Applied Superconductivity **3**, 367-381.

91. Tanaka, S., Itozaki, H. and Nagaishi, T. (1993) Properties of YBCO large washer SQUID, Jpn. J. Appl. Phys. **32**, L662-L664.

92. Itozaki, H., Tanaka, S., Nagaishi, T. and Kado, H. (1994) Multi-Channel High-T_c SQUID, IECE Trans. Electron. E77-C, 1185-1190.

93. Lee, L.P., Longo, J., Vinetskiy, V. and Cantor, R. (1995) Low-noise YBCO direct-

286

current SQUID magnetometer with direct signal injection, Appl. Phys. Lett. **66**, 1539-1541.
94. Lee, L.P. Histogram presented at the ASC'94 (unpublished).
95. Miklich, A.H., Koelle, D., Shaw, T.J., Ludwig, F., Nemeth, D.T., Dantsker, E., Alford, N.McN., Button, T.W. and Colclough, M.S. (1994) Low-frequency excess noise in YBCO dc SQUIDs cooled in static magnetic fields, Appl. Phys. Lett. **64**, 3494-3496.
96. Tavrin, Y., Zhang, Y., Mück, M., Braginski, A.I. and Heiden, C. (1993) YBCO thin film SQUID gradiometer for biomagnetic measurements, Appl. Phys. Lett. **62**, 1824-1826.
97. Tavrin, Y., Zhang, Y., Wolf, W. and Braginski, A.I. (1994) A second-order SQUID gradiometer operating at 77K, Supercond. Sci. Technol. **7**, 265-268.
98. Kirtley, J.R., Ketchen, M.B., Tsuei, C.C., Sun, J.Z., Gallagher, W.J., Yu-Jahnes, L.S., Gupta, A., Stawiasz, K.G. and Wind, S.J. (1996) Scanning SQUID microscopy. IBM J.of Research & Development **39**, 655-668.
99. Sun, J.Z., Gallagher, W.J. and Koch, R.H. (1994) Initial-vortex-entry-related magnetic hysteresis in thin-film SQUID magnetometers, Phys. Rev. B **50**, 13664-13673.
100. Sun, J. , Yu-Jahnes, L.S., Foglietti, V., Koch, R.H. and Gallagher, W.J. (1995) Properties of YBCO thin film single-level dc SQUIDs fabricated using step edge junctions, IEEE Trans. Appl. Supercond. **5**, 2107-2111.
101. Koch, R.H., Sun, J.Z., Foglietti, V. and Gallagher, W.J. (1995) Flux-dam, a method to reduce extra low frequency noise when a superconducting magnetometer is exposed to a magnetic field, Appl. Phys. Lett. **67**, 709-711.
102. Schöne, S., Mück, M. and Heiden, C. (1996) Reduction of low-frequency excess noise in SQUIDs by applying high-frequency magnetic fields, Appl. Phys. Lett. **68**, 859-861.
103. Glyantsev, V.N., Tavrin, Y., Zander, W., Schubert, J. and Siegel, M. (1996) Stability of dc and rf SQUID without magnetic shielding, Supercond. Sci. Technol. **9**, 105A-108A.
104. Ishikawa, N., Nagata, K., Kasai, N. and Kiryu, S. (1993) Effect of rf interference on characteristics of dc SQUID system, IEEE Trans. Appl. Supercond. **3**, 1910-1913.
105. Koch, R.H., Foglietti, V., Rozen, J.R., Stawiasz, K.G., Ketchen, M.B., Lathrop, D.K., Sun, J.Z. and Gallagher, W.J. (1994) Effects of radio frequency radiation on the dc SQUID, Appl. Phys. Lett. **65**, 100-102.
106. Knappe, S., Drung, D., Schurig, T., Koch, H., Klinger, M. and Hinken, J. (1992) A planar YBCO gradiometer at 77K, Cryogenics **32**, 881-884..
107. Daalmans, G.M., Bär, L., Kühnl, M., Uhl, D., Selent, M. and Ramos, J. (1995) Single layer YBCO gradiometer, IEEE Trans. Appl. Supercond. **5**, 3109-3112,
108. Matlashov, A.N., Koshelets, V.P., Kalashnikov, P.V., Zhuravlev, Yu.E., Slobodchikov, V.Yu., Kovtonyuk, S.A. and Filippenko, L.V. (1991) High sensitive magnetometers and gradiometers based on dc SQUIDs with flux focusers, IEEE Trans. on Magnetics, **27**, 2963-2966.
109. Erne, S.N., Hahlbohm, H.-D., Scheer, J. and Trontelj, Z. (1981), The Berlin

shielded room - performances, in: S.N. Erne, H.-D. Hahlbohm and H. Lübbig (editors) *Biomagnetism*, de Gruyter-Verlag, Berlin, 79-87.

110. Koelle, D., Miklich, A.H., Dantsker, E., Ludwig, F., Nemeth, D.T., Clarke, J., Ruby, W. and Char, K. (1993) High performance dc SQUID magnetometers with single layer YBCO flux transformers, Appl. Phys. Lett. **63**, 3630-3632.

111. Dantsker, E., Ludwig, F., Kleiner, R., Clarke, R., Teepe, M., Lee, L.P., McN. Alford, N. and Button, T. (1995) Addendum: "Low noise YBCO-SrTiO$_3$-YBCO multilayers for improved superconducting magnetometers", Appl. Phys. Lett. **67**, 725-726.

112. Drung, D., Ludwig, F., Müller, W., Steinhoff, U., Trahms, L., Koch, H., Shen, Y.G., Jensen, M.B., Vase, P., Holst, T., Freltoft, T. and Curio, G. (1996) Integrated YBCO magnetometer for biomagnetic measurements, Applied Physics Lett. **68**, 1421-1423.

113. Shen, Y.Q., Sun, Z.J., Kromann, R., Holst, T., Vase, P. and Freloft, T. (1995) Integrated high Tc superconducting magnetometer with multiturn input coil and grain boundary junctions, Appl. Phys. Lett. **67**, 2081-2083.

114. Kingston, J.J., Wellstood, F.C., Lerch, P. Miklich, A.H. and Clarke, J. (1990) Multilayer YBCO-SrTiO$_3$-YBCO films for insulating crossovers, Appl. Phys. Lett. **56**, 189-191.

115. Oh, B., Koch, R.H., Gallagher, W.J., Robertazzi, R.P. and Eidelloth, W. (1991) Multilevel YBCO flux transformers with high-T_c SQUIDs, Appl. Phys.Lett. **59**, 123-125.

116. Lee, L.P., Char, K., Colclough, M.S. and Zaharchuk, G. (1991) Monolithic 77K dc SQUID magnetometer, Appl. Phys. Lett. **59**, 3051-3053.

117. Grundler, D., David, B., Eckart, R. and Dössel, O. (1993) Highly sensitive YBCO dc SQUID magnetometer with thin-film flux transformer, Appl. Phys. Lett. **63**, 2700-2702.

118. Ludwig, F., Koelle, D., Dantsker, E., Nemeth, D.T., Miklich, A.H. and Clarke, J. (1995) Low noise YBCO-SrTiO$_3$-YBCO multilayers for improved super-conducting magnetometers, Appl. Phys. Lett. **66**, 373-375.

119. DiIorio M.S., Yang, K.-Y. and Yoshizumi, S. (1995) Biomagnetic measurements using low-noise integrated SQUID magnetometers operating in liquid nitrogen, Appl. Phys. Lett. **67**, 1926-1928.

120. Gassig, U., Schubert, J. and Zander, W. (1992), unpublished KFA results.

121. Ma, Q.Y. (1994) Multilayer HTS device processing with impurity ion implantation, Extended abstracts of HTSED '94 (Whistler, B.C.), **FED-136**, 210-213.

122. Keene, M.N., Satchell, J.S., Goodyear, S.W., Humphreys, R.G., Edwards, J.A., Chew, N.G. and Lander, K. (1995) Low noise HTS gradiometers and magnetometers constructed from YBCO/PBCO thin films, IEEE Trans. Appl. Supercond. **5**, 2923-2926 (CAM junctions used in that work are not discussed in this chapter).

123. Ockenfuss, G. (1995) Ph.D. Dissertation, Univ. Giessen, Germany (unpublished).

124. Zimmermann, J.E. (1971) Sensitivity enhancement of SQUIDs through the use of fractional-turn loops, J. Appl. Phys. **42**, 4483-4487.

125. Drung, D., Cantor, R., Peters, M., Scheer, H.J. and Koch, H. (1990) Low-noise high-speed dc SQUID magnetometer with simplified feedback electronics, Appl. Phys. Lett. **57**, 406-408.

126. Drung, D., Cantor, R., Peters, M., Ryhänen, T. and Koch, H. (1991) Integrated dc SQUID magnetometer with high dV/dB, IEEE Trans. on Magnetics, **27**, 3001-3004.

127. Drung, D., Knappe, S. and Koch, H. (1995) Theory of the multiloop dc SQUID magnetometer and experimental verification, J. Appl. Phys. **77**, 4088-4098.

128. Ludwig, F., Dantsker, E., Kleiner, R., Koelle, D., Clarke, J., Knappe, S., Drung, D., Koch, H., McN Alford, N. and Button, T.W. (1995) Integrated high-T_c multiloop magnetometer, Appl. Phys. Lett. **66**, 1418-1420.

129. David, B., Grundler, D., Krey, S., Doormann, V., Eckart, R., Krumme, J.P., Rabe, G. and Doessel, O. (1996) High-Tc SQUID magnetometers for biomagnetic measurements, Supercond. Sci. Technol. **9**, 96A-99A.

130. Drung, D., Dantsker, E., Ludwig, F., Koch, H., Kleiner, R., Clarke, J., Krey, S., Reimer, D., David, B. and Doessel, O. (1996) Low-noise YBCO SQUID magnetometers operated with additional positive feedback, Appl. Phys. Lett. **68** 1856-1858.

131. de Waal, V.J., van Nieuvenhuyzen, G.J. and Klapwijk, T.M. (1983) Design and performance of integrated dc SQUID gradiometers, IEEE Trans. on Magnetics **19**, 648-651.

132. Sweeny, M.F. (1985) An all-thin-film SQUID for ambient field operation, IEEE Trans. on Magnetics **21**, 656-657.

133. Masalov, V.V., Samoos, A.N., Matlashov, A.N., Slobodchikov, V.Y. and Maslennikov, Y.V. (1995) Multi-loop, self-shielded dc SQUID with meander-shaped input coil, IEEE Trans. Appl. Supercond. **5**, 3238-3240.

134. Mück, Diehl, D. and G., Heiden, C. (1990) Planar microwave rf SQUID gradiometer, Cryogenics **30**, 1149-1151.

135. Zhang, Y., Mück, M., Bode, M., Herrmann, K., Schubert, J., Zander, W., Braginski, A.I. and Heiden, C. (1992) Microwave rf SQUID integrated into a planar YBCO resonator, Appl. Phys. Lett. **60**, 2303-2305.

136. Zhang, Y., Mück, M., Braginski. A.I. and Töpfer, H. (1994) High-sensitivity microwave rf SQUID operating at 77K, Supercond. Sci. Technol. **7**, 269-272.

137. Gottschlich, M., Zhang, Y., Soltner, H., Zander, W., Schubert, J. and Braginski, A.I. (1995) Investigation of HTS rf SQUIDs with planar tank circuits, in D. Dew-Hughes (ed.) *Applied Superconductivity 1995*, **2**, 1553-1556.

138. Kornev, V.K., Likharev, K.K., Snigirev, O.V., Soldatov, Ye.S., Khanin, V.V., (1980) Radio Eng. Electronic Phys. **25**, 122-125.

139. Zhang, Y., Gottschlich, M., Soltner, H., Sodtke, E., Schubert, J., Zander, W. and Braginski, A.I. (1995) Operation of high-temperature rf SQUID magnetometers using dielectric $SrTiO_3$ resonators, Appl. Phys. Lett. **67**, 3183-3185.

PULSE TUBE REFRIGERATORS:
A COOLING OPTION FOR HIGH-T$_c$ SQUIDs

C. HEIDEN
University of Giessen
Institute of Applied Physics
Heinrich-Buff-Ring 16
D-35392 Giessen
Germany

KEYWORDS / ABSTRACT : HTS SQUID / YBCO SQUID / microwave SQUID / electronic SQUID gradiometer / SQUID cooling / pulse tube coolers / low noise cooling / mechanical vibrations / maintenance intervals / remote compressor

Recent developments concerning cooling options for SQUIDs in general and pulse tube refrigerators (PTR) in particular are reviewed, and a coaxial PTR is described which was used to operate a high-T$_c$ SQUID gradiometer. In order to reduce vibrations and interference noise at the cold head, the refrigerator inlet is connected by a flexible tube of several meters length to the compressor and its rotary valve. Two YBCO microwave SQUIDs, configured as an electronic gradiometer, were successfully operated without any shielding. External background noise level could be compensated to a level of the order of $6 \times 10^{-5} \phi_0 / \sqrt{Hz}$. Periodic interference signals from the cooler were of similar magnitude. Future improvements and developments are discussed including the possibility to cool SQUIDs made of conventional superconductors by multistage PTRs.

1. Introduction

Providing appropriate cooling for SQUIDs other than by liquid cryogen filled dewars has been a challenge now for over two decades [1, 2]. Although the situation is less severe due to the advent of SQUIDs operating at liquid nitrogen temperatures, the basic requirements that a cryocooler for SQUIDs has to meet are still the same: absence, as far as possible, of electromagnetic interference signals, and of mechanical vibrations, sufficient temperature stability, good reliability, and reasonable cost. Additional specifications, such as cooling power, size and weight, depend on the particular application and system.

Thermoelectric cooling devices, like Peltier columns, might be ideal for cooling SQUIDs, since they operate without any moving parts. It is conceivable that magnetic

H. Weinstock (ed.), SQUID Sensors: Fundamentals, Fabrication and Applications, 289–305.
© *1996 Kluwer Academic Publishers.*

fields associated with the electric current through the device can be nulled at the location of the SQUID. Unfortunately, up to now, such cooling devices do not reach the needed low temperatures [3]. This situation may change if suitable new materials with better thermoelectric conversion efficiencies can be found.

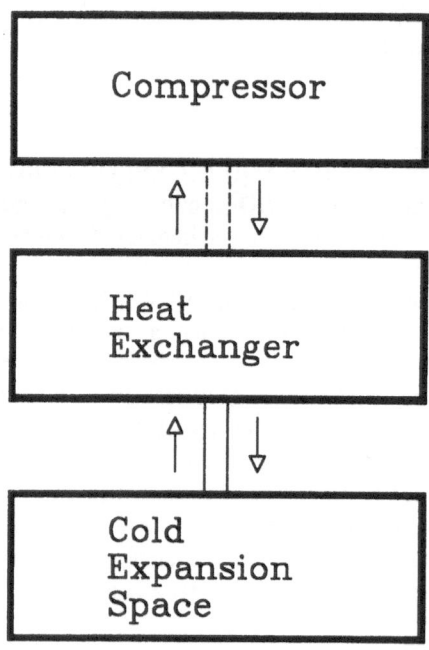

Figure 1. General scheme of refrigerators.

Major efforts towards developing suitable SQUID coolers therefore have been relying on thermodynamic cycles using appropriate gases or gas mixtures as the working fluid. The underlying principle of operation is simple (see Fig. 1): the gas, after being compressed, passes through a (warm) heat exchanger where it surrenders the heat of compression. It then is fed through an intermediate heat exchanger which connects the compressor to the cold expansion space. It serves as an efficient thermal isolation between the warm and cold part of the cooler, allowing at the same time a significant mass flow of the working fluid between these two parts. While passing through this heat exchanger, the gas is further cooled, either by exchanging heat with the porous matrix of a regenerative heat exchanger (the regenerator) or by exchanging heat in a recuperative heat exchanger with a counterflow gas stream. In the cold expansion space, the gas undergoes an expansion and absorbs heat from the surroundings, i.e., the cold heat exchanger. The whole cycle is closed by feeding the gas back to the compressor through the intermediate heat exchanger, where it picks up the heat which it gave away before. Different thermodynamic cycles can be realized using this scheme, depending on the features of the different components. A Joule-Thomson cooler will result, for instance, when using a continuous flow of high pressure gas passing through a recuperative heat exchanger and returning after isentropic expansion through a throttle valve. A Stirling refrigerator is realized by using a reciprocating compressor producing periodic pressure waves, a regenerative heat exchanger, and a thermodynamic cycle operating ideally under the restriction of constant gas volume and an isothermal expansion. Using a similar scheme, but replacing the valveless reciprocating compresssor by one with a steady output of pressurized gas, and producing the pressure waves instead by suitable valves, leads to the Gifford-McMahon refrigerator. A comprehensive survey of the different types of cryocoolers is given by Walker [4].

One of the first efforts towards developing a cryocooler for SQUIDs is the Stirling cooler by J. E. Zimmerman [5]. By using a multistage plastic displacer unit, temperatures well below the superconducting transition of niobium could be reached, whereby it was possible to demonstrate the operation of a Nb SQUID attached to the cooler. Although the feasibility of SQUID cooling by such schemes could be demonstrated, progress was slowed, if not interrupted, by the discovery of the high-T_c superconductors.

High-T_c SQUIDs that can be operated typically at temperatures in the range of liquid nitrogen are much easier to cool via cryocoolers. Nevertheless, little has been published so far concerning working high-T_c SQUID cryocooler combinations, a fact that may be due to some of the above mentioned requirements - above all, an essential reduction of electromagnetic interference and mechanical vibrations - that still are not satisfied when using commercially available refrigerators. Especially in the case of Stirling and Gifford-McMahon coolers, vibrations and magnetic interference signals are produced by an oscillating displacer in the cold expansion space.

Fortunately, there has been a development over the past decade that led to a new class of regenerative cryocoolers, the pulse tube refrigerators (PTR). These coolers - similar to the Joule-Thomson cooler - do not have moving parts in their cold space, but unlike the throttle valve in the Joule-Thomson cooler, they do not have a component that needs critical adjustment. This, in turn, is likely to have a positive impact on reliability and overall costs.

In the following, after a short outline of the working principle of pulse tube coolers, a laboratory version of such a refrigerator is described which has been used to cool a high-T_c SQUID gradiometer. Noise measurements have been carried out to characterize the performance of this SQUID-cooler combination. Possible further developements are described in the concluding section.

2. Working Principle of Pulse Tube Coolers

At present, there exist three different versions of the PTR. These are in historical order: the basic, the orifice and the double inlet type PTR, as illustrated schematically in Fig. 2. The pulse tube, consisting of a thin-walled cylinder with heat exchangers at both ends, is connected to a pressure oscillator via a regenerator. The pressure oscillation, p(t), can be generated by use of a valveless compressor like that of the Stirling refrigerator. Alternatively, the pressure oscillation can be provided with a Gifford-McMahon type compressor in combination with a flow-reversing valve which periodically connects the high- and low-pressure side of the compressor to the refrigerator. The regenerator is filled with a porous material of high heat capacity which provides good heat exchange with the working gas and has a low flow resistance. At temperatures above 30 K, a stack of metallic wire screens is commonly used as the regenerator matrix.

2.1. BASIC PULSE TUBE REFRIGERATOR

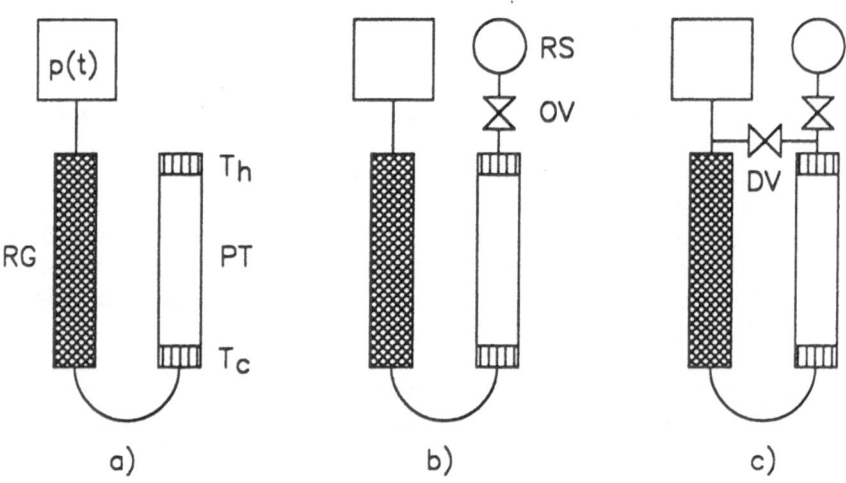

Figure 2. Schematic of the (a) basic, (b) orifice, and (c) double inlet pulse tube refrigerator. p(t): pressure oscillator; RG: regenerator; PT: pulse tube; T_h, T_c: temperatures of hot-end and cold-end heat exchanger, respectively; OV: orifice valve, DV: second-inlet valve.

The original version of the PTR, known today as the basic pulse tube refrigerator (BPTR), was invented and extensively studied by Gifford and Longsworth [6] in the early 1960s. In the BPTR the pulse tube is closed at the warm end (Fig. 2 a). In a simplified picture the working principle may be described as follows: during the compression part of the cycle, gas passing through the regenerator is cooled by heat transfer to the matrix. The gas entering the pulse tube then compresses adiabatically the gas already in the tube. The compression causes heating and motion of the gas toward the hot-end heat exchanger where some heat of compression is absorbed. In the following expansion part of the cycle, the gas in the tube is adiabatically cooled, and the temperature of the gas flowing out of the tube is lower than the temperature at which it entered. Therefore, heat is absorbed from the cold-end heat exchanger. Finally, the gas is warmed again while flowing in the regenerator toward the pressure oscillator.

Taking into account the typical high-to-low pressure ratios of less than four in a BPTR [7], shows that at high pressure the gas column in the pulse tube is compressed to only one half of its length at low pressure. This means that gas flowing into the tube through the cold heat exchanger will never reach the warm end. Therefore, an additional mechanism must be present which produces a net heat transport against the temperature gradient. This mechanism results from heat exchange between gas and tube wall, and is called surface heat pumping. Consider a small (disk-shaped) gas element in the cold part of the tube. During adiabatic compression this element is heated and simultaneously displaced toward the closed end. During the quiescent period at high pressure this gas element can give off heat to the tube wall, thereby cooling itself to some extent. During the subsequent expansion, the element is moved back to the starting position, reducing its temperature below the value it had at the start of the compression cycle. After coming to rest at low pressure, the cooled element can take up heat from the wall. For each given gas element (a) there is now a corresponding one (b) whose low pressure position coincides with the high pressure position of the given element. Therefore, element (b) can absorb the heat deposited on the wall by element (a) and transport it to a location closer to the warm end. The combined action of all pairs of gas elements in the tube results in a shuttle heat transfer to the warm heat exchanger.

Two important features of refrigeration by surface heat pumping have to be noted: (1) The surface heat exchange is confined to a thin gas layer of the order of the thermal penetration depth adjacent to the tube wall. A considerable fraction of the gas volume in the tube, therefore, does not contribute to the refrigeration effect. (2) When the temperature gradient along the tube is too steep, the surface heat pumping becomes ineffective or can even reverse its direction. Both features lead to a rather poor refrigeration efficiency of the BPTR, resulting in a lowest temperature of 124 K achieved thus far in a single-stage unit [7].

2.2. ORIFICE PULSE TUBE REFRIGERATOR

In 1984 the performance of the PTR has been considerably improved by Mikulin and co-workers [8] by connecting the warm end of the pulse tube via an orifice (usually realized by a needle valve) with a buffer reservoir. This modified arrangement, which is known as the orifice pulse tube refrigerator (OPTR), is shown in Fig. 2 b. Evidently, the presence of the orifice-reservoir assembly gives rise to an increased mass flow rate at the warm end. Thus, more compressed gas can transfer heat to the warm-end heat exchanger. In the same manner, more gas will pick up heat from the cold heat exchanger. Moreover, the orifice-reservoir assembly gives rise to refrigeration even in the ideal case of an adiabatic process within the pulse tube. It introduces a phase shift of less than 90 ⁰ between oscillating mass flow at the cold end and pressure in the tube. With respect to phase shift, the effect of the orifice is similar to that of the displacer piston in a Stirling cooler.

A more quantitative understanding of the ideal OPTR evolves from an enthalpy flow analysis elaborated by Radebaugh and co-workers [9, 10]. Application of the first law of thermodynamics to the cold and warm ends of the pulse tube shows that the time-averaged enthalpy flow <H> through the tube equals the gross refrigeration power of the cooler, which is rejected at the hot-end heat exchanger. For an ideal gas

$$< H >= (C_p/\tau) \oint m(t)T(t)dt \quad , \qquad (1)$$

where τ is the oscillation period, C_p is the specific heat of the gas at constant pressure, m(t) is the mass flow rate and T(t) is the gas temperature. In the case of small sinusoidal pressure and mass flow oscillations and adiabatic gas flow, Eq. (1) can be written as [10]

$$< H > = RT_c/(2 <p>)M_c P \cos\vartheta \quad , \qquad (2)$$

where R is the gas constant per unit mass, T_c is the cold-end temperature, M_c is the amplitude of the mass flow rate at the cold end, P is the pressure amplitude, <p> is the average pressure and ϑ is the phase angle between m_c and p. A relation between the mass flows at the cold and hot ends of the pulse tube is obtained from the continuity equation:

$$m_c = (T_h/T_c)m_o + V_t/(\gamma RT_c)dp/dt \quad . \qquad (3)$$

Here m_o is the mass flow rate through the orifice (neglecting the void volume of the hot heat exchanger), $T_h \approx 300$ K is the temperature of the hot heat exchanger, V_t is the pulse tube volume, and γ is the specific-heat ratio.

For the BPTR the mass flow m_o, of course, is zero. This leads to $\vartheta = 90$ ⁰ from Eq. (3), and thus from Eq. (2), to zero enthalpy flow under adiabatic conditions in the tube. In the OPTR, m_o is proportional to the dynamic pressure drop across the orifice, and, since the pressure in the reservoir is nearly constant, m_o is approximately in phase with the pressure oscillation. Therefore, in the OPTR, $\vartheta < 90$ ⁰, and an average enthalpy flow occurs from the cold to the hot end of the pulse tube.

According to Eq. (2) the ideal OPTR has a non-zero gross refrigeration power for all cold-end temperatures $T_c > 0$. This is in marked contrast to the BPTR, where, for a given tube length and pressure ratio, the refrigeration by surface heat pumping breaks down at a certain $T_c > 0$. The OPTR thus has a higher efficiency than the BPTR. A lowest temperature of 49 K has been reported for a single-stage OPTR [11].

2.3. DOUBLE INLET PULSE TUBE REFRIGERATOR

Still higher efficiencies can be achieved with the double inlet pulse tube refrigerator (DIPTR), which was introduced in 1990 by Zhu et al. [12]. The DIPTR is a modified version of the OPTR in which an additional flow impedance connects the pressure oscillator and the warm end of the pulse tube, as shown in Fig. 2 c.

The concept of the DIPTR has been developed by realizing that the enthalpy flow in an adiabatic pulse tube occurs in the form of work flow [12] as follows. In the middle section of the pulse tube there is a part of the gas which never leaves the tube. This "permanent" gas can be considered as a compressible piston. During the compression part of the cycle, the gas entering through the cold-end heat exchanger does work on the gas piston which, on the other hand, does work by displacing gas through the hot end and the orifice. Since the gas piston is compressible, a part of the gas flowing through the cold end is needed only to compress and expand the piston volume, and is not useful for obtaining refrigeration power. In the DIPTR configuration, this "useless" mass flow is supplied through the second inlet at the hot end, thereby increasing the refrigeration power per unit mass flow through the cold end.

A more quantitative understanding of the ideal DIPTR is possible by means of Eqs. (2) and (3). In Eq. (3) the useless mass flow through the cold end is represented by the second term, which is proportional to dp/dt (that leads p by 90 ^{0}) and which therefore does not contribute to the enthalpy flow. (This second term, in an electrical analogy, describes the charging and discharging of the pulse tube capacitor). For the DIPTR, the first term in Eq. (3) has to be modified by replacing m_o by m_o-m_2, where m_2 is the mass flow through the second inlet impedance into the hot end. The optimum effect of the second inlet is achieved when m_2 is in phase with dp/dt, because then the useless mass flow in Eq. (3) can be totally compensated by adjusting the second-inlet valve properly. In this case $\vartheta = 0$, and the enthalpy flow in Eq. (2) attains its maximum value for a given cold-end mass flow rate.

In comparison to the OPTR, the DIPTR needs a lower mass flow for the same gross refrigeration power. Moreover, the reduction of the mass flow through the regenerator also reduces the regenerator loss. Both effects of the second inlet are advantageous with respect to a higher efficiency of the cooler. Up to now, a lowest temperature of 28 K has been obtained with a single-stage DIPTR [13].

In a real DIPTR, m_2 is determined by the dynamic pressure drop across the second-inlet valve, and its phase therefore, in general, is not at its optimum value. This deficiency can be overcome when the mass flow rate at the hot end is independently controlled by use of an auxiliary piston [14] or, in the case of a valved compressor, by use of a second flow-reversing valve arrangement [15]. With an auxiliary piston at the warm end of a single-stage PTR, a minimum temperature as low as 23 K has been achieved [14].

296

The enthalpy flow model can explain the order of magnitude of the measured gross cooling power of an OPTR [16]. However, it does not take into account the various loss mechanisms which determine the net cooling power of a real cooler. One kind of loss is associated with the regenerator ineffectiveness and the thermal conductance along the tubes, as in other regenerative cryocoolers. Another kind of loss is connected with the dynamics of gas flow in the pulse tube: mixing of gas by turbulence and convective streaming can introduce additional thermal losses. More realistic models of the PTR, which incorporate the fluid dynamics on the basis of thermoacoustic theory, are currently being developed (see e.g., [17, 18]).

3. A Pulse Tube Cooler For High-T_c SQUIDs

3.1. DESIGN DETAILS

Practical operation of a SQUID gradiometer by use of a PTR requires a compact design of the cooler and of its vacuum jacket, and also a reduction of mechanical and electromagnetic noise from the compressor and the rotary valve used to create the pressure oscillations in the pulse tube.

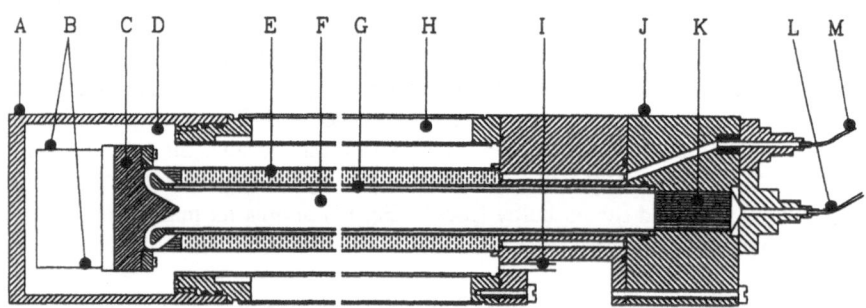

Figure 3. Cross section of coaxial pulse tube refrigerator. A: acrylic glass, B: SQUID mount, C: cold head, D: vacuum space, E: regenerator, F: pulse tube, G: thermal insulation, H: buffer volume (250 cm³), I: electrical feedthroughs, J: warm end cooled by convection, K: heat exchanger.

A cross section of the PTR designed for SQUID operation is shown in Fig. 3. In order to reduce the transverse dimension, a coaxial configuration of regenerator and pulse tube is used. Pulse tube and regenerator are made from stainless steel tubes with lengths of 250 and 205 mm, and inner diameters of 13.4 and 29 mm, respectively. The regenerator matrix consists of a stack of 220-mesh phosphorus bronze and stainless steel screens. The screens, which are annular in shape, are placed concentrically around the pulse tube.

One problem in the case of a coaxial configuration arises from the different temperature profiles along the regenerator and pulse tube, since these cause radial thermal conduction losses. In order to reduce such losses, a double wall structure is employed. This structure consists of an additional tube with inner diameter of 17 mm, which is welded to the ends of the pulse tube. The space between both tubes is filled with polyurethane foam. The part of the refrigerator which is located outside the vacuum insulation and to which the pulse tube and regenerator are attached, is fabricated of two cylindrical copper blocks containing the vacuum flange, gas inlets, the warm-end heat exchanger, and electrical feed-throughs. Axial grooves are machined into the surface of the copper blocks in order to enhance the cooling by air convection. The warm-end heat exchanger consists of a copper cylinder with thin axial flow channels providing heat exchange and flow straightening. The inner surface of the cold head (made of copper) is conically shaped, and serves for heat transfer and guidance of the gas flow. The design of the cold head is similar to that described in Ref. 14.

The vacuum vessel, with an outer diameter of 67 mm, consists of two pieces: the upper part, made from stainless steel, contains the reservoir in the form of an annular tank; the lower part, surrounding the cold head, is fabricated from acrylic glass. The total length of the cooler with vacuum vessel in Fig. 3 is 34 cm.

The entire cooler schematic is as shown in Fig. 4. A standard compressor for Gifford-McMahon coolers is used, and a motor-driven rotary valve serves to control the helium gas flow between the compressor and refrigerator inlets, as illustrated in Fig. 4, and is described in more detail in Ref.16.

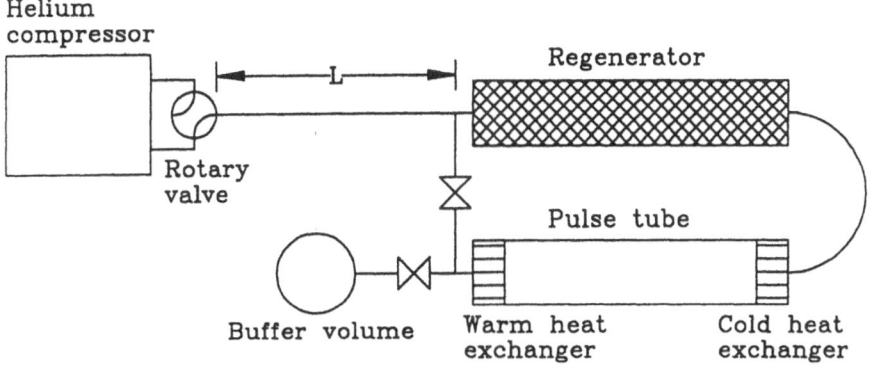

Figure 4. Schematic of double inlet type pulse tube cooler with all major components.

To reduce noise, the refrigerator is connected to the rotary valve by means of a flexible polyimide tube with inner diameter of 3 mm and with lengths that up to now have been chosen between $L = 1$ m and 4 m.

The two YBCO rf-SQUIDs, used in the configuration of an electronic gradiometer, are mounted at present on opposite sides of an aluminum block which is bolted to the cold head and is designed to give a base-line length of 45 mm - see Fig. 5. In order to reduce thermal noise originating from the entire structure, this aluminum stage can be replaced by a sapphire block of similar form. Each SQUID is placed in the center of an S-shaped microwave strip resonator, which is prepared from a laser-deposited YBCO film on a LaAlO$_3$ substrate. Coupling to the room temperature electronics is made by means of two semi-rigid coaxial cables (with outer diameter of 1.2 mm). The operating frequency is 3 GHz. A more detailed description of the SQUID preparation and the electronic readout system can be found in Refs. 19 and 20.

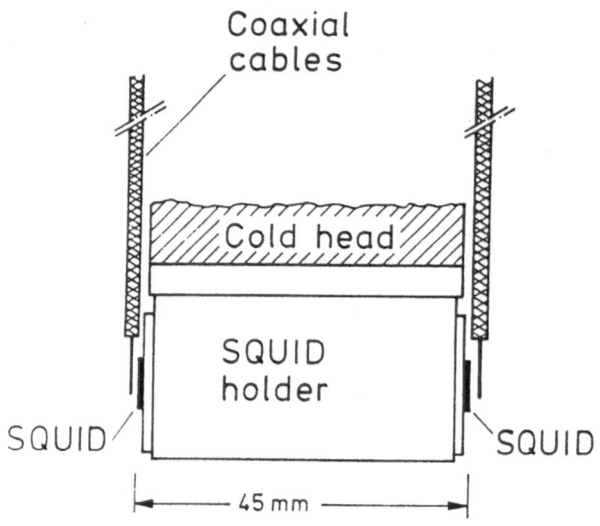

Figure 5. Configuration of sensor head with two microwave SQUIDs mounted on cooling block.

3.2. COOLER PERFORMANCE

The best performance of this coaxial PTR has been achieved in a double inlet configuration at a frequency near 2 Hz for the pressure wave. The average pressure in the cooler is typically 18 bar. The peak-to-peak pressure difference at the cooler inlet is $\Delta p = 7.8$ bar for a flexible tube length of $L = 1$ m and it decreases to $\Delta p = 7.2$ bar for $L = 4$ m.

Figure 6 shows the cool-down curves of the PTR with and without the SQUID gradiometer installed. In both cases no superinsulating foil was used in the acrylic glass region of the vacuum vessel. With the gradiometer installed, the lowest temperature increases from 48.6 K to 52 K, and a longer cool-down time of about 2.5 hours is needed. The rather long cool-down period in both cases is caused by the large heat capacity (≈ 125 J/K) of the copper cold head with the aluminum SQUID-holder attached. Increasing the length of the flexible tube in the range $L = 1 - 4$ m is found to increase the minimum temperature by about 0.4 K/m.

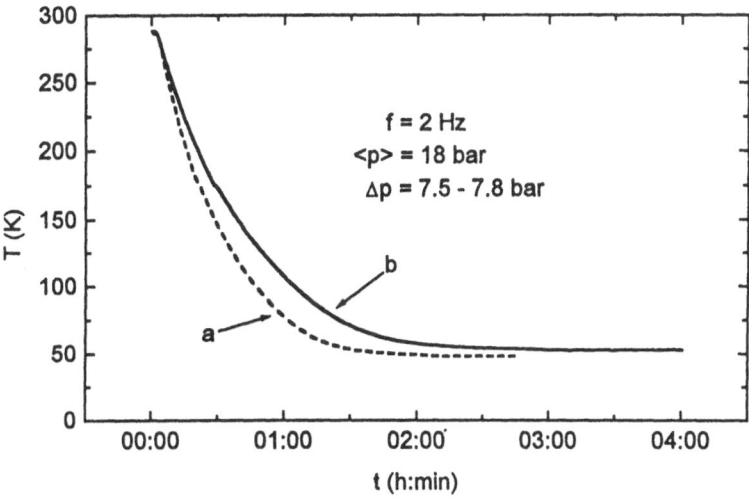

Figure 6. Cool-down behavior of coaxial PTR without (a) and with (b) attached SQUID gradiometer.

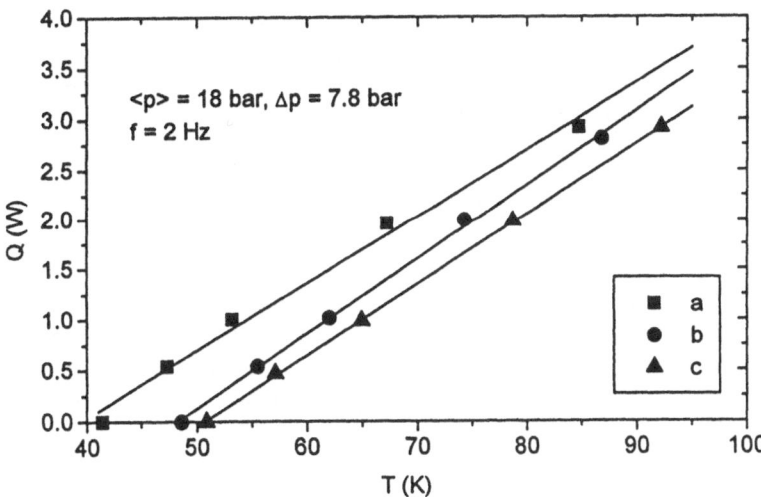

Figure 7. Temperature dependence of net cooling power (a) without gradiometer but with superinsulation around cold head, (b) without gradiometer and without superinsulation, and (c) with gradiometer but without superinsulation.

The measured net cooling power dQ/dt as a function of cold-end temperature for L = 1 m
with and without the gradiometer is displayed in Fig. 7. The thermal load was introduced by
a resistive heater fixed to the SQUID holder. The linear variation of dQ/dt with T is
explained in Ref. 16.

With proper radiation shielding of the cold head, a minimum no-load temperature of 41 K
and a net cooling power of dQ/dt = 2.8 W at 80 K are attained. Upon removal of the
superinsulation and installation of the gradiometer, the no-load temperatures increase to
49 K and 51 K, respectively. The slope of dQ/dt, however, remains nearly unaltered. The
vertical offset of $\approx$ 0.2 W between the two lower lines in Fig. 7, is due to the heat load from
the gradiometer, which is mainly due to thermal conductance along the microwave cables.
The net cooling power is found to decrease with increasing L by approximately 0.1 W/m.

The observed comparable weak dependence of net cooling power on the length of the
flexible plastic connecting line is a very attractive feature since it provides the possibility to
remove the sensor head sufficiently far from noise generating parts of the system, such as the
compressor and the rotating valve. It also provides the opportunity to build a system, e.g., for
nondestructive evaluation, which allows one to move the SQUID along 3-dimensional
trajectories.

4. Noise Characterization of the High-T_c SQUID, Pulse Tube Cooler Combination

For measurements of the flux noise, the SQUIDs were operated in a flux-locked loop in an
unshielded environment. The length of the flexible connecting tube to the rotary valve was
L = 2.2 m, resulting in a temperature of the cold head of 52 K. The compressor was placed
at a distance of 3 m from the rotary valve.

Figure 8 shows the measured equivalent flux noise at frequencies below 25 Hz. The data
were obtained for an electronic gradiometer configuration with three different distances
between the rotary valve and the cold head. At a distance of 1 m (Fig. 8 a) the flux noise
spectrum exhibits two peaks of the order of $3 \times 10^{-3} \phi_0 / \sqrt{Hz}$ located at 1 and 2 Hz. These
peaks can be related to the rotation of the valve shaft that occurs with a frequency of 1 Hz.
With increasing distance, these interference signals become weaker, and at 2 m they are
comparable in magnitude to the white background noise, which is of the order of 6×10^{-5}
$\phi_0 / \sqrt{Hz}$ -see Fig. 8 c. This white flux noise implies that the gradient noise of our laboratory
is equivalent to a field noise of a few pT/($\sqrt{Hz}$ cm). The observed reduction of the two
interference signals with increasing distance between SQUIDs and the rotary valve (at a
fixed flexible tube length) clearly shows that they are caused mainly by the rotary valve.

Increasing the distance to a value of 4 m, however, produced no further reduction of the
periodic interference signal, indicating a remaining noise source at the cold head. Various
possibilities were examined to account for this effect: (a) periodic temperature variations of
the sensor head with the frequency of the pressure waves; (b) periodic variations of the
diamagnetic moment of the helium gas in the pulse tube under the presence of a constant
background field (i. e., the earth's magnetic field), and (c) elastic mechanical oscillations of

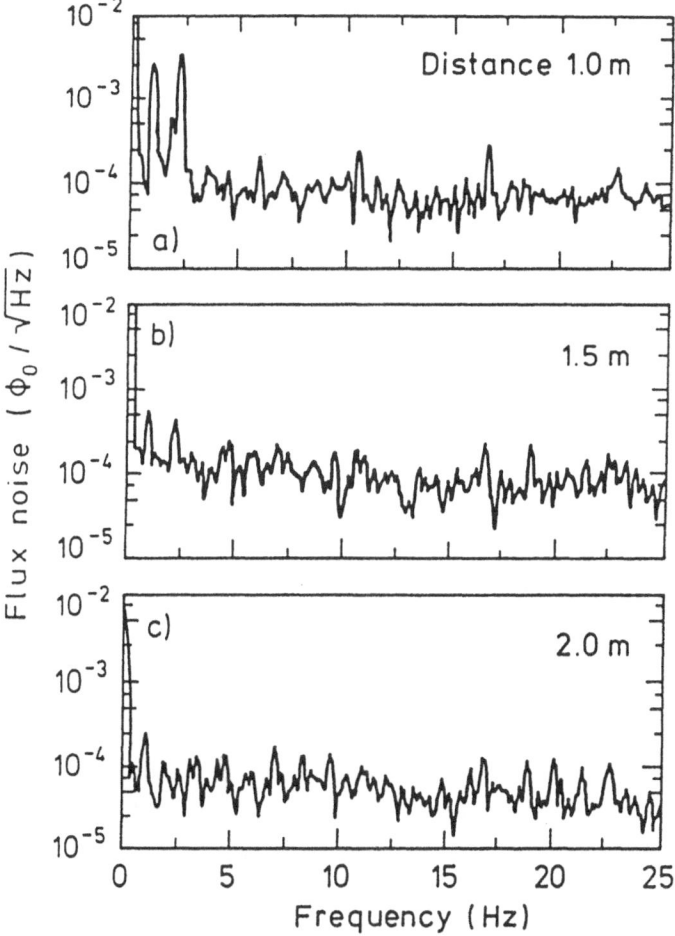

Figure 8. Flux noise of electronic SQUID gradiometer for three distances between sensor head and rotary valve.

the pulse tube due to the periodic pressure oscillations. With temperature variations of about 4 mK amplitude at the SQUID location, and an estimated magnetic helium dipole field signal of the order of 20 fT amplitude at the SQUIDs (calculated for a background field of 10^{-4} T), sources (a) and (b) are ruled out for the present experimental setup.

Pressure wave induced elastic oscillations of the pulse tube, therefore are considered as the cause of the remaining periodic interference signal. For the given pressure amplitude and pulse tube design, a length oscillation of 12 µm was calculated. This "breathing" of the structure at the cold head can, under the presence of a magnetic field gradient, produce the observed signal. As a possible source for the local magnetic field gradient, stainless steel components with a remanent magnetization could be identified. These components will no longer be used in a forthcoming improved construction of the cold head.

5. Future Developments

There are opportunities for further system improvements concerning different aspects, such as (a) further reduction of interference signals from the cooler itself, (b) improvements in system handling, and (c) reduction of operating temperature to a level appropriate for SQUIDs made of conventional superconductors such as niobium.

(a) For applications requiring higher sensitivity, it probably will be necessary to deal with the periodic elastic deformations of the pulse tube. One way to overcome this intrinsic problem is to mount the SQUID on a separate platform that is not mechanically anchored to the pulse tube. Flexible straps [21] of high thermal conductivity may be useful for heat transfer from this platform to the cold end of the pulse tube.

(b) Apart from being "quiet", ease of operation will be a decisive criterion for the acceptance of pulse tube refrigerators for cooling SQUIDs and SQUID systems. One step in this direction will be the use of smaller compressors with an input power of, say 0.5 kW, as used already in a commercial Joule-Thomson system [22]. Increasing the coefficient of performance, that is net cooling power divided by input power, is an important task in this context. In order to accomplish this, we must better understand and control the different loss mechanisms in the pulse tube cooler.

Appropriate vacuum technology for adequate thermal insulation (e.g., simultaneous use of the cold head for cryopumping) and an overall systems approach including SQUID electronics and the necessary electrical (or fiber optical) connections to the sensors, are further issues to be addressed.

Reduction of size also will be of interest for the proper pulse tube cooler. A reduction in volume by a factor of two appears to be feasible for the present design. It is doubtful, however, at the present time whether miniature pulse tube refrigerators, like those developed recently by TRW [23], can be used for SQUID cooling because of the inherent proximity of the compressor and its stray magnetic field.

(c) Recently it has been possible to achieve temperatures in the range of liquid ^{4}He by using multistage pulse tube refrigerators [24] with ErNi alloys as the regenerator material in the coldest stage. Such alloys [25] undergo a magnetic phase transition at temperatures in the vicinity of 10 K, resulting in a corresponding maximum in their specific heat as a function of temperature. The regenerator, therefore, can be operated to much lower temperatures before its heat storage function is exhausted. Figure 9 shows the net cooling power at the second stage of a liquid nitrogen precooled two-stage PTR for a total ^{4}He mass-flow input of about 1 g/s. Er_3Ni shot was used as the regenerator material in the second stage [26]. For optimum adjustment of the double inlet mode that was used for both stages, a lowest temperature of 2.2 K could be achieved. So far, this is the lowest temperature that has been reached using a multistage pulse tube cooler. It probably will be difficult to achieve lower temperatures with ⁴He as the working fluid because of its superfluid transition at the lambda line.

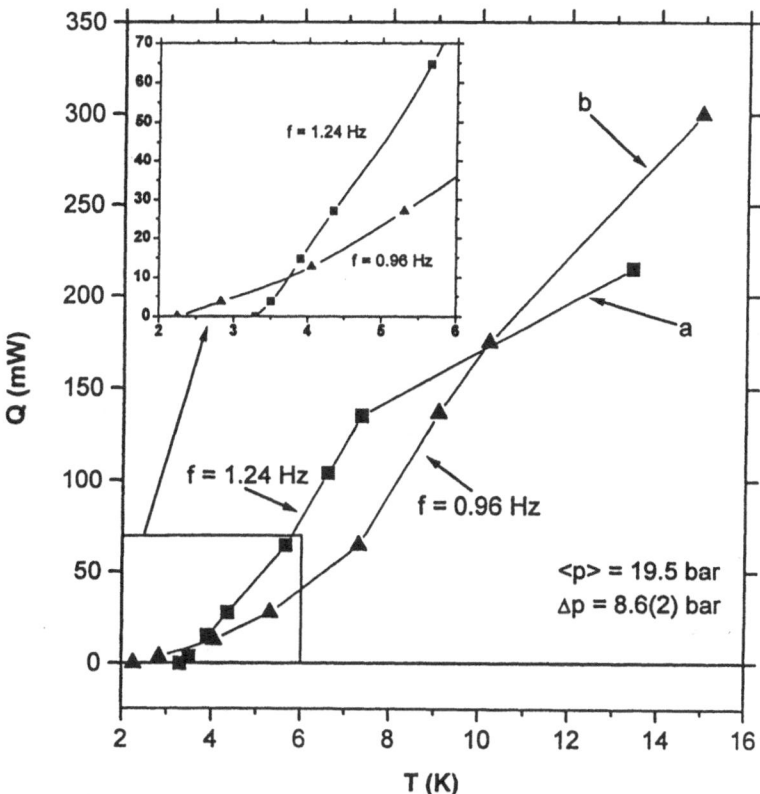

Figure 9. Net cooling power at the second stage of a liquid nitrogen precooled two-stage double inlet PTR. The temperature dependences shown for two pressure wave frequencies result from different settings of the orifice and double inlet valves: (a) for optimum net cooling power, and (b) for lowest temperature.

Reaching liquid ⁴He temperatures with pulse tube refrigerators offers the possibility of cooling SQUIDs made of conventional superconductors. Due to the absence of moving parts or components that need critical adjustments, this cooling technique may be of particular interest in cases where cooling for long uninterrupted periods is required. It remains to be seen whether the presence of magnetic material in the coldest stage of the refrigerator will be a serious problem for this PTR application.

304

Acknowledgement

The cooperative efforts of G. Thummes, M. Mueck, S. Bender, Y. Kuecuekkaplan, R. Landgraf and M. Wagner, leading to the design, construction and testing of the PTRs described in this report, are gratefully acknowledged. The YBCO SQUIDs were prepared at the Institute of Thin Film and Ion Technology of the Research Center Juelich (KFA) and kindly supplied by Y. Zhang. The project is supported by the German Ministry of Science and Technology under contract No. 13N6176.

6. References

1. Zimmerman, J. E., and Sullivan, D. B. (1982) A Study of Design Principles for Refrigerators for Low-Power Cryoelectronic Devices, NBS Technical Note 1049, U. S. Government Printing Office, Washington D. C. 20234.
2. Heiden, C. (1989) Cryogenics for Superconducting Electronics, in H. Weinstock and M. Nisenoff (eds.), Superconducting Electronics, Springer-Verlag Berlin, Heidelberg, pp. 409-429.
3. Buist, J. R., Fenton, J., Lichniak, G., and Norton, P. (1976) Low Temperature Thermoelectric Cooler for 145 K Detector Array Package, U. S. Army Night Vision Laboratories, Fort Belvoir, Virginia, Technical Report ADB008934 (Borg-Warner Thermoelectrics).
4. Walker, G., (1983) Cryocoolers, Part 1 and Part 2, Plenum Press, New York, USA.
5. Sullivan, D. B., Zimmerman, J. E., and Ives, J. T.(1981) Operation of a Practical SQUID Gradiometer in a Low-Power Stirling Cryocooler, in J. E. Zimmerman, D. B. Sullivan, and S. E. McCarthy (eds), Refrigeration for Cryogenic Sensors and Electronic Systems, NBS Special Publication SP-607, Superintendent of Documents, U. S. Government Printing office, Washington, D.C. 20234, pp. 186 - 194.
6. Gifford, W. E. and Longsworth, R. (1965) Pulse Tube Refrigeration Progress, Adv. Cryo. Eng., **10 B**, 69.
7. Longsworth, R. C. (1967) An Experimental Investigation of Pulse Tube Heat Pumping Rates, Adv. Cryo. Eng., **12**, 608.
8. Mikulin, E. I., Tasarov, A. A., and Shkrebyonock, M. P. (1984) Low-Temperature Expansion Pulse Tubes, Adv. Cryo. Eng., **29**, 629.
9. Storch, P. J. and Radebaugh, R. (1988) Development and Experimental Test of an Analytical Model of the Orifice Pulse Tube, Adv. Cryo. Eng., **33**, 851.
10. Radebaugh, R. (1990) A Review of Pulse Tube Refrigeration, Adv. Cryo. Eng., **35**, 1191.
11. Liang, J., Zhou, Y., and Zhu, W. (1990) Development of a single-stage pulse tube refrigerator capable of reaching 49 K, Cryogenics, **30**, 49.
12. Zhu, S. and Wu, P. (1990) Double inlet pulse tube refrigerators: an important improvement, Cryogenics, **30**, 514.
13. Ravex, A., Rolland, P., and Liang, J. (1992) Experimental study and modelisation of a pulse tube refrigerator, Cryogenics, **32**, ICEC 14 Supplement p. 9.
14. Ishizaki, Y. and Ishizaki, E. (1992) Experimental performance of modified pulse tube refrigerator below 80 K down to 23 K, Proc. 7th Int. Cryocooler Conference, Santa Fe, p. 140.
15. Matsubara, Y., Tanida, K. Gao, J. L., Hiresaki, Y., and Kaneko, M. (1993) Four valve pulse tube refrigerator, Proc. 4th Joint Sino-Japanese Seminar on Cryocoolers and Concerned Topics, Chinese Academy of Sciences, Beijing, p. 54.
16. Thummes, G., Giebeler, F., and Heiden, C. (1995) Effect of pressure wave form on pulse tube refrigerator performance, Cryocoolers 8, R.G. Ross, Jr. (ed.), Plenum Press, New York, p. 383.
17. Luo, E. R., Xiao, J. H., Zhou, Y. (1993) A simplified thermoacoustic modeling for pulse tube refrigerator, Proc. 4th Joint Sino-Japanese Seminar on Cryocoolers and Concerned Topics, Chinese Academy of Sciences, Beijing, p. 94.
18. Lee, J. M., Kittel, P., Timmerhaus, K. D., and Radebaugh, R. (1995) Steady secondary momentum and enthalpy streaming in the pulse tube refrigerator, Cryocoolers 8, R. G. Ross, Jr. (ed.), Plenum Press, New York, p. 359.
19. Mück, M. (1993) Progress in rf SQUIDs, IEEE Trans. Appl. Supercond. AS-3, 2003.
20. Mück, M. (1992) A Readout Electronics for 3 GHz rf SQUIDs, Rev. Sci. Instrum. **63**, 2268.
21. Sparr, L., Boyle, R., Loc, N., Frisch, H., Banks, S., and James, E. (1994) Design and Test of Potential Cryocooler Cold Finger Interfaces, Adv. Cryo. Eng. **39**, 1253.

22. A 0.5 kW compressor is used in the "CRYOTIGER" Cooling system of APD Cryogenics, Inc., Allentown, PA 18103, USA.

23. Tward, E., Chan, C. K., Raab, J., Orsini, R., Jaco, C., and Petach, M. (1995) Miniature Long-Life Space-Qualified Pulse Tube and Stirling Cryocoolers, Cryocoolers 8, R. G. Ross, Jr. (ed.), Plenum Press, New York, p. 329.

24. Matsubara, Y., and Gao, J. L. (1995) Multi-Stage Pulse Tube Refrigerator for Temperatures below 4 K, Cryocoolers 8, R. G. Ross, Jr. (ed.), Plenum Press, New York, p. 345.

25. Hashimoto, T., Yabuki, M., Eda, T., Kuriyama, T., and Nakagome, H. (1994) Effect of High Entropy Magnetic Regenerator Materials on Power of the GM Refrigerator, Adv. Cryog. Eng. **40A**, 665.

26. Thummes, G., Bender, S., and Heiden, C. (1996) Approaching the ^{4}He-Lambda Point with a Liquid Nitrogen Precooled Two Stage Pulse Tube Refrigerator, Cryogenics, submitted for publication.

HIGH-RESOLUTION MAGNETIC IMAGING: CELLULAR ACTION CURRENTS AND OTHER APPLICATIONS

J.P. WIKSWO, JR.
Vanderbilt University
Department of Physics and Astronomy
Box 1807 Station B
Nashville, TN 37235, USA

Abstract. The SQUID magnetometers used for biomagnetic studies typically have pickup coils with a 1- to 3-cm diameter that are at a comparable distance from the biological sample. While multiple measurements with such systems can be used to localize dipole-like sources with millimeter accuracy, the ability to resolve two adjacent dipoles, *i.e.*, the imaging resolution, is limited to approximately the source-to-coil separation of 1 to 2 cm. While such imaging resolution may be sufficient for many studies on humans, it is inadequate for studies of action currents at the cellular level, where the characteristic dimensions can be on the order of several hundred microns, and for many other applications, such as nondestructive evaluation (NDE). High resolution SQUID magnetometers, with miniature pickup coils a millimeter from the sample, have been used for a wide variety *of in vivo* and *in vitro* biomagnetic measurements, such as recording magnetic fields from nerves, skeletal muscle, cardiac tissue, intestinal smooth muscle, developing embryos, and the brain. Several of these studies and numerical simulations are presented to demonstrate the potential benefits of high-resolution magnetic imaging, and to provide the basis for understanding the factors that govern the spatial resolution of SQUID images of magnetic fields. The use of simple scaling arguments demonstrates that the performance enhancements achieved by minaturization of SQUID microscopes and arrays are governed not only by the sensitivity of the SQUID sensor, but also by the rate of fall-off of the field produced by the elements of a distributed source. Techniques are reviewed for the optimization of high-resolution SQUIDs for both biomagnetism and NDE. Because of the loss in sensitivity as pickup coils or SQUIDs are made smaller, high resolution SQUID imaging will definitely benefit from the development of lower noise SQUID sensors.

H. Weinstock (ed.), SQUID Sensors: Fundamentals, Fabrication and Applications, 307–360.
© *1996 Kluwer Academic Publishers.*

1. Introduction

SQUID systems developed for biomagnetic studies on humans have cryogenic pickup coils with a 1- to 3-cm diameter and a comparable distance
between the cryogenic pickup coils and the outside of the dewar. While
reducing the separation between the pickup coils and the subject might increase the signal strength somewhat, for these studies the thickness of the
skull or chest wall and the depth of the sources within the brain or heart
are the limiting factors. These biomedical SQUID systems cannot provide
either the sensitivity or spatial resolution required for detailed imaging of
action currents in isolated living tissue. Because the primary goal of the
research described in this chapter is to obtain high resolution (100 μm to
1 mm) images from biological samples, the engineering challenge, stated
most simply, is to keep the SQUID cold and the sample warm and wet.
This requires both good thermal insulation between the SQUID and the
sample, and the ability to keep the SQUID and/or the pickup coil below
its superconducting transition temperature.

In this chapter, we first demonstrate the need for higher resolution magnetometers for biomagnetic measurements. We then review the work to date
to obtain higher resolution for biological studies, including the work with
room-temperature toroidal current probes, the MicroSQUID magnetometers, and the recently introduced SQUID microscopes. We discuss the difference between imaging and localization resolution and show how they are
affected by pickup-coil diameter and coil-to-sample separation. Finally, we
examine how scaling laws apply to magnetic imaging and review techniques
to obtain the maximum possible spatial resolution.

While this chapter is directed towards the high-resolution imaging of
the magnetic fields produced by action currents, remanent magnetization,
and magnetic susceptibility in isolated living tissue and small experimental
animal preparations at the scale of 100 μm, the discussion of high resolution
SQUIDs is clearly applicable to the imaging of currents applied to metallic structures for nondestructive evaluation (NDE), and for determining
the distribution of diamagnetic, paramagnetic or ferromagnetic magnetization in inanimate objects. These applications are covered in detail in a
recent review [32]. This review draws heavily from several previous papers
[126, 122, 150, 78, 98, 97, 74, 144, 127, 147]. The more general problem of
mathematical deconvolution of magnetic field maps is treated in detail in
an accompanying chapter [129].

1.1. ACTION CURRENTS IN THE HEART

Before delving into the spatial resolution of SQUID magnetometers, it is
useful first to examine cardiac action currents as an example of electro-

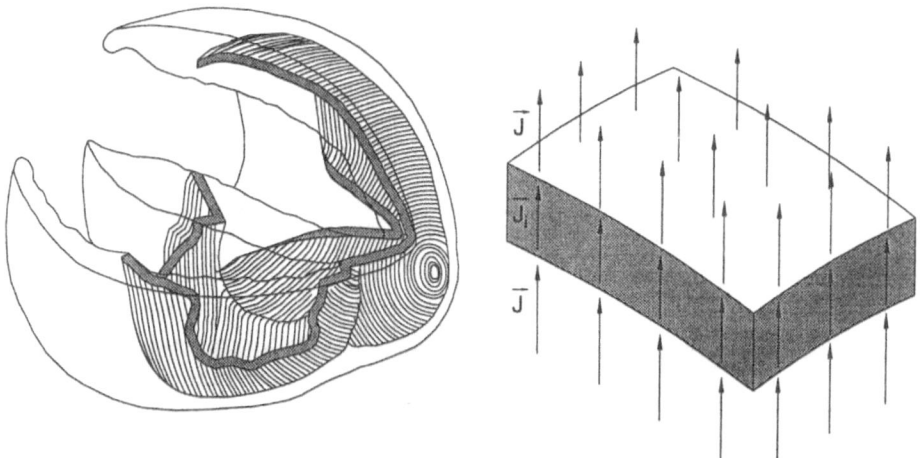

Figure 1. A depolarization wavefront propagating through the heart. (left) The wavefront 30 ms after the onset of ventricular depolarization. The near chamber is the right ventricle, and the far one is the left ventricle. Current leaves the outer surface, and reenters at the inner. (right) A representation of the current sources within the wavefront in terms of an impressed current density $\vec{J_i}$ and the Ohmic current $\vec{J}$. (From [126], with permission)

physiology and distributed bioelectric current sources that are of current interest and suitable for magnetic imaging.

1.1.1. *Propagating Action Potential Wavefronts in the Heart*

The heart is a tandem, two-stage, electrically-activated, metabolically-powered pump with a mean time to failure of 2×10^9 cycles. The contraction of the cardiac muscle is initiated by a spreading wave front of electrical activation in which the electrical potential across the cellular membrane changes in a millisecond or two from approximately -70 mV to +30 mV. This wavefront propagates with a velocity of up to 0.5 m/s. The activation of the ventricles of the heart, *i.e.*, the two high-pressure chambers, is synchronized by the nearly simultaneous activation of the epicardial (inner) surface of each chamber by a network of rapidly conducting Purkinje fibers that are connected by the atrioventricular node to the cardiac pacemaker in the sinoatrial node in the right atrium, one of the low pressure chambers. As a result of the combined effects of the simultaneous activation of the ventricular endocardium, the millisecond rise time of the transmembrane potential, and the rapid myocardial conduction velocity, the activation wavefront in the heart is only a millimeter or so in thickness. Figure 1 shows the wavefront 30 ms after the beginning of ventricular activation, when the wavefront has moved through half of the thickness of the ventricular wall. This wavefront can be best thought of as a thin, moving battery, with current leaving the outer, positive surface and entering the inner, negative one. Several very important observations can be made from this picture. First, while the potential and magnetic field distributions recorded at the surface of the chest may appear dipolar, the electrical source within the heart is clearly not a point electrical dipole. Furthermore, because of an approximate cancellation of the field contributions from sections of the wavefront on opposite sides of the heart, the determining factor for both the electrical potentials and the magnetic field outside of the heart, is the geometry of the edges of the wavefront[1]. One might think that by getting closer and closer to the heart, the magnetic fields would get stronger and stronger. While the current density within the wavefront can be as high as 20 to 50 A/m^2, the thickness and distributed nature of the wavefront results in the strongest magnetic fields occurring probably on the epicardial (outer) cardiac surface.

It would be ideal for cardiologists if either the electrocardiogram (ECG) or magnetocardiogram (MCG), alone or in combination, could be used to determine the locus of points that describes the wavefront at each instant in the cardiac cycle. Unfortunately, the lack of a unique solution to either the electric or magnetic inverse problem makes this impossible. This chapter presents a rationale not for solving this non-invasive clinical problem, but for the highly invasive experimental challenge of using magnetic measurements on isolated tissue to understand the factors that determine the shape, propagation velocity and strength of the activation wavefront.

Historically, the electrical behavior of the membrane of isolated cardiac myocytes has been typically studied using patch-clamp recordings and carefully designed current and voltage protocols [29]. The immediate response of cardiac tissue to electrical stimulation, as occurs during pacing and defibrillation, has been studied with micropipettes that measure the transmembrane potential, V_m, of a single cardiac cell, or dye/fluorescence techniques that record V_m from a number of adjacent cells, or macroscopic electrodes that are placed within the myocardium or on the epicardial, endocardial or torso surfaces, and record the extracellular potential. The models utilized to explain the resulting data have a similar span in spatial scales and describe, for example, the kinetics of single-ion channels, or the movement of activation wavefronts through the heart.

A microscopic understanding of the propagation of electrical activity though ventricular myocardium requires a knowledge of both the detailed electrical behavior of an individual cardiac cell, and the role of the cardiac syncytium that couples together the $\approx 10^9$ cells that form the ventricles. Over the past decade, patch-clamp techniques and the techniques of

---

[1]In the uniform double-layer approximation of the cardiac wavefront, the cancellation is perfect, and the potentials are determined by the solid angle subtended by the wavefront rim. For more information on this and the limitations of the uniform double-layer model, consult [144].

molecular biology are providing an increasingly clear picture of ion channel structure and the role of specific ion channels in the cardiac action potential. Progress has been slower in combining this knowledge with models of myocardial tissue, primarily because of the formidable computational challenge imposed by the requirements for 10-μm spatial discretization and 5-μs time steps in a simulated block of myocardium, no less the entire heart. Given the present impossibility of such a calculation, it is at present necessary to use simplified, homogenized models to link the sub-micron spatial scale associated with molecular electrophysiology to the 10-cm spatial scale of macroscopic electrical behavior of the intact heart.

The criteria that determine the threshold for fibrillation and subsequent defibrillation of the entire heart, which to date have been determined empirically, provide an example of how the successful coupling of microscopic and macroscopic descriptions of cardiac behavior would benefit clinical medicine [17]. Defibrillation has been explained with phenomenological theories including critical mass [152], upper and lower limits of vulnerability [9], extension of refractoriness [94], and synchronization of repolarization [15]. The term defibrillation threshold was established as an over-simplification to guide the design of implantable defibrillators [84]. The relationship between defibrillation energy and voltage or current, and the likelihood of success, has been found experimentally to be well-described by a sigmoid-shaped curve, reflecting a probabilistic function [14]. Actual defibrillation thresholds must be determined experimentally, and even mathematical models of defibrillation current distributions require empirical calibration [83]. Surprisingly, there is no theory or mathematical model that provides a first-principles connection between the vast knowledge of cellular cardiac electrophysiology and the growing understanding of fibrillation and defibrillation. Today, there are no models that can predict from a description of ion-channel kinetics the response of a fibrillating heart to a defibrillation shock. The role of tissue anisotropy in the distribution of defibrillation currents is unknown.

Similarly, the clinical cardiac literature is replete with observations of electrophysiological phenomena that have defied theoretical explanation, such as make-and-break stimulation with cathodal or anodal current, the strength-interval characteristics of two sequential stimuli, the ability of a pair of stimuli delivered to a single electrode to produce reentrant activation, and the differences in threshold for monophasic and biphasic stimuli. Some phenomena, such as the directional dependence of the rate of rise of V_m, have been addressed with models that may be unnecessarily complicated or whose wider implications are not fully understood, such as cardiac models with large numbers of discrete cells. As the role of individual ion channels is described in finer and finer detail, the severity of the gap be-

tween our understanding of the molecular electrophysiology of the heart
and our knowledge of how the $\approx 10^9$ cardiac cells interact to form the entire heart is becoming more pronounced. Reentrant phenomena that exist
at spatial scales of 1 mm to 1 cm will undoubtedly be governed by both
ion-channel kinetics and the nature of the three-dimensional cardiac cable,
which in some cases must include local heterogeneities.

1.1.2. *The Bidomain Model*

In the heart, the intracellular spaces of most adjacent cells are connected by
gap junctions that allow ions, and hence electrical current, to pass between
cells so that the heart behaves as a functional syncytium. As a result, the
cells of the ventricles form a three-dimensional cable, in which the intracellular spaces are all interconnected to form one of the conductors of the cable, while the cells share a common extracellular space that forms the other
conductor. The two conductors are separated by an insulating membrane
that has both capacitance and a time- and voltage-sensitive conductance.
The non-linear electrical properties of this membrane are responsible for
the ability of cardiac tissue to support a propagating activation wavefront.
The electrical behavior of such a system is best described with the bidomain
model.

In the bidomain model, cardiac tissue is represented by a homogeneous
three-dimensional electrical cable with distinct intracellular and extracellular spaces separated by cell membrane. In practice, the bidomain model
is represented by a pair of coupled partial differential equations governing
the intracellular, V_i, and extracellular, V_e, potentials

$$\nabla \cdot \widetilde{\sigma}_i \nabla V_i \;=\; \beta(C_m \partial V_m / \partial t + J_{ion}) - I_i \tag{1}$$

$$\nabla \cdot \widetilde{\sigma}_e \nabla V_e \;=\; -\beta(C_m \partial V_m / \partial t + J_{ion}) - I_e \;, \tag{2}$$

where $\widetilde{\sigma}_i$ and $\widetilde{\sigma}_e$ are the anisotropic electrical conductivities of the two
spaces (S/m), C_m is the membrane capacitance per unit area (F/m^2), β is
the ratio of cell membrane area to tissue volume (m^{-1}), J_{ion} is the membrane ionic current per unit area (A/m^2), and I_i and I_e are the external
current sources per unit volume (A/m^3) that are applied intracellularly or
extracellularly. The bidomain model simplifies the computations by treating
the macroscopic heart as a continuous, non-linear, three-dimensional cable,
with the effects of intercellular junctions incorporated into the anisotropic
intracellular conductivities. Because of the tissue architecture, the electrical
conductivities of the intracellular and extracellular spaces are directionally
dependent, *i.e.*, anisotropic, with differing anisotropies in the two spaces.
Tissue anisotropy, as described by the unequal-anisotropy bidomain model,
determines the spread of stimulus and action currents in a manner that affects the initiation and propagation of action potentials in both the normal

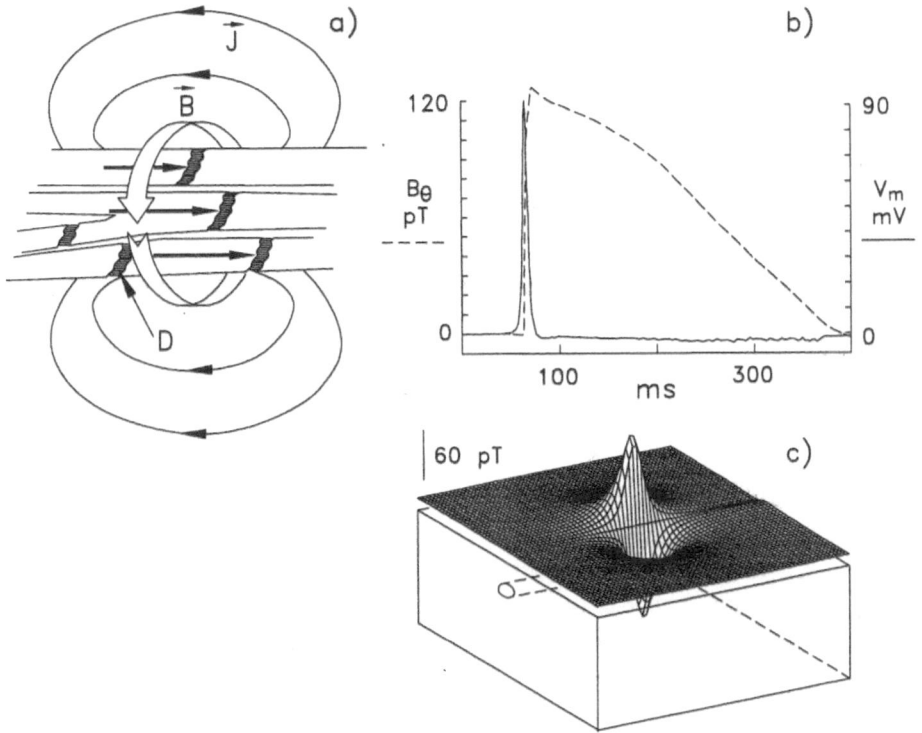

Figure 2. Cardiac action currents and their magnetic fields. (a) The equivalent current dipoles that represent the depolarization wavefront of an action potential as it propagates along a strand of cardiac tissue. The gap junctions within the intercalated disks, D, provide electrical connection between cells. (b) The transmembrane potential (dashed) and the axial action current (solid) for a one-dimensional strand of cardiac tissue. (c) A theoretical prediction of the output of a magnetometer as it is scanned above an active cardiac fiber immersed in a conducting bath. (From [126], with permission)

and abnormal heart. This, in turn, leads to the well-known anisotropic conduction velocity, as well as a variety of unexpected and interesting effects, many of which have been predicted and then confirmed experimentally [79, 81, 145, 89, 75, 128, 109, 136, 70, 80, 38, 37, 52].

1.2. MAGNETIC MEASUREMENTS IN CARDIAC ELECTROPHYSIOLOGY

Figure 2 illustrates one potential role of magnetic measurements in cardiac electrophysiology. Action currents deliver charge to the cell membrane. Magnetic fields are determined by the current $I(x, y, t)$. As this current flows through the membrane and extracellular space, there are associated with it voltage drops $V_m(x, y, t)$ and $V_e(x, y, t)$. In principle, magnetic measurements of $I(x, y, t)$ and electrical or optical measurements of $V_m(x, y, t)$

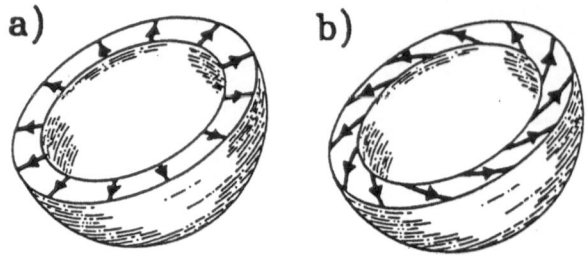

Figure 3. Hypothetical examples of cardiac activation wavefronts that have identical electric fields but differing magnetic fields. (a) A hemispherical uniform double-layer; the currents within the wave front are everywhere perpendicular to the wavefront. (b) A double-layer wavefront where the fiber architecture within the heart leads to a tangential, electrically-silent current circulating around the wavefront in addition to the radial currents in (a). (From [130], with permission)

and electrical measurements of $V_e(x, y, t)$ can be combined to determine tissue resistivities and anisotropies that are difficult to determine from electrical measurements alone. Further motivation for magnetic studies of cardiac tissue arises from the existence of phenomena in cardiac electrophysiology that are, as yet, poorly understood, but may be the result of unequal anisotropies, and from the possibility of applying SQUID magnetometry to problems in cardiac stimulation and defibrillation [76].

1.2.1. *The Information Content of the ECG and MCG*

Since the 1970's, magnetocardiographers have recognized the existence of an important, unanswered question of whether measurements of the magnetocardiogram (MCG) might provide information about cardiac electrical activity not available from the electrocardiogram (ECG). For historical reviews, see [144, 124]. The question of relative information content of the MCG and ECG has been addressed theoretically by devising a pair of hypothetical double-layer sources [130] that would produce identical ECGs but different MCGs (Fig. 3). The electrically-silent component did not appear to be present in the MCG of either normal subjects or patients with cardiac abnormalities, except for small but significant differences found using principal-component analysis [49].

The physiological basis of such silent sources was subsequently tied to the issue of tissue anisotropy [72] and then to the spiral architecture of the heart [77]. For example, the spiraling of fibers at the apex of the heart (Fig. 4a) could produce a magnetic field that would contain such new information [77] as a result of electrically-silent current loops flowing perpendicular to the direction of propagation (Fig. 4b). There are no reports in the literature of attempts to identify such magnetic fields.

The most recent development in the information-content question has

a)   b)

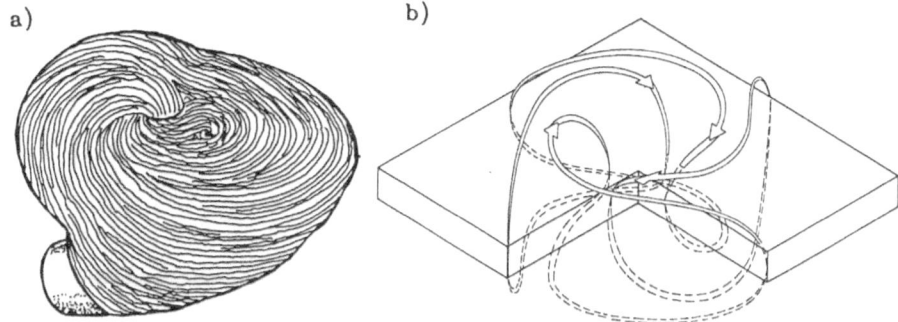

Figure 4.   The effects of spiral tissue architecture at the cardiac apex. (a) The fiber orientation at the apex [50]. (b) The apical magnetic field predicted for a bidomain model of a planar slab of tissue with spiral tissue architecture having a circularly symmetric cylindrical wavefront propagating outwards from the center. The field pattern is quite different from that produced by expanding wavefronts in tissue with parallel fibers, shown in Fig. 5. (From [77], with permission)

been the data obtained by the PTB group in Berlin during the R-wave and early in the T-wave at both resting and pharmacologically-induced high heart rates from a normal volunteer [107, 7]. At low heart rates and for both the ECG and MCG, the maps during the R and T waves are consistent with an electric dipole model of the heart. However, at a high heart rate, the MCG map during the T-wave undergoes a profound change and has a single central maximum; such a map could be produced by a magnetic dipole associated with a circulating current within the ventricular myocardium, shown in Fig. 3b. The magnetic data were used to determine the strength of the electric and magnetic dipole moments of the heart. Quantitative analysis of the data shows that these differences arise from a change in the sign of the z-component of the magnetic dipole in the T-wave at high heart rates; the electric dipole moment is unchanged.

In the absence of drugs or disease, the trajectory of propagation of cardiac depolarization does not differ substantially for low or high heart rates. At high heart rates, regional differences in action-potential duration may affect the relative contribution of timed and propagated components of repolarization. While depolarization is localized in a thin wavefront, re-polarization occurs over a large spatial extent. The role of tissue anisotropy and bidomain properties in repolarization is only now being examined [33] and is still poorly understood, yet could be significant because fiber orientation can change by 90° or more over the spatial extent of repolarization. The PTB data suggests that the MCG can provide a new means of studying repolarization phenomena at high heart rates that complements the ECG, even in normal subjects.

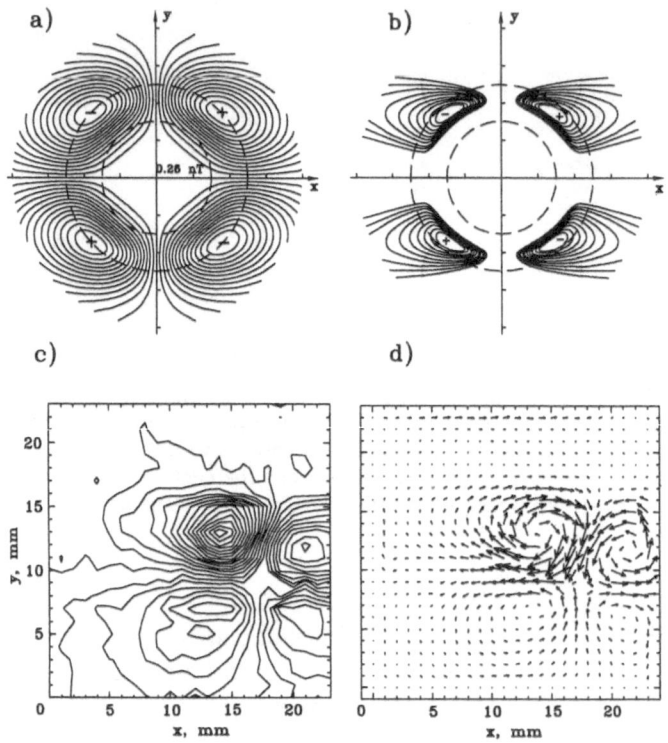

Figure 5. The calculated (a) magnetic field and (b) net current loops produced by an expanding wavefront in a two-dimensional bidomain [79]. The depolarization wavefront propagates outward and is confined to the region between the dashed circles. The myocardial fibers are oriented in the x-direction. In (a), the normal component of the magnetic field has a peak strength of 0.26 nT. (c) The magnetic field 2.5 mm above a slice of cardiac tissue. (d) The current density distribution calculated from the experimentally measured magnetic field. (From [89], with permission)

1.2.2. *The Quatrefoil Magnetic Field Produced by an Expanding Wavefront*

Another implication of the anisotropy of the cardiac bidomain, first proposed by Plonsey and Barr [64], is that the intracellular and extracellular action currents associated with an expanding wavefront are not equal and opposite. Rather, they form closed loops of net current with a four-fold (quatrefoil) symmetry (Fig. 5b). We realized that these current loops produce a magnetic field whose quatrefoil pattern (Fig. 5a) provides a unique signature of wavefront propagation [79], and that current injection produces a similar but quantitatively different pattern [81]. We have measured this magnetic field pattern using a high-spatial-resolution Superconducting QUantum Interference Device (SQUID) magnetometer scanned over a tissue slice from a dog heart [89] (Figs. 5c and d). A similar magnetic field pattern will exist when the cardiac surface is stimulated electrically to produce an expanding ellipsoidal wavefront. Measurement of the magnetic field is a particularly sensitive test of the bidomain model and unequal anisotropy ratios, because in the limit of equal anisotropy ratios, the magnetic field vanishes. Note that the spatial features have scales on the order of millimeters; our theoretical analysis [85] indicates that magnetic discrimination between bidomain and monodomain models is most accurate at spatial frequencies above 1 mm^{-1}. This spatial resolution would require sub-millimeter pickup coils; a 200-μm-diameter coil or SQUID that is 200 μm from the tissue would be ideal.

1.2.3. *Diastolic and Systolic Injury Currents*

The use of electrodes to record the very low frequency potentials associated with injury currents in biological systems is severely complicated by drifting electrode offset potentials. In a classic experiment, Cohen and Kaufman [11] used a SQUID magnetometer to detect these steady currents by means of their magnetic field. They demonstrated that reversible (acute) myocardial ischemia produced both systolic and diastolic injury currents that varied with time after ischemia. AC-coupled electrocardiograms are incapable of distinguishing between these two currents; they can exhibit ST segment elevation or depression depending upon a variety of cardiac or geometrical factors. Electrode artifacts and drift make DC-coupled ECG recording problematic. The Cohen and Kaufman experiment, while demonstrating the existence of the injury currents, did not examine in detail their time course. Visual inspection of the data showed that the observed systolic and diastolic shifts reached a stable maximum during the first fifteen minutes of the occlusion and that both shifts decreased thereafter. The data did not rule out the possibility that with the onset of irreversible infarction for occlusions longer than 15 minutes, there might be a new ST component due to the shortening of the transmembrane action potential, *i.e.*, a primary ST shift that was not accompanied by a diastolic current.

One of the first applications of the bidomain model was to calculate the response of the bidomain to injury currents [108]. Our modeling studies show that the diastolic and systolic injury currents observed by Cohen and Kaufman would produce measurable magnetic fields if the ischemic region borders on a passive, non-syncytial volume conductor, or if the tissue supporting these currents has unequal anisotropies. Otherwise, the intracellular and extracellular injury currents would be spatially superimposed and would produce no magnetic field. High resolution measurement of the temporal and spatial dependence of the injury currents at the border of an ischemic zone in an isolated rabbit heart should provide new insights into the sequence of physiological events during myocardial ischemia.

1.2.4. *Defibrillation Current Distributions*

At present, the distribution of defibrillation currents can only be inferred from electrode maps of extracellular potential distributions or optical images of transmembrane potential during defibrillation shocks. Hence, it is difficult to determine the role of tissue architecture and the associated anisotropy of the electrical resistivity in the distribution of defibrillation currents. The effects of the bidomain properties on the defibrillation currents is even less well understood. The distribution of potentials and currents and their gradients is of importance to understanding whether regions of low voltage gradient and defibrillation current are responsible for the reappearance of fibrillation after an unsuccessful shock [30]. SQUID magnetometers may be able to provide some of the required data, particularly on isolated hearts.

1.3. THE NEED FOR HIGHER RESOLUTION

By the mid 1980's, it had became clear that experiments to address the issues of information content in the MCG, identification of the sources of spontaneous activity in neural tissue, and studies of the relationship between cellular sources and their magnetic field, would require magnetometers with millimeter resolution. This was made particularly evident by the seminal studies by Barth *et al.* [4] in which a SQUID was used to map the epileptic spike activity of a rat: the pickup coil diameter was larger than the rat brain! The work at Vanderbilt at that time was directed towards understanding the magnetic field of a single nerve axon, a one-dimensional problem that was conceptually easier than mapping fields from the heart or brain, but which provided a new set of technical challenges.

1.3.1. *One-Dimensional Preparations*

The first measurements of the magnetic field of an isolated nerve demonstrated the advantages of getting close to the tissue under study. The distribution of electric and magnetic fields for an isolated nerve axon are shown in Fig. 6. The magnetic field falls off as $1/r$ close to the nerve, consistent with the finite spatial extent of the depolarization currents, $1/r^2$ further from the nerve, consistent with the dipole nature of the depolarization currents, and eventually as $1/r^3$, as expected for a quadrupolar current source [95]. Figure 7 shows the measured magnetic field pattern associated with a nerve impulse propagating along an isolated frog sciatic nerve. To measure this field, we placed an isolated frog sciatic nerve bundle next to the face of a SQUID magnetometer with a 1.5 cm × 2.7 cm elliptical pickup coil 15 mm from the nerve [135, 3]. The signal, shown in Fig. 7a, was barely detectable after averaging 1024 times.

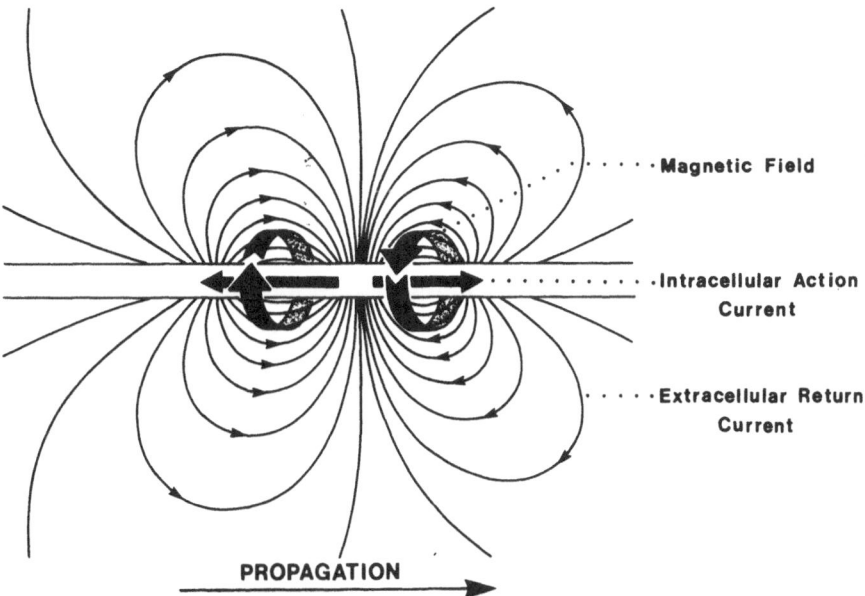

Figure 6. The electric and magnetic fields of a nerve impulse associated with a nerve action potential propagating from left to right. The dark bands represent the magnetic field and the thin lines the extracellular electric field. The effective current sources are represented by a pair of axial current dipoles, with the leading dipole depolarizing the nerve membrane and the oppositely-directed one repolarizing it.

As discussed in Ref. [125], the key to successful measurement of the magnetic field from such a weak and highly-localized current source is to place the magnetometer as close as possible to the tissue. Possibly the easiest way is to thread the nerve through a room-temperature, ferrite-cored, pickup coil so that the nerve serves as the primary to this transformer, and the copper wire wound on the core serves as the secondary. A number of different techniques can be used to sense the current or voltage induced in this coil by the magnetic field as it propagates through the toroid. Initially, we wrapped several turns of wire around the outer tail of our SQUID magnetometer dewar and connected this coil to the toroidal pickup coil [135, 3]. The SQUID was then used simply as an inductively-coupled ammeter, and we obtained the data in Fig. 7b. We subsequently improved the sensitivity and noise rejection by more closely matching the toroidal pickup coil to a SQUID that was housed in a cryogenic dip probe that could be operated in a liquid helium storage dewar [123].

The noise in Fig. 7b arises from three sources: the Johnson noise in the copper pickup coil, the Johnson noise from the rf-shielding transformer

320

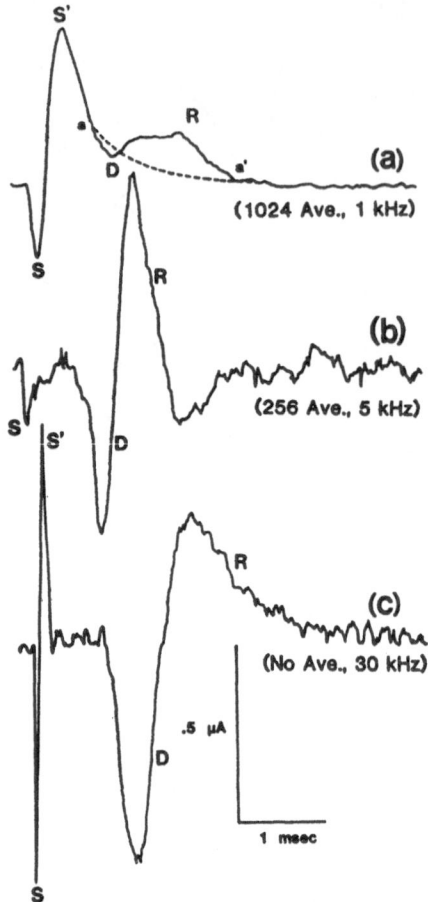

Figure 7. The magnetic field of an isolated frog sciatic nerve measured with different instruments. (a) The output of a SQUID optimized for recording the human vector MCG, with an isolated frog sciatic nerve in a conducting medium placed as close as possible to the pickup coil. S and S′ are the stimulus artifact, the deviation from the line aa′ is due to the magnetic field of the nerve, and D and R are the depolarization and repolarization phases [135, 3]. (b) The SQUID output when the frog nerve was coupled to the SQUID by a small, room-temperature toroidal coil [135]. (c) The corresponding output of the toroidal pickup-coil/semiconductor current-to-voltage converter [137, 23]. Note the differing number of averages and the bandwidths of the three recordings. (From [125], with permission)

within the probe, and the intrinsic SQUID noise. Since the Johnson noise in the toroidal pickup coil was dominant in our system, a properly-designed room-temperature amplifier should have sufficed instead of the quieter SQUID. However, since the toroidal pickup coils must be matched to the limited spatial extent of the propagating action potential, it is difficult to fabricate small toroids whose impedance is greater than 50 to 100 ohms. This, in turn, places a severe limitation on the use of conventional semiconductor amplifiers, since their voltage noise is usually excessive when the amplifier is connected to such a low source impedance. We overcame this limitation with a low-noise current-to-voltage converter with 10 parallel input stages designed specifically to reduce the input voltage noise of the amplifier [137]. Figure 7c shows the improved signal-to-noise ratio that was obtained with this instrument, which could achieve an effective input current noise of only 120 $pA/Hz^{1/2}$. These instruments, termed Toroidal Current Probes [123, 43, 137, 139, 112, 23], were used to make the first measurements of the magnetic fields from giant axons, in vivo human nerves, regenerating mammalian nerves [135, 141, 71, 115, 110, 111, 19, 44, 149, 134, 24, 39], the giant synapse of the squid *Loligo pealei* [133, 114], a single motor unit of skeletal muscle [23, 22, 20], a single muscle fiber [113], and cardiac Purkinje fibers and papillary muscles [140, 73, 131], as summarized in Fig. 8. The Toroidal Current Probe is a one-dimensional magnetometer with high sensitivity, bandwidth, and spatial resolution. The maximum sensitivity of 120 $pA/Hz^{1/2}$ corresponds to a magnetic field sensitivity of approximately 20 $fT/Hz^{1/2}$. The maximum spatial resolution is determined by the toroid thickness, which can be a fraction of a millimeter. Most of the toroidal pickup coils that have been used are a millimeter thick, which limits the spatial resolution. If thinner cores are used, it should be possible to achieve several-hundred-micron axial resolution, as would be required for measurements of the predicted reversal of action current at the intercellular junctions of septated nerves [2].

1.3.2. *Measurements in Two and Three Dimensions*

The need for high-resolution SQUIDs was demonstrated clearly in a series of simulations of the magnetic field of nerve action potentials propagating along the rat spinal cord [125]. The field pattern would have the same general shape as shown in Fig. 6. Figure 9 shows the calculated isofield contours for the vertical magnetic field component that would be recorded at several distances from an *in vivo*, horizontal rat spinal cord. The field measured at 16 mm from the cord is three orders of magnitude smaller than that recorded at 1 mm; the pattern at 16 mm is so spread out that it is difficult to imagine using it for localization of spinal cord damage or chronic pain foci.

It is important to recognize that the spatial spreading of the magnetic field with distance from the source also implies that for distributed sources, the contributions from adjacent regions have increasing overlap as the distance of the magnetometer from the source increases. If the source is also a current pattern that moves with time, for example a nerve signal propagating along a spinal cord, then the magnetic field will exhibit a distance-dependent spreading in time, as shown in Fig. 10.

322

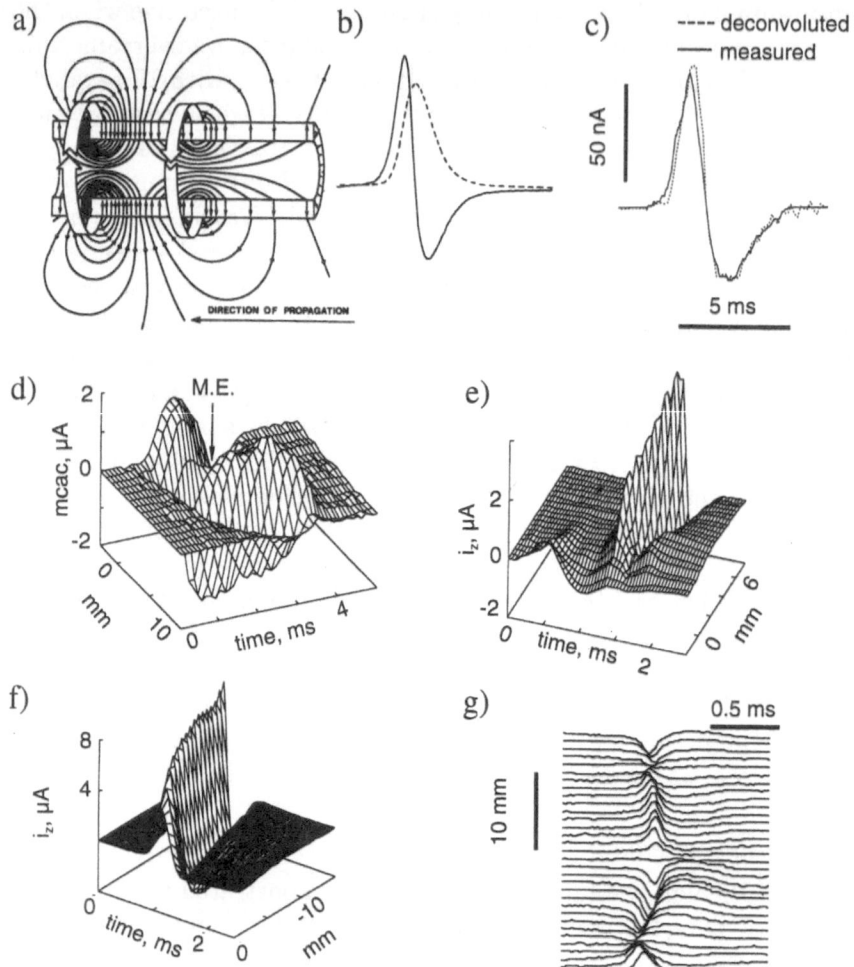

Figure 8. Representative data from magnetic measurements of action currents using the Vanderbilt Toroidal Current Probe. (a) A schematic representation of the electric (thin lines) and magnetic (broad bands) fields from a single axon [133]. (b) The transmembrane potential (solid) and magnetic field (dotted) recorded from a crayfish giant axon [71]. (c) Magnetic measurement of the compound action current (mcac) in a single frog skeletal muscle fiber [113]. (d) A scan of a toroidal pickup coil along a single motor unit of a rat EDL muscle, showing propagation out from the motor end-plate (ME) zone [133]. (e) Action currents along pre- and post-synaptic axons in the giant synapse of the squid *Loligo pealei* [111]. (f) Action currents as they propagate into the crushed end of a squid axon [111]. (g) The temporal and spatial variation of spontaneous oscillations in a squid axon [44].

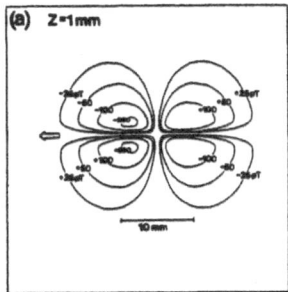 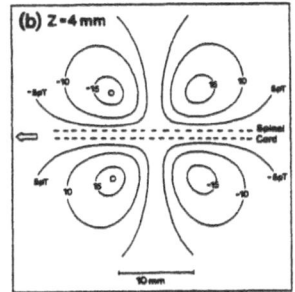 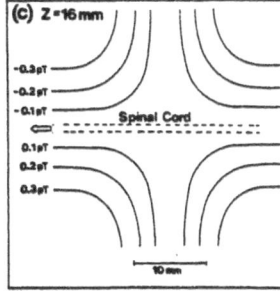

Figure 9. The calculated isofield contours for three distances above a rat spinal cord. The arrow shows the direction of propagation of the pattern. The field values are in picotesla. (From [125], with permission)

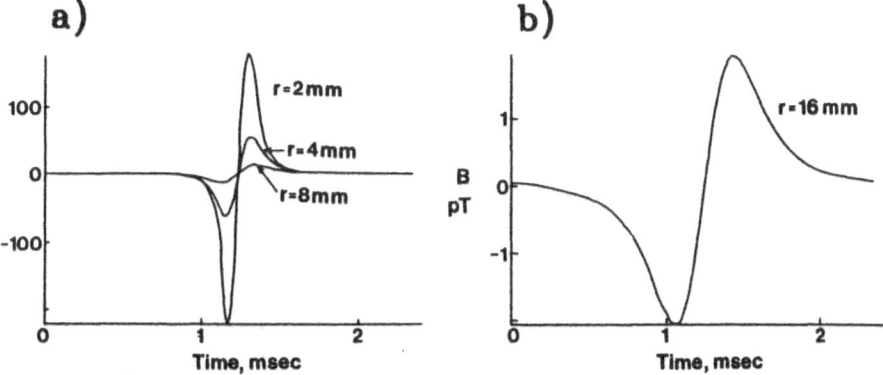

Figure 10. The time dependence of the magnetic field that would be recorded at several distances from the spinal cord of a rat as calculated with a volume-conductor model.

The requirement that the SQUID be placed near the nerve is emphasized in the plot of the peak-normal-field versus distance above the nerve in Fig. 11. The solid curve demonstrates the $1/r$ to $1/r^2$ to $1/r^3$ fall-off of the neuromagnetic field with distance, consistent with the calculation by Swinney and Wikswo [95]. The graph also shows the operating region of two SQUID systems with a 10-kHz bandwidth: a high-resolution system with coils 2 mm from the spinal cord and 50 fT/Hz$^{1/2}$ noise indicated by (a), and the other a standard biomedical SQUID system with pickup coils 10 mm from the cord and 20 fT/Hz$^{1/2}$ noise by (b). The closest possible coil-to-animal spacing locates the vertical line, while the system noise in the 10-kHz bandwidth required for recordings close to nerves determines the horizontal one. The high-resolution system would detect the spinal cord signal with a 20-to-1 signal-to-noise ratio, while the conventional one will have a 0.5-to-1

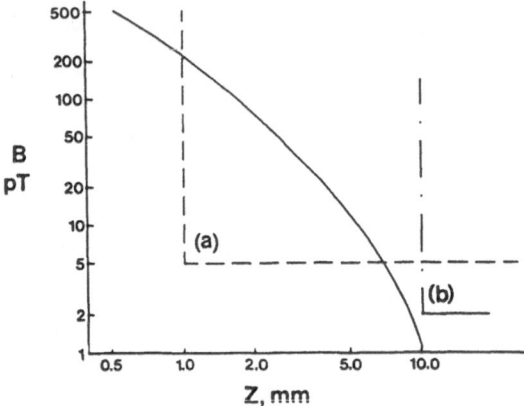

Figure 11. The calculated peak value of the normal component of the magnetic field from an intact rat spinal cord as a function of distance from the center of the spinal cord. The noise and spatial resolution limits of a MicroSQUID-class magnetometer are shown by corner (a) while those for conventional biomagnetic SQUIDs are shown by (b). (From [125], with permission)

signal-to-noise ratio. Reducing the SQUID noise to 10 fT/Hz$^{1/2}$ would still provide only a unity signal-to-noise ratio. This provides an example of how the loss in absolute field sensitivity accompanied by the use of small pickup coils can be compensated for by the large field strengths close to the source. As we will see later, this gain in signal strength depends upon the nature of the source.

The one-dimensional model has also been used to examine the spatial resolution of SQUIDs [132, 97]. The strong fall-off in spatial resolution with distance is seen in Fig. 12, in which the magnetic field is calculated for a volume conductor model of the brain containing a synchronous population of 10^5 apical dendrites of pyramidal cells in a volume of approximately 1 m × 1 mm × 3 mm. The lower row of the figure shows that the loss in spatial information with distance cannot be readily offset by a reduction in the magnetometer noise. Reducing the noise level in the large magnetometer by a factor of 100 raises the spatial frequency at which the noise dominates the signal to only 0.2 mm^{-1}. This analysis confirms that for mapping of the magnetic fields from localized sources, it is better to have a small, noisy magnetometer closer to the source than a larger, quieter one farther away.

The effect of coil-to-sample distance on the spatial resolution of a SQUID magnetometer can be quantified using the two-dimensional spatial-filtering approach [78], which is reviewed in more detail in the accompanying chapter [129]. In Fig. 7 of that chapter, we showed how the reconstruction of a current image from a field map recorded at $z = 0.1$ mm above a square

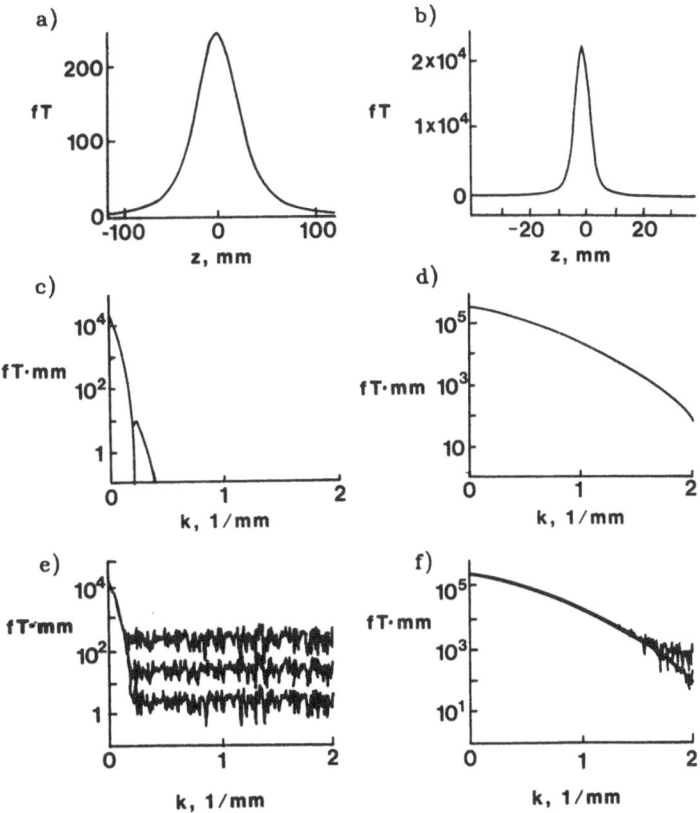

Figure 12. A simulation of the magnetic field from apical dendrites in the cortex of the brain that demonstrates the relation between coil-to-sample spacing and spatial frequencies. (left column) Magnetometer pickup coil 30 mm in diameter and 30 mm from the dendrites; (right column) 3 mm in diameter and 3 mm from the dendrites. (top row) The magnetic field as the magnetometer is scanned past the dendrites. Note the difference in the horizontal scales. (middle row) The spatial Fourier transform of the magnetic field. The closer the magnetometer to the source, the larger is the contribution from the higher spatial frequencies. (lower row) The spatial Fourier transforms in the presence of variant amounts of magnetic noise. For the 30-mm coil on the left, the noise is 15, 1.5 and 0.15 $fT/Hz^{1/2}$, while for the 3-mm coil on the right, the noise is 35, 3.5 and 0.35 $fT/Hz^{1/2}$. (Adapted from [132], with permission)

current loop would have a mean-square error of 0.03, while a reconstruction from a map recorded at 3 mm would result in an error of 0.95. Figure 13 compares two magnetometers, one that can be positioned only within 3 mm of the source (left column), and one that has a factor of 100 more noise, but can be placed within 1 mm of the source. As we might expect from Fig. 12, the magnetometer noise is an important factor in the quality of an image, but it is not as important as the spacing. Figure 13 shows the magnetic field maps (top row) and calculated images of the current density

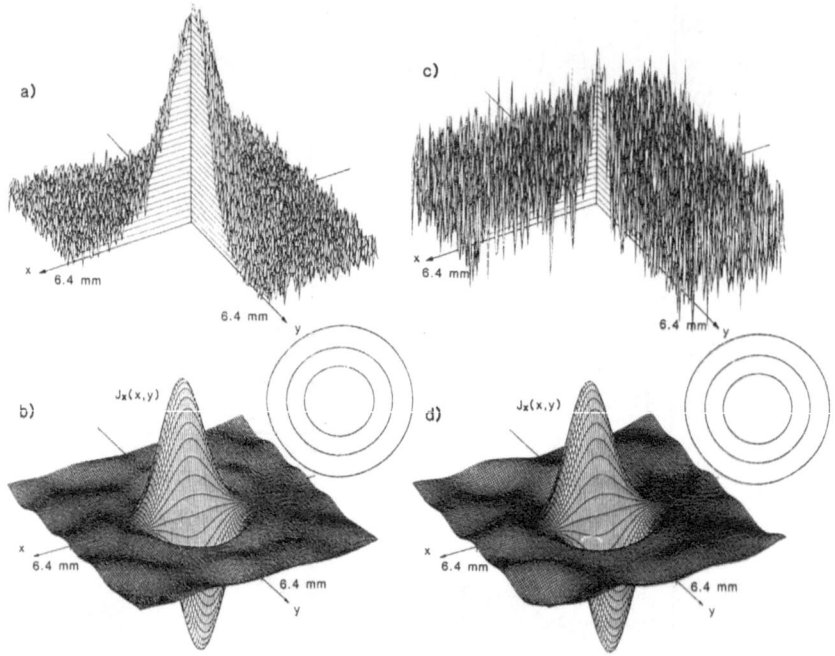

Figure 13. The magnetic field (a and c) and the image of the current density (b and d) for two different values of magnetometer noise and two different magnetometer distances. The current density is calculated from the z-component of the magnetic field produced by a square current loop. In (b) and (d), plots of both $J_x(x,y)$ and the current lines (upper right inset) are shown. In the magnetic field plots, the first quadrant is not shown so the signal can more easily be distinguished from the noise. The (a) magnetic field, 6.5 pT peak signal with 0.5 pT noise, and (b) image of the current density, 0.13 mA/mm^2, for a low-noise magnetometer 3 mm from the current. The (c) magnetic field, 126 pT, with 50 pT noise, and (d) image of the current density, 0.13 mA/mm^2, for a noisy magnetometer 1 mm from the current. (From [78], with permission)

(lower row) in these two cases. While the signal-to-noise ratio of the field mapped by the more distant magnetometer is better, the signal is much broader. As a result, the resulting current images exhibit nearly identical quality mean-square deviation (MSD) of 0.947. It follows that as long as the coil-to-current spacing is comparable to or larger than the characteristic scale length of the current source, improvement in the quality of the reconstructed current density is more sensitive to the coil-to-source distance than to the amount of noise in the magnetometer. If z is small, or if there is excessive noise in the magnetometer so that the signal-to-noise ratio is less than 1, then reducing the noise in the signal becomes increasingly important. Nevertheless, this analysis showed that in many applications of

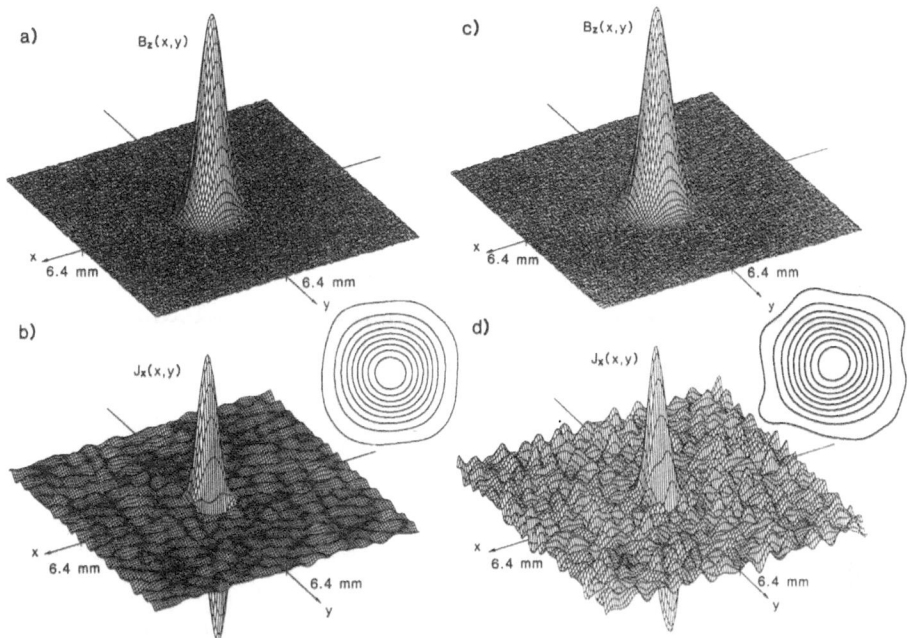

Figure 14. The magnetic field and the current density image calculated using two sizes of magnetometer pickup coils. (a) The magnetic field measured with a negligibly small coil radius, $z=1$ mm, noise = 0.5 pT, and (b) the current image, MSD=0.43. Both a plot of $J_x(x,y)$ and the current lines (upper right inset) are shown. The same (c) magnetic field, and (d) current image, except that the magnetometer pickup coil has a radius of 0.35 mm, MSD=0.54. (From [78], with permission)

SQUID magnetometry, dramatic increases in the spatial resolution of the imaged current might be obtained by decreasing the source-to-coil distance.

The size of the magnetometer pickup coil also plays a role in determining the quality of the image. Figure 14 shows, for two coil radii, the z-component of the magnetic field, with noise, measured 1 mm above the plane of the current, and the reconstructed image of the current density. In Fig. 14a, the magnetic field is calculated for the limiting case of a coil with zero radius, while in Fig. 14c, the field is averaged over a circular coil with a 0.35-mm radius. Although a coil of this size changes the shape of the magnetic field only slightly, it does degrade the current image.

2. SQUIDs for High Resolution Magnetic Imaging of Cellular Action Currents

In this section, we describe the first generation of high-resolution SQUIDs and the more recent SQUID microscopes, and then discuss in general terms several of the trade-offs that are important in the design of high resolution SQUIDs.

2.1. MICROSQUID

While the toroidal current probe described in the previous section is relatively inexpensive and simple to use, it is not suitable for two- or three-dimensional tissue preparations, nor can it detect steady or low-frequency magnetic fields. SQUIDLET, developed at the Open University in England, had 4-mm-diameter pickup coils that are 5.5 mm from room temperature [18]. The MicroSQUID magnetometer [8, 143, 150, 147], shown in Fig. 15a and b, was custom-built for Vanderbilt by Biomagnetic Technologies, Inc. (BTi) and Quantum Design (QD) in the 1980's. With this system, both the SQUID and its pickup coils are located in the vacuum space of the dewar wall. Because the inner portions of the dewar contract approximately several mm in length as the dewar is cooled from room temperature to 4 K, a close spacing between the room-temperature sample and the cryogenic pickup coils is achieved after cool down by subsequently lowering the pickup coils that are attached to the bottom of the helium reservoir, which forms the inner wall of the dewar, thereby providing close spacing during measurements and reduced helium boil-off when the system is not in use. The boil-off is 3.4 liters/day when the coils are in the close-spaced position, and 1.5 liters/day when the coils are retracted. Room- or body-temperature samples are placed outside a thin vacuum window. The pickup coils are only three millimeters in diameter and 1.4 mm from the sample, providing an order-of-magnitude higher spatial resolution as compared to typical biomedical SQUID magnetometers in conventional dewars.

Figure 15b shows the details of the tail of MicroSQUID. The system has 4 channels, each of which is a first-order differential magnetometer that has a 16-turn pickup coil with a 3-mm diameter and a single-turn balance coil with a 12-mm diameter. The pickup and balance coils are separated by 3 cm. External feedback minimizes crosstalk between the channels. The coils are hand-wound on a precision-ground sapphire substrate. The window is a 250-μm-thick fiberglass-epoxy composite. The frequency response of each SQUID is flat below 1 kHz. The noise in the four channels ranges from 90 to 120 fT/Hz$^{1/2}$ at 400 Hz. It should be possible to reduce this noise by 30% with improved coupling between the SQUIDs and the pickup coils.

When used with a high quality magnetic shield [47], and a non-magnetic scanning stage [150, 147], MicroSQUID has allowed our group to record a wide variety of two-dimensional magnetic images with millimeter resolution, shown in Fig. 16, and makes possible, for example, mapping of the temporal and spatial variation of action currents in isolated slices of cardiac tissue [89, 88, 87] and intestinal smooth muscle [93, 91, 92, 67, 21, 25], and of the developmental currents in the chick embryo [104]. Measurements on water, plexiglass and thin sections of rock demonstrated that it is possible

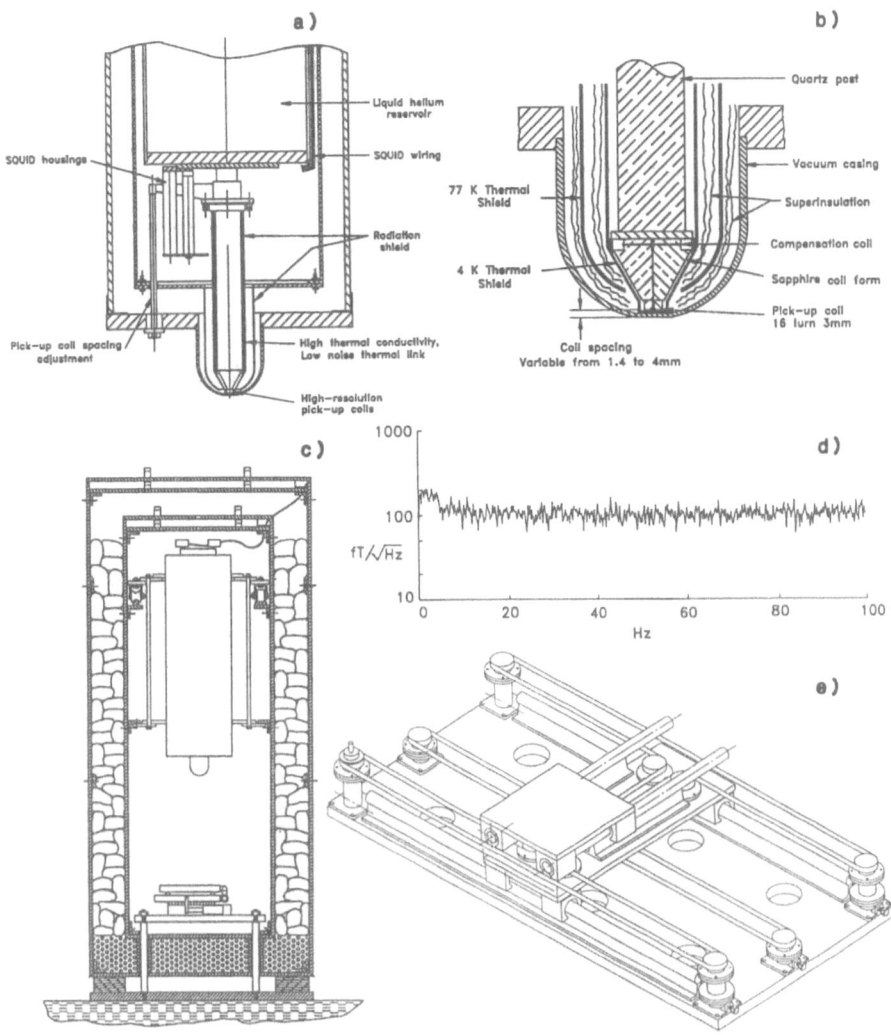

Figure 15. The MicroSQUID imaging facility. (a) The MicroSQUID magnetometer. (b) A close-up of the tail of MicroSQUID. (c) A cross-section of the magnetic shield that houses MicroSQUID [47]. (d) The noise spectrum of MicroSQUID recorded inside the magnetic shield. (e) The non-magnetic, high-speed scanning stage that can scan samples at up to 10 cm per second and is powered by servo motors outside the shield.

to image diamagnetic susceptibility distributions [82, 105, 48, 148, 138], as well as to detect the uptake of superparamagnetic microspheres in the rat liver [103, 100, 101]. It is important to recognize that the SQUID, shield, scanning stage and imaging software should be considered as a total system, since the quality of the magnetic maps and their interpretation can be degraded equally by inadequate performance of any of the components.

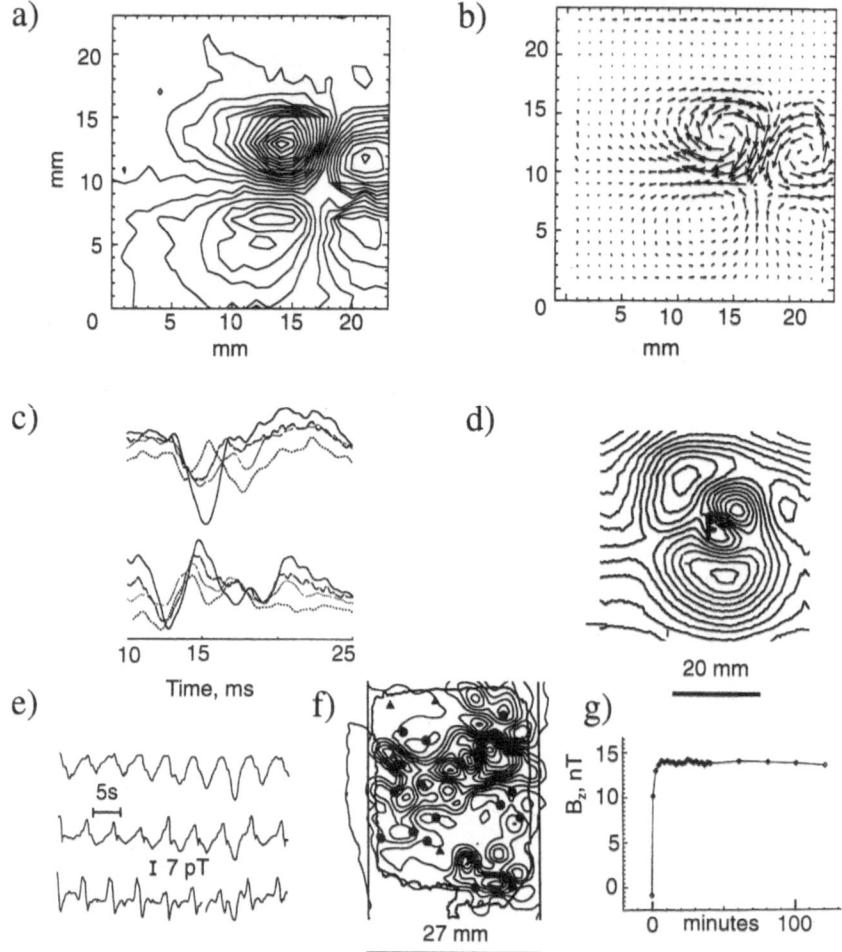

Figure 16. Representative data from the high resolution MicroSQUID magnetometer at Vanderbilt. (a) Magnetic field recorded 6 ms after the stimulus in a slice of canine cardiac tissue [89] (25 pT contours). (b) Current densities computed from the data in (a). (c) Simultaneous four-channel recordings of two different single-motor units in the human thumb [143]. (d) Magnetic field from a 94-hr chick embryo [104]. The embryo is the hook-shaped line. (e) Simultaneous, three-channel recording of the magnetic field from the basic electrical rhythm of isolated prairie dog small intestine [67]. (f) The induced magnetization distribution from a 50-μm-thick slice of pyroclastic rock in a 285 μT applied field with the locations of the magnetite (circles) and biotite (triangles) inclusions [102] (2 mA/m contour spacing). (g) The time-course of the uptake of a 900 μliter sample of superparamagnetic microspheres by a rat liver [101] in a 171 μT applied field. The applied field in (f) and (g) was provided by Helmholtz coils outside the cryostat.

A second MicroSQUID magnetometer with slightly higher sensitivity, but lower spatial resolution, has mapped magnetic fields evoked from turtle cerebellum, *in vivo* swine, slices of guinea pig hippocampus, and from spreading depression in the exposed avian retina [61, 56, 59, 54, 58, 60, 27, 40, 53, 55, 57].

Several high-resolution SQUID systems have been built for studying corrosion and for nondestructive evaluation of structural systems [28, 65, 10], including one utilizing a high-transition-temperature (HTS) SQUID [106] that has 1-mm spatial resolution and 2 pT/Hz$^{1/2}$ noise at 10 Hz.

2.2. SQUID MICROSCOPES

Scanning SQUID magnetometers have been used for over 25 years to map the distribution of trapped flux in superconducting thin films [26, 69]. Recently, scanning SQUID microscopes have been demonstrated with a spatial resolution of 10 or 20 μm for samples that are in the same cryogenic environment as the SQUID sensor [1, 5, 6, 35, 51, 117, 118, 36, 119]. While these microscopes are not useful for examining electric current distributions in living tissue, they provide an indication of the potential capabilities of SQUID microscopy: systems with both the sample and the SQUID immersed in liquid helium have 66 μm resolution and 5.2 pT/Hz$^{1/2}$ noise at 6 kHz [51], and 10 μm at 100 pT/Hz$^{1/2}$ [118]; the Conductus Scanning Magnetic Microscope provides micron resolution for helium-temperature samples. The IBM low-transition-temperature (LTS) SQUID microscope [36] has a 10 μm pickup coil and can map single magnetic flux quanta, ϕ_o, trapped in superconducting films, with a system noise of less than $2\phi_o$/Hz$^{1/2}$ or 40 pT/Hz$^{1/2}$. Wellstood and Black of Maryland built liquid nitrogen systems with a 100 mm^2 field of view, 80 μm spatial resolution, 80 pT/Hz$^{1/2}$ noise at 1 Hz [5] and 25 pT/Hz$^{1/2}$ above 500 Hz, and one with 15 μm resolution [6]. The Maryland group has just developed an HTS SQUID that is 40 μm from a room temperature sample [119], provides 50 μm spatial resolution, but has 34 pT/Hz$^{1/2}$ noise at 100 Hz and 260 pT/Hz$^{1/2}$ at 1 Hz (bias switching should improve the low-frequency performance). Lee, Dantsker, and Clarke at Berkeley have just fabricated an inverted HTS system designed for biological measurements [42].

2.3. DESIGN CONSIDERATIONS

2.3.1. *SQUIDs-in-the-Vacuum Technology*
The quantitative analyses of magnetic imaging reviewed above [78, 98, 96, 132, 97, 125, 85] have shown that a SQUID will have the best combination of spatial resolution and field sensitivity if the pickup-coil diameter is approximately the same as the spacing between the pickup coil and the

sample. In conventional dewars, the SQUIDs are located inside the helium reservoir, and therefore the coil-to-sample distance is rarely less than 1 cm. Several millimeters may be the practical limit for such systems. To attain the millimeter resolution required for MicroSQUID [8, 143], the coil-to-sample spacing had to be ≈1 mm. To achieve this, the SQUIDs and pickup coils of MicroSQUID were located in the vacuum beneath the helium reservoir, with a sapphire rod and bundles of copper wire used to cool the pickup coils, within 1.4 to 1.5 mm of room temperature, to approximately 5 K. The SQUID-in-the-vacuum approach has been so successful that BTi now uses this technology for their new 148-channel SQUID helmet. The Quantum Magnetics corrosion SQUID [28] utilizes this approach. The Maryland HTS SQUID microscope uses a sapphire rod in vacuum to support and cool the SQUID. The key point for high resolution SQUID magnetometers is that there is essentially no heat conduction through the vacuum, and only a small aperture is required in the thermal shielding for the SQUIDs and/or their pickup coils. Hence, the radiative load from room temperature to the SQUID is typically only a few milliwatts. In a later section, we discuss thermal loads and cooling.

2.3.2. *Window Design*

High resolution SQUIDs require as thin a window as possible that can support atmospheric pressure without excessive inward curvature or leaks. MicroSQUID has fiberglass/epoxy laminate windows with a 250 to 400 μm thickness and a 1- to 2-cm diameter. As the SQUIDs or their pickup coils are made smaller, the windows can be thinner. The Maryland group uses a 25 μm sapphire window. An alternative is ultrathin, single-crystal silicon or silicon-nitride wafers [116]. Single-crystal silicon wafers with specific crystal orientations can be produced (in 1- to 2-inch diameter) as thin as 2-4 μm. Such wafers are regularly supplied to commercial manufacturers of advanced, electronic sensors [12, 41], in which a 1 mm × 2 mm × 10 μm window has been shown to be vacuum-tight while supporting atmospheric pressure. Both the sapphire and the silicon windows are proven technologies; the primary challenge is to determine the optimum window parameters.

Single-crystal silicon is quite strong [62], with a Young's modulus of 1.31×10^{11} Pa and a Poisson's ratio of 0.29. Because the window will be thin and must support atmospheric pressure, any bowing will increase the separation between the SQUID and a planar sample. The thicker the window, the less bowing. We have calculated the optimum thickness for the Si windows with any particular aperture [31], *i.e.*, the thickness that results in the minimum sample-to-vacuum spacing (thickness plus deflection under vacuum). Our calculations were for a flat, clamped geometry [63] and did not allow for any prestressing that would further reduce the deflection under vacuum. We found that the minimum sample-to-vacuum spacing is proportional to the window diameter, and that even for a 1-cm^2 window, it is possible to obtain a spacing less than 200 μm. We also investigated rectangular windows, and found that a 5 mm × 10 mm window made from 100 μm <100> Si will have less than a 15 μm deflection under vacuum. In addition, silicon is a highly resistive substance with virtually no Johnson noise. It is also eminently coatable and can be easily metallized to provide rf shielding if the SQUID is to be used outside the confines of an rf-shielded room. The membranes can be purchased with a high polish, so that they can be aluminized to further reduce the thermal load on the SQUID. Sapphire windows have the advantage that they are optically transparent, so that the location of the SQUID with respect to the window can be determined visually. One design issue that has yet to be addressed fully is the requirement for some measurements that the window be immersed in saline; some epoxies are adversely affected by water, and silicon may be affected by the salt.

2.3.3. *Minimization of SQUID-to-Window Spacing*

A thin window is needed to obtain high spatial resolution, but it is also necessary to position the SQUID close to the window without allowing the SQUID to be pressed into the window when the cryostat is warmed. There are several methods of minimizing and/or adjusting the spacing between the outer surface of the window and the SQUID. With MicroSQUID, a lever mechanism actuated through a small room-temperature bellows seal is used to move the SQUID coils down toward the fixed window [143]; other approaches use a larger bellows or an O-ring/sealed-thread mechanism to move the window up toward a fixed SQUID array. The Quantum Magnetics corrosion SQUID system and the Maryland microscopes utilize metal bellows and an external adjustment mechanism to move the window as close as possible to the SQUID. An interesting approach, incorporated into the Maryland microscopes, utilizes differential contraction of dissimilar materials to construct a structure that either stays at a constant length or becomes shorter upon warming. An alternative design would have the SQUIDs located on the side of the dewar, so that the longitudinal contraction of the dewar upon cooling simply moves the SQUIDs parallel to the window. In any case, the goal is to adjust the SQUID-to-window spacing until the SQUID is heated above its critical temperature, and to then increase the spacing slightly to return the SQUID to its superconducting state.

2.3.4. *Thermal Loads*

Because vacuum is used as the primary insulation for separating the SQUID from the room-temperature window, the thickness of the region of vacuum is irrelevant. Except for where the SQUID approaches the window, all portions of the cryogenic region can be surrounded by the conventional configuration of superinsulation and intermediate-temperature thermal shields. By minimizing the area of the cryogenic regions exposed to 300 K thermal radiation, the heat load on the SQUID is minimal: the total heat load for a 0.5-cm^2 SQUID chip, shielded with only two layers of aluminized superinsulation, mounted 100 μm from a 1-cm^2, 300 K vacuum window is less than 5 mW. We have calculated [31], following the treatment of Richardson and Smith [68], the temperature distribution for the SQUID mounted in vacuum and cooled by a bundle of copper wires, as was done with MicroSQUID. A series thermal-resistance model was used which incorporated the material properties of the Si chip and the Cu wires, as well as an engineering estimate of the thermal resistance at the boundary [68]. With fewer than ten A.W.G. 30 wires, the temperature of the chip face can be maintained at 5.5 K, and the drop in temperature across the 1-mm-thick Si chip will be on the order of 1 mK. For single-SQUID designs, the cryogenic surface area adjacent to the window can be as small as 1 mm^2, so that the thermal problem is even easier. In MicroSQUID, with a less-than-ideal geometry and a larger exposed area, cooling is accomplished with bundles of hundreds of fine copper wires. Such high thermal conductivity copper links have been used routinely in SQUID systems developed by BTi, Conductus and SQM Technologies [65] over the past ten years. It is important to recognize that conduction cooling is more difficult at liquid nitrogen temperatures because of the reduced thermal conductivity of copper [119].

2.3.5. *Gradiometers*

The choice between building a magnetometer or gradiometer is governed to a large extent by the need for noise rejection [125], The complexity of fabricating a well-balanced second-order gradiometer may preclude its application to high-resolution measurements. The balance of a first-order, miniature gradiometer will depend strongly on the fabrication technique: thin-film deposition will lead to greater control than can be achieved for coils hand-wound from fine wire. An excellent magnetic shield might allow the use of simple pickup loops so that the system could operate as a magnetometer. Millimeter-diameter symmetric coils are difficult to balance adequately, whereas with an asymmetric design as was used in MicroSQUID [8], the upper coil can be 1 cm or larger in diameter so that permanent balance tabs can be installed and adjusted. The choice of turns and asymmetry ratios is governed by the need to match the coil to the SQUID sensor.

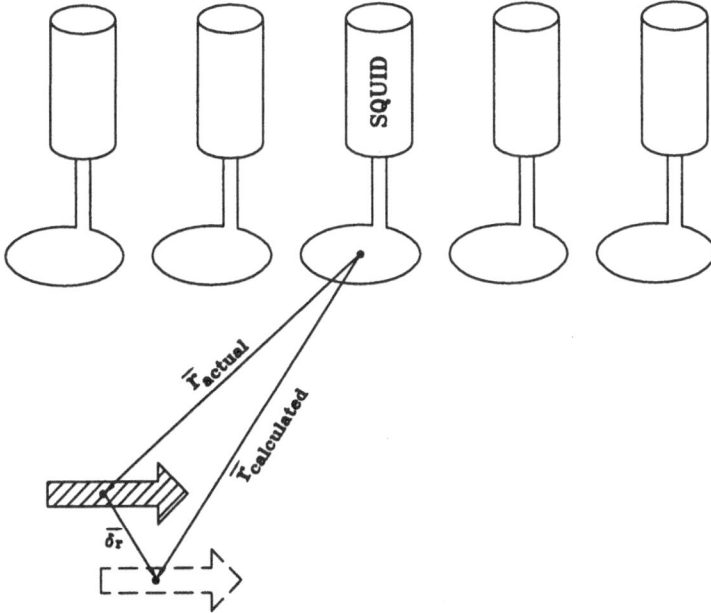

Figure 17. Localizing resolution: the ability of a magnetometer or a magnetometer array to determine the correct location of a source.

3. Resolution and Related Concepts

The spatial resolution typically quoted for biomagnetic measurements is the **localizing resolution**: the ability to determine the location of a spatially-constrained source of known geometry, *e.g.*, a current dipole. If the source is distributed or of unknown geometry, it may be more relevant to consider the **imaging resolution**: the ability to discriminate between two closely-spaced sources such as a pair of current dipoles. It is important to distinguish between these two, particularly in applications where imaging is becoming the norm, and to determine how they are affected by pickup-coil diameter and coil-to-sample separation.

3.1. LOCALIZING RESOLUTION

Localizing resolution is defined as the ability to determine the location of a spatially-constrained source of known geometry. Localizing resolution is often limited by the signal-to-noise ratio and uncertainties in geometrical and electromagnetic parameters, and can be improved through use of multiple magnetometers and least-squares fits. Figure 17   illustrates how localizing resolution can be measured – the displacement of a least-squares fit to the data relative to the known location of the source. The MEG literature has examined this problem in detail; at the scale of millimeters, the contribu-

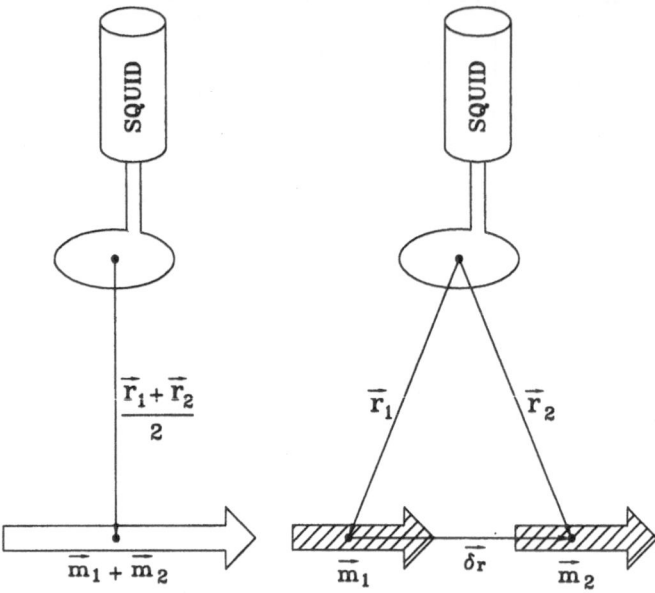

Figure 18. Imaging resolution: the ability of a magnetometer to distinguish between one source and a pair of closely spaced sources with the same total strength.

tion of the volume currents at the boundaries between regions of different conductivity at the scalp and within the head is a key factor in the accuracy of the localization. However, it is important to recognize that localization is a meaningful concept only if the source itself is localized, and many cortical and cardiac current sources are distributed over volumes of several cm^3. Localization algorithms can then identify the center of the distribution, if appropriate assumptions are made about the nature of the source. For high-resolution imaging of isolated tissue, the imaging resolution is often more important.

3.2. IMAGING RESOLUTION

Imaging resolution, shown in Fig. 18, is defined as the ability to discriminate between two closely-spaced sources, such as a pair of current dipoles, and a single source with the same total strength. Typically, imaging resolution is an order of magnitude worse than localizing resolution. Figure 19 provides a good example of limits of imaging resolution of a SQUID magnetometer [150]. The 25 mm × 25 mm square grid of conducting lines in Fig. 19a was formed on a printed circuit board and was imaged using an AC current and the lock-in algorithm. The large maximum and minimum at each side of

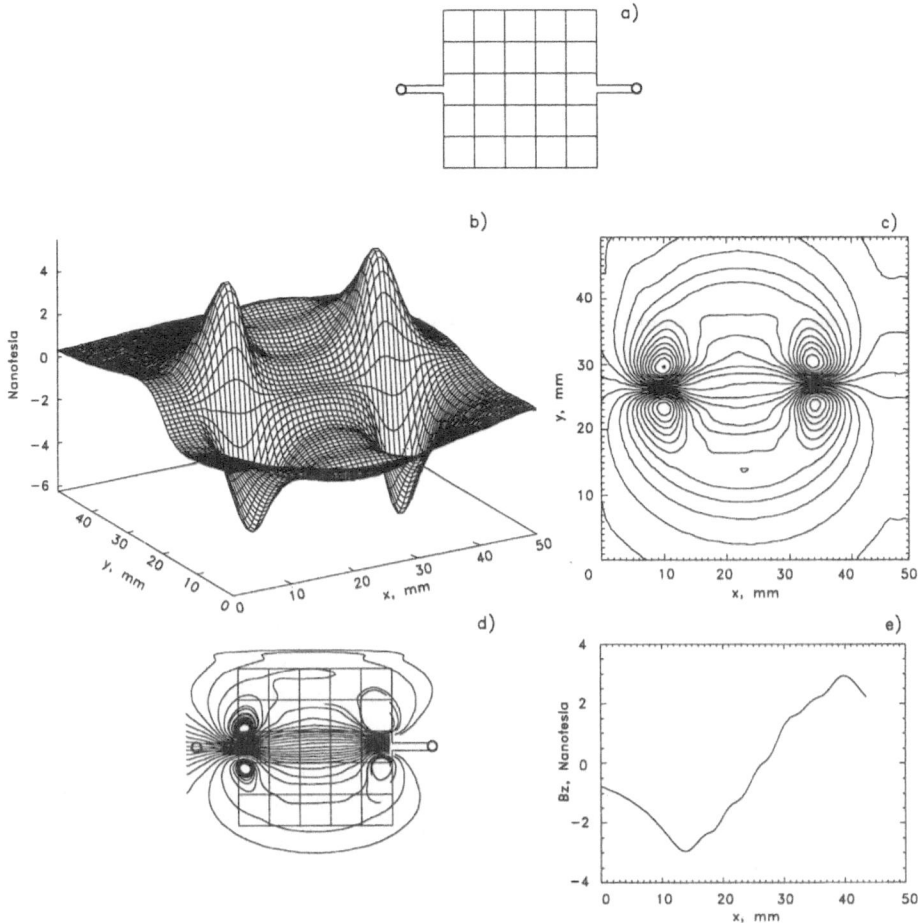

Figure 19. An example of limited imaging resolution. (a) A printed circuit current grid used to test the imaging resolution of MicroSQUID. (b) The magnetic field measured 2.5 mm above a square current grid. (c) Isofield contours for the magnetic field map (500 pT contour interval). (d) The current lines calculated from (b) and (c) using the FFT imaging algorithm [78] (5 μA per current line). The image of the grid is superimposed. (e) The z component of the magnetic field as a function of y for $x = 25$ mm, *i.e.*, a cross-section through the plot in (b) parallel to the y axis. (From [150], with permission)

the field map in Fig. 19b arise from both the current in the leads on either side of the grid and the spreading of the current in the x and y directions as the current from the single lead enters into the grid and then returns to the opposite lead. The six jumps in the field map in Fig. 19b are due to the individual horizontal lines in the grid. They are also evident in the graph of $B_z(y)$ at $x = 25$ mm (Fig. 19e), but are not fully resolved in either the isofield contour map (Fig. 19c) or the current image (Fig. 19d).

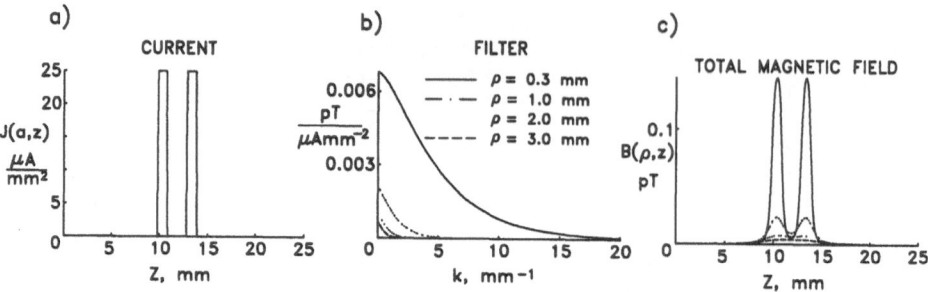

Figure 20. The effects of coil-to-source spacing on the magnetic field from two coaxial dipoles. (a) The intracellular current from two separated, but coaxial, cortical current sources lying along the z-axis of a fiber. Each source is 1 mm long, and they are separated by 3.0 mm. (b) The filter functions for calculating the magnetic field from the current for four different source-to-coil distances ρ. (c) The magnetic field produced by the two sources in (a) at different source-to-coil distances. (From [97], with permission)

The FFT imaging algorithm used to create Fig. 19d operates under the assumption that the currents are flowing in a homogeneous conducting plane of infinite extent, and, as a result, the image currents extend well beyond the boundaries of the grid. Techniques to constrain the current are described in the accompanying chapter [129]; more advanced algorithms may also improve the image quality and the resolution.

The one-dimensional nerve model can be used to examine imaging resolution in more detail [132, 97]. Figure 20 shows how the magnetic field of two dipoles blurs into a single peak as the magnetometer distance becomes equal to the dipole separation. The key point of this graph is that the magnetic field shows only a single broad peak when the dipole separation equals the source-to-coil distance (the dashed trace for $\rho = 3.0$ mm in Fig. 20c).

For quantitative studies of imaging resolution, it is convenient to adopt the rather conservative Rayleigh criterion [97] for judging whether the two peaks are resolved, in that if the peak-to-valley ratio is equal to 0.8, then the two peaks are just resolved. This is used to determine the critical distance ρ_c at which the peaks are defined as being resolved. Figure 21 shows how the imaging resolution degrades with distance from the source.

One of the challenges in defining imaging resolution is that the separation of features can be enhanced with image processing. Imaging algorithms can be viewed as inverse spatial filters, and hence are subject to noise-induced instabilities due to the amplification of high spatial frequencies that had been preferentially attenuated with distance from the source. In the accompanying chapter [129], we reviewed the Fourier inverse spatial filtering. Figure 22 shows the results of such an approach [97], where the magnetic field recorded at the Rayleigh limit $\rho_c = 1.7$ mm is deconvolved to reconstruct the current distribution. In the absence of noise, the spatial

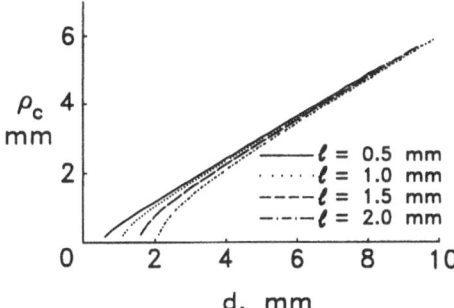

Figure 21. The source-to-coil distance ρ_c at which two sources are just resolvable in the magnetic measurement, *i.e.*, the Rayleigh criterion, as a function of the distance d between the sources. The different curves represent different lengths, ℓ, of the individual current elements. Note that the critical separation ρ_c is always a fraction of d. (Adapted from [97], with permission)

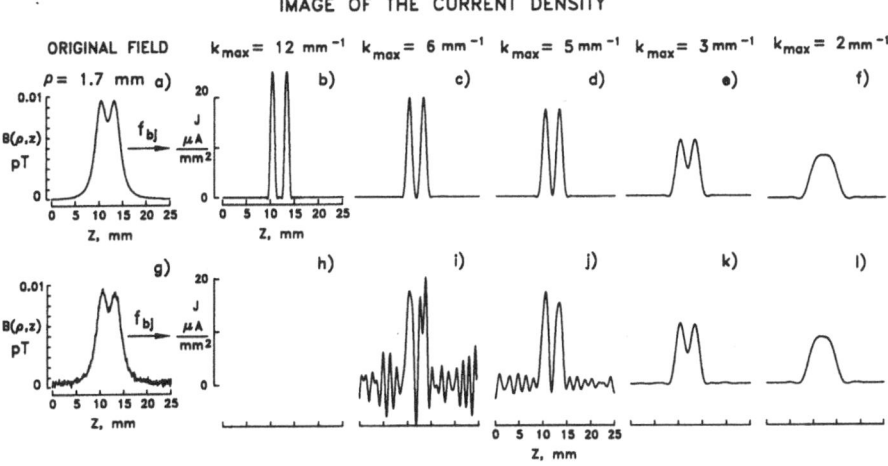

Figure 22. The effects of noise on the ability to reconstruct the current distribution from the magnetic field measured at the Rayleigh limit of resolution. (upper row): (a) Magnetic field without noise. (b)-(f) the current reconstructed with different windows. (lower row): The same calculations, but with a noisy magnetic field. The reconstruction in (h) was unstable. (From [97], with permission)

frequencies of the reconstruction have to be limited to those below 12 mm^{-1} to prevent numerical instabilities, but even so, the two sources are clearly separated. However, the addition of only a small amount of noise requires stronger windowing (low-pass filtering) of the data, and the reconstruction is severely degraded. While the imaging resolution is degraded, all of the nine current reconstructions in Fig. 22 have the same total dipole moment. The effects of SQUID and environmental noise, and techniques to minimize these effects, should be examined further.

3.3. SENSITIVITY

SQUID magnetometer sensitivity is generally specified in terms of the system noise in units of flux, energy or field. Some SQUID designers specify SQUID sensitivity in terms of an effective pickup-coil area. For imaging, it is useful to specify sensitivity in terms of a noise-equivalent source, which then determines the minimum detectable source in a unit bandwidth, *e.g.*, the minimum detectable current in a wire, I_{min}, the minimum detectable current dipole p_{min} in a conducting medium[2], or the minimum detectable magnetic dipole m_{min}. The advantage of this approach is that it is possible to compare widely different magnetometer configurations, both before and after image reconstruction, in terms of the sources that are being sought. It is functionally equivalent to referring an amplifier noise to the input; in this case the input to the SQUID is not the magnetic field, but rather the source that produces it. Figure 23 provides an example of this approach from a design study for a second-order SQUID magnetometer for magnetocardiography [122]. Such simulations can become challenging when there are a large number of degrees of freedom in the design, as when both the coil and SQUID parameters can be adjusted. In such cases, it may be most efficient to perform Monte Carlo analyses of thousands of designs to obtain the best for a particular application [121].

4. Scaling Laws and the Optimization of Resolution

In optimizing a high-resolution SQUID system, two of the most important questions that must be addressed are the desired size of the pickup coils, and how close the coils should be to the sample. Only then can one determine whether the desired sensitivity can be achieved without severely compromising spatial resolution. There is, of course, no single coil diameter

---

[2]There are, unfortunately, few suitable references that provide quantitative estimates of the strength of biological current sources. As can be inferred from Figs. 2 and 6, and can be derived from a one-dimensional cable model [23], the strength of the intracellular current density is simply the product of the intracellular electrical conductivity, $\widetilde{\sigma}_i$, and the longitudinal voltage gradient, $\Delta V / \Delta z$. From this, the net intracellular current along a fiber of radius a is given by

$$I_{axial} = \widetilde{\sigma}_i \, \pi a^2 \frac{\Delta V}{\Delta z} \tag{3}$$

$$= \frac{\widetilde{\sigma}_i \, \pi a^2}{u} \frac{\Delta V}{\Delta t} , \tag{4}$$

where u is the conduction velocity and $\Delta V / \Delta t$ is the time derivative of the transmembrane potential. Thus, small fibers with high conduction velocities will have weak currents and magnetic fields. Reference [23] provides data for various cardiac and nerve fibers; Okada [61] provides estimates of source strengths for cortical dendrites.

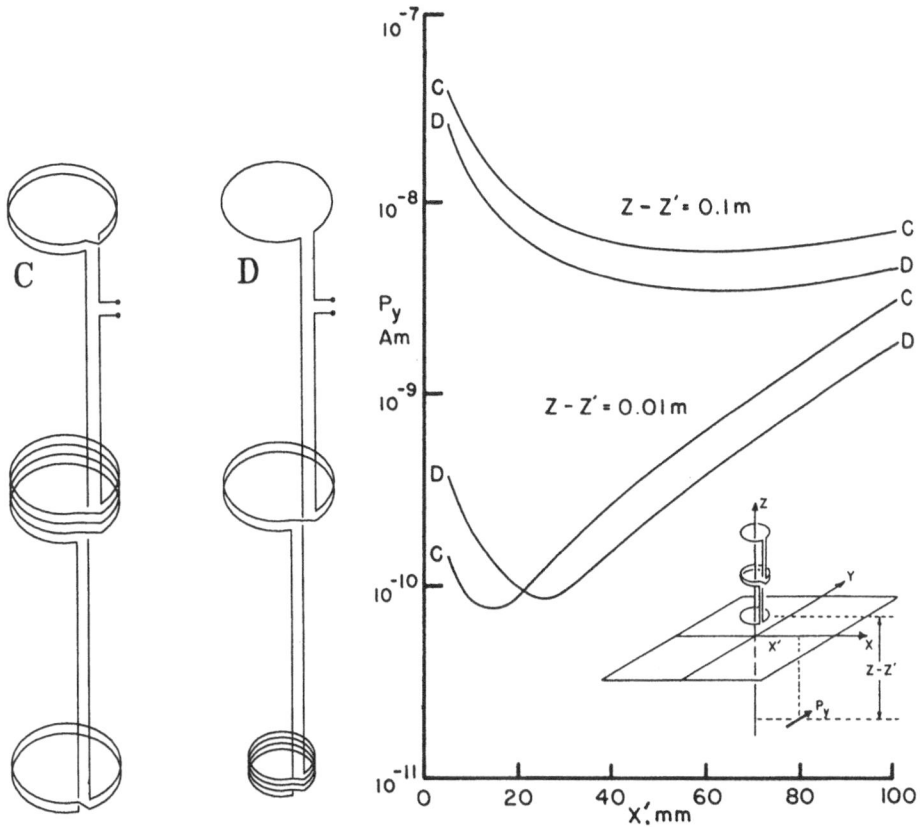

Figure 23. The optimization of a SQUID gradiometer using the minimum detectable current dipole. (left) Two second-order gradiometer configurations suitable for recording the MCG. The minimum detectable field for configuration C is 90 fT/Hz$^{1/2}$, while that for D is 140 fT/Hz$^{1/2}$. (right) The minimum detectable current dipole for configurations C and D at two different coil-to-source spacings as a function of lateral distance from the dipole. Note that C is more sensitive at greater distances, but D is both more sensitive and offers greater spatial resolution when close. (Adapted from [122], with permission)

that is best for general use. A large, closely-placed coil is not well suited for measuring fields from a compact source, since the average field over a large coil would be very low. The situation is even worse when a large coil is placed some distance from a compact source. Clearly, 1-mm or 3-mm diameter coils used to record the isofield contours in Fig. 9a or b will provide more accurate localization and a more detailed and accurate representation of the action currents than 15-mm coils used to record the field in Fig. 9c.

Pushing the state-of-the-art in high-resolution SQUID imaging for physiological samples requires that smaller magnetometers must be placed closer to the sample to realize the gain in resolution, but smaller SQUID magnetometers typically have lower sensitivities. While the design of an actual system would best be optimized by a series of detailed simulations of the images that would be obtained from the sources of interest, it is useful to look at a simpler conceptual approach that will demonstrate the trade-offs that will be encountered. Because SQUIDs measure magnetic flux, which is the product of field strength B and coil area A, the smaller the magnetometer, the lower will be its absolute field sensitivity. While this can be compensated in part by adding turns to the pickup coil, the ultimate limit is the energy density of the magnetic field. If a bare SQUID is used, the energy in the field is measured directly by the SQUID. If a flux transformer is used, the energy applied to the pickup coil is distributed between the face coil, any gradiometer balance coils and the SQUID coupling coil, so that only a fraction (less than half) of the flux that passes through the pickup coil is coupled to the SQUID sensor. Thus, the greatest energy sensitivity is achieved if the environmental noise can be reduced to the point at which bare SQUIDs can be used as magnetometers; for SQUID microscopes pushing the limits of sensitivity and spatial resolution, this factor of at least two can be important.

As we showed above, in general it is reasonable to assume that the pickup-coil diameter should be equal, or at least comparable, to the coil-to-sample spacing. In high-resolution SQUID systems with room- or biological-temperature samples and cryogenic SQUIDs, vacuum is the best thermal insulation, which in turn requires a thin vacuum window; the thickness of the window and the vacuum space are, of course, minimized. (In fully-cryogenic SQUID microscopes with cold samples, it is possible to have the coil-to-sample distance smaller than the minimum SQUID size, but this is not helpful for biological systems). To explore the trade-offs between sensitivity and resolution, let us initially assume that the spacing between the SQUID and the sample, r, is equal to the length of the side, a, of a square pickup coil or a square, bare SQUID; in a later section we will describe how coil apodization techniques can further improve resolution.

If the optimization is being performed around a commercial SQUID with fixed input inductance, the pickup coil optimization will then lead to a pickup coil of the same inductance. Within limits, the flux sensitivity of the SQUID will then be independent of the size of the pickup coil. Figure 24 shows an example of a graph used in such an optimization [125].

Given a fixed flux noise, ϕ_{noise}, and unity coupling between the SQUID sensor and a single-turn pickup coil, the sensitivity of the magnetometer in terms of the equivalent input noise, B_{noise}, is simply the flux noise divided by the area of the pickup coil, *i.e.*, $B_{noise} = \phi_{noise}/a^2$, implying that if our magnetometer size is reduced by a factor of 2, the field sensitivity of the magnetometer is reduced by a factor of 4.

A more rigorous analysis must assume that the SQUID has a certain en-

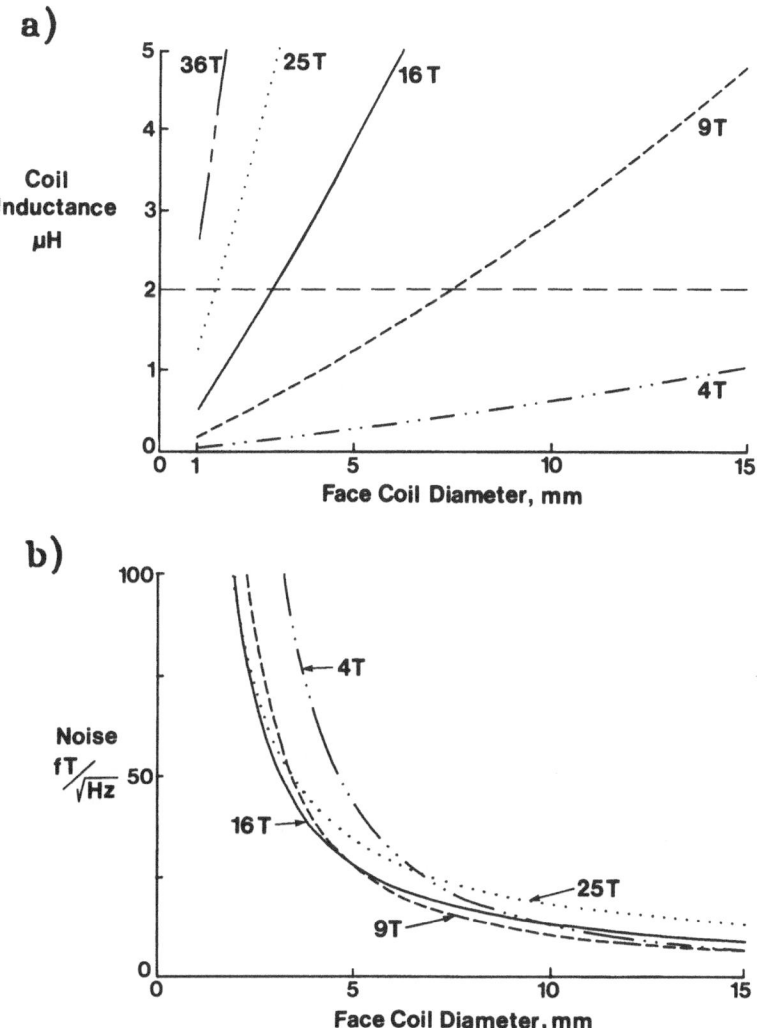

Figure 24. A design simulation for MicroSQUID showing the effects of various numbers of turns on the face coil of a first-order gradiometer upon (a) coil inductance and (b) noise as a function of face coil diameter. The horizontal dashed line in (a) shows the inductance of the SQUID sensor to which the pickup coil must be matched, which provides a constraint on the face coil diameter for a given number of turns. This diameter can then be used with the corresponding curve in (b) to determine the sensitivity of the system. (From [125], with permission)

ergy sensitivity, U_{noise}, that the flux sensitivity can be adjusted by changing the junction and washer geometries, and that the volume, V_{sense}, sensed by the pickup coil (proportional to the cube of the linear coil dimension, a) would contain a fixed-field energy which would then be optimally coupled to the SQUID by the correct choice of the numbers of turns in the pickup coil, *i.e.*,

$$\frac{1}{2\mu_o}(B_{noise})^2 V_{sense} = U_{noise} , \tag{5}$$

or

$$B_{noise} = (2\mu_o U_{noise}/V_{sense})^{1/2} \tag{6}$$

$$= (2\mu_o U_{noise})^{1/2}/K^{1/2}a^{3/2} , \tag{7}$$

where K is the proportionality constant between the coil dimension a and V_{sense}.

In either model, the smaller the magnetometer, the lower its absolute field sensitivity. However, because smaller magnetometers can be placed closer to the sample, there is an increase in the field produced by the source. If the source is a wire a distance a from the SQUID pickup coil, also of dimension a, the magnetic field strength is proportional to $1/a$. We can use Ampere's law for the magnetic field of a wire, $B_{wire} = \mu_o I/2\pi r$, to obtain the minimal current in a wire that can be detected by a magnetometer:

$$I_{min} = \frac{2\pi a}{\mu_o}B_{noise} = \pi\left(\frac{8U_{noise}}{\mu_o K a}\right)^{1/2} . \tag{8}$$

A smaller I_{min} means a better system. Hence, as the magnetometer is made smaller and closer, the field strength increases as $1/a$, and the sensitivity decreases as $a^{3/2}$, so that the minimum current detectable by the magnetometer is proportional to $a^{-1/2}$, *i.e.*, a smaller, closer magnetometer cannot detect as weak a current in a wire as can a larger, more distant magnetometer. If the sole object of an experiment is to detect the presence of a current-carrying wire, little is gained by miniaturizing the magnetometer, since the larger, more distant magnetometer will be able to detect weaker currents, and design compromises during the miniaturization process may lead to further reductions in sensitivity.

The situation is somewhat better if the magnetic field source is a current dipole immersed in a conducting medium, typical of many bioelectric sources. In this case, the magnetic field is proportional to $1/a^2$, so that the minimal detectable current dipole moment is proportional to $a^{1/2}$. There is a slight increase in detectability with smaller, closer magnetometers, assuming that the SQUID energy sensitivity is independent of a. On the other hand, if the object of the experiment is to locate the wire or image a distribution of current dipoles, it is clearly advantageous to use miniature magnetometers because of their increased spatial resolution. For our desired biological studies, we require both high sensitivity and high spatial resolution.

The advantages of magnetometer miniaturization are more obvious for magnetic fields that are produced by magnetic dipoles such as occur in magnetosomes or magnetic tracers. Since the magnetic field scales as $1/r^3$, while the sensitivity scales as $a^{3/2}$, the minimum detectable magnetic dipole, m_{min}, goes as $a^{3/2}$. Small a means a small, and hence good, m_{min}. This is why the relatively insensitive "magnetometer" used in a magnetic tape recorder can give very large signals: the spacing between the magnetic dipoles and the tape head can be on the order of 1 μm or less. Similarly, the magnetic force microscope is an intrinsically insensitive magnetometer, but for imaging the strong magnetic dipoles written on a computer disk, it can obtain a spatial resolution on the order of 10 nm with tip-to-surface spacings of 1 nm. However, the field sensitivity of this type of device is five orders of magnitude worse than what can be achieved with a SQUID microscope [118]. This simple analysis shows that the optimization of a SQUID microscope is governed not only by the nature of the source, but also by the intended measurement. In all cases, the more sensitive the SQUID, the better is the microscope. Table 1 summarizes these results.

TABLE 1. The relationship between SQUID dimension a, the signal strength S, the SQUID noise N, and the minimum detectable source (MDS) for various sources.

Source	S	N	S/N	MDS
Wire	a^{-1}	$a^{-3/2}$	$a^{1/2}$	$a^{-1/2}$
Electric Dipole	a^{-2}	$a^{-3/2}$	$a^{-1/2}$	$a^{1/2}$
Magnetic Dipole	a^{-3}	$a^{-3/2}$	$a^{-3/2}$	$a^{3/2}$

If the source is distributed, such as a current sheet or a cardiac depolarization wavefront, the field near the source may be, to first order, independent of distance, and hence there would be no increase in signal strength as the smaller magnetometer is placed closer, and therefore the minimum detectable currents will be larger than would be obtained for a single wire. It is important to realize from this, however, that a SQUID microscope which provides excellent images of distributions of ferromagnetic particles may not necessarily be capable of mapping the magnetic field from distributed, biological current sources.

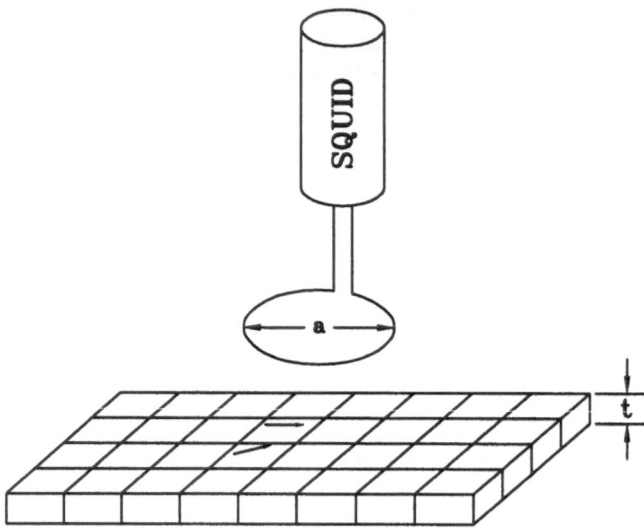

Figure 25. The relationship between a SQUID and a source that is discretized into volume elements, or voxels. Two current elements are shown.

4.1. PIXELIZATION

If the source is a current density J or a magnetization M distributed over a sample of thickness t, we need to discretize the source into voxels, as shown in Fig. 25. If we assume that the spatial resolution is determined by the coil-to-sample spacing and the pickup-coil diameter a, then the voxel size should be some fraction k of a. The volume assigned to each voxel of the source, and hence each pixel of the two-dimensional magnetic image, is thus equal to $k^2 a^2 t$, so that the pixel represents a current or magnetic dipole of strength $J_{min} k^2 a^2 t$ or $M_{min} k^2 a^2 t$. The minimum detectable current and magnetization densities are hence proportional to $k^{-2} a^{-3/2} t^{-1}$ and $k^{-2} a^{-1/2} t^{-1}$, respectively. If a is fixed, this can be used to adjust k to determine what pixel size is required to obtain a desired sensitivity to J or M, since larger pixels lead to increased sensitivity, but lower resolution.

4.2. CONTRAST

In a pixeled image, the minimum detectable contrast difference can be estimated from the ratio of the field due to the summed dipole moment of two adjacent pixels to that of the quadrupole moment formed by differences in the dipole moments of the two, as shown in Fig. 26. While the magnetic field of a current or magnetic quadrupole falls off as a power of a faster than that of a dipole, the quadrupole moment is also proportional to pixel size, so the contrast sensitivity will be independent of a, assuming k is fixed.

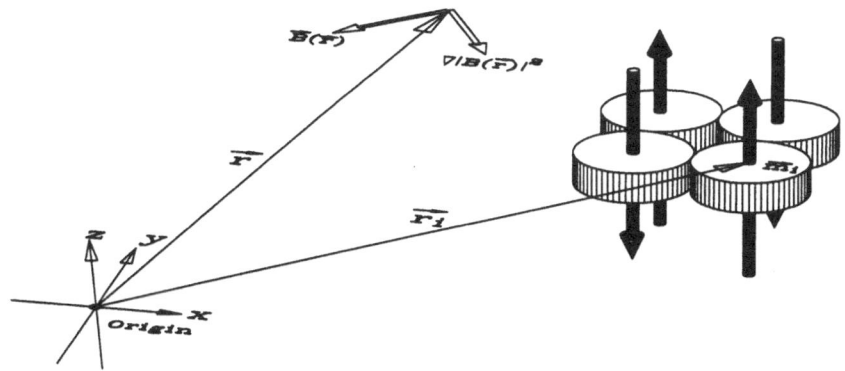

Figure 26. Four adjacent source voxels with differing normal magnetizations. The minimum detectable contrast difference along a particular line can be estimated by comparing the summed magnetic field from two adjacent dipoles to the difference of their fields.

In actual imaging situations, the simple power-law expressions for the field fall-off with distance is overly simplistic, particularly if only one component of the magnetic field is imaged, since there are additional geometric factors that come into play in the spatial variation of the field. In some situations, it may be necessary also to consider the finite size of the SQUID or the pickup coil.

5. Techniques for Improved Biomagnetic Current Imaging

The previous section delineates several considerations for the optimization of high resolution SQUIDs for biomagnetic imaging. Not surprisingly, the capabilities of SQUID imaging will be enhanced with quieter SQUIDs which provide a greater energy sensitivity that, in turn, will allow greater spatial resolution. For LTS SQUIDs, the sensitivity improvements are slow; for HTS SQUIDs, there are continuing improvements in sensitivity, but their performance does not approach that of the best LTS SQUIDs, particularly at low frequencies.

Great progress has been made on HTS SQUIDs that operate at liquid nitrogen temperatures. Sensitivities as good as 24 fT/Hz$^{1/2}$ have been achieved at 1 Hz with 4 cm × 4 cm flux concentrators [151], and 10 fT/Hz$^{1/2}$ at 100 Hz with a 7-mm diameter multiloop SQUID [16], but the performance of HTS SQUIDs is still inferior to that of the best LTS SQUIDs, particularly at frequencies below 10 Hz. It is at present difficult to fabricate complex pickup coil geometries without sacrificing sensitivity or introducing low-frequency drift. For measurements of strong, high-frequency bioelectric signals or fields from biogenic or injected magnetite, a liquid-nitrogen-based HTS SQUID microscope may be adequate, despite particularly bad noise at frequencies below 1 Hz. For measurements of weak or steady (<0.01 Hz) bioelectric currents from sources that fall off as only $1/r$ or $1/r^2$, the maximum possible SQUID sensitivity is required, indicating the need for liquid-helium-based LTS SQUIDs. Similarly, susceptibility measurements made in a strong applied magnetic field will for the present require 4 K, niobium SQUIDs.

Although HTS SQUIDs have been placed closer to room-temperature samples than have LTS SQUIDs, their reduced energy sensitivity translates to source sensitivities that are not yet sufficient for high-resolution imaging of the magnetic field from biological current sources. Given the limits of SQUID sensitivity and window-to-SQUID spacing imposed by cryogenic design, further enhancements to the performance of high-resolution SQUIDs may be achieved by developing specialized SQUID sensors and arrays, and magnetic-imaging algorithms.

5.1. APODIZATION

As was discussed above, the imaging resolution is limited in part by the diameter of the pickup coil, since the magnetic field B_z will be convolved with the spatial sampling (or turns) function H of the coil to give the detected flux Φ. In Fourier space, the magnetic field distribution from the sample $b(k_x, k_y)$ will be multiplied by the turns function $h(k_x, k_y)$ to give the flux $\phi(k_x, k_y)$. Our previous design work at Vanderbilt, presented in greater detail in another chapter in this volume [129], showed that it is theoretically possible to greatly enhance the effective spatial frequency bandwidth of a SQUID magnetometer by apodizing the pickup coils to move the zeros in the coil transfer function out of the range of spatial frequencies of interest [74]. With this approach, larger apodized coils can give better inverse images of currents than can smaller standard coils. The effect of the coil can be corrected by dividing $\phi(k_x, k_y)$ by $h(k_x, k_y)$ to obtain $B(k_x, k_y)$. An inverse FT gives the coil-corrected B_z. Unfortunately, for typical coils, $h(k_x, k_y)$ has zeros that limit the success of this approach. This is avoided by using apodized pickup coils [74, 129].

It also may be possible with thin-film photolithography techniques to construct a special planar gradiometer whose output is proportional to the curl of current directly below the center of the coil [90]. Computer models of such a spatially-extended curl coil suggest that it would be able to image the curl of current distributions far more accurately than can be accomplished with the present small coils. Since it is only the curl of currents which generates magnetic fields, we lose no information while greatly enhancing our ability to image the currents.

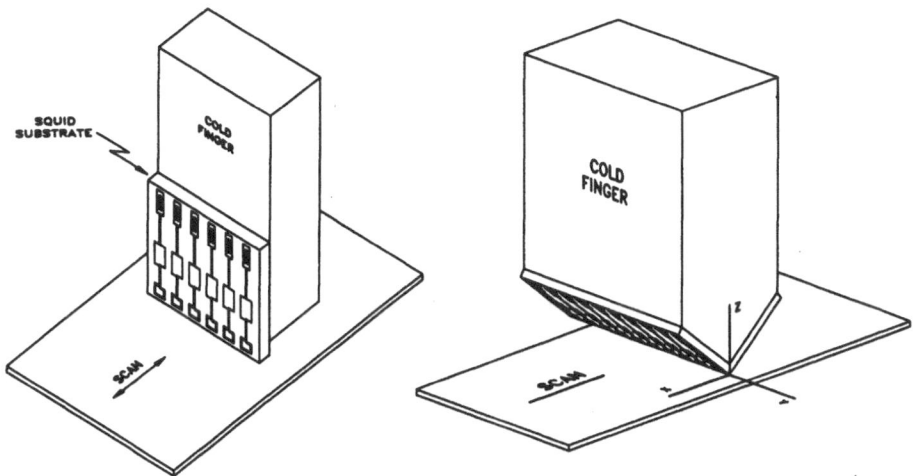

Figure 27. Two SQUID arrays, one that measures one diagonal tangential field component (left), and one that measures two orthogonal, diagonal components (right). While the second array is more difficult to fabricate, it has the advantage of being able, by computing the sum and differences of the two components, to determine both the normal and tangential field components [142].

5.2. SQUID ARRAYS

For many biological measurements, the time required to scan the magnetic field above the sample is a limiting factor. This can be ameliorated in part by using a SQUID array, as shown in Fig. 27, that samples the field at multiple locations at once. The scaling laws presented above apply equally well to SQUID arrays, for which the spacing of the individual SQUIDs and the array-to-source spacing should be comparable to the desired imaging resolution.

SQUID arrays may offer additional advantages. Our computer simulations [85] demonstrate that SQUID arrays that allow the simultaneous measurement of the magnetic field B_z and the partial derivatives $\partial B_z/\partial x$, $\partial B_z/\partial y$, and $\partial^2 B_z/\partial x\partial y$ will allow clear resolution of a pair of wires separated by 160 μm, with a measurement height of 300 μm.

For the planar, thin-film SQUIDs, it is often easier to measure the tangential field component than the radial one. For a current distribution on a closed surface, the tangential magnetic field components at the surface are simply μ_o times the orthogonal surface-current components, *i.e.*, $B_x = -\mu_o J_y^{eff}$ and $B_y = \mu_o J_x^{eff}$. Hence, the tangential magnetic field components measured over an arbitrary closed surface can be used to specify the effective current distribution on a surface that could reproduce the observed field without requiring any other sources within the bounding surface. The close relationships between B_x and J_y^{eff} (the y component of the effective surface current $\vec{J}^{eff}$) and the similar one between B_y and J_x^{eff} could simplify the immediate interpretation of the data when acquired, yet would still be suitable for off-line mathematical inversion to obtain quantitative source models.

Other considerations of great practical importance, but beyond the scope of this review, include the trade-offs between scanning speed, SQUID bandwidth and signal-to-noise ratio; possible development of multi-loop HTS SQUIDs [45, 46, 13, 16] for reducing microscope noise [120]; the possible use of flow cryostats to cool the SQUIDs [34, 31]; the need for magnetic shields to attenuate external noise; and the use of digital SQUIDs to obtain the dynamic range required for operation in magnetically-noisy environments and to allow the connection of a large number of SQUIDs to room temperature without large thermal loads [66].

6. Conclusions: High Resolution Magnetic Imaging

We conclude that the performance enhancements achieved by miniaturization of SQUID microscopes and arrays are governed not only by the sensitivity of the SQUID sensor, but also by the rate of fall-off of the field produced by the elements of a distributed source. The one-millimeter spatial resolution of MicroSQUID is inadequate to allow the study of either monolayers of cultured cardiac cells, or detailed studies on geometrically-simple yet physiologically-complex multicell preparations such as the nervous system of the leech. It is now possible, however, to utilize integrated circuit technology and advanced cryogenic design to construct ultrahigh-resolution SQUID magnetometers, used in a scanning magnetic microscope mode, whose pickup-coil dimensions and coil-to-sample spacing are comparable to the dimensions of cardiac myocytes and the squid giant synapse, i.e., an order of magnitude better than was achieved with MicroSQUID. The increases in spatial resolution that will be provided by such instruments, the capabilities provided by optical sensors of transmembrane potential, and the development of the requisite imaging algorithms and experimental techniques represent a major step toward the goal of using magnetometry to explore, at the cellular level, electrophysiological events such as: intercellular communication in cardiac muscle, smooth muscle and septated nerve axons; the effects of intercellular discontinuities and insulating islands of connective tissue on the propagation of cardiac action potentials and currents; the distribution of action currents in the giant synapse of the marine squid; the temporal and spatial dependence of injury currents in single axons and cardiac tissue; the role of axonal bifurcations in the stability of signal propagation in nerves; and the mechanism of action of various channel blockers on the propagating cardiac action potential. We anticipate that SQUID magnetometers with adequate sensitivity and spatial resolution will also be useful for the study of plasticity of neural networks, the physiological basis of spontaneous oscillations in neural structures, and the mechanisms of coupling of rhythmic activity in smooth muscle. SQUIDs may avoid the problems inherent with microelectrode recordings of action potentials in plants. SQUID magnetometers may be the optimum means to detect steady currents flowing inside of developing tissue, or determine the presence or distribution of biogenic or artificially-introduced magnetic particles.

The application of SQUID magnetometers to cellular electrophysiology may provide information not available with other techniques, but it is important to realize that in almost every case, SQUIDs will complement, but not replace the other techniques. For each of these phenomena, detailed knowledge of both action currents and action potentials will be required to discern the fundamental, governing mechanisms. We are presently working toward combining magnetic measurements of intracellular action currents, electrode measurement of extracellular potentials, and optical or microelectrode recordings of transmembrane potential to obtain improved measurements of the effects of the electrical anisotropy of both the intracellular and extracellular spaces of cardiac tissue [87, 86, 85]. We are also seeking a fundamental understanding of the relationship between biomagnetic fields, bioelectric fields, and their sources [130, 72, 77, 124]. As we explore this relationship in detail, our growing knowledge of the unique features, strengths and weaknesses of magnetic techniques is providing new insights regarding what data might best be obtained from magnetic techniques alone, electrical measurements alone, or the two in combination. One of the most important findings from our experimental and theoretical studies is that the ability to measure cellular action currents directly, without assumptions regarding tissue conductivities or anisotropies, can provide new and valuable insights into the interplay between tissue properties, electric fields and currents, and the propagation of electrical activity in multicellular systems [71, 87, 86, 85, 2].

While NDE and other applications of SQUID imaging can also benefit from increases in spatial resolution, it is important to remember that the imaging of weak sources that are deeper into the sample than the coil diameter would benefit from larger and more sensitive pickup coils. Hence SQUID microscopes, with 10 to 50 μm spatial resolution, are best suited for surface or two-dimensional studies, and systems with larger coils, still placed close to room temperature, would be better for the detection of deep flaws. For many NDE applications, the signal strength is determined by an applied field or current, and hence the signal-to-noise ratio of the image would be limited not by SQUID noise but by spatial variations in the material properties or structure of the sample [146].

The high-resolution magnetic imaging of cellular action currents remains as one of the most demanding appications of SQUIDs. The challenge to SQUID builders is to develop even better SQUIDs that have a spatial resolutions in the range of 100 μm to 1 mm, and are suited for *in vivo* measurements on small animals and *in vitro* measurements on isolated tissue. If they are successful, the combination of high sensitivity and hundred-micron spatial resolution should provide a whole new arena for the application of SQUIDs.

7. Acknowledgements

The preparation of this manuscript and much of the research described within it was funded by grants from the National Science Foundation and the National Institutes of Health. I am grateful to William Jenks for his effort towards developing our next generation of high resolution SQUIDs, and to Fred Wellstood and John Clarke for enjoyable discussions on SQUIDs and SQUID microscopes, to Bradley Roth for numerous discussions regarding cardiac electrophysiology and the bidomain model, and to Debra Echt for the information on cardiac defibrillation thresholds. I am indebted to Richard Friedman, Frans Gielen, William Jenks, Yu Pei Ma, Bradley Roth, Nestor Sepulveda, Daniel Staton, Ken Swinney, Shaofen Tan, Ian Thomas and the late Jan van Egeraat for their major contributions to the Vanderbilt magnetic-imaging effort, and their papers from which I have drawn heavily in preparing this chapter. I thank Margaret Khayat, Leonora Wikswo and Chris Bublitz for their comments on this manuscript, Licheng Li for her care in preparing the illustrations, and Cheryl Cosby for her assistance with the manuscript.

References

1. Anderberg, J. and Colclough, M.S. (1994) "Three channel SQUID-based magnetic microscope, utilizing 14 micrometer coils and cryogenic stepper motor driven stages," *Bull. Amer. Phys. Soc.*, **Vol. 39, no. 1**, p. 196.
2. Barach, J.P. and Wikswo, Jr., J.P. (1987) "Computer simulation of action potential propagation in septated nerve fibers," *Biophys. J.*, **Vol. 51**, pp. 177–183.
3. Barach, J.P., Freeman, J.A., and Wikswo, Jr., J.P. (1980) "Experiments on the magnetic field of nerve action potentials," *J. Appl. Phys.*, **Vol. 51**, pp. 4532–4538.
4. Barth, D.S., Sutherling, W., and Beatty, J. (1986) "Intracellular currents of interictal penicillin spikes: Evidence from neuromagnetic mapping," *Brain Res.*, **Vol. 368, no. 1**, pp. 36-48.
5. Black, R.C., Mathai, A., and Wellstood, F.C. (1993) "Magnetic microscopy using a liquid cooled $YBa_2Cu_3O_7$ superconducting quantum interference device," *Appl. Phys. Lett.*, **Vol. 62, no. 17**, pp. 2128–2130.

6. Black, R.C., Wellstood, F.C., Ludwig, F., Koelle, D., Nemeth, D.T., Miklich, A.H., and Clarke, J. (1994) "Eddy-current/RF/microwave magnetic microscopy using a liquid nitrogen cooled superconducting quantum interference device (SQUID)," *Bull. Amer. Phys. Soc.*, **Vol. 39, no. 1**, p. 523.

7. Brockmeier, K., Schmitz, L., Bobadilla Chavez, JdeJ, Burghoff, M., Koch, H., Zimmerman, R., and Trahms, L. "Magnetocardiography and 32-lead potential mapping: The repolarization in normal subjects during pharmacological stress," in preparation.

8. Buchanan, D.S., Crum, D.B., Cox, D., and Wikswo, Jr., J.P. (1990) "MicroSQUID: A close-spaced four channel magnetometer," in *Advances in Biomagnetism*, S.J. Williamson, M. Hoke, G. Stroink, and M. Kotani, Eds., Plenum, New York, pp. 677–679.

9. Chen, P.S., Shibata, N., Dixon, E.G., Martin, R.O., and Ideker, R.E. (1986) "Comparison of the defibrillation threshold and the upper limit of ventricular vulnerability," *Circ.*, **Vol. 73**, pp. 1022–1028.

10. Cochran, A., Donaldson, G.B. (1992) "Improved techniques for structural NDT using SQUIDs," in *Superconducting Devices and Their Applications*, H. Koch and H. Lubbig, Eds., Springer Verlag, Berlin, pp. 576–580.

11. Cohen, D. and Kaufman, A.L. (1975) "A magnetic determination of the relationship between the S-T segment and the injury produced by coronary occlusion," *Circ. Res.*, **Vol. 36**, pp. 414–424.

12. Cook, P., Virginia Semiconductor, Inc., Fredericksburg, VA, personal communation.

13. Dantsker, E., Ludwig, F., Kleiner, R., Clarke, J., Teepe, M., Lee, L.P., Alford, N. McN., and Button, T. (1995) "Addendum: Low noise $YBa_2Cu_3O_{7-x}$-$SrTiO_3$-$YBa_2Cu_3O_{7-x}$ multilayers for improved superconducting magnetometers," [Appl. Phys. Lett. 66, 373 (1995)] *Appl. Phys. Lett.*, **Vol. 67, no. 5**, pp. 725–726.

14. Davy, J.-M., Fain, E.S., Dorian, P., and Winkle, R.A. (1987) "The relationship between successful defibrillation and delivered energy in open-chest dogs: Reappraisal of the "defibrillation-threshold" concept," *Am. Heart J.*, **Vol. 113**, pp. 77–83.

15. Dillon, S.M. (1992) "Synchronized repolarization after defibrillation shocks: A possible component of the defibrillation process demonstrated by optical recordings in rabbit heart," *Circ.*, **Vol. 85**, pp. 1865–1878.

16. Drung, D., Dantsker, E., Ludwig, F., Koch, H., Kleiner, R., Clarke, J., Krey, S., Reimer, D., David, B., and Doessel, O. (1996) "Low noise $YBa_2Cu_3O_{7-x}$ SQUID magnetometers operated with additional positive feedback," *Appl. Phys. Lett.*, **Vol. 68, no. 13**, pp. 1856–1858.

17. Echt, D.S., personal communication.

18. Freake, S.M., Swithenby, S.J., and Thomas, I.M. (1988) "A miniature SQUID magnetometer for the detection of quasi-dc ionic current flow in developing organisms," in *Biomagnetism '87*, K. Atsumi, M. Kotani, S. Ueno, T.E. Katila, and S.J. Williamson, Eds., Tokyo Denki University, Tokyo, pp. 434–437.

19. Friedman, R.N., van Egeraat, J.M., and Wikswo, Jr., J.P. (1991) "Magnetic characterization of action currents at the primary branch of the squid giant axon," *Society for Neuroscience Abstracts*, **Vol. 17**, p. 605.

20. Friedman, R.N., Gielen, F.L.H., van Egeraat, J.M., Wijesinghe, R.S., and Wikswo, Jr., J.P. (1992) "Magnetic characterization of action currents in a slow-twitch mammalian skeletal muscle," *American Society for Gravitational and Space Biology*, **Vol. 6**, p. 94 (Abstract).

21. Garrard, C.L., Wikswo, Jr., J.P., Staton, D., Golzarian, J., Gallen, C., and Richards, W.O. (1994) "Noninvasive measurements of small bowel electrical activity," *Gastroenterology*, **Vol. 106, no. 4**, p. A502 (Abstract).

22. Gielen, F.L.H., Friedman, R.N., and Wikswo, Jr., J.P. (1991) "In vivo magnetic and electric recordings from nerve bundles and single motor units in mammalian skeletal muscle: Correlations with muscle force," *J. Gen. Physiol.*, **Vol. 98**, pp. 1043–1061.

23. Gielen, F.L.H., Roth, B.J., and Wikswo, Jr., J.P. (1986) "Capabilities of a toroid-amplifier system for magnetic measurement of current in tissue," *IEEE Trans. Biomed.*

354

Eng., **Vol. 33**, pp. 910–921.

24.   Gielen, F.L.H., Stasaski, R., and Wikswo, Jr., J.P. (1989) "Monitoring of peripheral nerve regeneration by means of a biomagnetic sensor," in *Images of the Twenty-First Century: Proc. 11th Annual International Conference of the IEEE Engineering in Medicine and Biology Society*, **Vol. 11**, pp. 997–998.

25.   Golzarian, J., Staton, D.J., Wikswo, Jr., J.P., Friedman, R.N., and Richards, W.O. (1994) "Diagnosing intestinal ischemia using a noncontact superconducting quantum interference device," *Am. J. Surgery*, **Vol. 167**, pp. 586–592.

26.   Goodman, W.L. and Deaver, B.S.D. (1970) "Detailed measurements of the quantized flux states of hollow superconducting cylinders," *Phys. Rev. Lett.*, **Vol. 24, no. 16**, pp. 870–873.

27.   Hashimoto, I., Papuashvili, N., Xu, C., and Okada, Y.C. (1996) "Neuronal activities from a deep subcortical structure can be detected magnetically outside the brain in the porcine preparation," *Neurosci. Lett.*, **Vol. 206**, pp. 25–28.

28.   Hibbs, A.D., Sager, R.E., Cox, D.W., Aukerman, T.H., Sage, T.A., Landis, R.S. (1992) "A high-resolution magnetic imaging system based on a SQUID magnetometer," *Rev. Sci. Instr.*, **Vol. 63**, pp. 3652–3658.

29.   Hille, B. (1992) *Ionic Channels of Excitable Membranes*, Sinauer Associates, Sunderland, MA.

30.   Idriss, S.F., Wolf, P.D., Oliver, L.M., Smith, W.M., and Ideker, R.E. (1996) "Ventricular fibrillation originates in weak potential gradient area during upper limit testing – regardless of S1 pacing site," *PACE*, **Vol. 19**, p. 603.

31.   Jenks, W.G. and Wikswo, J.P., Jr., in preparation.

32.   Jenks, W.G., Thomas, I.M., and Wikswo, Jr., J.P. "SQUIDs," in *Encyclopedia of Applied Physics*, G.L. Trigg, E.S. Vera, and W. Greulich, Eds., VCH Publishers, Inc., New York, NY, in press.

33.   Kanai, A. and Salama, G. (1995) "Optical mapping reveals that repolarization spreads anisotropically and is guided by fiber orientation in guinea pig hearts," *Circ. Res.*, **Vol. 77**, pp. 784–802.

34.   Kandori, A., Ueda, M., and Ogata, H. (1995) "Development of semi-portable DC-SQUID magnetometer," *IEEE Trans. Applied Supercond.*, **Vol. 5, no. 2**, pp. 2474–2477.

35.   Kirtley, J.R., Ketchen, M.B., Chaudhari, P., Khare, N., and Stawiasz, K.G. (1994) "Flux penetration and trapping in superconducting grain boundaries," *Bull. Amer. Phys. Soc.*, **Vol. 39, no. 1**, p. 222.

36.   Kirtley, J.R., Ketchen, M.B., Stawiasz, K.G., Sun, J.Z., Gallagher, W.J., Blanton, S.H., and Wind, S.J. (1995) "High-resolution scanning SQUID microscope," *Appl. Phys. Lett.*, **Vol. 66, no. 9**, pp. 1138–1140.

37.   Knisley, S.B. (1995) "Transmembrane voltage changes during unipolar stimulation of rabbit ventricle," *Circ. Res.*, **Vol. 77**, pp. 1229–1239.

38.   Knisley, S.B., Hill, B.C., and Ideker, R.E. (1994) "Virtual electrode effects in myocardial fibers," *Biophys. J.*, **Vol. 66**, pp. 719–728.

39.   Kuypers, P.D.L., Gielen, F.L.H., Wai, R.T.J., Hovius, S.E.R., Godschalk, M., and van Egeraat, J.M. (1993) "A comparison of electric and magnetic compound action signals as quantitative assays of peripheral nerve regeneration," *Muscle and Nerve*, **Vol. 16**, pp. 634–641.

40.   Kyuhou, S.-I. and Okada, Y.C. (1993) "Detection of magnetic evoked fields associated with synchronous population activities in the transverse CA1 slice of the guinea pig," *J. Neurophysiol.*, **Vol. 70, no. 6**, pp. 2665–2668.

41.   Lee, R., Lucas NovaSensor, Inc., Freemont CA, personal communication.

42.   Lee, T.S., Dantsker, E., and Clarke, J. (1996) "Scanning high-T_c SQUID system for room-temperature samples," *Bull. Am. Phys. Soc.*, **Vol. 41, no, 1**, p. 532 (Abstract).

43.   Leifer, M.C. and Wikswo, Jr., J.P. (1983) "Optimization of a SQUID clip-on current probe," *Rev. Sci. Instr.*, **Vol. 54**, pp. 1017–1022.

44.   Lin, S.F., Abbas, R.A., and Wikswo, Jr., J.P. (1993) "Magnetic localization of the

origins of self-sustained oscillation in SQUID giant axons," *Biol. Bull.*, **Vol. 185**, pp. 300–301.

45. Ludwig, F., Koelle, D., Dantsker, E., Nemeth, D.T., Miklich, A.H., Clarke, J., and Thomson, R.E. (1995) "Low noise $YBa_2Cu_3O_{7-x}$-$SrTiO_3$-$YBa_2Cu_3O_{7-x}$ multilayers for improved superconducting magnetometers," *Appl. Phys. Lett.*, **Vol. 66, no. 3**, pp. 373–375.

46. Ludwig, F., Dantsker, E., Kleiner, R., Koelle, D., Clarke, J., Knappe, S., Drung, D., Koch, H., Alford, N. McN., and Button, W. (1995) "Integrated high-T_c multiloop magnetometer," *Appl. Phys. Lett.*, **Vol. 66, no. 11**, pp. 1418–1420.

47. Ma, Y.P. and Wikswo, Jr., J.P. (1991) "Magnetic shield for wide-bandwidth magnetic measurements for nondestructive testing and biomagnetism," *Rev. Sci. Instru.*, **Vol. 62, no. 11**, pp. 2654–2661.

48. Ma, Y.P., Thomas, I.M., Lauder, A., and Wikswo, Jr., J.P. (1993) "A high resolution imaging susceptometer," *IEEE Trans. Applied Supercond.*, **Vol. 3, no. 1**, pp. 1941–1944.

49. MacAuley, C.E., Stroink, G., and Horacek, B.M. (1984) "Signal analysis of magnetocardiograms and electrocardiograms to test their independence," in *Biomagnetism: Applications and Theory*, H. Weinberg, G. Stroink, and T. Katila, Eds., Pergamon Press, New York, pp. 115–120.

50. Mall, F.P. (1911) "On the muscular architecture of the ventricles of the human heart," *J. Anat.*, **Vol. 11**, pp. 211–266.

51. Mathai, A., Song, D., Gin, Y., and Wellstood, F.C. (1993) "High resolution magnetic microscopy using a DC SQUID," *IEEE Trans. Applied Supercond.*, **Vol. 3, no. 1**, pp. 2609–2612.

52. Neunlist, M. and Tung, L. (1995) "Spatial distribution of cardiac transmembrane potentials around an extracellular electrode: Dependence on fiber orientation," *Biophys. J.*, **Vol. 68**, pp. 2310–2322.

53. Okada, Y.C. and Xu, C. (1994) "Detection of unaveraged spontaneous and event-related electrophysiological activities from focal regions of the cerebral cortex in the swine," in *Oscillatory Event-Related Brain Dynamics*, C. Pantev, B. Ltkenhoner, and T. Elbert, Eds., Plenum Press, pp. 389–400.

54. Okada, Y.C. and Xu, C. (1995) "Comparisons of MEG, EEG, ECoG and intracortical recordings for characterization of cortical tissues," in *Quantitative and Topological EEG and MEG Analysis*, M. Eiselt, U. Zwiener, and H. Witte, Eds., Universittsverlag Druckhaus-Mayer, Jena, pp. 109–116.

55. Okada, Y.C. and Xu, C. "Single-epoch neuromagnetic signals during epileptiform activities in guinea pig longitudinal CA3 slices," *Neurosci. Lett.*, in press.

56. Okada, Y.C., Kyuhou, S., and Xu, C. (1993) "Tissue currents associated with spreading depression inferred from magentic field recordings," in *Migrane: Basic Mechanisms and Treatment*, A. Lehmenkuhler, Ed., pp. 249–265.

57. Okada, Y.C., Papuashvili, N. and Xu, C. "What can we learn from MEG studies of the somatosensory system of the swine?" in *Visualization of Information Processing in the Human Brain: Recent Advances in MEG and Functional MRI*, I. Hashimoto, Y.C. Okada, and S. Ogawa, Eds., Elsevier, in press.

58. Okada, Y.C., Shah, B., and Huang, J.-C. (1994) "Ferromagnetic high-permeability alloy alone can provide sufficient low-frequency and eddy-current shieldings for biomagnetic measurements," *IEEE Trans. Biomed. Eng.*, **Vol. 41, no. 7**, pp. 688–697.

59. Okada, Y.C., Xu, C., and Kyuhou, S.-I. (1995) "Magnetic fields of synchronized population activities in the presence of picrotoxin in hippocampal slices of the guinea pig," in *Quantitative and Topological EEG and MEG Analysis*, M. Eiselt, U. Zwiener, and H. Witte, Eds., Universittsverlag Druckhaus-Mayer, Jena, pp. 43–50.

60. Okada, Y.C., do Carmo, R., Martins-Ferreira, H., and Nicholson, C. (1992) "Detection of magnetic fields from isolated avian retina during spreading depression," *Exper. Brain Res.*, **Vol. 23**, pp. 89–97.

61. Okada, Y.C., Kyuhou, S., Lähteenmäki, A., and Xu, C. (1992) "A high resolution

356

system for magnetophysiology and its applications," in *Biomagnetism: Clinical aspects*, M. Hoke, S.N. Ern, Y.C. Okada, and G.-L. Romani, Eds., Elsevier, pp. 375–383.

62. O'Mara, W.C., Herring, R.B., and Hunt, L.P. (1990) *Handbook of Semiconductor Silicon Technology*, Noyes Publications, Park Ridge, NJ.

63. Pilkey, W.D., (1994) *Formulas for Stress Strain and Structural Matrices*, John Wiley, New York.

64. Plonsey, R. and Barr, R.C. (1984) "Current flow patterns in two-dimensional anisotropic bisyncytia with normal and extreme conductivities," *Biophys. J.*, **Vol. 45**, pp. 557–571.

65. Podney, W.N. (1993) "Performance measurements of a superconductive micro-probe for eddy current evaluation of subsurface flaws," *IEEE Trans. Applied Supercond.*, **Vol. 3, no. 1**, pp. 1914–1917.

66. Radparvar, M. and Rylov, S. (1995) "An integrated digital SQUID magnetometer with high sensitivity input," *IEEE Trans. Applied Supercond.*, **Vol. 5, no. 2**, pp. 2142–2145.

67. Richards, W.O., Staton, D.J., Golzarian, J., Friedman, R.N., and Wikswo, Jr., J.P. (1995) "Non-invasive SQUID magnetometer measurement of human gastric and small bowel electrical activity," in *Biomagnetism: Fundamental Research and Clinical Applications - Proceedings of the 9th International Conference on Biomagnetism*, C. Baumgartner, L. Deecke, G. Stroink, and S.J. Williamson, Eds., IOS Press, Amsterdam, Netherlands, Vol. 7, pp. 743–747.

68. Richardson, R.C. and Smith, E.N. (1988) *Experimental Techniques in Condensed Matter Physics at Low Temperatures*, Addison-Wesley, Redwood City, CA.

69. Rogers, F.P. (1983), "A device for experimental observation of flux vortices trapped in superconducting thin films," Masters Thesis, Department of Electrical Engineering, MIT.

70. Roth, B.J. (1995) "A mathematical model of make and break electrical stimulation of cardiac tissue by a unipolar anode or cathode," *IEEE Trans. Biomed. Eng.*, **Vol. 42**, pp. 1174–1184.

71. Roth, B.J. and Wikswo, Jr., J.P. (1985) "The magnetic field of a single axon: A comparison of theory and experiment," *Biophys. J.*, **Vol. 48**, pp. 93–109.

72. Roth, B.J. and Wikswo, Jr., J.P. (1986) "Electrically-silent magnetic fields," *Biophys. J.*, **Vol. 50**, pp. 739–745.

73. Roth, B.J. and Wikswo, Jr., J.P. (1989) "Longitudinal resistance in cardiac muscle and its effects on propagation," in *Cell Interactions and Gap Junctions*, N. Sperelakis and W.C. Cole, Eds., CRC Press, Boca Raton, pp. 165–178.

74. Roth, B.J. and Wikswo, Jr., J.P. (1990) "Apodized pickup coils for improved spatial resolution of SQUID magnetometers," *Rev. Sci. Instr.*, **Vol. 61**, pp. 2439–2448.

75. Roth, B.J. and Wikswo, Jr., J.P. (1994) "Electrical stimulation of cardiac tissue: A bidomain model with active membrane properties," *IEEE Trans. Biomed. Eng.*, **Vol. 41**, pp. 232–240.

76. Roth, B.J. and Wikswo, Jr., J.P. (1996) "The effect of externally applied electrical fields on myocardial tissue," *Proceedings of the IEEE: Electrical Therapy of Cardiac Arrhythmias*, **Vol. 84, no. 3**, pp. 379–391.

77. Roth, B.J., Guo, W.-Q., and Wikswo, Jr., J.P. (1988) "The effects of spiral anisotropy on the electric potential and the magnetic field at the apex of the heart," *Mathemat. Biosci.*, **Vol. 88**, pp. 191–221.

78. Roth, B.J., Sepulveda, N.G., and Wikswo, Jr., J.P. (1989) "Using a magnetometer to image a two-dimensional current distribution," *J. Appl. Phys.*, **Vol. 65**, pp. 361–372.

79. Sepulveda, N.G. and Wikswo, Jr., J.P. (1987) "Electric and magnetic fields from two-dimensional anisotropic bisyncytia," *Biophys. J.*, **Vol. 51**, pp. 557–568.

80. Sepulveda, N.G. and Wikswo, Jr., J.P. (1994) "Bipolar stimulation of cardiac tissue using an anisotropic bidomain model," *J. Cardiovasc. Electrophysiol.*, **Vol. 15**, pp. 258–267.

81. Sepulveda, N.G., Roth, B.J., and Wikswo, Jr., J.P. (1989) "Current injection into

a two-dimensional anisotropic bidomain," *Biophys. J.*, **Vol. 55**, pp. 987–999.

82. Sepulveda, N.G., Thomas, I.M., and Wikswo, Jr., J.P. (1994) "Magnetic susceptibility tomography for three-dimensional imaging of diamagnetic and paramagnetic objects," *IEEE Trans. Mag.*, **Vol. 30, no. 6**, pp. 5062–5069.

83. Sepulveda, N.G., Wikswo, Jr., J.P., and Echt, D.S. (1990) "Finite element analysis of cardiac defibrillation current distributions," *IEEE Trans. Biomed. Eng.*, **Vol. 37**, pp. 354–365.

84. Singer, I. and Lang, D. (1992) "Defibrillation threshold: Clinical utility and theraputic implications," *PACE*, **Vol. 15**, pp. 932–949.

85. Staton, D.J. (1994) "Magnetic imaging of applied and propagating action currents in cardiac tissue slices: Determination of anisotropic electrical conductivities in a two-dimensional bidomain," Ph.D. Dissertation, Department of Physics and Astronomy, Vanderbilt University.

86. Staton, D.J. and Wikswo, Jr., J.P. (1993) "Magnetic determination of the anisotropic electrical conductivities in a two-dimensional cardiac bidomain," in *Engineering Solutions to Current Health Care Problems, Proceedings of the 15th Annual International Conference of the IEEE Engineering in Medicine and Biology Society*, **Vol. 15, Part II**, pp. 746–747.

87. Staton, D.J. and Wikswo, Jr., J.P. (1993) "Magnetic imaging of cardiac transmembrane potentials and measurement of anisotropic electrical conductivities," *Circ.*, **Vol. 88, no. 4**, p. I–623 (Abstract).

88. Staton, D.J., Friedman, R.N., and Wikswo, Jr., J.P. (1991) "High-resolution SQUID magnetocardiographic mapping of action currents in canine cardiac slices," *Circ.*, **Vol. 84, no. 4**, p. II–667 (Abstract).

89. Staton, D.J., Friedman, R.N., and Wikswo, Jr., J.P. (1993) "High resolution SQUID imaging of octupolar currents in anisotropic cardiac tissue," *IEEE Trans. Applied Supercond.*, **Vol. 3, no. 1**, pp. 1934–1936.

90. Staton D., Sepulveda, N.G., and Wikswo, Jr., J.P. "Curl and matched SQUID pickup coils for magnetic imaging," unpublished.

91. Staton, D.J., Wikswo, Jr., J.P., Friedman, R.N., and Richards, W.O. (1992) "First biomagnetic measurements of intestinal basic electrical rhythms (BER) in vivo using a high resolution magnetometer," *Gastroenterology*, **Vol. 103**, p. 1385 (Abstract).

92. Staton, D.J., Golzarian, J., Wikswo, Jr., J.P., Friedman, R.N., and Richards, W.O. (1993) "SQUID magnetometer diagnosis of experimental small bowel ischemia," in *Engineering Solutions to Current Health Care Problems, Proceedings of the 15th Annual International Conference of the IEEE Engineering in Medicine and Biology Society*, **Vol. 15, Part III**, pp. 1521–1522.

93. Staton, D.J., Soteriou, M.C., Friedman, R.N., Richards, W.O., and Wikswo, Jr., J.P. (1991) "First magnetic measurements of smooth muscle in vitro using a high-resolution DC-SQUID magnetometer," in *New Frontiers of Biomedical Engineering - Innovations from Nuclear to Space Technology (Proc. of the Annual International IEEE/EMBS Conf.)*, **Vol. 13**, pp. 550–551.

94. Sweeney, R.J., Gill, R.M., Steinberg, M.I., and Reid, P.R. (1990) "Ventricular refractory period extension caused by defibrillation shocks," *Circ.*, **Vol. 82**, pp. 965–972.

95. Swinney, K.R. and Wikswo Jr., J.P. (1980) "A calculation of the magnetic field of a nerve action potential," *Biophys. J.*, **Vol. 32**, pp. 719–732.

96. Tan, Shaofen, (1992) "Linear system imaging and its applications to magnetic measurements by SQUID magnetometers," Ph.D. Dissertation, Department of Physics and Astronomy, Vanderbilt University.

97. Tan, S., Roth, B.J., and Wikswo, Jr., J.P. (1990) "The magnetic field of cortical current sources: The application of a spatial filtering model to the forward and inverse problems," *Electroencep. and Clin. Neurophysiol.*, **Vol. 76**, pp. 73–85.

98. Tan, S., Ma, Y.P., Thomas, I.M., and Wikswo, Jr., J.P. (1993) "High resolution SQUID imaging of current and magnetization distribution," *IEEE Trans. Applied Supercond.*, **Vol. 3, no. 1**, pp. 1945–1948.

358

99. Tan, S., Ma, Y.P., Thomas, I.M., and Wikswo, Jr., J.P. (1996) "Reconstruction of a two-dimensional susceptibility distribution from the magnetic field of non-ferromagnetic materials," *IEEE Trans. Mag.*, **Vol. 32, No. 1**,. pp. 230–234.

100. Thomas, I.M. and Friedman, R.N. (1993) "Magnetic susceptibility imaging of macrophage activity in rat liver using intravenous superparamagnetic tracers," in *Engineering Solutions to Current Health Care Problems, Proceedings of the 15th Annual International Conference of the IEEE Engineering in Medicine and Biology Society*, **Vol. 15, Part I**, pp. 503–504.

101. Thomas, I.M. and Friedman, R.N. (1995) "Study of macrophage activity in rat liver using intravenous superparamagnetic tracers," in *Biomagnetism: Fundamental Research and Clinical Applications - Proceedings of the 9th International Conference on Biomagnetism*, C. Baumgartner, L. Deecke, G. Stroink, and S.J. Williamson, Eds., IOS Press, Amsterdam, Netherlands, Vol. 7, pp. 809–813.

102. Thomas, I.M., Moyer, T.C., and Wikswo, Jr., J.P. (1992) "High resolution magnetic susceptibility imaging of geological thin sections: Pilot study of a pyroclastic sample from the Bishop tuff," *Geophys. Res. Lett.*, **Vol. 19**, pp. 2139–2142.

103. Thomas, I.M., Sepulveda, N.G., and Wikswo, Jr., J.P. (1993) "Magnetic susceptibility tomography: a new modality for three-dimensional biomedical imaging," in *Engineering Solutions to Current Health Care Problems, Proceedings of the 15th Annual International Conference of the IEEE Engineering in Medicine and Biology Society*, **Vol. 15, Part I**, pp. 94–95.

104. Thomas, I.M., Freake, S.M., Swithenby, S.J., and Wikswo, Jr., J.P. (1993) "A distributed quasi-static ionic current source in the 3-4 day old chicken embryo," *Phys. Med. Biol.*, **Vol. 38**, pp. 1311–1328.

105. Thomas, I.M., Ma, Y.P., Tan, S., and Wikswo, Jr., J.P. (1993) "Spatial resolution and sensitivity of magnetic susceptibility imaging," *IEEE Trans. Applied Supercond.*, **Vol. 3, no. 1**, pp. 1937–1940.

106. Tinchev, S.S., Hinken, J.H., Stiller, M., Baranyak, A., and Hartmann, D. (1993) "High-Tc RF SQUID magnetometer system for high-resolution magnetic imaging," *IEEE Trans. Applied Supercond.*, **Vol. 3, no. 1**, pp. 2469–2471.

107. Trahms, L., Burghoff, M., Brockmeier, K., and Schmitz, L. (1996) "Vortex current detected by the magnetocardiogram," in *Proceedings of the Tenth International Biomagnetism Conference*, p. 247.

108. Tung, L. (1978) "A bidomain model for describing ischemic myocardial DC potentials," Ph.D Dissertation, MIT, Cambridge, MA.

109. Turgeon, J., Wisialowski, T.A., Wong, W., Altemeier, W.A., Wikswo, Jr., J.P., and Roden, D.M. (1992) "Suppression of longitudinal versus transverse conduction by sodium channel block: Effects of sodium bolus," *Circ.*, **Vol. 85**, pp. 2221–2226.

110. Van Egeraat, J.M. and Wikswo, Jr., J.P. (1990) "Application of a magnetic current probe to map axial inhomogeneities in a giant squid axon," *Biol. Bull.*, **Vol. 179**, p. 229 (Abstract).

111. Van Egeraat, J.M. and Wikswo, Jr., J.P. (1992) "Measurement of non-uniform propagation in the squid nervous system with a room-temperature magnetic current probe," in *Biomagnetism: Clinical aspects*, M. Hoke, S.N. Erné, Y.C. Okada, and G.-L. Romani, Eds., Elsevier, pp. 385–388.

112. Van Egeraat, J.M. and Wikswo, Jr., J.P. (1992) "A low-cost biomagnetic current probe system for the measurement of action currents in biological fibers," in *Biomagnetism: Clinical aspects*, M. Hoke, S.N. Erné, Y.C. Okada, and G.-L. Romani, Eds., Elsevier, pp. 895–899.

113. Van Egeraat, J.M., Friedman, R.N., and Wikswo, Jr., J.P. (1990) "Magnetic field of a single muscle fiber: First measurements and a core conductor model," *Biophys. J.*, **Vol. 57**, pp. 663–667.

114. Van Egeraat, J.M., Friedman, R.N., and Wikswo, Jr., J.P. (1991) "Characterization of non-uniform propagation in a squid giant axon and synapse," *Biophys. J.*, **Vol. 59**, p. 590a (Abstract).

115. Van Egeraat, J.M., Stasaski, R., Barach, J.P., Friedman, R.N., and Wikswo, Jr., J.P. (1993) "The biomagnetic signature of a crushed axon: A comparison of theory and experiment," *Biophys. J.*, **Vol. 64**, pp. 1299–1305.

116. Virginia Semiconductor, Inc., 1501 Powatan Street, Fredericksburg, VA 22401.

117. Vu, L.N. and Van Harlingen, D.J. (1993) "Design and implementation of a scanning SQUID microscope," *IEEE Trans. Applied Supercond.*, **Vol. 3, no. 1**, pp. 1918–1921.

118. Vu, L.N., Wistrom, M.S., and Van Harlingen, D.J. (1993) "Imaging of magnetic vortices in superconducting networks and clusters by scanning SQUID microscopy," *Appl. Phys. Lett.*, **Vol. 63, no. 12**, pp. 1693–1695.

119. Wellstood, F.C., personal communation.

120. Wellstood, F.C. "SQUID microscope design study for biological samples," unpublished.

121. Wijesinghe, R.S. and Wikswo, Jr., J.P. "Optimization of high-resolution SQUID magnetometers with rectangular pickup coils," unpublished.

122. Wikswo, Jr., J.P. (1978) "Optimization of SQUID differential magnetometers," *AIP Conference Proceedings*, **Vol. 44**, pp. 145–149.

123. Wikswo, Jr., J.P. (1982) "Improved instrumentation for measuring the magnetic field of cellular action currents," *Rev. Sci. Instr.*, **Vol. 53**, pp. 1846–1850.

124. Wikswo, Jr., J.P. (1983) "Theoretical aspects of the ECG-MCG relationship," in *Biomagnetism, An Interdisciplinary Approach*, S.J. Williamson, G.-L. Romani, L. Kaufman, and I. Modena, Eds., Plenum, New York, pp. 311–326.

125. Wikswo, Jr., J.P. (1988) "High-resolution measurements of biomagnetic fields," *Advances in Cryogenic Engineering*, **Vol. 33**, pp. 107–116.

126. Wikswo, Jr., J.P. (1990) "Biomagnetic sources and their models," in *Advances in Biomagnetism*, S.J. Williamson, M. Hoke, G. Stroink, and M. Kotani, Eds., Plenum, New York, pp. 1–18.

127. Wikswo, Jr., J.P., (1993) "Design considerations for magnetic imaging with SQUID microscopes and arrays," in *Proc. of the 4th International Superconductive Electronics Conference*, pp. 189–190.

128. Wikswo, Jr., J.P. (1994) "Tissue anisotropy, the cardiac bidomain, and the virtual cathode effect," in *Cardiac Electrophysiology, From Cell to Bedside, 2nd Ed.*, D.P. Zipes and J. Jalife, Eds., W.B. Saunders, Philadelphia, pp. 348–361.

129. Wikswo, Jr., J.P. "The magnetic inverse problem for NDE," this volume.

130. Wikswo, Jr., J.P. and Barach, J.P. (1982) "Possible sources of new information in the magnetocardiogram," *J. Theor. Biol.*, **Vol. 95**, pp. 721–729.

131. Wikswo, Jr., J.P. and Roth, B.J. (1985) "Magnetic measurement of propagating action potentials in isolated, one-dimensional cardiac tissue preparations," in *Biomagnetism: Applications and Theory*, H. Weinberg, G. Stroink, and K. Katila, Eds., Pergamon Press, pp. 121–125.

132. Wikswo, Jr., J.P. and Roth, B.J. (1988) "Magnetic determination of the spatial extent of a single cortical current source: A theoretical analysis," *Electroencep. and Clin. Neurophysiol.*, **Vol. 69**, pp. 266–276.

133. Wikswo, Jr., J.P. and van Egeraat, J.M. (1991) "Cellular magnetic fields: Fundamental and applied measurements on nerve axons, peripheral nerve bundles, and skeletal muscle," *J. Clin. Neurophys.*, **Vol. 8, no. 2**, pp. 170–188.

134. Wikswo, Jr., J.P., Abraham, G.S., and Hentz, V.R. (1985) "Magnetic assessment of regeneration across a nerve graft," in *Biomagnetism: Applications and Theory*, H. Weinberg, G. Stroink, and K. Katila, Eds., Pergamon Press, pp. 88–92.

135. Wikswo, Jr., J.P., Barach, J.P., and Freeman, J.A. (1980) "Magnetic field of a nerve impulse: First measurements," *Science*, **Vol. 208**, pp. 53–55.

136. Wikswo, Jr., J.P., Lin, S.-F., and Abbas, R.A. (1995) "Virtual electrodes in cardiac tissue: A common mechanism for anodal and cathodal stimulation," *Biophys. J.*, **Vol. 69**, pp. 2195–2210.

137. Wikswo, Jr., J.P., Samson, P.C., and Giffard, R.P. (1983) "A low-noise, low input impedance amplifier for magnetic measurements of nerve action currents," *IEEE*

Trans. Biomed. Eng., **Vol. 30**, pp. 215–221.

138. Wikswo, Jr., J.P., Sepulveda, N.G., and Thomas, I.M. (1995) "Three-dimensional biomagnetic imaging with magnetic susceptibility tomography," in *Biomagnetism: Fundamental Research and Clinical Applications*, C. Baumgartner, L. Deecke, G. Stroink, and S.J. Williamson, Eds., IOS Press, Amsterdam, Netherlands, Vol. 7, pp. 780–784.

139. Wikswo, Jr., J.P., Henry, W.P., Samson, P.C., and Giffard, R.P. (1985) "A current probe system for measuring cellular action currents," in *Biomagnetism: Applications and Theory*, H. Weinberg, G. Stroink, and K. Katila, Eds., Pergamon Press, pp. 83–87.

140. Wikswo, Jr., J.P., Barach, J.P., Gundersen, S.C., McLean, M.J., and Freeman, J.A. (1983) "First magnetic measurements of action currents in isolated cardiac Purkinje fibers," *Il Nuovo Cimento D*, **Vol. 2D**, pp. 368–378.

141. Wikswo, Jr., J.P., Barach, J.P., Gundersen, S.C., Palmer, J.O., and Freeman, J.A. (1983) "Magnetic measurements of action currents in an isolated lobster axon," *Il Nuovo Cimento D*, **Vol. 2D**, pp. 512–516.

142. Wikswo, Jr., J.P., Black, Jr., W.C., Hirschkoff, E.C., Marsden, J.R., and Paulson, D.N. (1995) "Magnetometer and method of measuring a magnetic field," *United States Patent* 5,444,372.

143. Wikswo, Jr., J.P., Friedman, R.N., Kilroy, A.W., van Egeraat, J.M., and Buchanan, D.S. (1990) "Preliminary measurements with MicroSQUID," in *Advances in Biomagnetism*, S.J. Williamson, M. Hoke, G. Stroink, and M. Kotani, Eds., Plenum, New York, pp. 681–684.

144. Wikswo, Jr., J.P., Malmivuo, J.A.V., Barry, W.H., Leifer, M.C., and Fairbank, W.M. (1979) "Theory and application of magnetocardiography," in *Cardiovascular Physics*, D.N. Ghista, E. van Vollenhoven, and W. Yang, Eds., Karger, Basil, pp. 1–67.

145. Wikswo, Jr., J.P., Altemeier, W., Balser, J.P., Kopelman, H.A., Wisialowski, T., and Roden, D.M. (1991) "Virtual cathode effects during stimulation of cardiac muscle: Two-dimensional in vivo measurements," *Circ. Res.*, **Vol. 68**, pp. 513–530.

146. Wikswo, Jr., J.P., Crum, D.B., Henry, W.P., Ma, Y.P., Sepulveda, N.G., and Staton, D.J. (1993) "An improved method for magnetic identification and localization of cracks in conductors," *J. Nondestr. Eval.*, **Vol. 12, no. 2**, pp. 109–119.

147. Wikswo, Jr., J.P., Ma, Y.P., Sepulveda, N.G., Staton, D.J., Tan, S., and Thomas, I.M. (1993) "Superconducting magnetometry: A possible technique for aircraft NDE," in *Nondestructive Inspection of Aging Aircraft*, M.T. Valley, N.K. Grande, and A.S. Kobayashi, Eds., SPIE Proceedings, Vol. 2001, pp. 164–190.

148. Wikswo, Jr., J.P., Ma, Y.P., Sepulveda, N.G., Tan, S., Thomas, I.M., and Lauder, A. (1993) "Magnetic susceptibility imaging for nondestructive evaluation," *IEEE Trans. Applied Supercond.*, **Vol. 3, no. 1**, pp. 1995–2002.

149. Wikswo, Jr., J.P., Henry, W.P., Friedman, R.N., Kilroy, A.W., Wijesinghe, R.S., van Egeraat, J.M., and Milek, M.A. (1990) "Intraoperative recording of the magnetic field of a human nerve," in *Advances in Biomagnetism*, S.J. Williamson, M. Hoke, G. Stroink, and M. Kotani, Eds., Plenum, New York, pp. 137–140.

150. Wikswo, Jr., J.P., van Egeraat, J.M., Ma, Y.P., Sepulveda, N.G., Staton, D.J., Tan, S., and Wijesinghe, R.S. (1990) "Instrumentation and techniques for high-resolution magnetic imaging," in *Digital Image Synthesis and Inverse Optics*, A.F. Gmitro, P.S. Idell, and I.J. LaHaie, Eds., SPIE Proceedings, Vol. 1351, pp. 438–470.

151. Zhang, Y., Krüger, U., Kutzner, R., Wördenweber, R., Schubert, J., Zander, W., Sodtke, E., and Braginski, A.I. (1994) "Single-layer $YB_2Cu_3O_7$ rf SQUID magnetometers with direct-coupled pickup coils and flip-chip flux transformers," *Appl. Phys. Lett.*, **Vol. 65, no. 26**, pp. 3380–3382.

152. Zipes, D.P., Fischer, J., King, R.M., Nicoll, A., and Jolly, W.W. (1975) "Termination of ventricular fibrillation in dogs by depolarizing a critical amount of myocardium," *Am. J. Cardiol.*, **Vol. 36**, pp. 37–44.

THE VOLUME CONDUCTOR PROBLEM IN BIOMAGNETISM

MARIA J. PETERS, S. P. VAN DEN BROEK AND F. ZANOW
University of Twente, Faculty of Physics
P.O. Box 217, 7500 AE Enschede, The Netherlands

Abstract. When biomagnetic and/or bioelectric measurements are used for the localization of electrical activity within the human body, mathematical models are needed that describe the electrical properties of the sources and the surrounding region, the so-called volume conductor. The sources can be described as arrangements of current dipoles. Commonly, the volume conductor is described by a compartment model, the compartments being piece-wise homogeneous. The volume conductor problem is illustrated for magnetoencephalography (MEG) and electroencephalography (EEG). First, the electrical conductivity of the tissues in the head is discussed. Subsequently, the basic theory, which is based on the quasi-static Maxwell equations, is presented. A set of concentric spheres is commonly used for modeling the head. It will be shown that radial dipoles within such a spherical volume conductor produce no magnetic field. The magnetic field generated by tangential dipoles is not influenced by the radii or conductivities. In contrast, the electrical potential is very dependent on these parameters. Since the shape of the human head differs from that of a sphere, realistically-shaped models are discussed. Such models can be taken into account using numerical methods, such as the boundary element method and the finite element method. Models will be described for MEG, EEG, MCG, and ECG (both for the adult and the fetus). The assessment of the adequacy of the models is the subject of the discussion.

Table of contents

1. INTRODUCTION
   1.1 General outline

2. THE CONDUCTIVITY

3. DUALITY
   3.1. The potential due to a current dipole in a homogeneous, isotropic conductor of infinite extent
   3.2. The magnetic field due to a current dipole in a homogeneous, isotropic medium of infinite extent

4. BASIC EQUATIONS

H. Weinstock (ed.), SQUID Sensors: Fundamentals, Fabrication and Applications, 361–394.
© *1996 Kluwer Academic Publishers.*

4.1. The quasi-static approximation
4.2. The quasi-static Maxwell equations
4.3. The Poisson equations
4.4. Secondary sources

5. A VOLUME CONDUCTOR WITH SPHERICAL SYMMETRY
    5.1. The potential
        5.1.1. *A dipole at the center of a homogeneous sphere*
        5.1.2. *A radial dipole in a homogeneous sphere*
        5.1.3. *A dipole in a set of concentric spheres*
    5.2. The magnetic field
        5.2.1. *A radial dipole in a set of concentric spheres*
        5.2.2. *A tangential dipole in a set of concentric spheres*

6. REALISTICALLY SHAPED MODELS
    6.1. The boundary element method
    6.2. The finite element method
    6.3. The generation of realistically shaped models from MRI

7. MODELS IN CARDIOGRAPHY
    7.1. The influence of the volume conductor on MCG and ECG
    7.2. The influence of the volume conductor on fetal MCG and ECG

8. DISCUSSION

9. REFERENCES

1. Introduction

Biomagnetic fields and bioelectric potentials are used to localize electrical active regions within the human body. The localization based on biomagnetic field measurements is a complementary technique to the one based on the potential measurements. The magnetic fields and the potentials are generated by the same sources that originate in the electrochemical activity of individual cells. These sources, in turn, give rise to ohmic currents in the surrounding conducting medium, the so-called volume currents, which also contribute to the magnetic field and the electrical potential. This constitutes the volume conductor problem in biomagnetism. The contribution by volume currents can be significant. We will show that the potential may be tripled due to the volume currents. The magnetic field may be nullified. A consequence of volume conduction is that the magnetic field mainly reflects activity from dipolar sources parallel to the body surface, being relatively insensitive to those sources perpendicular to the surface. This selective sensitivity of the magnetic field, together with the fact that the magnetic field generated by sources which are parallel to the surface is much less influenced by volume currents than the potentials, appears to give a magnetic-field distribution an advantage in source localization over the potential distribution.

The computation of the sources from the measured magnetic fields and/or electrical potential is called the inverse problem. Because the inverse solution does not lead to a unique solution [1], it is advantageous if MEG and EEG measurements are both used. If, within the error estimate, the same results are obtained, we can be more confident that this result is meaningful.

The recordings of the magnetic field near the human torso are called MCG's (magnetocardiograms), and of the difference of the electrical potential between two points at the torso surface, ECG's (electrocardiograms). Recordings of the magnetic field around the head are called MEG's (magnetoencephalograms), and of the potential at the scalp, EEG's (electroencephalograms). We will illustrate the volume conductor problem for the MEG and EEG. Both are used to localize electrical activity within the human brain, providing the possibility to relate mental faculties to specific parts of the brain. This is called functional imaging, as opposed to the anatomical imaging of x-rays, MRI (magnetic resonance imaging) or CT (computerized tomography). The time resolution of MEG and EEG imaging is a millisecond. Other techniques for functional imaging like PET (positron emission tomography), SPECT (single positron emission computerized tomography) and functional MRI have a time resolution which is at least a second. Using PET and SPECT imaging, the localities are identified by a radioactive tracer. In functional MRI, the changes of oxygen in the blood are mapped.

In this paper, we will discuss the volume conductor problem both for the biomagnetic field and the bioelectric potential. The reasons for including the study on the influence of the volume conductor on the electrical potential are:

- If both types of measurements are explained by the same result, one can feel more confident that the model used to obtain the result is justified.

- Measurements of the electrical potential provide complementary information.

- The solution of the localization procedure is more stable if it is based on a combination of magnetic and electric measurements.

- Magnetic field measurements must compete with electric measurements.

In other words, the confrontation of both techniques will help us to explain why so much effort is put into building elaborate instruments for the measurement of biomagnetic fields, when it naively appears that a simpler technique exists that can perform the same tasks.

1.1 GENERAL OUTLINE

About 10^4 neuronal cells must be synchronously active in order to generate a measurable potential or magnetic field at a distance of a centimeter or more. Such a patch of active neurons can be described by a discrete source, the so-called equivalent current dipole, which is a sort of battery, i.e., a current source, a current sink and an internal current from sink to source. Because the surrounding medium is conductive, ohmic currents will flow from source to sink. Since no current flows out of the head's surface, we have the boundary condition that the electric current must be tangential to the surface of the scalp, as is illustrated in Figure 1, where the head is modeled as a uniformly conducting sphere. Of course, the precise distribution of the volume currents, and hence the potential and magnetic field, depend on the conductivity distribution in the head. Unfortunately, our knowledge of the conductivity of the various tissues in the head is rather limited. The different conductivities will be discussed in section 2. The Poisson equation in electrostatics is quite similar to that in biomagnetism. Consequently, the solution is just like that for the electrostatic potential. These correspondences are discussed in section 3. The basic formulas are given in section 4.

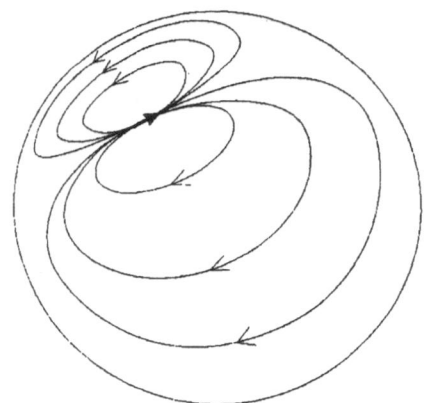

Figure 1. Current distribution due to a current dipole within a homogeneous sphere.

Usually the volume conductor is described by a so-called compartment model, the compartments being piecewise homogeneous and isotropic. As James Clerk Maxwell states: "In certain classes of cases, such as those relating to spheres, there are known mathematical methods by which we may proceed" [2]. If the head is described as a homogeneous isotropic conducting sphere, we can find a closed-form solution for the potential distribution at the surface as discussed in section 5.1. If the head is described by a set of concentric spheres, the analytic expression for the potential at the surface of the outer sphere is expressed as an infinite series of spherical harmonics. The magnetic field outside a homogeneous sphere is not influenced by the radius of the sphere if the source has a tangential orientation. This is true also for a model consisting of a set of concentric spheres, as shown in section 5.2.

If we have a radial source within a volume conductor that has spherical symmetry, the magnetic field outside is zero.

From simulation studies based on realistically-shaped multi-compartment models, it follows that the shape of the compartments can influence significantly the magnetic field and the electrical potential. These simulations are based on the boundary element method (BEM), where the contribution of the volume currents is represented by equivalent sources at the interfaces between regions with different conductivities. This will be discussed in section 6.1

In order to use the BEM, the compartments must be closed. The BEM has difficulty in dealing with a thin layer that has a much higher conductivity than the adjacent regions, such as the layer of cerebro-spinal fluid. It is found also that biological tissues have a conductivity which differs in the direction along the fibers from that perpendicular to the fibers. In other words, some compartments will have an anisotropic conductivity. A method which can deal with these phenomena, and which also can be used in the case that a compartment is not closed, is the finite element method (FEM) The FEM will be discussed in section 6.2. The geometry of the interfaces is derived from magnetic resonance images and will be discussed in section 6.3.

In section 7, volume conductor models for ECG and MCG (both for the adult and the fetus) will be presented.

A central problem in EEG and MEG research is to relate potentials measured at the scalp and magnetic fields measured near the head to the underlying physiological processes. The solution of this problem is called the inverse solution. The computation of the potential and/or the magnetic field distribution for the case in which the sources are known, is called the forward solution. Both forward and inverse solutions are used in simulations in order to assess the accuracy of a certain model of the volume conductor. This will be discussed in section 8. The optimal model must be simple enough to be handled mathematically, and at the same time realistic enough to enable source localization within the required precision limits. These limits are determined by the effect of noise on the measurements. So it makes no sense to improve the model when the background noise has a much larger effect on the localization than does the refinement of the volume conductor model.

2. The Conductivity

Impressed currents are produced by permeability changes in the cell membranes. These impressed currents generate currents in the active cells, in the interstitial space adjacent to the cell walls, in the non-active cells, and at larger distances in the volume conductor. We are interested in the potential and magnetic-field distribution measured at points which are at a distance of centimeters from the active cells. The surrounding medium is assumed to be linear. In other words, the conductivity σ does not depend upon the electric field $\vec{E}$, so that the relation between the current density $\vec{j}$ and $\vec{E}$ is a linear one. The different biological tissues which constitute the conductor are treated macroscopically, and inhomogeneities on a cellular level (e.g., a neuron surrounded by extra-cellular fluid) are neglected.

It is convenient to write

$$\vec{j} = \vec{j}_p + \sigma \vec{E}, \qquad (1)$$

where $\vec{j}_p$ is the primary current density, which is non-zero in the source region, and $\sigma \vec{E}$ expresses the ohmic currents in the surrounding medium.

Typical values for the conductivity of various biological materials at low frequencies (< 1000 Hz) are listed in Table 1. For comparative purposes a few physical materials are also included. It is clear that biological tissues are not "good" conductors, in comparison, for example, to copper.

ideal conductor	∞	skull	0.0042
copper	5.5×10^7	scalp tissue	0.33
blood	1.5	water	2×10^{-4}
heart muscle	0.2	vernix caseosa	3.8×10^{-6}
lung tissue	0.06	ideal insulator	0
liver	0.14		
fat	0.04		
mean value of body tissues	0.22		
brain tissue	0.33		
cerebro-spinal fluid	1.0		

Table 1. Typical values for the conductivity in S/m, these values are taken from [3], [4], and [5].

Most values given in Table 1 are not very precise because our knowledge of the conductivity in the human body is rather limited. Measurements of this parameter are difficult to carry out, because in post-mortem or biopsy materials, they will be influenced by necrosis and fluid loss. Most investigators use values given by Geddes and Baker [3], who collected values of the specific conductivities of biological materials. The values given for a macroscopic mass of brain tissue are measured in animals under various conditions. The conductivity of the cortex of the rabbit measured at 5 Hz is 0.33 Sm^{-1}, and at 5 kHz it is 0.43 Sm^{-1}. The conductivity of the white matter of the rabbit is given only at 5 kHz and is a factor 4.1 larger than σ of the cortex at that frequency. Many biological tissues have some degree of directional organization and therefore it is expected that the electrical conductivity also will depend on the direction.

For instance, the white matter, which consists of fiber tracts shows anisotropy. Ratios of longitudinal to transverse conductivities are found to vary between 5.7 and 9.4. Nevertheless, it is common practice to take 0.33 Sm^{-1} as a global value for the conductivity of the brain. The range of global conductivities for the skull presented in the literature is large, with a factor of 10 between extremal values. According to Rush and Driscoll [6], the skull has a smaller conductivity transversely than in directions parallel to its surface. This is due to the fact that the skull consists of three layers, two of them highly ohmic and one conducting alveolar layer. Moreover, the conductivity of a certain tissue may be non-uniform. For instance, Law [7] drilled twenty samples from a skull and measured the resistivity of these samples. He found that the conductivity of the skull varied widely over the surface of the human head. The values ranged from 1,360 ohm-cm to 21,400 ohm-cm. The four lowest resistivity values were from samples that contained suture lines along the bone plug. The highest resistivity values were from temporal bone.

Magnetic properties shall not be considered since the magnetic permeability of biological matter is for present purposes equal to that of free space, that is $\mu = 4\pi.10^{-7}$.

3. Duality

The mathematics of current flow in a volume conductor correspond closely with that of charges in a dielectric. In order to show these correspondences we start with some known formulas. In electrostatics, the potential due to a charge density ρ in a homogeneous dielectric of infinite extent, with permittivity ε, obeys Poisson's equation:

$$\nabla^2 V = -\frac{\rho}{\varepsilon} \tag{2}$$

with solution

$$V(\vec{r}) = \frac{1}{4\pi\varepsilon} \int \frac{\rho(\vec{r'})}{R} dv' \tag{3}$$

according to the law of Coulomb. The volume v' encloses all charges, and $\vec{R} = \vec{r} - \vec{r'}$.

This leads, for a point charge Q at the origin, to

$$\vec{E}(\vec{r}) = \frac{Q}{4\pi\varepsilon r^2} \vec{e}_r \tag{4}$$

An electric dipole, which consists of two charges of equal magnitude and opposite charge which appear close together for an observer some distance away, gives the following expression for the potential:

$$V(\vec{r}) = \frac{\vec{p} \cdot \vec{r}}{4\pi\varepsilon r^2} \qquad (5)$$

where $\vec{p} = Q\vec{s}$ is the dipole moment, $\vec{s}$ defines the distance between the opposite charges, and the dipole is at the origin.

The equations used to compute bioelectric potentials and magnetic fields are similar to those used in electrostatics. These correspondences are applications of "the principle of duality" [8] and will be used in the next sections.

3.1 THE POTENTIAL DUE TO A CURRENT DIPOLE IN A HOMOGENEOUS, ISOTROPIC CONDUCTOR OF INFINITE EXTENT

Okada and Nicholson [9] studied the neural basis of magnetic evoked fields of the brain with an isolated turtle cerebellum. It followed from their measurements that the field is dipolar at a distance which is on the order of the dimensions of the active region or larger. The dipole is oriented along the longitudinal axis of the active cells, and a synchronous activity in 1 mm^3 of nerve tissue gives rise to a magnetic field of 0.1 pT at a source-to-field distance of 2 cm.

In order to study the influence of the volume conductor on the EEG and MEG, sources will be described by current dipoles. Since all other source configurations (e.g., higher-order multipoles, dipole layers) are a combination of dipoles, this choice does not limit the generality of the results. A current dipole is a current element. Because this current element is situated within a conducting medium, one of the terminals of the element will act as a current source and the other as a current sink. For the case of a current source with strength I, situated in a homogeneous, isotropic conductor of infinite extent, we find that $\vec{j}$ is radial, and I is the same through each sphere with the source as center - see Figure 2. Thus, for a source at the origin,

$$\vec{j}(\vec{r}) = \frac{I}{4\pi\sigma r^2}\vec{e}_r \qquad (6)$$

and because $\vec{j} = \sigma\vec{E}$, it is found that the electric field pattern is identical to that of a point charge as given by Equation 4, provided we replace ε by σ, and I by Q.

As a consequence, the potential of a current dipole embedded in a homogeneous isotropic medium of infinite extent can be found from the expression for V of a charge dipole as given by equation 5 by replacing ε by σ and the charge dipole by a current dipole. This yields

$$V_\infty(\vec{r}) = \frac{\vec{p} \cdot \vec{r}}{4\pi\sigma r^3} \tag{7}$$

where $\vec{p} = I\,\vec{s}$ is the current dipole moment, $\vec{s}$ is the distance between the sink and source, and the dipole is at the origin.

3.2 THE MAGNETIC FIELD DUE TO A CURRENT DIPOLE IN A HOMOGENEOUS, ISOTROPIC CONDUCTOR OF INFINITE EXTENT

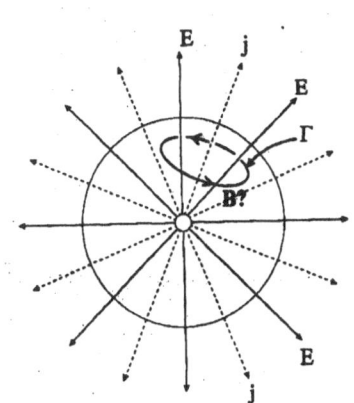

Figure 2. Current distribution and field lines due to a current source (current mono-pole) within a homogeneous, isotropic conductor of infinite extent.

The resultant magnetic flux density $\vec{B}$ of a current dipole is the vector sum of the field generated by the volume currents from the source and those from the sink, and the field caused by the current element. The volume currents due to a source in a homogeneous, isotropic conductor of infinite extent are drawn in Figure 2. If we draw a loop Γ on a sphere of radius r, as shown in the figure, then some current will flow through the loop, so we can expect to find a magnetic field circulating in the direction shown. A different choice of Γ would allow us to conclude that its direction is exactly opposite to that shown. Thus, the solution is that $\vec{B}$ due to the volume currents from the source is zero. The same is true for $\vec{B}$ due to the volume currents from the sink. Consequently, we find that the magnetic field is due only to the current element. The magnetic flux density generated by a current element $\vec{p} = I\,\vec{s}$ at the origin is given by the law of Biot and Savart:

$$\vec{B}_\infty(\vec{r}) = \mu_0 \frac{\vec{p} \times \vec{r}}{4\pi r^3} \tag{8}$$

4. Basic Equations

4.1 THE QUASI-STATIC APPROXIMATION

Bioelectric sources are time-varying, however the highest frequency component of interest will be on the order of 1000 Hz. As argued by Plonsey and Heppner [10], the Maxwell equations can be used in quasi-static approximation. In other words, capacitive, inductive, and propagation effects can be neglected. For example, if in the Maxwell equation

$$\nabla \times \vec{B} = \mu_0 (\vec{j} + \frac{\partial \vec{D}}{\partial t}) \qquad (9)$$

$\partial \vec{D}/\partial t$ is small compared to the $\vec{j}$ term, the ohmic currents will be much more important than the displacement currents. For a sinuoidal wave with angular frequency ω, we find that

$$\frac{\partial \vec{E}}{\partial t} = i\omega \vec{E} \qquad \text{and} \qquad \vec{j} = \sigma \vec{E}$$

Thus,

$$\left| \frac{\frac{\partial \vec{D}}{\partial t}}{\vec{j}} \right| = \frac{\varepsilon \omega}{\sigma} \qquad (10)$$

For most living tissue the dielectric constant ε is on the order of $500 \, \varepsilon_0$. For example, the dielectric constants of tissues in the torso at 1000 Hz are in the range of 50 (fat) - 320 (heart muscle); at 100 Hz these values are 150 - 820 [4]. The mean conductivity of human tissue is 0.2 S/m. It follows that for most human tissues $\varepsilon \omega / \sigma \ll 1$ for frequencies below 1000 Hz, and as a result, capacitive effects can be neglected.

According to the law of conservation of charge,

$$\nabla \cdot (\vec{j} + i\omega\varepsilon \vec{E}) = 0 \qquad (11)$$

which leads, at an interface between two regions 1 and 2, with conductivities of respectively. σ_1 and σ_2 and dielectric constants ε_1 and ε_2 to the boundary condition for the normal component of the electric field:

$$\sigma_1(1 + i\varpi\varepsilon_1)E_{1n} = \sigma_2(1 + i\varpi\varepsilon_2)E_{2n} \tag{12}$$

which reduces to

$$\sigma_1 E_{1n} = \sigma_2 E_{2n} \tag{13}$$

in those cases for which $\omega\varepsilon/\sigma \ll 1$, i.e., the case for most biological tissues as stated before.

Outside the head we find $\sigma = 0$ and $\varepsilon = \varepsilon_0$. Consequently, the boundary condition at the surface of the head is

$$\sigma_1(1 + i\varpi\varepsilon_1)E_{1n} = i\varpi\varepsilon_0 E_{2n} \tag{14}$$

This reduces to

$$E_{1n} = j_{1n} = 0 \tag{15}$$

because $\omega\varepsilon_0/\sigma_1 \ll 1$.

4.2 THE QUASI-STATIC MAXWELL EQUATIONS

As mentioned previously, bioelectric phenomena can be described by the Maxwell equations in the quasi-static approximation.
The quasi-static Maxwell equations are

$$\nabla \cdot \vec{D} = \rho \tag{16}$$

$$\nabla \cdot \vec{B} = 0 \tag{17}$$

$$\nabla \times \vec{E} = 0 \text{, from which follows } \vec{E} = -\nabla V \tag{18}$$

$$\nabla \times \vec{B} = \mu_0 \vec{j} \tag{19}$$

4.3 THE POISSON EQUATIONS

Applying the law of conservation of charge and equation 1 yields

$$\nabla \cdot \vec{j} = \nabla \cdot \vec{j}_p + \nabla \cdot \sigma \vec{E} = \nabla \cdot \vec{j}_p - \nabla \cdot \sigma \nabla V = 0 \tag{20}$$

For a homogeneous, isotropic conductor of infinite extent, σ is a constant. It follows that

$$\nabla^2 V = \frac{\nabla \cdot \vec{j}_p}{\sigma}. \tag{21}$$

The solution corresponds to Equation 3 being the solution of Equation 2, except for changing ε to σ, and ρ to the term $-\nabla \cdot \vec{j}_p$; thus, we obtain

$$V(\vec{r}) = -\frac{1}{4\pi\sigma} \int \frac{\nabla' \cdot \vec{j}_p(\vec{r}')}{R} dv' \tag{22}$$

which is identical to Equation 7.

Using a well-known vector identity for ∇^2, and applying Equations 17 and 19 gives

$$\nabla^2 \vec{B} = \nabla(\nabla \cdot \vec{B}) - \nabla \times \nabla \times \vec{B} = -\nabla \times \nabla \times \vec{B} = -\mu_0 \nabla \times \vec{j} \tag{23}$$

Application of Equation 1 gives

$$\nabla^2 \vec{B} = -\mu_0 (\nabla \times \vec{j}_p) - \mu_0 (\nabla \times \sigma \vec{E}) \tag{24}$$

In a medium which is homogeneous, isotropic and of infinite extent, σ is constant. Application of the vector identity $\nabla \times \nabla V = 0$, and using the principle of duality again, yields in this case

$$\vec{B}_\infty(\vec{r}) = \frac{\mu_0}{4\pi} \int \frac{\nabla \times \vec{j}_p(\vec{r}')}{R} dv', \tag{25}$$

which can be rewritten, applying $\nabla \times (\dfrac{\vec{j}_p}{R}) = -\vec{j}_p \times \nabla(\dfrac{1}{R}) + \dfrac{1}{R}\nabla \times \vec{j}_p$.

First one integrates both sides over a volume that contains all sources $\vec{j}_p$. Applying the theorem

$$\int \nabla \times \vec{a}\,dv = -\int \vec{a} \times d\vec{S}$$

to the left-hand side shows that this term equals zero, since $\vec{j}_p$ is zero over the bounding surface. Hence, Equation 25 becomes

$$\vec{B}_\infty(\vec{r}) = \dfrac{\mu_0}{4\pi}\int \vec{j}_p(\vec{r}') \times \nabla'(\dfrac{1}{R})dv' = \dfrac{\mu_0}{4\pi}\int \vec{j}_p(\vec{r}') \times \dfrac{\vec{R}}{R^3}dv' \qquad (26)$$

which is identical to Equation 8.

4.4 SECONDARY SOURCES

Assuming the conductivity is not constant over all space, implies that in Equation 24 the term $\nabla \times \sigma \vec{E}$ is not zero. Hence, the magnetic field is determined by both the primary sources and the currents in the volume conductor. We assume that the volume conductor is piece-wise homogeneous. The solution of Equation 24 is [11]

$$\vec{B}(\vec{r}) = \vec{B}_\infty(\vec{r}) - \dfrac{\mu_0}{4\pi}\sum_k \int \sigma_k \nabla V(\vec{r}') \times \dfrac{\vec{R}}{R^3}dv' \qquad (27)$$

$\displaystyle\sum_k$ implies the summation over all compartments.

Applying the vector identities $\nabla \times a\vec{A} = a\nabla \times \vec{A} - \vec{A}\nabla a$ and $\int \nabla \times a\vec{A}\,dv = -\int a\vec{A} \times d\vec{S}$ and $\nabla \times \nabla a = 0$ to the second term on the right-hand side gives

$$\vec{B}(\vec{r}) = \vec{B}_\infty(\vec{r}) - \frac{\mu_0}{4\pi} \sum_i \int_{S_i} \Delta\sigma_i V(\vec{r}') \frac{\vec{R}}{R^3} \times d\vec{S}_i \tag{28}$$

where $\Delta\sigma_i$ is the conductivity in the region inside boundary S_i minus that outside boundary S_i, and $\sum_i$ is summation over all interfaces between the compartments, the outer surface included.

For the electric potential, the compartment model gives a rather similar formula. The derivation is not given here, but is based on Green's theorem [12, 13]:

$$V(\vec{r}) = V_\infty(\vec{r}) - \frac{1}{4\pi\sigma} \sum_i \int_{S_i} \Delta\sigma_i V(\vec{r}') \frac{\vec{R}}{R^3} \cdot d\vec{S}_i \tag{29}$$

Equations (28) and (29) can be interpreted as yielding the magnetic field and electric potential as if the entire space were uniform, but in addition secondary sources are set up at the interfaces between regions of different conductivity. The secondary sources are not a physical reality. It is clear, for example, that they are not in accord with the law of conservation of charge. These fictitious secondary sources ensure that the boundary conditions are obeyed. The secondary sources have a current dipole density $\Delta\sigma_i V$ (where V is the local potential) and their orientations are in the direction of the local normal vector.

The primary current density as given in Equation 1 has two components, namely the impressed currents in the cell membrane and an ohmic part. In an analogous way, we can describe the magnetic field of the ohmic part in $\vec{j}_p$ by the magnetic field caused by secondary sources at the boundaries between membranes and the intracellular or extracellular media. Estimations by Plonsey [14] showed that the contributions to the magnetic field and electrical potential by the impressed currents are overshadowed by the contribution of the secondary sources.

5. A Volume Conductor with Spherical Symmetry

Usually the head is described by a set of four concentric spheres describing the brain, the cerebro-spinal fluid, the skull and the scalp. While this model may appear to be a gross oversimplification of the actual state of affairs, it provides a rather useful description of the observed potential distributions on the surface of the head and the observed magnetic field distributions near the surface of the head. For a spherically symmetric model, analytical expressions can be found both for the magnetic field and the electric potential. To find a solution, it is required that the solution in the first place satisfies the boundary conditions at the surface of the head and at the interfaces between the various compartments and secondly satisfies Laplace's or Poisson's equation everywhere. According to the uniqueness theorem, there can not exist more than one potential that satisfies these requirements.

5.1. THE POTENTIAL

The boundary conditions are:

- The normal components of the current densities are continuous:

$$\sigma_i \left(\frac{\partial V_i}{\partial r} \right)_{r=R_i} = \sigma_j \left(\frac{\partial V_j}{\partial r} \right)_{r=R_i} \tag{31}$$

Since no current flows out of the body's surface, we have at the body's surface the boundary condition that the normal component of the current density is zero. In other words, the electric current must be tangential to the body's surface.

- $V = 0$ for $r \to \infty$.

- V is finite for $r \to 0$.

5.1.1. *A Dipole at the Center of a Homogeneous Sphere*
As an example, we consider the case where a current dipole is placed at the center of a homogeneous sphere with conductivity σ and radius R_0. The region outside the sphere is non-conducting. A homogeneous sphere can be considered as the simplest model for the head.

A current dipole in the center of a sphere (which is chosen as the origin of the coordinate system) is symmetrical around the z-axis, so the potential is independent of the azimuthal angle ϕ, and therefore it will be a function of r and θ only. In this case Laplace's equation reduces to:

$$\frac{1}{r^2} \frac{\partial}{\partial r} \left(r^2 \frac{\partial V}{\partial r} \right) + \frac{1}{r^2 \sin\theta} \frac{\partial}{\partial \theta} \left(\sin\theta \frac{\partial V}{\partial \theta} \right) = 0 \tag{32}$$

The general solution is

$$V(r,\theta) = \sum_{n=0}^{\infty} (A_n r^n + C_n r^{-(n+1)}) P_n(\cos\theta) \tag{33}$$

The $P_n(\cos\theta)$ terms are called Legendre polynomials [15].

The general solution must be added to the particular solution of Poisson's equation:

$$V(r,\theta) = V_{particular} + V_{general} = \frac{\vec{p}\cdot\vec{r}}{4\pi\sigma r^3} + \sum_{n=0}^{\infty} (A_n r^n + C_n r^{-(n+1)}) P_n(\cos\theta) \tag{34}$$

The solution of this problem is the one which accounts for the boundary conditions. The potential should be finite for $r \to 0$, which requires $C_n = 0$ for $n \geq 0$. The values of the coefficients A_n follow from the boundary condition that for $r = R_0$,

$$\frac{\partial V}{\partial r} = 0$$

because the normal component of the current density is continuous, and outside the sphere the current density is zero.

The application of this boundary condition to Equation 34 states that for any θ,

$$\frac{-2p\cos\theta}{4\pi\sigma R_0^3} + \sum_{n=0}^{\infty} nA_n R_0^{n-1} P_n(\cos\theta) = 0 \tag{35}$$

This gives

$$A_1 = \frac{2p}{4\pi\sigma R_0}$$

where $A_n = 0$ for $n \geq 2$.

A_0 is undetermined. Because the potential V is not uniquely defined, one can add any constant. $A_0 = 0$ is chosen.

Thus, at $r = R_0$ the potential of the sphere is

$$V = \frac{3p\cos\theta}{4\pi\sigma R_0^{\ 3}} = 3V_\infty \tag{36}$$

In other words, the potential due to a current dipole in the center of a homogeneous isotropic conducting sphere gives rise to a potential at the boundary which is enhanced by a factor of three because of the volume conductor effects.

5.1.2 A Radial Dipole in a Homogeneous Sphere

In general, the current dipole will not be in the center. The method of harmonics also is used in this case. The particular solution for the potential at the surface of a sphere with radius R_0 caused by a radial dipole at a distance r' from the center, is

$$V_{particular}(\vec{r}) = \frac{\vec{p}\cdot\vec{R}}{4\pi\sigma R^3} = \frac{\vec{p}\cdot(\vec{R}_0 - \vec{r}')}{\left|\vec{R}_0 - \vec{r}'\right|^3}, \tag{37}$$

which is expressed in spherical coordinates and chosen such that the radial dipole is centered at the z-axis. Therefore,

$$V_{particular}(\vec{r}) = \frac{p_z}{4\pi\sigma}\frac{R_0\cos\theta - r'}{R^3}, \tag{38}$$

where R can be expanded according to [15]:

$$\frac{1}{R} = \frac{1}{R_0}\sum_{n=0}^{\infty}\left(\frac{r'}{R_0}\right)^n P_n(\cos\theta) \tag{39}$$

$V_{general}$ is given by Equation 33, because we still have azimuthal symmetry. Application of the boundary conditions leads to the following expression for the potential at the surface of the sphere:

$$V = \sum_{n=0}^{\infty}(A_n R_0^{\ n} + \frac{p_z}{4\pi\sigma}n\frac{(r')^{n-1}}{R_0^{\ n+1}})P_n(\cos\theta), \tag{40}$$

where

$$A_n = \frac{p_z}{4\pi\sigma}(n+1)\frac{(r')^{n-1}}{R_0^{2n+1}}.$$

It is clear that if r' is not much smaller than R_0, that is, if the dipole is near the surface of the sphere, the number of terms needed to obtain a precise solution may be quite large. If the dipole is not radially oriented the appropriate solution for the Laplace equation should be taken. The expression for a dipole in a homogeneous sphere can be rewritten in closed form [16].

5.1.3. A Dipole in a Set of Concentric Spheres

Usually the head is described by three or four concentric spheres. There are two main reasons for its application. First, the implementation is easy because there is an analytical expression for the electric potential generated by a current dipole. This expression can be derived in a manner similar to that described in the previous section and can be found in the literature [17,18], even for the case where the conductivity is anisotropic in the restricted sense that the conductivity in the radial direction differs from that in the tangential directions[19,20]. Second, there are only a few parameters to be adjusted, namely the common center of the spheres, and the radius and conductivity value for each of the three or four layers. The geometry parameters are often obtained by taking the best-fitting sphere with respect to the electrodes used to measure the EEG's or with respect to the scalp surface.

Stok [21] studied the influence of the radii and the conducivities of the various compartments systematically using a four-sphere model. In order to do so, the potential due to a dipole was calculated, and the simulated data were used to obtain the current dipole by solving the inverse problem. In order to make it possible to establish the influence of one parameter, the inverse solution was based on a reference model, which differed only in one aspect from that used in the forward computations. He showed that 20% changes in the values of the conductivities could cause changes of over 60% in the potential values. The changes in the location of the dipole were about 1 to 3.5 mm. The influence of the radii on the location of the current dipole was shown to be on the order of 0.5 to 3.5 mm when the radii were varied up to 2.5 %. De Munck [17] showed that the potential distribution also is influenced by anisotropy.

5.2. THE MAGNETIC FIELD

5.2.1. *A Radial Dipole in a Set of Concentric Spheres*

We shall use the method of axial expansion for determining the external magnetic field generated by a radial current dipole in a set of homogeneous concentric spheres [22]. Because the current density is zero outside the volume conductor, Equation 19 becomes

$$\nabla \times \vec{B} = 0 \tag{41}$$

Therefore, we can express the magnetic field outside the volume conductor as

$$\vec{B} = -\nabla\Phi \tag{42}$$

where Φ is the magnetic scalar potential.

The external potential is independent of the azimuthal angle ϕ and thus - see Equation 34 - can be represented by spherical harmonics:

$$\Phi(r,\theta) = \sum_{n=0}^{\infty}(A_n r^n + C_n r^{-(n+1)})P_n(\cos\theta) \tag{43}$$

where θ is measured with respect to the symmetry axis.

Since for points on the axis, $\theta = 0$, it follows that for any point on the axis, $P_n(\cos\theta) = 1$ for all values of n [15]. Consequently, the potential on the axis reduces to

$$\Phi(z) = \sum_{n=0}^{\infty}(A_n z^n + C_n z^{-(n+1)}), \tag{44}$$

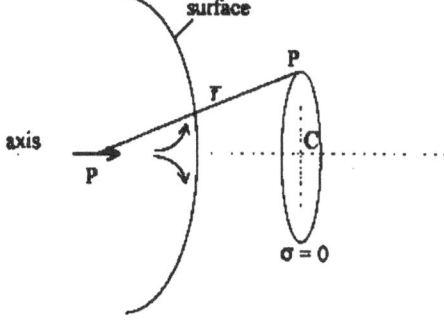

Figure 3. A volume conductor with axial symmetry and a current dipole on the axis.

where z is the distance from the center of the spheres along the axis. The z-component of the magnetic flux density on the symmetry axis (i.e., the z-axis in Figure 3) is $B_z = 0$ for any z. It follows from Biot and Savart's law: (1) the current dipole causes a magnetic field which is perpendicular to the z-axis; and (2) the components of the volume currents are parallel to the z-axis. The components of the volume currents perpendicular to the axis are schematically given in Figure 4. Current lines **a** and **b** give a contribution to B_z which cancel each other, as do all other pairs.

Consequently, the potential at any point on the axis outside the volume conductor is

$$\Phi(z) = \int_{\infty}^{z} B_z(z')dz' = 0. \tag{45}$$

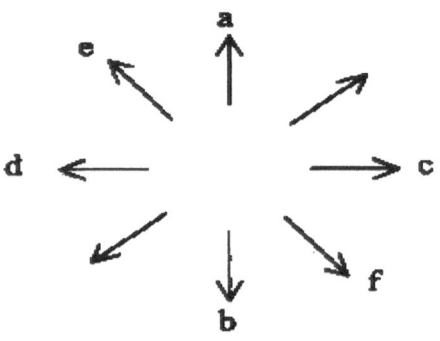

Figure 4. The components of the volume current lines perpendicular to the axis.

The coefficients A_n and B_n in Equation 44 must then be zero and can be substituted in Equation 43. So, from the fact that the potential is zero on the axis, it follows that the potential is zero for all points outside the volume conductor. It follows from Equation 42 that for a radial dipole, $\vec{B} = 0$ for all points outside a volume conductor which has spherical symmetry. In other words, the magnetic field generated by the source is nullified by that generated by the volume currents. This is true also for any current dipole which is on an axis of symmetry.

When a focal portion of cortex is activated, it produces a current dipole that is oriented perpendicular to the cortical surface. The brain's surface is folded into a hills-and-valleys arrangement, the so-called gyri and sulci. Because a spherically-symmetric model of the head is a rather good description for certain sections of the head, we may conclude that the MEG will not reveal activity arising from gyral sources.

5.2.2. A Tangential Dipole in a Set of Concentric Spheres

Electrical active patches of neurons which are located in the sulci of the cortex can be described as tangential dipoles. Let us consider a tangential dipole within a spherical symmetric volume conductor. This is equivalent to a tangential dipole in an infinite homogeneous medium plus secondary sources at the interfaces. The latter are all radially orientated. Thus, according to the Biot-Savart law, the radial component

$\vec{B}_r(\vec{r})$ at points outside the volume conductor generated by the secondary sources (at position r') of a surface element (with orientation $\vec{e}_n$) is linearly proportional to $(\vec{e}_n \times (\vec{r} - \vec{r}')) \cdot \vec{e}_r = 0$, where $\vec{e}_r$ is the unit vector in the radial direction $\vec{r}$. Thus, the radial component of the magnetic field outside the volume conductor due to the secondary sources of any surface element is zero. Consequently, the resultant radial component of the magnetic field is $\vec{B}_\infty \cdot \vec{e}_r$, where $\vec{B}_\infty$ is given by Equation 8.

This radial component for a dipole $\vec{p} = p\vec{e}_y$, located at $\vec{r}' = (0,0,b)$ is given by

$$B_r(\vec{r}) = \frac{\mu_0}{4\pi} \left(\frac{\vec{p} \times (\vec{r} - \vec{r}')}{|\vec{r} - \vec{r}'|^3} \right) \cdot \vec{e}_r \qquad (46)$$

Rewriting this expression in spherical coordinates yields:

$$B_r(\vec{r}) = \frac{\mu_0 p_y b \sin\theta \cos\varphi}{4\pi(r^2 - 2br\cos\theta + b^2)^{3/2}} \qquad (47)$$

Contour plots of the distribution of B_r on a concentric sphere (with a radius which is larger than the radius of the outer sphere of the volume conductor) are shown in Figure 5b. From this dipole pattern, the angle θ at the extremes can be extracted. The distance between the extremes is found for

$$\frac{\partial B_r}{\partial \theta} = 0,$$

yielding

$$\frac{r}{b} = \frac{(3 - \cos^2\theta) \pm \sqrt{((\cos^2\theta - 3)^2 - 4\cos^2\theta)}}{2\cos\theta} \qquad (48)$$

In Equation 47 the conductivity of the spheres, as well as their radii, are not present; hence, they do not influence the magnetic field.

An expression for the scalar magnetic potential Φ generated by the tangential dipole $\vec{p}_y$ at $(0,0,b)$ within a set of concentric spheres, is derived by Grynszpan and Geselowitz [23] and by de Munck [18]:

$$\Phi(\vec{r}) = \frac{\mu_0 p_y \cos\varphi}{4\pi b \sin\theta} \left(\frac{b\cos\theta - r}{\sqrt{(r^2 - 2br\cos\theta + b^2)}} + 1 \right) \tag{49}$$

The tangential components can be computed from Equation 49, since $\vec{B} = -\nabla\Phi$. Hence, we can conclude that they too are not influenced by the radii and conductivities.

Maps depicting the distributions of $|B|$, B_r, B_θ and B_φ on a concentric sphere with a radius larger than that of the outermost sphere are presented in Figure 5.

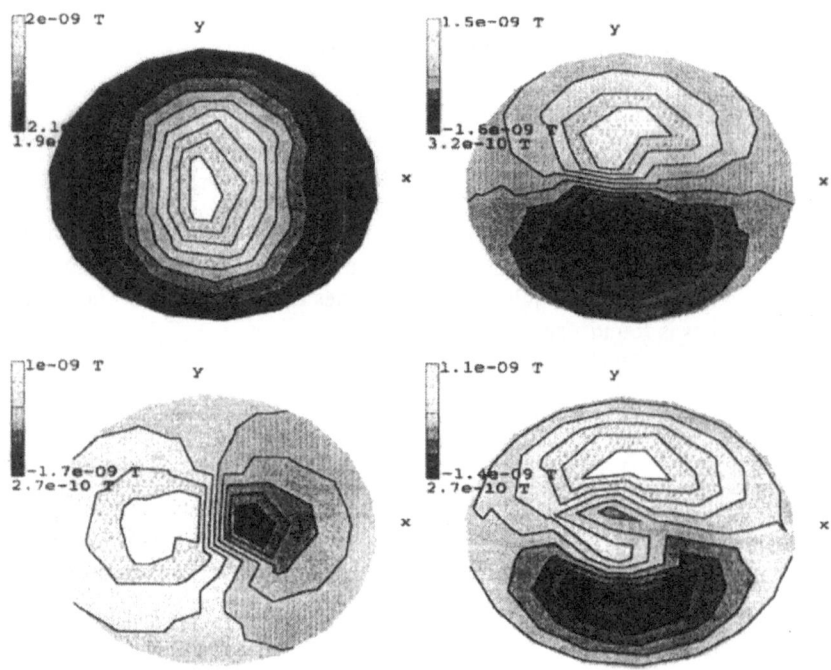

Figure 5. The magnetic flux density generated by a tangential current dipole in a homogeneous sphere with radius $R_0 = 10.0$ cm. The dipole is positioned at $z = 8$ cm and orientated in the x-direction. The arrows depict the magnetic flux density as a vector. The magnetic flux density is computed at 59 points on a concentric hemisphere with a radius of 12.0 cm.
(a) strength; (b) radial component; (c) φ-component; (d) θ-component.

6. Realistically-Shaped Models

As argued above, both the conductivities and the radii of the successive spheres do not have any influence on the MEG. However, these values are of importance for the EEG. Bearing in mind the paucity of knowledge of these parameters, it confirms the opinion that it is easier and more accurate to use MEG's instead of EEG's to localize sources. Yet, the symmetry condition involved is stringent, since all surfaces across which the conductivity is discontinuous should form concentric spherical surfaces. Moreover, all fibers should point in the radial direction or be parallel to a surface. The advantage of a model consisting of a set of concentric spheres is that an analytical expression is known for the potential and magnetic field as discussed above. Nevertheless, this involves some serious problems. As a sphere is a poor approximation of the shape of the head, the localization errors may be substantial. Furthermore, the choice of the geometrical reference is subjective. Consequently, the parameters of the sphere model are subjective also. The geometrical parameters are often obtained by taking the best-fitting sphere with respect to the electrodes used to measure EEG's, or with respect to the scalp surface. Another problem is that the source space is limited to the innermost sphere, which does not necessarily include the entire brain. The projection of the electrode positions onto the outer sphere also may cause some errors because their distance to the sphere surface can be several centimeters.

When a sphere fitting is executed for four tissue boundaries individually, a set of eccentric spheres is found. The potential distribution in an eccentric-sphere model can be found analytically [24]. Also for a spheroidal model with a layered structure, an analytical expression is derived [19]. More realistically-shaped models can be studied by numerical methods such as the boundary element and the finite element method [25]. Several research groups have constructed individual realistically-shaped volume conductor models to account for the variations in the size and shape of the head and its internal inhomogeneities.

6.1 THE BOUNDARY ELEMENT METHOD

The application of the Boundary Element Method (BEM) enables us to use realistically shaped models of the volume conductor boundaries. Three-compartment models describing the brain, the skull and the scalp are being used to solve the inverse problem [27, 28, 29]. The cerebrospinal fluid is included in the brain compartment.

The BEM is based on a discrete version of Equations 28 and 29. These equations imply that for both EEG's and MEG's the influence of the volume conductor can be described by the so-called secondary sources situated on the interfaces between regions of different conductivity, the outer surface included. The secondary sources are linearly proportional to the local potential. Normally, the interfaces are discretized by triangulation. The potential on each triangle is assumed to be constant or to vary linearly over a triangle. An analytical expression for the approximation of the integral of

$$\alpha_i^N = V(\vec{r'})\frac{\vec{R}\times \vec{e}_n}{R^3} \tag{50}$$

over a triangle N at surface i, is available [26]. Consequently, the numerical approximation of Equation 28 is

$$\vec{B}(\vec{r}) = \vec{B}_\infty - \frac{\mu_0}{4\pi}\sum_i\left(\sum_N \alpha_i^N S_i^N\right). \tag{51}$$

We can write this as a system of linear equations. The resulting matrix equation can be solved directly by applying matrix inversion [29]. For the potential a similar matrix equation is found. The numerical accuracy is limited if the skull conductivity is set at a low value compared to the brain and scalp conductivities. A method that accounts for this problem, the so-called isolated-problem approach, is developed by Hämäläinen and Sarvas [30].

Since the matrix is exclusively determined by the properties of the volume conductor, the magnetic field and/or potential can be found at little computational cost once the inverse of the matrix has been obtained. This is particularly useful when many sources must be localized within the same volume conductor. Hence, the question arises: if and to what extent, is it possible to use a standard boundary element model for the head? For MEG, a homogeneous conductor with the shape of the inner surface of the skull is (within a 5% error) equivalent to a model consisting of three realistically-shaped homogeneous, isotropic compartments describing the brain, the skull and the scalp, as argued by Hämäläinen and Sarvas [30]. Simulations were carried out using realistically-shaped models with the shape of the inner skull surface of three persons [31]. An impression of such a model is shown in Figure 6. One of the models was used to compute the magnetic field distribution caused by a single current dipole at different positions. For the subsequent dipole localizations, these data served as input information. The other two models were used as volume conductor models in the inverse procedure. In order to fit the different sizes of the models, these models were scaled in three different directions. It was found that a standard head may cause

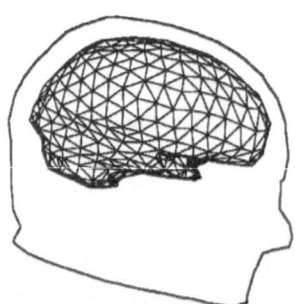

localization errors that are even larger than those of the sphere model. The errors are strongly dependent on the source location and the particular model. One could conclude that the application of standard head models must take into account the individual differences in geometrical shape. A sufficiently large model library might be of advantage if the adequacy of the model can be estimated in advance.

Figure 6. A triangulated model of the head with the shape of the inner surface of the skull.

6.2 THE FINITE-ELEMENT METHOD (FEM)

Complex-shaped compartments with an inhomogeneous and anisotropic conductivity can be studied by means of the FEM. As shown in section 4, the physical problem to be solved is a Poisson equation (Equation 20) with boundary conditions. From the Poisson equation an integral equation is derived, in which only first derivatives of the potential appear. Thus, the numerical version of this equation must obey the condition that the first derivative of the approximated potential exists and that it is continuous. The volume is divided into small elements, such as tetrahedrons or hexahedrons. The potential is calculated at the corner points of the elements. Within an element the values are obtained by linear interpolation, which is in accordance with the condition mentioned above. The tensor describing the conductivity is allowed to differ for all these elements [32]. A hole in the skull can be modeled simply, for instance, by giving some elements belonging to the skull layer the same conductivity as the scalp region. Simulations of the electric potential using the FEM [33] have indicated a noticeable influence of such holes.

Since the dipole is a discrete source, the potential distribution will be singular at the dipole position. This singularity can be removed by introducing a function Ψ defined by

$$\Psi = V - V_\infty,$$ (52)

where V_∞ is the solution of the problem for the case in which the volume conductor is a homogeneous conductor of infinite extent, the solution for which is given by Equation 7. Now ψ is the solution of the problem:

$$\nabla \cdot (\sigma \nabla \Psi) = 0.$$ (53)

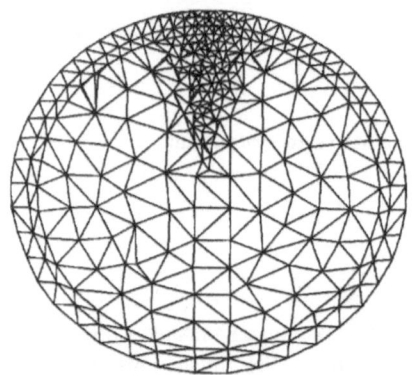

Figure 7. Cross-section of a three-sphere model of the head. The mesh, consisting of tetrahedrons, is refined in the region near the dipole.

The removal of the singularity at the source location results in an additional term in the boundary condition. The solution of the potential V is obtained by first solving ψ and then adding V_{∞}. Refinement of the mesh is required in regions where the gradient of Ψ changes rapidly. An example of such a refined mesh is given in Figure 7.

Subsequently, the magnetic induction can be computed from the calculated potential distribution, using the Biot-Savart law as given in Equation 27. The volume integral over the volume conductor is computed as the sum of the volume integrals over the single elements. This method to compute the magnetic field where the volume conductor consists of non-closed regions with inhomogeneous and anisotropic conductivities, without the need to mesh the space surrounding the head, is described by van den Broek [34]. The FEM was used to simulate holes in the skull. It was found that these holes were of little influence on the MEG. An impression of a skull with a hole is depicted in Figure 8.

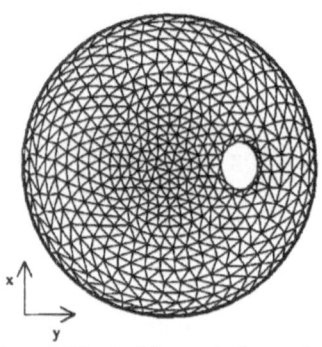

Figure 8. The skull layer of a three-sphere model of the head. Some of the tetrahedrons are attributed to the scalp region in order to model a hole.

As argued previously, the magnetic field of a radial dipole in a set of concentric spheres is zero. Since the shape of the human head is obviously quite different from a perfect sphere, it is to be expected that the field of a radial dipole in an actual head is suppressed when compared to that of a tangential one. Detailed FEM models of the head were constructed by Haueisen et al. [35], and the suppression ratio was computed. An average suppression ratio of 0.19 ± 0.07 was found for different regions and depths in the gray matter. It was found that the computed magnetic field of radial sources varied significantly with the conductivities of the surrounding tissues where the dipole was located.

6.3 THE GENERATION OF REALISTICALLY-SHAPED MODELS FROM MRI

Most numerical methods for solving bioelectric field problems require that the continuous domain be broken into discrete elements, the so-called mesh. Because of the complex geometries, construction of the mesh can become one of the most time-consuming aspects of the modeling process. The structural information needed to create models can be obtained from MRI. First, the MR scans are segmentated, i.e., the information in the MRI is classified according to its type of tissue. This step is necessary to enable the generation of the surface of the head and the interfaces between tissues. A data set of an MRI head scan consists of voxels which represent the different tissues by means of varying gray values. A method for the automatic segmentation of the head, where it is automatically determined which voxels are part of the head and which are not, and where the brain, skull and scalp regions are discriminated, is described by Wieringa and Peters [36]. Second, a sufficient number of points must be selected on each boundary. These points form the basis for subsequent interpolation, e.g., by splines [29] or a surface harmonic expansion [37]. This interpolation offers the possibility of generating new points on the surface. For the FEM, a volume element mesh is created by adding points within the volume. If the gradient of the potential differs more than a given value between two adjacent surface or volume elements, these elements should be made smaller by inserting a new point. Such a regional refinement is advantageous because an overall refinement strongly increases the calculation time if applied to inverse-solution methods.

7. Models in Cardiology

The volume conductor problem plays a role whenever measurements of the potential or magnetic fields are used to estimate sources. These sources may be situated in muscles, nerves, the spine, etc. We will give some examples of volume conductors used to localize sources within the adult and fetal heart. Both the adult MCG and the fetal MCG may be measurable with high-T_c magnetometer systems. MCG's have a diagnostic value, and satisfactory models for the torso are available. Fetal MCG is a new tool. Models which explain the MCG are important in order to estimate the direction of the sources in the fetal heart.

7.1. THE INFLUENCE OF THE VOLUME CONDUCTOR ON MCG AND ECG

A simple model of the torso is a homogeneous isotropic half space (i.e., a semi-infinite conductor). This conductor is equivalent to a sphere with a radius of infinite extent. Consequently, the volume currents will not contribute to the component of the magnetic field which is perpendicular to the surface, and a source which is perpendicular to the

388

surface will give a zero magnetic field. The potential on the surface will be doubled by the volume currents - this can be shown by the method of images, see e.g., [38]. However, such a model gives rise to substantial localization errors. A homogeneous,

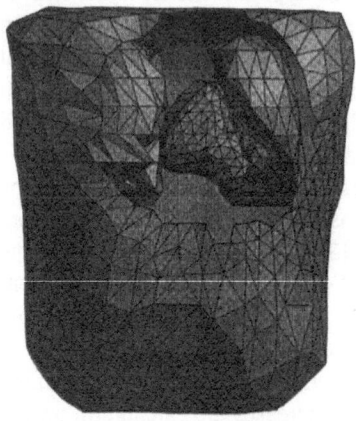

Figure 9. Model of a torso, displaying the boundaries of the lungs and the blood masses in the heart. Part of the surfaces are removed to reveal the inside. (Courtesy of Dr. T.F. Oostendorp, Department of Medical Physics and Biophysics, Catholic University of Nijmegen, The Netherlands).

standard-size model with the shape of the outer torso surface gives a better accuracy. Such a model has been applied to localize the ventricular pre-excitation sites in patients from MCG measurements. The average accuracy was found to be about 2 cm when compared with invasive localization results obtained during surgery [39]. Internal inhomogeneities that have a pronounced effect are the lungs and the heart's cavities. Such a model is depicted in Figure 9. It is to be expected that the most accurate inverse solutions will be obtained using individualized torso models based on MRI scans of a given subject [40]. The appropriate method for solving the inverse problem for these models is the BEM. The effect of the anisotropic skeletal muscle layer on the ECG is studied using the FEM [41]. Results showed that the magnitude of the surface potentials decreased by the introduction of the anisotropic skeletal muscle layer, while the shape of the potentials remained almost unaffected.

The most extensive model described in the literature [42], the Utah Torso model, contains detailed anatomical structures of bone, including the sternum, 12 ribs at both sides, the spine, and the left and right clavicle. The blood vessels, like the aorta, pulmonary veins, etc. also are included, as are the myocardium, the blood-filled chambers of the heart, the lungs, skeletal muscles and a subcutaneous fat region. The model consists of about 1 million tetrahedrons. It appeared to be difficult to draw conclusions from simulations with this model. Simulations starting with a homogeneous model and adding an inhomogeneity, showed that the structures closer to the torso surface tend to have the largest impact on the ECG. Starting with the full model and removing one inhomogeneity led to the conclusion that the structures closest to the heart have the largest impact. These detailed models have been developed to study the influence of parameters on the model in forward solutions. Using them in inverse solutions would be too time-consuming.

7.2. THE INFLUENCE OF THE VOLUME CONDUCTOR ON FETAL MCG AND ECG

During the entire period of pregnancy, the recording of a fetal ECG using abdominal leads is a troublesome technique because the amplitude is small and the signal-to-noise ratio is low. The amplitude of the fetal ECG recorded at the maternal abdomen varies during pregnancy. Starting from a low value at the beginning of pregnancy it reaches its maximal value, on the order of 15 µV, at about the 27th week of gestation. Afterwards, it drops drastically to below the noise level. After the 32nd week, the magnitude of the fetal ECG increases again. However, the distribution of the potential at the maternal abdomen does not correspond to the actual equivalent dipole. In order to explain these phenomena, models of the volume conductor were developed by Oostendorp [43].

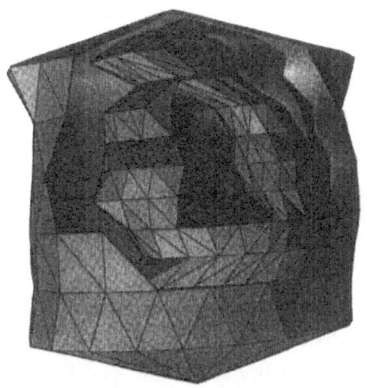

Figure 10. Model of the maternal abdomen, displaying the boundaries of the fetus and the uterus. Part of the surfaces are removed to reveal the inside. (Courtesy of Dr. T.F. Oostendorp, Department of Medical Physics and Biophysics, Catholic University of Nijmegen, The Netherlands).

He arrived at the following hypothesis concerning the conduction of the current generated by the fetal heart at the surface of the maternal abdomen. As the fetus grows, the strength of the current generated by the heart increases. At about 20 weeks, the current becomes strong enough for the ECG to be measured. From that moment on until 28 weeks, the conduction is best described by a model consisting of three compartments, i.e., the maternal abdomen, the uterus filled with amniotic fluid and the fetal body - see Figure 10. Apart from a scaling factor it also can adequately be described by a homogeneous model with the shape of the abdomen of the mother. These models do not apply well later in the gestation period. Around 28 weeks, the layer of vernix caseosa, which has a resistivity of about a factor 10^6 higher than that of the surrounding tissue and amniotic fluids, starts to develop. The fetus is almost completely electrically isolated. As pregnancy continues open patches at unpredictable positions appear in the layer of vernix. To simulate the current which is escaping from two holes in an electrically-isolating layer, the source description of two monopoles is used, each having the same strength, but opposite polarity. The volume conductor model consists of two compartments, i.e., the maternal abdomen (0.22 Sm^{-1}) and the fetal body surrounded by an isolating layer of vernix (0 Sm^{-1}). This model was found to be inadequate because the positions of the monopoles were not always found near the fetus.

The fetal MCG, on the other hand, can be measured during a long period of pregnancy, beginning with the 13th week of gestation. The amplitudes of the QRS complexes are roughly proportional to the fetal age. A decrease of the amplitude is not observed in the

390

period when the amplitude of the fetal ECG shows a strong decrease. Oostendorp [44] used two different volume conductor models to simulate the fetal MCG. The first one consisted of the fetal body, the uterus filled with fluid, and the maternal abdomen. This model corresponds to the situation in early pregnancy. In the second one, the volume conductor consisted of the fetal body only. This corresponds to the situation in which the fetus is covered by an isolating layer of vernix. Measurements of the magnetic field distribution with respect to the abdomen will help to decide if these models are adequate.

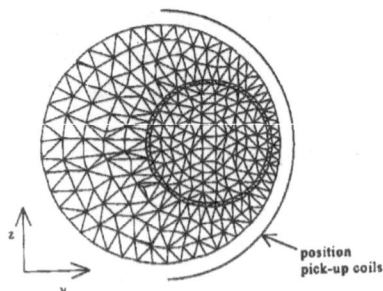

Figure 11. Cross-section of a finite-element model of the maternal abdomen, which consists of two concentric spheres representing the fetus and the layer of vernix, and an eccentric sphere which represents the maternal abdomen.

In order to study the volume conductor problem for the period after the 32nd week of gestation, we constructed a model which consisted of two concentric spheres, representing the fetus and the vernix layer, and an eccentric sphere which represents the maternal abdomen. This model is depicted in Figure 11. The conductivities assigned to the fetus and the mother were taken to be 1 (in arbitrary units) and that of the vernix, 10^{-6}. Two holes were made in the isolating layer. Simulations were carried out using the FEM in combination with the Biot-Savart law. The results showed that the influence of the holes on the potential distribution is enormous. The distribution depends strongly on the positions of the holes, while the orientation of the dipole, representing the fetal heart activity, plays a minor role. The influence of the holes in the vernix layer on the MCG is rather small, although it is seen easily in Figure 12. The contour maps still reflect the orientation of the dipole. However, the shape of these contours depends on the orientation with respect to the holes.

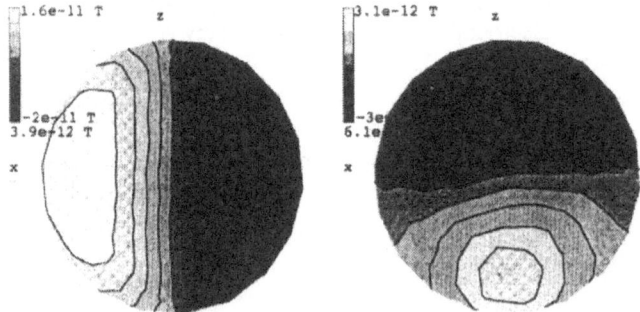

Figure 12. Simulations based on the model depicted in Figure 11. At the z-axis, two holes in the layer of vernix are modeled. The dipole representing the fetal heart activity is in the center of the sphere describing the fetus. The magnetic flux density is calculated at 76 points on a hemisphere, as shown in Figure 11. The contours display the distribution of the radial component of the field due to: (a) a dipole in the x-direction, (b) a dipole in the z-direction.

8. Discussion

Source localization is always based on assumptions concerning the geometry of the volume conductor, the conductivities of the tissues and the type of source. These assumptions are a compromise between the desire to localize the source as accurately as possible and to use a method which is as simple as possible. In order to study the influence of the model parameters, two different procedures can be used.

First, the potential and/or the magnetic field distribution of a current dipole in a certain model of the volume conductor are calculated, and the simulated data are used for the inverse algorithm to calculate the equivalent dipole. The inverse algorithm is based on a model that differs only in one aspect from that used in the forward calculations. The difference in location is a measure of the influence of the aspect studied.

Second, forward calculations are carried out for two different models and the resulting distributions of the field or potential are compared. As a measure, the relative difference between the two solutions can be computed, defined as [45]

$$
\mathrm{RDM} = \sqrt{\frac{\sum_i (V_{i1} - V_{i2})^2}{\sum_i V_{i1}^2}} \, ,
$$

where V_{i1} and V_{i2} are the calculated values of the potential or a component of the magnetic field at measurement point i.

The experimental verification of the adequacy of a volume conductor model is difficult. Methods to assess the merits of MEG- and EEG-based localization procedures are:

1. Phantoms can be used. For instance, phantoms have been used to compare the influence of a spherical model on source localization with that of a skull-shaped model. Henderson et al. [46] compared electric source localization of artificial dipoles positioned in a sphere with those in a specially prepared human skull. A saline-soaked woolen cloth simulated the scalp. EEG electrodes were stitched to this 'scalp'. A three-sphere model was used for the inverse procedure in both cases. The maximum observed distance between the actual and calculated dipole was 9.4 mm for the skull-shaped model, as opposed to 5.2 mm for a homogeneous sphere. Menninghaus and Lütkenhöner [47] compared magnetic source localization with the help of a skull filled with a saline solution. They found that the magnetic field of a radial dipole in a skull-shaped conductor was suppressed by a factor of about 6 (dependent on the depth) compared with that of a tangential dipole.

2. Electrodes implanted in patients being evaluated for epileptic surgery as artificial dipoles also can be used to test the localization procedure with a chosen model. A weak current is passed between two electrodes to act as a current dipole. For example, localization errors for EEG using a three-shell model are reported on the order of 1 cm for dipoles in the frontal part of the brain [48], and 2 cm for dipoles in the hippocampus and amygdala [49]. The same type of experiment has been used to study the adequacy of localizations based on MEG {50]. Homma et al. [51] used the method to study the localization error for EEG with a realistically-shaped, three-compartment model. The error was minimized by changing the conductivity of the skull with respect to the conductivities of the scalp and brain tissue. When using the relative conductivity of 1/80 of the brain conductivity, the estimated errors became minimal.

3. MEG- and EEG-based localizations are used to localize the site of an epileptic focus for pre-surgical planning, and to localize the source of slow-wave and beta-wave activity associated with brain lesions. The success of the operation is a measure of the adequacy of the localization procedure. MCG- and ECG-based localizations are used to localize the arrythmogenic zone in the heart, an accessory pathway in WPW syndrome, or a Kent bundle. The results of catheter ablation or surgery will help us to estimate the efficacy of the model used.

4. MEG- and EEG-based localizations can be compared with results obtained with other methods for functional imaging, such as PET, SPECT and functional MRI. PET and SPECT provide a measure of chemical specificity, while functional MRI shows the changes in blood flow. All of these methods have different time resolutions. It is difficult at this moment to say to what extent the results from these different modalities are comparable.

5. Last, but not least, combining MEG and EEG localization procedures will lead to more precise results.

9. References

1. Helmholtz, H. (1853) Über einige Gesetze der Verteilung elektrischer Ströme in körperlichen Leitern mit Anwendung auf die Tierisch-elektrischen Versuche, *Ann. Phys. Chem.*, 89, 211-233, 353-377.
2. Maxwell, J.C. (1873) *Treatise on Electricity and Magnetism*, Dover Publications, ING, New York
3. Geddes, L.A. and Baker, L.E. (1967) The specific resistance of biological material - A compendium of data for the biomedical engineer and physiologist, *Med. & Biol. Eng.*, 5, 271-293.
4. Schwan, H.P. and Kay, C.F. (1957) The conductivity of living tissue, *Ann.N.Y.Acad.Sci*, 65, 1007-1013.
5. Oostendorp, T.F.(1989) *Modelling the fetal ECG*, PhD thesis, Catholic University of Nijmegen, The Netherlands.
6. Rush, S. and Driscoll, D.A. (1969) EEG electrode sensitivity: an application of reciprocity, *IEEE Trans. on Biomed. Eng.*, BME-16, 15-22.
7. Law, S.K. (1993) Thickness and resistivity variations over the upper surface of the human skull, *Brain Topography*, 6, 99-109.

8. Plonsey, R. and Barr, R.C. (1991) *Bioelectricity-A quantitative approach*, Plenum Press, New York and London.

9. Okada, Y.C. and Nicholson, C. (1988) Magnetis evoked field associated with transcortical current in turtle cerebellum, *Biophys. J.*, 53, 723-731.

10. Plonsey, R. and Heppner, D.B. (1967) Considerations of quasi-stationarity in electrophysiological systems, *Bulletin of Mathemaical Biophysics*, 29, 657-664.

11. Geselowitz, D.B. (1970) On the magnetic field generated outside an inhomogeneous volume conductor by internal current sources, *IEEE Trans Magn* MAG-6, 346-347.

12. Barnard, A.C.L., Duck, I.M., Lynn, M.S. and Timlake, W.T. (1966), The application of electromagnetic theory to electrocardiology II, *Biophys. J.*, 7, 463-491,

13. Geselowitz, D.B. (1967) On bioelectric potentials in an inhomogeneous conductor, *Biophys. J.*, 7, 1-11.

14. Plonsey, R. (1982), The nature of sources of bioelectric and biomagnetic fields, *Biophys. J.*, 39, 309-312.

15. Lorrain, P., Corson, D.R. and Lorrain, F., *Electromagnetic Fields and Waves*, W.H.Freeman and Company, New York, page 92.

16. Frank, E. (1952) Electric potential produced by two point current sources in a homogeneous conducting sphere, *J. of Applied Physics*, 23, 1225-1228.

17. Zhang, Z. and Jewett, D.J. (1993), Insidious errors in dipole localization parameters at a single time-point due to model misspecification of number of shells. *Electroenceph. and Clin. Neurophys.*, 88, 1-11.

18. De Munck, J.C. (!989) *A mathematical and physical interpretation of the electromagnetic field of the brain*, PhD thesis, University of Amsterdam, The Netherlands.

19. De Munck, J.C. (!988) The potential distribution in a layered anisotropic spheroidal volume conductor, *J.Appl. Phys.*, 64, 766-778.

20. Zhou, H. and van Oosterom, A. (1992), Computation of the potential distribution in a four-layer anisotropic concentric spherical volume conductor, *IEEE Trans. on Biomed. Eng.*, BME-34, 154-158.

21. Stok, C.J., (1987) The influence of model parameters on EEG/MEG single dipole source estimation, *IEEE Trans. on Biomed. Eng.*, BME-34, 289-296.

22. Jefimenko, O.D. (1966) *Electricity and Magnetism*, Appleton-Century-Crofts, New York .

23. Grynszpan, F. and Geselowitz, D.B.(1973), Model studies of the magnetocardiogram. *Biophys. J.*, 13, 911-925.

24. Cuffin, B.N. (1991) Eccentric spheres models of the head, *IEEE Trans. on Biomed. Eng.*, 38, 871-878.

25. Pruis, G W., Gilding, B.H. and Peters, M.J. (1993), A comparison of different numerical methods for solving the forward problem in EEG and MEG, *Physiol. Meas.*, 14, A1-A9.

26. De Munck, J.C. (1992), A linear discretization of the volume conductor boundary integral equation using analytically integrated elements, *IEEE Trans. on Biomed Eng.*, 39, 986-990.

27. He, B., Ye, W., Okamoto, Y. and Musha, T. (1987), Inhomogeneous head model for dipole tracing in the brain, *IEEE, Ninth Annual Conf. of the Eng. in Medicine and Biology Soc.*, 939-940.

28. Meijs, J.W.H. (1988) *The influence of head geometries on electro- and magnetoencephalograms*, PhD thesis, Twente, The Netherlands.

29. Zanow, F. and Peters, M.J. (1995), Indivdually shaped volume conductor models of the head in EEG source localization, *Med. & Biol. Eng. & Comput.*, 33, 582-588.

30. Hämäläinen, M.S. and Sarvas, J. (1989) Realistic conductivity geometry model of the human head for the interpretation of neuromagnetic data, *IEEE Trans. on Biomed. Eng.*, BME-36, 165-172.

31. Zanow, F., Verhoef, R., Knoesche, T. and Peters, M.J. (1995) Are standard head models superior to the sphere model in MEG source localizations?, In: *Biomagnetism: Fundamental Research and Clinical Applications*, C. Baumgartner et al. (Eds), Elsevier Science, IOS Press, 450-454.

32. Bertrand, O., Thevenet, M. and Perrin, F. (1991) 3D Finite element method in brain electrical activity studies, In: J. Nenonen, H.M. Rajala and T. Katila (Eds.), *Biomagnetic localization and 3D modeling*, Helsinki University of Technology, Dep. of Techn. Phys., report TKK-F-A689, 154-171.

33. Bertrand, O., Thevenet, M., Perrin, F. and Pernier, J. (1992), Effects of skull holes on the scalp potential distribution evaluated with a finite element model, *Satellite Symp. on Neuroscience and Technology, 14th An. Int. Conf. of the IEEE Engin. in Medicine and Biology Soc..*, Lyon, 42-45.

34. van den Broek, S.P., Zhou, H and Peters, M.J. (1995) Computation of neuromagnetic fields using the finite element method and the Biot-Savart law, accepted for publication in: *Med. & Biol. Eng. & Comp.*

35. Haueisen, J., Ramon, C., Czapski, P. and Eiselt, M. (1995) Realistic volume conductor modelling, Accepted for publication in: *Annals of Biomedical Engineering*

36. Wieringa, H.J. and Peters, M.J. (1993) Processing MRI data for electromagnetic source imaging, *Med. & Biol. Eng. & Comput* 31, 600-606.

37. Hren, R., Vardy, D., Miller, R., Stewart Ferguson, A. and Stroink, G. (1995) Approximation of torso geometries by surface harmonic expansion, in: Biomagnetism: Fundamental Research and Clinical Applications, Baumgartner, C., Deecke, L., Stroink, G., Williamson, S.J. eds, IOS press, Elsevier, 668-670.

38. Nunez, P.L. (1981), *Electric fields of the brain,* Oxford University Press.

39. Nenonen, J.,(1991) Realistic torso models in the MCG, In: J. Nenonen, H.M. Rajala and T. Katila (Eds.), *Biomagnetic localization and 3D modeling,* Helsinki University of Technology, Dep. of Techn. Phys., report TKK-F-A689, 62-70.

40. Huiskamp, G and van Oosterom, A. (1989) Tailored versus realistic geometry in the inverse problem of electrocardiography, *IEEE Trans. on Biomed. Eng.,* 36, 827-835.

41. Zhou, H. (1994), *Anisotropic volume conduction in biological media,* PhD hesis, Catholic University of Nijmegen, The Netherlands.

42. Klepfer, R.N., Johnson, C.R. and Mac Leod, R.S. (1995) The effects of inhomogeneities and anisotropies on electrocardiographic fields: A three-dimensional finite element study, *IEEE EMBS,* to appear.

43. Oostendorp, T.F. (1989) *Modelling the fetal ECG,* PhD thesis, Catholic University of Nijmegen, The Netherlands.

44. Oostendorp, T.F. and van Oosterom, A. (1991) Modelling the fetal magnetocardiogram, *Clin. Phys. Physiol. Meas.,* 12, Suppl. A, 15-18.

45. Meijs, J.W.H., Weier, O., Peters, M.J. and Van Oosterom, A. (1989) On the numerical accuracy of the boundary element method. *IEEE Trans. On Biomed. Eng.,* BME-36, 1038-1049.

46. Henderson, C.J., Butler, S.R. and Glass, A. (1975) The localization of equivalent dipoles of EEG sources by the application of electric field theory, *Electroenceph. and Clin. Neurophysiol.,* 39, 117-130.

47. Menninghaus, E. and Lütkenhöner, B. (1995) How silent are deep and radial sources in neuromagnetic measurements, in: *Biomagnetism: Fundamental Research and Clinical Application,* Baumgartner, C., Deecke, L., Stroink, G., Williamson, S.J. eds, IOS press, Elsevier, 352-356.

48. Cuffin, B.N., Cohen, D., Yunokochi, K., Maniewski, R., Cosgrove, G.R., Ives, J., Kennedy, J. and Schoner, D. (1991) Tests of EEG localization accuracy using implanted sources in the human brain, *Annals of Neurology,* 29, 132-138.

49. Smith, D.B., Sidman, R.D., Flanigin, H., Henke, J. and Labiner, D. (1985) A reliable method for localizing deep intracranial sources of the EEG, *Neurology,* 36, 1702-1707.

50. Balish, M., Sato, S., Connaughton, P. and Kufta, C., (1991) Localization of implanted dipoles by magnetoencephalography, Neurology, 41, 1072-1076.

51. Homma, S, Musha, T., Nakajima, Y., Okamoto, Y., Blom, S., Flink, R., Hagbarth, K.E. and Moström , U. (1994) Location of electric current sources in the human brain estimated by the dipole tracing method of the scalp-skull-brain (SBB) model, *Electroenceph. and Clin. Neurophysiol.,* 91, 374-382.

Magnetocardiography, an introduction

S.N. Erné and J. Lehmann
Zentralinstitut für Biomedizinische Technik
Universität ULM, 89069 Ulm, Germany

Abstract- As introduction to magnetocardiography, first a general view of the cardiovascular system is given with special emphasis on the anatomy and physiology of the heart. A short overview of the pathophysiology of the heart so far relevant for the understanding of the possible applications of magnetocardiography is followed by an historical survey of the magnetocardiography development in the last 20 years. The main applications of magnetic measurements in cardiology are presented.

1. - Introduction to the cardiovascular system

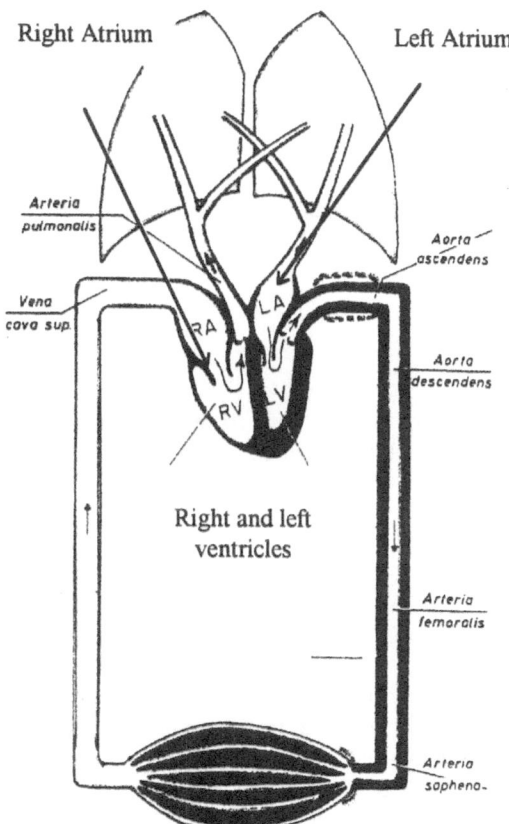

Right Atrium

Left Atrium

Arteria pulmonalis

Aorta ascendens

Vena cava sup.

Aorta descendens

Right and left ventricles

Arteria femoralis

Arteria saphena-

Figure 1 - Schematic diagram of the blood circulation in the human body

Blood can fulfill it's goal of transporting hemoglobin only if it is continuously circulating through the body. The pump providing the mechanical energy for this circulation is the heart. The blood circulation system in humans consists of two completely separated loops: a peripheral loop to provide oxygen-rich blood to all muscles of the body and a second loop to regenerate blood in the lungs (eliminating CO_2 and acquiring new O_2). The heart is the connection point between the two circulation paths.

From the standpoint of mechanical engineering the heart consist of two complete adjacent pumps, located physically in parallel, but functionally operating in series. The left half of the heart (pump number one) pumps the blood in the artery system of the body; the right half of the heart (pump number two) pumps the blood coming back from the body to the lungs in order to collect new oxygen.

395

H. Weinstock (ed.), SQUID Sensors: Fundamentals, Fabrication and Applications, 395–412.
© *1996 Kluwer Academic Publishers.*

From the lungs the bloods come back to the left half of the heart through the pulmonary veins. The left and right halves of the heart consist of two muscular cavities named *atrium* and *ventricle*, that are separated by valves.

The blood coming from the veins flows into the right atrium, from the atrium into the right ventricle; from the right ventricle into the lungs through the *arteria pulmonalis*, from the lungs into the left atrium, and from there into the left ventricle. Eventually the blood is pumped from the left ventricle through the *aorta* into the body periphery. At the transition between the ventricles and arteries there are again valves; these are shaped in a suitable way to establish the direction of the unidirectional blood flow.

Due to the fact that both of the heart halves are bound together, they always work synchronously, and at each heart beat the same blood volume is ejected: for this reason under physiological conditions there is no liquid transfer between the two partial circulation paths.

2. Heart mechanics

Similar to the well-known piston motor in a car, the heart operates in a four-step cycle. In the following, the description of a single phase refers to the left half of the heart: this does not reduce the generality of the description, due to the fact that, as mentioned above, the two heart halves work synchronously and on the same volume. The only difference is the fact that the left ventricle must produce a higher pressure and for this reason consists of a stronger muscle.

2.1 SYSTOLE

Systole is the name of the active phase of the heart activity. At the begining we assume that the left ventricle is almost full with blood, with a volume of about 140 ml.

2.1.1.Tension Phase (ca. 50 ms)

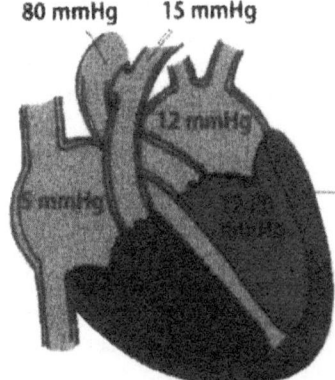

80 mmHg   15 mmHg

12 mmHg

5 mmHg

Figure 2 - First systolic phase: the tension phase

During this part of the contraction (about 70 ms), the ventricular muscles change shape around the liquid volume, as consequence the blood pressure increase but the volume remains constant (isovolumetric contraction). With the increase of pressure the valves between atrium and ventricle remain closed. Soon the pressure in the ventricle surpasses the pressure in the aorta the outlet valve opens, starting the next phase:

2.1.2 Ejection Phase (ca. 250 ms)

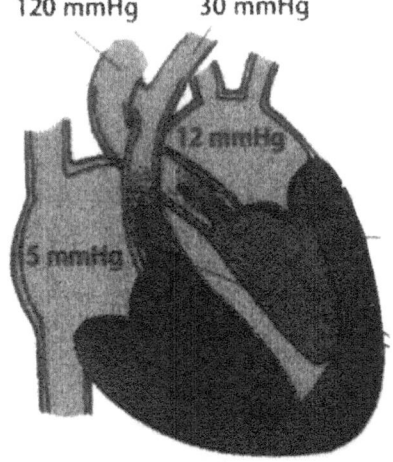

120 mmHg   30 mmHg

Figure 3 - Second systolic phase: the ejection phase

Now the pressure increases further to its maximum value and simultaneously about 50% of the blood present in the ventricle is ejected into the aorta. After the maximum has been reached the blood pressure slowly decreases; when the value in the ventricle is lower than the value in the aorta, the aorta valve closes and the systole is completed.

2.2.   DIASTOLE

The diastole can be considered as the rest phase of the heart cycle.

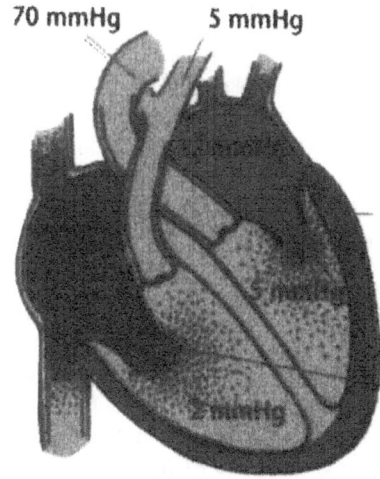

70 mmHg   5 mmHg

Figure 4 - Diastolic phase

2.2.1   Relaxing phase (ca. 100 ms)

In this phase the blood pressure continues to diminish, although the volume of the ventricle remains constant. After the pressure is decreased below the value in the atrium (5 mmHg), the valves between atrium and ventricle open and the next phase begins.

2.2.2   Filling phase (600 ms)

The filling procedure is initially fast, and is performed without muscular work; about 95% of the required blood amount flows into the ventricle. Thereafter, there is an important phase called diastase, in which the entire heart muscle is relaxed. After the diastase, the atrium starts contracting and presses the remaining 5% of the blood into the ventricle. During the diastase, the heart muscle has its best blood perfusion through the coronary arteries.

With increasing heart frequency, the diastole, and especially the diastase, are shortened. At high beat rates the perfusion of the heart becomes insufficient.

3. Electrophysiology of the heart.

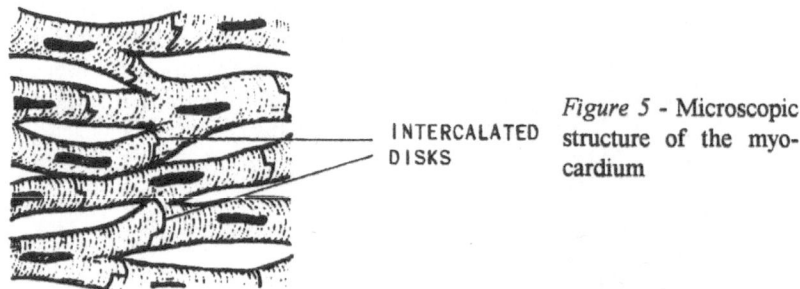

INTERCALATED DISKS

Figure 5 - Microscopic structure of the myocardium

The basic elements for the heart work are the myocardium fibers, which consist of a series of cells interconnected via the so called *intercalated discs*. The intercalated discs provide a good connection so that the activation can propagate from one cell to the next. From the point of view of functionality, two families of cardiac fibers exist:

- the fibers of the myocardium, that perform the real mechanical work described above and

- the specific fibers of the activation and conduction system, that control the timing of the pump.

The rhythmic activity of the heart is controlled by a pacemaker and timing system inside the heart. An isolated heart can, under suitable conditions, continue to beat with a constant frequency. This is called *automaticity*. The region of the heart containing cells, that are able to provide such an excitation, is called *pacemaker*. Under physiological conditions the heart beat starts from the *sinus node*. The sinus node is located where the upper vein (vena cava superior) enters the atrium. In a normal quiet regime, the sinus paces the heart with a rhythm between 60 and 70 beats per minute

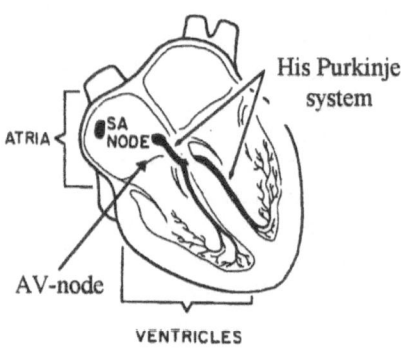

His Purkinje system

ATRIA

SA NODE

AV-node

VENTRICLES

Figure 6 - Heart Conduction System

The excitation passes through the atria using as a conduction medium the myocardial tissue. The region between the atria and the ventricles is called the annulus fibrosus, and it is basically non conducting. Only the Atrium-Ventricular node (AV-node) and the following His bundle provide as part of the specific conduction system a means for the further propagation of the excitation toward the ventricles.

To guarantee that the atrial excitation is completed, the AV node provides a short time delay before passing the conduction signal to the His bundle.

The His bundle and the following, more and more tree like, bifurcating bundles consist of Purkinje fibers with a fast conduction velocity, the result is that the time between the activation of the His Bundle and the onset of ventricular activation is about 40 ms.

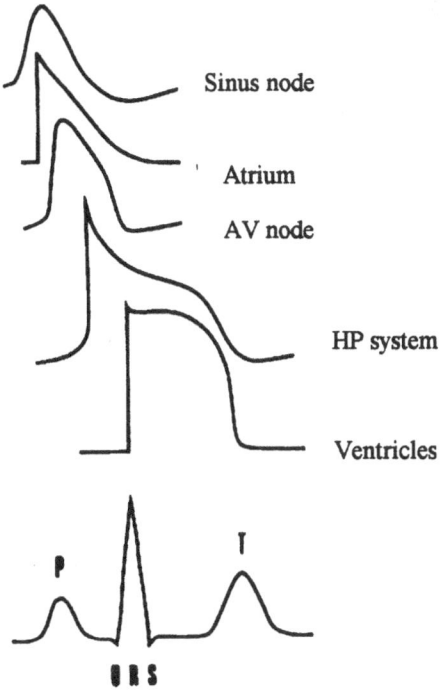

Sinus node

Atrium

AV node

HP system

Ventricles

P

T

Q R S

Electrophysiologically the different components contributing to the pacing and the regulation of the heart can be identified by their electrical signature, the monophasic action potential, which characterizes the electrical activation sequence. In Figure 7 the monophasic action potentials of the five relevant structures are summarized. At the bottom of the figure a standard limb ECG provides the time scale and a macroscopic reference. In the ECG signals, the letters P, Q, R, S and T indicate the prominent signal components.

Figure 7 - Monophasic action potentials in different heart tissues

For a better understanding of this nomenclature, in Fig. 8 the development of the ECG trace is shown in parallel with the schematic representations of the progress of the excitation in the heart: at the time corresponding to the maximum of the P wave, the excitation of the atria is almost completed; during the PQ interval, the atrial excitation is completed, and the His-Purkinje system is activated.

At Q the myocardial tissue near the apex of the ventricle starts becoming activated. At R, the largest signal in ECG, the heart activation extends to the total septum and a large part of the ventricles; in S and ST the excitation of the ventricles is completed. During T the ventricular depolarization is ongoing, and at the end of the beat (post T) the ECG goes back to zero, and the heart is at rest.

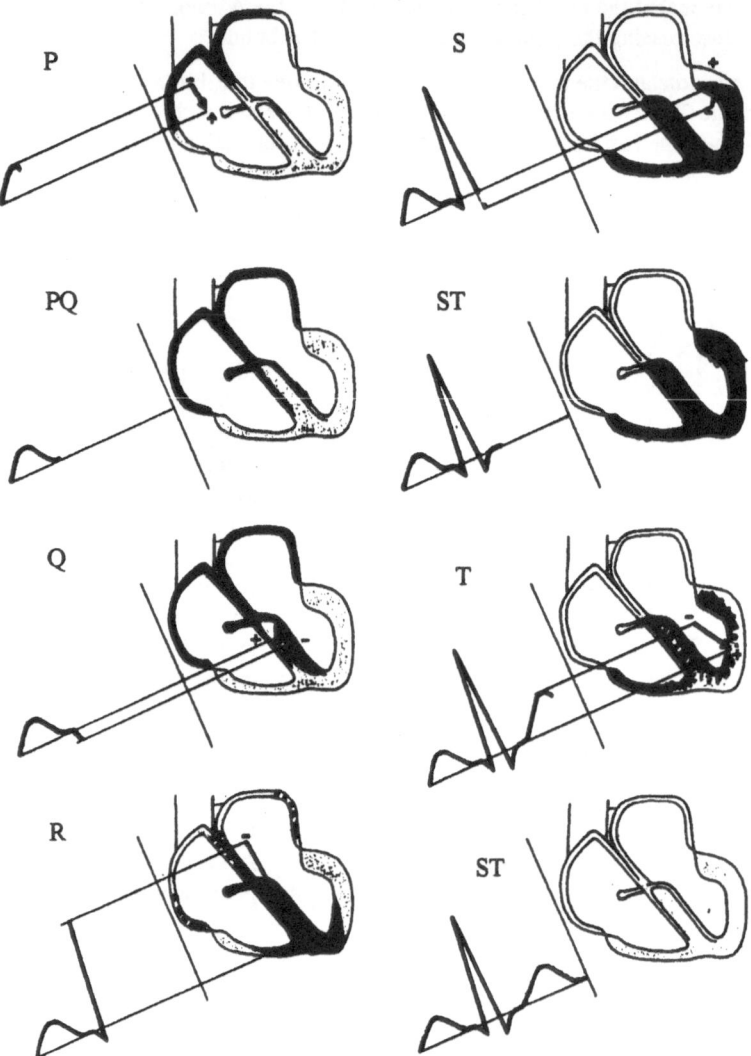

Figure 8 - The ECG and the heart activation sequence

4. Pathophysiology of the Heart

The reduction of the functionality can be due to changes in morphology (hypertrophy or dilatation of the ventricles) as well as due to reduced ability to perform mechanical work (reduced blood perfusion with reduction of oxygenation, reduced wall motility).

4.1. REDUCTION OF THE PUMPING FUNCTION

This general reduction of functionality follows an adaptation process to a chronic overload of the heart.

Pressure overload: the most important cause is an increased flow resistance in the blood flow (arterial or pulmonary hypertension). The pressure overload concerns mainly the ventricles.

Volume overload: the most common origin is an improper operation of the heart valves, leading to a situation where blood can oscillate between atria and ventricles.

The system has three ways to compensate the overload:

> 1 - increasing the amount of blood filling the heart
>
> 2 - change the contractility
>
> 3 - increase the muscle mass (hypertrophy)

1 -- An increased filling has a positive effect, shifting the working cycle for the myocardial fibers to a better working point; the increased amount of blood in the ventricle is pumped out, however, at the price of an increased amount of blood remaining in the ventricle.

2 -- an increase of the contractility is induced (in case of acute need) via the sympathicus system.

3 -- under pressure overload the mass of the myocardial muscle increases, maintaining the heart volume unchanged. Under volume overload the wall thickness increases together with the heart volume, while the geometrical relation between wall thickness and ventricle volume remains constant.

However, independent of the origin of the reduced functionality of the heart, one can observe a characteristic change in the geometry of the heart, called dilatation. Dilatation is characterized by a strong increase of the heart volume, whereas the wall thickness remains constant or decreases. The consequences of such generic reduction in heart functionality are different depending on the involved part of the heart.

Reduced functionality of the left ventricle:

> Reduced functionality of the left ventricle leads to a reduction of cerebral and muscular efficiency, to a reduction of the kidney functionality, to a reduction of resorption in the stomach and intestine, to accumulation of blood in the lung's vessels, and to lung's edema, eventually to an increasing load of the right ventricle.

Reduced functionality of the right ventricle:

> Reduced functionality of the right ventricle leads to a reduction of the perfusion of the lungs, blood accumulation in the body circulation,a nd slower blood flow velocity with increased tendency to thrombosis.

4.2 MYOCARDISCHEMIA

The myocardium is in an ischaemic condition if, as a consequence of inadequate blood perfusion of the myocardium, the available oxygen is not sufficient to cover the needs of the tissue. Microscopic structure changes following an ischemia develop within the first few minutes; the degenerative process develops rapidly and reaches the maximum after ca. 120 minutes. Under ischemic conditions the pumping function of the heart decays very rapidly, in a few heart beats.

Repetitive, in intensity and duration varying ischemic attacks are known as angina pectoris. Angina pectoris is in some way a premonition of a myocardial infarction and can be the first step in the generation of the infarction.

During the myocardial infarction a thrombus occludes one or more vessels that normally provide the heart with blood. The occlusion of these vessels produces an ischemic region with a series of irreversible modifications of the cardiac tissue. The chronic consequences of an infarction are a scar in the myocardial tissue; these scars influence not only the mechanical action of the heart (reduced wall motility, aneurysm) but also have a negative influence on the electrical control mechanism of the heart. In many cases the heart becomes prone to life-threatening tachyarrhythmias.

4.3 DISTURBANCES OF THE HEART EXCITATION, ARRHYTHMIAS.

The first regular pacemaker in the heart is, as seen before, the sinus node. Disturbances in the functionality of the sinus node generally influence directly the heart rate. However, due to the action of the AV node and of the following conduction system, very often the action is limited to the atria.

A much more widespread family of arrhythmias is due to disturbances of the conduction system; in this case two mechanisms are the basis:

■ a delay or a block of the conduction somewhere in the system, generally resulting in a reduction of the beat rate. Simple conduction blocks or delays can be located everywhere in the propagation path. However, blocks located functionally 'after' the AV node are true pathological situations due to an organic failure of parts of the conduction system.

■ a combination of unidirectional conduction and reentry, resulting in an increased beat rate until possible tachycardia and fibrillation. Reentry is a phenomenon similar to a positive feedback; it requires the existence of two pathways for the excitation, one of them being capable of retrograde conduction and both with an associated conduction velocity reduction. Under these circumstances, when the conduction time around the reentry loop exceeds the refractory time of the tissue involved, a self-sustained oscillation can be generated. Reentry loops can exist at a macroscopic level (involving large parts of the myocardium and of the conduction system) or at a microscopic level around scar tissue.

5. Magnetocardiography

Magnetocardiography is basically an electrophysiological study of the heart, since it concentrates on measurements of the electrical activity of the heart via the detection of the magnetic field generated by the intracellular current. On this one has the following possible applications of magnetocardiography to cardioelectrophysiological problems:

-- as an alternative to the standard ECG;

-- as an alternative to the so-called High Resolution ECG;

    ::   diagnostic of the heart conduction system

    ::   detection of the magnetic equivalent of 'late potentials'

-- as an alternative to Body Surface Potential Mapping in the preoperative localization of arrhythmogenic foci, especially for catheter ablation;

-- as an alternative to fetal ECG.

5.1. MAGNETOCARDIOGRAPY AS AN ALTERNATIVE TO THE STANDARD ECG

Although the official start of MCG was in the early sixties with the pioneering work of Baule and McFee, MCG started being a real tool of possible interest in cardiology only with the introduction of the SQUID magnetometer. The early studies, performed with single-channel instruments using second-order gradiometers in an unshielded environment were concentrated in exploiting the possibilities of the new technique in comparison with standard ECG. The major early work in this field had been done in the 1970's by the Finnish group led by Siltanen and Katila [1 to 4], and a substantial contribution was made in Japan.

All of these studies have demonstrated that MCG is practical and provides at least the same information that usually is obtained by the ECG; perhaps MCG has been shown to be superior in the diagnosis of special pathologies, such as ventricle hypertrophy. However, after some time period, all groups had the feeling, that independent of the small advantages compared to ECG the MCG did not have any chance of being considered as an alternative to ECG, due to the obvious increased major costs and major technical efforts necessary to perform MCG recordings compared with the standard ECG. Nowadays, with the introduction of high-T_c sensors, there are expectations that the cost of simple MCG systems can drop significantly. However, the MCG alternative to the standard ECG remains rather questionable.

5.2. MAGNETOCARDIOGRAPHY AS AN ALTERNATIVE TO HR- ECG

The pessimistic attitude towards MCG changed at the beginning of the 1980's when several groups, in Rome, Cleveland and Berlin began to consider High Resolution MCG as an alternative to HR-ECG, a modification of the ECG, that was then becoming increasingly important, thanks to a newly available very low-noise preamplifier, that has allowed recordings limited primarily only by the patient intrinsic noise. The HR-ECG concentrated mainly in the study of conduction disturbances (measurement of the His bundle activity, of the AV-node) and of fractionated activity in the late QRS end or ST segment (late potentials).

5.2.1. His bundle detection.

The classical measurement of the His bundle activity is an invasive measurement, performed with an electrical catheter during an electrophysiological study. Locating the tip of the electrode in proximity to the His bundle, it is possible to record a spike, a bump about 40 ms prior to the onset of ventricular activation (in a heart operating under normal conditions).

The first studies performed in Rome and in Cleveland were controversial: in Rome Barbanera *et al.* [5] recorded a very weak bump in the right temporal sequence followed by a ramp, a signal increasing almost steadily until the onset of ventricular activation; in Cleveland Farrell *et al.* [6] were able to record only the ramp like signals. Early attempts to model the generator of the signal including the complete conduction system (His bundle, left and right bundles and Purkinje fibers) have shown that the ramps were surely generated by the part of the conduction system subsequent to the actual His bundle [7].

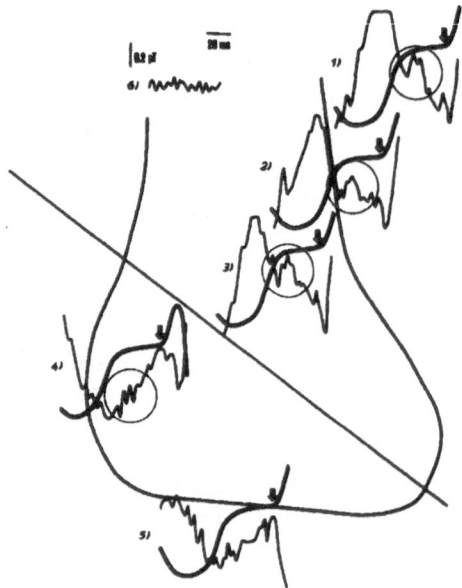

Figure 9 - First recording of 'bumps' in the PR segment, that could be associated with the activation of the His Bundle (after Fenici et al.)

In a series of studies, Erné *et al.* [8] later provided a clear cut answer to this problem. The results can be summarized as follows:

a - in the last part (lasting 30 to 40 ms) of the PR-segment, two types of signals are appreciable, ramps and bumps, after having performed the best possible subtraction of the atrial repolarization.

b - the ramps can be explained from their time evolution, as well as on the basis of simple or more complicated models as the signal generated by the propagation of the excitation along the conduction system.

c - the bumps are of more difficult interpretation, due to the fact that often more than one bump can be detected, not only at the beginning of the ramp but also during the ramp. A detailed model study by Erné *at al.* [9] has demonstrated the origin of the bump can be seen not only in the His bundle, but also in other anatomical structures, such as the bifurcation of the left and right bundle, strongly depending on the actual anatomical and physiological situation.

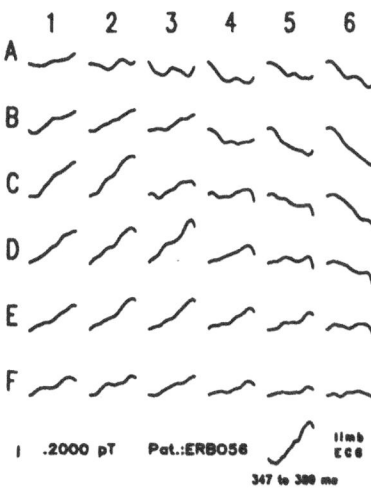

Figure 10 - MCG recording of the PR interval showing 'ramps' and 'bumps'

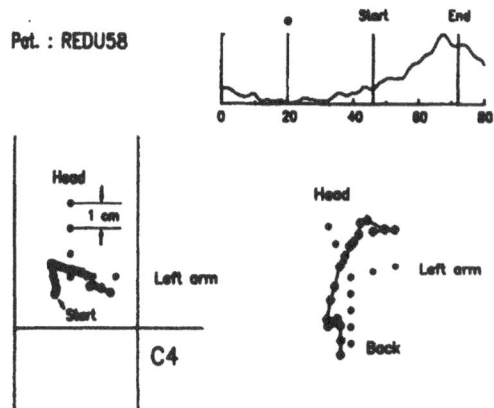

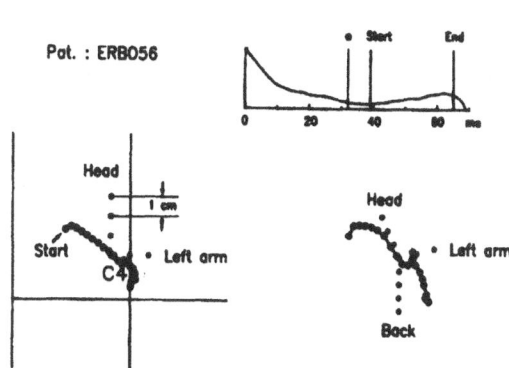

Figure 11 - Reconstruction from MCG data of the activation path of the His-Purkinje system in a normal subject.

On the basis of these results and on the basis of an experience with about 100 subject studied Erné *et al.* [8] arrived at the conclusion, that only in 67% of the subjects is it possible to obtain definitive information about the timing and functionality of the heart conduction system non-invasively via HR-MCG. However, they also demonstrated that in a large number of cases it was possible not only to localize the site of the His bundle, but also to reconstruct the excitation pathway.

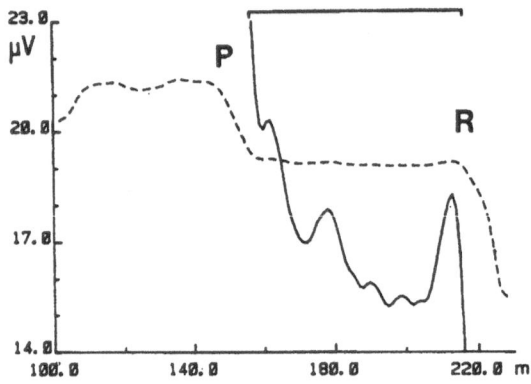

Figure 12 - High Resolution ECG recording (after averaging) of His bundle activity recorded from the body surface.

This was, however, only a demonstration of the sensitivity of magnetocardiography at that time: comparable results obtained with High Resolution Electrocardiographic recordings - see Fig. 12 - have precluded large-scale application of MCG in this field.

5.2.2 Detection of late fields

It is generally accepted that in patients affected by recurrent ventricular tachyarrhythmias, often a postexcitation syndrome occurs. It consists of a delayed activation of one or more regions of the ventricular myocardium, resulting in a fragmented activity appreciable after the end of the QRS complex. For these reasons the related signals are named late potentials in ECG and late fields in MCG: this phenomenon is important in cardiology because it is considered to be correlated to a susceptibility to life-threatening ventricular arrhythmias.

Late potentials can be recorded invasively from the epicardium or endocardium and non-invasively via body-surface recordings. However the measurement of such signals remains only a partially-solved problem due to the fact that intrinsically the late potentials can vary in amplitude, morphology and time correlation with respect to the preceding QRS complex. In 1983 Erné *et al.* [10] demonstrated for the first time that the magnetic equivalent of the late potentials can be successfully recorded.

The results of these first measurements, confirmed also by later findings, do not provide a clear cut answer for the role of MCG in recording late potentials: one aspect is positive, the ability to detect activity over a longer time interval than in the surface ECG (as shown in Fig 13), the other is negative, namely due to the strong dependence on the localization of the sensor, the quality of the recording is very sensitive to the position of the sensor. Extensive studies should be done with large multichannel systems, since in this case the localization problem becomes less relevant. However, experience shows that by considering more global effects such the spatio-temporal evolution of the heart field, one should have a better chance of success.

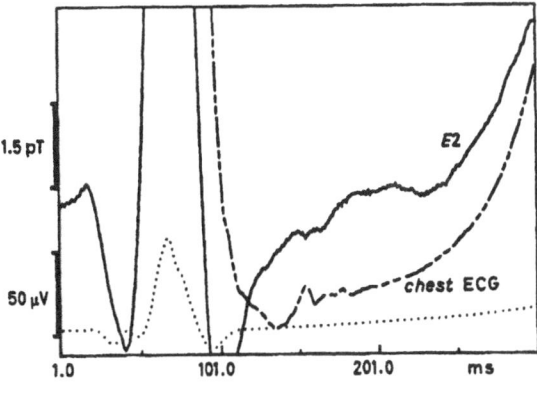

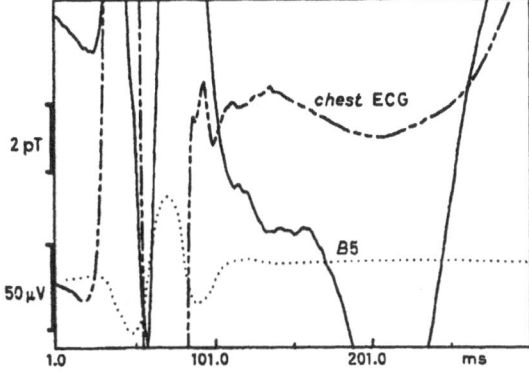

Figure 13 - MCG and ECG recordings of late fields and potentials in a patient with documented ventricular tach-yarrhythmias

5.3. PREOPERATIVE LOCALIZATION OF ARRHYTHMOGENIC FOCI

On the basis of the success in reconstructing the excitation pathway it was natural to seek other clinical uses of MCG to localize a signal source in a particular section of the heart provided the activated region is sufficiently small to be represented by a simple model. Acting on a suggestion by R.Fenici, Erné *et al.* [11] demonstrated in 1985 the feasibility of the localization of a special pathway involved in the generation of cardiac arrhythmias.

The basic idea utilized had been to avoid following the heart activity through the ar-rhythmic cycle and through the reentry path, but instead to concentrate on detecting and localizing a part of the abnormal conducting tissue (possibly at the beginning of the cycle). In this way the problem is reduced to the following question: can the current distribution at the beginning of an ectopic excitation be represented by a simple model, such as a current dipole? This question has been analyzed using a simplified model of the excitation wave front (specifically a discretized representation of a dipole layer) and a complex model of the myocardium including the anisotropy in propagation velocity and conductivity due to the layered structure of the tissue.

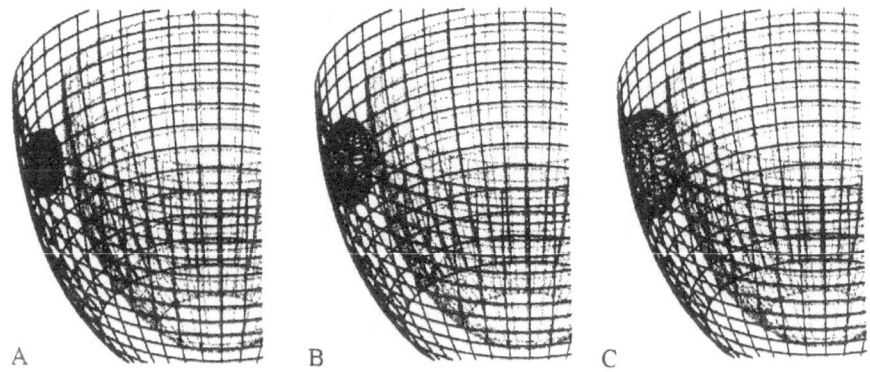

A   B   C

Figure 14 - Evolution of the excitation wave front in the myocardial tissue after an ectopic excitation: A 9 ms, B 12 ms and C 15 ms after beginning of the excitation

The model calculations have shown that generally, due to the tissue anisotropy, the total field measurable outside the thorax is almost dipolar, allowing the use of simple models to calculate the site of the start of the excitation. However, under certain conditions, especially when the wave front reaches the wall causing symmetry breaking quadrupolar components also can be observed.

Almost all groups involved in the development of preoperative MCG localization of arrhythmogenic foci started using Wolff-Parkinson-White (WPW) patients as a model. The Wolff-Parkinson-White syndrome is a congenital heart malformation. Due to a 'leak' in the insulating annulus fibrosus, which normally separate the atria from the ventricles, the ventricles are not only excited by the regular mechanism of ventricular excitation via conduction system, but they show also a 'preexcitation', due to an accessory pathway in the myocardial tissue. In most cases the existence of such an accessory pathway, also called a Kent bundle, can easily be detected in the ECG appearing as a kind of 'shoulder' at the onset of QRS, the so-called delta wave. Under normal circumstances the beat rate is not influenced by the presence of the excitation. However, under special conditions, the Kent bundle can start conduction in a retrograde fashion; in this case the normal excitation of the ventricle can pass through the atria and return to the ventricle via the normal conduction pathways. In such a case a macro reentry circuit is created with positive feedback, which generates a sustained asynchronous arrhythmia (decoupled to the heart pacemaker). Such an arrhythmia can become life threatening, so that WPW patients are candidates for ablation.

Ablation is a recent form of semi-invasive therapeutic intervention: using a specialized electrical catheter positioned in the vicinity of the end of the Kent bundle, electrical energy in the form of radio frequency current can be applied to destroy the tissue.

In the last decade all groups active in magnetocardiography have been involved in studies on the preoperative localization of arrhythmogenic foci, especially in patients with WPW syndrome. In this way a large number of successful localizations and ablations have been reported in the literature. However, due to the differences in the instrumentation used and in the different approaches, as well as large-scale validation, remain important goals of magnetocardiography.

5.4. FETAL MAGNETOCARDIOGRAPHY

The fetal magnetocardiogram is one of the first biomagnetic signals reported: Kariniemi *et al.* reported in 1974 the first successful measurements of the magnetic signals related to the cardiac activity of the fetal heart [11]. Fetal MCG has been considered from the beginning an interesting application for biomagnetic instrumentation because it presents some advantages compared to electrical measurements. Fetal ECG is noisy, with a strong superimposition of the maternal ECG, and, especially in a 'critical' period between the 25 and 36 week, is of very limited amplitude due to the isolating action of the vernix around the fetus. Although the MCG cannot be called noise free, its spatial resolution power permits the measurement of the fetal MCG with a minimal influence of the maternal MCG. As the MCG is not strongly influenced by conduction problems in the tissue surrounding the source, the fetal MCG can be recorded easily between week 25 and week 36 of the pregnancy.

The instantaneous heart rate is used as a tool for fetal monitoring during pregnancy and for neonatal intensive care. Accelerations of the fetal heart rate due to sympathetic reflexes are considered signs of fetal well being. Conversely deceleration of the fetal heart rate due to the vagal reflex is considered a sign of imminent fetal jeopardy.

Another parameter is the heart rate variability, which is an expression of the functioning of the central nervous system. The variability appears around the 17th gestation week and seems to grow during gestation.

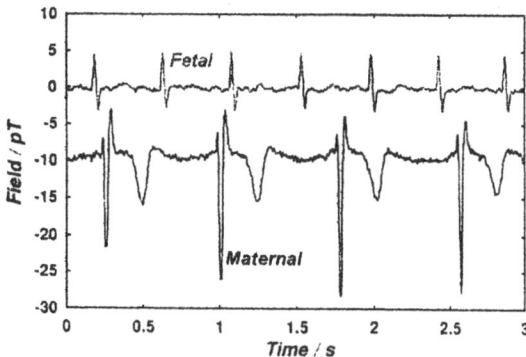

Figure 15 - Fetal and maternal MCG recorded with a high T_c sensor [12]

410

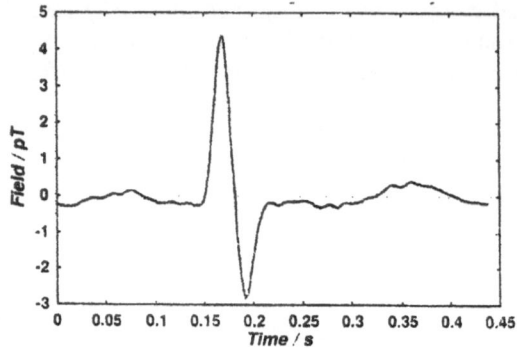

Figure 16 - Averaged fetal MCG recorded with a high T_c sensor [12]

Considerable pionieering work was performed by the Finnish group on this topic [e.g. 13]. However, fetal magnetocardiography, despite of such successful work, never became an accepted tool in clinical practice. Many factors were probably responsible: the instrumentation was complex, noisy and required skilled personnel. Furthermore it was not trivial to find a way to integrate the new information into the clinical management of critical phases of the pregnancy or of the perinatal phase.

A new start in the field was made about 15 years later by the group of Wakai [14]. Using modern instrumentation, they were able to offer to clinicians not only beat rate and beat rate variability information, but also information about the morphology of the fetal MCG [15]. In the last time, the introduction of High Tc sensors has provide one step more to bring fetal MCG into the clinic.

To characterize the state of the art in Fig. 15 and 16 illustrate examples of the work of Donaldson's group at the University of Strathclyde are shown [more in the same volume]. In both cases the measurements have been performed using the new generation of SQUID sensors fabricated using high-T_c materials: In Fig. 15 it is possible to appreciate the clear separation between the maternal and fetal signals, in Fig. 16 it is evident that the quality of the fetal MCG allows also for a morphological interpretation.

6. Conclusion

After reviewing the developments in magnetocardiography over the last two decades, one can conclude that there are interesting application possibilities despite the fact that until now the MCG has generally not been considered a clinical tool. The main reason, in the authors opinion, is the lack of instrumentation developed for clinical application: it is not a question of number of sensors, mapped surface and noise (although all these parameters play an important role), but it is much more a matter of patient handling, and measurement and analysis time. Only when it is possible to perform complete patient measurements and data analysis at a rate of few patients per hour, will one be able to achieve the patient statistics required to validate this new cardiological methodology. With the availability of high quality SQUID's, this situation is becoming a distinct possibility.

7. References

1. Saarinen, M., Karp, P.J., Katila, T.E. and Siltanen, P. (1974) The magnetocardiogram in cardiac disorders. *Cardiovascular Res.* **8**, 820-834

2. Katila, T.E., Karp, P.J. (1982) Magnetocardiography: morphology and multipole presentation, in Williamson, S.J., Romani, G.L., Kaufman, L., Modena, I., (eds.), Biomagnetism: An Interdisciplinary Approach, Plenum Press, New York and London, 237-261

3. Karp, P.J., Katila, T.E., Saarinen, M., Siltanen, P. and Varpula, T.T., (1978) Etude comparative des magnetocardiogrammes normaux et pathologiques, *Ann. Cardiol. Ageiol.* **27**, 65-70

4. Karp, P.J., Katila, T.E., Saarinen, M., Siltanen, P. and Varpula, T.T., (1980) The normal human magnetocardiogramm: A multipole analysis. *Circ. Res.* **47**, 117-130

5. Barbanera, S:, Carelli, P., Leoni, R., Romani, G.L., Bordoni, F., Modena, I., Fenici, R., Teppilli, P., (1981) Magnetocardiographic study of some human cardiac electrophysiological phenomena: preliminary observations, in Erné, S.N., Hahlbohm, H.-D., Lübbig, H. (eds.), Biomagnetism, Walter de Gruyter, Berlin and New York, 282-290

6. Farrell, D.E., Tripp, J.H. and VanDoren, C.L. (1981) High resolution cardiomagnetism, in Erné, S.N., Hahlbohm, H.-D., Lübbig, H. (eds.), Biomagnetism, Walter de Gruyter, Berlin and New York, 273-281

7. Fenici, R.R., Romani, G.L., Leoni, R., (1983) Magnetic Measurements and Modeling for the Investigation of the Human-Heart Conduction System, *Il Nuovo Cimento* **2D**, 280-290

8. Erné, S.N., Fenici, R.R., Hahlbohm, H.-D., Korsukewitz, J., Lehmann, H.P., Uchikawa, Y., (1984) Magnetocardiographic study of the PR segment of normals, in Weinberg, H., Stroink, G. and Katila, T. (eds.), Biomagnetism: Application & Theory, Pergamon Press, New-York, 132-136

9. Erné, S.N., Lehmann, H.P., Masselli, M., Uchikawa, Y., (1984) Modelling of the His-Purkinje heart conduction system, in Weinberg, H., Stroink, G. and Katila, T. (eds.), Biomagnetism: Application & Theory, Pergamon Press, New-York, 126-131

10. Erné, S.N., Fenici, R.R., Hahlbohm, H.-D., Jaszczuk, W., Lehmann, H.P., Masselli, M., (1983) High-Resolution Magnetocardiographic Recordings of the St segment in Patients with Late Potentials, *Il Nuovo Cimento* **2D**, 340-345

11. Kariniemi, V., Ahopelto, J., Karp, P.J., Katila, T.E., (1974) The fetal magnetocardiogram, *J. Perinat. Med.* **2**, 214-216

412

12. Donaldson, G., private communication

13. Kariniemi, V. and Hukkinen, K. (1977) Quantification of fetal heart rate variability by magnetocardiography and direct electrocardiography, *Am. J. Obstet. Gynecol.* **128**, 526-530

14. Wakai, R.T., Wang, M., Pedron, S.L., Reid, D.L., Martin, C.B. (1993) Spectral analysis of antepartum fetal heart rate variability from fetal magnetocardiogram recordings, *Early Human Development* **35**, 15-24

15. Wakai, R.T., Wang, M., Martin, C.B. (1994) Spatiotemporal properties of the fetal magnetocardiogram, *Am. J. Obstet. Gynecol.* **170**, 770-776

MAGNETOCARDIOGRAPHIC AND ELECTROCARDIOGRAPHIC MAPPING STUDIES

G. STROINK[1,2], M.J.R. LAMOTHE[1] and M.J. GARDNER[2,3]
Department of Physics[1], Physiology and Biophysics[2] and Medicine[3]
Dalhousie University
Halifax NS, Canada, B3H-3J5

Abstract The electrical activity of the heart can be monitored electrically with electrodes or magnetically using SQUIDs. With multiple measuring sites, covering a significant portion of the upper torso, Body Surface Potential Maps (BSPMs) or Magnetic Field Maps (MFMs) can be constructed every 1 or 2 ms, providing detailed temporal and spatial information about cardiac electrical activity. Several methods are available to extract clinically useful parameters from this wealth of information. Using inverse solutions, cardiac function can be assessed, and cardiac events located. When such an event is implicated in arrhythmia, knowledge of the location of this site can be used to guide the catheter toward it for possible ablation. Lately, the BSPM technique has been used to record maps that result from endocardial catheter pacing. The resulting BSPM is characteristic for the pacing site, and when similar to the surface maps obtained during spontaneous arrhythmogenic events, the pacing catheter is assumed to be close to the cardiac tissue initiating the arrhythmia. This method of localization provides an alternative to the traditional inverse solutions based on numerical methods. A similar technique of matching patterns also can be used with MFMs. We review the different localization techniques that use MFMs and/or BSPMs. Such techniques, together with MRI, are now under development to provide the clinician with electrical images of the heart surface for the assessment of cardiac function. We also summarize results of the analysis of MFMs and BSPMs of the same patient or patient group with an emphasis on finding landmarks in such maps that are predictors of clinical cardiac events. The results obtained so far are encouraging for both BSPM and MFM. Systematic multichannel MFM studies with substantial patient populations are needed to demonstrate the clinical importance of cardiac magnetic field mapping. This new mapping method, made possible by recent developments in SQUID technology, could provide, by itself, or together with BSPM, a powerful, quick, non-invasive method to image electrical activity of the heart to assist in clinical diagnosis.

1. Introduction

The periodic contractions of cardiac muscle are triggered by depolarization of the membrane potential of cardiac muscle cells. These changes in membrane potential generate currents in the surrounding tissue. The currents flowing through the resistive body volume (volume currents) create potential differences on the body surface which can be displayed as electrocardiograms (ECGs). The currents also contribute to magnetic fields measured near the torso surface, recorded as magnetocardiograms (MCG). These potentials or magnetic fields can be recorded by an array of electrodes or SQUIDs, covering a large portion of the human thorax. From such multiple ECG and MCG recordings we can construct time sequences of Body Surface Potential Maps (BSPM) or Magnetic Field Maps (MFM), respectively. During a typical 1 second heart cycle, such maps can be produced every 1 or 2 ms. Both types of displays are used to assess cardiac

413

H. Weinstock (ed.), SQUID Sensors: Fundamentals, Fabrication and Applications, 413–444.
© *1996 Kluwer Academic Publishers.*

function, diagnose cardiac abnormalities and even locate the region of origin of such abnormalities.

Possibly the first BSPM was obtained by Nahum *et al.* in 1951 [1]. They used a single electrode to sample each point on the torso sequentially. In 1969, using a similar, single-probe technique, Cohen and Chandler [2] published temporal MCG traces displayed on a grid of measurement locations near the torso. The first MFM, displaying contour lines, based on single probe measurements, was published in 1981 [3], and an analysis of measured BSPMs *and* MFMs on the same patients was published by us around 1984 [4]. Results based on maps obtained with a truly multichannel (larger than 30 leads measured simultaneously) system appear for BSPM in 1978 [5] and for MFM in 1990 [6]. Both, the Physikalish-Technische Bundesanstalt at Berlin and the BioMag Laboratory at the Helsinki University Central Hospital have now the capability to routinely measure both BSPM and MFM on the same patient with multichannel systems [7].

From this short history it is clear that body surface mapping of both the potential and magnetic field distributions of the same electrocardiological events in the same patient with ms resolution is a recent development. The availability of multichannel SQUID systems and the ability to collect and analyze large amounts of data have made such detailed measurements possible.

We shall provide a theoretical background for the potential and magnetic fields observed, compare the different mapping techniques, and report on studies that use similar patient groups to extract information from the two data sets. Recent reviews on MCG, by itself, have been written by Siltanen [8], Fenici *et al.* [9], Stroink [10] and Nakaya and Mori [11] and on BSPM by Flowers and Horan [12], as well as by Green and Abildskov [13].

2. The Forward Solution

The forward solution calculates, at the body surface, the potential or magnetic field produced by the cardiac sources. Forward solutions are essential for our understanding of the relationship between the cardiac electrophysiology and the measured maps. As we shall see later, accurate theoretical models also are needed to use such maps to obtain information about the cardiac generators (the inverse solution). Any such mathematical model of cardiac activity requires a source model (the cardiac generators) and a volume conductor model (volume in which these sources operate).

2.1. SINGLE CURRENT DIPOLE IN A HOMOGENEOUS CONDUCTIVE MEDIUM

The simplest forward model assumes that, at a particular moment during the heart cycle, the complex and often distributed cardiac sources can be lumped into a single equivalent dipole located at the center of the heart. It also assumes that the body can be represented by an infinite or semi-infinite homogeneous volume conductor. This source and volume conductor model [14] has been extremely useful in predicting the main features of the MCG and ECG at specific lead locations; it also provides a general understanding of the BSPM and MFM measured.

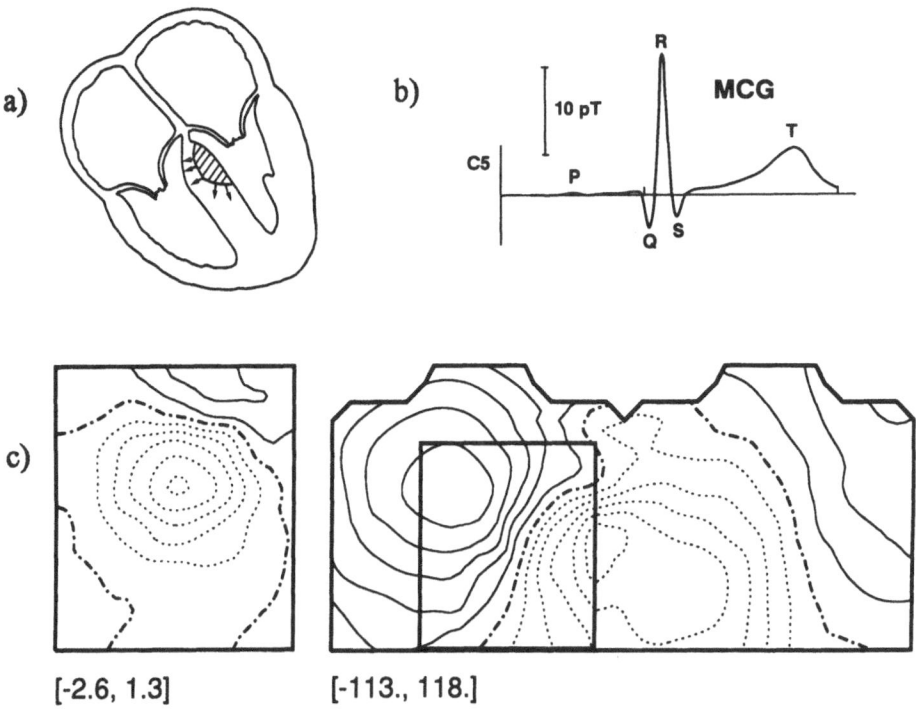

Figure 1: The onset of the ventricular depolarisation wavefront at the interventricular septum (a) produces the Q-wave in the ECG and MCG (b). Combining all magnetic field values, B_z, at 56 MCG measuring locations in a plane near the anterior surface at this moment, produces the Q-wave Magnetic Field Map (MFM) shown on the left hand side of panel (c) as contour plots with equal steps. The numbers below the map give the extrema in pT. Also shown is the 117-lead Body Surface Potential Map (BSPM) on the right hand side of panel (c) measured at this same time. The numbers below that map give the extrema in µV. The BSPM is a projection on a cylinder of the potential measured on the body surface. The cylinder is unwrapped by cutting along a line from the right shoulder to the right hip. Consequently, the left part of this map represents the anterior torso. The positive region (solid lines) at the right of this map continues as the positive region on the left. The rectangular area in the BSPM is the area where the MFM (left) is measured. Details of the map format also are given in Figure 9.

Figure 1a is a simplified picture of the onset of the left ventricular depolarization wavefront originating at the interventricular septum. Because of the relatively small amount of tissue involved, it is not unreasonable to represent this wavefront by a single equivalent current dipole which points to the right side of the body. This dipole creates a potential difference between the wrists which, by convention, is negative. It also creates a magnetic field that points into the body (defined as positive) above the location of this current dipole and out of the body below it. This onset of ventricular depolarisation is labeled Q in the MCG shown in Figure 1b. The magnetic field at this instant is displayed as negative because it is measured near the torso below the level of the current dipole and consequently points out of the body. A typical MFM and BSPM measured during this onset of ventricular depolarisation is shown in Figure 1c. The MFM is measured in a plane near and parallel to the chest; the BSPM covers both the anterior and posterior body surface. Both maps show a dipolar pattern (one maximum and one minimum) typical for a single current dipole vector that represents the activity shown in Figure 1a.

Figure 2a shows the result of an analytical solution [14] for the potential V and the magnetic field B_z [1] due to a current dipole in an infinite homogeneous conductive medium. The dipole ($-p_x$) points to the right side of the body. Comparing this figure with the measurements on a subject (Figure 1c), one sees that in the case of this localized, small wavefront, the maps predicted by this simple model show similarities to the maps actually measured. However, the zero contour line (dash-dotted line) of the measured maps is not at the same location as that of the calculated maps, and the relative values of the maximum and minimum are different. So even for this relatively localized event, the onset of ventricular depolarization, the model predicts only the general features. Improvements in this simple model are needed to obtain better agreement with the measurements.

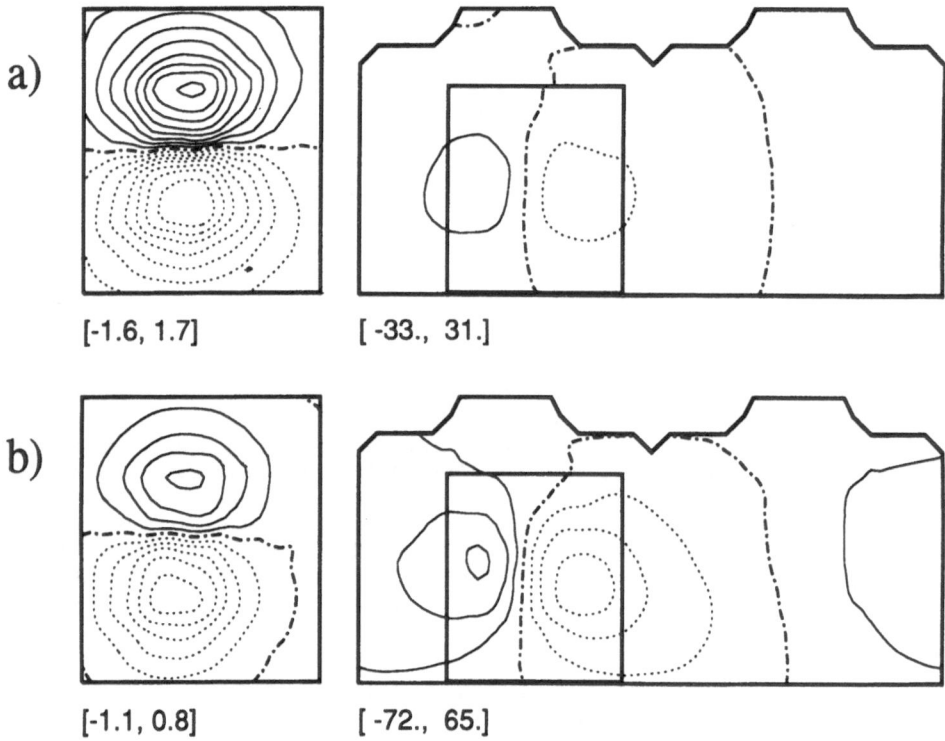

a) [-1.6, 1.7]   [-33., 31.]

b) [-1.1, 0.8]   [-72., 65.]

Fig. 2 The magnetic field B_z (left) and the potential V (right) due to a current dipole of 1 μA·m located at the interventricular septum (Q onset) pointing in the - x direction (towards the right side of the body). In the upper figure (a) the torso, the torso compartments and the region outside the torso all have the same conductivities. That is, the dipole is in an infinite homogeneous conductive medium. In the bottom figure (b) the torso, lungs and heart have different, realistic conductivities. The conductivity outside the torso is zero. The rectangular box on the anterior surface of the potential map (right side of the figure) indicates where the magnetic field, shown at the left side of this figure, is calculated. The values under the maps are the extrema, in pT (left figure) and μV (right figure).

---

[1] The z -axis is pointing out of the body, perpendicular to the anterior plane. The y-axis points from toe to head, the x-axis from the right to the left arm.

2.2. SINGLE CURRENT DIPOLE IN A PIECEWISE HOMOGENEOUS CONDUC-
TIVE MEDIUM

In Figure 2b we show also the theoretical results for the same current dipole at the same location, but now placed in a realistic, male torso model with different conductivities for the torso, lungs and blood masses. These different conductivities will distort the volume currents and consequently influence the potentials on and the magnetic fields near the torso. No analytical method exists to calculate the potential and magnetic field of a piecewise homogeneous torso model, and we need to use a numerical technique like the Boundary Element Method (BEM) [15, 16]. Improvements to this technique as it applies to the calculation of cardiac surface maps have recently been published [17]. Figure 2b shows that the resulting patterns are less symmetric than those obtained in Figure 2a. Taking into account that the dipole direction is somewhat different and that the torso used in the calculations is not that of the subject measured, one observes that the patterns start to resemble the real measurements shown in Figure 1c. It is interesting to note that even with the different conductivities, the patterns keep their dipolar features.

Experience shows that under the special circumstance that the cardiac sources can be represented by a single current dipole in the x-y plane, and provided that this dipole is relatively near the anterior surface, the maps calculated with the piecewise homogeneous torso model show similarities to those obtained with the infinite homogeneous volume conductor.

The situation is different for a dipole pointing out of the body (p_z). Such a source, in an infinite homogeneous volume conductor, has no magnetic field component in the z direction, B_z, is zero. Figure 3a shows that B_z has non-zero, but relatively small values in a realistic torso model. Figure 3a also shows that in a BSPM, p_z produces large potentials in the direction it is pointing, the anterior surface. A more detailed study of the potential and magnetic fields of a current dipole at many locations in the heart shows that, on average, the magnetic field maximum due to p_z dipoles is about 1/4 that of the tangential sources (p_x and p_y). In contrast, the potential maxima for p_z dipoles are approximately 1.5 times greater than that of the tangential sources. Based on this, one can conclude that the MFM samples primarily the p_x and p_y components, and the BSPM samples primarily the p_z component [10, 18, 19]. This is further illustrated in Figure 3b where we have modeled a dipole pointing between the -x and z direction. As can be seen the magnetic map emphasizes the x component (shown in Figure 2b), whereas the BSPM emphasizes the z component of the dipole (shown in Figure 3a). The data suggest that to determine the dipole components from maps, it is advantageous to have access to both the MFM and BSPM.

Numerical models, such as the BEM or the Finite Element Method (FEM) are essential not only for modeling p_z sources, but also when one wants to model sources that are located posteriorly in the heart (deep sources). The large changes in the conductivity caused by the proximity of the poorly-conducting lung and the highly-conductive bloodmass result in dramatic changes in the volume currents that influence both the potentials and the magnetic fields at the surface. The infinite homogeneous volume conductor model, which ignores all this, does not suffice here.

418

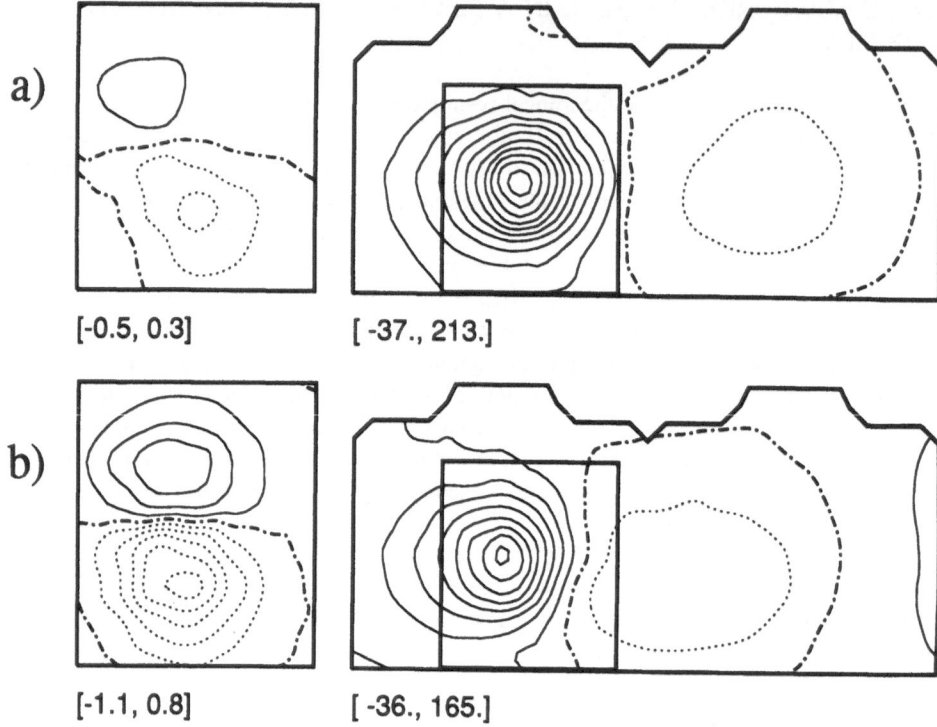

Figure 3: The magnetic field B$_z$ (left) and the potential V (right) due to a current dipole of 1 µA·m located at the interventricular septum in a realistic torso, pointing in the z-direction (a) and midway between the -x and z direction (b). In this forward model, torso, lungs and heart have realistic conductivities, the same as used to calculate the maps in Figure 2b. The map layout is the same as in Figure 2.

Several authors [19, 20] have compared the sensitivity of the forward solution of electric potentials and magnetic fields to the inclusion of internal organs in a realistic torso model. If either the BSPM or MFM is less sensitive to the presence of inner organs then this would indicate less inter-individual variability of the map measured by that technique. There is general agreement that when using a dipolar source model and a realistic torso model, on average there is no clear advantage to measuring either the electric potential or magnetic field with respect to their sensitivity to the different conductivities of the organ. In fact, several authors strongly recommend using individualized torsos that include the lung and heart boundaries (as derived from the MRI of that person) to obtain accurate forward solutions for the BSPM and MFM [19, 21].

2.3. MORE COMPLICATED SOURCE MODELS

Several earlier studies [22] characterize the heart's electrical activity by numerous electrical dipole sources which are activated in sequence based on measured activation wavefronts in isolated human hearts. Such models also use realistic torso models with a homogeneous conductivity to determine the potential and magnetic fields [23] at the torso surface. The agreement with the measured data throughout the complex ventricular depolarization (QRS) is only superficially good. Van Oosterom *et al.* [24] used the classical uniform double-layer model for ventricular depolarization. This source model consists of a surface of uniform current-dipole density at the boundary between the activated part of the ventricles and the remainder. This activation wavefront propagates in time over the ventricles. Potentials and magnetic fields calculated with this source model generate realistic ECG and MCG QRS waveforms on a torso surface. Using multiple lead ECG subject data and this source model, they concluded that the QRS MCG can be predicted to a large degree by the ECG measurements, and therefore, that the MCG contains little information that can not be obtained from the ECG [24]. Using more general arguments this conclusion was also reached earlier by Plonsey [25] based on the same source model.

However, there is growing experimental evidence that the isotropic conduction volumes and the source models considered so far are inadequate to explain the complex activation wavefronts. Macroscopic [26] and microscopic models of cardiac tissue with anisotropic conductivity (allowing the source currents to follow the fiber direction in the ventricles) [27] have illustrated the importance of anisotropy in the genesis of the ECG and MCG, and consequently in the BSPM and MFM. Such simulations are based on anisotropic bidomain theory, and the interaction between neighboring cells is managed by cellular automata [28]. As an illustration of such detailed models, we show the BSPM results by Hren *et al.* [29] in Figure 4. The activation sequence is initiated at a distinct endocardial site in the left ventricle and allowed to propagate by means of cellular automata in a three-dimensional ventricular model containing about 1.7 million cells. Each cell has a characteristic action potential and fiber direction assigned to it. Similarly, MFMs can be calculated and will soon appear in the literature. Initial results show that the magnetic field is more sensitive than the electrical potential to the underlying fiber orientation [30]. Anisotropic conduction shows up most clearly in high-resolution measurements with a MicroSQUID positioned within millimeters of 2-D slices of cardiac tissue [31, 38].

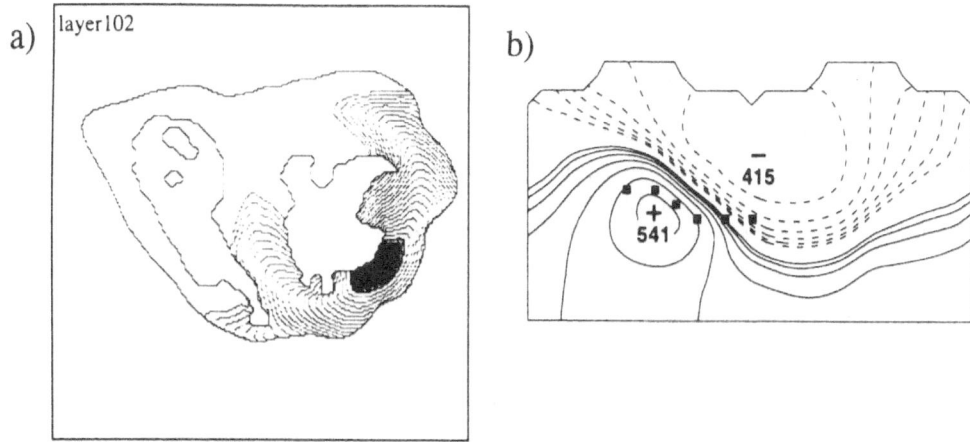

Figure 4. a) Isochrones of the simulated activation sequence at a left lateral endocardial pacing site. Isochrones are displayed at 2 ms intervals in a cross-section through the heart about 5 cm above the apex. b) Body surface potential maps simulated at 50 ms after the ectopic activation started. The extrema are in μV.

The detailed forward models help us understand the BSPM and MFM obtained, for example, during endocardial pacing, crucial in the treatment of patients with ventricular tachycardia [32]. Endocardial pacing makes use of catheters to start the depolarization at the pacing site; it provides information about the local wavefront propagation properties. Such models also give us insight into the fundamental differences between BSPM and MFM, and how to make optimal use of the information (resulting from the different sensitivities of these methods to propagating currents) in the assessment of cardiac disease.

That such differences exist also has been demonstrated by our analysis of BSPM and MFM subject data, which shows quantitative evidence for information content in MFM which is independent of BSPM [4]. The nature of this ECG-independent component in the MCG can be explored with such detailed models.

3. Inverse Solutions Applied to Model BSPM and MFM Data.

In an inverse solution one uses the measured maps to characterize and locate the cardiac sources that generate the maps. We have addressed in several ways the question whether the BSPMs or MFMs, or perhaps both, will yield the most accurate localization. Good accuracy is required if one wants to use the results of such a localization to guide a catheter as close as possible to the source of the arrhythmogenic signals on the endocardium for possible ablation. Another potential application of this inverse solution is to use the information from the BSPM and MFM to display the activation wavefront on the epicardial surface. The epicardial surface can be obtained from MRI. This imaging of the electrical potential or current density on this surface during the heart cycle would be a powerful tool in assessing cardiac function. In one example, a lack of such currents is expected in cardiac regions that have suffered an infarction. Determining non-invasively the location, size and electrical properties of the infarcted region and the ischemic region immediately surrounding it, would have clinical applications.

To test and improve the inverse solution algorithm, one can use as input, model data of calculated MFM or BSPM as described in the previous sections. The advantage of this method is that the sources, torso boundaries and lead positions, as well as other "experimental" parameters, are well defined. The disadvantage of using model data is obvious as well. For measured body surface maps, the actual source characteristics, the volume conductor, the exact location of the sensors and the noise profile of the measurement are not always accurately known. Consequently, any conclusion based on the accuracy of such modeling with the forward and inverse solution that ignores some or all these factors, is limited in scope. However, such studies make us aware of the intrinsic limitations of such inverse solutions for both BSPM and MFM.

In one such study [33] we calculated, using the Boundary Element Method (BEM), the BSPM and MFM due to a single dipole, located at one of 17 different locations in the heart mass; ten of these locations are in the plane of the atrio-ventricular (AV) ring and pointing perpendicular to this plane; five sources are "deep sources" located near the posterior area of the heart. The ten sites selected on the AV ring are those originally proposed as possible epicardial preexcitation sites in Wolff-Parkinson-White Syndrome patients. Their location is well documented [34]. The potentials and normal component of the magnetic field (B_n) were calculated at each of 117 lead positions on the anterior and posterior body surface of the upper torso. We also calculated the MFM (B_z) at 56 positions in a plane near the anterior surface to simulate more conventional measurements as can be obtained with a multi-channel SQUID system. The realistic torso model used in the forward calculations included both lungs and blood masses. The inverse solution uses a variety of volume conductor models. In addition, in one series of tests, noise, 10% of the peak values, was added to the BSPM and MFM model data. Some of the results are shown in Table 1.

Table 1: The average localization error (in cm) and standard deviation (in brackets) of 17 dipole sources in the cardiac region (all), with five dipole sources in the chest (deep). The forward solution uses a torso model that included both lungs and blood masses; the inverse solution was calculated with the infinite homogeneous model (i), torso with lungs (tl) and blood (tlb) [33].

	source	i (deep)	tl (deep)	tl (all)	tlb (all+noise)
BSPM	p_x	1.00(0.20)	0.13(0.06)	0.20(0.10)	0.52(0.38)
	p_y	0.84(0.55)	0.32(0.10)	0.25(0.09)	0.49(0.33)
	p_z	0.96(0.65)	0.24(0.14)	0.23(0.13)	0.41(0.26)
MFM	p_x	1.87(0.37)	0.95(0.30)	0.53(0.35)	0.51(0.45)
	p_y	2.02(0.89)	0.65(0.24)	0.43(0.23)	0.52(0.49)
	p_z	2.85(0.74)	0.66(0.38)	0.60(0.34)	0.85(0.59)

The results show that for noise-free BSPM and MFM data, substantial improvements in dipole localization, particularly for deep sources, can be achieved by using a realistic volume conductor (as opposed to an infinite homogeneous model) in the inverse solution. This conclusion also was reached previously with similar calculations using MFM model data for measurements in the anterior plane [35, 36]. As can be anticipated from the differences in signal strength of different dipole components (discussed earlier), the BSPM is more accurate for localizing p_z dipoles than is the MFM. With noisy data the accuracies of the BSPM and MFM are comparable for p_x and p_y sources. Reducing the MFM measuring points from 117 to 56, all in the anterior plane, increases the localization error in the second-to-last column by about 0.2 cm [35]. In all cases the average localization error, when using realistic torsos, is well below 1 cm, as compared typically to 1 cm or greater for the infinite homogeneous model. In practice, a localization error of less than 1 cm is probably the best that can be achieved when imaging current sources because of errors in postioning the subject relative to the measuring probes or leads, and the uncertainty in the position of the heart and other organs relative to such leads.

A more sophisticated approach to test localization accuracy and imaging capabilities of the BSPM and MFM, is to use different source models in the forward and inverse solutions. Detailed maps, both MFM (anterior area, 56 positions) and BSPM (full torso, 117 leads), were calculated [37] based on an anisotropic propagation model of the human ventricular myocardium, embedded in a homogeneous volume-conductor model of the human torso. The model simulates the activation wavefront near an accessory pathway (AP) across the atrio-ventricular (AV) ring 40 ms after onset for seven (of the ten earlier mentioned) different locations. No noise was added. Using an inverse solution that uses a current dipole as the source, and a piecewise homogeneous torso (with lungs and blood masses) as part of the inverse solution, we calculated [18] the dipole position and compared its location with that of the known position of the accessory pathway. The results are shown in Table 2.

Table 2: Average inverse solution localization errors (in cm) using an infinite medium (i) and homogeneous torso (ht) in the inverse solution [18].

	i	ht
BSPM	1.72(0.49)	1.42(0.45)
MFM	2.55(0.98)	1.04(0.37)

Comparing Tables 1 and 2 we see that an inverse solution that fits a dipole source to a map calculated with a more realistic source model results in larger localization errors. It is interesting to note that under these circumstances the MFM inverse solution, although covering a smaller map area involving fewer leads, gives (on average) more accurate results than the BSPM.

So far we have considered inverse solutions which represent the total activity of the heart at a given moment by a single equivalent current dipole. It is clear that this representation is inadequate in describing the complex source configurations found in an active heart. Several techniques have been proposed to find the current distribution on the epicardial surface using the BSPM or MFM maps [38]. In one such approach we represented the epicardial surface, obtained from MR images of a subject, by 176 triangles. The forward and inverse solutions used the individualized homogeneous torso of the subject, which also were obtained from MR images [39]. Gaussian noise with a standard deviation of 10% and 50% of the peak sensor signal was added to the forward calculated 117-lead BSPM (posterior and anterior body surface) and 56-lead MFM (anterior surface). For test purposes this forward solution was based on a single dipole located at any of 10 AP locations on the AV-ring discussed earlier. The inverse solution, based on the minimum-norm method [40], allows for two orthogonal dipoles to lie in each triangle. Consequently, this imaging program allows for 352 unknown dipoles. The torso and epicardial boundaries provide the physiological and anatomical constraints needed to solve the underdetermined system. While the minimum-norm method does an adequate job in recovering such dipoles using either MFM and BSPM, substantial improvements can be achieved by using a probability-based approach. This method weighs the best fitted dipole at each epicardial triangle by the probability that that dipole generates the measured field or potential map. The results for a particular accessory pathway (left anterior) is shown in Figure 5 [41]. Similar algorithms applied to MCG also have been presented by Graumann et al. [42] and Fuchs et al. [43]. It is anticipated that this type of current-density imaging, displaying complicated current configurations on the epicardium, can provide sufficient detail to characterize cardiac dysfunction.

In the BSPM literature substantial progress is reported on inverse solutions which do not use a particular source model. Several researchers have shown that reliable isochrones on the epicardium can be calculated directly from BSPM data (sometimes referred to as myocardial activation imaging). Knowing how the activation wavefront propagates in time over the epicardial surface would reveal the location of the onset of tachycardias. However, the mathematical methods for relating heart and body-surface potentials through transfer functions, are unstable when small amounts of noise are added and often lack adequate spatial resolution. Despite the difficulties, good results have been obtained

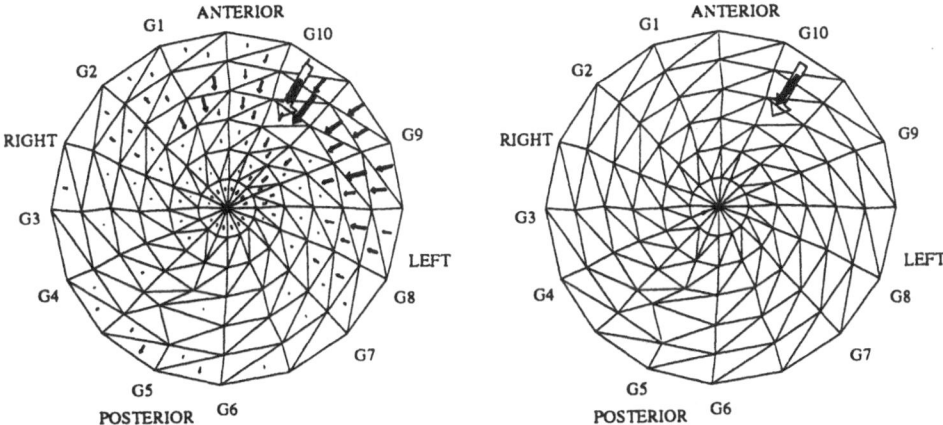

Figure 5. Distributed current solutions (solid arrows) calculated by using the MFM generated near the anterior surface by a source dipole (large arrow) with 10% random noise added. The apex of the heart was positioned at the origin of the this polar projection of the epicardial ventricular surface. The circumference represents the AV-ring. a) a "fiber constrained solution" using a single dipole on each epicardial triangle oriented parallel to the heart axis. b) a current probability solution with 10 repeat measurements at each of 56 anterior measuring points of the MFM resulting in an accurate reproduction of the original dipole that generate the MFM [41].

as was demonstrated with animal and tank models under well-defined measuring conditions [44, 45]. The images obtained generally resemble the features found in low-resolution epicardial activation maps, obtained invasively. So far, MFM has not been used to calculate isochrones directly. However, such developments can be anticipated, and it is hoped that MFM, in combination with BSPM, can enhance the spatial resolution and accuracy currently obtained in BSPM myocardial activation imaging.

The non-invasive localization and visualization of cardiac electrical activity-current density imaging- has clinical and research applications. It is, therefore, expected that this area of research, which involves powerful mathematical techniques [40, 41, 44], will grow rapidly. Key to its success is the amount and accuracy of the information (data, geometry and probe characteristics) available. Because the inverse solutions are ill-behaved it is essential that maps are constructed from many data points covering a large area. We have shown [46] that to locate the simplest cardiac source, a current dipole, themagnetic field at at least 30 locations within a area of approximately 24 by 28 cm should be known. This study also implies that more SQUID channels are needed when the sources are more complicated, as is usually the case. In addition, detailed torso geometries, with each compartment having its own clearly-defined conductivity, are needed. This requires the use of individualized torsos that can be obtained with MRI. MRI-based geometrical information also can be incorporated into the inverse solution by forcing the sources to lie on particular surfaces, as dictated by physiology. Probe and electrode positions relative to the torso should be well defined, and the inverse solution needs to be stable when measuring conditions (lead placement, torso geometry and signal-to-noise ratio) are varied to some degree, as can be expected during real measurements. Finally, the inverse solution has to be validated in a number of ways. Ideally, this involves accurate epicardial and body surface measurements on a variety of subjects with different cardiac conditions. Before reviewing localization results on patient data, we shall first describe in more detail the experimental techniques of BSPM and MFM measurements. We shall emphasize the special noise-reduction techniques used in MFM measurements.

4. Measuring and Analysis Techniques.

4.1. DATA ACQUISITION AND SENSOR NOISE LEVEL

The techniques for measuring multi-lead ECG (BSPM) or multiprobe MCG (MFM) are very similar. The electronics beyond the initial sensor and pre-amplifier stage must be identical. This avoids recording differences in the data sets due to a different design of the hardware alone. A combined single-probe MCG/multi-lead BSPM system has been described by us in detail [47]. Typically, data in a 0.025 - 250 Hz bandwidth is collected for 30 s to several minutes at a sampling rate of 1000 Hz. Recommended standards for data-acquisition and analysis of high resolution ECG measurements are given by Breithardt et al. [48]. Most of these standards also apply to high resolution MCG. To obtain meaningful comparisons between BSPM and MFM, it is essential that all software used to handle and analyze the MFM and BSPM data sets also is identical. Frequently, the MCG and ECG data are averaged to enhance the signal-to-noise ratio of each heart-beat trace. Provided environmental noise can be greatly reduced (see below) averaging MCG data is, in our experience, less labour intensive than averaging ECG data. Electrode artifacts (such as short and long term drifts) can hamper the averaging process.

Placing many leads on the body and providing good contact by removing hair and even lightly abrading the skin at each electrode location can be time consuming and, at least in our lab, not all subjects agree to undergo this process. Because, normally, no leads are needed, measurements with a multiprobe MCG system are fast, which eliminates much of the resistance from clinicians, who view BSPM as a complicated, time-consuming procedure. However, the acquisition cost of a SQUID multiprobe system is prohibitive for most centers.

As is clear from the maps displayed in Figures 1 to 3, the body surface area covered by the MFM and BSPM differs. The optimal area to sample, the number of leads, and their position, all have been debated extensively in the literature [46, 49]. Many different schemes for the placement of BSPM electrode and MFM probe positions are presently in use. There is some advantage of being able to measure on the back, for this will help stabilize the inverse solution [19]. This is possible, but not very practical with the liquid helium cryostats used to house the SQUID probes. The anterior area that needs to be covered by the probes is, to a large degree, determined by the locations of the larger MFM extrema, important parameters in defining map characteristics. Our fixed grid area of 24 x 28 cm with 4 cm probe spacing (Figure 9), includes approximately 90% of the extrema generated at the anterior surface during the heart cycle in a sample of 60 patients and normals. The separation between the probes is determined by the spatial sampling frequency required. Eigenfunction analysis of maps [63] obtained with probe separation of 2 cm and 4 cm on the same adult male shows that the larger separation does not result in loss of information [4]. Based on this discussion and comments made earlier on the minimum number of probes needed to accurately locate a current dipole with an inverse solution, one can conclude that a multichannel system, covering a circular area with a diameter of 30 cm and containing about 50 SQUID probes with a probe separation of about 4 cm, can measure, instantaneously, most magnetic cardiac activity in adults that can be used in a detailed analysis of spatial or temporal features or for current density imaging. If the cardiac signal is repetitive, a valuable MFM can be constructed from sequential measurements with a single probe, or perhaps a few, covering the same or smaller area, as determined by the question that is being addressed, e.g., in vector MCG the 3 vector components of the magnetic field at a single location are measured [50]. Only a few leads are needed to determine the spectral content or the existence of "late-fields" at the end of the QRS in a particular patient population (see section 5.1).

One aspect that needs discussion is the desired signal-to-noise ratio in BSPM and MFM measuring systems. It is sometimes stated that the noise requirements for MCG measurements are not so severe as is the case for Magnetoencephelograms (MEG) and that, therefore, a high T_C-SQUID system, which typically has larger noise levels than low T_C-SQUIDs, should be adequate. This is correct for determining some MCG parameters, such as heart rates, where one detects the peak of the QRS complex or for monitoring map extrema, as in done in some applications [78]. However, for most clinical applications and particularly applications that involve an inverse solution (imaging), one needs the best possible signal-to-noise ratio. Also, when comparing the diagnostic value of high-resolution BSPM and MFM data, one needs similar signal-to-noise ratios in both data sets. For BSPM and MFM measurements we typically collect about 30 beats before averaging the signals. Of these 30, only 20 to 25 beats are sufficiently similar to be accepted by our averaging program. Based on the results of a group of 27 normal subjects, we find that the averaged ECG data has, in a bandwidth of 0.05-125 Hz, a noise level of about 1.3 μV and a peak value of 1.8 mV during the QRS complex. The MCG data of this subject group show average QRS peak values of 16 pT. To achieve a similar signal-to-noise ratio as in the ECG signal, we deduce from these data that the noise level in the averaged MCG data should not exceed 12 fT in a bandwidth of about 125 Hz, which translates to about 1.2 fT/√Hz. This means that the noise level in the unaveraged data should be about 6 fT/√Hz. Currently, only the intrinsic noise level of low-T_C SQUID systems is low enough to achieve this, and consequently, low T_C-SQUIDs should be used to obtain a meaningful MFM/BSPM comparison.

High T_c-SQUIDs have the obvious advantage that they do not require liquid helium, a commodity not always available and relatively expensive. The cryostats needed to cool with liquid nitrogen can be simpler than those used for liquid helium, making innovative designs, advantageous for MCG measurements, possible. One can visualize with such nitrogen cryostats, probes near the left side and back of the body, measuring the normal field component, B_n, or cryostats with built-in sensors that adjust to the shape of the body. In addition, the distance between the probe and the outside of the cryostat can be reduced. This would increase the signal strength and consequently the signal-to-noise ratio, compensating, at least in part, for the higher intrinsic noise level of the high T_c-SQUID systems.

Of central importance is the signal-to-noise ratio. In addition to the intrinsic SQUID noise, noise sources external to the SQUID system contribute to the overall MCG noise level. In the next sections we shall identify such noise sources and describe some of the techniques available to reduce their influence.

4.2. EXTERNAL MAGNETIC NOISE SOURCES.

The pick-up coils connected to the SQUID system will, in addition to the cardiac signals, also measure other magnetic fields generated by the subject, as well as magnetic noise generated by equipment, machinery, power lines or moving metallic objects nearby -see Figure 6.

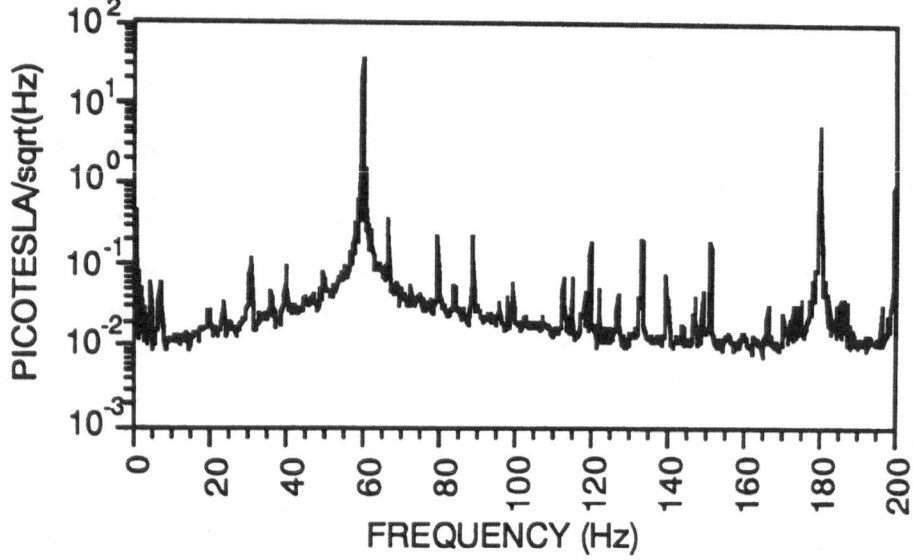

Figure 6. Typical magnetic noise spectrum obtained with a 2nd-order gradiometer coil system connected to a SQUID in a non-shielded environment [51]. The peaks originate from equipment operating at distinct frequencies.

Patient noise originates from a variety of sources. The stomach produces magnetic signals at regular intervals, known as magnetoenterograms (MEnG), in the frequency range of 40 - 60 mHz. Although these signals near the abdominal area, can be quite large (about 10 pT) compared to the MCG signals (on average 16 pT), their frequency is too low to seriously affect the MCG. When lungs are contaminated by magnetic dust, as is the case of subjects working in a mine or technicians using grinding tools, breathing also can show up as a regular, slowly-varying background. This breathing signal can be monitored separately and subtracted through adaptive filtering, or a curve can be fitted to magnetic field values at equivalent moments in successive heart beats and subtracted. Occasionally, one also observes magnetic signals arising from dental work. These show up as a series of spikes during the heart beat and are due to cardiac ballistic effects. The change of impulse of the blood during the cardiac cycle causes the body and consequently metallic objects on the body to move with the same frequency as the heart beat. These series of sharp spikes superimposed on the MCG can be readily recognized and eliminated by moving a magnetic tape eraser near the mouth. Finally, both the ECG and MCG can sense, in principle, the thermal noise of the patient. This is generated by the random movement of electrons in the conductive torso. This magnetic noise is estimated to be about 0.1 fT/$\sqrt{Hz}$ [52], below the intrinsic noise of a state-of-the-art SQUID sensor. In summary, patient noise sources are important, in that their proximity to the detector makes it difficult to shield against such interference. However, with relatively simple means, measures can be taken to eliminate or reduce their effect on the measurements.

The environmental noise can be four to five orders of magnitude larger than the peak value of the MCG in the frequency band of interest (0-250 Hz). For an overview of such noise sources see the article in this volume by J. Vrba [53]. The temporal or frequency spectrum display of the measured noise provides clues to the nature of the noise source. Monitors, oscilloscopes, copying machines, moving elevators and cars, all provide their own characteristic signatures. Magnetic signals at the line frequency and its harmonics normally dominate all other noise components. Fortunately, many techniques are available to reduce the magnetic interference of such sources. They vary from such simple solutions as moving the source away or plugging the equipment into a different, better grounded outlet, to more sophisticated techniques such as advanced electronic or software filtering, sensing the noise fields and using feedback techniques and/or building a shielded room. Often, several of these techniques are used simultaneously.

Most important in reducing the effect of the external noise sources on the measurements is the use of gradiometers. The simplest system, a magnetometer, consists of a single superconducting loop, for cardiac studies typically 2 cm in diameter, connected to the SQUID. A 1st order gradiometer consists of two such loops wound in series opposition, typically 6 to 7 cm apart from each other. If the "far-away" noise source can be represented by a magnetic dipole, then a single loop senses a magnetic field that falls off with distance as $1/r^3$. A 1st-order gradiometer would measure the first spatial derivative of this source. That is, it senses a source that falls off as $1/r^4$. Consequently, it is less sensitive to this noise source than a magnetometer. This arrangement also ensures that the system is insensitive to uniform fields, like the Earth's magnetic field. A 2nd-order gradiometer consisting of two 1st-order gradiometers back to back, measures the 2nd spatial derivative of the "far-away" noise source ($1/r^5$). It also is insensitive to uniform fields and the field gradient along the axis of the gradiometer. The frontal loop still catches almost all the flux generated by the biomagnetic source directly underneath the probe. The total effect of gradiometers is that they preferentially measure the nearby biological sources and are relatively insensitive to "far-away" noise sources.

MFMs are generated by a multitude of sources active at the same time. Sources nearby or further away, will be sensed differently by different gradiometer systems positioned at the same torso location, resulting in somewhat different MFMs. This makes a detailed comparison of such maps, obtained by different research groups using different gradiometers, difficult. Algorithms exists to convert between maps [54]. To facilitate such comparison, maps should be converted to that measured with a magnetometer.

Multichannel systems use a multitude of SQUID sensors to measure all magnetic field components (B_x, B_y, B_z) and first- and second-order gradients of the noise field with coil systems located higher up in the dewar, away from the cardiac sources. This information on the noise field is then used to subtract these magnetic field components from the sensing coils located as close as possible to the heart. These sensing coils can be 1st-order gradiometers or, ideally, just magnetometers. This form of noise rejection, involving many sensors, is discussed in detail by Vrba [53]. Finally, the signal common to all measuring coils near the chest area can be found by summing the output of all channels together and then subtracting the average from the signal in each channel. This again leads to noise reduction. It is based on the assumption that the noise fields are common to all channels and the biomagnetic signals of interest different in each channel.

Of the many methods used to reduce external magnetic noise fields, the shielded room is possibly the most controversial because of the space it occupies, the limitations it imposes in providing access for and to the patient and its expense. Attempts to measure biomagnetic signals without such rooms, using techniques described above, have met with moderate success [55, 56]. Because noise reduction technology in biomagnetic measurement is a rapidly growing field, it is likely that the importance of shielded rooms will diminish. However, if the aim is to obtain data with good signal-to-noise ratio, uninterrupted, at any time of the day in an active research or clinical environment, and view the signal of interest at the same time it is measured, then shielded rooms are still the most effective method to reduce magnetic noise.

4.3. SHIELDED ROOMS

Shielded rooms reduce the environmental noise directly over all or most of the frequency range of interest. Reducing the peak amplitudes in Figure 6 by a factor of 100 would eliminate most interference, except for the line frequencies (60 and 180 Hz), which can then be filtered. Such shielding allows direct observation of the cardiac signal, not normally possible given the noise spectrum of Figure 6. Two types of shielding are used: eddy current shielding and μ-metal shielding.

4.3.1. *Eddy-Current Shielding*

The walls of an eddy-current shielded room are made of highly conductive material (aluminium or copper). The time-varying environmental magnetic fields induce eddy currents in these walls that produce a magnetic field that opposes these external fields. Consequently, the field inside this room is reduced (Figure 7a). The crucial parameter that determines the performance of this room is the skin depth δ. This is the depth in the metal where the magnetic field is reduced to $1/e$ of its value just outside the material. δ is given by:

$$\delta = \sqrt{(\pi \mu_o \sigma f)^{-1}} \tag{1}$$

where σ is the conductivity of the material, f is the frequency of the external magnetic field and μ_o is the permeability of free space. Equation (1) shows that δ will decrease for increasing frequencies. For aluminum $\delta \cong 30$ mm at 10 Hz; 10 mm at 50-60 Hz. For the frequency range of interest in biomagnetic measurements, typically $\delta > t$, the thickness of the conductive shield. In this limit the shielding factor S, the ratio of the uniform magnetic field outside the room, H_e, to that inside the room, H_i, is to first order given by [57]:

$$S = \sqrt{1 + (2\pi f \tau)^2} \tag{2}$$

For a cubic enclosure,

$$\tau = \frac{1}{4} \mu_o w t \sigma \tag{3}$$

where w is the length of the room. It has been shown that this equation predicts the shielding factor for uniform fields and field gradients accurately, provided the entrance is covered by a door or an extra wall to provide a corridor [58]. This equation says that the larger the room, the larger the shielding factor, suggesting that, at least in principle, one can construct an effective eddy-current shielded room within the walls of an existing room, providing easy access to the subject. For an aluminum room with $t = 2$ cm and large enough to contain a bed ($w = 2.4$ m), $S \cong 200$ (50 dB) at 60 Hz. However, at dc ($f = 0$) the room is transparent for magnetic fields -see bottom solid curve in Figure 8. Such an aluminum shielded room is relatively inexpensive (approximately \$50,000 when done by local contractors). It provides, at most locations, the necessary shielding to observe, with a 2nd- or 3rd-order gradiometer, the cardiac signal directly.

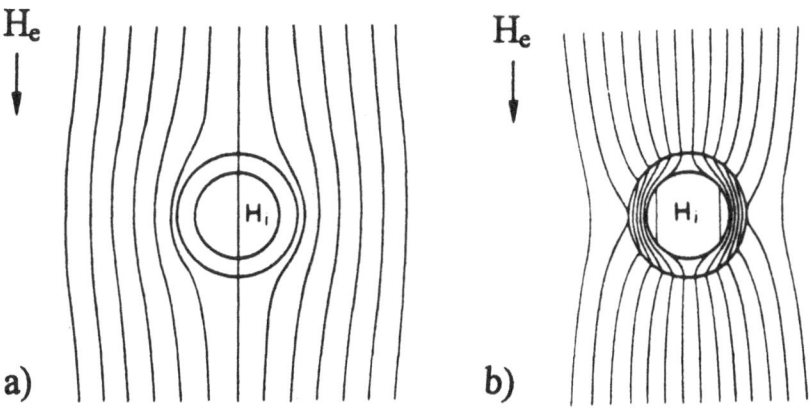

Figure 7. Schematic diagram of the effect of a shielded room in an uniform field H_e, on the internal field, H_i, when this room is constructed of a) highly conductive material and b) ferromagnetic material.

A possible concern when using an eddy-current shielded room is the thermal magnetic noise of the conductive materials surrounding the detector. This thermal noise, proportional to $\sqrt{\sigma}$, can be observed by placing the SQUID sensor near the wall [59]. However, the strength of this noise field reduces rapidly with distance away from the conductive walls. At the center of the room, where the SQUID is usually located, this noise is well below that of the detector.

It is now possible to construct a shielded room of high-T_C superconducting material [60]. Several such experimental enclosures, operating at liquid-nitrogen temperatures (77K) and sufficiently large to contain a subject, have been built. Shielding factors as large as 10^7 (140 dB) have been measured. MCGs have been recorded in such rooms.

4.3.2. μ-Metal Shielded Room.

The large μ value of some ferromagnetic materials ensures that the magnetic-field line density, and consequently B, is extremely large inside the shielding material relative to the field line density directly adjacent to the materials (Figure 7b). In contrast to the eddy-current shielded room, this type of shielding also is effective at very low frequencies, including dc magnetic fields. To first approximation the shielding factor S, for a cubic enclosure with cube edge w, can be written as:

$$S = 1 + 0.7\mu_r t / w \qquad (4)$$

430

where t is the thickness of the shield and μ_r is the relative permeability. Note that the shielding is inversely proportional to the size of the room. Using typical values for μ_r (2 x 10^4) and t (1.5 mm), yields for a room with w = 2.4 m a shielding factor for dc fields of 13.5 (23 dB). The shielding factor at higher frequencies is more difficult to predict. It increases slowly with frequency, but, for the larger rooms, not as fast as predicted by theory.

In many shielded rooms for biomagnetic measurements one uses several layers of μ metal and aluminum to make optimal use of the different frequency and shielding characteristics of the two different types of shielding. Several research groups have attached μ-metal to their, several cm thick, aluminum shielded room [61, 62]. The total shielding factor is approximately the product of the shielding factors for aluminum and μ-metal. From the numbers given above one can estimate that adding a 1.5-mm layer of μ-metal to the inside and outside of a 2 cm thick, 2.4-m long aluminum room would add another 40 dB to the shielding factors for a total of about 40 dB at dc and 90 dB at 60 Hz. The μ-metal alone would cost about $ 80,000 , resulting in a total cost for this Al/ μ-metal room of about $ 130,000. when constructed by local contractors; about $ 300,000.- when constructed by companies specializing in this field.

If the above measures to reduce the environmental noise fields are not sufficient more layers of highly conductive material and μ-metal can be added. Rooms with even more

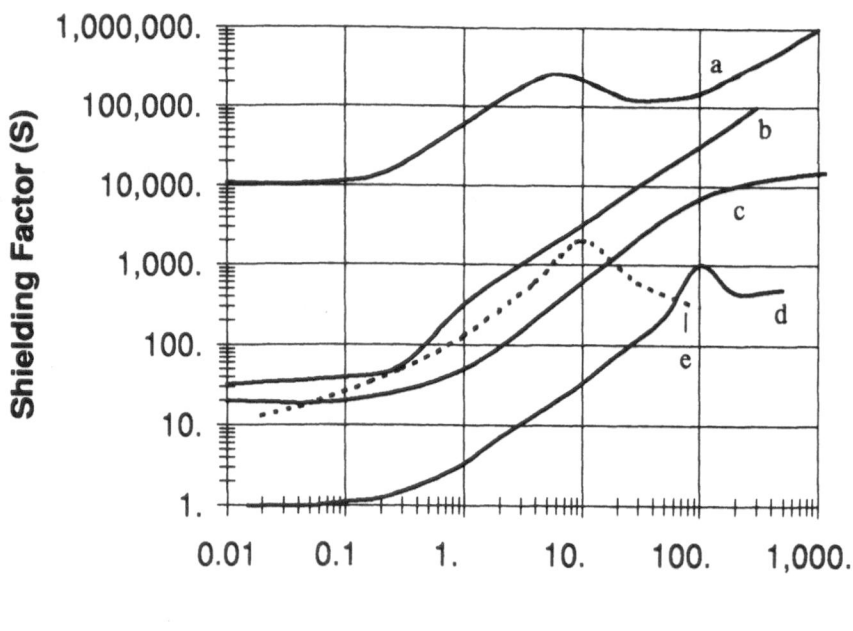

Figure 8. The shielding factor as function of frequency for several types of "walk-in" shielded rooms currently in use for biomagnetic measurements. From top to bottom, solid lines a) the Berlin shielded room (6 layers of μ-metal; 1 inner layer of copper) [64], b) a commercial 3-layer shielded room [63] (2 layers of a μ-metal, one of aluminum), c) a commercial 2-layer (μ-metal/aluminum) shielded room [64], d) a thick aluminum shielded room [58], and e) a thick aluminum shielded room covered on the inside with μ-metal [62] (dotted line).

layers and other special features have been build in Berlin, Helsinki and recently in Tokyo at great expense (exceeding $ 1,000,000). For such rooms shielding factors for uniform fields of 10^4 (80dB) at dc and 10^5 (100dB) at 50 Hz have been achieved (see Figure 8).

To determine optimal shielding for a particular environment one usually obtains a Fast Fourier Transform of the magnetic field measured with the available SQUID system in the frequency band of interest (Figure 6) and determines from this noise spectrum the shielding factors needed to reduce the environmental noise to below the intrinsic SQUID noise. Although the expressions for shielding factors discussed so far are for uniform fields, they give a good estimate of what is required even if higher-order magnetic field components are present. Economics then dictate what level of magnetic shielding in addition to other measures, such as hardware and software filtering, are required to achieve the desired noise level. To estimate from the desired shielding factors what type of shielded room is needed, we have plotted, in Figure 8, the frequency dependence of the shielding factor of some of the biomagnetic shielded rooms presently in use.

4.4. MAP ANALYSIS

The large amount of information contained in a time series of high resolution BSPMs or MFMs created every 1 or 2 ms during the heart cycle, is difficult to display or analyze in a form that is useful or even acceptable to a clinician [13]. Some data compression is required to cast the data in a form that can result in a quick, reliable diagnosis. The BSPM literature contains different solutions to the comprehensive presentation of the essential information. For example, insight into differences between measured map sequences of patients with different types of myocardial infarctions (MI), can be obtained by calculating time integrals over specific time segments during the heart cycle. In such a representation, the sum value for each measuring location over the selected time segment is obtained and multiplied by the segment duration. This value is then displayed directly as a *time-integral map* or averaged for the patient group under study. As an example, in Figure 9, we show the T-wave time integral MFM and BSPM of a normal subject. Such maps look dramatically different for the different MI groups [65].

In *extrema trajectory plots* one searches for the location of the extrema in the spatial maps, and plots this position in the measurement area as a function of time during a preselected time interval of the cardiac complex. The path of these trajectories is a good indicator of normal vs. abnormal cardiac function. For a normal cardiac sequence, these trajectories follow a continous smooth path; for an abnormal heart such trajectories are fragmented and spread over a larger map area.

Other useful displays that involve data compression are moving-dipole trajectory plotting and eigenfunction analyses. In the *moving dipole trajectory plots* one analyses the trajectory of the position of the origin of the Equivalent Current Dipole (ECD) as determined by an inverse solution. The ECD is the dipole moment and location that best fits the recorded maps. This procedure requires that one specifies the volume conductor. Often, one selects the infinite homogeneous volume conductor. However, with a realistic

432

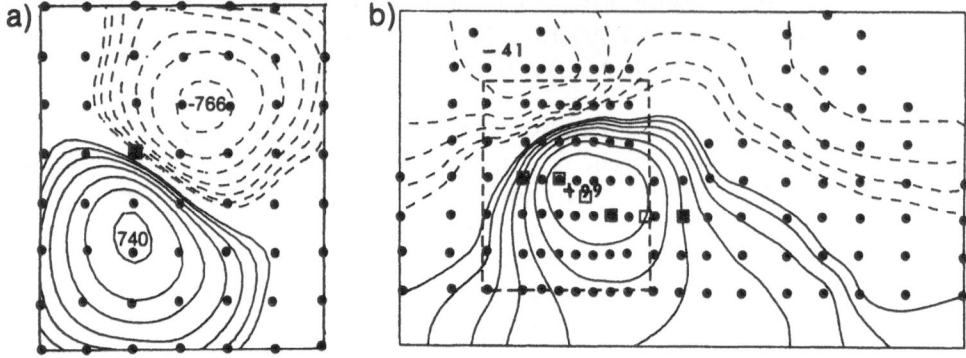

Figure 9. a) T-wave isointegral magnetic map for a normal subject. Dots indicate the locations, 4 cm apart, at which field measurements were made. The extrema are expressed in pT· ms. The black square gives the location of the junction between sternum and fourth intercostal space. b) T-wave isointegral body surface potential map from a normal subject. Dots indicate the location of the electrodes, squares the location of the standard V1 - V6 electrodes. The extrema are expressed in mV· ms. The rectangular area enclosed by the dotted line represents the area covered by a) [65].

torso model, more reliable conclusions can be drawn about the underlying physiology [66]. Any temporal or spatial distribution of the measured potentials or magnetic fields can be expanded in a set of orthogonal *temporal or spatial eigenfunctions*, respectively [67]. Such an orthogonal expansions of the temporal or spatial information can be based on a singular value decomposition (SVD) or the related Karhunen-Loeve transformation (KLT) [68]. The number of such eigenfunctions which are needed to completely represent the data depends on the complexity of the map and the noise level. The eigenfunction maps with the largest values represent the most dominant features. Such eigenfunction maps can then be used for further analysis [69].

These data reduction methods and others [12] that focus on e.g. the frequency content of the temporal information, are regularly used in the BSPM community and have found applications in the analysis of MFM as well. With the relatively few parameters derived from such data compression techniques, one can quantitatively compare the information contained in the BSPM and MFM of the same patient or, when dealing with a patient group, use statistics to calculate how well BSPM and MFM characterize particular heart diseases.

5. BSPM and MFM of Patients with Arrhythmias

The largest effort in BSPM and MFM analysis deals with identifying patients at risk of suffering from arrhythmias. It is believed that in the US alone more than 400,000 people (about 50% of all cardiovascular deaths) die suddenly each year from sustained ventricular tachy-cardia (VT) or ventricular fibrillation (VF). Most have coronary artery disease and have suffered a myocardial infarction (MI) in the past. Diseased cells surrounding the infarcted cells may conduct slowly, and depolarization is delayed. In about 10% of the MI patients, this slow-moving wave may find a circular path back to its origin where repolarization already has taken place. It can then depolarize this region again (and from there the rest of the ventricular muscle), and the process repeats itself, bypassing the normal conduction process. The result is a fast and ineffective heart beat (tachycardia). It is recognized that among individual patients recovering from myocardial infarction, finding improved methods for detecting those at risk for VT or VF is essential for reducing mortality in this patient group. When identified, drug treatment is prescribed. However, if this is not effective, the re-entrant circuit needs to be interrupted by ablation of the heart tissue involved. The ablation, usually done with a radio-frequency (rf) catheter, requires that the location of this re-entrant circuit be found accurately.

A second group of patients that has received considerable attention in the research literature, are those suffering from Wolff-Parkinson-White (WPW) syndrome. In such patients one or several accessory pathways (AP) between the atria and ventricles can, in addition to the His-Purkinje System, conduct the electrical signals from the atria to the ventricles. If the signal enters the ventricles and finds a conductive path back to the atria, the resulting re-entrant circuit bypasses the heart's normal method of regulating the cardiac cycle, allowing the atria and ventricles to contract very rapidly. Such arrhythmias can be disabling. The incidence of people suffering from WPW syndrome, manifested as abnormalities in the ECG at sinus rhythm is small (1-3 per 1000 people). For this group, the risk of sudden cardiac death is less than 1%. Treatment can consist of drug therapies. However, the long-term side effects of drug treatments, particularly in younger patients, is not always desirable, and rf ablation of the AP is presented as an option. This requires a localization of the accessory pathway.

5.1. MAP FEATURE ANALYSIS BASED ON HIGH-RESOLUTION MEASURE-MENTS.

Over the past decades many investigators have recorded high-frequency, low-level signals in the averaged ECG of patients prone to sustained ventricular tachycardia (VT) or ventricular fibrillation (VF). These abnormal signals have been associated with the delayed ventricular activation in border zones of damaged myocardium discussed above. The presence of such low-level signals at the end of the QRS, that is, after the normal depolarization process has taken place, have been used to identify patients at risk of sudden death. About 80% of patients with VT demonstrate such late potentials (LP). However, the predictive value of the existence of late potentials at the end of the QRS or other ECG variables as an indicator of arrhythmic events within one year is relatively low (5-30%) [71, 72]. It is very high in predicting a good outcome, showing a 95% event-free survival if no late potentials are detected. Predictive accuracy can be enhanced if other, independent diagnostic indices that measure cardiac function are used as well [70], or when other ECG parameters, in particular the total duration of the QRS complex, are included [71, 72, 73]. Several researchers have shown that late, high-frequency activity at the end of the QRS also exists in high-resolution MCGs of VT patients [74, 75, 76, 77, 78]. Montonen [79] recently reviewed the methods used to analyze MCGs to identify patients prone to malignant arrhythmias.

The BSPM community has explored spatial features to extract abnormal, VT-specific information. The main analysis techniques used are: trajectory plotting [80], time-integral

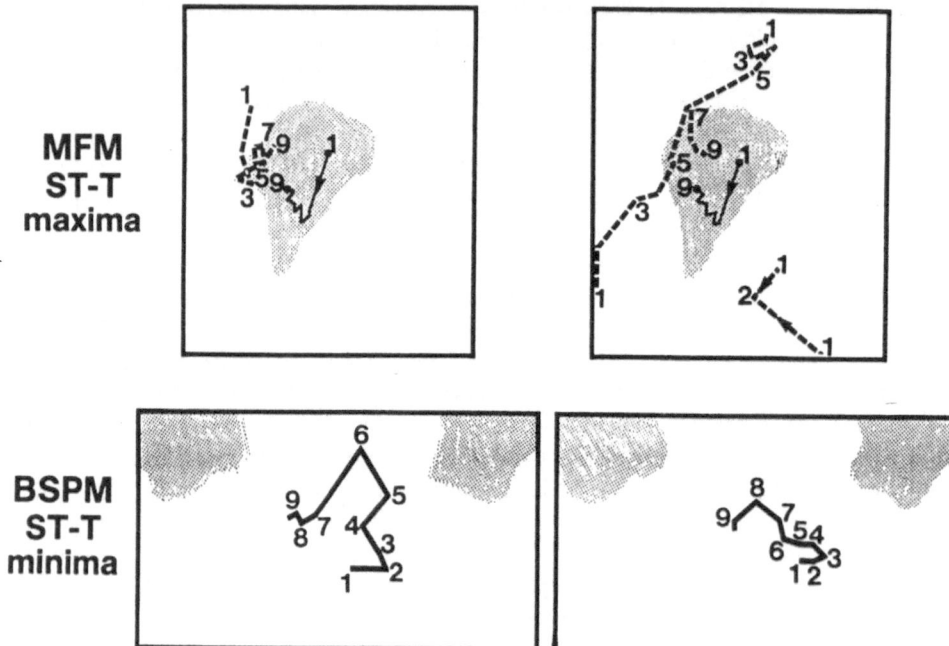

Figure 10. MFM (top) and BSPM (bottom) trajectory plots of extrema during the ST-T interval for a patient with Anterior Myocardial Infarction (AMI) (left) and a patient with an AMI and episodes of sustained VT (right). The data are obtained with the patient in sinus rhythm. The range of the trajectories for normal subjects is indicated by the shaded area. Patient with VT show a larger proportion of trajectories that split into simultaneously occurring, multiple trajectories (dashed lines).

maps [81] and eigenfunction analysis [82]. We have used both MFM [83] and BSPM integral maps and trajectory plots to discriminate MI patients suffering episodes of sustained VT from MI patients without such episodes, and from normal subjects [84]. The study group of 60 subjects comprised three sex- and age-matched groups: 30 normals, 15 patients who had previously suffered an MI and 15 patients with sustained VT of which 11 had previously suffered an MI. Figure 10 shows the BSPM and MFM extrema trajectory plot of a patient with MI and that of a patient with MI and VT. The figure shows that the patient with VT has more trajectories, and that these trajectories are more fragmented and widespread than that of the MI patient without VT. This trend is consistent with the fragmented electrical activity known to exist in patients prone to VT or VF. In an attempt to quantify these results, we defined T as the number of discrete trajectories in the plot and F as the number of time instants at which two or more discrete trajectories coexist. Using both these criteria in the ST-T interval for MFM and BSPM trajectory plots, enabled separation of MI and VT patients with accuracies of about 80% for each method. This means that it also is possible to find in a training group, parameters in MFM that are accurate predictors of patients prone to VT. However it is not clear how these predictors work for a large test group. This points to a major weakness in much of the present MCG work. Unless large groups of between 100-200 patients are investigated, as has been done in ECG and BSPM research, it is extremely difficult to validate such risk factors. In order to investigate such large numbers, multichannel MCG systems, which allow measurements that take only a few minutes per patient, are essential.

5.2. INVERSE SOLUTIONS AS APPLIED TO PATIENT MAPS

5.2.1. *Ventricular Arrhythmias*
Both MFMs and BSPMs have been used to determine the location of the site of origin of ventricular arrhythmias. This work is stimulated by the development of surgical and catheter-mediated ablation therapies to treat sustained VT. Knowledge of the precise location of the site of VT origin is imperative to destroy the arrhythmogenic tissue without causing significant damage to the normal myocardium. It is estimated that the lesion produced by the application of the rf ablation current is < 6 mm in diameter and < 3 mm in depth [85, 86]. Currently, the most reliable methods for precise localization are invasive activation sequence mapping and/or pace-mapping. They involve introducing several catheters through arteries or veins into the heart chambers and mapping the cardiac activation sequences at or near the endocardium, a time-consuming procedure.

SippensGroenewegen et al. [32] recorded the BSPM during endocardial pacing in patients with normal cardiac anatomy. Pacing creates an ectopic beat that starts at the pacing site. For each pacing site, the resulting surface potentials were recorded and displayed as a BSPM integral map. They found 38 distinct patterns, originating from well-defined, unique areas of the endocardium. These maps can then be used as a reference data base for comparison with the maps obtained during the onset of ectopic ventricular beats in VT patients. It is assumed that the paced BSPMs in normal subjects and the ectopic-beat BSPMs of VT patients, which appear similar in topography, originate from the same location. It also is assumed that this site of origin of the ectopic beat in VT patients also is the site of the onset of the VT. The same researchers validated this procedure by comparing the BSPM obtained during endocardial stimulation in VT patients with the BSPM morphology of spontaneous or induced VT in the same patient. Using this procedure they were able to locate accurately (i.e., within 2 cm of the endo-cardial site of VT origin) 80% of the VT morphologies, indeed a substantial improvement on 12-lead ECG localization procedures [86]. This inverse solution method relies on matching BSPM patterns. No mathematical solution involving a detailed source and volume model is necessary.

The first MCG localization of the site of origin of a sustained VT was obtained in 1986 with a single probe MCG in a hospital environment by Fenici et al. [9]. Maps were obtained during the onset of the induced VT complex and used in the inverse solution containing a current dipole in an infinite homogeneous medium. The work was repeated a year later on the same patient during sinus rhythm using the onset of ectopic beats. The inverse solution results were similar, indicating that the ectopic beats found their origin at the same site as the arrhythmias, as anticipated. Similar localization techniques were used by Moshage et al. [87, 88] and Weismuller et al. [89], who used multi-channel systems and the anatomical information as derived from MRI. Fenici et al. have consistently used magnetic mapping, pre-operatively, to determine the site of the arrhythmogenic area. They showed a localization accuracy of about 2 cm, sufficiently accurate to shorten the ablation procedure. Both this group [9] and Moshage et al. [87] have developed non-magnetic catheters to determine localization accuracy by comparing the catheter tip location, as determined radiographically, with the inverse solution of the MFM at the onset of the ventricular excitation induced with this catheter. This way, the localization accuracy of less than 2 cm could be validated [89]. Weismuller *et al.* [91] also used multichannel MFM of the late fields at the end of the QRS complex to locate the origin of this delayed activity in 8 VT patients. They found that this activity was located within the border zone of the infarction in six of the eight patients. Weismuller [92] recently reviewed some of the multichannel work on VT patients and showed that the MFM/MRI localization procedure based on the onset of ectopic activity or VT, could be fast (30 seconds to a few minutes), accurate (2-3 cm) and easily visualized.

Nomura *et al* [93] used both BSPMs and MFMs of ectopic beats of 10 patients with VT. They obtained from the MFM, recorded 20 ms after the beat onset, an inverse solution with the single current dipole (and infinite homogeneous) model. The results were compared with the classification scheme of Ushijima *et al.* [94] that connects the topology of ectopic BSPM maps with the site of origin of such ectopic beats. They found good agreement between these two different localization procedures in almost all patients.

From the above it is clear that the technology exists to obtain MFMs at the onset of a VT sequence or ectopic beat, and to calculate the location of this onset with inverse solutions or, as an alternative localization technique, to compare the topology of such maps with the endocardially-paced MFM maps. As has been shown by SippensGroenewegen *et al.* [32] for BSPM, this latter method can lead to an accurate localization of ventricular foci in VT patients and avoids the use of the inverse modeling procedure. In the final analysis, accuracy, speed and cost will decide which one of these two mapping technologies will be used clinically.

5.2.2. *WPW Patients*

The majority of the arrhythmias in WPW patients can be cured by catheter ablation of the accessory pathway (AP) across the AV ring after the site has been located by invasive electrophysiological studies (EPS). EPS uses pacing catheters to determine the location of the earliest ventricular depolarisation. The general location of the AP site can be determined non-invasively from ECG or MCG recordings of the delta-wave, observed with the patient in sinus rhythm as a ramp just before the QRS (Figure 11). Many criteria for localizing the AP from the polarity of this ramp in 12-lead ECG have been published [95]. Such localization schemes assist the physician in determining, roughly, the AP location before starting EPS. Such information can help in planning the ablation procedure; i.e., the presence of either a left- or right- sided AP requires entry of the catheter through different veins or arteries.

There have been several BSPM studies of the non-invasive localisation of the AP's [12, 13, 96, 97]. Such studies show that the BSPM topology (e.g., the location of the map extrema on the torso) 20 - 40 ms after the onset of the delta wave, is a sensitive indicator of the location of the AP, as verified by electrophysiological studies (EPS). Accuracies of 2 cm or less are quoted. This non-invasive localisation procedure can be done well in advance of the catheterisation. In addition to saving time during the normally lengthy search for the AP with EPS, it also enables the physician to anticipate complicated conduction pathways. We recently showed [98] that the large information content of the BSPM can be used to successfully identify the existence of two AP's with the additional pathology of two distinct atrial pacemaker sites. One AP was found to conduct continuously; the second AP, intermittently. Depending on which sites were active, this resulted in 4 different map sequences. Data analysis techniques allowed us to extract the maps belonging to each delta wave and use its topology to determine each AP location. Twelve-lead ECG can not provide this detailed information. Such results emphasize the need for either BSPM or multichannel MFM as a preoperative analysis tool.

Dubuc *et al.* [99] recorded BSPMs during endocardial ventricular pacing at the AP near the AV-ring and compared it with the BSPM delta-wave map obtained during sinus rhythm. A precise match indicated that the pacing catheter is positioned at the site of the AP. This procedure is similar to that discussed earlier for the localization of VT foci. With the pacing catheter in the right location, ablation has a high success rate (93%). This localization technique, which involves obtaining BSPMs in the catheter lab, results in considerable time savings during the ablation procedure. The authors showed that a 5-mm change in the position of the pacing catheter produced a significantly different BSPM. They also demonstrated that the time of the average ablation procedure (about 4.5 hours) was reduced to less than one hour.

The biomagnetism community has used the numerical inverse solution applied to MFMs at the onset of the delta wave, as the primary route to locate the AP non-invasively. There is a relatively large number of publications that base this localization on the current-dipole model in an infinite or semi-infinite homogeneous volume conductor [9, 10, 11]. In general, good agreement is found between the localization results of this inverse solution

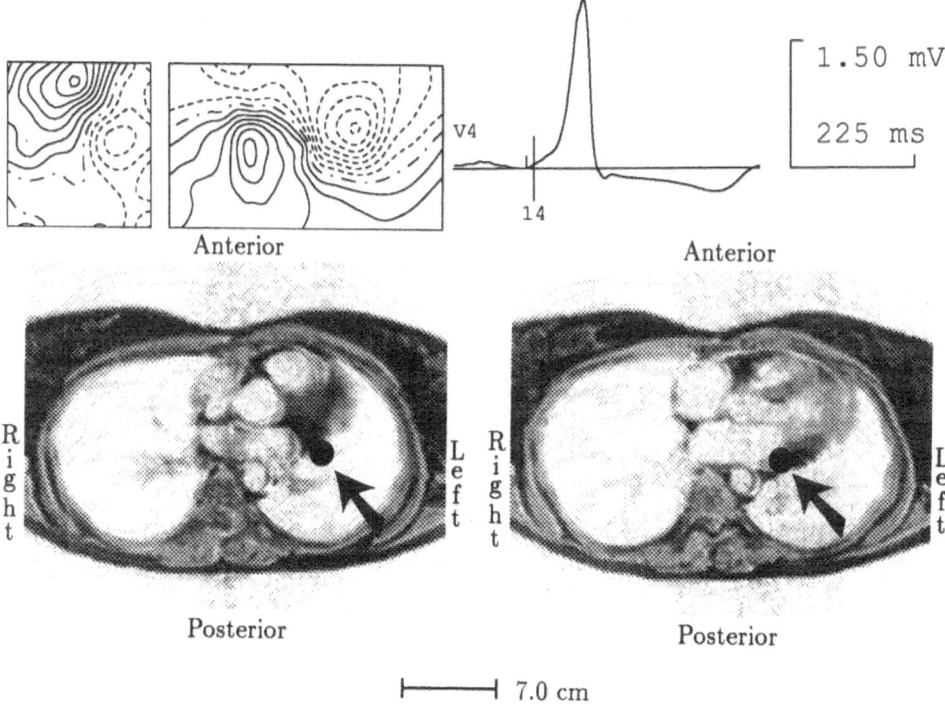

Figure 11. Current dipole inverse solution representing the location of the accessory pathway, projected with a dot (•) onto MR images of a WPW patient. The inverse solutions, at 14 ms after the onset of the δ-wave are obtained from the maps in the left hand corner, using an inhomogeneous volume conductor. ECG lead V4, shown in the upper right corner, shows at which time the delta-wave maps are plotted [18].

and EPS [90, 92]. Nakaya *et al.* [100] measured both BSPM and MFM of 19 patients with WPW. Using a BSPM classification scheme based on the position of the extrema, they obtained good agreement between this type of localization of the AP and the MFM inverse solution, and EPS when available. We explored the MFM inverse solution based on the homogeneous realistic torso [101]. Several researchers used this homogeneous realistic torso model for AP localization [102, 103]. By using this volume conductor model (scaled to fit the individual torso), the accuracy, expressed as the average distance between the results of the inverse solution and the EPS results, improved from 8.5 cm, for the infinite homogeneous model, to 2.8 cm [103]. In our group, Lamothe [18] also used this conductor model on both BSPM and MFM data of 16 WPW patients. Although the inverse solutions (see Figure 11) do not nearly produce such accuracy in locating the AP as is reported above, it was shown that the BSPM-determined locations are, on average, closer to the EPS-determined AP locations than are the MFM-determined locations. This discrepancy can partly be explained by the larger signal-to-noise ratio in our BSPM data and the fact that the MFMs were obtained with a single channel. Above all, these results reflect the intrinsic difficulties in obtaining reliable results of the desired accuracy. In addition to approximations made in the inverse solution, there are localization errors associated with the measurement itself. The error in the location of the measuring grid relative to the torso and the torso relative to the heart are of the order of centimeters. The location of the AP site as determined by EPS, relative to the measuring grid, is accurate to within several cm-only after substantial effort can catheter location accuracies of 0.7 cm be achieved [32]. Using individualized (as opposed to just realistic) torsos in the inverse solution, produces localization results based on the MFM and BSPM that are substantially closer to each other than when only homogeneous torsos are used [18]. However, the effort in producing such torsos with lungs and heart from MRI is considerable [39, 104], although faster routines are now becoming available.

We also have measured and analysed the MFMs and BSPMs of 16 patients suffering from WPW to obtain delta-wave maps for common AP sites around the AV ring [18, 105]. Overall, the changes in the topology of both sets of maps going from site to site are substantial, as was noted before for the BSPM and as also was shown from forward calculations of such maps [30, 37]. For some sections around the AV ring, the BSPM shows slow variations, hence further discrimination might be achieved using the MFM. The opposite is true as well, suggesting that both techniques should be used on the same patient to improve the non-invasive, pre-EPS, localization based on map topology. We estimate that using both type of maps, localization accuracies of 1.5 cm can be achieved.

In the end, the localization process must lead to a successful ablation of the AP. The method of Dubuc *et al.* [99] described earlier, which uses catheter ablation by comparing endocardial BSPM pace mapping with delta-wave maps, is very promising. There is every reason to expect that this map sensitivity to catheter placement also can be achieved with MFM.

6. Conclusions

Both MFM and BSPM provide a detailed record of complex cardiac electrical activity. Both can be used as the starting point for inverse solutions in the traditional sense, that is, finding the dipole or dipoles that best fit the observed maps. Recently, inverse solutions that determine isochrones from the BSPM or current-density distributions from MFM have given promising results and could assist in determining non-invasively, anomalies in the activation sequence. However, such methods are mathematically demanding and require detailed knowledge of individual anatomical features. Obtaining an estimate of the location of tissue implicated in arrhythmias by comparing patient BSPMs or MFMs characteristic of such activity with catalogued standard maps of activation sites, is more practical in a clinical setting. It gives a quick, first-order estimate of the location, which is helpful in planning treatment, such as ablation. Such maps can be obtained well in advance of the required catheterization.

There are initial reports in the BSPM literature which suggest that searching with the pacing catheter until the pacing BSPM is similar to that of the BSPM characteristic of the arrhythmia, increases the accuracy and shortens the time of the procedure substantially. Although the volume of BSPM research in this area is growing quickly, the number of such studies in MFM and the number of subjects participating in such studies is still small. It is hoped that the recent appearance of several multichannel systems in a hospital setting, capable of covering the whole anterior torso will change this situation. Based on the limited data available, it appears that the spatial resolution of such maps is different for different endocardial areas and probably different for each technique (BSPM or MFM). It also appears that MFM is more senstive to anisotropic conduction processes. Ideally, both techniques should be used simultaneously to make optimal use of the intrinsically different sensitivities.

The debate about which one is the better method for the clinical, non-invasive evaluation of cardiac function and which one provides the better predictive value for severe heart conditions is still wide open. Lacking sufficient clinical data to make such comparisons, the arguments center around technical issues such as expense and convenience. Research, involving either method, has demonstrated that they both contribute to our understanding of the mathematical, physiological, clinical and technical issues involved in the non-invasive mapping and imaging, and the subsequent diagnosis of cardiac dysfunction in different patient groups.

7. Acknowledgment

This work has been funded through the support of the Nova Scotia Heart and Stroke Foundation, and the Natural Science and Engineering Research Council of Canada. We thank R. Meeder, S. Ritcey and A. Adams for their valuable comments and assistance during the preparation of this manuscript.

8. References

1. Nahum, L.H., Mauro, A., Chernoff, H. and Sikand, R.S. (1951) Instantaneous equipotential distribution on surface of the human body for various instants in the cardiac cycle, *J. Appl. Physiol.* **3**, 454-464.
2. Cohen, D. and Chandler, L. (1969) Measurements and a simplified interpretation of magnetocardiograms from humans, *Circulation* **39**, 395-402.
3. Peters, M.J., van de Graaf, A.P. and van Oosterom, A. (1981) The influence of the torso on the magnetic field of a current dipole, in S.N. Erne, H.D. Hahlbom and H. Lubbig (eds.), *Biomagnetism*, de Gruyter, Berlin, pp. 337-342.

440

4. MacAulay, C.E., Stroink, G. and Horacek, B.M. (1985) Signal analysis of magnetocardiograms to test their independence, in H. Weinberg, G. Stroink and T. Katila (eds.), *Biomagnetism: Applications and Theory*. Pergamon Press, New York, pp. 115-120.

5. Spach, M.S., Barr, R.C., Lanning, C.F. (1978) Experimental basis for QRS and T wave potentials in WPW syndrome, *Circulation* 62, 103-118.

6. Schneider, S., Hoenig, E., Reichenberger, H., Abraham-Fuchs, K., Moshage, W., Oppelt, A., Stefan, H., Weikl, A. and Wirth A. (1990) Multichannel biomagnetic system for study of electrical activity of the brain and heart, *Radiology* 176, 825-830.

7. Simelius, K., Ahonen, A., Huotilainen, M., Ilmoniemi, R., et al. (1995) BioMag: Functional brain and heart research in clinical environment, *Proceedings 17th Ann. Int. conf. IEEE Eng. in Med. Biol. Soc.*, Montreal.

8. Siltanen, P. (1989) Magnetocardiography, in P.W. MacFarlane and T.D. Veitch Lawrie (eds.), *Comprehensive Electrocardiology*, Pergamon Press, New York, pp. 1405-1438.

9. Fenici, R.R., Melillio, G. and Maselli, M. (1991) Clinical magnetocardiography, *Int. J. Cardiac Imag.* 7, 151-167.

10. Stroink, G. (1993) Cardiomagnetic Imaging, in B.L. Zaret, L. Kaufman, A.S. Berson and R.A. Dunn (eds.), *Frontiers in Cardiovascular Imaging*, Raven Press, New York, pp. 161-177.

11. Nakaya, Y. and Mori, H. (1992) Magnetocardiography, *Clin. Phys. Physiol. Meas.* 13, no.3, 191-229.

12. Flowers, N.C. and Horan, L.G. (1995) Body surface potential mapping, in D.P. Zipes and J. Jalife (eds.), *Cardiac electrophysiology: from cell to bedside*, W.B. Saunders Comp., Philadelphia, pp.1049-1067.

13. Green, L.S. and Abildskov, J.A. (1995) Clinical applications of body surface potential mapping, *Clin. Cardiol.* 18, 245-249.

14. Williamson, S.J. and Kaufman, L. (1981) Magnetic fields of the cerebal cortex, in S.N. Erne, H.D. Hahlbom and H. Lubbig (eds.), *Biomagnetism*, de Gruyter, Berlin, pp. 353-402.

15. Gulrajani, R.M. (1988) Models of the electrical activity of the heart and computer simulation of the electrogram, *CRC Crit. Rev. Biomed. Eng.* 16, no.1, 1-66.

16. Ferguson, A.S., Zhang, X. and Stroink, G. (1994) A complete linear discretization for calculating the magnetic field using the Boundary Element Method, *IEEE Trans. Biomed. Eng.* BME-41, no. 5, 455-460.

17. Ferguson, A.S. and Stroink, G. (1996) Factors affecting the accuracy of the Boundary Element Method in the forward problem. Part I: Calculating surface potentials. To be published.

18. Lamothe, M.J.R. (1993) The feasibility of separating concurrent sources in cardiac magnetic field and body surface potential maps, Ph.D. thesis, Dalhousie University.

19. Purcell, C., Stroink, G. and Horacek, B.M. (1988) Effect of torso boundaries on electrical potential and magnetic field of a dipole, *IEEE Trans. Biomed. Eng.* BME-35, 671-678.

20. Rudy, Y. (1987) The effects of the thoracic volume conductor (inhomogeneities) on the electro-cardiogram, in J. Liebman, R. Plonsey and Y. Rudy (eds.), *Pediatrics and fundamental electro-cardiography*. Martinus Nijhoff, Boston. pp. 49-74.

21. Van Oosterom, A. and Huiskamp, G.J. (1989) The effect of torso inhomogeneities on body surface potentials quantified using "tailored" geometries, *J. of Electrocard.* 22, no. 1, 53-72.

22. Geselowitz, D.B. and Miller, W.T. (1973). Extra-corporal magnetic fields generated by internal biomagnetic sources, *IEEE Trans. Mag.* Mag-9 (3), 392-398.

23. Horacek, B.M. (1973) Digital model for studies in magnetocardiography, *IEEE Trans. Magn.* MAG-6, 346-347.

24. Van Oosterom, A., Oostendorp, T.F., Huiskamp, G., and ter Brake, H.J.M. (1990) The magnetocardiogram as derived from electrocardiographic data, *Circulation Research*, 67, 1503-1509.

25. Plonsey, R. (1982) The nature of sources of bioelectric and biomagnetic fields, *Biophys. J.*, 39, 309-312.

26. Sepulveda, N.G. and Wikswo, J.P. (1987) Electric and magnetic fields from two-dimensional anisotropic bisyncytia, *Biophys. J.*, 51, 55-568.

27. Nenonen, J., Horacek, B.M. and Katila, T. (1992), Torso and heart models in magnetocardiology, in M. Hoke, S.N. Erne, Y.C. Okada and G.L. Romani (eds.), *Biomagnetism, Clinical Aspects*, Elsevier, Amsterdam, pp. 417-425.

28. Leon, L.J. and Horacek, B.M. (1991) A computer model of excitation and recovery in the anisotropic myocardium, *J. Electrocardiol.*, 24, 1-41.

29. Hren, R., Nenonen, J. MacInnes, P. and Horacek, B.M. (1995) Simulated body-surface potential maps for paced activation sequences in human ventricles, in A. Murray, R. Arzbaecher (eds.), *Computers in Cardiology*, IEEE Press, Piscataway, pp. 95--98

30. Nenonen, J. and Horacek, B.M. (1995). Comparison of electric and magnetic fields of anisotropic myocardium, to be published.

31. Staton, D.J., Friedman, R.N. and Wikswo, Jr. J.P. (1993) High resolution SQUID imaging of octupolar currents in anisotropic tissue. *IEEE Trans. Appl. Supercond.* **3**, no. 1, 1934-1936.

32. SippensGroenewegen, A., Spekhorst H., van Hemel N.M., Herre Kingma J., Hauer R.N.W., de Bakker, J.M.T., Grimbergen, C.A., Janse, M.J. and Dunning A.J. (1993) Localization of the site of origin of postinfarction ventricular tachycardia by endocardial pace mapping, *Circ.* **88**, no. 5, 2290-2306.

33. Hren, R., Zhang, X. and Stroink, G. (1996) Comparison between electro-cardiographic and magnetocardiographic inverse solutions using the boundary element method, *Med. & Biol. Eng. & Comp.* **34**, no.2, 110-114.

34. Startt-Selvester, R.H. (1992) Nomina Anatomica Contradicta, *J. Electrocardiol.* **25**, 157-159.

35. Hren, R. (1993) The effect of inhomogeneities on electrocardiographic and magneto-cardiographic inverse solutions. M.Sc. Thesis, Dalhousie University.

36. Forsman K., Nenonen, J., Purcell, C. and Stroink, G. (1992) Biomagnetic inverse solution with a realistic torso, in M. Hoke, S.N. Erne, Y.C. Okada and G.L. Romani (eds.), *Biomagnetism; Clinical Aspects,* Elsevier, Amsterdam, pp. 819-823.

37. Nenonen, J., Edens, J.A., Leon, L.J. and Horacek, B.M. (1991) Computer model of propagation excitation in the anisotropic human heart, in K. Ripley and A. Murray (eds.), *Computers in Cardiology,* IEEE Computer Society Press, Los Alamitos, CA, 217-220.

38. Wikswo, J.P., this volume.

39. Stroink, G., Greek, L.S., Elliott, P., Nenonen, J. and MacGregor, J.H. (1992) Is there a need for individualized homogeneous torso models for magnetic inverse solutions?, in M. Hoke, S.N. Erne, Y.C. Okada and G.L. Romani (eds.), *Biomagnetism, Clinical Aspects,* Elsevier, Amsterdam, pp. 813-817.

40. Hamalainen, M.S. and Ilmoniemi, R.J. (1984) Interpreting measured magnetic fields of the brain: estimates of current distributions, Helsinki University of Technology report TKK-F-A559.

41. Ferguson, A.S. and Stroink, G. (1995) Localization of epicardial sources using magnetic and potential maps, in L. Deecke, C. Baumgartner, G. Stroink, and S.J. Williamson (eds.), *Biomagnetism: Fundamental research and clinical applications,* Elsevier, Amsterdam, pp. 641-646.

42. Graumann, R., Abraham-Fuchs, K., Moshage, W. and Schneider, S. (1992) Reconstruction of current densities with anatomical constraints, in M. Hoke, S.N. Erne, Y.C. Okada and G.L. Romani (eds.), *Biomagnetism, Clinical Aspects,* Elsevier, Amsterdam, pp. 813-818.

43. Fuchs, M., Wagner, M., Wischman, H.-A., Dossel, O. (1995) Cortical current imaging by morphologically constrained reconstructions, in L. Deecke, C. Baumgartner, G. Stroink and S.J. Williamson (eds.), *Biomagnetism: Fundamental research and clinical applications,* Elsevier, Amsterdam, pp. 320-325.

44. Gulrajani, R.M., Roberge, F.A. and Savard, P. (1989) The inverse problem of electrocardiography, in P.W. MacFarlane and T. D. Veitch Lawrie (eds.), *Comprehensive Electrocardiology,* Pergamon Press, New York, pp. 327-284.

45. Oster, H.S. and Rudy, Y. (1992) The use of temporal information in the regularization of the inverse problem in electrocardiology, *IEEE Trans. Biomed. Eng.* **BME-39**, pp. 65-75.

46. Tan, G.A., Brauer, F., Stroink, G. and Purcell, C.J. (1992) The effect of measuring conditions on MCG inverse solution, *IEEE Trans. Biomed. Eng.* **BME-39**, pp. 921-927.

47. Stroink, G., Purcell, C., Lamothe, R., Merritt, R., Horacek, B.M. and ten Voorde, B.J. (1988) Body surface potential and magnetic mapping, in K. Atsumi, M. Kotani, S. Ueno, T. Katila, and S.J. Williamson (eds.), *Biomagnetism '87,* Tokyo Denki Univ. Press, Tokyo, pp.74-81.

48. Breithardt, G., Cain, M.E., El-Sherif, N., Flowers, N.C., Hombach V., Janse, M., Simson, M.B. and Steinbeck, G. (1991) Standards for analysis of ventricular late potentials using high-resolution or signal-averaged electrocardiography. *Circulation,* **83**, no.4, 1481-1488.

49. Barr, R.C., Spach, M.S. and Herman-Giddens, G.S. (1971) Selection of the number and position of measuring locations for electrocardiography. *IEEE Trans. Biomed. Eng.* **BME-18,** 125-138.

50. Nousiainen, J.J., Oja, O.S. and Malmivuo, J.A. (1994) Normal vector magnetocardiogram. I. Correlation with the normal vector ECG. *J. Electrocardiol.* **27**, no. 3, 221-231.

51. Bateman, G. (1993) Magnetocardiographic measurements in a magnetically noisy environment. M.Sc. thesis, Dalhousie University.

52. Varpula, T. and Poutanen, T. (1984) Magnetic field fluctuations arising from thermal motion of electric charge in conductors, *J. Appl. Phys.* **55**, 4015-4021.

53. Vrba, J., this volume.

54. Numminen, J., Ahlfors, S., Ilmoniemi, R., Montonen, J. and Nenonen, J. (1995) Transformation of multi-channel magnetocardiographic signals to standard grid form, *IEEE Trans. Biomed. Eng.* **BME-42,** 72-78.

442

55. Vrba, J., Betts, K., Burbank, M. Chueng, T. *et al.* (1995) Whole cortex 64 channel system for shielded and unshielded environments, in C. Baumgartner, L. Deecke, G. Stroink, S.J. Williamson (eds.), *Biomagnetism: Fundamental research and clinical applications*, Elsevier, Amsterdam, pp.521-525.

56. Tavrin, Y., Zhang, Y., Mock, M.A. and Braginski, A.I. (1994) A second order SQUID gradiometer operating at 77 K, *Supercond. Sci. Technol.*, **7**, 265-268.

57. Zimmerman, J.E. (1977) SQUID instruments and shielding for low-level magnetic measurements, *J. Appl. Phys.* **48**, 702-710.

58. Stroink, G., Purcell, C., Brauer, F. and Blackford, B. (1983) An eddy current shielded room with partially closed entrance, *Il Nuovo Cimento* **2D**, no.2, 195-202.

59. Stroink, G. and MacAulay, C. (1986) Thermal magnetic noise generated by an eddy current shielded room, *Rev. Sci. Instrum.* **57** (4), 658-660.

60. Matsuba, H., Shintomi, K., Yahara, A., Irisawa, D., Ima, K., Yoshida, H. and Seike, S. (1995) Superconducting shield enclosing a human body for biomagnetic measurements, in C. Baumgartner, L. Deecke, G. Stroink, S.J. Williamson (eds.), *Biomagnetism: Fundamental research and clinical applications*, Elsevier, Amsterdam, pp.483-489.

61. Ma, Y.P. and Wikswo, Jr., J.P. (1991) Magnetic shield for wide-bandwidth magnetic measurements for nondestructive testing and biomagnetism, *Rev. Sci. Instrum.* **62** (11), 2654-2661.

62. Sullivan, G.W., Lewis, P.S., George, J.S. and Flynn, E.R. (1989) A magnetic shielded room designed for magnetoencephalography, *Rev. Sci. Instrum* **60** (4), 765-770.

63. Amuneal Manufacturing Corp., Philadelphia, PA 19124, USA.

64. Vacuumschmelze GMBH, D-6450 Hanau, Germany.

65. Lant, J., Stroink, G., Montague, T.J., Gardner, M.J. and Mieszkowski, M. (1990) Discrimination between myocardial infarct groups through the use of iso-integral magnetic field maps, *Am. J. Noninv. Cardiol.* **5**, 215-222.

66. Purcell, C., Stroink, G. and Montague, T.J. (1989) Classification of infarcts using electric and magnetic inverse solutions, in S.J. Williamson, M. Hoke, G. Stroink and M. Kotani (eds.), *Advances in Biomagnetism*, Plenum Press, New York, pp. 429-432.

67. Lux, R.L., Evans, A.K., Burgess, A.K., Wyatt, R.F. and Abildskov, J.A. (1981) Redundancy reduction for improved display and analysis of body surface potential maps, I: Spatial compression, *Circ. Res.* **49** (1), 186-196.

68. Press, W.H., Flannery, B.P., Teukolsky, S.A. and Vetterling, W.T. (1992) *Numerical recipes: The art of scientific computing*, Cambridge University Press.

69. Lamothe, R. and Stroink, G. (1991) Orthogonal expansions: their applicability to signal extraction in electrophysiological mapping data, *Med. & Biol. Eng.& Comp.* **29**, 522-528.

70. Vester, E.G. and Strauer, B.E. (1994) Ventricular late potentials: state of art and future perspectives, *Europ. Heart J.*, **15** (Suppl. C), 34-48.

71. Cain, M.E., Ambos, H.D., Arthur, R.M. and Lindsay, B.D. (1992). Signal-averaged electrocardiography: methods of analysis and clinical impact, in W.W. Parmley and K. Chatterjee (eds.), *Cardiology: Physiology, Pharmacology, Diagnosis*, JB Lippincott Co., Philadelphia, pp.1-20.

72. Savard, P., Davies, R.F., Dupuis, R., Ferguson , J., Gardner, M., Lauzon, C., Morel, P., Poitras, N., Stewart, D.J., Sussex, B., Talajic, M., Warnica, W.J. and Rouleau, J.L. (1996) Risk stratification after myocardial infarction using signal-averaged electrocardiographic criteria adjusted for sex, age and myocardial infarction type. To be published.

73. Makijarvi, M., Montonen, J., Toivonen, L., Leini, M., Siltanen, P., and Katila, T. (1994) High-resolution and signal-averaged electrocardiography to separate post-myocardial infarction patients with and without ventricular tachycardia, *Europ. Heart J.* **15**, 189-199.

74. Erne, S.N., Fenici, R.R., Hahlbohm, H.-D., Jaszczuk, W., Lehmann, H.P. and Masselli, M., (1983) High resolution magnetocardiographic recordings of the ST segment in patients with electrical late potentials, *Il Nuovo Cimento*, **2D**: 340-345.

75. Stroink, G., Vardy, D., Lamothe, R. and Gardner, M. (1989) Magnetocardiographic and electro-cardiographic recordings of patients with ventricular tachycardia, in S.J. Williamson, M. Hoke, G. Stroink and M. Kotani (eds.), *Advances in Biomagnetism*, Plenum Press, New York, pp. 437-440.

76. Weismuller, P., Richter, P., Abraham-Fuchs, K., Harer, W., Schneider, S., Hoher, M., Kochs, M., Edrich, J. and Hombach, V. (1993) Spatial differences of the duration of ventricular late potentials in the signal-averaged magnetocardiogram in patients with ventricular late potentials, *Clin. Electrophysiol.* **16** (1): 70-79.

77. Makijarvi, M., Montonen, J., Toivonen, L., Leinio, M., Siltanen, P. and Katila, T. (1992) High-resolution magnetocardiography can identify ventricular tachycardia patients after myocardial infarction, in M.

443

Hoke, S.N. Erne, Y.C. Okada and G.L. Romani (eds.), *Biomagnetism, Clinical Aspects,* Elsevier, Amsterdam, pp. 483-486.

78. Erne, S.N., this volume.

79. Montonen, J. (1995) Magnetocardiography in identification of patients prone to malignant arrhythmias, in C. Baumgartner, L. Deecke, G. Stroink, S.J. Williamson (eds.), *Biomagnetism: Fundamental research and clinical applications,* Elsevier, Amsterdam, pp. 606-611.

80. Smith, E.R., Gardner, M.J., Montague, T.J. and Horacek, B.M. (1985) Sudden cardiac death: the search for a non-invasive means to detect the electrical substrate for the development of life-threatening cardiac arrhythmias, *Clin. and Invest. Med.* **8**, No.1, 41-47.

81. Gardner, M.J., Montague, T.J., Armstrong, C.S., Horacek, B.M. and Smith, E.R. (1986) Vulnerability to ventricular arrhythmias: assessment by mapping of body surface potentials, *Circulation* **73**: 684-692.

82. Hubley-Kozey, C.L., Mitchell, L.B., Gardner, M.J., Warren, J.W., Penny, C.J., Smith, E.R., Horacek, B.M. (1995) Spatial features in Body-Surface Potential Maps can identify patients with a history of sustained Ventricular Tachycardia. *Circulation,* **92**, 1825-1838.

83. Stroink, G., Lant, J., Elliott, P., Charlebois, P. and Gardner, M.J. (1992) Discrimination between myocardial infarct and ventricular tachycardia patients using magnetocardiographic trajectory plots and iso-integral maps, *J. Electrocard.* **25**, 129-142.

84. Stroink, G., Lant, J., Elliott, P., Lamothe, R. and Gardner, M. (1992) Magnetic field and body surface potential mapping of patients with ventricular tachycardia, in M. Hoke, S.N. Erne, Y.C. Okada and G.L. Romani (eds.), *Biomagnetism, Clinical Aspects,* Elsevier, Amsterdam, pp. 471-475.

85. Huang, S.K.S. (1987) Use of radiofrequency energy for catheter ablation of the endomyocardium: a prospective energy source, *J. Electrophysiol.* **1**, 78-91.

86. SippensGroenewegen, A., Spekhorst, H., van Hemel, N.M., Herre Kingma, J., Hauer, R.N.W., Janse, M.J. and Dunning, A.J. (1990) Body surface mapping of ectopic left and right ventricular activation, *Circulation,* **82**, 879-896.

87. Moshage, W., Achenbach, S., Gohl, K., Harer, W., Schneider, S. and Bachman, K. (1992) Magnetocardiography in combination with MRI: Verification of localization accuracy with a nonmagnetic pacing catheter, in M. Hoke, S.N. Erne, Y.C. Okada and G.L. Romani (eds.), *Biomagnetism, Clinical Aspects,* Elsevier, Amsterdam, pp. 447-451.

88. Moshage, W., Achenbach, S., Schneider, S., Gohl, K., Abraham-Fuchs, K., Graumann, R. and Bachmann, K. (1992) Application of multichannel systems in magneto-cardiography, in M. Hoke, S.N. Erne, Y.C. Okada and G.L. Romani (eds.), *Biomagnetism, Clinical Aspects,* Elsevier, Amsterdam, pp. 439-446.

89. Weismuller, P., Abraham-Fuchs, K., Schneider, S., Richter, P., Harer, W., Kochs, M., Edrich, J. and Hombach, V. (1992) Magnetocardiographic localization of ventricular tachycardias with a multichannel system, in M. Hoke, S.N. Erne, Y.C. Okada and G.L. Romani (eds.), *Biomagnetism, Clinical Aspects,* Elsevier, Amsterdam, pp. 465-469.

90. Moshage, W., and Achenbach, S. (1995) Functional localization in cardiology with MCG, in C. Baumgartner, L. Deecke, G. Stroink, S.J. Williamson (eds.), *Biomagnetism: Fundamental research and clinical applications,* Elsevier, Amsterdam, pp. 552-556.

91. Weismuller, P., Abraham-Fuchs, K., Killmann, R., Richter, P., Harer, W., Hoher, M., Kochs, M., Eggeling, Th. and Hombach, V. (1995) Localization of the site of origin of ventricular late fields in the signal averaged magnetocardiogram in patients with ventricular late potentials, in C. Baumgartner, L. Deecke, G. Stroink, S.J. Williamson (eds.), *Biomagnetism: Fundamental research and clinical applications,* Elsevier, Amsterdam, pp. 566-570.

92. Weismuller, P. (1992) Role of magnetocardiography (MCG) in cardiology, in C. Baumgartner, L. Deecke, G. Stroink, S.J. Williamson (eds.), *Biomagnetism: Fundamental research and clinical applications,* Elsevier, Amsterdam, pp.542-545.

93. Nomura, M., Nakaya, Y., Saito, K., Kishi, F., Miyoshi, H., Ito, S, Wada, M., Fujita, S., Takae, T., Tamura, I. (1995) Localization of the focus in ventricular tachycardia by magnetocardiogram, in C. Baumgartner, L. Deecke, G. Stroink, S.J. Williamson (eds.), *Biomagnetism: Fundamental research and clinical applications,* Elsevier, Amsterdam, pp. 571-575.

94. Ushijima, S., Magara, T., Kawasuji, M. et al. (1985) Diagnosis of the origin of ventricular tachycardia by body surface maps-evaluation of QRS wave mapping and T wave mapping, *Jpn. J. Electrocardiol* **5**: 190-197.

95. Yuan, S., Blomstrom, P., Pehrson, S. and Olsson, S.B. (1991) Localization of cardiac arrhythmias: conventional noninvasive methods, *Int. J. Cardiac Imag.* **7**, 193-205.

96. Iwa, T. and Magara, T. (1981) Correlation between localization of accessory conduction pathway and body surface maps in WPW syndrome, *Jpn. Circ. J.* **45**, 1192-1198.

444

97. Benson, D.W., Sterba, R., Gallagher, J.J., Waltson, A.I.I., Spach, M.S. (1982) Localization of the site of ventricular preexcitation with body surface potential maps in patients with WPW syndrome, *Circulation* **65**, 1259-1268.

98. Lamothe, R.M.J., Stroink, G. and Gardner, M.J. (1996) BSPM recording of a WPW patient with two accessory pathways and two atrial complexes, *J. Electrocard.* **129** (2), 139-147.

99. Dubuc M., Nadeau, R., Tremblay, G., Kus, T., Molin F., Savard, P. (1993) Pace mapping using body surface maps to guide catheter ablation of accessory pathways in patients with WPW syndrome, *Circulation* **87**, 135-143.

100. Nakaya, Y., Nomura, M., Saito, K., Kishi, F., Miyoshi, H., Nishikado, A., Bando, S. and Nishitani, H. (1995) Comparative studies of magnetocardiographic and electrocardiographic mappings for the localization of accessory pathway in WPW syndrome, in C. Baumgartner, L. Deecke, G. Stroink, S.J. Williamson (eds.), *Biomagnetism: Fundamental research and clinical applications*, Elsevier, Amsterdam, pp. 580-585.

101. Purcell, C., Stroink G. and Horacek, B.M. (1987) Magnetic inverse solution using a homogeneous torso model. Proc. 9th Ann. Conf. of IEEE Eng. in Med. and Biol., 214-215.

102. Makijarvi, M., Nenonen, J., Leinio, M., Montonen, J., Toivonen, L., Nieminen, M.S., Katila, T. and Siltanen, P. (1992) Localization of accessory pathways in WPW syndrome by high resolution magnetocardiographic mapping, *J. Electrocard.* **25**, 143-155.

103. Nenonen, J., Purcell, C.J., Horacek, B.M., Stroink, G. and Katila, T. (1991) Magnetocardiographic functional localization using a current dipole in a realistic torso. *IEEE Trans. Biomed. Eng.* **38**, 658-664.

104. Hren, R. and Stroink, G. (1995) Application of the surface harmonic expansions for modeling the human torso, *IEEE Trans. Biomed. Eng.* **42**, 521-524.

105. Lamothe, R., Stroink, G. and Gardner, M.J. (1995) Body surface potential and magnetic field maps of WPW syndrome patients, in L. Deecke, C. Baumgartner, G. Stroink, and S.J. Williamson (eds.), *Biomagnetism: Fundamental research and clinical applications*, Elsevier, Amsterdam, pp. 591-594.

NEUROMAGNETISM AND ITS CLINICAL APPLICATIONS

G.L. ROMANI, C. DEL GRATTA
Institute of Medical Physics, and
Institute of Advanced Biomedical Technologies
Gabriele D'Annunzio University
Via dei Vestini 33, 66013, Chieti, ITALY

V. PIZZELLA
Institute of Solid State Electronics - CNR
Via Cineto Romano 42, 00156, Rome, ITALY

1. Introduction

This contribution is devoted to the presentation of the latest results in the field of neuromagnetism, namely the application of SQUIDs to the patho-physiological study of the human brain. For completeness sake, these results are preceded by an account, in the form of a tutorial, of the theoretical and experimental aspects of neuromagnetism: the methods of field calculation and source estimation, and an overview of the state-of-the-art instrumentation. A recent review of neuromagnetism, that covers theoretical, experimental and clinical fields, can be found in [1]. Complete and more advanced treatments of the volume conductor effect, of distributed source modelling, and of integration of biomagnetism with Magnetic Resonance Imaging (MRI) and other techniques for medical imaging are published elsewhere in this volume.

2. Generation of biomagnetic fields

In this section, the bases of neuromagnetic field generation will be assessed, stating the physical framework of the theory and connecting elementary physiological processes to basic current patterns.

2.1. THE QUASISTATIC APPROXIMATION OF MAXWELL'S EQUATIONS

The time variability of bioelectric phenomena corresponds to a frequency range extending from dc to a few hundred hertz - typically below 1 kHz. Inductive and displacement effects at these relatively low frequencies are negligible. Indeed, if we consider a sinusoidal component of the signal at frequency f, it can be seen that the displacement-current term and the real- current term in Maxwell's equations are in the

H. Weinstock (ed.), SQUID Sensors: Fundamentals, Fabrication and Applications, 445–490.
© 1996 *Kluwer Academic Publishers.*

ratio of $2\pi\varepsilon f/\sigma$ [2]. Typical values of $\sigma \cong 0.25\ \Omega^{-1}m^{-1}$ and $\varepsilon \cong 10^{-10}\ Fm^{-1}$ yield for this ratio a value of about $10^{-9}\ f$. This quantity is negligible in the frequency range of interest. The induction term can be shown to be negligible as well. Therefore, explicit time-derivative terms in Maxwell's equations may be neglected. We consider here for simplicity a non-polarised, non-magnetised medium, for which the equations are:

$$\text{div } \mathbf{E} = \rho / \varepsilon_0 \tag{1}$$

$$\text{curl } \mathbf{E} = 0 \tag{2}$$

$$\text{div } \mathbf{B} = 0 \tag{3}$$

$$\text{curl } \mathbf{B} = \mu_0 \mathbf{J} \tag{4}$$

To these we must add the equation stating the law of charge conservation, which, in the quasi-static approximation, is given by:

$$\text{div } \mathbf{J} = 0 \tag{5}$$

and the equation stating the differential form of Ohm's law:

$$\mathbf{J} = \sigma \mathbf{E} \tag{6}$$

where σ represents the conductivity distribution of the medium, hereafter considered isotropic for simplicity. The above equations (1-6) rule all bioelectric phenomena.

In describing bioelectric currents we will distinguish between active currents, which take place inside the cell and sustain a potential difference along the cell, and passive currents, which take place in the extracellular medium. The former are called *impressed currents*, and are denoted by $\mathbf{J}^i$, while the latter are called *volume currents*. Therefore, at any point inside the conductor, the total current density is the sum of the impressed currents and the volume current density:

$$\mathbf{J} = \mathbf{J}^i + \sigma \mathbf{E} \tag{7}$$

The latter is the quantity that appears in Equations 4 and 5.
Let us now consider an infinite homogeneously conducting medium. Taking the divergence of Equation 7, and using Equation 5, we obtain:

$$\text{div } \mathbf{E} = - \text{div } \mathbf{J}^i / \sigma \tag{8}$$

which gives, using Equation 2:

$$\nabla^2 \Phi = \text{div } \mathbf{J}^i / \sigma \tag{9}$$

where Φ is the electric potential. Substituting Equation 7 into Equation 4 and taking the curl gives, after taking Equations 2 and 3 into account:

$$\nabla^2 \mathbf{B} = - \mu_0 \text{ curl } \mathbf{J}^i \tag{10}$$

Equations 9 and 10 show that both the electric potential and the magnetic field at any point of an infinite homogeneous conductor depend only on the impressed currents and not on the volume currents. Moreover, not the currents themselves, but their divergence and curl, are the mathematical sources of the field. Thus, for instance, a divergence-free current, like a current loop, would produce no electric potential, and a curl-free current, like a radial current, would produce no magnetic field.

In general, the electric potential and magnetic field are the solutions of Equations 9 and 10, respectively:

$$\Phi_\infty(\mathbf{x}) = \frac{1}{4\pi\sigma} \int \frac{\text{div}' \mathbf{J}^i(\mathbf{x}')}{|\mathbf{x} - \mathbf{x}'|^3} dv' \tag{11}$$

$$\mathbf{B}_\infty(\mathbf{x}) = \frac{\mu_0}{4\pi} \int \frac{\text{curl}' \mathbf{J}^i(\mathbf{x}')}{|\mathbf{x} - \mathbf{x}'|^3} dv' \tag{12}$$

where the integration is extended to a volume containing the sources.

The latter expressions may be transformed in the following forms showing the direct dependence on the sources [3]:

$$\Phi_\infty(\mathbf{x}) = \frac{1}{4\pi\sigma} \int \mathbf{J}^i(\mathbf{x}') \cdot \frac{(\mathbf{x} - \mathbf{x}')}{|\mathbf{x} - \mathbf{x}'|^3} dv' \tag{13}$$

$$\mathbf{B}_\infty(\mathbf{x}) = \frac{\mu_0}{4\pi} \int \mathbf{J}^i(\mathbf{x}') \times \frac{(\mathbf{x} - \mathbf{x}')}{|\mathbf{x} - \mathbf{x}'|^3} dv' \tag{14}$$

The ∞ subscript in the above Equations 11 to 14 indicates that they are valid for an unbounded homogeneous medium.

2.2. ELECTROPHYSIOLOGY OF EXCITABLE CELLS

2.2.1. *The cell membrane*
In electrically-active cells, the membrane plays a very important role in the generation and propagation of electric signals. The cell membrane consists of a double layer of phospholipids spanned by proteins. The hydrophylic head groups of the phospholipids are in contact with water in the intracellular medium on one side and the extracellular medium on the other side, whereas the hydrophobic tails of one layer are in contact

with those of the other layer in the middle of the membrane. This configuration gives the membrane a high electrical resistance: for example, the membrane of a 1-mm-long segment of a 100-μm-diameter nerve axon has a resistance of 300 kΩ [4]. The proteins have an axial channel that allows certain ions to cross the cell membrane, thus making the membrane selectively permeable to several ion species. Some of the proteins are gated, i.e., voltage dependent, which means that these proteins allow passage of ions only when they are in a specific configuration, depending on the transmembrane potential. In its resting state, the membrane sustains a negative potential difference between the inside of the cell and the outside: in typical nerve cells this potential difference ranges from - 45 mV to -70 mV. This resting transmembrane potential is due to several simultaneous mechanisms: (1) there is a high concentration gradient of ion species across the membrane, specifically the K^+ ion concentration inside the cell is greater than outside, whereas the Na^+ ion concentration outside the cell is greater than inside; (2) the Na-K pump drives Na^+ ions out of the cell and K^+ ions into the cell; (3) the membrane is selectively permeable to different ion species, i.e., permeability to K^+ ions is greater than that to Na^+ ions. Let us first consider the case in which the membrane is permeable to only one ion species, namely K^+ ions. Diffusion of ions through the membrane protein channels builds up a potential difference across the membrane that settles at a value at which the ion current due to the concentration gradient balances the ion current due to potential difference. The equilibrium is stable and no energy is required to maintain it.

If we consider next the simultaneous action of two ion species, such as K^+ and Na^+, with opposed concentration gradients, the outflux of K+ ions must be balanced by the influx of Na^+ ions in order to leave the potential difference across the membrane undisturbed. However, if this were the only mechanism at play, currents due to concentration gradients would not be counterbalanced by currents of the corresponding ion species due to potential difference, and the diffusion process would continue undisturbed, leading to a depletion of K^+ ions inside the cell and of Na^+ ions outside the cell. In fact, the concentration gradient is maintained by the action of the Na-K pump, which uses metabolic energy derived from ATP. Since the membrane is much more permeable to K^+ ions, the transmembrane potential is close to the Nernst potential of K^+ ions.

The membrane is stable with respect to small variations in the transmembrane potential: for example, a small influx of Na^+ ions would make the inside of the cell less negative with respect to the outside, thereby increasing the outflow of K^+ ions due to potential difference, and thus compensating the former potential imbalance. Such a decrease in transmembrane potential due to an inward current is called a *depolarisation* and the membrane reaction an *electrotonic potential*. If the depolarisation is large enough to exceed a threshold value, usually about 15 mV, the Na^+ channels of gated proteins open, thus allowing a large influx of Na^+ ions that further depolarises the cell. This depolarisation continues until the potential of the inside of the cell becomes 30 to 50 mV positive with respect to the outside. At this

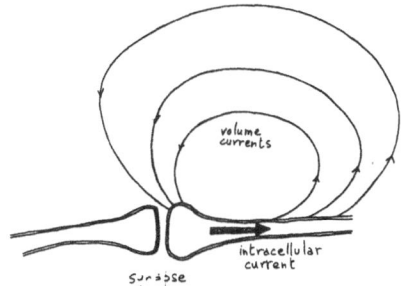

a) postsynaptic potential

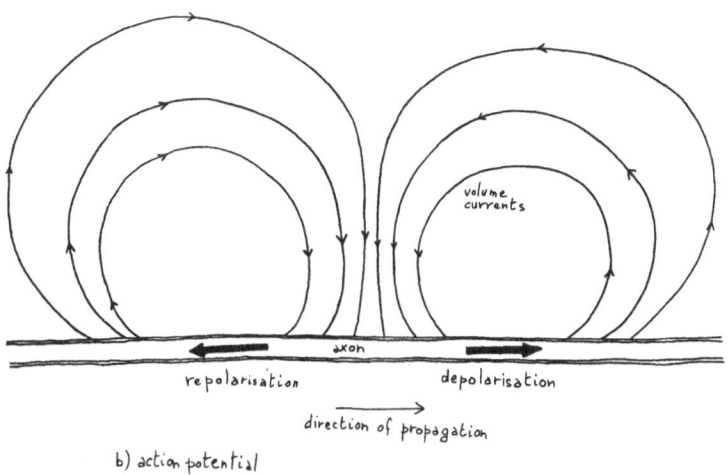

b) action potential

Figure 1. Action potential and postsynaptic potential

voltage the Na$^+$ channels close, reducing the Na$^+$ ion influx, and the K$^+$ channels open, allowing a large outflux of K$^+$ ions. The latter phenomenon is called *repolarisation* because it restores the original transmembrane potential. Since the number of ions involved in the depolarisation as well as in the repolarisation of the membrane is small, the concentration gradients of K$^+$ and Na$^+$ ions across the membrane are not significantly altered by these processes [5].

2.2.2. *Action potential and postsynaptic potential*

The presence of gated proteins and the high resistance of the membrane allow the propagation of a perturbation such as a depolarisation followed by a repolarisation. Let us imagine that the membrane is locally depolarised over a small area. When, at the climax of Na^+ influx, the inside of the cell is positive with respect to the outside, the Na^+ ions are localised near the open Na^+ channels. While the open Na^+ channels close again and the K^+ channels open, the Na^+ ions diffuse inside the cell making a nearby region positive. This determines the opening of other Na^+ channels and the local depolarisation of the membrane. Thus, the depolarisation propagates along the cell, closely followed by a repolarisation.

Typical nerve cells consist of a soma surrounding the nucleus, several terminations called dendrites, and an additional longer termination called the axon. The dendrites establish connections with other neurons, whereas the axons connect the cell to peripheral muscles and organs. Both the axon and the dendrites terminate in a *synapse* that allows signal transmission by electrical or chemical means. When the intracellular potential in the soma, at the base of the axon, falls below a certain threshold value, a depolarisation wave is triggered that can travel along the axon, followed by a repolarisation wave. The close succession of these two waves is called an *action potential*. The duration of an action potential is of a few milliseconds.

The electric potential in the soma is, in turn, determined by the impulses coming from the dendrites of the cell. These impulses are triggered at the synapse and are therefore called *postsynaptic potentials*. They consist of a slow depolarisation wave followed by a much slower repolarisation. In comparison with the action potential, the postsynaptic potential may last several tens of milliseconds [6].

2.3. A SIMPLE MODEL OF BIOELECTRIC CURRENTS: THE CURRENT DIPOLE

These physiological processes now must be given a mathematical description in order to develop a model of the sources of the magnetic field. We shall separate the currents at play into intracellular currents, transmembrane currents and extracellular currents. Intracellular currents flow from a depolarising region to a region in the resting state, or from a region in the resting state to a repolarising region. Transmembrane currents flow radially out of the cell in repolarising regions or into the cell in depolarising regions. They contribute negligibly to the magnetic field because they are directed radially and because they are very short, due to the thinness of the membrane [6]. Finally, extracellular currents close the current loop from a depolarising region to a repolarising region, spreading in the extracellular medium. Neglecting transmembrane currents, we can identify intracellular currents with the active or impressed currents $\mathbf{J}^i$, and extracellular currents with the passive currents $\sigma \mathbf{E}$ of Equation 7.

The intracellular current may be represented by a straight and short current element along the axis of the dendrite or axon, characterised by its length and its intensity.

Such a current element is called a current dipole because the current pattern it determines in the extracellular medium is equal to that of two pointlike, closely spaced, positive and negative current sources - source and sink, or equivalently that of a time varying electric dipole. The moment Q of a current dipole is a vector and is equal to the product of current intensity and length [6]:

$$Q = IL \tag{15}$$

The current density function corresponding to a current dipole is given by

$$J^i(x) = Q \, \delta(x\text{-}x_0) \tag{16}$$

where δ is the Dirac delta function and x_0 is the position vector of the current dipole. Inserting this current-density function in Equations 13 and 14 gives the electric potential and magnetic field for a dipole immersed in an infinite homogeneously-conducting medium:

$$\Phi_\infty(x) \quad = \quad \frac{1}{4\pi\sigma} \; Q \; \bullet \; \frac{(x-x_0)}{|x-x_0|^3} \tag{17}$$

$$B_\infty(x) \quad = \quad \frac{\mu_0}{4\pi} \; Q \; \times \; \frac{(x-x_0)}{|x-x_0|^3} \tag{18}$$

The current dipole can be used as a model of elementary cellular events. For example, a postsynaptic potential may be represented by a current dipole, whereas an action potential may be represented by two closely spaced and oppositely-directed current dipoles [6]. The latter source configuration has a vanishing current dipole moment and is called a current quadrupole. The field of a quadrupole falls off as the inverse cubic distance. Therefore, quadrupole fields are much weaker than dipole fields. For this reason magnetic signals from the brain are believed to originate mainly from postsynaptic potentials rather than from action potentials.

When large assemblies of cells are involved in a cooperative process, as in brain function, it is not possible to take into account every cell, and the current distribution must be integrated over the active area [7]. Thus, a macroscopic description of the sources is adopted, in accordance with the experimental fact that macroscopic fields are measured. When the integral of the microscopic current distribution over the active area is non- vanishing, the current distribution has a leading first-order term called the *equivalent current dipole*:

$$Q \;\; = \;\; \int J^i(x') \, dv' \tag{19}$$

In the case of brain activity, the size of an active cell group may be estimated by comparing the magnetic signal of a single neuron to those actually measured. The field due to a postsynaptic potential may be evaluated using a simple model [6]. Let us consider the dendrite as a cylinder with diameter d, length L, and conductivity σ; let V be the potential difference along the dendrite. The current in the dendrite is given by V / R, where R is the dendrite resistance, equal to $4 L / (\sigma \pi d^2)$. Therefore, the magnitude of the corresponding current dipole moment is, according to Equation 15:

$$Q = V \sigma \pi d^2 / 4 \qquad (20)$$

Substituting for typical values of $\sigma \cong 0.25 \ \Omega^{-1} m^{-1}$, $d \cong 4 \ \mu m$, $V \cong 10$ mV, we obtain Q $\cong 3 \times 10^{-14}$ Am. The corresponding field intensity at a 4-cm distance is, according to Equation 18, about 2×10^{-18} T. This value is well below the sensitivity limits of biomagnetic instrumentation and also below the thermal noise of tissues. For comparison, typical field values measured over the scalp in response to somatosensory or auditory stimulation are on the order of 10^{-13} T. Therefore, about 50,000 synapses are needed to build typical fields. Fields over the scalp mainly originate from these postsynaptic potentials, rather than from action potentials. The latter are, indeed, much faster and their field is much weaker.

In conclusion, the equivalent current dipole can be used as a model of active sources. It must be considered as a macroscopic current dipole, in the sense that it results from the summation over a large number of microscopic dipoles, even though it is small compared to the size of body organs.

3. Solution of the direct problem

3.1. OVERVIEW

The direct, or forward, problem consists of the calculation of the magnetic field distribution, given the current distribution inside the body. This calculation requires a model for the currents as well as for the surrounding conductor. We shall consider here only the equivalent current-dipole model. In principle, any current distribution may be viewed upon as made of a large number of current dipoles.

Modelling of the conductor may take into account organ features of increasing complexity. The simplest requirement is that the conductor be bounded rather than infinite, and this implies that (in general) volume current contribution may not be neglected. Then, the shape of the conductor should approximate that of the investigated organ. We shall see that simple geometrical shapes, for instance the sphere, permit one to obtain an analytical expression for the field [8]. More accurate modelling should take into account dishomogeneities in tissue conductivity; this is done in multi-compartment models [9, 10]. Finally, modelling the precise shape of the organs requires numerical calculations, as well as a numerical representation of tissue boundaries [11,12]; this technique recently has benefited from medical imaging tools such as MRI [13]. In addition, some features of the organs, such as tissue anisotropy, require sophisticated modelling and lengthy calculations [14, 15].

3.2. FIELD OF A CURRENT DIPOLE IN A BOUNDED MEDIUM

We now shall solve the biomagnetic direct problem for a case that is general enough to be useful in many applications, but still not so general as to be exceedingly complex from a numerical point of view. We will assume that tissue conductivity is isotropic and piecewise homogeneous, i.e., tissue dishomogeneities are concentrated at the interfaces between different tissues [3].

Moreover, we shall assume that any region of uniform conductivity is surrounded by another region of uniform conductivity, while currents are confined to the innermost region. The field is to be calculated outside the outermost region. Let V_1, V_2, ... , V_n be n regions such that

$$V_1 \subset V_2 \subset ... \subset V_n \tag{21}$$

and let S_1, S_2, ..., S_n be their respective boundary surfaces, so that S_1 is the innermost interface and S_n is the outermost interface. For each surface S_j , with $j = 1,...,$ n, let σ_{j+}, and σ_{j-} be the outer and inner conductivity, respectively, and let **n** be the unit vector normal to the surface and directed outwards. Substituting Equation 7 into Equation 4, and taking the curl gives

$$\nabla^2 \mathbf{B} = - \mu_0 (\text{curl } \mathbf{J}^i + \text{curl } (\sigma \mathbf{E})) \tag{22}$$

since the conductivity distribution σ is no longer uniform. The second term on the right hand side of Equation 22 takes into account the contribution of volume currents to the field. Here (and in the following), in accordance with our discussion on the equivalent current dipole, we consider $\mathbf{J}^i$ to be a macroscopic current distribution. Taking Equation 2 into account we have

$$\text{curl } (\sigma \mathbf{E}) = \text{grad } \Phi \times \text{grad } \sigma \tag{23}$$

We can see that this term is nonvanishing only at the interfaces between regions with different conductivities. At interface S_j we have

$$\text{grad } \sigma = (\sigma_{j+} - \sigma_{j-}) \, \delta(s_j(\mathbf{x})) \, \mathbf{n} \tag{24}$$

where $s_j(\mathbf{x}) = 0$ is an implicit representation of surface S_j. Using Equations 23 and 24, Equation 22 becomes

$$\nabla^2 \mathbf{B} = - \mu_0 \text{ curl } \mathbf{J}^i - \mu_0 \sum_j (\sigma_{j+} - \sigma_{j-}) \, \delta(s_j(\mathbf{x})) \text{ grad } \Phi \times \mathbf{n} \tag{25}$$

This is the same as Equation 10 for an infinite homogeneously-conducting medium, except for an extra source term distributed over the interfaces between regions of different conductivities. The solution of Equation 25 is

$$\mathbf{B}(\mathbf{x}) = \frac{\mu_0}{4\pi} \int \frac{\text{curl}' \, \mathbf{J}^i(\mathbf{x}')}{|\mathbf{x} - \mathbf{x}'|} dv' + \sum_j (\sigma_{j+} - \sigma_{j-}) \frac{\mu_0}{4\pi} \int_{S_j} \frac{\text{grad}' \, \Phi \times \mathbf{n}'}{|\mathbf{x} - \mathbf{x}'|} da' \tag{26}$$

This equation can be rewritten in a form analogous to that of Equation 14 by using the formula for the gradient of a product and Stokes' theorem [16]:

$$\mathbf{B}(\mathbf{x}) = \frac{\mu_0}{4\pi} \int \mathbf{J}^i(\mathbf{x}') \times \frac{(\mathbf{x} - \mathbf{x}')}{|\mathbf{x} - \mathbf{x}'|^3} dv' + \frac{\mu_0}{4\pi} \sum_j (\sigma_{j+} - \sigma_{j-}) \int_{S_j} \Phi \mathbf{n}' \times \frac{(\mathbf{x} - \mathbf{x}')}{|\mathbf{x} - \mathbf{x}'|^3} da' \tag{27}$$

We thus see that, in a bounded piecewise homogeneous medium, the contribution of volume currents may be described as due to a fictitious current distribution restricted to the interfaces between regions with different conductivities. These fictitious currents also are called *secondary currents*, whereas the impressed currents are called *primary currents*.

Note that the secondary currents do not flow on the surfaces, but are orthogonal to them. Rather than a surface current density, the secondary sources constitute a current dipole distribution with dipole moment per unit area equal to ($\sigma_{j+} - \sigma_{j-}$) Φ. Therefore, to calculate the volume current contribution to the magnetic field, the electric potential over the surfaces S_j must be known.

Taking the divergence of Equation 7, and using Equations 5 and 2, we obtain

$$\sigma \nabla^2 \Phi = \text{div } \mathbf{J}^i - \sum_j (\sigma_{j+} - \sigma_{j-}) \, \delta(s_j(\mathbf{x})) \text{ grad } \Phi \cdot \mathbf{n} \tag{28}$$

where $s_j(\mathbf{x}) = 0$ is an implicit representation of surface S_j. Let us consider a point $\mathbf{x}$ inside the volume V_n but not on any surface S_j. We can use Equation 28 to derive an integral form similar to that of Equation 26:

$$\sigma(\mathbf{x})\Phi(\mathbf{x}) = \frac{1}{4\pi} \int \frac{\text{div}' \, \mathbf{J}^i(\mathbf{x}')}{|\mathbf{x} - \mathbf{x}'|} dv' + \sum (\sigma_{j+} - \sigma_{j-}) \frac{1}{4\pi} \int_{S_j} \frac{\text{grad}' \, \Phi \cdot \mathbf{n}'}{|\mathbf{x} - \mathbf{x}'|} da' \tag{29}$$

This equation can, in turn, be transformed to yeld a form similar to Equation 27 [17]:

$$\sigma(\mathbf{x})\Phi(\mathbf{x}) = \frac{1}{4\pi} \int J^i(\mathbf{x}') \cdot \frac{(\mathbf{x} - \mathbf{x}')}{|\mathbf{x} - \mathbf{x}'|^3} \, dv' + \frac{1}{4\pi} \Sigma(\sigma_{j+} - \sigma_{j-}) \int_{Sj} \Phi \, \mathbf{n}' \cdot \frac{(\mathbf{x} - \mathbf{x}')}{|\mathbf{x} - \mathbf{x}'|^3} \, da'$$

(30)

From this equation an integral equation can be derived for the potential at any point on the surfaces S_j, taking into account the discontinuity of the integral over S_j as the point $\mathbf{x}$ goes through the surface [16]. Thus, for a point $\mathbf{x}$ on surface S_k, the following equation holds:

$$\frac{(\sigma_{k+} + \sigma_{k-})}{2} \Phi(\mathbf{x}) = \frac{1}{4\pi} \int J^i(\mathbf{x}') \cdot \frac{(\mathbf{x} - \mathbf{x}')}{|\mathbf{x} - \mathbf{x}'|^3} \, dv' + \frac{1}{4\pi} \Sigma(\sigma_{j+} - \sigma_{j-}) \int_{Sj} \Phi \, \mathbf{n}' \cdot \frac{(\mathbf{x} - \mathbf{x}')}{|\mathbf{x} - \mathbf{x}'|^3} \, da'$$

(31)

This integral equation must be solved for the electric potential Φ over all the surfaces S_j. Then the magnetic field may be calculated using Equation 27. Note that

$$d\Omega' = \mathbf{n}' \, (\mathbf{x} - \mathbf{x}') / |\mathbf{x} - \mathbf{x}'|^3 \, da'$$ (32)

is the solid angle subtended at point $\mathbf{x}$ by an infinitesimal surface element da' at $\mathbf{x}'$. Note also that the first term in the right hand side of Equation 31 is the potential at point $\mathbf{x}$ for an infinite homogeneous medium with unit conductivity.

In fact, Equation 31 is never solved analytically. When the forward problem is solved using an analytical conductor surface, such as a sphere, it is possible to calculate the magnetic field directly by using the magnetic scalar potential, as will be outlined in the following. When the boundary surfaces are more complicated, then Equation 31 must be solved numerically. This will be described in a later paragraph.

3.2.1. *Spherical conductor*
If the boundaries of the conductor are spherical, the volume-current contribution can be shown to be tangential to the conductor surface. In the following we shall take the origin of the coordinate system at the centre of the sphere. We now shall calculate the radial component of the magnetic field at a point $\mathbf{x}$. Let $\mathbf{e}_x = \mathbf{x} / |\mathbf{x}|$ the radial unit vector at point $\mathbf{x}$. The unit vector normal to the surface is also radial and $\mathbf{n}' = \mathbf{x}' / |\mathbf{x}'|$. The radial component of the field is then given by

$$\mathbf{B}(\mathbf{x}){\cdot}\mathbf{e}_x = \frac{\mu_0}{4\pi} \int \mathbf{J}^i(\mathbf{x}'){\times}\frac{(\mathbf{x}-\mathbf{x}')}{|\mathbf{x}-\mathbf{x}'|^3}{\cdot}\mathbf{e}_x\,dv' + \frac{\mu_0}{4\pi} \Sigma(\sigma_{j+}{-}\sigma_{j-}) \int_{S_j} \Phi\mathbf{n}'{\times}\frac{(\mathbf{x}-\mathbf{x}')}{|\mathbf{x}-\mathbf{x}'|^3}{\cdot}\mathbf{e}_x\,da'$$

$$(33)$$

In the integrand of the second term on the right side of this equation, the vectors $\mathbf{n}'$, $(\mathbf{x}-\mathbf{x}')$ and $\mathbf{e}_x$ lie in a plane so that their mixed product vanishes. Therefore, the second term vanishes identically, and the radial component of the field is equal to that obtainable from Equation 14 for an infinite medium. If the source is a current dipole of moment $\mathbf{Q}$, the radial component of the field is given by

$$\mathbf{B}_r(\mathbf{x}) = \frac{\mu_0}{4\pi}\,\mathbf{Q} \times \frac{(\mathbf{x}-\mathbf{x}_0)}{|\mathbf{x}-\mathbf{x}_0|^3}{\cdot}\mathbf{e}_x \tag{34}$$

We can show further that, in the case of a piecewise homogeneous conductor, where the compartments are concentric spherical shells, the field $\mathbf{B}$ does not depend at all on secondary currents, i.e., it is independent of the conductivity distribution. Outside the conductor, the field $\mathbf{B}$ satisfies, according to Equation 4,

$$\text{curl}\,\mathbf{B} = 0 \tag{35}$$

Therefore, $\mathbf{B}$ is the gradient of a scalar magnetic potential:

$$\mathbf{B} = -\,\text{grad}\,\Phi_m \tag{36}$$

Since the scalar magnetic potential may be taken to vanish at infinity, it may be calculated at a point $\mathbf{x}$ by integrating the field along a radial line starting from $\mathbf{x}$ and extending to infinity:

$$\Phi_m(\mathbf{x}) = -\int_{\mathbf{x}}^{\infty} \mathbf{B}_r(\mathbf{x}')\,d\mathbf{x}' \tag{37}$$

We can see from the above expression and from Equation 34 that Φ_m, and hence $\mathbf{B}$, does not depend on the conductivity distribution. This same procedure may be used to obtain the expression for the field of a current dipole of moment $\mathbf{Q}$ in a piecewise homogeneous sphere, including the volume-current contribution [3, 18, 19]:

$$\mathbf{B(x)} = \frac{\mu_0}{4\pi F^2} (F\mathbf{Q} \times \mathbf{x}_0 - \mathbf{Q} \times \mathbf{x}_0 \bullet \mathbf{x} \text{ grad } F) \tag{38}$$

where

$$F = a(a x + \mathbf{a} \bullet \mathbf{x}), \ \mathbf{a} = \mathbf{x} - \mathbf{x}_0,$$

and

$$\text{grad } F = (x^{-1} a^2 + a^{-1} \mathbf{a} \bullet \mathbf{x} + 2a + 2x)\mathbf{x} - (a + 2x + a^{-1} \mathbf{a} \bullet \mathbf{x})\mathbf{x}_0$$

Finally, from Equation 38 we can see that the field of a radial current dipole in a piecewise homogeneous sphere vanishes identically outside the sphere.

The piecewise homogeneous sphere model has proven useful for the qualitative analysis of neuromagnetic fields as well as for the interpretation of real neuromagnetic data. We shall say more about data analysis in a later section. Although the brain shape is not spherical, some qualitative conclusions may be drawn from the above results: (1) when a field component is measured that is almost perpendicular to the head surface, its dependence on volume currents is very weak; (2) if a source is almost radial, its field is very weak, therefore, assuming that the dendrites are mostly directed perpendicularly to the cortex - as occurs for the apical dendrites of pyramidal neurons, magnetic measurements are most sensitive to the activity of neurons located in the fissures of the cortex; (3) the radial field pattern of a current dipole in a sphere gives good insight to the field that is measured outside the head when the source has a leading dipolar moment. As an example, let us consider the radial component of the field at the surface of a sphere of radius R for a tangentially-oriented dipole of moment Q at depth d below the surface. Transforming Equation 34 to polar coordinates yields

$$B_r(\vartheta, \varphi) = \frac{\mu_0 Q [(R/d) - 1] \sin\vartheta \cos\vartheta \sin\varphi}{4\pi d^2 [((R/d)^2 - 1) \sin2\vartheta + 1]^{3/2}} \tag{39}$$

where ϑ and φ are the usual spherical polar angles, and the dipole is located on the z-axis and directed along the x-axis. The origin is at the centre of the sphere. From this equation we can see that the field is antisymmetric with respect to the xz-plane that contains the dipole and is zero in that plane. If we consider the variations of B_r in the plane orthogonal to the dipole, we can see that it is extremal at an angle ϑ_m which does not depend on the dipole-moment intensity. Therefore, the field has a maximum and a minimum, the dipole lying midway between them. The value of ϑ_m may be found by differentiating the field expression, Equation 39, and it turns out that it has a roughly linear dependence on dipole depth for shallow dipoles, increasing more slowly as the dipole moves deeper [8].

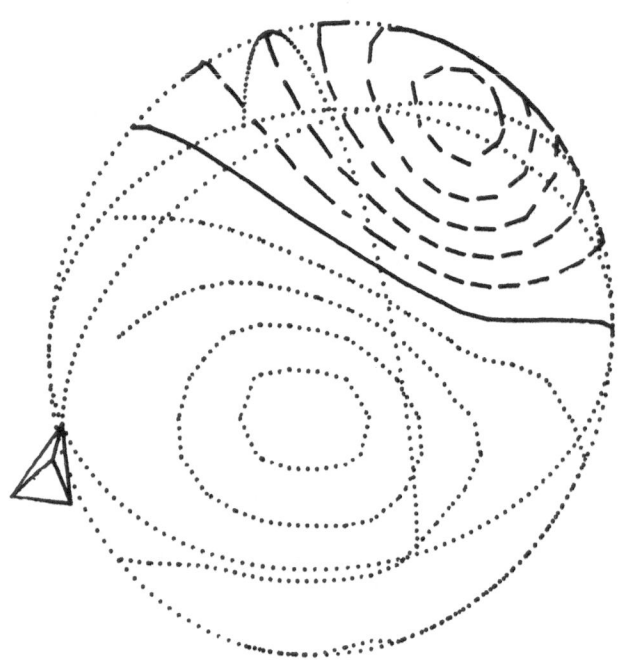

Figure 2. Distribution of the radial field component for a dipolar source in a spherical head

3.3. NUMERICAL CALCULATION OF THE MAGNETIC FIELD

In order to take into account the precise shape of the organ boundaries, Equation 31 and the surface integrals in Equation 27 must be solved using a numerical representation of real surfaces. Such a representation is a list of a large number of points lying on the surface.

3.3.1. *The Boundary Element Method*

In the Boundary Element Method, the integrals are calculated by triangulating the surfaces: the latter are approximated by a large number of adjacent triangles, the vertices of which are the surface points. The integrals then are calculated over the triangulated surfaces. Each integral over a surface therefore can be approximated by a sum of integrals over triangles. In turn, each integral may be approximated by the integrand mean value over the triangle [11, 12, 20, 21, 22]. In this way, the integral equation, Equation 31, can be discretised in the form:

$$\Phi_j = 2\ \Phi_{j\infty} + (1/2\pi)\ \Sigma_{jk}\ \omega_{jk}\ \Phi_k \tag{43}$$

where the indices j and k run over all triangles, Φ_j is the potential value at the centroid of triangle j, $\Phi_{j\infty}$ is the potential value at the centroid of triangle j calculated as though the medium were infinite, homogeneous and with unit conductivity, and ω_{jk} is the solid angle subtended by triangle k at the centroid of triangle j if j ≠ k, and is equal to unity if j = k. Therefore, the potential Φ_j is the solution of a linear system:

$$\Sigma_k\ A_{jk}\ \Phi_k = \Phi_{j\infty} \tag{44}$$

the coefficients of which depend only on conductor geometry and not on the current distribution. The matrix A_{jk} is singular, corresponding to the fact that the potential is determined only to within an additive constant. The singularity may be removed by deflation [23], which in practice means that A_{jk} is replaced by $A_{jk} + 1/N$, where N is the total number of triangles. Solving this new linear system yelds the potential at each triangle:

$$\Phi_j = \Sigma_k\ M_{jk}\ \Phi_{k\infty} \tag{45}$$

where $\Phi_{k\infty}$ can be calculated by means of Equation 13 at any surface point.

Finally, we can calculate the field according to Equation 27. After substituting for the potential at each triangle according to Equation 44, we have an equation of the type

$$\mathbf{B}(\mathbf{x}) = \mathbf{B}_\infty(\mathbf{x}) + \Sigma_j\ \mathbf{P}_j\ \Sigma_k\ M_{jk}\ \Phi_{k\infty} \tag{46}$$

where $P_j = (\sigma_{j+} - \sigma_{j-})\ a_j\ n_j$, a_j is the surface of the j-th triangle and n_j is a unit vector orthogonal to the j-th triangle.

This method has been applied to the direct problem solution in magnetoencephalography. The homogeneous conductivity compartments are: the brain, the cerebrospinal fluid, the skull and the scalp. A simplified version is the homogeneous head model that neglects the skull and the scalp volume-current contribution, thereby reducing the calculation burden. It has been shown [24] that the latter model gives satisfactory results when calculating the magnetic field. This is due to the poor conductivity of the skull, which has the effect of confining the most intense currents to the brain volume.

3.3.2. *Need for a realistic model*
The solution of the direct problem by numerical techniques usually requires costly computations. Therefore, the question that must be addressed here is: when is the use of a numerical model necessary, and when, to the contrary, is an analytical model suitable? Generally, the answer to this question depends on the location of the sources. In recent years the effect of organ boundaries on the magnetic field distribution has been extensively studied by computer simulations [25, 26] using a current dipole as a source and realistic conductor shapes. Results show that the use of a realistic model may in many instances considerably improve the field representation. Non-sphericity in the temporal or frontotemporal regions may result in large deviations from the prediction of the sperical model. The deeper the sources are located, the larger the deviations.

4. Solution of the inverse problem

4.1. OVERVIEW

The retrieval of current sources inside a conductor from measurements of the external magnetic field falls into the broader class of inverse problems. The biomagnetic inverse problem is an ill-posed problem, in the sense that it does not have a unique solution. Indeed, there are sources that produce zero magnetic field outside the conductor (e.g., a radially-oriented current dipole inside a homogeneous sphere) and, due to the superposition principle, adding such a source to a solution for a specific field distribution gives another solution for the same field. Although a solution of this problem is not possible in a mathematical sense, a meaningful estimate of the currents may still be undertaken. To this end it is necessary to constrain the problem in order that it have a unique solution. This is done by means of a model, describing the source and the conductor, which does not contain silent sources. Due to the availability of only a finite and noisy data set, such a model must be characterised by a limited number of parameters. It must be stressed that any information on electrophysiological sources obtained by this method, does not come from a direct measurement, but instead comes from a parameter estimation. Therefore, any conclusion about the sources is valid only to the extent that the model assumptions are valid. This aspect is very important in clinical applications. Another possibility is to evaluate the response of a magnetometer to an arbitrary source distribution, thus ruling out inaccessible information. This is done in the lead-field approach.

4.2. ESTIMATE OF SOURCE PARAMETERS

4.2.1. *Choice of a model*

The first step in the interpretation of biomagnetic data is the choice of a model for the equivalent source, as well as for the volume conductor. The simplest model of a bioelectric source is the equivalent-current dipole. However, it is a suitable model only when the currents are confined to a single small region. In the presence of several simultaneously active regions, multiple dipole models may be used, thereby increasing the flexibility of the model, but also increasing the number of parameters. When the current distribution has no net dipolar moment, a current quadrupole may be used. Higher order multipolar terms may be used in addition to a current dipole. A current loop may be modelled by an equivalent magnetic dipole. The choice of the conductor model is primarily dictated by the measurement conditions. Generally, the smaller the area of field measurement and the shallower the source, the better is the approximation obtained by means of analytical models. The spherical model was described in the previous section in relation to the forward problem solution. While the source-model parameters are the object of the biomagnetic investigation, the volume-conductor-model parameters are obtained by means of an anatomical investigation. The latter may be very simple, such as the measurement of typical head dimensions, or more sophisticated, such as a series of MRI slices of the head for the reconstruction of brain, skull and scalp boundaries. In these examples, the experimenter would, in the first case, find the centre and the radius of the best-fitting sphere to the area of measurement, and in the second case, build triangulated surfaces representing the conductor boundaries. The choice of a volume-conductor model also is a compromise between accuracy and computational costs. The sphere and half-space models have been the most widely used models during the early years of biomagnetism due to the simplicity of the calculations. It may be noted that in some neuromagnetic investigations, important results were obtained using the half-space model [27]. The development in recent times of realistic models responds to an increased need for accuracy. Magnetoencephalography has evolved towards the use of large multichannel instruments allowing simultaneous whole- head mapping of the magnetic field. Under these conditions, volume current contributions are detected coming from several parts of the brain, which are modelled as a secondary source distribution over the brain surface. The latter shows different curvature radii in different regions and therefore cannot be represented by a sphere.

4.2.2. *Least squares search*

A practical method for the estimation of the model parameters is the use of the least-squares estimator. This is a maximum likelihood estimator, i.e., it yields parameter estimates for which the measured data values have the largest probability. In addition, the least-squares estimator gives a statistical criterion for the acceptance or the rejection of the model and confidence intervals for the parameters [28]. The parameter values obtained by this method are those that minimise the quantity

$$\chi^2 = \sum_j \frac{(B'_j - B(x_j; a))^2}{\sigma_j^2} \tag{47}$$

where B'_j is the measured field component at point x_j, $B(x_j; a)$ is the value predicted by the model, which depends on the parameter vector a, and σ_j^2 is the expected variance of B'_j. When B is a linear function of a, the parameter values a_m that minimise Equation 47 may be found by solving a linear system. This is the case for a fixed dipole, i.e., a dipole the location of which is assumed *a priori*. This method may be useful in reducing the number of parameters when the active region is expected to have one or more precise anatomical locations. Generally, however, the dipole location cannot be assumed *a priori*, and its three space coordinates must be considered as free parameters. In this latter case, the field dependence is nonlinear and the minimisation of Equation 47 must be performed by an iterative method. The most widely used one is the Levenberg-Marquardt algorithm [29], which combines a steepest descent, far from the minimum, with a Hessian matrix inversion near the minimum. Note that this algorithm requires that the forward solution corresponding to the current parameter values be calculated many times. Therefore, it is important that the forward calculation be as time efficient as possible. Once the best estimate parameters are found, confidence regions and confidence limits may be obtained by using a linear approximation of the function $\chi^2(a)$ near the minimum. The usual percentage levels corresponding to $\Delta\chi^2$ values are valid when the errors σ_j are normally distributed [28]. This does not seem to be the case in biomagnetic measurements. Therefore, Monte Carlo techniques for the estimate of parameter uncertainty are to be preferred. In the latter case, appropriate noise simulation must be provided, taking into account various noise sources. The uncertainty in a biomagnetic measurement arises mainly from instrument noise, sensor positioning errors and subject noise. The first contribution is small in state-of-the-art instruments that feature a field-noise density below 5 fT/$\sqrt{Hz}$, compared to a signal level of several tens of fT. The second contribution is, in large multichannel systems, a systematic error, and hence shows coherence among different channels. This means that these errors may be taken into account using a model with a small number of parameters. Finally, subject noise shows correlation among channels as well [30]. Subject noise comes from thermal noise in the body, as well as from physiological activity which the experimenter is not interested in. A typical example of this situation is background brain activity when measuring specific evoked or spontaneous signals.

4.3. INTEGRATION WITH MAGNETIC RESONANCE IMAGES

We have mentioned already that conductor boundary surfaces may be extracted from magnetic resonance images. We now shall discuss the relationship between biomagnetism and anatomical investigation tools, and in particular MRI. This aspect in the evolution of biomagnetism is gaining increasing importance and deserves being mentioned. Principally, the use of sophisticated medical imaging was motivated by the need for numerical representation of boundary surfaces. These can be obtained, for instance, from magnetic resonance contiguous cross-sections by means of automatic segmentation techniques. Voxels with similar grey values are assumed to belong to the same conducting region and therefore are grouped together. Then, the boundaries of these regions are found by continuity, starting from a single voxel inside a region; this technique is called region growing. Control and smoothing processes may be used to improve the surface quality. Then points on the surfaces are obtained by tracing lines from the center of each slice and finding their intersection with the surfaces. Finally, triangulation is performed using algorithms that assure the homogeneity of triangle dimensions over the surface [31].

In addition to providing numerical representation of the conductor boundary surfaces, anatomical imaging provides an efficient means of displaying the data that combines anatomical and functional investigation. Once the source coordinates are found by the inverse procedure, they can be shown at their anatomical location superimposed on the appropriate magnetic-resonance slices.

Note that, in order to perform the forward calculation correctly, as well as to locate the sources on magnetic-resonance slices, the sensor positions with respect to the body surface must be measured. This is done by placing small current-carrying coils on the skin, usually at precise anatomical landmarks that behave mathematically like magnetic dipoles. These coils are then localised by means of an inverse procedure, similar to that for a current dipole. If three or more coils are localised for the same magnetometer position, then the position and orientation of the body with respect to the magnetometer may be retrieved [32, 33]. Since the coils define a coordinate system attached to the body, the source coordinates may be given in this system. When an MRI investigation is performed, the same anatomical landmarks are localised using markers visible in the scans, thus finding the coordinates of each voxel in the previous system. This allows the superposition of the current-source image and the anatomical image.

4.4. LEAD FIELD THEORY

In the above mentioned techniques for neural-current estimation a specific model of the source was used, like for example the equivalent current dipole. On the contrary, in the lead field approach no a priori assumption is made on the source. This method is based on the evaluation of the spatial sensitivity function of each sensor [34, 3]. Since the component B_j of the field measured by the j-th sensor depends linearly on the current distribution $J(x)$, it can be expressed as follows:

$$B_j = \int L_j(x) \cdot J(x) \, dv \qquad (48)$$

where the integral is extended to the conducting region. $L_j(x)$ is a vector function called the *lead field* of the j-th sensor. Mathematically, $L_j(x)$ can be expressed in terms of the sensor response to a pointlike source. If the current distribution is a current dipole at point x_0, it is given by Equation 16, and the measured field value from Equation 48 becomes

$$B_j = L_j(x_0) \cdot Q \qquad (49)$$

We can calculate the lead field of a pointlike magnetometer at point x measuring the radial component $B(x) \cdot e_x$ of the magnetic field outside a homogeneous sphere. By comparing Equation 49 with Equation 34 we obtain

$$L_j(x) = - \frac{\mu_0}{4\pi} \, e_x \times \frac{(x - x_0)}{|x - x_0|^3} \qquad (50)$$

The lead-field functions depend on sensor position and conductivity distribution.

Let us now consider the set of all current distribution functions defined over the conducting region. This is a vector space where an inner product is defined:

$$(J_1, J_2) = \int J_1(x) \cdot J_2(x) \, dv \qquad (51)$$

From the above inner product, a norm can be defined by

$$\|J\| = \sqrt{(J, J)} \qquad (52)$$

From Equation 48 it can be seen that

$$B_j = (\, L_j \,, J \,) \qquad (53)$$

Thus, the magnetic field measurements give only information on the current distributions that are a linear combination of the lead-field functions L_j. Therefore, we must look for a solution in the form

$$J = \Sigma_j \, b_j \, L_j \qquad (54)$$

where b_j are numerical coefficients. Note that this relation is not the expression for the most general current distribution, since the lead fields L_j do not span the vector space of all current distributions. Such a solution is called a *minimum norm estimate,* since it is the one with the smallest norm among all possible current distributions that give rise to the set of measurements given by Equation 53. Inserting Equation 54 in Equation 53, we obtain a set of linear equations:

$$\Sigma_k \, A_{jk} \, b_k = B_j \qquad (55)$$

where the coefficients A_{jk} are the inner products (L_j , L_k). The matrix A_{jk} is non singular if the lead fields L_j are linearly independent. However, A_{jk} may possess very small eigenvalues if in practice the lead fields L_j are only "weakly" linearly independent. This leads to large errors in the computation of the coefficients b_j. To remove this numerical instability, the solution of Equation 55 requires regularisation [3], the principles of which are outlined in the following.

A small eigenvalue (and its corresponding eigenvector) identifies a direction in the subspace spanned by the lead fields, { L_j }, that is "weakly coupled" to the sensor. This means that a current distribution along this direction generates comparatively smaller magnetic signals than a current distribution along other directions in the subspace. Depending on the value of the signal-to-noise ratio, current distributions along directions corresponding to small eigenvalues may generate magnetic signals that are strongly degraded by the noise. In the regularisation procedure, directions corresponding to small eigenvalues are simply suppressed. The threshold level for the eigenvalues below which a direction should be discarded depends on the signal-to-noise ratio. Thus, the subspace in which the solution is sought is actually smaller than { L_j } and the minimum norm estimate resulting from regularisation reproduces the measured signals only approximately, but still within the measurement errors.

Although lead-field analysis does not impose the restrictions of a pointlike model, as, for example, the current dipole, it still requires that the sources be *a priori* constrained in a region where the solution is unique. Therefore, the sources are commonly constrained to lie in a given plane. The choice of this plane is made arbitrarily on anatomical bases or, alternately, by means of a previous estimate of the equivalent current dipole of the current distribution.

5. Biomagnetic Instrumentation

The challenge for biomagnetic instrumentation is the detection of extremely weak magnetic signals (1 fT to 100 pT) in the presence of a very noisy background (~ 10 μT and above). Properly designed instrumentation must, therefore, combine high sensitivity with the ability to reject unwanted signals (or noise). Another cause for concern relates to the requirements for clinical applications, where the necessity for measuring the magnetic field pattern over the entire area of interest without repositioning the system is mandatory if this instrumentation is to receive widespread use.

Modern biomagnetic instrumentation is based on SQUIDs that provide the sensitivity to detect magnetic fields as weak as 1 fT. The fundamental concepts of SQUID devices are described in Chapter 1 and elsewhere in this volume. For our purposes, the SQUID, and its related electronics, may be considered as a black box with a low-noise magnetic-flux-to-voltage converter in it. Reviews of biomagnetic instrumentation may be found in [1, 35, 36].

5.1 THE SINGLE CHANNEL BIOMAGNETIC SYSTEM

To understand how a large multichannel biomagnetic system is assembled, it is useful to look at the different parts of a single "unit" channel of a biomagnetic system. The main components are: the detection coil with the flux transformer, the SQUID with its related electronics and the dewar. These components always are present in a biomagnetic system, although their characteristics may vary according to specific needs.

5.1.1 *Detection coil*
The detection coil is actually the sensor of the magnetic field. It is convenient to use a separate detection coil apart from that of the SQUID loop, because it is possible to change the field spatial sensitivity of the device without affecting the SQUID performance. The magnetic flux sensed by the detection coil is transferred to the SQUID through an input coil which is often built directly onto the SQUID chip. It is worth noting that the entire flux transformer - detection coil and input coil - is a superconducting loop. Thus, the external magnetic field induces a current in this loop which is proportional to the field itself:

$$I = \frac{\Phi}{L} = \frac{(\mathbf{B} \cdot \hat{\imath}) S}{L} \tag{56}$$

where $\mathbf{B}$ is the magnetic field, $\hat{\imath}$ is the normal to the coil, S is the area of the coil, and L is the coil inductance. This is different from the usual case of resistive coils where only the derivative of the field over time induces a current in the loop. Finally, it must be remembered that a single sensor measures only a component of the magnetic field. In the following section we shall refer to "the magnetic field sensed by a detection coil at one point in space," although it really senses only one component of the magnetic field at that point. A detailed discussion of different types of detection coils and their properties is presented elsewhere in this volume; here, only some helpful concepts will be reviewed.

The simplest detection coil is made by a single turn (or few turns) of superconducting wire (magnetometer). If the magnetic field is constant over the coil area, the current flowing in the loop is simply proportional to the field intensity. The important advantage of this simple type of detection coil is that it is easy to integrate within the SQUID chip, thus simplifying the construction of complex biomagnetic systems.

The multiturn coil is the best type of detection coil from the sensitivity point of view. Specific geometries for the detection coil may reduce conveniently the sensitivity to noise sources, with little loss of sensitivity for the biomagnetic sources of interest. The most used detection coil of this type is the first-order gradiometer. The first-order gradiometer is made by adding two magnetometer coils wound in an opposite sense. These coils may be displaced along their common axis (producing an axial gradiometer) or in their common plane (producing a planar gradiometer). The magnetic field sensed by this type of coil is, therefore, the difference of the field between the pick-up and the compensation coil. These gradiometer-type detection coils are the most widely used in the currently-available multichannel biomagnetic systems.

Another type of detection-coil system is the second-order axial gradiometer. This device couples two first-order gradiometers end-to-end yielding a detector which is insensitive to spatially uniform fields and linear gradients. This type of coil system was once very popular when single-channel or small multichannel systems were built for operation in unshielded environments. Today, the complexity of these coils, which require a large amount of machine and hand work and a relatively large volume inside the dewar (e.g., a 5-cm baseline results in a 10-cm device), limits their use.

5.1.2 *Dewar*
The dewar used in biomagnetic systems is a critical part of the instrument and must satisfy severe requirements: (1) the distance of the detection coils from the head of the

subject must be as small as possible - usually less than 20 mm, typically 15 mm - since the field intensity decreases as $1/r^2$; (2) the magnetic noise must be less than the noise of the sensors. Usually fiberglass is used as a material to build the dewar. Fiberglass has optimal magnetic properties. However, its use poses some problems in the mechanical stability of the dewar, especially in the tail section. To guarantee the safety of the subject, thick fiberglass layers are used, especially in flat-bottomed systems, but this causes the detection coils to be far from the subject. Moreover fiberglass does not provide any shield against radiation, and therefore, many (50 - 100) layers of mylar must be wrapped around the inner portion of the dewar for radiation shielding. However, these shields increase the magnetic noise of the dewar, since eddy currents are allowed to flow on the aluminated side of the mylar and generate magnetic field noise [37]. A compromise must be made to obtain the best overall performance. Typical biomagnetic dewars exhibit noise below 5 fT/Hz$^{1/2}$. Last, when designing clinical systems, it is important to consider practical issues for system operation and maintainance. For instance, the helium reservoir should be large enough to ensure several days between helium refills, yet without increasing by too much the external dimensions and weight of the dewar. Typical whole-head neuromagnetic dewars have a helium capacity of ~30 liters with a helium refill interval of 2-4 days, whereas for smaller dewars, helium may last for 7-8 days.

5.1.3 *Shielded room*

Nowadays, biomagnetic multichannel systems always operate in magnetically-shielded rooms. These rooms are relatively large - typical inner dimensions are 3×4×2.5 meters - and comfortable. The shielded rooms are built in accordance with two physical phenomena: eddy-current shielding (provided by layers of highly-conducting materials, copper or aluminium), and magnetic shielding (provided by layers of high-magnetic-permeability materials, Fe-Ni alloys, typically with small percentages of Mo, Cu, Cr and Al). A description of the different techniques used to build these rooms may be found in [38].

Eddy-current shielding works as follows: time-varying magnetic fields induce electrical currents circulating in the conducting layer in such a way that their magnetic field tends to cancel the changes of the applied magnetic field by virtue of Lenz's law. The efficiency of the shield depends on the layer thickness, on the conductivity of the materials and on the frequency of the external magnetic field. While it is relatively simple to shield against radiofrequency interference ($f > 500$ kHz) using conducting layers with less than 1-mm thickness, the efficiency of eddy-current shielding drops dramatically at low frequencies, i.e. in the range of the power-line frequency and below. Usually aluminum layers, 20 mm or more thick, are used. At very low frequencies, where the eddy-current shielding is no longer effective, the only effective shielding method is to use low hysteresis ferromagnetic materials with high magnetic permeability μ, the so-called soft magnetic materials. Pure iron ("soft" iron with a carbon content less then 100 ppm) and several alloys based on iron and nickel show μ values larger than 10^3 and are suitable for shielding applications. However, for practical reasons, and especially to obtain a reasonable price-to-performance ratio, a value of $\mu > 10^4$ is mandatory.

Today, typical medium-quality shielded rooms are built using two layers of high-μ materials, with a thick layer of high-conductivity material (usually aluminium) in between. The performances of two shielded rooms installed at CNR-IESS (Rome) and at the ITAB at the University of Chieti are shown in Figure 3.

5.2 MULTICHANNEL BIOMAGNETIC SYSTEMS

Several multichannel systems (some of which are commercial units) now are operating in laboratories and hospitals in several countries. Two different kinds of systems are in use: flat-bottomed systems, mainly devoted to magnetocardiographic measurements, but also used for the study of the peripheral nervous system, and spherical or "head-shaped" systems for neuromagnetic measurements. Most of these systems operate in a medium- quality shielded room. The most important systems are:

− The 122-channel system of Neuromag Ltd, Helsinki (Finland), for brain measurements [39]. This system is made of 61 pairs of orthogonal planar gradiometers arranged on a helmet-like dewar tail to cover the entire cortex of the subject. The average distance between the gradiometers is 44 mm. Each gradiometer has an area of 5.3 cm^2 with a baseline of 16.5 mm. The noise level is less than 5 fT/(cmHz$^{1/2}$).

− The 64-channel system of CTF Systems Inc, Port Coquitlam (Canada), for brain measurements [40]. This total-head system uses wire-wound first-order axial gradiometers with a 5-cm baseline and a 2-cm diameter. Appropriate algorithms simulate 2nd- and 3rd-gradiometer response for each channel, thus allowing one to work in unshielded environments with an overall noise below 10 fT/Hz$^{1/2}$. If a shielded room is used, and if hardware first-order gradiometers are used, the noise is approximately 4 fT/Hz$^{1/2}$. This system can be customised to improve the spatial resolution in particular areas of interest with the addition of extra sensors up to the limit of 100 channels (including reference channels). A new system with 143 MEG sensors and 26 reference channels has been announced - see Chapter 3 in these proceedings.

− The 37-channel system of BTi, San Diego (USA), devoted to magnetoencephalography [41]. The basic 37 channels use first-order gradiometers. This instrument can be paired with a second one specifically designed to operate up-side-down for simultaneous recordings over the two hemispheres of the head, thus doubling the total number of channels. Each concave tail covers a circular region of about 14 cm in diameter; the white-noise level is less than 10 fT/Hz$^{1/2}$.

- The 31-channel system of Philips AG, Hamburg (Germany) [42], for brain measurements, assembled using individual probes and with a spherical dewar bottom. The detection coils are first-order gradiometers with a 20-mm diameter and 70-mm baseline. The white-noise level is below 10 fT/Hz½ for each channel.

- The 128-channel system, Chiba and ETL, Tsukuba (Japan) [43]. By virtue of its unique 3/4-helmet shape, this system can measure the brain activity of a subject lying on a bed, regardless of head shape or size. It allows simultaneous recording of the parietal, occipital and one side of the temporal regions. The sensors are coupled to first-order gradiometers, and the noise level is less than 10 fT/Hz½ above 5 Hz.

- The 256-channel system of the ETL and SSL (Japan) [44]. This system is being developed as a national project of the Superconducting Sensor Laboratory, and it

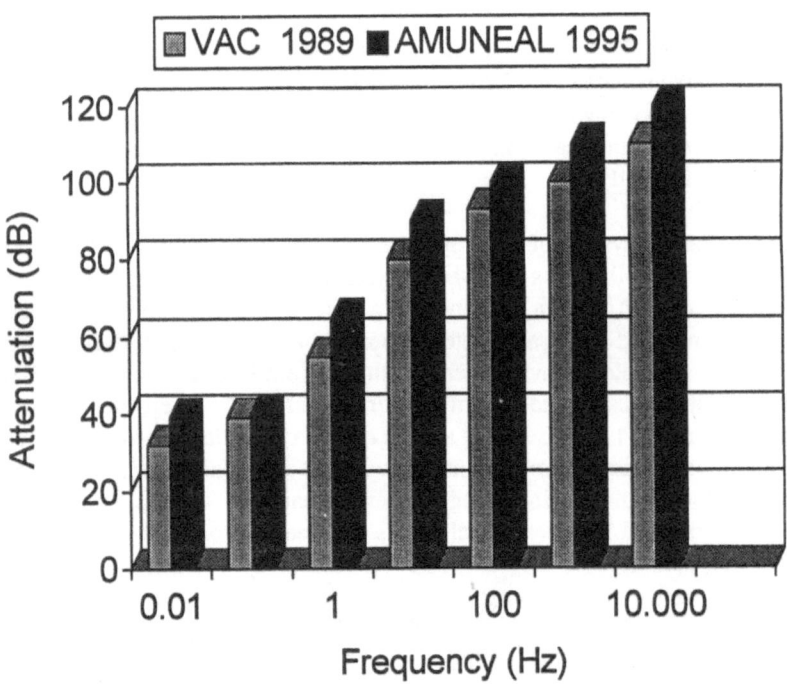

Figure 3. Performances of two shielded rooms installed at CNR-IESS (Rome) (VAC 1989) and at the ITAB at the University of Chieti (AMUNEAL 1995). Attenuation factor is governed by μ-metal at frequencies below 10 Hz, while it is governed by aluminium at frequencies greater than 10 Hz.

includes a high-quality shielded room, liquid helium cooler and other related technologies.

– The 83-SQUID system of the PTB, Berlin (Germany) [45]. This system features an almost circular, flat sensitive area of 21-cm average diameter. It employs on-chip integrated magnetometers, with 49 primary sensors which are sensitive to the normal component of the field B_z, 14 sensors which are sensitive to the tangential components B_x and B_y, and 20 additional sensors used to form first- and/or second-order electronic gradiometers for noise cancellation. The typical white-noise level is 2.5 fT/Hz$^{1/2}$ in the first-order gradiometer mode. The system operates in a medium-quality shielded room.

– The 28-channel system of the IESS - CNR, Rome (Italy) [46], devoted to brain measurements. It covers an area of 16 cm in diameter and consists of an array of 16 axial gradiometers surrounded by an array of 9 simple magnetometers; 3 additional magnetometers are used to perform 3-axis noise cancellation on the magnetometer channels. The white-noise level is less than 5 fT/Hz$^{1/2}$ inside a medium-quality shielded room. This system is an intermediate step toward the building of a 162 channel, total-head instrument completely based on SQUID integrated magnetometers and vectorial noise cancellation [47]. Furthermore, a flat system with 74 channels (including 19 channels for noise compensation) also is under development.

Other systems are operating in laboratories around the world. The current trend favors systems capable of performing measurements with a single positioning and completing data analysis within 30-60 minutes. Such operation is uniquely required to achieve the goal of routine clinical application. For this reason great effort must be focused on data acquisition systems and data analysis algorithms.

5.3 SIGNAL CONDITIONING AND DATA ACQUISITION

All biomagnetic measurements are performed using computerised data acquisition systems, therefore signal conditioning is required to best match signal characteristics in terms of amplitude and bandwidth with the dynamic range, the resolution and the sampling rate of the analog-to- digital converter (A/DC) in use. Indeed, thanks to digital signal processors (DSPs), the implementation of digital filters, featuring much better performance than their analog counterparts, now is possible at very reasonable prices. Therefore, the use of analog filters is limited to DC filtering to avoid offset problems and anti-aliasing low-pass filters prior to A/D conversion.

5.3.1 *Bandpass filtering*
Requirements for the recording bandpass are determined by the A/D process. The high-pass filter must ensure that there is no DC offset in the signal to properly utilise the A/DC resolution. Usually a first-order filter is employed with cut-off frequency as

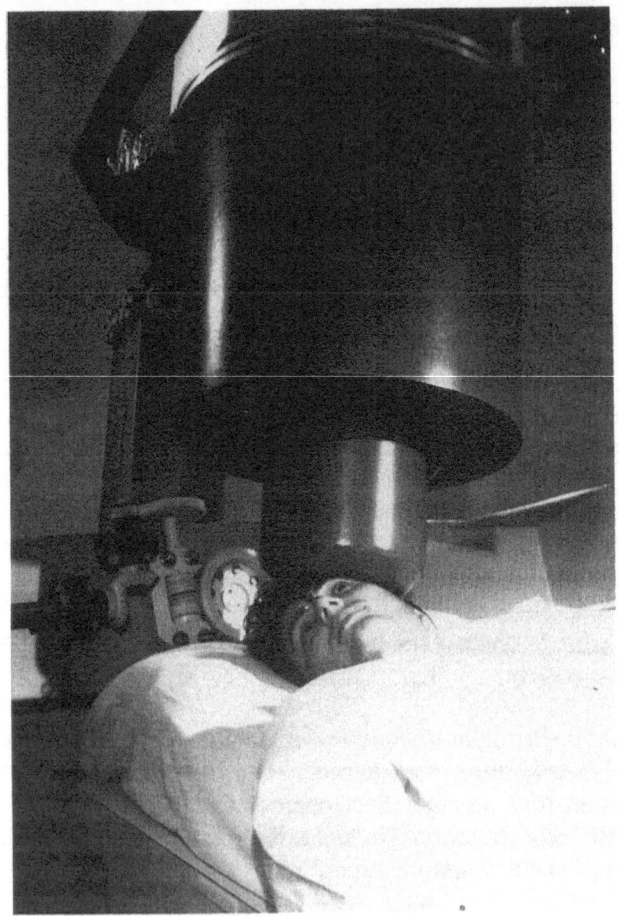

Figure 4. The 28-channel system operating at IESS-CNR, Rome Italy.

low as possible according to the A/DC dynamic range. With regard to the low-pass filter, the sampling theorem requires that, if the sampling frequency is f_s, the signal to be sampled must have components at frequencies higher than $f_s/2$ less than the sensitivity of the A/D converter. If this is not true, the components at frequency greater than $f_s/2$ are transformed into low-frequency components, thus resulting in unwanted noise. This phenomenon is known as "aliasing." To avoid aliasing, an analog filter is used prior to A/D conversion. Usually 48 dB/octave or even 96 dB/octave Butterworth filters are used, and to ensure a linear phase shift in the signal bandwidth, the cutoff frequency is usually set at 1/3 or 1/4 of the sampling frequency.

5.3.2 *Line rejection filtering*
Currently the use of dedicated analog filters for rejection of the magnetic signal due to power lines is limited to special applications with small systems operating in

unshielded environments. As a matter of fact, the use of a shielded room, and often electronic noise suppression, greatly reduces this noise, and residual noise is either negligible or comfortably treatable with digital filtering after the A/D conversion.

If measurements are carried out in an unshielded environment without special electronic noise cancellation, it is mandatory to filter out this noise prior to the A/D conversion to properly use the A/DC resolution. Notch filters are not used for this purpose because of their poor performance. Two other types of filters, namely the comb filter and the hybrid filter, although complex and expensive, are used to fulfill this task [48].

5.3.3 *Analog-to-digital conversion*

After analog conditioning, the signal is A/D converted. Usually a 12-bit A/DC provides enough dynamic range. Only if no analog high-pass filter is used, or power-line noise is large, 14- or 16-bit converters are required. Since modern A/DCs are much faster and less expensive than in the past, it is now favorable to perform the A/D conversion by "oversampling" at a much higher frequency (32 to 64 times) than the signal bandwidth. In this way, to avoid aliasing, a simple first-order 6 dB/octave analog filter may be used, with low-pass frequency set at $f_s/32$ or $f_s/64$.

In multichannel systems it is very important that all channels are sampled simultaneously: for this reason every channel must be connected to its own sample-and-hold circuit, followed by an analog multiplexer and an A/D converter. If oversampling is used, an A/D converter is dedicated to each channel, thus avoiding multiplexing.

5.3.4 *Digital signal processing*

In modern systems most of the on-line data treatment, e.g., done by bandpass filters and line-frequency filters, is done after A/D conversion. For instance, for the case of oversampling, data should be low-pass filtered and downsampled to the desired sampling rate before being written to disk. There are many available DSPs that can be used to accomplish this task. The main advantages in using digital filters for the low-pass and high-pass filters are the identical behaviour of these filters across different channels, and the absence of thermal drifts, ageing and noise pick-up. Additionally, it is possible to filter out environmental noise from data using suitable data processing algorithms. An example of noise cancellation is shown in Figure 5.

The use of DSPs allows one also to simulate different types of detection coils from actual measurements as discussed in Chapter 3 of this volume.

5.3.5 *Experimental control*

If a biomagnetic system is to be used in a clinical environment, the experimental control features must meet the demands of the medical staff. Usually the system hardware, including data acquisition, may be controlled through a high-end PC or,

474

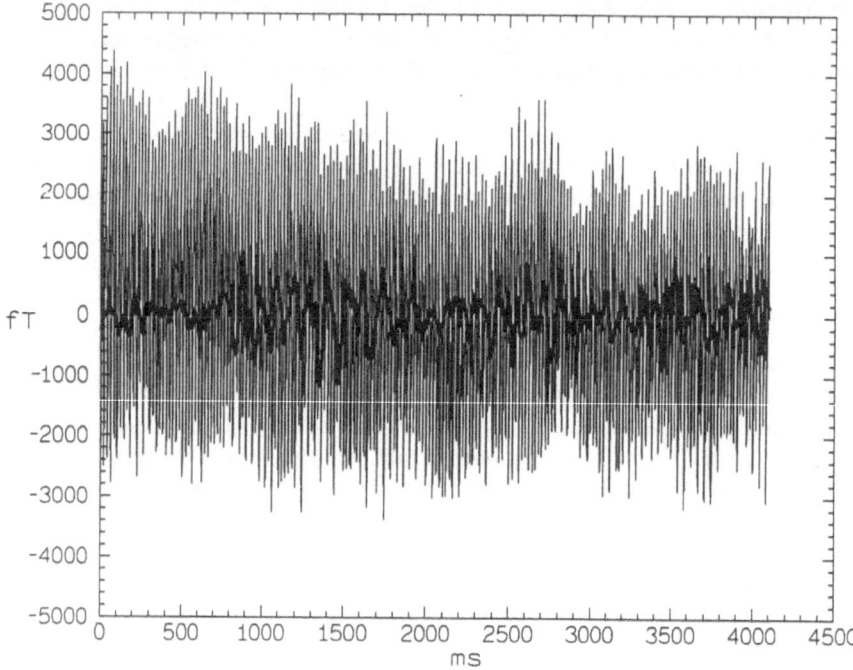

Figure 5. Example of magnetometer data output before and after noise cancellation.

preferably, a workstation. Highly graphic performance is needed in order to monitor all measuring channels during data acquisition. An example of such an environment may be seen in Figure 6. From different windows it is possible to set SQUID operating parameters and data acquisition parameters such as sampling frequency and bandpass filters, as well as to monitor incoming data.

6. Experimental set-up and data analysis

Before proceeding with a more detailed description of the results obtained by the biomagnetic technique, it is useful to examine a typical set-up required to run successfully a neuromagnetic experimental session. The experimental set-up plays an important role in the design of the experiment, and must be carefully considered to obtain the best results from this technique.

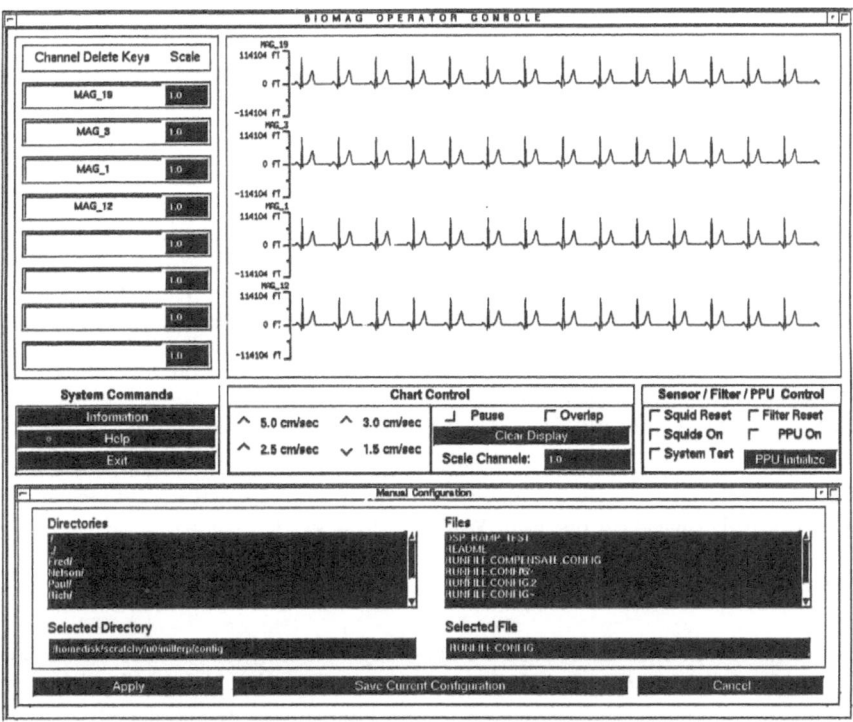

Figure 6. Layout of a high-end user interface for biomagnetic system control.

6.1 SET-UP OF A NEUROMAGNETIC MEASUREMENT

First, almost all multichannel systems are designed to operate inside a magnetically-shielded room. These rooms are quite large and comfortable, but nevertheless impose some restrictions on additional instrumentation that may be required in studying special patients. Inside the shielded room all ferromagnetic material must be avoided, and therefore, instruments for sensory stimulation such as TV pattern generators and current-pulse generators must be placed outside the shielded room. Before entering the shielded room, the subject must be carefully demagnetised,e.g., with a TV monitor degausser, since small magnetic particles are always present on the hair and the skin, and a net magnetisation of these particles may increase the overall noise of the measurement. Once the subject is ready, he is introduced into the shielded room and the sensor is positioned close to his head. Data may be acquired for several minutes, depending on the specific measurement, but usually 5-10 minutes are sufficient. If the sensor is able to record magnetic data over the entire scalp, the experimental session is over. If a smaller multichannel system is used, it is often necessary to reposition the sensor at several places over the head in order to obtain the complete magnetic field distribution over the scalp. This necessity to move the sensor, which was inevitable until a few years ago, gave rise to a major limitation of this method and prevented this biomagnetic methodology from being adopted in clinical practice.

Great care also must be taken in the determination of sensor position with respect to the head. If several sensor positions are necessary, it is mandatory to know exactly the coordinates of each position in order to reconstruct the spatial magnetic field distribution. If a single position of the sensor is sufficient, its coordinates are needed to successfully integrate electrophysiological data with anatomical data taken from MRI or CT. The accuracy of the spatial location of the active parts of the cortex is dependent on many specific characteristics of the activation itself, and this is still a subject of discussion [49-51]. If the Equivalent Current Dipole model is used, the best accuracy is for tangential dipoles located at a maximum depth of 3-4 cm inside the head. For ordinary conditions, an accuracy of 4-5 mm is achieved, while it may be as good as 2-3 mm under the most favourable conditions. Most of the errors occur are in locating precisely the sensor positions with respect to brain anatomy.

6.2 DATA ANALYSIS

The biomagnetic method is focused on the study of biological functions through the study of the intra- and extra-cellular currents inside the biological tissue under examination. Therefore, the final goal of the biomagnetic technique is to find the current-density distribution inside the examined volume. This task is generally impossible, as there is no unique solution to the inverse problem of finding the sources generating a known field. However, if some assumptions about the sources are made, some characteristics of the current distribution may be obtained. Information on sources generating the biomagnetic field may be obtained also by measuring the electric potential on the scalp. However, in this section we will restrict ourselves to the analysis of magnetic field data.

Usually data analysis is divided into a two-step process: first, the distribution of the magnetic field generated by the particular source under study is found; second, the source characteristics are gathered from the magnetic field values.

6.2.1 *From raw magnetic field values to processed magnetic field values*
This step, sometimes labelled pre-processing, is devoted to the extraction of the signal of the source under study from the whole population of magnetic sources, including also intrument noise sources. Often all "unwanted" magnetic fields, i.e., the magnetic field not generated by the source under study, are called "noise", while the magnetic field generated by the source under study is called the "signal." Several tasks are to be completed during this step:

- Magnetic field values are examined in order to exclude that portion of the data corrupted by A/DC overflows, large external noise or possible instrument failures. This task is sometimes done during data acquisition before data is written on disk.
- Data contaminated by head or eye movements are removed from the data set.

These two tasks are not related to the particular source under study; they just improve signal-to-noise ratio by excluding portions of high noise data. The next two tasks maximize the signal-to-noise ratio while trying to separate, as much as possible, the signal from the noise using information on the specific source under study.

- Data coming from the particular source under study is extracted from the remaining data set, that is the signal is extracted from the noise. This task is the most critical one, and the success of the source characterization procedure strongly depends on this task. We shall come back to this point later.
- Last, the selected data are averaged (if desired), and the final topography of the magnetic field is found.

To clarify the different tasks, let us describe a simple but typical session of evoked-field measurement. After data collection, data is automatically checked for identification of spikes, large drifts and other typical signals clearly not generated by the examined source. After removal of these components, data is averaged over all the stimuli presented to the subject. The averaging procedure cancels out all (noise) signals not phase-locked to the stimulus, including the magnetic field generated by intracerebral currents, but not related to the source activated by the stimulus. In this way the magnetic field values reflect only the source under study, i.e., the one activated by the stimulus.

In this simple case, the averaging procedure takes care also of data extraction, since all unwanted data averages to zero. However, there are other cases when there is a need for specific data selection and extraction. For instance, when recording the magnetic field generated by spontaneous activity of an epileptic patient, the field values are sampled during a period of several minutes. Only during specific time windows the pathological activity is present, and these epochs must be correctly identified in order to proceed to source characterization. Several tools are used in order to successfully reach this identification: correlation with a template, fourier analysis, principal component analysis, etc. Bandpass filtering also may be included in these tools as a procedure to discard noise from data.

6.2.2 *From processed magnetic field values to source characteristics*
Once data have been properly processed, the source characteristics may be inferred from magnetic field topography. The simplest source model used in neuromagnetic data analysis is the current dipole as described in section 2.3. In this simple case the source characteristics are completely determined by the spatial position and strength of the dipole. Although this is a very simple model, it is useful to condense source

information: regardless of the real source nature, it is always possible to use the abstraction of the Equivalent Current Dipole (ECD) to condense field spatial distribution information into the parameters used to characterize the ECD. Finally, this model may be extended including two or more ECDs to more accurately match the different activation sites of the brain [52].

A different approach is focused on the reconstruction of the current density distribution inside the head. As described in section 4.4, it is possible to determine the current density J flowing in a two-dimensional plane from the measurements of the magnetic field distribution over the plane, and by imposing that the overall amplitude of the current distribution $\|J\|$ is minimum [53].

7. Applications and examples : Basic neurophysiology and clinical investigations

7.1. FUNCTIONAL AND TOPOGRAPHIC STUDIES OF SENSORY AREAS

In recent years much attention has been devoted to the topographic mapping of the sensory areas in order to reproduce and confirm what was already known from the pioneering work by Penfield [54]. For example, in addition to the neuromagnetic characterisation of the tonotopic organisation of the auditory cortex [55, 56], some other interesting properties have been identified, such as the characterisation of an amplitopic map in the auditory cortex [57], or the localisation of the missing fundamental [58].

A new thrust in the area of evoked activities, more concerned with recognition and attention, has been given by Näätänen and his co-workers, who first described the so-called mismatch negativity (MMN), an electrically negative component that could be detected when a rare different stimulus was randomly inserted in a series of equal stimuli [59]. The same group has devoted particular attention over the past years to the magnetic counterpart of MMN - see [60] for a review of their work. These experiments now are performed in many research and clinical centers, including MEG laboratories, and their importance for cognitive functions, learning, memory and language development is generally accepted.

In the area of sensory-motor integration, Kristeva and co-workers [61] have performed recordings of the motor evoked field (MEF) due to movement of the second finger as compared to stimulation of the same finger. The aim of this study was to investigate whether it is possible to discriminate non-invasively in humans the contributions of mainly proprioceptive and cutaneous areas in the primary somatosensory area. The results show that the dipole locations for the first component of the motor evoked field, namely MEF I or MEFI, were significantly deeper than the early somatosensory evoked field (SEF) components, consistently with a position on the floor of the central sulcus. The first two SEF components did not show significant location differences, and their position was consistent with a generator located in area 3b that occupies most of the posterior bank of the central sulcus. This study together with the studies by Hari and co-workers [62, 63] is a further demonstration of MEG capability of localizing activity in sub-areas of the primary cortices in man.

Kristeva and co-workers also have carried out a study dealing with the phenomenon of sensory gating [64]. Here again the stability of the responses over time is important. The experiment protocol involved electric sensory stimulation, self-paced movements and sensory stimulation triggered by a self-paced movement. In all of the investigated subjects the overlay of somatosensory stimulation on movement provoked a decrement of the brain responsiveness as determined by the SEF amplitude. Gating was found not to be restricted to the SEF's components within 50 ms after movement onset, but it affected the later components as well, showing that the gating phenomenon persists during the entire movement.

7.2. EPILEPSY

The dipole localisation procedure finds an ideal field for clinical application in the study of epilepsy, and namely, the presurgical diagnosis and detection of epilepic foci. There has been impressive development in this area by many laboratories, and the current trend relies strongly on the combination of MEG with other techniques, both invasive and non-invasive. As an example, we would like to quote the work by Nakasato and co-workers [65], where the non-invasive source analysis obtained with MEG and EEG was compared with ECoG localizations, surgical outcome and the presence of lesions suggested by MRI. The authors show that whenever a good dipole fit could be obtained with satisfactory reproducibility, there was "excellent agreement between MEG dipole sources and ECoG sources, as well as surgical outcome and presence of MRI lesions" while EEG source analysis, though satisfactory, was not as consistent.

7.3. OTHER CLINICAL STUDIES

Another interesting study was performed several years ago by Pelizzone and co-workers [66] on patients with a cochlear implant. In this case the major advantage of the neuromagnetic approach (with respect to a more traditional technique) is related to the possibility of evaluating, from the intensity of the magnetic evoked response, the number of residual fibers that still contribute to the analysis of auditory input at a cortical level. Some authors have measured auditory evoked fields from both hemispheres of aphasic patients: normal responses were measured from the intact hemisphere, whereas the 100m component, i.e. the component occurring 100 ms after the stimulus in the magnetic evoked response, was totally absent over the injured one [67]. The CT examination confirmed the presence of ischemic lesions in the primary auditory cortex and in the associative areas immediately posterior to Brodmann's areas 41 and 42. Karhu and co-workers [68] first have measured the SEF due to median- and

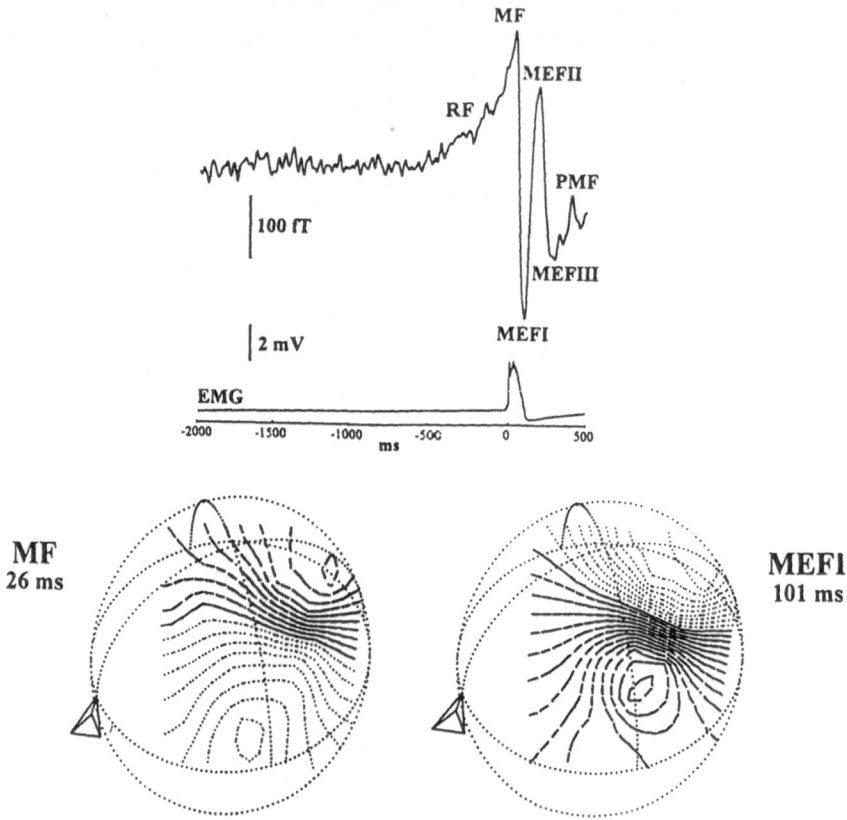

Figure 7. Motor Evoked Fields. Top: time trace recorded during index finger movement from a lateral position of the left scalp, together with the nerve electromyographic signal. The six different components are indicated on the MEG trace: RF (readiness field), MF (motor field), MEF I, MEF II, MEF III (three movement evoked fields), PMF (post movement field). Bottom: magnetic field distribution during MF and MEF I (isofield step 30 fT). The field patterns suggest anteriorly-directed current flow for the MF and posteriorly directed current flow for the MEF I.

ulnar-nerve stimulation in patients affected by multiple sclerosis. These authors observed an increase in latency of the N20m response, similarly to what happens to the electric N20. Surprisingly, three of the subjects studied showed an impressive enhancement of the 60m component. This finding, not yet observed in electric recording, was interpreted as possibly due to the effect of prolonged action potentials, known to appear in demyelinated nerves, that could increase postsynaptic excitation and thereby enhance the amplitude of middle latency SEFs.

7.4. HEMISPHERIC ASYMMETRIES

A systematic investigation of hemispheric asymmetries in a population of patients with focal hemispheric lesions due to vascular problems has been performed by Rossini and co-workers [69]. The aim of the study was at establish whether or not the neuromagnetic method was suitable for identifying some parameter associated with cerebral *plasticity*, during recovery of sensory motor functions after a stroke. The authors studied the location and strength of the equivalent sources activated in the primary somatosensory cortex after left and right median nerve stimulation. No statistically significant differences between the spatial coordinates of the equivalent sources in the two hemispheres were observed. Minor differences were shown to be related to interhemispheric asymetries in the position of the central sulcus as revealed by MRI. Finally, the authors observed that, when comparing the location of the generators across individuals, interhemispheric differences fluctuated less than absolute values, thus suggesting the use of the former as a discriminative parameter in clinical studies on hemispheric lesions and subsequent cortical reorganization.

7.5. CEREBRAL PLASTICITY

On the basis of the preceding example it might be inferred that the neuromagnetic method is suitable for identifying and monitoring plasticity phenomena. In order to validate this hypothesis, the same group has recently performed a neuromagnetic study [70] aimed at investigating the cortical rearrangement of one-finger representation in the primary somatosensory cortex transiently induced by an anaesthetic block of the sensory information from adjacent fingers. The basis for this experiment was the observation that the neuromagnetic system used could identify well the ECD accounting for the sources elicited by stimulation of individual fingers (the 1st, 3rd, and 5th fingers), as illustrated in Figure 6. In order to investigate a possible transitory cortical reorganisation, similar to what had been observed in animals [70], Rossini and co-workers used a recording protocol based on three stages: (A) control condition, in which a specific finger was stimulated using ring electrodes, and the magnetic response was measured over the contralateral hemisphere; (B) ischemic condition, where the measurement was performed following complete anaesthesia of the non-test fingers after about 20 minutes of ischemia obtained by tightly fastening rubber bands at the base of the non-test fingers, and an anaesthetic cream also was applied to eliminate the burning pain induced by the ischemia alone; (C) replication of the control condition, after reopening of blood perfusion and regaining of sensation. It was observed that the ECDs responsible for the cortical responses N20m and P30m from an un-anaesthetised finger were significantly changed after a relatively brief period of sensory deprivation from the adjacent fingers. Such changes of the ECDs with respect to the control conditions, were characterised by an increase in strength in all cases, by

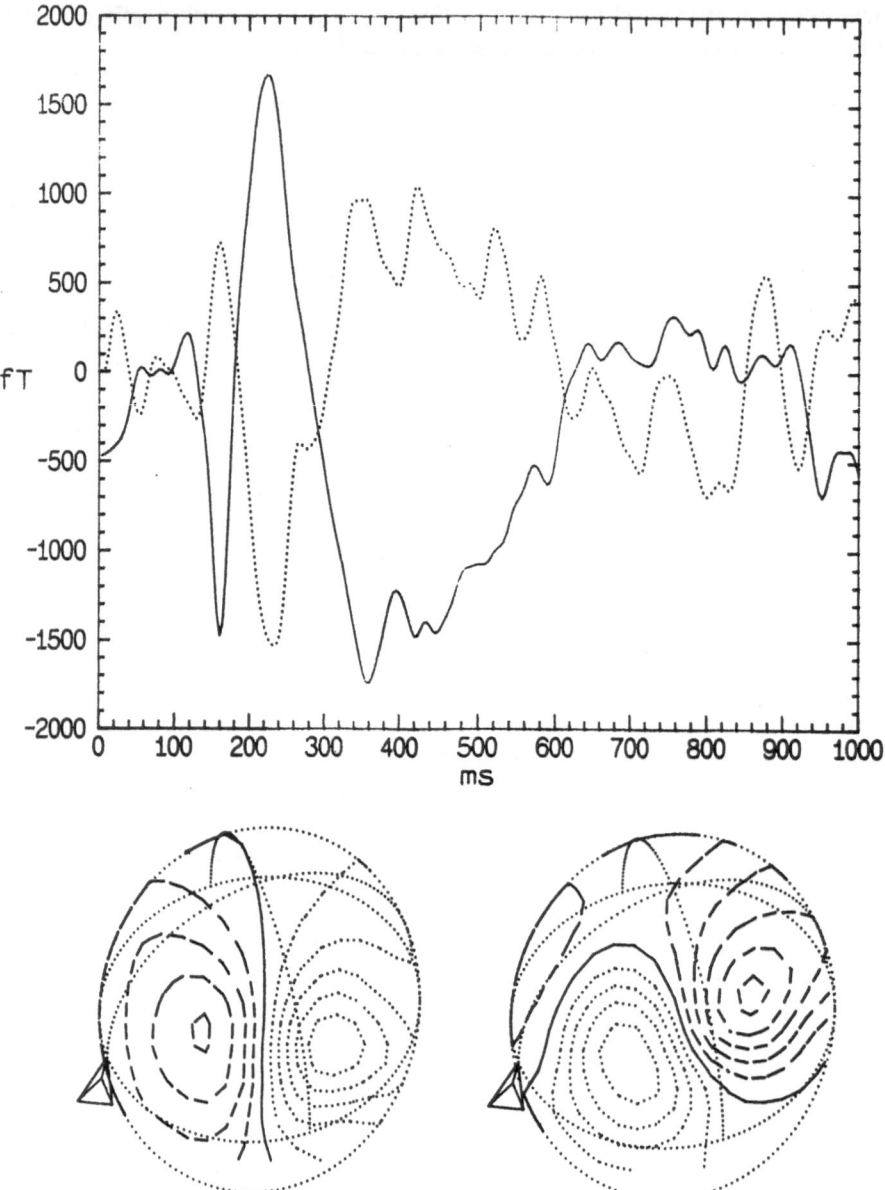

Figure 8. Epilepsy. Top: time traces representing a spike-and-wave complex measured over the left temporal lobe of a patient affected by partial epilepsy. The solid and dotted lines are measured over the maxima of opposite polarity. Bottom left: distribution of the radial field at 160 ms (spike); isofield step is 250 fT. Bottom right: distribution of the radial field at 215 ms (wave); isofield step is 300 fT. Note the different field distribution suggesting the presence of two different generators.

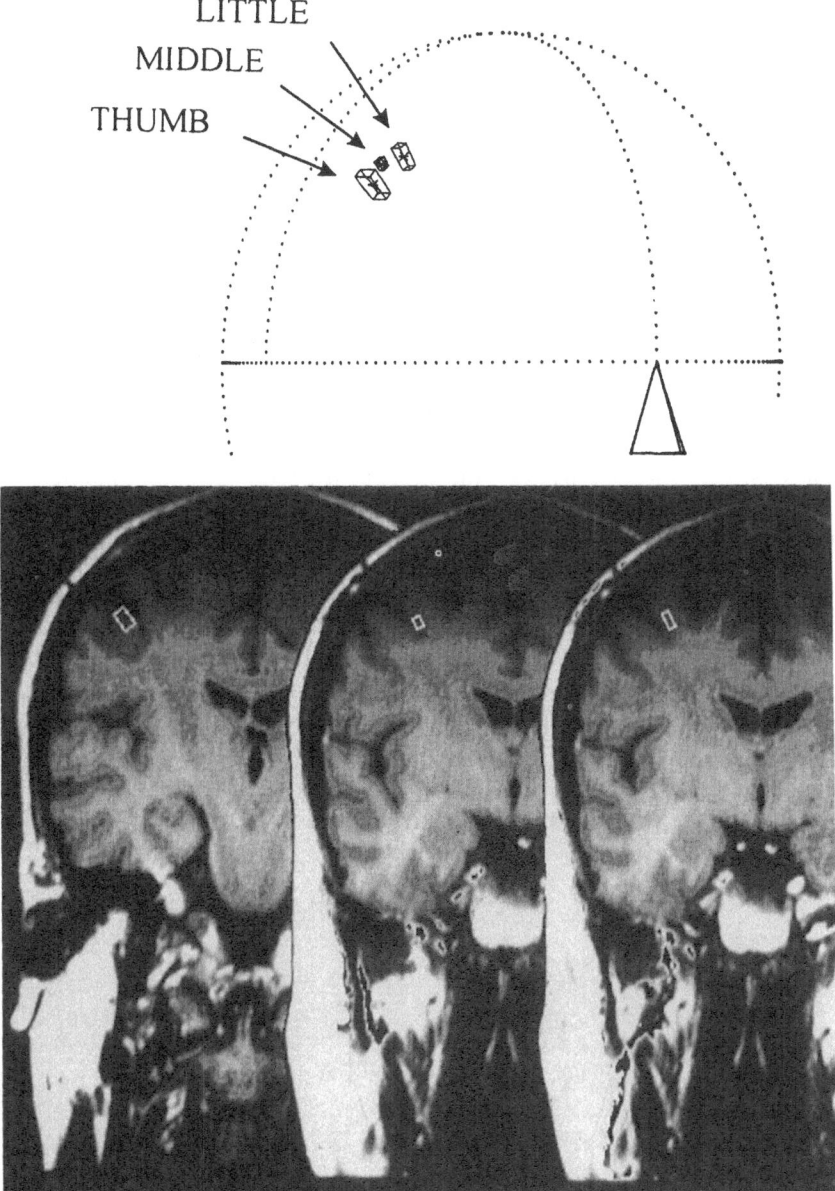

Figure 9. Plasticity. Top: location of the ECDs elicited by successive stimulation of thumb, middle and little finger. The three ECDs follow the finger somatotopy of the primary sensory cortex, the little finger being most medial and high, and the thumb most lateral and low. Bottom: the ECD locations, transformed into the MRI coordinate system, are shown on the appropriate MRI slice.

a deepening of the ECD for the middle finger, and by a shift on the coronal plane of the ECDs for the thumb and the little finger (medial for the former, lateral for the latter, as illustrated in Figure 7). Such modifications became evident in stage B, but still persisted in stage C for some of the subjects studied. The authors concluded that the proposed protocol was suitable for further development of monitoring methods for the evaluation of pharmacological and rehabilitative therapeutic protocols, which are aimed at the restoration of a lost function due to a brain lesion.

A brilliant confirmation of the capabilities of the neuromagnetic method in the investigation of cerebral plasticity was recently provided by the work of Flor and co-workers [71]. These authors studied the modification of cortical representation in the somatosensory cortex of subjects suffering of the phantom-limb syndrome. In this pathology, an amputated limb may produce pain or other feelings that can, in turn, result in severe impairment. The results of the neuromagnetic investigation demonstrated that a strong correlation existed between cortical reorganization and phantom-limb pain, in the sense that the stronger the reorganization, the greater the pain. The authors concluded by suggesting "that phantom-limb pain is related to, and may be a consequence of, plastic changes in the primary somatosensory cortex."

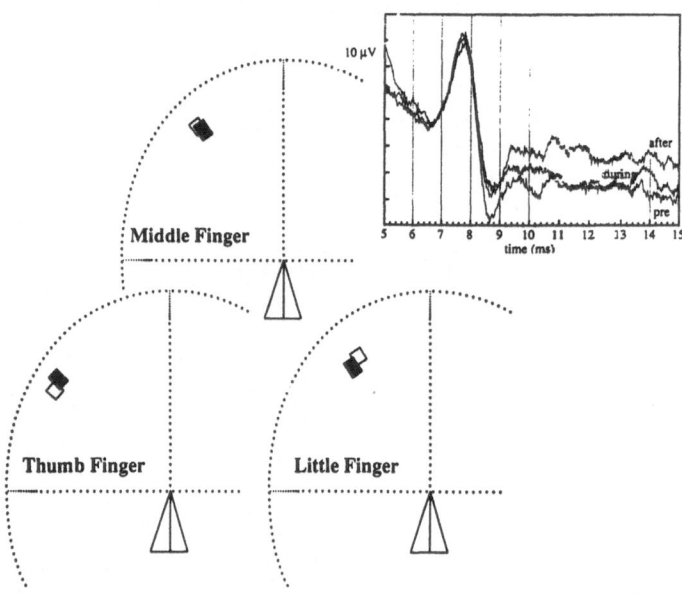

Figure 10. Plasticity. Antero-posterior view of the location of the ECDs elicited by the stimulation of the test fingers in the control stage (empty symbols) and in the ischemisation stage (filled symbols). The thumb projection shifts toward a more medial position, while the little-finger projection shifts toward a more lateral position. No appreciable shift is visible for the little finger. Box size corresponds to the 95% confidence interval. Insert: electric signal from the nerve recorded at the elbow during little finger stimulation ; notice that both latency and amplitude responses do not vary during the different experimental stages.

8. Future developments

The areas that will deserve major attention in basic and clinical research will certainly involve dynamic properties of nervous cells, neuro-pharmacological studies and applications of evoked responses to short term variations of neurophysiological and pathological parameters, or to early diagnosis of degenerative and vascular diseases.

Concerning the technology, three major issues to be addressed in the next years are: i) the introduction of high critical temperature (HTc) SQUIDs; ii) the implementation of new high speed analysis software; iii) the integration with other biomedical imaging techniques.
Regarding the introduction of HTS SQUIDs [72] for neuromagnetic measurements, cryogenic requirements would be significantly simplified, with attendant reduction in operating costs. Unfortunately, so far, the performance of these devices has not been sufficiently satisfying. Indeed, only recently an impressive reduction of their noise level, particularly in the frequency range below 100 Hz, has made possible their use for neuromagnetic measurements. Preliminary results will be made in the near future in several laboratories.

An important aspect of systems devoted to clinical research is the analysis software. This is a point that so far has not been adequately considered and that requires future attention. When clinical measurements are considered, there is an objective need to consider a large number of patients, thus requiring short measurement time, fast data analysis and large data-storage capability. The analysis console should be user friendly, and direct integration with CT or MRI data should be available in order to present the results in the frame of a multimodal imaging procedure.

Finally, the availability of large systems now provides a variety of results both for fundamental studies and in clinical applications. This will strengthen the interest of researchers and clinicians, and lead to widespread use of neuromagnetometry. Particular attention should be paid to the integration of the results of neuromagnetic investigation with those of PET and functional MRI, since multimodal images will provide the most powerful diagnostic tool for the near future.

9. References

1. Hämäläinen, M., Hari, R., Ilmoniemi, R., Knuutila, J., and Lounasmaa, O. (1993) Magnetoencephalography - theory, instrumentation, and applications to noninvasive studies of the working human brain, *Rev. Mod. Phys.* **65**, 413-497.
2. Tripp, J.H. (1983) Physical concepts and mathematical models, in S.J. Williamson, G.L. Romani, L. Kaufman, and I. Modena (eds.), *Biomagnetism: An Interdisciplinary Approach*, Plenum Press, New York, pp. 101-139.

3. Sarvas, J. (1987) Basic mathematical and electromagnetic concepts of the biomagnetic inverse problem, *Phys. Med. Biol.* **32**, 11-22.

4. Wikswo, J.P. (1989) Biomagnetic sources and their models, in S.J. Williamson, M. Hoke, G. Stroink, and M. Kotani (eds.), *Advances in Biomagnetism*, Plenum Press, New York, pp. 1-18.

5. Koester, J. (1985) Resting membrane potential and action potential, in E.R. Kandel and H.J. Schwartz (eds.), *Principles of Neural Science*, second edition, Elsevier, Amsterdam, pp.49-57.

6. Williamson S.J., and Kaufman L. (1990) Theory of neuroelectric and neuromagnetic fields, in F. Grandori, M. Hoke, and G.L. Romani (eds.), *Auditory Evoked Magnetic Fields and Electric Potentials*, Adv. Audiol., Karger, Basel, pp. 1-39.

7. Nenonen, J. and Katila, T. (1991) Noninvasive functional localisation by biomagnetic methods - Part I, *J. Clin. Eng.* **16**, 423-434.

8. Williamson S.J., and Kaufman L. (1981) Magnetic fields of the cerebral cortex, in S.N. Erné, H.D. Hahlbohm, and H. Lübbig (eds.), *Biomagnetism*, Walter de Gruyter, Berlin. pp. 353-402.

9. Stok, C.J. (1986) *The inverse problem in EEG and MEG with application to visual evoked responses*, PhD thesis, University of Twente, Enschede, The Netherlands.

10. Meijs, J.W.H., Peters, M. and Oosterom, A. van (1985) Computation of EEGs and MEGs using a realistically shaped multi-compartment model of the head, *Med. Biol. Eng. Comput.* **23** suppl. part 1, 36-37.

11. Hämäläinen, M., and Sarvas, J. (1989) Realistic conductivity geometry model of the human head for the interpretation of neuromagnetic data, *IEEE Trans. Biomed. Eng.* **36**, 165-172.

12. Nenonen, J., Purcell, C.J., and Horacek, B.M. (1991) Magnetocardiographic functional localization using a current dipole in a realistic torso, *IEEE Trans. Biomed. Eng.* **38**, 658-664.

13. Wieringa H.J. (1993) *MEG, EEG and the integration with magnetic resonance images*, PhD thesis, University of Twente, Enschede, The Netherlands.

14. Munk, J. C. de (1989) *A mathematical and physical interpretation of the electromagnetic field of the brain*, PhD thesis, University of Twente, Enschede, The Netherlands.

15. Peters, M. and Munk, J. C. de (1990) On the forward and the inverse problem for EEG and MEG, in F. Grandori, M. Hoke, and G.L. Romani (eds.), *Auditory Evoked Magnetic Fields and Electric Potentials*, Adv. Audiol., Karger, Basel, pp. 70-102.

16. Geselowitz, D.B. (1970) On the magnetic field generated outside an inhomogeneous volume conductor by internal current sources, *IEEE Trans. Magn.* **6**, 346-347.

17. Geselowitz, D.B. (1967) On bioelectric potentials in an inhomogeneous volume conductor, *Biophys. J.* **7**, 1-11.

18. Ilmoniemi, R., Hämäläinen, M., and Knuutila, J. (1985) The forward and the inverse problems in the spherical model, in H. Weinberg, G. Stroink, and T. Katila (eds.), *Biomagnetism: Applications and Theory*, Pergamon, Oxford, pp. 278-282.

19. Grynszpan, F., and Geselowitz, D.B. (1973) Model studies of the magneto-cardiogram, *Biophys. J.* **13**, 911-925.

20. Oostendrop, T.F., and Oosterom, A. van (1989) Source parameter estimation in inhomogeneous volume conductors of arbitrary shape, *IEEE Trans. Biomed. Eng.* **36**, 382-391.

21. Oostendrop, T.F., and Oosterom, A. van (1991) Source parameter estimation using realistic geometry in bioelectricity and biomagnetism, in J. Nenonen, H.-M. Rajala, and T. Katila (eds.), *Biomagnetic Localisation and 3D Modelling*, Report TKK-F-A689, Helsinki University of Technology, pp. 71-86.

22. Barnard, A.C.L., Duck, I.M., Lynn, M.S., and Timlake, W.P. (1967) The application of electromagnetic theory to electrocardiology II, *Biophys. J.* **7**, 463-491.

23. Lynn, M.S., and Timlake, W. P. (1968) The use of multiple deflations in the numerical solution of singular systems of equations to potential theory, *SIAM J. Numer. Anal.* **5**, 303-322.

24. Hämäläinen, M., and Sarvas, J. (1987) Feasibility of the homogeneous head model in the interpretation of neuromagnetic data, *Phys. Med. Biol.* **32**, 91-92.

25. C. J. Stok (1987) The influence of model parameters on EEG/MEG single dipole source estimation, *IEEE Trans. Biomed. Eng.* **34**, 289-296.

26. Meijs, J.W.H. (1988) *The influence of head geometries on electro- and magnetoencephalograms*, PhD thesis, University of Twente, Enschede, The Netherlands.

27. Romani, G.L., Williamson, S.J., and Kaufman, L. (1982) Tonotopic organization of the human auditory cortex, *Science* **216**, 1339-1340.

28. Press, W.H., Flannery, B.P., Teukolski, S.A., and Vetterling, W.T. (1986) *Numerical Recipes - The Art of Scientific Computing*, Cambridge University Press.

29. Marquardt, D.W. (1963) An algorithm for least-squares estimation of nonlinear parameters, *J. Soc. Ind. Appl. Math.* **11**, 431-441.

30. Knuutila, J., and Hämäläinen, M. (1988) Characterization of brain noise using a high sensitivity 7-channel magnetometer, in K. Atsumi, M. Kotani, S. Ueno, T. Katila, and S.J. Williamson (eds.), *Biomagnetism '87*, Tokyo Denki University Press, pp. 186-189.

31. Wieringa, H.J., and Peters, M. (1991), MRI and MEG, 3D models and display, in J. Nenonen, H.-M. Rajala, and T. Katila (eds.), *Biomagnetic Localisation and 3D Modelling*, Report TKK-F-A689, Helsinki University of Technology, pp. 10-22.

32. Knuutila, J., Ahonen, A.I., Hämäläinen, M., Ilmoniemi, R., and Kajola, M.J. (1985) Design considerations for multichannel SQUID magnetometers, in H.D. Hahlbohm and H. Lübbig (eds.), *SQUID 85 - Superconducting Quantum*

Interference Devices and their Applications, Walter de Gruyter, Berlin, pp. 939-944.

33. Erné, S.N., Narici, L., Pizzella, V., and Romani, G.L. (1987) The positioning problem in biomagnetic measurements: a solution for arrays of superconducting sensors, *IEEE Trans. Magn.* **23**, 1319-1322.

34. Hämäläinen, M., and Ilmoniemi, R. (1984) *Interpreting magnetic fields of the brain: estimates of current distributions*, Report TKK-F-A559, Helsinki University of Technology.

35. Romani, G.L., Williamson. S.J., Kaufman. L. (1982) Biomagnetic Instrumentation, *Rev. Sci. Instrum.* **53**, 1815-1845.

36. Ryhänen, T., Seppä, H., Ilmoniemi, R., Knuutila J. (1989) SQUID magnetometers for low-frequency applications, *J. Low Temp. Phys.* **76**, 287-386.

37. Nenonen, J., Katila, T., and Montonen J. (1989) Thermal noise of a biomagnetic measurement dewar, in S.J. Williamson, M. Hoke, G. Stroink, M. Kotani (eds.), *Advances in Biomagnetism*, Plenum Press, New York, pp.729-732.

38. Erné, S.N. (1983) Shielded rooms, in S.J. Williamson, G.L. Romani, L. Kaufman, I. Modena (eds), *Biomagnetism: An Interdisciplinary Approach*, Plenum Press, New York, pp.569-578.

39. Knuutila, J., Ahonen, A., Hämäläinen, M., Kajola, M., Laine, P., Lounasmaa, O., Parkkonen, L., Simola, J. and Tesche, C. (1993) A 122-channel whole-cortex SQUID system for measuring the brain's magnetic fields, *IEEE Trans. Magn.* **29**, 3315-3320.

40. Vrba, J., Betts, K., Burbank, M., Cheung, T., Cheyne, D., Fife, A.A., Haid, G., Kubik, P.R., Lee, S., McCubbin, J., McKay, J., McKenzie, D., Mori, K., Spear, P., Taylor, B., Tillotson, M. and Xu, G. (1995) Whole cortex 64 channel system for shielded and unshielded environments, in C. Baumgartner, L. Deecke, G. Stroink, S.J. Williamson (eds.), *Biomagnetism: Fundamental Research and Clinical Applications*, Elsevier, Amsterdam, pp.521-525.

41. Benzel, E.C., Lewine, J.D., Bucholz, R.D., Orrison, W.W. (1993) Magnetic Source Imaging: a review of the Magnes system of Biomagnetic Technologies Incorporated, *Neurosurgery* **33**, 252-259.

42. Dössel, O., David, B., Fuchs, M., Krüger, J., Lüdeke, K.M., Wischmann, H.A. (1993) A Modular 31-channel SQUID system for biomagnetic measurements, *IEEE Trans. Appl. Supercond.* **3**, 1883-1886.

43. Ueda, M., Kandori, H., Ogata, H., Takada, Y., Komuro, T., Kazami, K., Ito, T., Kado, H., (1995) Development of a biomagnetic measurement system for brain research, *IEEE Trans. Appl. Supercond.* **5**, 2465-2469

44. Kado, H., Uehara, G., (1994) Multi-channel SQUID system, *FED Journal* **5**, 12-19.

45. Drung, D. (1995), The PTB 83-SQUID system for biomagnetic applications in a clinic, *IEEE Trans. Appl. Supercond.* **5**, 2112-2117.

46. Foglietti, V., Del Gratta, C., Pasquarelli, A., Pizzella, V., Torrioli, G., Romani, G.L., Gallagher, W.J., Ketchen, M.B., Kleinsasser, A.W., Sandstrom, R.L.,

(1991), 28-channel hybrid system for neuromagnetic measurements, *IEEE Trans. Magn.*, **27**, 2959-2963.

47. Pizzella, V. (1995) The Italian Biomagnetic Projects, in C. Baumgartner, L. Deecke, G. Stroink, S.J. Williamson (eds.), *Biomagnetism: Fundamental Research and Clinical Applications*, Elsevier, Amsterdam, pp.476-482.

48. Erné, S.N. (1983) Analog filtering techniques, in S.J. Williamson, G.L. Romani, L. Kaufman, I. Modena (eds), *Biomagnetism: An Interdisciplinary Approach*, Plenum Press, New York, pp.579-589.

49. Cohen, D., Cuffin, B.N., Yunokuchi, K., Maniewski, R., Purcell, C., Cosgrove, G.R, Ives, J., Kennedy, J. G. and Schomer, D.L. (1990), MEG versus EEG localization test using implanted sources in the human brain, *Annals of Neurology* **28**, 811-817.

50. Anogianakis, G., Badier, J.M., Barrett, G., Erné, S.N., Fenici, R., Fenwick, P., Grandori, F., Hari, R., Ilmoniemi, R., Mauguiere, F., Lehmann, D., Perrin, F., Peters, M., Romani, G.L., Rossini, P.M., (1992) A consensus statement on relative merits of EEG and MEG, *Electroencephalogr. clin. Neurophysiol.* **82**, 317-319.

51. Wikswo, J. P., Gevins, A., Williamson, S. J., (1993), The future of the EEG and MEG, *Electroencephalogr. clin. Neurophysiol.* **87**, 1-9

52. Mosher J. C., Lewis, P. S., Leahy R. (1992), Multiple dipole modelling and localisation from spatio-temporal MEG data, *IEEE Trans. Biomed. Eng.* **39**, 541-557.

53. Ioannides, A. A., Bolton, J. P. R., Clarke, C. J. S. (1990), Continuous probabilistic solutions to the biomagnetic inverse problem, *Inverse Probl.* **6**, 523-542.

54. Penfield, W. and Rasmussen T. (1950) *The cerebral cortex of man.* Macmillan, New York.

55. Pantev, C., Hoke, M., Lehnertz, K., Lütkenhöner, B., Anogianakis, G., and Wittkowski, W. (1988) Tonotopic organization of the human auditory cortex revealed by transient auditory evoked magnetic fields, *Electroencephalography and clinical Neurophysiology* **69**, 160-170.

56. Romani, G. L., Williamson, S. J., and Kaufman, L. (1982) Tonotopic organization of the human auditory cortex. *Science* **216**, 1339-1340.

57. Pantev, C., Hoke, M., Lehnertz, K., and Lütkenhöner, B. (1989) Neuromagnetic evidence of an amplitopic organization of the human auditory cortex, *Electroencephalography and clinical Neurophysiology* **72**, 225-231.

58. Pantev, C., Hoke, M., Lütkenhöner, B., and Lehnertz, K., (1989) Tonotopic organization of the auditory cortex: pitch versus frequency representation, *Science* **246**, 486-488.

59. Näätänen, R., Simpson, M., and Loveless, N. E. (1982) Stimulus deviance and evoked potentials, *Biological Psychology* **14**, 53-98.

60. Näätänen, R., Paavilainen, P., Tiitinen, H., Jiang, D., and Alho, K. (1993) Attention and mismatch negativity, *Psychophysiology* **30**(5), 436-450.

61. Kristeva-Feige, R., Rossi, S., Pizzella, V., Tecchio, F., Romani, G. L., Erné, S., Edrich, J., Orlacchio, A., and Rossini, P. M. (1995) Neuromagnetic fields of the brain evoked by voluntary movement and electrical stimulation of the index finger, *Brain Research* **682**, 22-28.

62. Hari, R., Karhu, J., Hämäläinen, M., Knuttila, J., Salonen, O., Sams, M., and Vilkman V. (1994) Functional organisation of the human first and second somatosensory cortices: a neuromagnetic study, *European Journal of Neuroscience* **5**, 724-734.

63. Kaukoranta, E., Hämäläinen, M., Sarvas, J., and Hari R. (1986) Mixed and sensory nerve stimulations activate different cytoarchitectonic areas in human primary somatosensory cortex SI, *Experimental Brain Research* **63**, 60-66.

64. Kristeva-Feige, R., S. Rossi, V. Pizzella, L. Lopez, S. Erné, J. Edrich and P. M. Rossini (1995) A neuromagnetic study of movement-related somatosensory gating in the human brain, *Experimental Brain Research* **107**, 504-514.

65. Nakasato, N., Levesque, M. F., Barth, D., Baumgartner, C., Rogers, R. L., and Sutherling, W. W. (1994) Comparisons of MEG, EEG, and ECoG source localization in neocortical partial epilepsy in humans, *Electroencephalography and Clinical Neurophysiology* **91**, 171-178.

66. Pelizzone, M., Hari, R., Mäkelä, J., Kaukoranta, E., and Montandon, M. (1987) Cortical activity evoked by a multichannel cochlear prosthesis, *Acta Otolaryngol.* **103**, 632-636.

67. Leinonen, L., and Joutsiniemi S., L., (1989) Auditory evoked potentials and magnetic fields in patients with lesions of the auditory cortex, *Acta Neurol. Scand.* **79**, 316-325.

68. Karhu, J., Hari, R., Mäkelä, J., Huttunen, J., and Knuutila, J. (1992) Somatosensory evoked magnetic fields in multiple sclerosis, *Electroencephalography and Clinical Neurophysiology* **83**, 192-200.

69. Rossini, P. M., Martino, G., Narici, L., Pasquarelli, A., Peresson, M., Pizzella, V., Tecchio, F., Torrioli, G., and Romani, G. L., (1994) Short-term brain 'plasticity' in humans: transient finger representation changes in sensory cortex somatotopy following ischemic anaesthesia, *Brain Research* **642**, 169-177.

70. Nicolelis, M. A. L., Lin, R. C. S., Woodward, D. J., and Chapin, J. K. (1993) Introduction of immediate spatiotemporal changes in thalamic networks by peripheral block of ascending cutaneous information, *Nature* **361**, 533-536.

71. Flor, H., Elbert, T., Knecht, S., Wienbruch, C., Pantev, C., Birbaumer, N., Larbig, W., and Taub, E. (1995) Phantom-limb pain as a perceptual correlate of cortical reorganization following arm amputation, *Nature* **375**, 482-484.

72. Braginski, A. I., this volume.

INTEGRATING COMPETING TECHNOLOGIES WITH MEG

M.C.GILARDI, G.RIZZO, G.LUCIGNANI and F.FAZIO
INB-CNR, Università di Milano, Istituto Scientifico H S.Raffaele
via Olgettina 60, 20132 Milano, Italy

Abstract. Alternative and/or complementary information are provided by functional imaging techniques. Physiological, biochemical, pharmacological variables can be assessed by positron emission tomography (PET) and single photon emission tomography (SPET). Recent technological advances, in particular the implementation of fast acquisition sequences, such as echo-planar technique, allow functional investigation by Magnetic Resonance Imaging (fMRI). For each modality the physical principles and research/clinical applications are presented, relative merits and limitations in assessing functional activity in the brain are discussed in relation to magneto-encephalography (MEG). The combined use of the different information provided by multi-modal bio-imaging and bio-signal methods is enhanced by registration techniques, allowing for a spatial correspondence of the multi-modal information. Different approaches to the registration problem are described. Integration of functional imaging techniques and MEG provides an accurate description of cerebral functions with high spatial and temporal resolution, this representing a powerful probe for brain investigation both in basic and clinical neuroscience.

1. Introduction

Several techniques are currently available for the examination of the brain under normal and pathological conditions. A number of non-invasive neuroimaging methods, such as Magnetic Resonance Imaging (MRI) and x-ray Computed Tomography (CT), have been developed which supply information about the brain structure. Other techniques such as Positron Emission Tomography (PET), Single Photon Emission Tomography (SPET) and, more recently, functional MRI (fMRI) permit the investigation of the functional activity of the human brain by measurements of blood flow, metabolism and receptor distribution [1,2]. Moreover, a non-invasive three-dimensional localization of the bio-electric sources of the brain's electric and magnetic fields at the scalp surface can be achieved by electro-encephalographic (EEG) and magneto-encephalographic (MEG) signal analysis [3].

Each of these techniques presents relative merits and constraints in the ability to temporally and spatially resolve cerebral processes and structures. They do not compete with each other, since their integrated use can help to achieve better comprehension of the human brain. The combined use of multi-modal imaging and bio-signal methods represents a new and powerful strategy for investigating the brain in basic and clinical

H. Weinstock (ed.), SQUID Sensors: Fundamentals, Fabrication and Applications, 491–516.
© *1996 Kluwer Academic Publishers.*

neuroscience. Indeed, the integration of multi-modal information helps in overcoming the limitations relative to each technique, thus enhancing the specific value of each.

In this chapter the basic principles of functional neuroimaging methods, emission tomography (PET and SPET) and fMRI will be reviewed briefly. The second part of the chapter will focus on methods for the spatial integration of multi-modal information obtained by the different bioimaging methods and MEG. Finally, the application of this integrated approach to the interpretation of anatomical and functional brain mapping will be discussed.

2. Functional imaging methods

2.1. POSITRON EMISSION TOMOGRAPHY (PET)

Positron Emission Tomography produces images of radioactivity distributed in selected body sections following the administration of a radiopharmaceutical to the patient under examination. By measuring radioactivity concentrated in the brain, PET permits the study *in vivo* of biochemical and physiological processes [4].

The isotopes most commonly used in PET are listed in Table 1.

TABLE 1. Positron emitting isotopes

Isotope	Half life (min)
^{11}C	20.4
^{13}N	10.0
^{15}O	2.0
^{18}F	109.8

Besides being positron emitters, they present two main characteristics:
  - each has a short half-life, on the order of minutes. This implies the need for a particle accelerator, a cyclotron, in proximity to the PET centre. The costs related to such instrumentation and its complexity explain the limited use of PET with respect to other functional imaging techniques;
  - they are isotopes of elements which are constituents of the biological matter (oxygen, carbon, nitrogen and fluorine, a substitute for hydrogen). In principle, all molecules present in the human body can be labelled by ^{15}O, ^{11}C, ^{13}N and ^{18}F; physiological radioactive molecules are thus available. They trace physiological and biochemical processes, which can be assessed by external detection of the emitted radiation. The availability of physiological tracers represents one of the main advantages of PET, allowing not only for the qualitative assessment of functional parameters, but also for their absolute quantitative estimation [5 to 7]. Images shown in Figure 1 are representative of the regional cerebral glucose metabolic rates, assessed following the

administration of ^{18}F-labelled 2-fluoro-2-deoxy-D-glucose ([^{18}F]FDG), an analogue of glucose.

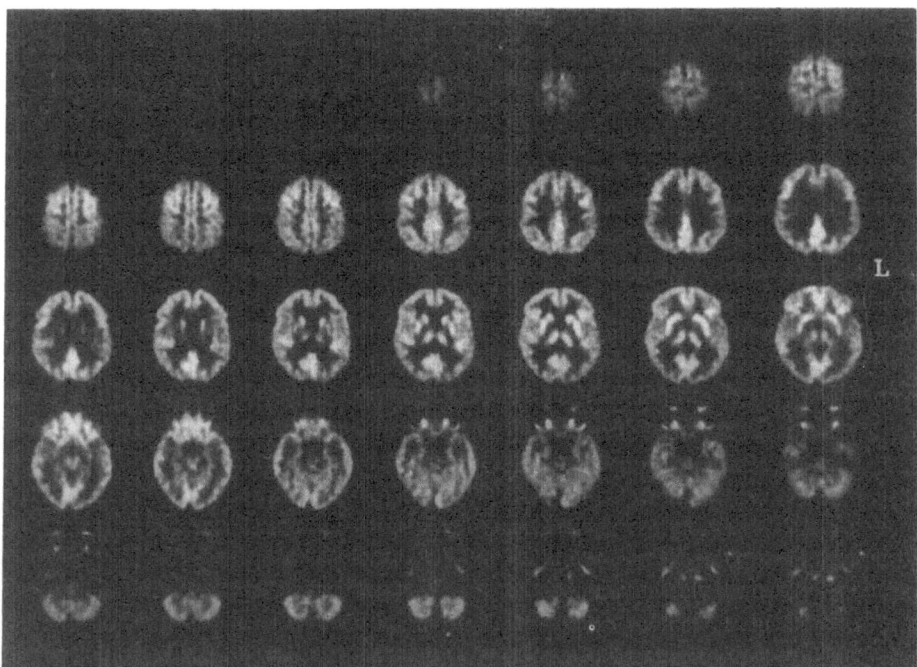

Figure 1. [^{18}F]FDG PET images of regional cerebral glucose metabolic rates in a normal volunteer.

2.1.1. *Technological aspects*

When a positron is emitted by a radioactive nucleus in the human body following the administration of a radiotracer, it travels a short distance by interacting with the surrounding matter and losing its kinetic energy. It then interacts with an electron in an annihilation event in which the mass of the two particles (electron and positron) disappears to generate electromagnetic radiation in the form of two 511 keV gamma rays emitted 180° to each other. These gamma rays are detected in PET, indirectly allowing for the detection of the positron decay.

A PET tomograph generally consists of a ring of detectors surrounding the organ under study and connected via temporal coincident circuitry. One event is recognized when two gamma rays are detected simultaneously or, better, within a very short time interval (on the order of nsec) by two detectors in the ring. The line connecting the centres of the two detectors involved in a coincident event is recognised as the line along which the annihilation event has occurred. During PET scanning, radioactivity in the body section under study is sampled at different linear and angular positions to generate a set of activity profiles at different projection angles [8]. These are the input for the reconstruction algorithm, which generates an image representing the radioactivity distributed in the body section of interest. The reconstruction algorithms are similar to those used for other tomographic techniques, e.g., CT [9].

Current generation PET tomographs consist, in fact, of several adjacent rings generating multiple slices and permitting a simultaneous three-dimensional (3D) representation of the radioactivity distributed in the organ under study.

Without presenting details of PET instrumentation (as it is beyond the scope of this paper), the principal features of current-generation PET scanners are [10 to 13]:
- Significant improvement has been obtained in spatial resolution, which is on the order of 4 mm in commercial scanners. Advances in detector technology allows for the use of small bismuth germanate (BGO) detectors with mm-size dimensions. The good definition of anatomical details in a PET study of regional cerebral glucose metabolism performed in a normal volunteer by a state-of-the-art PET tomograph can be observed in Figure 1.
- A second technological achievement is the extension of the axial field of view (FOV). Several detector rings can be assembled providing multiple transaxial slices of the organ under study for a 3D simultaneous assessment of the radioactivity distribution. Current generation tomographs are multiple ring systems characterised by an axial FOV of 15 cm or more, allowing the entire brain to be covered in a single scan. In Figure 1, a sequence of 35 images is shown, each image being 4.5 mm thick.
- Finally, a real 3D PET imaging technique has been developed. In fact, conventional bi-dimensional (2D) PET scanners are equipped with inter-plane shields, positioned between two adjacent rings to avoid the detection of oblique radiation. A set of 2D transaxial images is then obtained, each image being generated by the detection of coincidence events between detectors within the same ring or in adjacent rings. However, by removing the shields, the detection of all coincidences, even the very oblique ones, is possible. A real 3D sampling of the radioactivity distribution is thus performed with a significant increase in the number of detected events and of the sensitivity of the technique (by a factor 5-10) [14 to 16]. The drawback of this approach is the need for handling a much larger amount of data with respect to the conventional 2D mode and the need for more sophisticated reconstruction and image-processing algorithms [14]. However, powerful and fast computers are increasingly available at cheaper cost. Significant advantages of the improved sensitivity are found both in clinical and research applications, allowing for either an improvement in image quality in terms of signal-to-noise ratio or a reduction of the radiation dose administered to patients. This approach has proven to be very important in low-count-rate studies (e.g., receptor studies or pediatric studies) or when repeat PET measurements are desired, as in PET activation studies [15, 17].

In the study of the functional activity of the brain, a limiting factor in PET is represented by temporal resolution. Temporal resolution can be defined as the minimum time to detect a signal, being on the order of minutes in PET. Shorter scanning time would, in fact, generate excessively noisy data. However, in a different interpretation, temporal resolution represents the minimum time between two subsequent scans, being dependent in PET on the time for dispersal of the radioactivity distributed in the human body. In this sense, temporal resolution depends both on the physical half-life of the radioisotope and on the kinetics of the molecule in the body. In many cases the physical radioactive decay represents the limiting factor to temporal resolution. As an example, in the case of repeated PET blood flow studies with $H_2^{15}O$, the time resolution is on the order of 10 min because before performing a second scan and injecting a new dose of radioactivity, the previous activity must have decayed, and this takes approximately 5-6 half-lives (i.e., 10-12 min for ^{15}O).

2.2. SINGLE PHOTON EMISSION TOMOGRAPHY (SPET)

Single photon emission tomography is a technique very similar to PET. A radiotracer is administered to the patient, and it is distributed in the patient 's body. Similarly to PET, SPET provides images of radioactivity distributed in body sections by the detection of the emitted radiation [18]. However, while positron-emitting isotopes are used in PET studies, in SPET, single photon emitters are employed. The most used isotopes in SPET are shown in Table 2.

TABLE 2. Single photon emitting isotopes

Isotope	Half life (d)
^{99m}Tc	0.25
^{123}I	0.55
^{131}I	8.0
^{111}In	2.8
^{201}Tl	3.0

They have longer half-lives than PET isotopes, thus they do not require an on-site cyclotron, but can be delivered to the users from elsewhere. This is one of the main advantages of SPET over PET, and explains the greater use of this technique. However, these isotopes are not "physiological", meaning that they can be labelled to simulate, but not directly trace biochemical and physiological processes in the body, as in PET. The availability of physiological tracers, and thus the ability to quantify *in vivo* functional parameters, is the main advantage of PET over SPET.

2.2.1. Technological aspects
From the technological point of view SPET, as for PET, requires sampling of the radioactivity distributed in the section under examination at different linear and angular positions. The most diffuse SPET tomograph consists of a gamma camera, the basic instrumentation of scintigraphic imaging, equipped with a large detector of thallium activated sodium iodide [NaI(Tl)]. In order to obtain tomographic images, the gamma camera must rotate 360° around the patient. These gamma-camera-based SPET systems are advantageous because they are cheap and can be used for imaging of various organs. However, in order to improve detection efficiency, more detection volume around the patient would be needed. For this, multiple-head cameras [19] or dedicated tomographs

now preferred for SPET [19, 20]. The technological advances achieved in these SPET systems permit significantly improved image quality, and SPET is now competitive to PET, at least in clinical applications, because of its more widespread availability and flexibility. SPET spatial resolution is on the order of 8-10 mm in state-of-the-art tomographs. PET performance remains superior particularly in terms of detection efficiency, which is by a factor of 20 lower in SPET than in 2D PET, and more than 100 times lower than in 3D PET [21]. The long half-life of radioisotopes used in SPET (on the order of hours) is the determining factor in limiting the temporal resolution of the technique.

It also should be noted that equipment is on the market for the assessment of the distribution of positron-emitting radiotracers by modified SPET cameras. In particular, SPET cameras can be equipped with high-energy collimators designed to detect 511 keV gamma rays [22]. More recently, a dual-head camera has been equipped with coincidence circuitry yielding spatial resolution on the order of 5 mm. The possibility of imaging PET radiotracers by low-cost SPET instrumentation might become a winning strategy in the future.

2.3. APPLICATIONS OF PET AND SPET

Applications of both PET and SPET in neurology are related to the assessment of regional alterations of vascular and/or metabolic activity in the brain induced by pathology [23, 24]. Some of these (clinical and research) applications are described in this section.

In the evaluation of patients with cerebral-vascular disorders, PET and SPET studies of perfusion and/or metabolism can be related to CT, which is the reference technique for the diagnosis of this pathology [25]. When morphological brain damage has not yet occurred, as in the case of a transient ischemic attack, PET and SPET permit the localisation of the ischemic areas, and thus an early diagnosis. At the time of the stroke onset, the functional variables assessed by PET and SPET may be the only useful measurable markers of this pathology [26]. The predictive value of SPET perfusion studies on patient outcome and the relation between the results of SPET perfusion studies, CT, and cerebral angiography have been demonstrated [27]. This SPET merit can have a direct impact on therapeutic planning, one of the most relevant issues in acute-stroke patients. Furthermore, PET and SPET are unique in assessing cortical changes in areas distant from the ischemic regions, changes due to interruption of neuronal connections, a phenomenon known as diaschisis [28]. Follow-up PET and SPET studies of the cerebral hemodynamics involving the monitoring of the reperfusion mechanisms with time have prognostic value.

Partial epilepsy may be the consequence of a brain lesion, but, in many cases, the assessment of brain morphology with CT or MRI does not demonstrate a detectable lesion, even though structural alterations below the resolution of the instrumentation may be present. In patients with partial seizures, the assessment of perfusion and metabolism by emission tomography (PET and SPET) may allow one to observe areas of interictal hypoperfusion and hypometabolism, and/or increases of perfusion and metabolism during the ictal phase even without evidence of morphological lesions [29, 30]. For the routine clinical evaluation of patients with epilepsy, the combination of interictal and ictal studies of perfusion provides the most complete information [31]. Assessment of perfusion and metabolism is used by some surgeons in the presurgical evaluation of patients with medically refractory partial epilepsy to confirm the uniqueness of the focus and its location. This practice, however, has not yet gained widespread acceptance and is considered in many cases insufficient for surgical intervention without stereo-EEG with depth electrodes. PET and SPET studies in epileptic patients have shown that the area of altered flow and metabolism is often wider than the epileptogenic zone, and that there are within the hypometabolic and underperfused areas, various zones with different electrical activity: epileptogenic, irritative and lesional. Thus, the relation between electrical activity and perfusion/metabolic alterations needs further clarification in order to achieve a full exploitation of the PET and SPET information in planning surgical therapy.

The assessment of cerebral metabolism and blood flow with emission tomography has been shown to be useful in the evaluation of demented patients, particularly in the early phase of probable Alzheimer's disease (AD). A pattern of bilateral hypometabolism has been consistently demonstrated by PET in the parietal and temporal cortex [32 to 34] with high prognostic value for the development of AD. Similar results have been obtained by SPET in combination with blood-flow tracers [35 to 37]. A systematic comparison of the sensitivity of PET and SPET techniques in the same series of patients with AD has been performed by state-of-the-art PET and SPET instrumentation. Significant differences between AD patients and controls were found both with PET and SPET in the parietal and temporal cortices, and differences were inconsistently observed only in other associative areas [38].

In oncology, the possibility of assessing one or more different biochemical and physiological variables of a tumour is unique to PET and SPET. An increased glucose metabolism assessed by PET and $[^{18}F]FDG$ or an increased uptake of ^{201}Tl assessed by SPET in the tumour lesion can be associated with the degree of malignancy [39, 40]. Furthermore, other imaging modalities with SPET and monoclonal antibodies are attracting increasing interest, in particular those aimed at signal amplification by tumour pre-targeting techniques [41].

The number of radioactively-labelled compounds produced for the assessment of several neuropharmacological variables has increased continuously over the last decade, a period when images of brain receptors were produced first with SPET and then with PET [42 to 45]. The tracers developed for SPET, all labelled with iodine, include substances for the assay of dopaminergic, benzodiazepine, serotoninergic, muscarinic and opiate receptors. Some of these tracers have been used not only for research purposes but also for some clinical applications. However, it is in the area of PET, rather than SPET, that most neuropharmacological research with emission tomography has been concentrated. Not only have many of the tracers developed for PET been tested, but they also have been used for the evaluation of patients with neurological and psychiatric disorders. PET is at present unique for quantification of the dynamic processes (uptake, utilisation and excretion) involved in kinetics of radioactively-labelled chemicals entering the metabolic pathways of the living human being.

Finally, the relation between neuronal activity and cerebral blood flow has become the basis for the assessment by PET and SPET of cerebral functional activity during the execution of defined tasks (known as activation studies) [46 to 49]. Cerebral activation is, in fact, accompanied by a regional increase of blood flow, detectable by emission tomography. The use of tracers labelled with short-lived isotopes allows one to perform a complete activation study, i.e., baseline and activations, within the same study session by use of PET, e.g., with ^{15}O-labelled water or carbon dioxide. Alternate techniques based on the use of SPET, originally with radioactive xenon, and more recently with perfusion tracers labelled with ^{99m}Tc, have been developed. However, SPET activation studies present serious limitations, in particular due to poor temporal resolution of the technique, making repeat base-line and activation studies difficult to perform in the same day. PET activation studies have made possible the definition of neural circuits and pathways involved in sensory-motor activity, attention, memory, word processing and visual recognition. The functional reorganization of the human brain after focal injury has been investigated with emission tomography in patients with stroke, and the recruitment of additional motor areas has been reported, as well as the reorganization of the brain, in different individual functional patterns [50].

2.4.   FUNCTIONAL MRI (fMRI)

High resolution images of the brain anatomy are obtained by MRI, showing high contrast differentiation between different tissue cerebral components - white matter, gray matter, cerebrospinal fluid. Contrast is based on specific characteristics of the cerebral tissues, such as the time evolution of the resonant phenomenon, described by the two time-relaxation constants T_1 and T_2, and the density of protons [51].

However, recent developments in MRI have demonstrated the possibility of non-invasive mapping of the human brain's functional activity with good spatial and temporal resolution. This is known as functional MRI, or fMRI for short [52, 53].

There are different approaches to fMRI, generally based on the use of contrast agents. Contrast agents are represented by paramagnetic compounds which do not cross the blood brain barrier in healthy brain, remaining in the intravascular space. They create magnetic field gradients around the blood vessels, resulting in a reduction of the MR signal (in T_2 weighted images). Contrast agents, such as gadolinium or dysprosium chelates, are used in conventional morphological MRI, for example, in the diagnosis of tumours to enhance contrast between pathological and normal tissue. In tumours the blood brain barrier can be damaged by the pathology, and the contrast agent is not limited to the intravascular space, but it can expand to the tumour mass.

fMRI methods involving the administration of an exogenous contrast agent (e.g., Gd-DTPA) have been used to assess the regional increase of blood flow following a photic stimulus, such an experiment representing one of the first fMRI applications [54].

However, a totally non-invasive fMRI method, which does not require the administration of contrast agents, has been proposed based on the Blood Oxygenation Level Dependent (BOLD) principle. The basis for this method is that deoxygenated hemoglobin (deoxyHb) is paramagnetic, while oxygenated hemoglobin (oxyHb) is diamagnetic. Each of these has a different value of susceptibility, and an increase in the concentration of deoxyHb increases the susceptibility of blood. DeoxyHb may act as an endogenous contrast agent. Thus, variation in the level of oxygenation of the blood induces variations in the MRI signals (in T_2 weighted images). In particular, an increase of deoxyHb concentration would result in a drop of the MRI signal [55, 56].

The capability of the BOLD method to map hemodynamic processes in the brain is used to assess cerebral functional activity, as for PET in activation studies. As already discussed for PET, during stimulation, blood flow increases in the brain region involved. However, this blood flow increase is accompanied by an increase, but to a lower degree, of the extraction of oxygen. During activation blood flow vasodilatation is not associated with a significant increase of oxidative metabolism. This results in a global increase of oxyHb concentration in a voxel, in a drop in deoxyHb concentration, thus in a regional increase of the MRI signal [57].

An issue raised on the interpretation of the fMRI signal is the following: the detected signal might not only be generated at the capillary level in the activated brain region, but also in large venous vessels draining it. Thus, quantification of fMRI images is complicated by the presence in the detected signal of such an artifact.

2.4.1. Technological aspects

The following issues regarding fMRI imaging technology are worthy of consideration [52]:
1.   Signal intensity depends on the intensity of the magnetic field. Many fMRI studies have been performed at 1.5 T by commercial scanners. Improvement of signal-to-noise ratio and, thus, of the possibility to detect small signal changes following activation have been observed at higher magnetic field (e.g., 4 T MRI scanners).
2.   Spatial resolution on the order of 3 mm has been obtained by fMRI methods, worse than for conventional MRI, as a consequence of a compromise between spatial and temporal sampling.
3.   The first fMRI studies were performed using the echoplanar imaging (EPI) technique, a recently developed method characterized by very fast sampling of the MRI signal. EPI allows for a bidimensional image to be recorded in a time on the order of 40 msec, and three-dimensional images of the brain can be obtained in 1-3 sec [53, 58, 59]. EPI requires the implementation of additional expensive equipment, and, at the present time, its utilization is limited, although EPI systems now are commercially available.

More recently, fMRI studies have been performed by commercial conventional scanners using fast pulse acquisition sequences which yield an image in a time on the order of 1-2 sec.

Short acquisition time for the EPI technique is an important advantage in fMRI studies in which a significant loss of signal can occur due to motion artifacts produced by both physiological motion originated by the brain's pulsation with the cardiac cycle and involuntary head motion.

Furthermore, as mentioned for PET, the brain response to a stimulus ideally should be monitored by very fast dynamic scans in order to follow the physiological process. Thus, temporal resolution plays a relevant role in activation studies by fMRI. However, although acquisition times on the order of 40 msec are possible using EPI technology, temporal resolution, defined as the time interval between two subsequent scans, presents further limitations. The intensity of the MRI signal versus time is, in fact, dependent on spin saturation, and this effect has a characteristic time on the order of 1-2 sec. Furthermore, the characteristic time for deoxyHb concentration changes is on the order of 2 sec. Based on this, the physiological process under study can be measured by fMRI at a sampling rate of not less than approximately 2 sec.

Most applications of fMRI have been addressed to the assessment of the functional activity of the brain when an external stimulus is applied or while a specific cognitive task is performed - these are known as activation studies [60]. This approach has been widely used for neurophysiological and neuropsychological studies in normal subjects. More recently, clinical use of this fMRI method has been reported in pathology. In particular, fMRI has been used in the surgical planning of brain tumours to identify and preserve functional areas surrounding the lesion [60].

2.5. COMPARATIVE CONSIDERATIONS

In the evaluation of the potentials of each imaging modality, whether for scientific research or in the process of defining an optimal diagnostic protocol, some essential physical features of the instruments available must be considered. The following considerations compare schematically the characteristics of functional-imaging modalities to those of MEG.

The possibility of detecting concurrent events occurring in distant cerebral regions is an important issue. The entire brain can be simultaneously studied by all these techniques due to the large axial field of view in current-generation PET and SPET tomographs, the availability of multi-slice acquisition sequences in fMRI, and helmet systems in MEG.

Spatial resolution is of obvious importance in the examination of any organ, but in the examination of brain morphology and function, it becomes essential to differentiate contiguous, yet distinct, anatomical-functional structures. At the present time, the highest spatial resolution is provided by CT and MRI - less than 1 mm. Functional imaging techniques (PET, SPET, fMRI) are characterized by spatial resolutions on the order of a few mm. Similar precision is obtained in MEG in the localization of magnetic dipoles in cortical regions. However, PET, SPET and fMRI allow for a 3D representation of the functional activity in the brain in terms of hemodynamic and metabolic variables, while MEG maps the electrical cerebral activity at the scalp surface. The identification and localization of the source dipoles in the brain by MEG can not be determined unequivocally since there is not a unique solution to the inverse problem.

Limitations to temporal resolution in PET, SPET and fMRI already have been discussed in the previous sections on instrumentation performance, physical characteristics of the measured signal and the physiological process under examination. It should be obvious then that MEG, with a temporal resolution of 1 msec or less, is superior in its ability to follow temporal changes, even in fast physiological processes.

Quantification of functional variables describing the process under study might be crucial in several applications. The availability of physiological tracers whose behaviour in the human body is known and can be mathematically modelled makes PET a quantitative technique. On the other hand, quantification in SPET is limited by the non-physiological behaviour of the available tracers. In fMRI, the physical and physiological interpretation of the signal is not completely clear, and additional investigation is required for data quantification.

Another important issue is the invasiveness of the various techniques. PET, SPET and CT are limited by the problem of radiation exposure. However, the increased detection efficiency achieved in current-generation 3D PET and SPET technology permits a reduction of the amount of radioactivity administered to the patient. Fundamental studies on the safety of MRI are still very limited. However, the results of studies to date do not indicate any limit in the repeatability of the MRI examination. fMRI based on the BOLD principle does not require the administration of exogenous contrast agents, being then, as is MEG, a totally non invasive method.

3. Integration of multi-modal information on the brain

Biomedical images and signals (from PET, SPET, fMRI, MRI, CT, EEG and MEG) represent different information describing the morphology and the functions of the brain. Correlation of such different information can support the comprehension of the human brain in normal and pathological conditions [61]. However, combining independent data obtained by the different imaging and signal modalities can be a cumbersome procedure for the physicians, who must interpret all the information obtained to achieve a unique description of the process under examination. To help forming a correct interpretation of the relation among the different modalities, a unique spatial reference system is defined, into which one must represent images and signals relative to the same subject. Within this spatial reference system, the head and the brain of the subject represent the anatomical model for which multi-modal information can be mapped. Integration of multi-modal information regarding the brain may then be utilized to find the point-by-point correspondence of the same anatomical structures in the different modalities, or, for example, localisation of an EEG/MEG dipole within the brain anatomy.

3.1. INTEGRATION OF MULTI-MODAL BIOMEDICAL IMAGES

Tomographic images (such as PET, SPET, MRI, fMRI and CT) describe the brain as a set of contiguous bidimensional slices, thus providing a three-dimensional representation of the organ under study. This three-dimensional representation is consistent with the spatial reference system specific for each tomographic imaging device.

Image integration requires both the identification of the spatial reference systems in which each volumetric representation is seen and the determination of the geometrical transformation parameters - translation, rotation and scaling factors - to remap one study into the other. This task can be performed by registration techniques in which direct comparison of the same anatomical-functional structures in the different multi-modal studies can be established.
In the following sub-section, a review of the principal approaches to the registration problem is presented.

3.1.1. Fiducial markers

A common strategy for image registration of two assumed tomographic studies is based on the use of fiducial markers, whose coordinates recognized allow for the determination of the registration parameters [62 to 64]. The markers are, in general, point or line sources, comprised of selected radioactive or contrast agents which are visible in the different imaging modalities. They are located either directly on the patient's skin or fixed to a device used for patient positioning. The coordinates of the markers in the two studies to be registered are used as input to a minimization algorithm which estimates the best geometric transformation to overlay the markers and achieve spatial correspondence of the two image volumes.

Registration methods based on this approach have widespread application, being independent of the imaging modalities to be matched and of the alteration in the image pattern induced by the pathology. The registration accuracy is strictly related to the precision by which markers are repositioned in the different tomographic studies. The main limitation of this approach is the need for a predefined acquisition protocol, requiring special care in the fixation and filling of the markers.

Skin markers. External skin markers are usually small spherical or cylindrical sources, directly positioned on the patient's head. They can be filled at the beginning of each study with a different component appropriate for each specific examination. Alternately, they can be constructed to be simultaneously visible by more than one imaging modality. For example, a solution of water and a radioactive isotope, such as the positron emitter ^{68}Ge (physical half-life: 287 days), can be visualized by both MRI and PET. By choosing long-lived radioisotopes, as in the case of ^{68}Ge, markers very seldom require filling. Markers appear as spread points in the axial tomographic images. Their position, calculated by the coordinates of their centroids, represent the spatial reference points to be used by the registration process. As an example of a registration algorithm based on point markers, the mathematical method proposed by Hawkes *et al.* [62] is briefly reported here. In this approach, the parameters of the geometrical transformation (translation, rotation and scaling factors) are decoupled and separately calculated. The algorithm uses a least-squares-fitting procedure based on singular-value-decomposition which minimizes the root-mean-square distance between marker-pair positions [65]. In general, the accuracy of the registration procedure is strictly related to the number of markers used. A miminum of three markers is required to identify the three-dimensional geometrical transformations. A greater number of points improves the precision of the registration by minimizing the uncertainty in the marker coordinates. However, the accurate positioning of the markers on the head is time consuming, and the use of several markers can be uncomfortable for patients. Thus, in clinical situations, no more than 4 to 5 markers are generally employed. The anatomical positions of the markers also is crucial for registration accuracy. Anatomical points on the head, where the skin does not easily slide with respect to the skull, should be selected in order to avoid spatial variations of the marker positions. Moreover, it is not always possible to perform different examinations keeping the markers fixed to the patient. Errors in the reproducibility of the markers' positions can occur during the repositioning procedure between one study and another. In order to improve the accuracy of repositioning, anatomical points which can be localized easily should be chosen. Anatomical positions, such as for the mastoid processes of the temporal bones and the zygomatic processes of the frontal bone, seem to be suitable both for small skin movements associated with these areas and for their identification reproducibility.

Positioning devices. Especially for integration of brain studies, a great variety of positioning devices has been proposed. A special category of these systems is represented by stereotactic frames which are attached to the patient's head by means of screws directly fixed into the skull [66 to 68]. The use of these 'traumatic' positioning devices is generally limited to the study of patients scheduled for neurosurgery, for which the stereotactic frame is required. Stereotactic holders guarantee accurate repositioning of the patient's head with respect to the holder in both therapeutic and diagnostic procedures. A stereotactic frame is always provided with external markers, generally V- or N-shaped, describing straight lines in space. The markers appear as points in the tomographic image planes. The positions of these points in the image matrices permit the determination of the orientation and position of the stereotactic frame in three-dimensional space. The geometry of the markers is chosen in order to analytically compute the spatial transformation without applying more complex minimization algorithms. Other positioning devices have been developed to be used in less critical applications where the invasiveness of a stereotactic holder is not justified. In this case, the fixation is reached using patient-personalized tools such as special masks modeling the face of the subject under examination, or individual dental molds of the patient [69, 70]. These more patient-friendly approaches maintain almost all the advantages offered by the stereotactic technique (i.e., simple marker geometry, reduction of head movements during acquisition, accurate repositioning of the patient for follow-up and multi-modal studies), except for a less accurate fixation of the head. This fact influences the accuracy of the registration procedure (less than 1 mm for stereotactic frames vs. 2-3 mm for non-stereotactic head holders). However, it should be noted that for many clinical uses, an accuracy of a few mm is considered adequate.

3.1.2. Image feature matching
A different approach to the registration problem relies on the consideration that in two multi-modal studies of the same organ it is possible to extract homologous geometrical features relative to the anatomical region under examination. If these features contain the spatial information needed for correctly identifying the spatial reference systems in which the tomographic images are represented, matching of the equivalent features results in matching the multi-modal image volumes [71 to 74].

Based on this principle, several methods have been proposed, differing from each other principally in the class of homologous features chosen for the extraction and/or for the registration algorithm, by which the features themselves are overlapped. Thus, these different methods present different performances in terms of registration accuracy, computer time, suitability for clinical use and dependence on the image alterations induced by pathology.

In general, these techniques are based on image-processing methods which can be performed off-line, and they do not present constraints during acquisition. For this reason, they have the common advantage, with respect to the use of external markers, of allowing retrospective registrations, which is a relevant feature for wide clinical use.

On the other hand, a limitation of this approach to image registration is that these methods can be applied only to those imaging modalities in which several homologous anatomical structures can consistently be identified in the studies to be correlated.

Matching of homologous points. A method assuming anatomical point markers as matching features has been proposed by Evans *et al.* [71] to correlate PET and MRI cerebral images in normal subjects. In the images to be correlated, some homologous cortical and subcortical landmarks are recognized in order to identify two sets of corresponding 3D points. From the computational point of view, this approach is very similar to that based on external point markers. A linear least-squares optimization technique is used to calculate the geometrical transformation which minimizes the root-mean-square distance between the two sets of equivalent points. The registration algorithm is based on the Procrustes algorithm, which estimates in two successive steps the translation and rotation components. From simulation studies it was estimated that at least 15 points should be used for a registration accuracy of less than 1 mm, assuming that an uncertainty of 5 mm or less can affect the position identification of the landmarks. It is important to note that strong user interaction, supervised by an expert in neuroanatomy, is needed for properly locating the corresponding point pairs in the two studies. In general, the application of this method depends on the identification reliability of a consistent number of anatomical structures in the two image volumes to be correlated. This criterion can limit the extension of this approach to other imaging modalities. For example, the limited spatial resolution of SPET, and the apparent resulting mixture of etherogeneous tissues, might mask some anatomical landmarks, affecting the feasibility of the method. In low-count statistics for SPET or PET images, statistical noise might prevent the recognition of the landmarks. Furthermore, for pathological conditions, the functional cerebral alterations can make some markers invisible, complicating the application of this approach to registration.

Principal axes matching. Registration algorithms, based on matching spatial moments of corresponding objects, such as the centroids or the principal axes, have been proposed. From the classical theory of rigid bodies, the spatial position of an object is uniquely defined by knowledge of its center of mass and its orientation with respect to its center of mass, described by its inertial principal axes. Alpert *et al.* [75] applied this approach to the registration of cerebral PET, MRI and CT image volumes, considering the organ under examination as a rigid body with uniform mass-density functions. The registration procedure extracts from the two image volumes equivalent surfaces and transforms the principal axes coordinate system of one study into the reference study system. This registration procedure is fast and simple because it requires only analytic computation of moments and matrix inversion. However, a high degree of symmetry in the objects to be matched can result in eigenvector solutions which are not unique. The major practical limitations of this approach are related to the limited three-dimensional sampling of the volume and to the difficulty in exactly delineating corresponding volumes in the two image sets.

Surface matching. A different strategy that allows a retrospective registration of MRI, CT and PET cerebral images has been developed by Pelizzari *et al.* [72]. In this method, equivalent surfaces corresponding to the external surface of the head are identified in each of the two studies. One surface, named "head" and represented as a set of contours, is used as a reference; the second one, named "hat" and represented by a set of points, is fitted to the first one. For each hat point, the intersection with the head surface of a ray from the transformed hat point to the centroid of the head model is evaluated. A non-linear least-squares algorithm is used, which minimizes a cost function defined by the mean square of such "hat-head" distances. The allowed class of transformations includes rigid-body rotation and translation and linear scaling along three orthogonal axes, which accounts for uncertainty in the scan pixel sizes and for possible linear distortion in MRI. This method assumes that the cost function presents a single global minimum, no severe distortion is present between the two studies and every surface point in the "hat" has a corresponding point in the "head". For this last requirement to be accomplished, the axial extension in the reference study must be greater than in the study to be matched. The uniqueness of the solution is not ensured due to the intrinsic limitation of non-linear minimization algorithms. Thus, some user interaction is required to drive the minimization algorithm in choosing a good search starting point and in modifying the transformation parameters, if necessary. The limiting accuracy of this method is determined by the pixels with the lowest spatial resolution. Other algorithms based on the Pelizzari "head-hat" approach have been proposed to improve the computational performance. Registration is obtained by a matching algorithm in which a generalized average distance, calculated via a distance map between the equivalent surfaces, is used as the measure of correspondence between the two studies. This more sophisticated approach results in better registration accuracy and in a smoothed function to be minimized [76, 77].

3.1.3. Pixel-by-pixel analysis

Fully automated registration techniques that are based on the pixel-by-pixel analysis of the image characteristics also have been proposed [78 to 80]. These techniques rely on the hypothesis that the organ under examination shows similar characteristics in two independent tomographic scans. This means that pixel values in the two studies to be matched are strongly correlated. For this reason, these methods seem to be particularly powerful for the registration of studies acquired with the same acquisition modality (i.e., in PET follow-up, test-retest and activation studies).

An extension of this approach to the registration of multi-modal studies, in particular MRI and PET, also has been implemented [81]. The hypothesis of pixel value similarity is not verified in this case, and some modifications must be introduced. Specifically, the MRI study is pre-processed by deleting the structures which are not imaged by PET (e.g., scalp, skull and meninges) and by segmenting MRI data into separate components on a thresholding basis. The algorithm minimizes the uniformity of the PET pixel values within each partition by iteratively minimizing a weighted average of the standard deviation of the PET pixels. The computation time is quite variable and depends on a number of parameters, such as pixel size, number of slices and counting statistics. Usually a time of 30-40 min is required for PET/MRI registration with a standard setting of the parameters. Extra time must be considered in the manual editing of the MRI study to eliminate the structure external to the brain.

The procedure described above is able to register MRI and PET studies with a residual error estimated to be less than 3 mm. However, in the presence of pathology resulting in an altered image pattern, this method must be used carefully by verifying that the presence of a structural lesion does not cause residual misalignment after registration of the two studies.

3.2. INTEGRATION OF MEG TO MULTI-MODAL BIOMEDICAL IMAGES

The anatomical localization of electrical sources estimated by MEG measurements as current dipoles is usually performed by representing the head as a homogeneously conducting sphere [82]. In principle, individual anatomical knowledge is not required, restricting the interpretation of MEG data to a coarse localization of the MEG sources.

This limitation can be overcome by integrating MEG measurements with the three-dimensional description of the real head of the subject under examination as provided by high resolution MRI. In such a way, the neuromagnetic localization in the individual head allows for the direct association of the electrical phenomena to the involved cerebral areas. Furthermore, MRI-based morphological information can be included in the estimate of the MEG dipole, resulting in a more accurate anatomical localization.

In analogy to the registration of multi-modal images, matching of the spatial reference systems consistent with the MEG probe and MRI volumetric data can be performed by identifying at least three corresponding points, by which the registration parameters can be estimated.

Head fixation systems have been proposed to reproduce the same head position in MEG and MRI devices. Head fixation has been obtained by a personalized mold of a patient's teeth. However, this approach to the registration of MEG and MRI data has not had widespread use due to the difficulty in fitting the head fixation system and the MEG probe during MEG measurements.

Another approach to MEG/MRI registration uses, as corresponding points, anatomical landmarks recognized in MRI images and external MEG markers fixed in the same anatomical positions. MEG markers can be made by small radiofrequency transmitters whose spatial position can be revealed by appropriate radiofrequency receivers. Alternately, small copper coils can be adopted as MEG markers, generating an alternate current revealed (simultaneously with the MEG data) by the same superconducting array [83]. Methods based on external MEG markers and anatomical MRI points do not require a predefined MRI acquisition protocol. A major advantage of this approach is that it allows retrospective integration to MRI. Thus, MRI can be performed at any time and without special care in patient positioning. Accuracy is dependent on the accuracy of recognition of anatomical points in MRI. This requires expertise in neuroanatomy and high-quality, high-definition MRI images.

A different registration approach consists in the use of external markers in both MRI and MEG positioned at the same anatomical points. MRI markers are generally small tubes filled with water or copper-sulphate solution visible to MRI. This technique does not require knowledge of brain anatomy, since external markers are easily recognized as the positions of relevant signals. Moreover, this approach can be extended to the direct correlation of MEG to other imaging modalities such as PET and SPET, independent of the localization in MRI. On the other hand, the use of external markers recalls the previously mentioned limitations on accuracy due to marker repositioning and a need for a pre-defined acquisition protocol, which prevents retrospective analysis.

An example of integration of anatomical (MRI) and functional (PET and MEG) information in a normal subject during visual stimulation is shown in Figure 2. Integration of MRI, PET and MEG was performed by use of external markers positioned in the three studies in the same positions on the patient's skin.

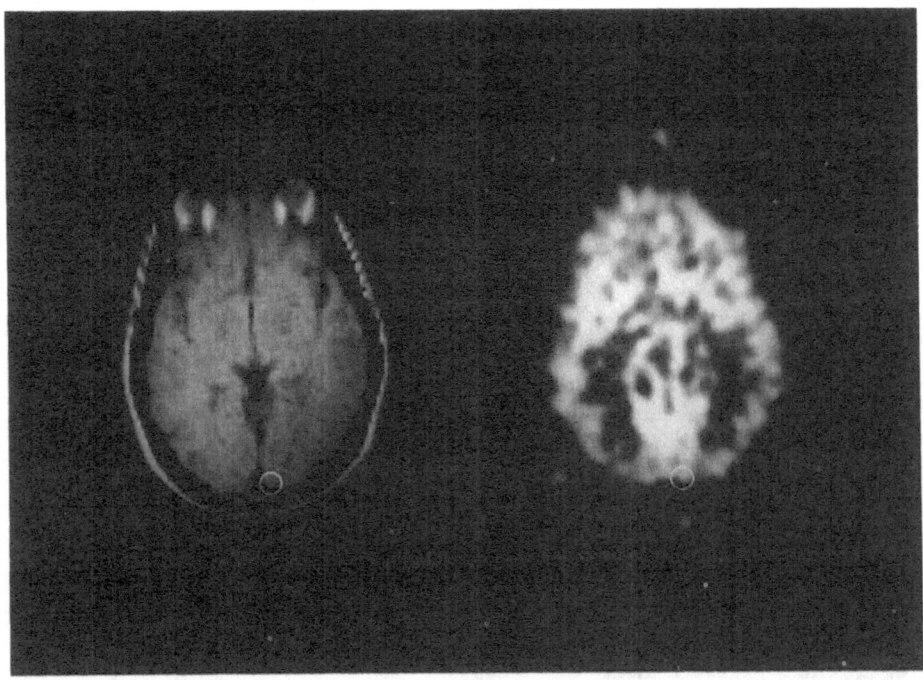

Figure 2. Integration of MRI, PET and MEG information in a normal subject during visual stimulation. MRI (left) and PET (right) images are shown. The MEG dipole (o) is localized in the calcarine cortex.

4. Application of the multi-modal approach for the analysis of brain function

Images generated by the various acquisition techniques contain different and often complementary information whose synergistic use can improve the analysis of brain function. An important aspect of this multi-modal approach is represented by the possibility of localizing functional information on individual brain anatomy. The analysis of functional images and signals obtained by PET, SPET, fMRI, EEG and MEG is, in fact, enhanced by taking into account morphological information as provided by CT or MRI.

Functional images represent physiological and biochemical processes, and they do not intrinsically show the underlying morphology. Extreme situations of such sparse

508

morphological information are found in PET/SPET neuroreceptor studies and in tumour detection with monoclonal antibodies, where radioactivity is distributed in isolated spots without the appearance of related anatomical structure. The point-by-point correspondence of anatomical and functional images, as obtained by a registration technique, allows for the unambiguous association of functional areas with morphological districts. In Figure 3, anatomical-functional correlation is applied to the evaluation of a patient with a meningioma. The lesion detected by MRI in the right ventricular region corresponds in the SPET study to an area of high tracer uptake.

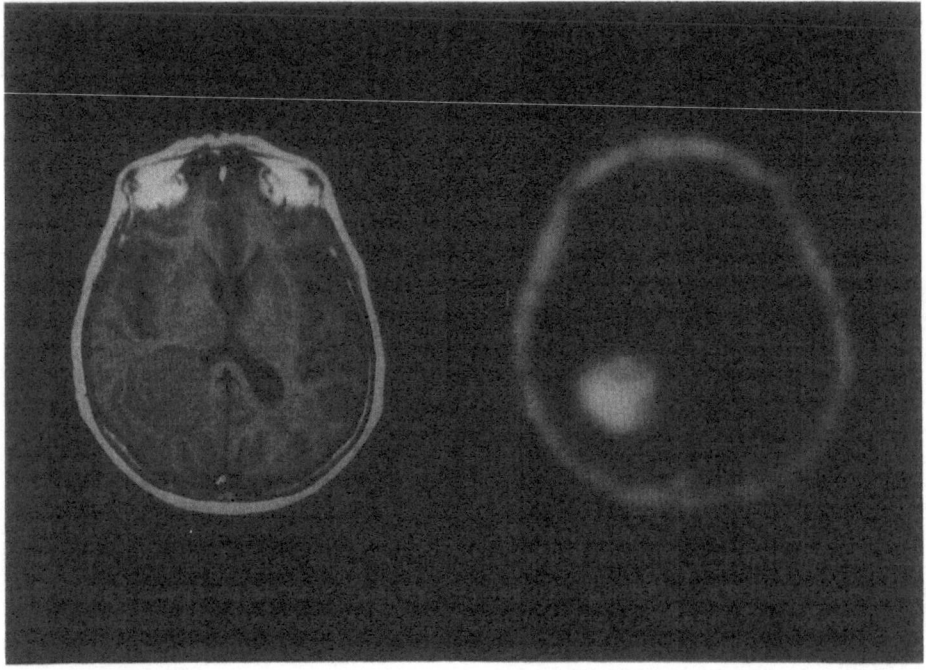

Figure 3. Integration of MRI (left) and monoclonal antibodies SPET (right) in a patient with a meningioma.

An immediate benefit of this approach is in providing assistance in the quantitative and semiquantitative analysis of nuclear medicine images (PET and SPET), carried out by the definition of regions of interest. Anatomical-functional correlation also is crucial in activation studies to properly localize the activation areas in individual brain anatomy.

Furthermore, anatomical-functional correlation can be applied to more accurate quantification of functional parameters. Morphological data can be included within the tomographic reconstruction process to improve signal-to-noise ratio and spatial resolution [84]. Underestimation of radioactive concentration in small anatomical structures, as a result of the limited system spatial resolution, effect known as partial volume effect, could be compensated by knowledge of the shape and size of the particular anatomical region under examination [85, 86]. **Quantification of functional** images is affected by the presence of etherogeneous tissues characterized by different metabolic or flow rate (e.g., gray matter, white matter and cerebrospinal fluid) [87]. Quantification accuracy can be improved by knowledge of the regional tissue composition. This information is achieved by segmentation of high spatial definition and contrast anatomical images, such as MRI, into homogeneous tissue components. As an example, in Alzheimer's disease the measurement of the degree of cerebral atrophy assessed by MRI could be used for correcting the related apparent reduction of functional activity (such as blood flow and metabolism) assessed by PET or SPET [88, 89].

By analogy, the integration of MEG data to morphological MRI images permits the anatomical localization of the MEG electrical sources in the individual brain. This approach improves the analysis of MEG data recorded during stimulation by providing a direct and accurate identification of the cortical area responsible for the measured electrical activity. In pathology, brain mapping of the electrical activity obtained by this integrated MEG-MRI approach provides good support for a diagnosis. For instance, anatomical localization of the electrical dipole estimated by MEG in epileptics guides the identification of the epileptic focus inducing electrical abnormalities [90, 91]. Furthermore, knowledge of brain anatomy can improve the procedure of spatial localization of the MEG dipole, accounting for the real shape of the individual brain instead of the spherical head model.

From the clinical diagnostic point of view, the synergistic use of the information specific to each imaging modality should produce a more complete comprehension of pathophysiological mechanisms and generate a more accurate definition of the clinical picture. Although the choice of specific methods to be used in the investigation of patients with neurological diseases must remain (in principle) selective and sequential, new information can be obtained without any further patient involvement at little additional cost, by the fusion of individual images, integrating and correlating information obtained with different imaging devices and methods. As this synergistic strategy allows one to gain a more complete understanding of any given physiological, pharmacological or pathological process, it could represent the basis for the modification of current diagnostic procedures. The synergistic use of the available biomedical technologies also is crucial for an effective diagnostic process and economic use of the available technologies. Therefore, this is an issue that can have a relevant economic and social impact.

The ultimate goal of diagnostic imaging procedures performed on patients with neurological symptoms is the identification of the abnormal morphological and functional patterns that correlate with the symptoms. Currently, CT and MRI are the diagnostic methods of first choice in all neurological diseases [92]. Both CT and MRI are very sensitive techniques. However, in numerous cases, signs and symptoms cannot be completely explained by methods of morphological imaging. Functional disturbances can, in fact, precede detectable morphological damage. Therefore, only functional imaging may permit (a) the early assessment of the biochemical damage that precedes the morphological damage, (b) the correlation of functional alterations and clinical symptoms in the absence of detectable morphological damage, and eventually (c) the explanation of symptoms due to functional abnormalities that occur in areas proximal and remote from the site of the primary lesion due to neuronal disconnection.

Thus, in several cases it is only with the integration of morphological and functional information that it is possible to achieve a full appreciation of the nature and extent of the damage that produces the clinical symptoms.

The value of this approach of integrated diagnosis can be appreciated in the evaluation of all neurological diseases, as described in section 2.3, where PET and SPET are clinically used in conjuction with MRI or CT. Figure 4 illustrates a case of cerebral-vascular ischemia. The estimate of the extent of functional damage is improved by correlation with the cerebral anatomy.

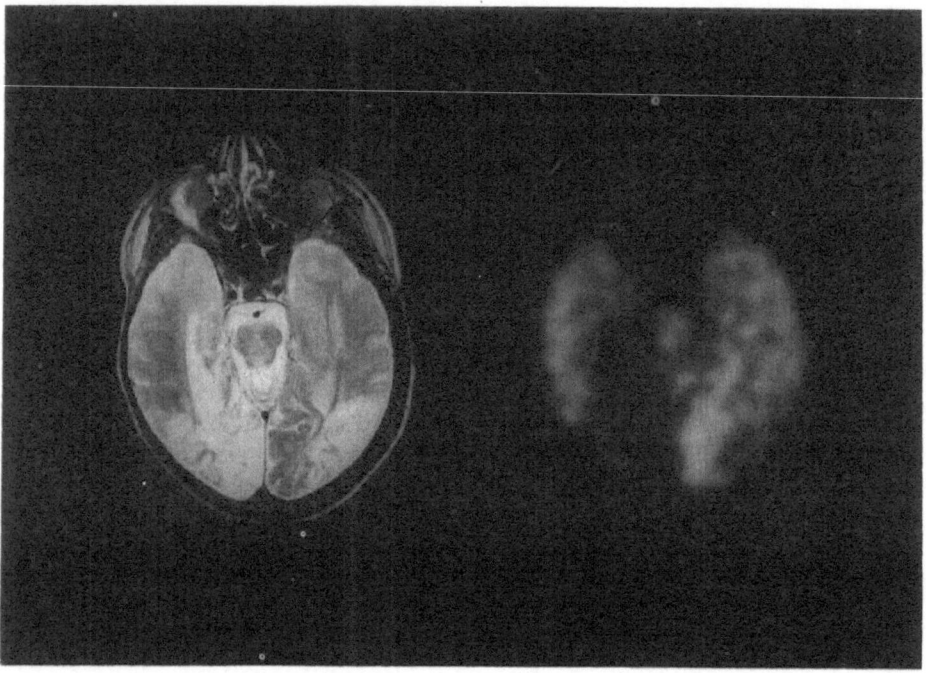

Figure 4. Integration of MRI (left) and [18F]FDG PET (right) in a case of cerebral-vascular ischemia in the occipital lobe.

Oncology represents another clinical situation where a single diagnostic modality might not provide sufficient information for a complete definition of the pathology. The diagnostic work-up of brain tumours is based on morphological imaging, first with CT and more recently also with MRI. However, CT and MRI show an altered signal in the tumour region, but they might not be able to differentiate between tumour recurrence and necrotic tissues induced by radiotherapy. The *in vivo* biochemical imaging of tumours is best achieved by emission tomography. An increased level of glucose consumption in the presence of high-grade tumour recurrence has been, in fact, observed by PET and [18F]FDG. It is only the conjunction of anatomical information by MRI and functional information by PET that makes the differential diagnosis of tumour recurrence possible [93].

Another important aspect is the temporal evolution of the process under study, crucial for complete comprehension of the brain functional activity in normal and pathological conditions. In functional imaging techniques (PET, SPET and fMRI) temporal resolution is often insufficient to describe the rapidity of functional changes. An example is found in the study of epilepsy. PET, SPET and MRI permit the localization of the source of the seizure, but only EEG and MEG provide temporal analysis of the electrical activity [94, 95]. Similarly, in activation studies, both PET and fMRI map functional activity in the brain showing one or more regions of increased blood flow without information on their temporal relations. Neuronal activation events occur in time on the order of msec, and only EEG and MEG techniques are adequate to describe such phenomena with sufficient temporal discrimination [96 to 98]. The assessment of neural networks is a challenging issue in neurophysiology which only can be approached by the combined spatial-temporal information provided by the integration of multi-modal biomedical images and signals.

In conclusion, none of the available biomedical imaging and signal techniques is unique at the present time in describing completely the functional activity of the brain. Each presents relative merits and limitations. However, it is not a competition, but rather the combination of the specific values of each that might bring about a deeper comprehension of brain function.

Acknowledgement.

This work was partially financed by the Progetto Finalizzato ACRO/1915, CNR.

References.

1. Young, I.R. (1994) Review of modalities with potential future in radiology, *Radiology* **192**, 307-317.

2. Raichle, M.E. (1994) Visualizing the mind, *Scientific American*, **April**, 36-44.

3. S.Sato (ed.) (1990) *Advances in Neurology: Magnetoencephalography.Vol.54*, Raven Press, New York:

4. Phelps, M.E. and Mazziotta, J.C. (1985) Positron Emission Tomography: Human Brain Function and Biochemistry, *Science* **228**, 799-809.

5. Huang, S.C. and Phelps, M.E. (1986) Principles of tracer kinetic modeling in positron emission tomography and autoradiography, in M. Phelps, J. Mazziotta, and H. Schelbert (eds), *Positron emission tomography and autoradiography: principles and application for the brain and heart*, Raven Press, New York, pp. 287-346.

6. Frackowiak, R.S.J., Lenzi, G.L., Jones, T. and Heather, J.D. (1980) Quantitative measurement of regional cerebral blood flow and oxygen metabolism in man using ^{15}O and positron emission tomography. Theory, procedure and normal values, *J.Comput.Assist.Tomogr.* **4**, 727-736.

7. Reivich, M., Kuhl, D., Wolf, A., Greenberg, J., Phelps, M., Ido, T., Casella, V., Fowler, J., Hoffman, E., Alavi, A., Som, P. and Sokoloff, L. (1979) The [^{18}F]fluorodeoxyglucose method for the measurement of local cerebral glucose utilization in man, *Circ. Res.* **44**, 127-137.

512

8. Hoffman, E.J. and Phelps, M.E. (1986) Positron emission tomography principles and quantitation, in M. Phelps, J. Mazziotta, and H. Schelbert (eds.), *Positron emission tomography and autoradiography: principles and application for the brain and heart*, Raven Press, New York, pp. 237-286.

9. Brooks, R. and Di Chiro, G. (1976) Principles of computer assisted tomography (CAT) in radiographic and radioisotopic imaging, *Phys.Med.Biol.* 21, 689-732.

10. Koeppe, R.A. and Hutchins, G.D. (1992) Instrumentation for positron emission tomography: tomographs and data processing and display systems, *Semin.Nucl.Med.* 22, 162-181.

11. Wiehenard, K., Dahlbom, M., Eriksson, L., Michel, C., Bruckbauer, T., Pietrzyk, U. and Heiss W-D. (1994) The ECAT EXACT HR: performance of a new high resolution positron scanner, *J.Comput.Assist.Tomogr.* 18, 110-118.

12. DeGrado, T.R., Turkington, T.G., Williams, J.J., Stearns, C.W., Hoffman, J.M. and Coleman R.E. (1994) Performance characteristics of a whole-body PET scanner, *J.Nucl.Med.* 35, 1398-1406.

13. Spinks, T.J., Jones, T., Bailey, D.L., Townsend, D.W., Grootoonk, S., Bloomfield, P.M., Gilardi, M.C., Casey, M.E., Sipe, B. and Reed, J. (1992) Physical performance of a positron tomograph for brain imaging with retractable septa, *Phys.Med.Biol.* 37, 1637-1655.

14. Townsend, D.W., Geissbuhler, A., Defrise, M., Hoffman, E.J., Spinks, T.J., Bailey, D.L., Gilardi, M.C. and Jones, T. (1991) Fully three dimensional reconstruction for a PET camera with retractable septa, *IEEE Trans. Med. Imag.* 10, 505-512.

15. Bailey, D., Jones, T., Spinks, T.J., Gilardi, M.C. and Townsend, D.W. (1991) Noise equivalent count measurements in a Neuro-PET Scanner with retractable septa, *IEEE Trans. Med. Imag.* 10, 256-260.

16. Cherry, S.R., Dahlbom, M. and Hoffman, E.J. (1991) 3D PET using conventional multislice tomograph without septa, *J.Assist.Comput. Tomogr.* 15, 655-668.

17. Cherry, S.R., Woods, R.P., Hoffman, E.J. and Mazziotta, J.C. (1993) Improved detection of focal cerebral blood flow changes using three-dimensional positron emission tomography, *J.Cereb.Blood Flow Metab.* 13, 630-638.

18. Rosenthal, M.S., Cullom, J., Hawkins, W., Moore, S.C., Tsui, B.M.W. and Yester, M. (1995) Quantitative SPECT imaging: a review and recommendations by the Focus Committee of the Society of Nuclear Medicine Computer and Instrumentation Council, *J.Nucl.Med.* 36, 1489-1513.

19. Kouris, K., Jarritt, P.H., Costa, D.C., Ell, P.J. (1992) Physical assessment of the GE/CGR Neurocam and comparison with a single rotating gamma-camera, *Eur.J.Nucl.Med.* 19, 236-242.

20. Genna, S. and Smith, A.P. (1988) The development of ASPECT, an annular single crystal brain camera for high efficiency SPECT, *IEEE Trans.Nucl.Sci.* NS-35, 654-658.

21. Bailey, D.L., Zito, F., Gilardi, M.C., Savi, A.R., Fazio, F. and Jones, T. (1994) Performance comparison of a state of the art neuro-SPET scanner and a dedicated neuro-PET scanner, *Eur.J.Nucl.Med.* 21, 381-387.

22. Leichner, P.K., Morgan, H.T., Holdeman, K.P., Valentino, F., Lexa, R., Kelly, R.F., Hawkins, W.G. and Dalrymple, G.V. (1995) SPECT imaging of Fluorine-18, *J.Nucl.Med.* 36, 1472-1475.

23. Ell, P.J. (1992) Mapping cerebral blood flow, *J.Nucl.Med.* 33, 1843-1845.

24. Messa, C., Fazio, F., Costa, D.C. and Ell, P.J. (1995) Clinical brain radionuclide imaging studies, *Semin.Nucl.Med.* 25, 111-143.

25. Jarenwattananon, A., Khandji, A. and Brust, J.C.M. (1988) Diagnostic Neuroimaging in Stroke, in W.H. Theodore (ed.), *Clinical Neuroimaging*, Alan R. Liss inc., New York, pp. 11-47.

26. Bogousslawsky, J., Delaloye-Bishof, A., Regli, F. and Delaloye, B. (1990) Prolonged hypoperfusion and early stroke after transient ischemic attack, *Stroke* 21, 40-46.

27. Giubilei, F., Lenzi, G.L., Di Piero, V., Pozzili, C., Pantano, P., Bastianello, S., Argentino, C. and Fieschi, C. (1990) Predictive value of brain perfusion single-photon emission computed tomography in acute ischemic stroke, *Stroke* 21, 895-900.

28. Pantano, P., Baron, J.C., Samson, Y., Bousser, M.G., Derouesné, C. and Comar, D. (1986) Crossed cerebellar diaschisis: further studies, *Brain* 109, 677-694.

29. Baldy-Moulinier, M., Lassen, N.A., Engel, J., Askienazy S. (eds.) (1989) *Focal epilepsy; clinical use of emission tomography*, John Libbey, London, Paris, Rome.

30. Theodore W. H. (1988) Epilepsy, in W.H. Theodore (ed.), *Clinical Neuroimaging*, Alan R. Liss inc., New York, pp. 183-210.

31. Rowe, C.C., Berkovic, F., Benjamin Sia, St., Austin, M., Mc Kay, J.W., Kalnins, R.M. and Bladin, P. (1989) Localization of epileptic foci with postictal single photon emission tomography, *Ann.Neurol.* 26, 660-668.

32. Alavi, A., Dann, R., Chawluk, J., Alavi, J., Kushner, M. and Reivich, M. (1986) Positron emission tomography imaging of regional cerebral glucose metabolism, *Semin. Nucl. Med.* 16, 2-34.

33. Haxby, J.V., Grady, C.L., Duara, R., Sclageter, N., Berg, G. and Rapoport, S.I. (1986) Neocortical metabolic abnormalities precede nonmemory cognitive defects in early Alzheimer's type dementia, *Arch.Neurol.* 43, 882-885.

34. Duara, R., Grady, C., Haxby, J., Sundaram, M., Cutler, N.R., Heston, L., Moore, A., Sclageter, N., Larson, S. and Rapoport, S.I. (1986) Positron emission tomography in Alzheimer's disease, *Neurology* 36, 879-887.

35. Johnson, K.A., Mueller, S.T., Walsche, T.M., English, R.J. and Holman, B.L. (1987) Cerebral perfusion imaging in Alzheimer's disease: use of single photon emission computed tomography and Iofetamine Hydrochloride I-123, *Arch.Neurol.* 44, 165-168.

36. Perani, D., Di Piero, V., Vallar, G., Cappa, S., Messa, C., Bottini, G., Berti, A., Passafiume, D., Scarlato, G., Gerundini, P., Lenzi, G.L. and Fazio, F. (1988) Technetium-99m HM-PAO-SPECT study of regional cerebral perfusion in early Alzheimer's disease, *J.Nucl.Med.* 29, 1507-1514.

37. Burns, A., Philpot, M.P., Costa, D.C., Ell, P.J. and Levy, R. (1989) The investigation of Alzheimer's disease with single photon emission tomography, *J.Neurol.Neurosur.Psych.* 52, 248-253.

38. Messa, C., Perani, D., Lucignani, G., Zenorini, A., Zito, F., Rizzo, G., Grassi, F., DelSole, A., Franceschi, M., Gilardi, M.C. and Fazio, F. (1994) High-resolution Technetium-99m-HMPAO SPECT in patients with probable Alzheimer's disease: comparison with fluorine-18-FDG PET, *J.Nucl.Med.* 35, 210-216.

39. Di Chiro, G., De LaPaz, R.L., Brooks, R.A., Sokoloff, L., Kornblith, P.L., Smith, B.H., Patronas, N.J., Kufta, C.V., Kessler, R.M., Johnson, G.S., Manning, R.G. and Wolf, A.P. (1982) Glucose utilization of cerebral gliomas measured by [^{18}F] fluorodeoxyglucose and positron emission tomography, *Neurology* 32, 1323-1329.

40. Kim, K.T., Black, K.L., Marciano, D., Mazziotta, J.C., Guze, B.H., Grafton, S., Hawkins, R.A. and Becker, D.P. (1990) Thallium-201 SPECT imaging of brain tumours: methods and results, *J.Nucl.Med.* 31, 965-969

41. Paganelli, G., Magnani, P., Zito, F., Villa, E,. Stella, M., Lopalco, L., Siccardi, A.G. and Fazio, F. (1991) Antibody guided tumor detection in CEA positive patients using the avidin-biotin system, *Cancer Res.* 51, 5960-5966.

42. Brucke, T., Podreka, I., Angelberger, P., Wenger, S., Topitz, A., Kufferle, B., Muller, Ch. and Deecke, L. (1991) Dopamine D2 receptor imaging with SPECT: studies in dfferent neuropsychiatric disorders, *J.Cereb.Blood Flow Metab.* 11, 220-228.

43. Stöcklin, G. (1992) Tracers for metabolic imaging of brain and heart, *Eur.J.Nucl.Med.* 19, 527-551.

44. Lucignani, G., Moresco, R.M. and Fazio, F. (1989) PET-Based Neuropharmacology: State of the Art, *Cerebrovasc.Brain Metab.Rev.* 1, 271-287.

45. Young, A.B., Frey, K.A. and Agranoff, B.W. (1986) Receptor assays; in vitro and in vivo, in M. Phelps, J. Mazziotta, H. Schelbert (eds.), *Positron emission tomography and autoradiography: principles and application for the brain and heart*, Raven Press, New York, pp. 73-111.

514

46. Frackowiak, R.S.J. and Friston, K.J. (1994) Functional neuroanatomy of the human brain: positron emission tomography - a new neuro anatomical technique, *J.Anat.* **184**, 211-225.

47. Perani, D., Gilardi, M.C., Cappa, S.F. and Fazio, F. (1993) PET studies of cognitive functions: a review, *J. Nucl. Med. Biol.* **36**, 324-336.

48. Pantano, P., Di Piero, V., Ricci, M., Fieschi, C., Bozzao, L. and Lenzi, G.L. (1992) Motor stimulation response by technetium-99m hexamethylpropylene amine oxime split-dose method and single photon emission tomography, *Eur.J.Nucl.Med.* **19**, 939-945.

49. George, M.S., Ring, H.A., Costa, D.C., Ell, P.J., Kouris, K. and Jarrit, P.H. (1991) *Neuroactivation and neuroimaging with SPECT*, Springer-Verlag, Londra.

50. Weiller, C., Ramsay, S.C., Wise, R.J.S., Friston K.J. and Frackowiack, R.S.J. (1993) Individual patterns of functional reorganization in the human cerebral cortex after capsular infarction, *Ann.Neurol.* **33**, 181-189.

51. Chakeres, D.W. and Schmalbrock, P. (1992) *Fundamentals of Magnetic Resonance Imaging*, Williams & Wilkins, Baltimore.

52. Cohen, M.S. and Bookheimer, S.H. (1994) Localization of brain function using magnetic resonance imaging, *TINS* **17**, 268-277.

53. Prichard, J.W. and Rosen, B.R. (1994) Functional study of the brain by NMR, *J.Cereb.Blood Flow Metab.* **14**, 365-372.

54. Belliveau, J.W., Kennedy, D.N., McKinstry, R.C., Buchbinder, B.R., Weisskoff, R.M., Cohen, M.S., Vevea, J.M., Brady, T.J. and Rosen, B.R.(1991) Functional mapping of the human visual cortex by magnetic resonance imaging, *Science* **254**, 716-719.

55. Ogawa, S., Lee, T-M., Nayak, A.S. and Glynn, P. (1990) Oxygenation-sensitive contrast in magnetic resonance image of rodent brain at high magnetic fields, *Magn.Reson.Med.* **14**, 68-78.

56. Turner, R., Le Bihan, D., Moonen, C.T., Despres, D. and Frank, J. (1991) Echo-Planar Time course of MRI cat brain oxygenation changes, *Magn.Reson.Med.* **22**, 159-166.

57. Moseley, M.E. and Glover, G.H. (1995) Functional MR imaging: capabilities and limitations, in B.P. Drayer (ed.), *Functional Neuroimaging*, W.B. Saunders Company, Philadelphia, pp. 161-192.

58. Stehling, M.K., Turner, R. and Mansfield, P. (1991) Echo-planar imaging: magnetic resonance imaging in a fraction of a second, *Science* **254**, 43-54.

59. Turner, R., Jezzard, P., Wen, H., Kwong, K.K., Le Bihan, D., Zeffiro, T. and Balaban, R.S. (1993) Functional mapping of the human visual cortex at 4 tesla and 1.5 tesla using deoxygenation contrast EPI. Magn. Reson. Med. **29**, 281-283.

60. Latchaw, R.E., Ugurbil, K. and Hu, X. (1995) Functional MR imaging of perceptual and cognitive functions, in B.P. Drayer (ed.), *Functional Neuroimaging*, W.B. Saunders Company, Philadelphia, pp. 193-206.

61. Maisey, M.N., Hawkes, D.J. and Lukawiecki-Vydelingum, A.M. (1992) Synergistic imaging, *Eur.J.Nucl.Med.* **19**, 1002-1005.

62. Hawkes, D.J., Hill, D.L.G., Lehmann, E.D., Robinson, G.P., Maisey, M.N. and Colchester, A.C.F. (1990) Preliminary work on the interpretation of SPECT images with the aid of registered MR images and an MR derived 3D neuro-anatomical atlas, in H. Hoene, S. M. Pizer and H. Fuchs (eds.), *3D Imaging in Medicine*, Nato ASI Series F 60 K, Springer-Verlag, Berlin, pp. 241-252.

63. Mandava, V.R., Fitzpatrick, J.M., Maurer, C.R.jr, Maciunas, R.J. and Allen, G.S. (1992) Registration of multimodal volume head images via attached markers, in *Proceedings of SPIE Medical Imaging VI: Image processing*, SPIE Press, Bellingham W.A. **1652**, pp. 271-282.

64. Rizzo, G., Gilardi, M.C., Prinster, A., Lucignani, G., Bettinardi, V., Triulzi, F., Cardaioli, A., Cerutti, S. and Fazio, F. (1994) A bioimaging integration system implemented for neurological application, *J.Nucl.Biol.Med.* **38**, 566-572.

65. Arun, K.S., Huang, T.S. and Blostein, S.D. (1987) Least-squares fitting of two 3D point sets, *IEEE Trans.PAMI* **9**, 698-700.

515

66. Henry, C.J., Collins, D.L. and Peters, T.M. (1991) Multimodality image integration for stereotactic surgical planning, *Med.Phys.* **18**, 167-177.

67. Peters, T.M., Clark, J.A., Olivier, A., Marchand, E.P, Mawko, G., Dieumegarde, M., Muresan, L.V. and Ethier, R. (1986) Integrated stereotaxic imaging with CT, MR imaging and digital subtraction angiography, *Radiology* **161**, 821-826.

68. Zhang, J., Levesque, M.F., Wilson, C.L., Harper, R.M., Engel, J., Lufkin, R., Behnke, E.J. (1990) Multimodality imaging of brain structures for stereotactic surgery, *Radiology* **175**, 435-441.

69. Bettinardi, V., Scardaoni, R., Gilardi, M.C., Rizzo, G., Perani, D., Paulesu, E., Striano, G., Triulzi, F. and Fazio, F. (1991) A new head-holder for patient positioning repositioning and fixation in PET, MR and CT devices, *J.Comp.Assist. Tomogr.* **15**, 886-892.

70. Schad, L.R., Boesecke, R., Schlegel, W., Hartmann, G.H., Sturm, V., Strauss, L.G. and Lorentz, W.J. (1987) Three-dimensional image correlation of CT, MR and PET studies in radiotherapy treatment planning of brain tumors, *J.Comput.Assist.Tomogr.* **11**, 948- 954.

71. Evans, A.C., Marrett, S., Collins, L. and Peters, T.M. (1989) Anatomical-functional correlative analysis of the human brain using three dimensional imaging system, in R.H.Schneider, S.J. Dwyer III, R.Gilbert Jost (eds.), *Proceedings of SPIE Medical Imaging III: Image processing*, SPIE Press, Bellingham W.A. **1092**, pp. 264-274.

72. Pelizzari, C.A., Chen, G.T.Y., Spelbring, D.R., Weichselbraum, R.R. and Chen C.T. (1989) Accurate Three-Dimensional Registration of CT, PET and/or MR Images of the Brain, *J.Comput. Assist. Tomogr.* **13**, 20-26.

73. Steinmetz, H., Huang, Y., Seitz, R.J., Knorr, U., Schlaug, G., Herzog, H., Haclander, T. and Freund, H.(1992) Individual integration of positron emission tomography and high-resolution magnetic resonance imaging, *J. Cereb. Blood Flow Metab.* **12**, 919-926.

74. Pietrzyk, U., Herholz, K. and Heiss, W.D. (1990) Three-dimensional alignment of functional and morphological tomograms, *J. Comp. Assist. Tomogr.* **14**, 51-59.

75. Alpert, N.M., Bradshaw, J.F., Kennedy, D. and Correia, J.A. (1990) The principal axes transformation - a method for image registration, *J.Nucl. Med.* **31**, 1717-1722.

76. Mangin, J.F., Frouin, V., Bloch I., Bendriem, B. and Lopez-Krahe, J. (1994) Fast non supervised 3D registration of PET and MR images of the brain, *J.Cereb.Blood Flow Metab.* **14**, 749-762.

77. Jiang, H., Holton, K. and Robb, R. (1992) Image registration of multimodality 3-D medical images by chamfer matching, in *Proceedings of SPIE Biomedical image processing and three-dimensional microscopy*, SPIE Press, Bellingham W.A. **1660**, pp. 356-366.

78. Woods, R.P., Cherry, S.R. and Mazziotta, J.C. (1992) Rapid automated algorithm for aligning and reslicing PET images, *J.Comput.Assist.Tomogr.* **16**, 620-633.

79. Hoh, H.K., Dahlbom, M., Harris, G., Choi, Y., Hawkins, R.A., Phelps, M.E. and Maddahi, J. (1993) Automated iterative three-dimensional registration of positron emission tomography images, *J.Nucl.Med.* **34**, 2009-2018.

80. Apicella, A., Kippenhan, J.S. and Nagel, J.H. (1988) Fast multimodality image matching, in R.H.Schneider, S.J. Dwyer III, R.Gilbert Jost (eds.), *Proceedings of SPIE Medical Imaging III: Image processing*, SPIE Press, Bellingham W.A. **1092**, pp. 252-263.

81. Woods, R.P., Mazziotta, J.C. and Cherry, S.R. (1993) MRI-PET registration with automated algorithm, *J.Comput.Assist.Tomogr.* **17**, 536-546.

82. Clarke, C.J.S., Ioannides, A.A. and Bolton, J.P.R. (1990) Localized and distributed source solutions for the biomagnetic inverse problem I., in S.J. Williamson et al. (eds.), *Advances in Biomagnetism*, Plenum, New York, pp. 587-590.

83. Romani, G.L. and Pizzella, V. (1990) Localization of brain activity with magnetoencephalography, in S. Sato (ed.), *Advances in Neurology: Magnetoencephalography* Vol.54, Raven Press, New York, pp. 67-78 .

84. Chen, C., Ouyang, X., Wong, W.H., Hu, X., Johnson, V.E., Ordonez, C. and Metz, C.E. (1991) Sensor fusion in image reconstruction, *IEEE Trans.Nucl.Sci.* **NS-38**, 687-692.

85. Hoffman, E.J., Huang, S.C. and Phelps, M.E. (1979) Quantitation in Positron Emission Computed Tomography: 1. Effect of object size, *J. Comput.Assis.Tomogr.* **3**, 299-308.

86. Muller-Gartner, H.W., Links, M.J., Prince, J.L., Bryan, R.N., McVeigh, E., Leal, J.P., Davatzikos, C. and Frost J.J. (1992) Measurement of radiotracer concentration in brain gray matter using positron emission tomography: MRI-based correction for partial volume effect, *J.Cereb.Blood Flow Metab.* **12**, 571-583.

87. Schmidt, K., Lucignani, G., Moresco,.R.M., Rizzo, G., Gilardi, M.C., Messa, C., Colombo, F., Fazio, F. and Sokoloff, L. (1992) Errors introduced by tissue heterogeneity in estimation of local cerebral glucose utilization with current kinetic models of the [^{18}F] fluorodeoxyglucose method, *J. Cereb. Blood Flow Metab.* **12**, 823-834.

88. Herscovitch, P., Auchus, A.P., Gado, M., Chi, D. and Raichle, M.E. (1986) Correction of Positron Emission Tomography data for cerebral atroph, *J.Cereb.Blood Flow Metab.* **6**, 120-124.

89. Videen, T.O., Perlmutter, J.S., Mintun, M.A. and Raichle, M.E. (1988) Regional correction of positron emission tomography data for the effects of cerebral activity, *J.Cereb.Blood Flow Metab.* **8**, 662-670.

90. Stefan, H., Shneider, S., Abraham-Fuchs, K., Bauer, J., Feistel, H., Pawlik, G., Neubauer, U., Rohrlein, G. and Huk, W.J. (1990) Magnetic source localization in focal epilepsy, *Brain* **113**, 1347-1359.

91. Paetau, R., Kayola, M., Karhu J., Nousiainen, U., Partanen, J., Tiihonen, J., Vapalahti, M. and Hari, R. (1992) Magnetoencephalographic localization of epileptic cortex - impact on surgical treatments, *Ann.Neurol.* **32**, 106-109.

92. Theodore, W H. (ed.) (1988) *Clinical Neuroimaging*, Alan R. Liss inc.,New York.

93. Di Chiro, G., Oldfield, E., Wright, D.C., De Michele, G., Patronas, N., Doppman, J.L., Larson, S.M., Masanori, I. and Kufta, C.V. (1988) Cerebral necrosis after radiotherapy and/or intraarterial chemotherapy for brain tumors: PET and neuropathologic studies, *Am.J.Roentgenol.* **150**, 189-197.

94. Stefan, H., Schneider, S., Feistel, H., Pawlik, G., Schuler, P., Abraham-Fuchs, K., Schlegel, T., Neubauer, U. and Huk, W.J. (1992) Ictal and interictal activity in partial epilepsy recorded with multichannel magnetoencephalography: correlation of electroencephalography/ electrocorticography, magnetic resonance imaging, single photon emission computed tomography and positron emission tomography findings, *Epilepsia* **33**, 874-887.

95. Pantev, C., Hoke, M., Lehnertz, K., Lutkenhoner, B., Fahrendorf, G. and Stober, U. (1990) Identification of sources of brain neuronal activity with high spatio-temporal resolution through combination of neuromagnetic source localization (NMSL) and magnteic resonance imaging (MRI), *Electroencephalogr. Clin.Neurophysiol.* **75**, 173-184.

96. Gevins, A., Cutillo, B., Durousseau, D., Le, J., Leong, H., Martin, N., Smith, M.E., Bressler, S., Brickett, P., McLaughlin, J., Barbero, N. and Laxer, K. (1994) Imaging the spatio-temporal dynamics of cognition with high resolution evoked potential methods, *Human Brain Mapping* **1**, 101-116.

97. Tatcher, R.W. (1995) Tomographic electroencephalography-magnetoencephalography: dynamics of human neural network switching, *J.Neuroimag.* **5**, 35-45.

98. Walter, H. Kristeva, R., Knorr, U. , Schlaug, G., Huang, Y., Steinmetz, H., Nebeling, B., Herzog, H. and Seitz, R.J. (1992) Individual somatotopy of primary sensorimotor cortex revealed by intermodal matching of MEG, PET, and MRI, *Brain Topography*, **5**, 183-187.

SUPERCONDUCTING MAGNETIC GRADIOMETERS FOR MOBILE APPLICATIONS WITH AN EMPHASIS ON ORDNANCE DETECTION

T. R. CLEM, G. J. KeKELIS, J. D. LATHROP, D. J. OVERWAY, and W. M. WYNN
Coastal Systems Station, Dahlgren Division, Naval Surface Warfare Center
Panama City, FL, USA. 32407-7001

Abstract. Passive magnetic sensors provide one means to conduct mobile area surveys and search operations, useful for a number of applications, including sea mine countermeasures and the detection of unexploded ordnance and packaged biological, chemical and radioactive waste for environmental cleanup. To date, the generally accepted method for such detection involves the generation of two- or three-dimensional magnetic anomaly field maps, using primarily total-field magnetometers. Sensor configurations measuring spatial gradients of magnetic field offer a new opportunity for better localization and classification. Sensors incorporating Superconducting Quantum Interference Devices (SQUIDs) provide the greatest sensitivity available with current technology for magnetic anomaly detection. During the late 1970's and early 1980's, the Naval Surface Warfare Center Coastal Systems Station (CSS) developed the Superconducting Gradiometer/Magnetometer Sensor (SGMS) specifically for mobile operations outside the laboratory environment. This sensor technology utilized niobium superconducting components cooled by liquid helium. The SGMS has demonstrated rugged, reliable performance even onboard airborne and undersea towed platforms. In this article, a general perspective for the use of passive magnetic sensors for mobile operations will be established. The SGMS design will be described in some detail. General design principles underlying its mobile application, fundamental sensor and environmental noise issues, and approaches to compensate for them, will be presented. The magnetic sensor detection and classification concept developed for sea mine countermeasures and results from that demonstration will be discussed. Recent developments and future opportunities, especially encompassing the use of high temperature (high-T_c) superconducting components cooled by liquid nitrogen, will be addressed.

Table of Contents.   1. Introduction

2. Prospectus on Magnetic Sensors for Mobile Operations
   2.1. Field and Gradient from a Magnetic Dipole
   2.2. Sensor Types
      2.2.1. Magnetometers
      2.2.2. Gradiometers
      2.2.3. Superconducting Gradiometers
      2.2.4. Quantitative Comparison of Sensor Types
   2.3. Operational Scenarios for Gradiometers in Motion

H. Weinstock (ed.), SQUID Sensors: Fundamentals, Fabrication and Applications, 517–568.
© *1996 Kluwer Academic Publishers.*

518

2.4. Early Sensor Development at CSS

3. The Superconducting Gradiometer/ Magnetometer Sensor
    3.1. Sensor Description
        3.1.1. Dewar
        3.1.2. Cryogenic Probe Assembly
        3.1.3 Flux Transformer Assembly
        3.1.4. Room Temperature SQUID and Datalink Electronics
    3.2. Basic Gradiometer Performance Parameters
        3.2.1. Sensor Sensitivity
        3.2.2. Dynamic Range

4. Noise Sources in Mobile Operation and Their Compensation
    4.1. Primary Issues with the Terrestrial Magnetic Field
        4.1.1. Balance and Gradient Admixtures
        4.1.2. Eddy Current Compensation
    4.2. Secondary Issues with the Terrestrial Magnetic Field
        4.2.1. Geomagnetic Noise
        4.2.2. The Terrestrial Magnetic Gradient
        4.2.3. Geologic Noise
        4.2.4. Ocean Magnetohydrodynamic Wave Noise
    4.3. Sensor-Intrinsic Noise Sources
        4.3.1. A General Perspective
        4.3.2. SGMS Enhancements
    4.4. Electromagnetic Interference
    4.5. Platform Noise

5. Demonstrations of a Gradiometer in a Mobile Operation
    5.1. The Magnetic and Acoustic Detection of Mines (MADOM)
        5.1.1. System Concept
        5.1.2. Parallel Array Signal and Image Processing System
        5.1.3. Model to Detect, Classify and Localize Targets
        5.1.4. Results from System Sea Tests
        5.1.5. Clutter Rejection
    5.2. The Mobile Underwater Survey System

6. Recent Developments and Future Trends for Mobile Applications
    6.1. Development of a High Performance Thin Film Gradiometer
    6.2. Opportunities with High-T_c Technology
        6.2.1. Benefits from Nitrogen Cooling for Naval Operations
        6.2.2. Advanced Refrigeration Concepts
    6.3. High-T_c Sensor Development

7. Conclusions and Summary

1. Introduction

Magnetic sensors have proven merit for mobile area surveys and search operations conducted from air, land or sea. Two applications which will be the focus of this paper include (1) the environmental cleanup of munitions and chemical, biological and nuclear waste and (2) sea mine countermeasures. Similarly, magnetic sensors can be operated from mobile platforms for geophysical surveys, for locating underground or undersea telephone cables, for antisubmarine warfare, for the detection of underground hidden facilities and tunnels, and for archeology and treasure hunting. Magnetic sensors are used by civil engineers to locate pipelines, old foundations, buried gas tanks, and well heads. They have been utilized in police operations to locate buried or sunken cars, planes and boats.

As a result of war, military training, and weapons testing, unexploded ordnance (UXO) contaminates millions of acres of land and coastal areas throughout the world. Examples of such UXO are displayed in Figure 1.1. In order to obtain a limited perspective on the magnitude of the

105 mm HOWITZER   1000 LB BOMB

60 mm MORTAR   500 LB BOMB

Figure 1.1. Examples of munitions to be removed in the cleanup of formerly-used defense sites.

problem, the Army Corps of Engineers has the current operational mission for munitions remediation and cleanup for approximately 900 formerly-used defense sites. Since the housing for ordnance is generally made of steel, passive magnetic sensors can be used to locate the ordnance [1]. In fact, in an assessment conducted by the Jet Propulsion Laboratory for the Army Corps of Engineers, magnetic sensors in general were identified among the most useful sensors for munitions detection and localization, and superconducting gradiometers were specifically identified as the most useful tool in a class by themselves [2].

Sea mine countermeasures represents another area of interest for the application of magnetic sensors. Sea mines represent a low cost and effective means to hinder ship movements and to deter amphibious assaults. While acoustic approaches to mine detection and classification are most successful in deeper waters, sonar capabilities and ranges diminish in shallower waters, and mine burial of bottom mines represents a limitation to sonar effectiveness. Sea mines generally are housed in steel casings so that passive magnetic sensors provide a useful means for detection and localization. In the shallower waters, the fusion of magnetic and acoustic sensor data provides a promising approach for mine reconnaissance and hunting [3].

2. A Prospectus on the Application of Magnetic Sensors for Mobile Operations

The selection of magnetic sensor type will largely depend on the operational requirements, determined primarily by the desired detection range, which is, in turn, a function of the magnetic moment of the targets and of sensor sensitivity. The selection also depends on such factors as financial budgets, logistical support and technical expertise of the operators. To date, magnetic sensor approaches have provided limited localization and mapping capabilities. To gain widespread acceptance, approaches must be introduced which provide accurate localization and target classification, and which lend themselves to straightforward interpretation and minimal training. Performance must not be limited by magnetic noise from the host platform and other subsystems. For land-based operations, the system must be capable of operating over rough, overgrown terrain. The sensor and associated signal processing also must deal effectively with environmental noise. Geomagnetic temporal variations of 0.1 nT are common, and geologic spatial variations cover a wide range of length scales. For undersea operations, ocean waves generate significant magnetic signals through magnetohydrodynamic effects arising from the motion of the salt ions in the sea water relative to the Earth's magnetic field.

In this paper we shall describe two types of sensors which detect magnetic anomalies: sensors which detect changes in the local magnetic field, *magnetometers;* and sensors which measure the spatial derivatives of magnetic field, *first order gradiometers*. Gradiometers measuring higher than first-order spatial derivatives of magnetic field have been investigated and are especially useful for the magnetic detection of nearby magnetic sources for applications such as neuromagnetism, but only a first-order gradiometer (hereinafter simply gradiometer) is expected to have merit for detecting targets at a distance for applications such as ordnance detection. For the interested reader, a detailed discussion of different classes of gradiometers is given in Ref. 4, Chapter 3 in this volume.

2.1 FIELD AND GRADIENTS FROM A MAGNETIC DIPOLE

The performance of a gradiometer in a mobile search mode is measured by its detection range, which is a function of the noises of the sensor channels in motion (including environmental noises) and the magnetic moment of the targets of interest. In the far field, a target can be well approximated as a magnetic dipole. In this approximation, relatively simple, analytic expressions can be written to relate sensor sensitivity requirements in terms of nominal values for target magnetic moment and range.

For the special case in which a circular, connected conducting loop with area A is carrying an electrical current I, we can define the magnetic moment of the loop as the vector $\mathbf{m} = \text{IA}\,\hat{n}$ where $\hat{n}$ is the normal to the loop in the direction defined by the right-hand rule for positive current. The International System (SI) unit for magnetic moment is ampere-meter-squared (A-m²). The SI system also contains a related quantity called the magnetic dipole moment $\mathbf{M} = \mu_o \mathbf{m}$ given in units of weber-meter.

For a general, bounded distribution of current described by the current density $\mathbf{J(r)}$, the magnetic vector potential $\mathbf{A(r)}$ is given exactly by

$$\mathbf{A(r)} = \frac{\mu_o}{4\pi} \int \frac{dV' \mathbf{J(r')}}{|\mathbf{r} - \mathbf{r'}|} \tag{2.1}$$

The denominator of the integrand can be expanded in a Taylor series and the leading term in the resulting series for $\mathbf{A}$ is given by

$$\mathbf{A(r)} \cong \frac{\mu_o}{4\pi} \frac{\mathbf{m} \times \mathbf{r}}{r^3} \tag{2.2}$$

where $\mathbf{m}$ is defined by the volume integral

$$\mathbf{m} = 1/2 \int dV' \ \mathbf{r'} \times \mathbf{J(r')} \tag{2.3}$$

Here, $\mathbf{m}$ is defined as the magnetic moment of the distribution, and reduces exactly to $\text{IA}\,\hat{n}$ for an elemental current loop. The vector potential given by Equation (2.2) is called the magnetic dipole vector potential. From the general equation $\mathbf{B} = \nabla \times \mathbf{A}$, the field for a magnetic dipole can be written as

$$\mathbf{B} = \frac{\mu_o}{4\pi} \left[\frac{3(\mathbf{m} \cdot \mathbf{r})\mathbf{r}}{r^5} - \frac{\mathbf{m}}{r^3} \right] \tag{2.4}$$

We note in passing that it is common to see Equation (2.4) written in the form

$$\mathbf{B} = \frac{3(\mathbf{m} \cdot \mathbf{r})\mathbf{r}}{r^5} - \frac{\mathbf{m}}{r^3} \tag{2.5}$$

where the factor of $\mu_o/4\pi$ is absorbed into the moment vector. This is not consistent with SI units, but has been used practically, especially in older U.S. Navy literature. There, the moment $\mathbf{m}$ would be expressed in gamma-foot-cubed (1 gamma = 1nT). In air, the magnetic induction $\mathbf{B}$ is related to the magnetic field $\mathbf{H}$ via $\mathbf{B} = \mu_o\mathbf{H}$. Hereafter, $\mathbf{B}$ will be used exclusively and will be referred to as the magnetic field. The following units for magnetic field will frequently be used to represent sensor sensitivities: nanotesla (1 nT $=10^9$T), picotesla (1 pT$=10^{12}$ T), and femtotesla (1 fT$=10^{15}$T).

A number of sensors, notably fluxgate magnetometers and superconducting SQUID magnetometers, measure the individual vector components of field. (In fact, SQUID magnetometers only measure changes of field components referenced to a baseline value.) Other magnetometers, notably those based on nuclear or atomic resonance processes, measure the magnitude of the total magnetic field and are known as total-field magnetometers. Let $\mathbf{B}_0$ denote the induction field of the Earth, and let $\mathbf{b}$ denote the induction generated by an anomaly. If $|\mathbf{b}| \ll |\mathbf{B}_0|$, then the signal observed by a total field sensor (referenced to the baseline earth field) is

$$b_m \equiv |\mathbf{B}_0 + \mathbf{b}| - |\mathbf{B}_0| = \sqrt{B_0^2 + 2\mathbf{B}_0 \cdot \mathbf{b} + b^2} - B_0 \approx \frac{\mathbf{B}_0 \cdot \mathbf{b}}{B_0} \tag{2.6}$$

As a result of the right-hand approximation above, a total-field magnetometer does not simply measure the magnitude of the magnetic-field anomaly, but measures instead the projection of that anomalous signal onto the earth's field. The basic idea underlying the detection of total-field and vector magnetometer signals is depicted in Figure 2.1.

The gradient of the magnetic field (in standard MKS units of T/m) is a second-rank tensor with components given in Ref. 5 by

$$G_{ij} \equiv \frac{\partial B_i}{\partial x_j} = -\frac{3}{r^7}\frac{\mu_o}{4\pi}\{\mathbf{m} \cdot \mathbf{r}\,(r_i\,r_j - r^2\delta_{ij}) - r^2(r_i m_j + r_j m_i)\} \tag{2.7}$$

As a result of Maxwell's equations in free space, only 5 of these 9 tensor elements are independent. For this reason, gradiometers are typically designed with 5 independent gradient channels, using the minimum number which permits characterization of the local tensor gradient field. It is feasible to determine the bearing vector and the magnetic-moment vector direction of the dipole by inversion of the gradient equations at a single point only [5]. More recently it has been shown that the addition of gradient rate information at a single point leads to a unique solution for dipole position and moment vector [6].

An example of one configuration to measure a single-gradient tensor component and a simple configuration to measure 5 independent gradient components are displayed in Figure 2.2. Each

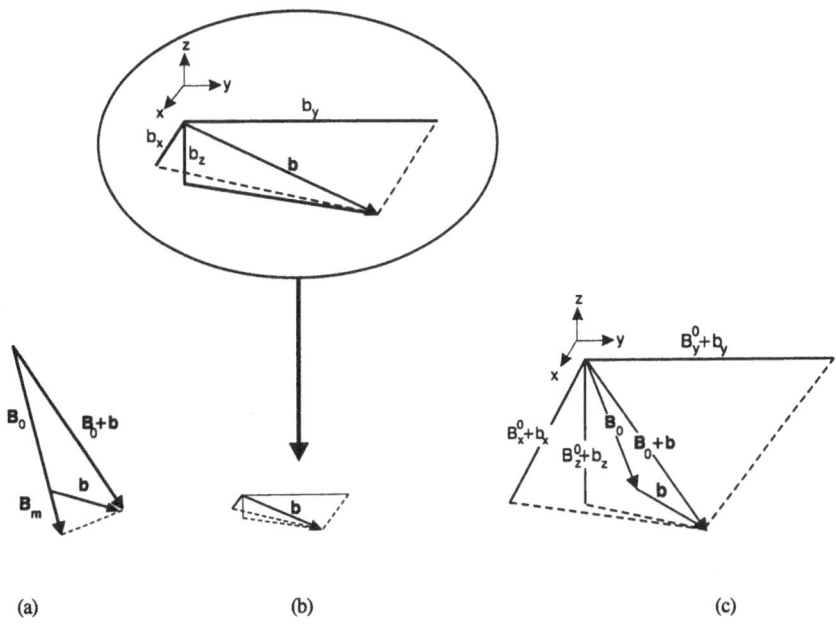

(a)   (b)   (c)

Figure 2.1. Target signals detected from (a) a total-field magnetometer, (b) a 3-channel SQUID vector magnetometer, and (c) a 3-channel fluxgate vector magnetometer.

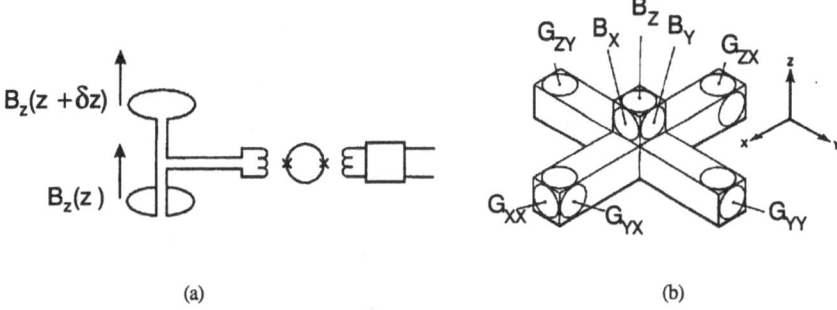

(a)   (b)

Figure 2.2. Gradiometer sensor concepts including (a) a single SQUID-based gradiometer channel and (b) a conceptually simple 5-channel gradiometer configuration capable of magnetic dipole localization and moment classification (with 3 orthogonal magnetometers for motion compensation).

gradient tensor component is measured by a spatially separated loop pair connected in a common-mode rejection configuration. Observe, from Equations (2.4) and (2.7), that the field for a dipole diminishes in the far field with the third power of separation distance, while the gradient diminishes with the fourth power of separation distance.

The contraction of the gradient tensor defined by

$$G \equiv \sqrt{\sum_{i=1}^{3} \sum_{j=1}^{3} G_i G_{ij}} \qquad (2.8)$$

is a rotational invariant associated with the gradient tensor analogous to the magnitude of the field vector. This quantity may prove very useful for applications in which a gradiometer is subjected to large rotations during the period of measurement, e.g., hand-held operation (in contrast to straight runs onboard stabilized platforms) [6].

2.2 SENSOR TYPES

In this section, we shall describe the basic features of magnetometers and gradiometers, and compare these two sensor types.

2.2.1. Magnetometers

The generally accepted method for ordnance detection involves the generation of two- or three--dimensional magnetic field maps. To date this has been made possible using total-field magnetometers. Technologies available for the total-field measurements have been based on nuclear and atomic resonance processes, with a change in magnetic field measured as a shift in the resonance frequency. A major advantage of this type of sensor is its insensitivity to rotation in the Earth's background field of 50,000 nT (since total field is a rotational invariant).

Because total field magnetometers measure the projection of the anomalous magnetic field vector onto the Earth's magnetic field, these maps are more difficult to interpret than, say, a map of the magnitude of the anomaly field vector. Interpretation often requires an experienced operator, and precise anomaly locations are difficult to obtain. Since total field magnetometers provide only one channel of information, they lack valuable target vector information. In particular, they provide very limited localization and little capability for anomaly classification through moment determination. Moreover, these sensors are limited in field operation to sensitivities at levels approaching 0.1 nT as a result of geomagnetic noise, i.e., temporal variations in the Earth's field, without the use of very sophisticated compensation schemes.

A three-axis vector magnetometer, likely using fluxgate or SQUID sensors, is far more useful for localization (providing three channels of information). For stationary applications in geophysics and barrier defense, such sensors are effective. However, to date, they have not proven to be an effective approach for mobile applications, since means to compensate the anomalous signals arising from rotations in the Earth's magnetic field have not been devised.

2.2.2. Gradiometers

Gradiometers offer the potential to remove many of the limitations associated with magnetometers because the output of a gradiometer is typically produced by twin magnetometers operating in differential mode. In particular, this configuration provides common-mode rejection of the nominal 0.1-nT temporal variations in the Earth's field and of the nominal 1000-nT field changes arising from typical 1 degree sensor rotations while in towed motion.

Expanding on this last point, very large signals are generated in a vector magnetometer from sensor rotation in the Earth's field. A tilt of 1 degree produces a magnetic field signal of up to 1000 nT in the Earth's background field, assuming a nominal value of 50,000 nT (Figure 2.3.a). The relative importance of motion noise may be substantially reduced by measuring the spatial gradient with a magnetic field gradiometer. By careful matching and balancing of the two magnetometer elements in the gradiometer and by signal processing compensation using auxiliary magnetometers, the ratio of the motion noise in a gradiometer to sensor noise can be reduced by a factor on the order of 10^{-7}/m as compared to a vector magnetometer. In this case, the 1000-nT field change from nominal rotations will generate an erroneous gradient signal of magnitude 1×10^{-4} nT/m. In conclusion, a gradiometer can be made insensitive to the very large signals generated by sensor rotations in the Earth's field.

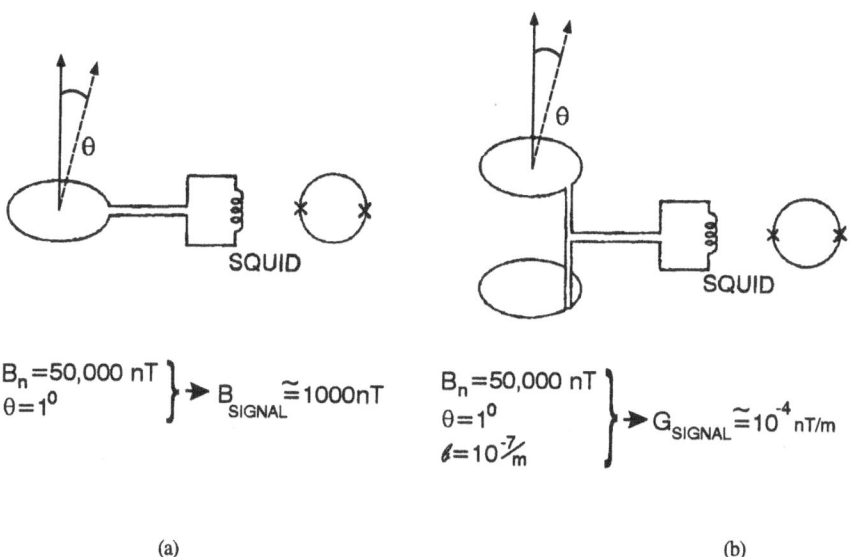

Figure 2.3. Magnetic signals generated by rotation in the Earth's magnetic field for (a) a vector magnetometer and (b) a first-order gradiometer. In this figure, the magnetic field normal to the sense loops is 50,000 nT, and the sense loops are rotated 1 degree. In this example, gradiometer balance, a measure of the gradiometer's common-mode rejection of magnetic field (Section 4.1.1), is assumed to be 10^{-7}/m.

Gradiometers may be fabricated using many available magnetometer technologies. Available fluxgate and total-field magnetometers can perform at levels approaching 1-10 pT, while superconducting magnetometers can perform at levels on the order of 1-10 fT. Short gradiometer baselines i.e., sizes compact enough to permit packaging within one host platform (typically on the order of 0.3-1.0 m), are important in order to maintain coherence between elements. In a short-baseline configuration of 0.3 m, only superconducting technology currently can exceed 0.1 pT/m. (It should be noted here that this ideal stationary performance may not be realized in mobile applications because of a host of noise issues introduced by the sensor motion.)

A package consisting of 5 gradient channels (sufficient to characterize the local gradient tensor field) and 3 magnetometers for motion compensation provides complete characterization of magnetic dipole sources, obtaining localization and classification information not possible by the single-channel total-field magnetometers. It should be noted that total-field magnetometers can be operated in gradiometer mode, providing the spatial gradient of the total field along one direction in space. A configuration with four such sensors aligned along three orthogonal axes will provide substantially more classification and localization information than a single total-field magnetometer. This capability falls short of the capability afforded by the 5-channel vector gradiometers and the use of the total-field sensors is complicated by the Earth's field projection issue.

2.2.3. Superconducting Gradiometers
Superconducting gradiometers are attractive for large area searches because such sensors provide the greatest sensitivity available today for magnetic anomaly detection. Almost all of the efforts with SQUID sensor technology have dealt with sensors inside a very controlled laboratory environment, and to a more limited extent outside the laboratory at stationary locations, notably for geophysical measurements. As shown by CSS, superconducting magnetic sensors can be efficiently and accurately operated onboard a moving platform to determine the range and bearing of a magnetic target, such as a sea mine or a submarine, and to classify the target in terms of its magnetic characteristics. Extreme performance available from SQUID magnetometers means short baselines can be implemented in gradiometers. The short baseline provides a compact package with extreme coherence between magnetometers. The coherence is required to maintain high performance when in motion, and the compact package is amenable to implementation aboard one vehicle. In addition, with the compact sizes, a large number of gradient channels can be packed into a small package to obtain complete position and moment determination of a magnetic dipole target at a single point in space.

Under some field conditions, CSS has demonstrated a superconducting gradiometer which has attained sensitivities on the order of $1 \text{pT/m-}\sqrt{Hz}$ at 0.1 Hz. In comparison total-field gradiometers (with a 0.3-meter baseline) specifically designed for operation onboard a moving platform have achieved sensitivities on the order of $30 \text{ pT/m-}\sqrt{Hz}$ at 0.1 Hz.

2.2.4. Quantitative Comparison of Sensor Types

The signal strength of a magnetic dipole decreases as the third power of the range for magnetic fields, and as the fourth power of the range for magnetic field gradients. The approximate ranges of magnetometers and gradiometers are displayed in Figure 2.4 as functions of dipole strength and sensor sensitivity.

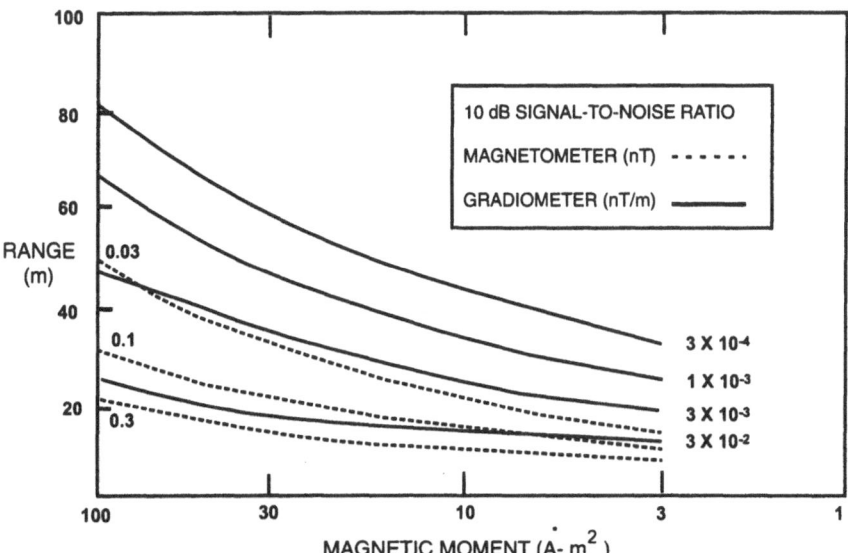

Figure 2.4. Approximate ranges (in meters) of magnetometers and gradiometers as a function of target strength in terms of magnetic moment (in units of A-m^2). Curves are given for different sensor sensitivities (in units of nT for magnetometers and nT/m for gradiometers) assuming a 10-dB signal-to-noise ratio.

It can be shown that the sensitivity requirements for a magnetometer and a gradiometer, respectively, to have the same detection range r against a given dipole target, is given by the approximate relation $N_g/N_m \sim 3/r$. The 10-dB signal-to-noise ratio detection of a mortar shell with a magnetic moment of 0.1 A-m^2 at a range of 3 m requires a 0.36 nT magnetometer or a 0.36 nT/m gradiometer. A 3 pT/m gradiometer can detect an individual 500-pound bomb (with a moment of 30 A-m^2) at a range of 33 m and a 60-mm mortar shell (with a moment of 0.1 A-m^2) at a range of 8 m. It should be noted that the high rate of signal reduction with distance in the case of a gradiometer represents a shortcoming for the gradiometer configuration. However, a gradiometer does have merit at large distances because it is relatively insensitive to target moment, since detection range drops off only as the fourth root of the target's magnetic moment.

528

An example of the ease of interpretation for 5-channel gradiometer data compared to single-channel total-field magnetometer data is displayed in Figure 2.5. Magnetic profiles have been generated for a 60-mm mortar shell buried 1 meter under ground for two different orientations with respect to the Earth's background field. In this example, the magnetometer profiles and gradiometer profiles are given by the anomalous total field (Eqn (2.6)) and the corresponding changes in magnitude of the gradient tensor (Eqn (2.8)), respectively. The complex total field profiles require precision data and critical interpretation to localize dipole sources. The symmetric gradiometer profile leads to straightforward interpretation convenient for gradient searches for dipole localization.

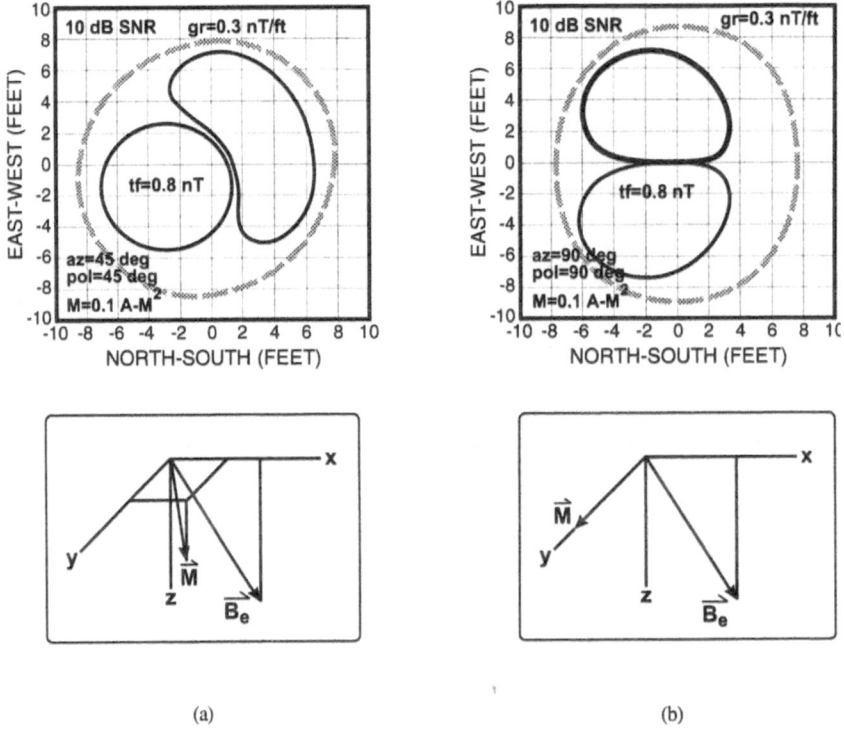

(a)   (b)

Figure 2.5. A comparison of magnetometer and gradiometer capabilities for target localization. The profiles measure tf = total field (Equation (2.6)) and gr = gradient magnitude (Equation (2.8)). Profiles are given for a magnetic dipole centered at origin with (a) a general orientation and (b) orientation in an east-west direction. Observe that the gradiometer profiles are approximately circularly symmetric about the dipole's location so that "gradient" searches normal to the gradient magnitude profiles are meaningful for the gradiometer. In contrast, the magnetometer profiles are not amenable to such straightforward interpretation.

2.3 SCENARIOS FOR GRADIOMETER MOBILE OPERATION

Selection of gradiometer sensor technology will largely depend on the operational scenario (folded in with cost considerations). We can envision three general types of operational scenarios: relatively long-range searches for ordnance clusters, more moderate-range searches against individual targets, and detailed close-in surveys. High sensitivity will be critical for the long-range and moderate-range scenarios, but such sensitivity will likely be sacrificed for the close-in surveys at ground level.

A perspective on the role of higher sensitivity gradiometers used for wide-area searches and lower sensitivity gradiometers for close-in surveys can be obtained from the following example (displayed in Figure 2.6). A 3 pT/m gradiometer can detect a grouping of twenty 500-pound bombs (clustered in a circle several meters in diameter) at a range of 46 meters. An area search rate of 1 km^2/hr can be obtained when the sensor's altitude is 15 meters moving at a forward speed of 16 km/hr. When deployed from the ground, a less sensitive 300 pT/m gradiometer would provide detection ranges of 10 meters and 2.5 meters for the detection of a 500-pound bomb and a 60-mm mortar shell, respectively.

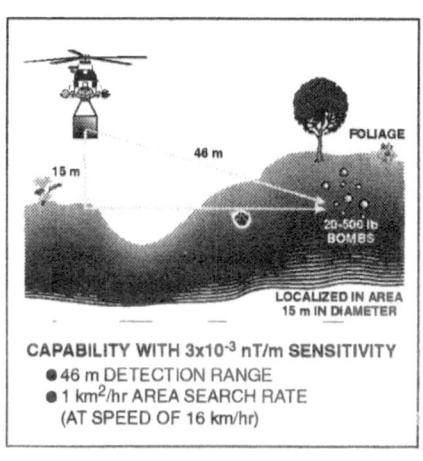

CAPABILITY WITH 3x10^{-3} nT/m SENSITIVITY
● 46 m DETECTION RANGE
● 1 km^2/hr AREA SEARCH RATE
   (AT SPEED OF 16 km/hr)

(a)

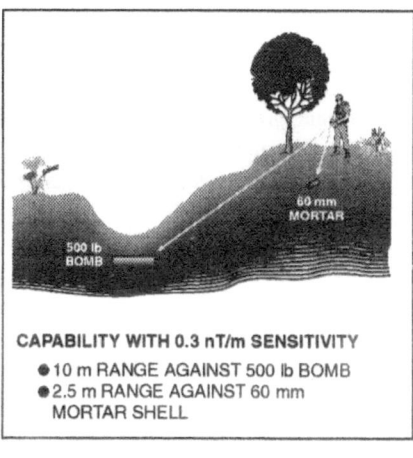

CAPABILITY WITH 0.3 nT/m SENSITIVITY
● 10 m RANGE AGAINST 500 lb BOMB
● 2.5 m RANGE AGAINST 60 mm
   MORTAR SHELL

(b)

Figure 2.6. Two modes of operation for 5-gradient channel sensors: (a) wide-area surveys using high sensitivity sensors and (b) short-range searches for single targets using less sensitive sensors.

2.4. EARLY SENSOR DEVELOPMENT AT CSS

In the late 1960's the CSS embarked on a program to develop superconducting magnetic sensors for anti-submarine warfare (ASW) and sea mine countermeasures (MCM) applications. In the early years of this program, much of the effort was aimed at optimizing sensor design parameters and configurations. From those early research efforts, it was determined that the measurement of five independent spatial magnetic field gradients (3-off axis, 2 on-axis) was sufficient to determine the bearing vector, vector direction of magnetic moment, and the ratio of moment magnitude to the fourth power of the range of a dipole target. In addition, it was determined that information from the three magnetic field components could be effectively utilized to compensate for motion and eddy-current noise in the gradiometer channels produced by motion of the sensor. Algorithms were developed and tested to implement these capabilities.

A 5-channel gradiometer using fluxgate technology was developed in 1972. Individual channels were constructed from 2 individual fluxgate sensors mounted wit h 1 meter separation, and the gradients were obtained by subtraction of signals from the two individual sensors. Good stationary performance was obtained at that time, but there was insufficient dynamic range in the processing of the differential signals to operate the sensor successfully in motion.

In 1975 a superconducting gradiometer was developed by CSS and Superconducting Technology, Inc. (ScT) [7]. This sensor, displayed in Figure 2.7a, is believed to be the first five-gradient-component superconducting sensor ever built, and hence was the first superconducting gradiometer capable of magnetic dipole localization and moment classification. In contrast to the fluxgate gradiometer, signal subtraction occurs within the loop structure prior to the SQUID stage, thereby intrinsically extending the dynamic range of the sensor. This sensor featured bulk and wire superconducting technology, used point contact rf SQUIDs, adjustable vane balancing, and a vertical dewar system which cooled the superconducting components by immersion. This sensor was deployed on a number of occasions to demonstrate the utility and feasibility of superconducting gradiometer detection systems. While the sensor was bulky, was difficult to set up and tune, was subject to frequent circuit failure, and was only moderately sensitive, it proved invaluable in providing information for the design of the next generation of sensors.

The early successes with this sensor led to the development of a series of gradiometers with the investigation of a wide variety of nonacoustic phenomena. In 1978 a single-axis prototype superconducting gradiometer was developed by ScT (Figure 2.7.b). In particular, it featured for the first time a liquid helium dewar designed to operate in a horizontal position for optimal motion stability in air-tow operations. This sensor was flight tested in 1978, but unfortunately was lost during preliminary flights prior to the collection of high quality data. However, preliminary evaluation did provide proof-of-principle of key concepts for the next generation of sensors.

The Superconducting Gradiometer/Magnetometer System (SGMS), displayed in Figure 2.7.c, was developed in the late 1970's and early 1980's by the CSS in conjunction with Unisys [8]. Unisys delivered the first 2 SGMS sensors to CSS in 1982 to provide the U.S. Navy with a magnetic sensor for mine countermeasures, specifically designed for deployment in an undersea towed vehicle. The sensor employs largely niobium bulk and wire superconducting technology (with thin-film Josephson Junctions), features dc SQUIDs, and is cooled by cyrocooled helium gas evaporating from a liquid helium reservoir. More recently in 1989 one of the SGMS sensors was refurbished by Loral [9], IBM [10] and Quantum Magnetics [11], leading to impressive improvement in sensor reliability and operability.

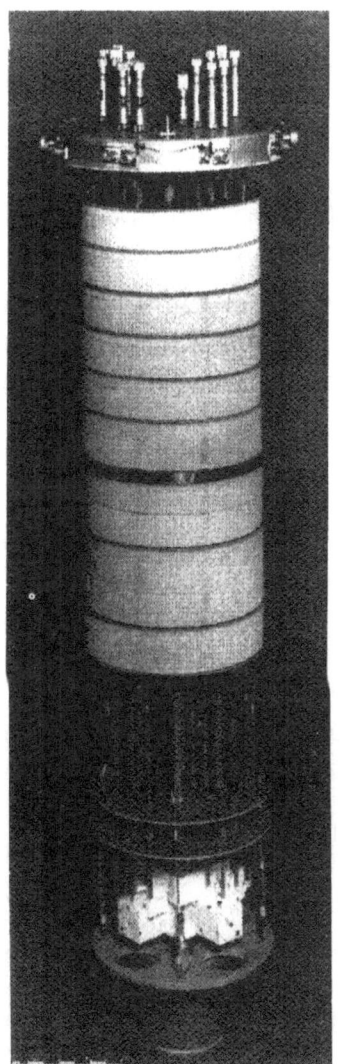

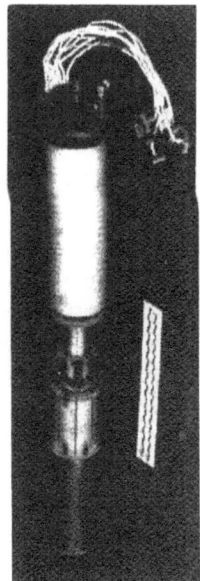

Figure 2.7. The evolution of superconducting gradiometers developed by the U.S. Navy, cryogenic probe assemblies for :
(a) the first 5-channel gradiometer permitting complete determination of dipole position and moment, (b) the first horizontal gradiometer for mobile tow operation, and (c) the Superconducting Gradiometer/Magnetometer Sensor. The sizes of these probe assemblies are displayed on approximately the same scale, with the lengths ranging from 1.5m down to 1m for the probe assemblies in (a) and (c), respectively.

In 1983, after a series of successful technical tests, internal sensor noise sources were identified that limited gradiometer performance when it was placed in motions typical of mobile deployment. As a result of dedicated experimentation, the system was modified in 1984, leading to a dramatic increase in performance. Subsequently, the sensor demonstrated, for the first time, high sensitivity and rugged, robust, reliable performance in sea testing outside the laboratory environment. Under operational conditions, the sensor has attained sensitivities on the order of 1 pT/m $\sqrt{Hz}$ at 0.1 Hz. Following this improvement in sensor performance, the SGMS has been successfully field deployed in sea mine countermeasure demonstrations (Section 5.1) and will be utilized for locating unexploded ordnance in coastal regions (Section 5.2).

Efforts in progress to advance the technology of superconducting gradiometers are discussed in Section 6. These include development of low-T_c thin-film sensors with increased sensitivity and the development of high-T_c sensors for compact packaging and reduced support requirements.

3. The Superconducting Gradiometer/Magnetometer Sensor (SGMS)

In this section, we shall describe the SGMS sensor (including its fundamental features, subassemblies and critical components) and fundamental performance parameters essential to characterize sensor performance.

3.1. SENSOR DESCRIPTION

The SGMS incorporates conventional low-T_c superconducting materials, such as niobium, and is cooled by liquid helium. An artist's conception of this system is shown in Figure 3.1.a. It has three major subassemblies: (1) a passive cooling unit referred to as a *dewar*, (2) a cryogenic probe assembly, and (3) a room-temperature sensor electronics package.

3.1.1. Dewar

The SGMS dewar is designed to operate horizontally, distinguishing it from standard laboratory vertical dewar designs (Figure 3.1.b). The probe assembly is cooled by evaporated helium gas, which is subsequently vented to the atmosphere. Special designs and stringent materials selection (not necessary for laboratory applications) are required to assure sensor performance in motion. The dewar is 45" long and 13" in diameter. The capacity of the dewar is 15 liters of liquid helium with a hold time of 3 days. The dewar has three concentric chambers: the inner most chamber housing the probe assembly, the next chamber holding liquid helium, and the outermost evacuated chamber containing thermal radiation shielding. The helium chamber contains baffling which minimizes helium sloshing when the sensor is in motion. G-10 fiberglass is used for the outer wall, and quartz-epoxy composite materials are used for the inner construction.

3.1.2. Cryogenic Probe Assembly

The probe structure fits into the innermost chamber of the dewar and is composed of two structural components, a neckplug and a gradiometer support structure (Figure 3.1.c). The neckplug provides the mount for the probe structure into the dewar and provides thermal insulation for the open end of the dewar. The gradiometer support structure is a combination of structural cylinders

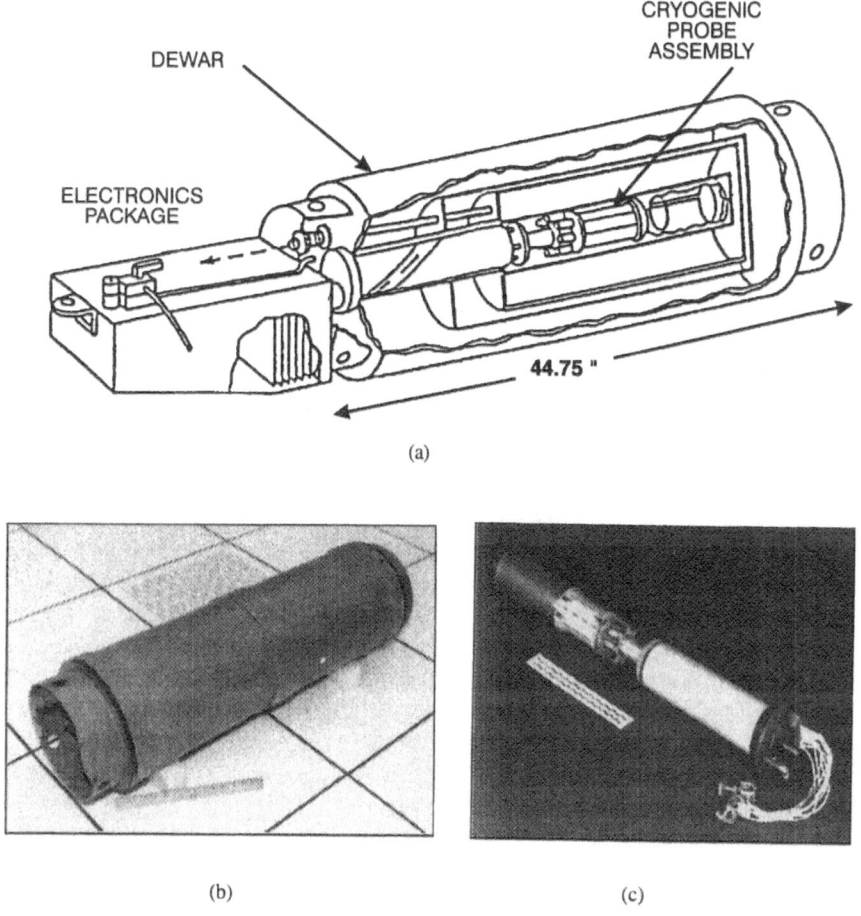

Figure 3.1. The Superconducting Gradiometer/Magnetometer Sensor: (a) artist's conception, (b) dewar, and (c) cryogenic probe assembly.

and flanges on which the SQUID assemblies, known as *Flux Transformer Assemblies* (FTAs), and the sense-loop array are mounted. The SGMS incorporates 5 gradient sense loops (to determine all components of the local tensor gradient field) and 3 additional magnetic field sense loops (for motion compensation of the gradiometer signals). These sense loops are formed with niobium wire hand laid in grooves on a fuzed quartz cylinder. In contrast to the simple gradient loop layout depicted in Fig. 2.2.b, the SGMS gradiometer and magnetometer sense loops employ a less intuitive, more intricate layout on a cylindrical substrate (Figure 3.2).

The substrate supports a wire-in-groove sense loop array. The neckplug and support structure are fabricated with quartz epoxy composite. The probe assembly is cooled by evaporated helium gas. This gas is vented to atmosphere through a spiral vent tube (fabricated into that part of the inner

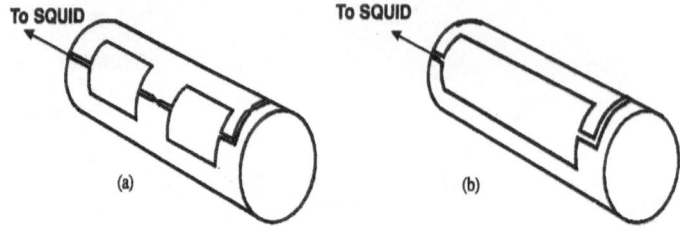

(a)

(b)

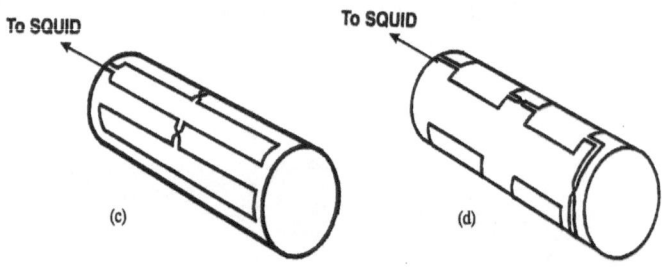

(c)

(d)

(e)

Figure 3.2. Artist's conception of the gradiometer loops in the SGMS to measure (a) G_{yx}, (b) G_{yy}, (c) G_{yz} (d) G_{zx}, and (e) G_{zz}.

chamber housing the neck plug) which serves to cool the neckplug via conduction.

3.1.3   Flux Transformer Assembly

The Flux Transformer Assembly (FTA) is the heart of the sensor [12]. It houses the SQUID and the transformers that interface the SQUID to its sense loop and to its warm-side electronics. The FTA is similar in design to other SQUID packages developed in the early 1980's. Figure 3.3 shows the details of the FTA.

The SQUID is dc-biased and non-hysteretic. It is a hybrid design, featuring a bulk toroidal inductive element and thin-film Josephson junctions. The transformers include an rf-isolation transformer (superconducting primary and secondary windings) which connects the SQUID to the sense loops, and acts as a low-pass filter with a cutoff at 100 kHz; a feedback transformer (normal primary, superconducting secondary) which provides closed-loop operation for the SQUID; a modulation transformer; and a signal transformer (normal primary and secondary with a permalloy core) which couples the output of the SQUID to the warm-side electronics.

The FTA components are mounted on a small circuit board, and the entire assembly is mounted in a niobium or lead canister for shielding against rf interference and external magnetic fields. The signal lines connecting the FTA to the sense loops are shielded in niobium tubes.

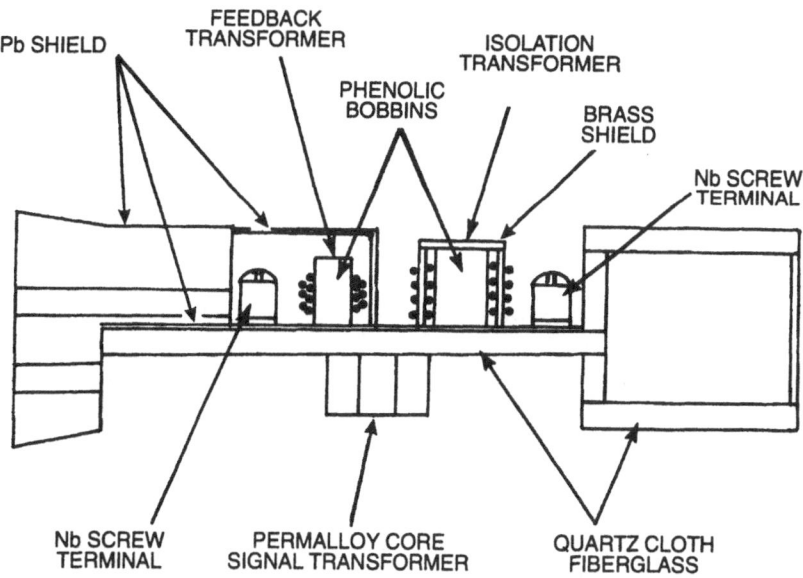

Figure 3.3. Sketch of the SGMS Flux Transformer Assembly.

536

Modulated analog SQUID output signals are coupled electrically via transmission lines to the room-temperature electronics package outside the probe/refrigeration assembly. Electrical leads connecting the FTA to the warm-side electronics are shielded in CuNi tubes.

3.1.4. Room Temperature SQUID and Datalink Electronics

The electronics package for the SGMS permits signal readout and circuit control for each of the eight channels. The package consists of SQUID electronics, anti-aliasing filters, analog-to-digital converters, and control circuits for remote tune, gain and offset control. A standard flux-locked loop modulation-demodulation technique is utilized for electronic readout because of the low voltage output of the SQUID and the necessity to match the low impedance of the SQUID with the high impedance of the preamplifier. A block diagram of the analog electronics is shown in Figure 3.4.

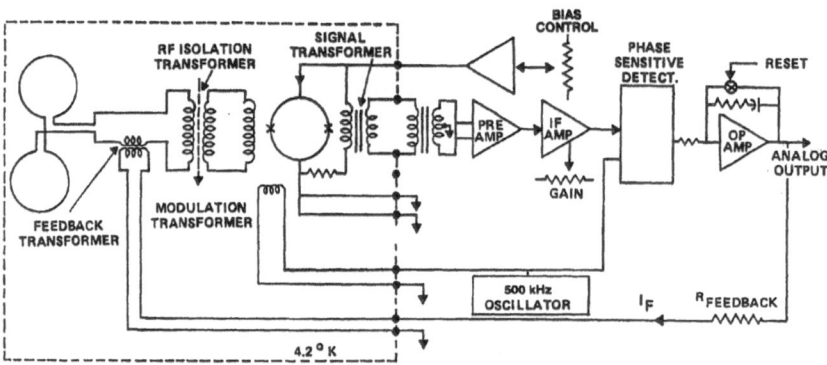

Figure 3.4. Schematic of the SGMS flux-locked loop feedback electronics.

The SGMS analog electronics are typical of feedback systems utilizing a 500-kHz square-wave modulation signal inductively coupled to the SQUID input. The magnitude of the modulation signal is selected so that, in the absence of an input signal, the operating point of the SQUID will switch between points of equal voltage on the V-Φ curve. In that case, the output of the SQUID will not change, and the electronics will produce no feedback signal. However, if there is a change in the input flux, the SQUID will switch between two different voltage points, with the resulting square-wave output proportional to the input signal and with the phase relationship (0 or 180 degrees) between the signal and modulation indicating the polarity of the input signal. This output signal is transformer-coupled (for impedance matching) into a low-noise pre-amplifier and then into a high-gain intermediate frequency (IF) amplifier. Subsequently, the signal is demodulated with a phase-sensitive detector (which preserves polarity information), and finally the resulting signal is integrated to filter out the modulation and to generate a low frequency voltage output proportional to the gradient input signal. Negative feedback is utilized to make the SQUID a null detector; i.e., to lock the SQUID onto equal voltage operating points on the V-Φ curve indicative of a zero input signal. To this end, a feedback current proportional to the demodulated SQUID voltage output signal is generated across the feedback resistor and is inductively coupled back into the gradiometer loop circuit via the feedback transformer.

The outputs of the SQUID system electronics are processed and digitized by the datalink package. Upon entering the datalink, the analog signals are amplified via programmable gain amplifiers and are low-pass filtered to eliminate aliasing using 4-pole filters with a cut-off frequency at 8 Hz. Programmable gain amplifiers are used to maximize the dynamic range of the datalink. The data are sampled at a rate of 20 Hz.

3.2. BASIC GRADIOMETER PERFORMANCE PARAMETERS

In this section, we shall establish terminology to describe baseline gradiometer performance and establish a context for aspects of performance peculiar to mobile applications. Performance figures for the SGMS will be included.

3.2.1. Sensor Sensitivity

For a given SQUID system there are a number of interrelated measures used to describe the sensitivity of the system. One measured parameter often quoted is the flux noise, denoted by ϕ_n, typically in units of $\phi_o / \sqrt{Hz}$. Related to the flux noise is the "flux noise power spectral density" given by $S_\phi = \phi_n^2$.

The noise power spectral density of SQUIDs is typified by white noise for all but low frequencies and a frequency-dependent component at the lowest frequencies for which the flux noise typically has a 1/f spectrum. The white noise can be explained largely in terms of Johnson noise in the resistive shunts. The 1/f spectrum for thin-film dc SQUIDs has received considerable attention and definitive results have been obtained only recently. For a number of low-T_c junction types, notably aluminum-oxide trilayer junctions, the 1/f nose has been attributed to variations in junction critical current (arising from electron trapping in the junctions).

SQUID performance also is represented in terms of energy sensitivity. For a SQUID itself, intrinsic energy sensitivity is $\varepsilon = S_\phi/2L$ where L is the SQUID inductance. For the case in which the SQUID is coupled to a sense loop structure, the appropriate energy figure is the coupled energy sensitivity $\varepsilon_\kappa = \varepsilon/k^2$, where k is the coupling factor between the input coil and the SQUID. (Note that $k^2 = M/LL_i$, where L_i is the inductance of the input coil, and M is the mutual inductance between the SQUID loop and the input coil.) See Reference 13 in this volume for an in-depth discussion of this topic and for a complete set of references.

For gradiometer operation, gradient sensitivity denoted as G_n (in units of nT/m-$\sqrt{Hz}$, pT/m-$\sqrt{Hz}$ or fT/m-$\sqrt{Hz}$) represents the ultimate figure of merit for sensor performance. G_n is usually represented as a function of frequency f in terms of the flux noise and the transfer function $\partial G/\partial \phi$ by

$$G_n(f) = \frac{\partial G}{\partial \phi}(f) \cdot \phi_n(f) \tag{3.1}$$

In order to understand the fundamental mechanisms underlying SQUID intrinsic noise, such as sensor white noise and 1/f noise, SQUID circuits are typically operated in a laboratory-controlled low-field environment shielded from the Earth's magnetic field and from external electromagnetic interference. This has been the focal point in much of the research and development for SQUID technology. For applications outside of such laboratory environments, the effect of magnetic field and electromagnetic interference will significantly impact SQUID performance. From the perspective of this article, the intrinsic sensitivity of a SQUID circuit is defined to be its sensitivity in an environment well-shielded from external magnetic field and electromagnetic interference. This represents a baseline from which to assess the impact from noise sources arising from field operation, in general, and mobile operation, in particular. It should not be at all surprising that these noise sources may raise the noise floor of the sensor. The object of the research described in the following sections is to mitigate the effects of such noise sources.

The intrinsic stationary sensitivity of the SQUIDs in the SGMS in terms of flux noise is nominally 200 $\mu\phi_0/\sqrt{Hz}$ at 0.1 Hz. The 1/f noise knee for these SQUIDs is approximately 10 Hz. The stationary gradient sensitivity of the SGMS is 0.2 pT/m-$\sqrt{Hz}$ at 0.1 Hz. In motions with accelerations on the order of 0.02 g and rotations on the order of 0.02 radians at 0.1 Hz, the SGMS displays sensitivities from 1-3 pT/m-$\sqrt{Hz}$ at 0.1 Hz.

3.2.2. Dynamic Range
A performance parameter less often cited is the dynamic range of the SQUID system which measures the operational range of the SQUID electronics, i.e., the ratio of the magnitude of the signal processed by the electronics to the magnitude of signals which can be resolved from the signal processed. This factor is especially important for mobile applications in order to maintain sensitivity in the face of the large field variations on the order 1000 nT arising from rotations in the Earth's magnetic field. Dynamic range is usually referenced at the SQUID output so that the dynamic range is the ratio of full-scale output ϕ_{max} to the flux sensitivity ϕ_n. This quantity is frequently represented in decibels (dB) by the relation

$$20\log_{10}\left(\frac{\phi_{max}/\sqrt{Hz}}{\phi_n}\right) \tag{3.2}$$

Dynamic range is largely determined by the design of the SQUID feedback loop. A typical dynamic range for a SGMS channel is 126 dB.

Once digital signal processing is incorporated with the analog electronics, the dynamic range of the analog-to-digital converters (ADCs) must be taken into account. In a straightforward approach, it is desirable that the ADC dynamic range be commensurate with that of the analog feedback electronics. For current ADC technology, figures of 80-100 dB are common.

4. Noise Sources in Mobile Operation and Compensation Approaches

CSS has actively investigated sources of noise which could potentially deteriorate gradiometer performance during mobile operations. Some noise sources occur when the sensor is operated unshielded, even when the sensor is stationary, e.g., temporal changes in the Earth's magnetic field, magnetohydrodynamic signals associated with ocean waves, and electromagnetic interference. Other noise sources are associated with the sensor's motion, e.g., relative motion of the sensor with respect to the Earth's magnetic field, to the underlying geological structure, and to the sensor platform. In addition, motion can induce signals internal to the sensor, e.g., magnetic signals and temperature-induced effects associated with the sloshing cryogen, signals associated with the motion of magnetic components with respect to the sense loops, eddy-current signals, and field-induced magnetization effects associated with changing sensor orientation in the earth's magnetic field. Compensation approaches, including specific sensor design features, electronic feedback and software subtraction of signals detected by auxiliary sensors, modeling of phenomena, and deliberate platform maneuvers have been investigated and implemented in order to minimize the effects of such noise sources.

In Section 4.1 we shall address the most signficant noise sources associated with the Earth's magnetic field alone, namely, the admixture into each gradiometer channel of signals from the magnetic field vector components and from the eddy-current signals. A discussion of affiliated compensation approaches is included. Additional issues associated with the Earth's magnetic field are discussed in Section 4.2. A limited discussion on sources intrinsic to the sensor, on electromagnetic interference, and on sources associated with the platform of operation will be given in Sections 4.3 through 4.5, respectively. More detailed treatments of these subjects will be the topic of subsequent papers to be published.

4.1. THE LOW FREQUENCY TERRESTRIAL MAGNETIC ENVIRONMENT

To first order, the Earth's magnetic field is locally uniform, with the magnitude of total field nominally 50,000 nT. The field lies in a vertical North-South plane, with vertical and horizontal components varying with latitude. Sensors operating in motion must be able to discriminate relatively small signals from the field variations encountered coincidentally with platform motions. This presents stringent demands on the dynamic range of vector magnetometers and gradiometers and necessitates the requirement for precision balance for gradiometers.

4.1.1. Balance and Gradient Admixtures
Balance is the common-mode rejection of a uniform magnetic field by a gradiometer, i.e., a measure of the extent to which a gradiometer is ideal without admixture of magnetometer components. In practice, a gradiometer signal will have magnetometer components (Figure 4.1).

540

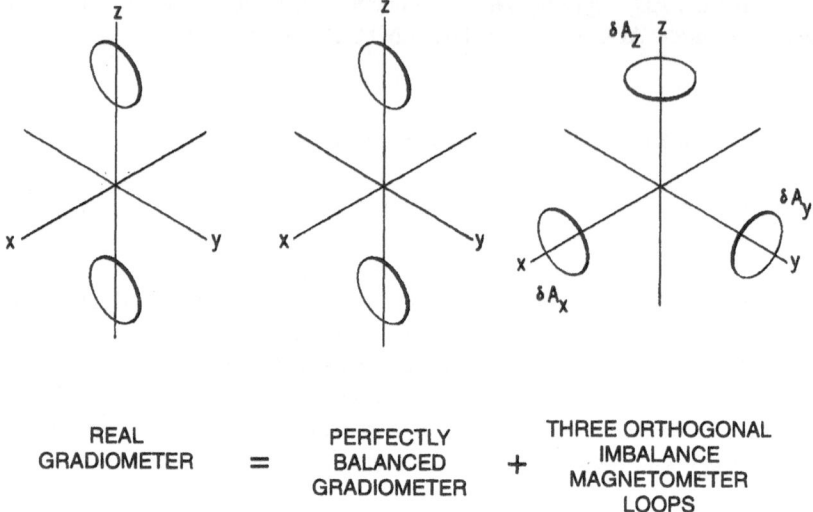

| REAL
GRADIOMETER | = | PERFECTLY
BALANCED
GRADIOMETER | + | THREE ORTHOGONAL
IMBALANCE
MAGNETOMETER
LOOPS |

Figure 4.1. Decomposition of a real gradiometer (with manufacturing imperfections) into a perfectly balanced gradiometer component and three orthogonal imbalance magnetometer components.

Gradiometer imbalance can be represented in terms of a tensor b, with the error signal G_{ij} resulting from the application of the uniform magnetic field components B_k given by

$$G_{ij} = \sum_{k=1}^{3} b_{ij}^k \cdot B_k \qquad (4.1)$$

where the "imbalance" tensor b is often loosely referred to as the "imbalance vector." The imbalance signal is typically the largest noise signal and often is larger than the target signals seen in mobile operations.

Approaches to enhance balance include the following:

1. *Design and fabrication:* Loops are designed to be matched as closely as possible in effective area and alignment. Perturbing effects due to diamagnetic SQUID shield cans and diamagnetic stripline shields are minimized by optimization of symmetry, separation, size, etc.

2. *Fixed vane trimming:* Superconducting (diamagnetic) discs are placed near various loops to compensate for asymmetries and field distortions arising from magnetic components. This basic approach has been implemented in the SGMS by controlled deformation of the sense loops.

3. *Potentiometer feedback:* Magnetometer feedback can be introduced through electronic coupling with manual preset prior to sensor operation. Potentiometer feedback is sometimes called "fixed potentiometer feedback" or "screwdriver-adjusted feedback".

4. *Multiplying digital-to-analog converter (MDAC) feedback:* Electronic feedback can be used to dynamically change the balance by means of multiplying digital-to-analog converters (MDACs) in the data link by an operator or by automation. MDAC feedback is similar to fixed potentiometer feedback, except that the "screwdriver" is "turned" by software.

5. *Postprocessing noise cancellation:* Signal post-processing can be employed for compensation of imbalance signals. This approach has the distinct advantage of providing frequency-dependent compensation.

For the SGMS gradiometer (utilizing wire-in-groove pickup-loop geometry), the magnitude of the imbalance tensor components are typically 2×10^{-3} m^{-1} using fixed vane trimming. For "straight line" tows with field changes of approximately 2000 nT/$\sqrt{Hz}$, the imbalance signals will be approximately 4 nT/m-$\sqrt{Hz}$.

The admixture of other gradient components into a pure gradient channel will arise from fabrication and assembly imperfections as in the case of magnetometer imbalance components. In addition, admixtures will be signifigant in all but the simplest geometries. For example, the SGMS calibration matrix relating the voltage signals S_{ij} (in units of V/$\sqrt{Hz}$) for the 5 gradiometer channels to the specified pure gradient components G_{ij} (in units of nT/m-$\sqrt{Hz}$) is given by

$$
\begin{bmatrix} G_{xx} \\ G_{yx} \\ G_{zx} \\ G_{zy} \\ G_{zz} \end{bmatrix} = \begin{bmatrix} .16 & 12.2 & -.13 & -.30 & -13.5 \\ -14.9 & .59 & -.82 & .43 & -.10 \\ -.66 & -.26 & 16.1 & -.10 & -.46 \\ .07 & -.52 & -.56 & 11.5 & -.16 \\ .26 & .92 & .13 & .98 & 15.2 \end{bmatrix} \begin{bmatrix} S_{yx} \\ S_{yy} \\ S_{zx} \\ S_{zy} \\ S_{zz} \end{bmatrix}
$$

Another example (displayed in Figure 4.2) shows the admixtures required to reconstruct certain gradients for configurations in which planar gradiometers are employed. Planar gradiometers will naturally provide information about off-diagonal gradients, e.g., G_{zx} or G_{zy}. However, by appropriate orientation of planar gradiometers, admixtures of both on-diagonal and off-diagonal components can be used to construct on-diagonal components, e.g., G_{xx}, G_{yy} and G_{zz}, can be determined. For the coordinate system in Figure 4.2, G_{zz} can be determined from appropriate addition of the two admixtures displayed for G_{xx} and G_{zz}, and for G_{yy} and G_{zz}. For any given configuration, a calibration matrix measuring the gradient admixtures must be measured.

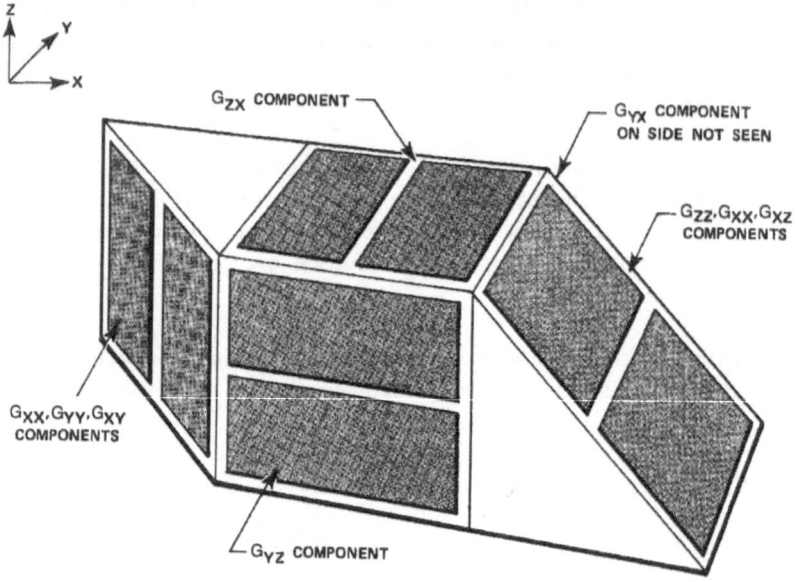

Figure 4.2. A 5-channel planar gradiometer configuration. A planar gradiometer channel intrinsically measures the spatial variation of the field component normal to the circuit's planes along a direction in the plane, e.g., G_{xy}, G_{xx}, and G_{yz}. The axial gradients G_{xx}, G_{yy}, and G_{zz} can be reconstructed by superposition as suggested in this figure.

4.1.2. Eddy-Current Compensation

Eddy currents originating from metallic components within the sensor and from other metallic components in the overall system represent another major motion-induced noise source. Compensation approaches are applied similar to those used for balance. Consider a circular conductor of resistance R and inductance L, and a sinusoidal magnetic field applied in a direction normal to the loop $B_a = B'e^{i\omega t}$. Under the condition that the resistance of the normal conductor is much greater than the inductive reactance, i.e., $R >> \omega L$, the impact of transient effects can be ignored. Then the well known result

$$I = \frac{-\omega AB_a e^{i\delta}}{\sqrt{\omega^2 L^2 + R^2}} \qquad (4.2)$$

applies with $\delta = tan^{-1}(R/\omega L)$. For the case $R >> \omega L$, the magnetic signals (from eddy currents induced by sensor rotations) which are sensed by the gradiometer channels, are proportional to frequency and 90° out of phase with the applied field. These basic conclusions from the analysis for a disc or a loop apply to more general cases, although the integrals may be difficult in all but the simplest cases. In fact, the eddy-current sources can be considered as lumped LR cirucuits [4].

Reasonable care in construction of the sensor, and the fact that the power in the rotational spectrum of the sensor is limited to low frequencies, will generally insure that the condition $R \gg \omega L$ applies at all frequencies of interest.

As a result of analysis, there exists an "eddy-current" tensor e, analogous to the imbalance tensor b, which relates the error signals G_{ij} arising from eddy-current sources to the time changes in field:

$$G_{ij} = \sum_{k=1}^{3} e_{ij}^k \frac{dB_k}{dt} \qquad (4.3)$$

A representative magnitude of an eddy-current tensor component for the SGMS is 2×10^{-5} sec/m, with corresponding error signals on the order of 0.04 nT/m-$\sqrt{Hz}$ at 0.1Hz.

As in the case of balance, eddy-current gradients can be partly compensated using the time derivatives of the field components in conjunction with software signal processing and electronic feedback. Eddy-current signals also can be reduced by minimizing the quantity of normal conductors, increasing resistivity, minimizing current path dimensions, symmetry, etc. Eddy-current effects, in principle, also could be compensated by use of normal-metal vanes, much as the effect of superconducting material is compensated by using superconducting trim vanes.

4.2. SECONDARY ISSUES WITH THE TERRESTRIAL MAGNETIC FIELD

In this section we shall discuss secondary issues arising from the Earth's magnetic field including its temporal variations defined as geomagnetic noise, its gross spatial gradients, the finer local geologic features, and magnetohydrodynamic noise arising from ocean waves and currents in oceanic applications.

4.2.1. Geomagnetic Noise
Geomagnetic noise refers to the temporal variations in the Earth's field arising from ionospheric currents. The interested reader is referred to Reference 14 for a general reference on this topic. These variations occur nominally at levels from 10^{-3} - 10^{-1} nT $/\sqrt{Hz}$ at 0.1 Hz, depending on magnetic latitude and atmospheric electrical activity (Figure 4.3). The spatial scales are on the order of 100 km or greater. This represents a major limitation in the use of magnetometers, thus providing an impetus for the use of gradiometer configurations which naturally provide a high degree of common-mode rejection to such changes.

4.2.2. The Terrestrial Magnetic Gradient
The Earth has a magnetic gradient on the order of $1-3 \times 10^{-2}$ nT/m, arising from sources inside the Earth which are closer to the Earth's center than to the surface. Just as rotations in the Earth's field will generate imbalance signals in gradiometer loops, anomalous signals will arise from gradiometer rotations in the Earth's magnetic gradient. These gradients are directional, so that large signals may be obtained in some channels, while the signals in the remaining channels may

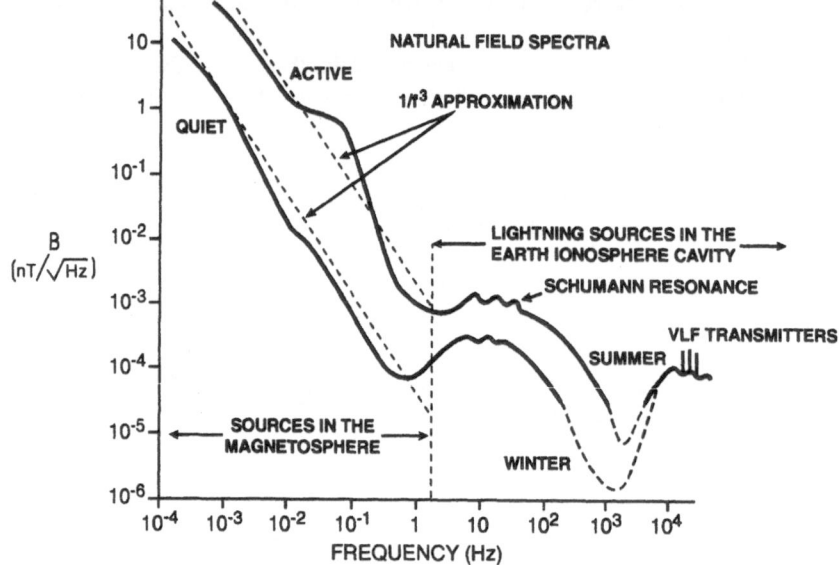

Figure 4.3. Power spectral density of temporal variations in the Earth's magnetic field geomagnetic noise. Typical spectra are displayed for quiet atmospheric electrical activity in winter and active atmospheres in summer (after Cladis et al. [15] and Campbell [16]).

be two or three orders of magnitude smaller. The interested reader is referred to Reference 17 for a discussion of this topic.

The order of magnitude of the noise signals, ΔG, is given by

$$\Delta G = 2\Delta\Theta \cdot G_E \tag{4.4}$$

where G_E is the Earth's magnetic field gradient, and $\Delta\theta$ is the angle through which the sensor is rotated. Then for $G_E = 0.03$ nT/m and $\Delta\theta = 0.01$ radians, $\Delta G = 6\times10^{-4}$ nT/m. Hence, typical rotations in the Earth's gradient induce signal changes in gradiometer channels on the order of a percent or two of the gradient magnitude.

Detailed maps of the Earth's gradient field are not currently made. It should be possible to model the Earth's field well enough to obtain a good estimate of the gradient for many regions (noting that local variations from this estimate do arise in areas with significant magnetic signature from crustal magnetic structure, as discussed below in Section 4.2.3 on geology).

Compensation of the noise from the gradient field is possible using either gradient maps or gradient modeling in conjunction with measurement of the sensor's instantaneous orientations, following an approach analogous to the magnetometer compensation for imbalance. Another approach is to use rotational invariants which can be constructed from the gradiometer and magnetometer sensors. With this approach, multiple signals from sensor channels are added to create a rotationally invariant signal. To the extent that it can mitigate the problem, it is a desirable approach because it circumvents the requirement for complex and costly hardware approaches; however, this approach greatly reduces the number of independent signal channels.

4.2.3. Geologic Noise

Magnetic signals due to geology stem from sources within the Earth's outer crust (in contrast to the more uniform gradient field arising from sources deep inside the Earth as described in Section 4.2.2) and from local surface features. The magnitude of geologic signals may greatly exceed the Earth's field gradient in certain locales, notably in mountainous and rocky areas. Typical values of geologic noise range from 0.05nT/m in low gradient areas to in excess of 10nT/m in high gradient areas. Large-scale geologic noise arises from distributed magnetic sources on the sea floor in bands approximately 20 km in width. The impact of geology will depend on the spatial scale of the geology compared to the scales of the targets of interest. Geologic noise may be the source of false alarms if the spatial scales of the local geologic features are commensurate with the scales of targets of interest.

The only effective method for mitigating severe geological noise is by means of geological maps combined with high quality spatial registration. In the absence of adequate spatial registration, the maps can still be used tactically to select measurement traverses that place the geological noise outside of the signal bandpass range of interest.

4.2.4. Ocean Magnetohydrodynamic Wave Noise

When seawater, a conductor, moves in the Earth's magnetic field, electric currents are generated which, in turn, produce an anomalous magnetic fields and gradients. Such magnetohydrodynamic noise can arise from wind-driven waves, swell from distant sources and internal waves. For ocean-based operations these effects must be considered.

The theory for surface wave noise is well developed and has been experimentally verified for a variety of conditions and frequencies. Not as much is known about internal waves due to the difficulty in observing them in the open ocean. It is known that they exhibit large temporal variability and length scales varying from many kilometers to meters. Although the velocities associated with internal waves are relatively small, their ubiquity may provide a limiting noise source for magnetic measurements above and within the open ocean. The importance of internal wave activity in shallow coastal waters is more problematical, but their noise contribution will probably be less significant than, say, turbulence due to coastal currents and surface wave-bottom interactions. Other sources including tidal currents and currents such as the Gulf Stream are present. Prediction of the magnetic signals from such sources has not received the same degree of attention as surface or even internal waves. They may be too infrequent, out of the relevant target bandpass, very low in power, or so poorly understood that nothing can be done about them presently.

4.3. SENSOR-INTRINSIC NOISE SOURCES

A superconducting gradiometer such as the SGMS in mobile applications is operated in the magnetic and gravitational fields of the Earth. When towed, the sensor undergoes small angle rotations with corresponding changes in the magnetic field, ΔB, and the sensor experiences both linear and angular accelerations, Δg. The magnitudes of ΔB and Δg are determined by the dynamics of the sensor platform. Typical values for stabilized underwater tow systems investigated by the CSS are $0.02 g_E$ and $0.02 B_E$ (where g_E and B_E are respectively the gravitational acceleration and magnetic field of the Earth). Due to both the tow system dynamics and the signal physics of interesting phenomena/targets, the frequency band of the signals of interest extends from approximately 10 mHz up to 1 Hz. This is also the frequency region for which the power spectrum of the accelerations Δg and the field changes ΔB are typically greatest.

Field changes and accelerations might be expected to induce noise signals intrinsic to the sensor that do not correlate simply with the imbalance and eddy-current compensation. In this section, a general perspective for such sensor-intrinsic noise sources and one explicit example encountered in the SGMS development will be discussed.

4.3.1. A General Perspective
The accelerations will cause substrate bending or flexure that can change the balance of the gradiometer sense loops and that will generate signals arising from the relative motion between the substrate and magnetic inclusions in the sensor system within the frequency range of interest. Cryogen sloshing arising from the acceleration will generate error signals as a result of the cryogen's magnetic susceptibility. Slosh baffling is used to diminish this effect.

The field changes arising in sensor operation can potentially result in magnetization changes in both the normal and superconducting materials within the sensor. These changes can be made negligible for the normal materials by using good engineering practices and judicious material selection. There has been concern about flux-trapping effects in the superconducting materials. Investigations conducted in conjunction with the bulk technology used in the SGMS indicate that this is not a significant factor for the SGMS, provided that careful operational procedures are followed, e.g., avoiding sensor exposure to fields much greater than B_E. This issue is more relevant for low-T_c thin-film technology than for the low-T_c bulk technology and gives rise to major research activity in the area of high-T_c thin-film materials and devices.

Linear and rotational acceleration changes also drive temperature fluctuations, ΔT, as a result of liquid cryogen slosh and gas convection. ΔT is on the order of a millikelvin at 0.1 Hz for the SGMS dewar for typical motions observed in tow bodies of interest. These temperature changes can cause changes in the susceptibility of the paramagnetic materials in the cold space of the sensor or change the dimensions of the loops through thermal expansion. Temperature-induced penetration-depth changes proved to be an important factor impacting the performing of the first generation of the SGMS. For the advanced liquid helium dewar described in Section 6.1, ΔT is only on the order of 1 microkelvin when subjected to comparable motions, essentially eliminating this as a noise source.

Sensor-intrinsic noise sources arising from accelerations, field change and temperature change, and sensor design principles to mitigate them, represent major issues for mobile applications. These issues will be the subject of subsequent papers. In this paper, we shall limit discussion to one example concerning modifications in the SGMS to reduce its thermal sensitivity in motion.

4.3.2. SGMS Enhancements

In 1983, after a series of successful technical tests, noise sources that reduced gradiometer sensitivity in motions typical of field deployment were identified. The SGMS and other sensors were characterized under controlled conditions, and the thermal sensitivities of the SQUID circuits were identified as a significant noise source. Research was conducted to investigate the origin of the high thermal sensitivities.

Temperature-induced penetration-depth changes were identified as one major noise source. From the standpoint of sensor operation, penetration-depth changes can change (1) the inductance of a circuit which, in turn, changes the gradient signal transfer function or (2) the magnetic moment of a circuit element which, in turn, is detected as a field change by the pickup loops. As a result of these investigations, niobium has been identified as the most desirable of the more commonly-used materials for both bulk components and for wire. In addition to material selection, surface quality has been identified as an important factor with respect to this penetration-depth effect. Electropolishing bulk components and wires has been demonstrated to reduce this effect. Circulating currents in the superconducting circuits represent one important factor which enhances the penetration-depth effect. In order to reduce circulating currents, thermal switches to break these currents were implemented in the 1989 refurbished SGMS sensor. This work is summarized in Reference 18.

Three versions of the FTA package were manufactured by Unisys to improve sensitivity. In addition to modifications in the selection of superconducting materials to address the penetration-depth effect, material changes were made to reduce the magnetic signature of normal-metal components in the FTA package, including the RFI shielding sleeve, the ground plane, wire for electrical circuitry and solder for interconnects. In the original version, brass was used for the RFI shielding sleeve, and BeCu was used for the ground plane. NbTi wire was used not only for the essential superconducting circuitry, but also for the SQUID modulation coil. PbSn solder was used for interconnection of non-superconducting wire elements.

In the second version (MOD 1), manganin wire was used for the modulation coil in place of NbTi wire. Susceptibility measurements on FTA packages made by the Naval Research Laboratory indicated that the PbSn solder exhibited a broad superconducting transition and hence made a significant contribution to the magnetic signature of the FTA in the range of temperatures of interest in sensor operation. Hence, PbSn solder was replaced by In-Ag solder, because it would remain normal in the range of operational temperatures.

In the third version (MOD 2), the brass and BeCu components were replaced by #220 bronze counterparts. The #220 bronze was selected because it has no lead contamination, and bronze was used throughout to avoid problems with dissimilar metals. Phosphor-bronze wire was used in place of manganin wire in order to reduce ferromagnetic contamination. In addition, electropolished Nb bulk material and wire were utilized in place of NbTi and Pb to address the issue of penetration-depth changes.

With these modifications, an overall 25 *dB* increase in sensitivity at 0.1 Hz for components in the SGMS sensors has been demonstrated (Figure 4.4). The original performance, as measured by noise-canceled power spectral densities of gradiometer data taken during motion tests, was 20 to 35 *dB* above the desired sensitivity goal. A 15 *dB* increase for the sensor in motion has been realized by the simple retrofit of components in MOD 1 (Figure 4.4).

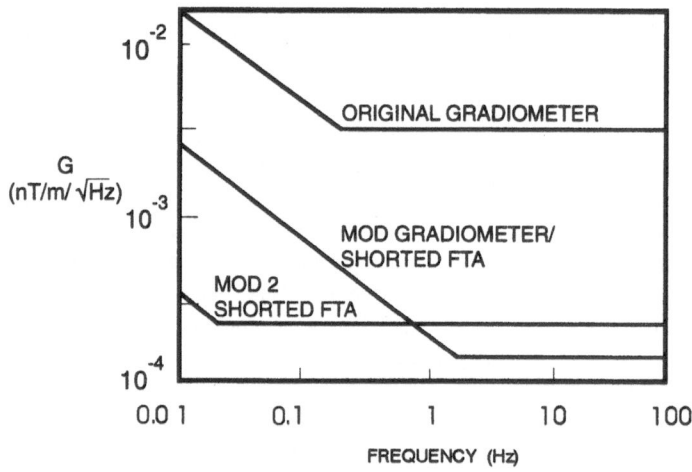

MATERIALS FOR SQUID ASSEMBLIES			
COMPONENTS	ORIGINAL	MOD 1	MOD 2
SQUID	NbTi	–	Nb (ELECTROPOLISHED)
SHIELD	Pb	–	Nb (ELECTROPOLISHED)
MODULATION COIL	NbTi	MANGANIN	PHOSPHOR BRONZE
SUPER WIRE	NbTi	–	Nb (ELECTROPOLISHED)
NORMAL METAL	BRASS	–	BRONZE
SOLDER	PbSn	In	In

Figure 4.4. Power spectral density of the average response of the SGMS in motion: (a) original sensor performance in 1983, (b) performance following the first modification in 1984 (Mod 1), and (c) performance using new SQUID assemblies in 1985 (Mod 2). Materials changes in these modifications are tabulated.

In contrast to the retrofit approach, the more dedicated changes of MOD 2 led to a substantual improvement in SQUID performance. The noise in the new sensors was reduced an additional 10 dB at 0.1 Hz compared to results for MOD 1, and the frequency break point for $1/f$ noise was reduced to 0.02 Hz (Figure 4.4). The stability of the new SQUID sensors against operating temperature fluctuations was improved by a factor of at least three as compared to the earlier sensors.

4.4. ELECTROMAGNETIC INTERFERENCE

Electromagnetic interference (EMI) can be considered to arise from electromagnetic radiation at any frequency. However, within the context established above where natural terrestrial magnetic sources dominate the low-frequency spectrum, it is convenient to consider electromagnetic interference to begin at 50 and 60 Hz arising from low-frequency power-line transmissions. (Note that high-energy dc power systems are used in some areas of Europe.) Other predominant sources include atmospherics, notably lightning, and manmade transmissions with frequencies ranging from hundreds of kilohertz and extending up to hundreds of megahertz. (It is worth noting that higher frequency elecromagnetic signals are greatly attenuated in sea water so that EMI is not an issue for undersea tow operations except from the electrical and electronic subsystems internal to the host platform.)

A SQUID-based gradiometer provides the greatest sensitivity of any low-frequency magnetic sensor currently available. Target-signal bandpasses for mobile operations generally will fall in the range from 0.01 to 1 Hz. The gradiometer, however, is inherently a broadband sensor, sensitive to signals from frequencies less than 0.001 Hz up to frequencies in the gigahertz range. The signals outside the target signal bandpass can limit the sensor's performance in the bandpass by such nonlinear mechanisms as flux popping, transfer-function washout, and aliasing, which are described below.

As discussed in Section 3.1.4, SQUID flux-locked-loop electronics operate by "locking on" to an operating point on the SQUID V-ϕ curve; i.e., the electronics circuit attempts to drive the flux error signal at the SQUID to zero by means of negative feedback. The circuit requires a finite time to react to a changing applied flux and generate the appropriate feedback. So long as changes in the flux error signal are less than $\phi_o/4$ in magnitude, the circuit will be able to track the input signals. If the rate of change of the EMI input signals is large enough to create a flux error signal with magnitude greater than $\phi_o/4$, the SQUID current will exceed the critical current of a junction, and one or more flux quanta will enter or leave the SQUID, an effect commonly described as *flux popping*. As a consequence, the feedback circuit will "lose lock." Once the large EMI input signals subside, the feedback electronics will stabilize and "lock on" to a new operating point and the flux-locked loop will apply feedback to maintain the new flux operational point of the SQUID V-ϕ curve. Information about the low-frequency signals of interest is lost when the feedback circuit loses lock.

Slew rate is the parameter defined to quantify this capability for the feedback electronics to maintain lock. Slew rate measures the rate at which a SQUID system can track an incoming signal and is especially important when the sensor must operate in an active electromagnetic environment when the sensor is unshielded and exposed to atmospherics (lightning) and lower-frequency radio emissions. The circuit will remain locked so long as the flux error signal Φ_e does not exceed $\Phi_o/4$. Hence, the (maximum) slew rate of the system is given quantitatively by the relation

$$\frac{d\Phi_e}{dt} = \omega\Phi_o/4 \tag{4.7}$$

Because the rate of change of flux is proportional to the frequency, higher-frequency signals cause a greater problem for a given amplitude. In particular, the slewing capability of the current generation of circuits is effective against electromagnetic emissions with frequencies less than 1 MHz. Slew rate is primarily a function of the preamplifier circuit (reflecting the requirement that the circuit must respond rapidly to the input signal to maintain a small flux error signal), and to a lesser extent, the configuration of the system. High slew rates are essential to gain acceptance of this technology for reliable high-sensitivity field operation. The SQUIDs in the SGMS have a nominal slew rate of $2.5 \times 10^4 \, \Phi_o$/sec. The advanced (optimized) thin film SQUID system developed by Wellstood et al. has achieved slew rates of $3 \times 10^6 \, \Phi_o$/sec [19].

In addition to the slew-rate issue, EMI may cause transfer-function washout and aliasing. Three ways in which EMI can deteriorate sensor performance have been identified and investigated: (a) CW EMI will raise the white noise floor, (b) digital noise also will raise the low-frequency noise floor, and (c) modulated CW will be aliased down to a low-frequency noise [20]. Signals also can be aliased by conditioning of the SQUID output signal, e.g., filtering and digitial sampling.

For the shielded SQUID operation utilized in the SGMS, the primary entry path of EMI into the SQUID is via the pickup loops, with the EMI magnetic field detected as an imbalance signal. In addition to the use of high slew-rate circuits, two general approaches to improve EMI immunity are to shield the pickup loops with a Faraday cage or use low-pass filtering between the loops and the SQUID. Loop shielding is successfully accomplished in laboratory environments by the use of EMI shielded rooms. EMI shields can be built into the sensor for stationary field operation, but their performance has been less than satisfactory in our experience and the loop shield becomes a major source of eddy-current problems in mobile operation.

RF filtering has been implemented in SGMS by means of an isolation transformer between the pickup loops and SQUID, with primary and secondary separated by a normal metal rf shield. The filters currently used introduce a frequency cutoff at 100 kHz. It should be noted that the RFI isolation transformer causes a loss in sensitivity through the addition of Johnson noise generated by the rf shield.

4.5. PLATFORM NOISE

The ideal situation for operating sensitive magnetic sensors onboard moving platforms is to provide an environment in proximity to the sensor which is ferromagnetically clean, free of other metallic eddy-current sources, and which is distant from electrical and electronic equipment, power plants, electrical distribution paths, and propulsion systems. In reality, we only can design a system to reduce the impact of such factors and to devise schemes to compensate for such noise sources.

To the extent feasible, this philosophy leads to system configurations in which the sensor operates in a magnetically-sterile package at a sufficient standoff from a primary vehicle (such as a truck on land, a helicopter in the air, or a boat in water) towing the sensor package. In the MADOM program (described in Section 5.1) approaches were developed so that the SGMS was operated successfully in a tow configuration within an underwater fiberglass body working in proximity to a sonar sensor.

In order to gain widespread acceptance of magnetic sensors in general, especially high-performance sensors, it is very desirable to provide system concepts in which the magnetic sensor can be operated onboard the primary vehicle, be it a tractor, airplane or boat. This represents a major challenge using current technologies. Such efforts involve significant costs in making the subsystems magnetically and electrically quiet, introduce significant complexity in order to mount auxiliary sensors for motion compensation, and require sophisticated signal processing. Currently, towing configurations are most attractive to avoid unnecessary complexity for one-vehicle system concepts. Decisions on the use of a high-performance magnetic sensor over alternate sensors, relate to the value of a magnetic sensor to satisfy system detection, classification and localization requirements in comparison to cost and inconvenience (as determined in system design tradeoffs).

5. Demonstrations of Gradiometer Operations

Here we discuss the operation of the SGMS in a mine countermeasures and environmental cleanup advanced technology demonstrations for the U.S. Navy.

5.1 THE MAGNETIC AND ACOUSTIC DETECTION OF MINES (MADOM)

Starting in 1984, the SGMS was operated onboard an undersea vehicle towed by a surface ship. Following this initial success, the CSS initiated the MADOM project in 1985. This project successfully demonstrated the value of magnetic and acoustic sensor fusion for mine reconnaissance. The SGMS showed for the first time, high sensitivity and rugged, reliable, fully-automated performance of a superconducting gradiometer operating onboard a moving, towed underwater vehicle outside a laboratory environment. In fact, the sensor was utilized in sea testing for a period of 7 years.

5.1.1 *System Concept*
A system concept depicted in Figure 5.1 was developed incorporating a sensor triad, including the SGMS, a synthetic-aperture sonar (SAS) and a high-frequency sidescan sonar, operating from towed underwater vehicles.

552

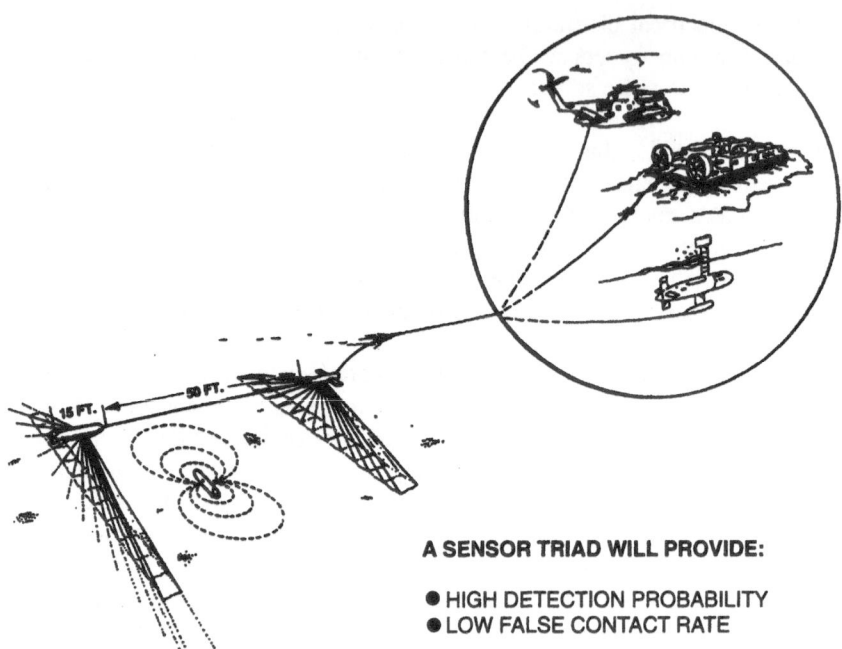

A SENSOR TRIAD WILL PROVIDE:

● HIGH DETECTION PROBABILITY
● LOW FALSE CONTACT RATE

Figure 5.1. The MADOM system concept.

The sensor triad, a high-speed towed vehicle system and a real-time processing system were developed under the project. The experimental configuration utilized in the demonstration, consisting of a float, a dead-weight depressor, and a second tow body is depicted in Fig. 5.2. The sidescan sonar was housed in the dead-weight depressor, while the SGMS and the SAS were housed in the aft tow body. Operation of the SGMS and the SAS in the aft tow body provided better motion stability, a factor essential to the performance of these two sensors. A photograph of the tow body housing the SGMS and the SAS is given in Figure 5.3.

5.1.2. Parallel Array Signal and Image Processing System

The Parallel Signal and Image Processing System (PARSIPS) was developed to control the sensor suite and to process data collected (Figure 5.4). The heart of the PARSIPS computer system is a bank of eight array processors, four of which are dedicated to sonar data conditioning, motion compensation and beamforming, and four of which are dedicated to gradiometer operation data conditioning, motion compensation and target localization. Of particular interest in this article, PARSIPS processing of gradiometer data provides magnetic target detection and classification, as well as localization, fully automated effectively in real time.

Through PARSIPS, the sensor triad is highly automated and is easily managed by a single person. In particular, gradiometer operation consists of turning on the power to the gradiometer and to the helium pump, tuning the gradiometer (a three-minute semi-automated process typically executed at the start of operation), inputting towing vessel speed, and executing keystrokes to start and stop the automated gradiometer data recording and processing.

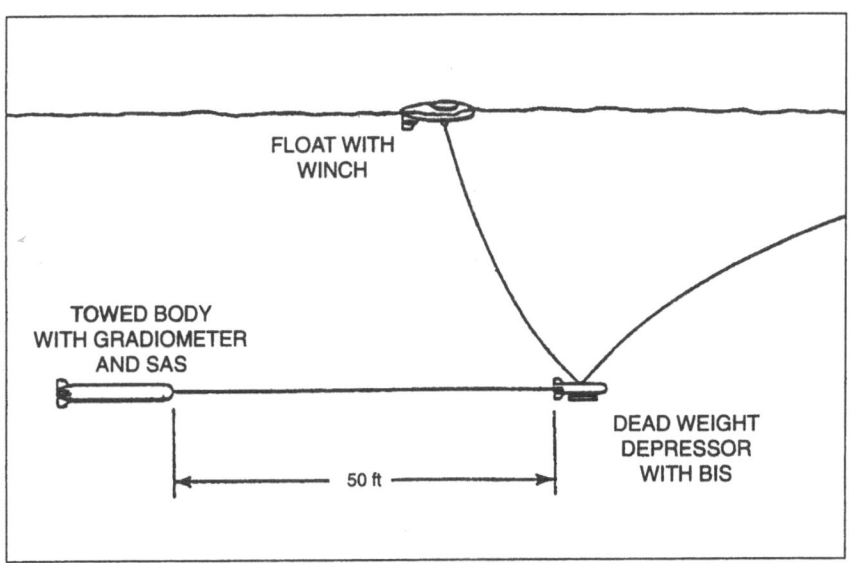

Figure 5.2. Experimental towing configuration utilized in the MADOM demonstrations. The configuration features a float with winch, a dead weight depressor housing a Bottom Imaging Sensor (BIS), and a second trailing tow body housing the SGMS and a Synthetic Aperture Sonar (SAS).

Figure 5.3. The undersea tow vehicle for the SGMS and the SAS.

554

Figure 5.4. The PARSIPS sonar/gradiometer display.

The gradiometer collects the time history of the gradient of the magnetic field in its frame of reference as it moves along its trajectory through the magnetic field. From this collected data, the gradiometer analysis algorithm provides real-time determinations of the locations and moments of the magnetic dipoles. This algorithm operates on overlapping segments of the gradiometer data, compensating for the motion of the sensor and iteratively solving for the three-dimensional position and moment vectors of the source dipoles until a minimal set of dipoles is found that is sufficient to describe the measured signal. This magnitude of the magnetic moment of a target is utilized to classify that target as either a magnetically mine-like or nonmine-like object. The algorithm requires no inputs of any kind (such as thresholds or instrument noise levels) except for the tow speed.

The images of the sea floor obtained from the SAS and sidescan sonars are displayed "waterfalling" in real time on the left and right portions of the high-resolution PARSIPS monitor. The locations of magnetic targets found by the automated gradiometer processing algorithm are displayed as icons superimposed on the sonar images of the sea bottom. The shape and color of these icons denote the size of the target's magnetic moment and the relative confidence of the algorithm in the target's localization. The icons are mapped onto their locations on the screen after a delay following the point of closest approach in order to assure sufficient data for the analysis.

5.1.3. Model to Detect, Classify and Localize Multiple Dipole Targets

A mathematical model for detection, classification, and localization of multiple stationary magnetic dipole targets using a gradiometer (with 5 independent gradiometer channels appropriately selected and 3 orthogonal magnetometer channels) moving in a straight line trajectory past the targets at a constant speed has been developed. This model explicitly compensates for the imbalance and eddy-current signals of the 5 primary gradiometer channels.

In the model, the 5 output signals $S_i(t)$ (i=1,2,....5) from the 5 independent gradiometer channels and the 3 magnetic field components $B_l(t)$ (l =1,2,3) are measured as a function of time as the sensor moves past the targets. The time derivatives of field $dB_l(t)/dt$ are calculated from the $B_l(t)$ for eddy-current compensation. The object in this problem is to extract the dipole signals $G_{ij}^{(k)}(t)$ (k=1,2,...,n) for the unknown number n of dipole targets and then to determine the magnetic moments and the positions of the n targets.

The model describes the signals S_i in the 5 gradiometer channels by the equations

$$S_i(t) = \sum_{j=1}^{5} c_{ij}\left\{\sum_{k=1}^{n} G_{ij}^{(k)}(t)\right\} + \alpha_i + \sum_{l=1}^{3} \beta_{il}B_l(t) + \sum_{l=1}^{3} \gamma_{il}\dot{B}_l(t) + v_i(t) \quad (5.1)$$

where c is the calibration matrix for the gradiometer, $v_i(t)$ and $\alpha_i(t)$ are uncompensated noise (setting the noise floor of the channel compensation parameters per channel) and channel biases, respectively, and β_{il} and γ_{il} are the balance and eddy-current vectors for channel i. An iterative analysis first estimates the α's, β's and γ's (a total of 35 parameters) and then executes a gradient search for the location of the single target that best fits the residual signal. The α's, β's and γ's, and target location and moment are then optimized, and the procedure is repeated for a second target. Targets continue to be added until finally no target can be found whose signal contributes substantially to a reduction in total signal power, at which point the algorithm terminates.

5.1.4. Results from System Sea Tests

In order to demonstrate the effectiveness of the MADOM concept, tests were conducted to measure the probability of detection and the probability of classification of buried mines in a drill mine field and to measure the incidence of magnetic and acoustic bottom debris (clutter). The tests showed that the MADOM concept of fusing magnetic and acoustic sensor data, does provide a powerful approach to buried mine detection and classification, and to clutter rejection in the shallow-water environment. It is expected that this approach will be even more important in the very shallow water environment where sonar performance is likely to deteriorate.

With regard to the magnetic sensor capability demonstrated, the SGMS did provide long-range detection compared to conventional, non-superconducting magnetic sensors. The sensor is capable of detecting 100% fully-buried mines without diminished performance. As a result of the multi-channel approach, the sensor also provided accurate moment and localization determination. The magnetic detection and classification signal processing was demonstrated to be effective, providing a fully-automated, real-time capability. This high-tech sensor did provide reliable, rugged performance in undersea tows conducted over a period of seven years. In these tests, the SGMS was operated adjacent to the SAS in the same tow vehicle without a loss in performance.

5.1.5. Clutter Rejection

As indicated in the preceding section, sensor fusion is essential for the reduction of false alarms. The sea bottom, especially in shallow water environments, may be cluttered with debris that might be falsely classified as minelike with a single sensor. Figure 5.5 depicts the environment in which a mine reconnaissance or hunting system may operate, displaying bottom mines and bottom debris. In coastal regions, the density of such clutter may lead to a high false-alarm rate using conventional imaging sonar approaches alone.

For effective clutter rejection, it is very desirable to use distinctly different sensor approaches. The application of two or more collocated sensors operating simultaneously has the potential to reduce false alarms and provide robust detection in a wide variety of background conditions. Use of two sonars operating at different frequencies offers one such approach. A better approach would be to use sonars operating in different modes, e.g., imaging and resonant backscatter modes. Better yet is the use of sensors which sense totally different target characteristics. For mine reconnaissance and hunting, the combination of magnetic sensors with sonars provides such an alternative.

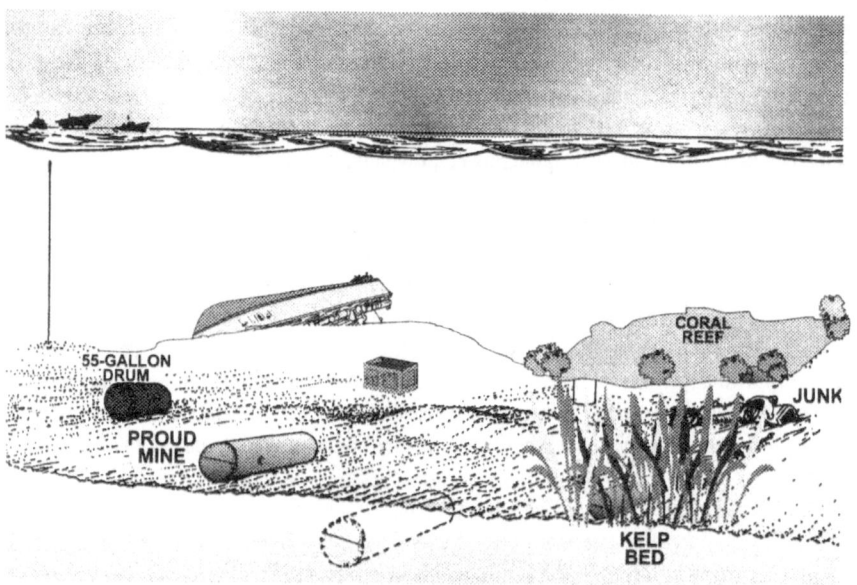

Figure 5.5. A typical environment encountered in a shallow water bottom mine-hunting operation. The mine-hunting system must distinguish between mines, debris and bottom features and must provide bottom penetration.

A magnetic sensor will identify a target as mine-like if the magnitude of the target's moment falls within a certain range of acceptable values. A target's magnetic moment is principally a function of the amount of steel in the mine. Hence, harbor debris such as 55-gallon drums and sunken automobiles will be classified not to be mine-like if their magnetic moments fall outside the acceptable range. However, other harbor debris such as truck axles, engine blocks and refrigerators will be magnetically indistinguishable from a ferrous mine if they have magnetic moments comparable to that of a ferrous mine. Conversely, the acoustic sensors rely on the shape of an acoustic target in order to classify that target as mine-like or not. However, a mine's shape may be acoustically indistinguishable from the shapes of non-mine objects. Thus, the acoustic sensors would classify 55-gallon oil drums as mine-like, but not truck axles.

Fusion of the magnetic and acoustic data provides the means to distinguish both a 55-gallon oil drum and a truck axle from mines. From this line of reasoning, one can see that the fusion of magnetic and acoustic data may provide a powerful tool for discriminating between mines and clutter. This premise has been successfully validated during the MADOM sea testing, which showed that more than 90% of the acoustically mine-like clutter was not magnetically mine-like.

5.2. THE MOBILE UNDERWATER SURVEY SYSTEM

Of the 900 Formerly Used Defense Sites targeted for cleanup by the Army Corp of Engineers, about 50 are under water. Many of the sites with beach or near-shore acreage are of significant commercial, recreational and/or cultural importance to their respective communities, and the risk to the public is increasing each year. At the present time there are no established approaches viable to locate such ordnance in coastal regions.

As described in Section 2, a project recently has been initiated under the sponsorship of the U.S. Strategic Environmental Research and Development Program (SERDP) to develop and evaluate a Mobile Underwater Debris Survey System (MUDSS) for locating UneXploded Ordnance (UXO) in coastal regions at closing U.S. military installations. The effort involves a collaboration between CSS and the Jet Propulsion Laboratory (JPL). This system will extend the sensor fusion approach developed in MADOM to incorporate not only magnetic and acoustic sensors, but also electro-optic and chemical sensors. The system initially will incorporate the SGMS with replacement by an advanced superconducting sensor projected for subsequent testing. Figure 5.6 depicts the first version of MUDSS which was fielded in the summer of 1995.

A successful demonstration will prove the concept of a trailerable, low-maintenance, catamaran-based system capable of finding and accurately mapping the locations of UXO ranging from small shells to large bombs in water depths of from 4 to approximately 100 feet. Besides UXO, other similar underwater debris (such as oil drums or chemical, biological or radioactive waste) can be surveyed using the MUDSS technologies. MUDSS technologies also may be applicable to land-based cleanup efforts.

Using the experience that JPL has developed for 3-D visualization in conjunction with space exploration, JPL will develop and demonstrate a 3-D visualization tool to assist in interpreting the

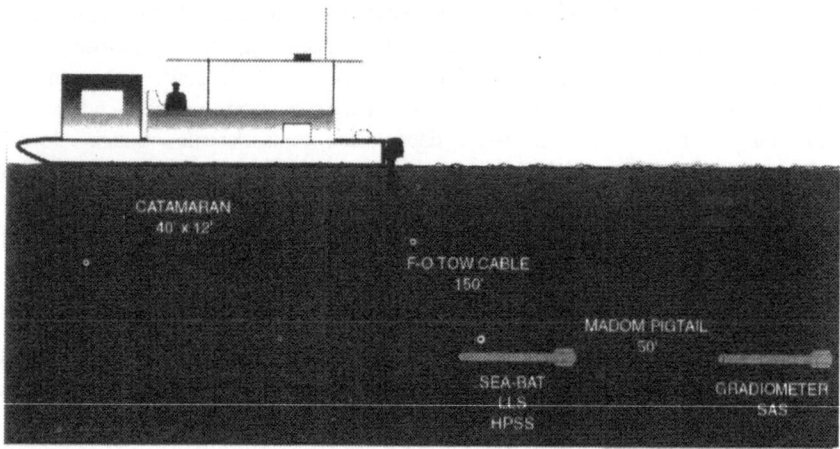

Figure 5.6. An artist's conception of the first experimental configuration of the Mobile Underwater Debris Survey System (MUDSS) to be demonstrated in 1995. This system features a trailerable, low-maintenance, catamaran-based system. The tow-body configuration is similar to the MADOM configuration. The sensor suite in MUDSS enhances capabilities in MADOM. The suite includes a forward looking sonar (the Sea Bat), a Laser Line Scan Electro-optic Sensor (LLS), and the High Performance Side Scan Sonar (HPSS) in a forward tow-body/dead-weight depressor. The suite also includes the Synthetic Aperature Sonar (SAS) and the SGMS in a trailing tow body.

results and to help if divers are called upon to make a first-hand inspection. The visualization will show the lay of the underwater terrain, the positions of the targets and (through an icon-driven selection) the targets' optical or sonar images and their magnetic moments. This will enhance greatly the imaging and data fusion approaches developed in MADOM.

6. Recent Developments and Future Trends

The search for enhanced performance continues. We describe below the development of low-T_c thin-film gradiometers and superior dewar approaches to provide greater sensitivity than afforded by the earlier SGMS technology. The advent of high-T_c sensors will lead to smaller sensor packaging, reduced cryogenics support, significant reduction in logistics and some reduction in operational costs which will expand the use of superconducting sensors. These enhancements will be critical to obtain widespread acceptance of this technology for mobile surveys and searches.

6.1.   DEVELOPMENT OF A HIGH PERFORMANCE THIN-FILM GRADIOMETER

We believe that the current technology, represented by the SGMS sensor, is reaching its performance limit.   This technology is largely characterized by the use of bulk and wire superconducting components.   More advanced approaches are required to obtain greater sensitivity in motion.   Following a technology evaluation and concept study to establish a preliminary concept design and recommendations for development of an advanced all thin-film (niobium) sensor prototype, CSS awarded a contract in 1991 to Loral [9] with IBM Research [10] as a subcontractor to develop an all thin-film gradiometer sensor probe assembly.   An artist's conception of this sensor concept, referred to as the Thin-Film Gradiometer (TFG), is displayed in Figure 6.1.

Major advances being explored in the TFG development include: 5" micron-precision planarized niobium processing, wide-band flux-lock-loop electronics with a modulation frequency of 16 MHz, dynamic balancing techniques and detailed noise modeling.   A test article has been fabricated to validate the thin-film sensor concept.   This test article has provided *the successful demonstration of a high sensitivity, all thin-film sensor to operate totally unshielded in a harsh magnetic and electromagnetic environment.*   Based on the evaluation of that test article, a final sensor prototype is currently being fabricated in order to demonstrate a significant increase in performance over the older SGMS technology.

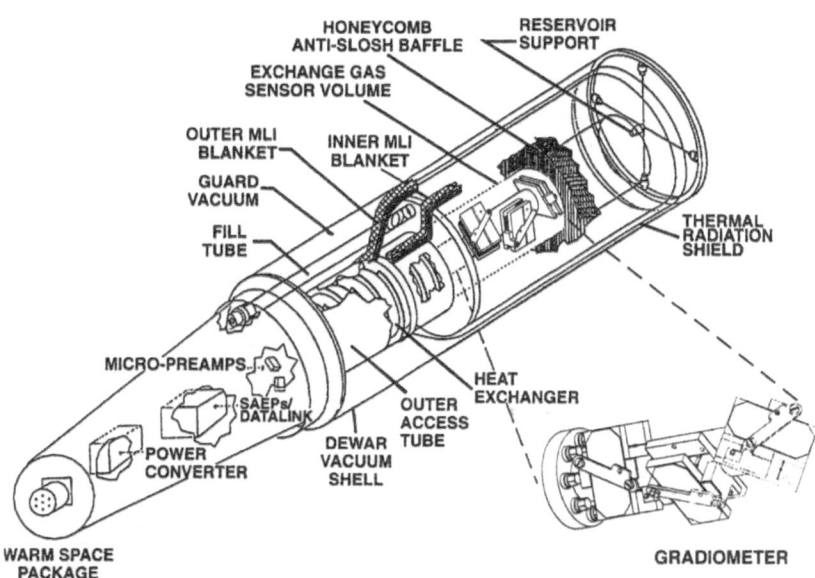

Figure 6.1. An artist's conception of the Thin-Film Gradiometer.

CSS, pursuing development of an advanced dewar concept to remove the dewar as a limiting noise source for gradiometer performance, awarded a contract to Ball Aerospace [21] in 1989 for an advanced dewar. A flexible design approach supported by detailed thermal, mechanical and field calculations, and stringent thermal and magnetic budgets, was established to assure this goal. Fabrication and acceptance testing of the dewar prototype were completed in 1992. *We believe that this dewar provides the most stable passive cooling of a large volume that is available today, providing temperature fluctuations on the order of 1 $\mu K/\sqrt{Hz}$ at 0.1 Hz in motions typical of tow operations, a factor three orders of magnitude better than for the SGMS dewar.*

6.2. OPPORTUNITIES WITH HIGH-T_c TECHNOLOGY

High-T_c technology has matured (since its inception in 1986) to the extent that practical magnetic sensors are now feasible [13, 22, 23]. Several distinct classes of materials, most notably yttrium-barium-copper oxide ($YBa_2Cu_3O_{7-x}$) are currently being explored for commercial applications of magnetic sensors. There has been excellent progress by numerous groups in the development of high-T_c Josephson junctions, SQUIDs and magnetic sensing circuits using $YBa_2Cu_3O_{7-x}$. Since 1993, there have been a number of results reported on magnetometer prototypes with white noise better than 200 fT/$\sqrt{Hz}$, including recent reports of 39 and 20 fT/$\sqrt{Hz}$ at 1 Hertz [24, 25]. As a result of nitrogen cooling, the development of sensors utilizing the high-T_c materials provides an opportunity for significant size reduction, an ease of maintainability and convenience in comparison to the low-T_c technology using helium cooling, factors critical to gain widespread acceptance of the superconducting technology over other magnetic sensor approaches.

6.2.1. Benefits from Nitrogen Cooling for Naval Operations

The naval application of sea-mine countermeasures provides a specific interesting perspective to assess the benefits which can be obtained from nitrogen cooling. The major enhancement offered by the new technology arises from the opportunity to replace the helium dewars with nitrogen cooling units, ranging from dewars to active cryocoolers. A high-T_c sensor with liquid nitrogen refrigeration offers the opportunity to implement into small underwater vehicles, the capabilities already established in the MADOM Project using the low-T_c technology (Figure 6.2).

Based on an assessment of refrigeration technology conducted for the CSS by Ball Aerospace [21], liquid nitrogen dewars represent the best choice for near-term development. Results of the concept analysis indicate that an open (vented) liquid nitrogen dewar with dimensions of 17" in diameter and 30" in length could be built which would satisfy a sensitivity goal of 300fT/m-$\sqrt{Hz}$ at 0.1 Hz. and which would have a hold time of approximately 33 days (Figure 6.3).

A design in which the dewar is an integral part of the body section effectively expands the dewar's diameter to 21", thus providing a substantial increase in hold time. Such long hold-time nitrogen dewars would eliminate the need for cryogen support during active operations for a number of mission scenarios with cryogen replenishment performed only during general ship resupply at

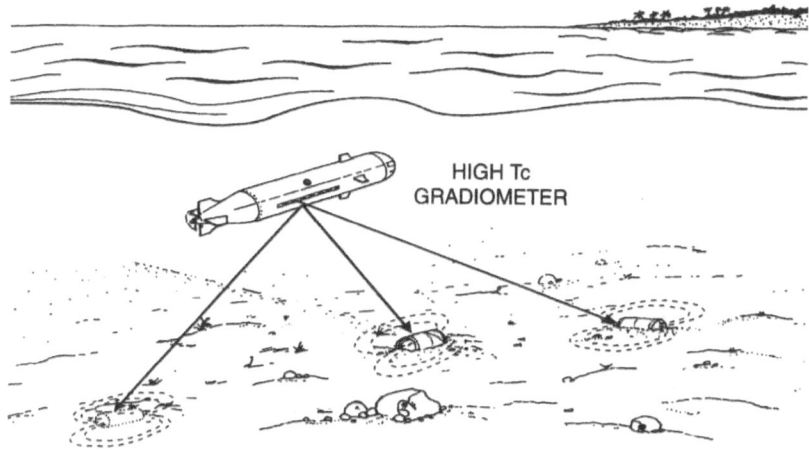

Figure 6.2. A multi-sensor autonomous mine-hunting system concept including a magnetic detection/classification capability. Integration of a long range magnetic sensor into an autonomous underwater vehicle is now possible with the advent of high-T_c sensors operating in compact nitrogen dewars less than one half the volume of conventional helium-cooled superconducting gradiometers.

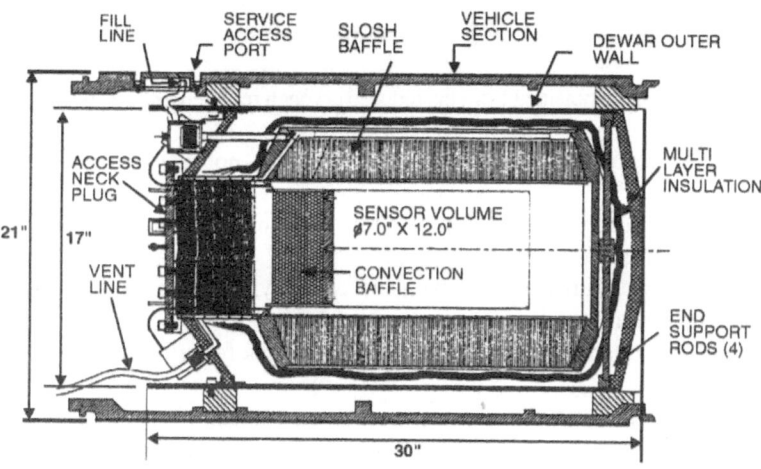

Figure 6.3. Liquid nitrogen dewar concept to provide the volume to house 5-channel gradiometers with a sensitivity design goal of 300 fT/m-Hz$^{1/2}$ (using the Three Sensor Gradiometer (TSG) concept described in Section 6.3) and to provide a hold time approximated at 33 days. Variations in the dewar dimensions could be considered to accommodate requirements in size and hold time for different systems and operational scenarios.

dock or via a ship tender. Reductions in diameter and length are possible for operation in smaller vehicles. In particular, the length of such a sensor could be reduced to 18" without a loss of sensitivity, although there will obviously be a reduction in hold time nominally to 10 days. As the performance of the high-T_c devices improves, the dimensions of the sensor could be further reduced maintaining the same gradient sensitivity. Such size reductions are possible within the current technology if reduced sensitivity (still better than that of existing conventional nonsuperconducting technologies) can be tolerated.

If liquid nitrogen dewars are utilized in place of liquid helium dewars, increases in hold times by a factor of 5 or even 10 are projected, reducing the quantity of cryogen that must be carried on ship (in comparison to liquid helium) and possibly eliminating the need to carry cryogen onboard mine countermeasure ships. For scenarios in which dewar hold time exceeds the duration of on-site operation without resupply (say less than 30 days), the mission could be conducted without replenishment during the mission, thus eliminating the requirement for onboard ship cryogen support. Cryogen replenishment could then be conducted during resupply from a ship tender or at port.

For scenarios in which the on-site operation time without resupply exceeds the hold time of the nitrogen dewar, and the ship must store cryogen, nitrogen cooling still has an advantage over helium cooling in that nitrogen refills would be performed only on the order of every 20 to 30 days instead of the 2- to 3-day refilling required in the case of liquid helium. As a result of the smaller quantities of cryogen required, the nitrogen storage dewars can be stored beneath the underwater vehicle without adding any new footprint onboard the ship. In comparison, the large sizes of the helium storage dewars required for helium replenishment will likely add a substantial footprint on the ship deck on the order of 70 square feet for one mission scenario analyzed.

The benefits which can be obtained from the reduced cryogen requirements include:

1. a significant reduction in down time, or possibly no sensor down time, during operations;
2. affiliated reductions or elimination in labor requirements for cryogen support during active operations;
3. greatly reduced failures in the cryogenic circuitry or in the dewar (such failures typically occur during cryogen recycling); and
4. the elimination of an additional footprint on ship deck for cryogen storage.

In addition, the use of liquid nitrogen in place of liquid helium significantly reduces supply logistics as a result of the wide availability of nitrogen on the market at domestic and most foreign ports and the availability of a large number of liquifiers in the U.S. Fleet, with at least 54 units identified. Significant cost savings are expected from reduced costs for cryogen supply. For commercial purchases, cryogen costs per month per unit are estimated at $25 for a liquid nitrogen dewar compared to $1000 for a liquid helium dewar.

6.2.2. Advanced Refrigeration Concepts for Mobile Operations

In the future, a wide range of new refrigeration approaches may become available to offer more attractive concepts for mobile operation. Alternate cryogens (notably hydrogen with an atmospheric boiling point of 20 K, neon with a boiling point of 27 K, argon with a boiling point of 87 K, and oxygen with a boiling point of 90 K) have been considered. However, with the successful demonstration of high-T_c magnetic sensors in liquid nitrogen (with a boiling point of 77 K) by a large number of groups, liquid nitrogen is the preferred candidate because of its widespread availability, ease of handling, low cost and the relative simplicity of nitrogen refrigeration approaches.

Vented solid nitrogen dewars operating at 63 K offer some benefit over liquid nitrogen units as a result of reduced thermal noise at the lower operating temperature and by the elimination of liquid slosh in moving applications. Sealed (nonventing) solid nitrogen units operating at 63 K offer an interesting approach useful for some applications where cryogen storage is a significant burden. Such units would not require nitrogen replenishment (with recooling using an auxiliary active cryocooler when the sensor is not in operation). Although this approach would permit only relatively short operation periods compared to other approaches, say on the order of 14 hours before cryogen refreezing is required, this constraint may be acceptable if cryogen storage is not feasible.

The use of an active cryocooler approach in place of the passive cooling approaches described above has the appeal of totally eliminating any cryogen support issues. A wide range of approaches including Joule-Thomson, pulse-tube, Stirling-cycle, Gifford-McMahon, magnetic and thermo-electric cryocoolers have been developed for cryogenic electronic applications including microwave circuitry, infrared detectors or digital circuitry. None of these electronic applications require the extreme noise-free environment at low frequencies necessary for magnetic sensors. These cryocoolers will typically introduce some combination of thermal, vibrational, magnetic or electrical noise. The majority of these refrigerator types have substantial magnetic signatures at the sensor volume arising from their electrical power requirements. The Joule-Thomson approach, which is based on the cooling effect of an expanding gas, probably provides the best opportunity for an actively-cooled magnetic sensor. As noted in Reference 26 (in this volume), pulse-tube coolers also can be considered. Both of these cooler types avoid requirements for electrical pumps in proximity to the sensor; however, issues of vibrational and thermal stability remain. In any case, substantial developmental investment will be required before the cryocooler technologies will be satisfactory for high sensitivity, low frequency magnetic sensor applications.

6.3. HIGH-T_c SENSOR DEVELOPMENT

A project to develop a compact high-T_c superconducting gradiometer for mobile deployment was initiated by the Office of Naval Research in 1993. This project includes the efforts of the CSS, the Naval Research Laboratory, Loral [9], IBM Research [10], Quantum Magnetics [11], Ball Aerospace [21] and Conductus [27].

A technology evaluation was initiated in FY 1993 and completed in FY 1994 to assess current industrial capabilities and preferred design concepts for a field-deployable High-T_c Superconducting Gradiometer (HTSG). This included analysis, component fabrication and experimentation. Based on the conclusions from this evaluation, a HTSG preliminary design concept has been established and its performance in motion has been modelled. The HTSG is expected to surpass significantly the motion performance of any conventional non-superconducting magnetic sensor technology and is projected to have a sensitivity better than that of the low-T_c SGMS.

This sensor concept features a novel gradiometer circuit design in order to circumvent current limitations in high-T_c fabrication technology (available in 1995) for both wires and for thin films. In order to manage linearity and dynamic range requirements in low-T_c niobium superconducting gradiometer circuits, signal subtraction is generally performed prior to the SQUID gain stage. This subtraction approach has been implemented in the niobium technology by multipli-connected, counterwound wire gradiometer loops or counterwound thin-film loops requiring multilayer crossovers. Such a thin-film concept is depicted in Figure 6.4.a. In the high-T_c setting, neither wire-wound loop structures with resistanceless joints nor multilayer thin-film structures on 2- to 3-inch scales have been manufactured to date that will provide the required performance.

In order to circumvent these fabrication limitations, a novel approach patented by IBM Research has been incorporated into the HTSG concept [28]. This approach is based on the concept that a gradiometer can be configured using two independent magnetometers with signal subtraction performed at the output of the two magnetometers, following the approach using fluxgate magnetometers pursued by CSS in the early 1970's. Good stationary performance was obtained at that time, but there was insufficient dynamic range in the processing of the differential signals for operation in motion. In order to circumvent the limitation in dynamic range, a third sensor (either a high-T_c SQUID or a fluxgate magnetometer) is used for common-mode rejection, feeding back a signal to the two primary SQUID magnetometers in order to null out the ambient background field (Figure 6.4.b). This Three-Sensor Gradiometer (TSG) approach circumvents the requirement for refined high-T_c multilayer processing on 2- to 3-inch scales (which has not been demonstrated to date) providing long baselines with normal metal wire connecting the independent SQUID magnetometers.

This three-sensor concept recently has been pursued successfully using room-temperature fluxgate magnetometers (in place of the SQUID magnetometers) demonstrating a sensitivity better than 0.3 nT/m at 0.1 Hz in motion [29]. Although this fluxgate technology will not provide the long detection ranges afforded by the superconducting technology, it does provide the same detection/classification capability for short-range applications where factors such as cost, portability and convenience outweigh the need for greater sensitivity.

Further improvement in performance is expected using the high-T_c SQUID magnetometers in place of fluxgate magnetometers as a result of the greater sensitivity afforded by the high-T_c sensors. A laboratory test article is being developed and will be evaluated late in 1996 to validate the high-T_c sensor concept. An artist's conception of the test article is given in Figure 6.5.

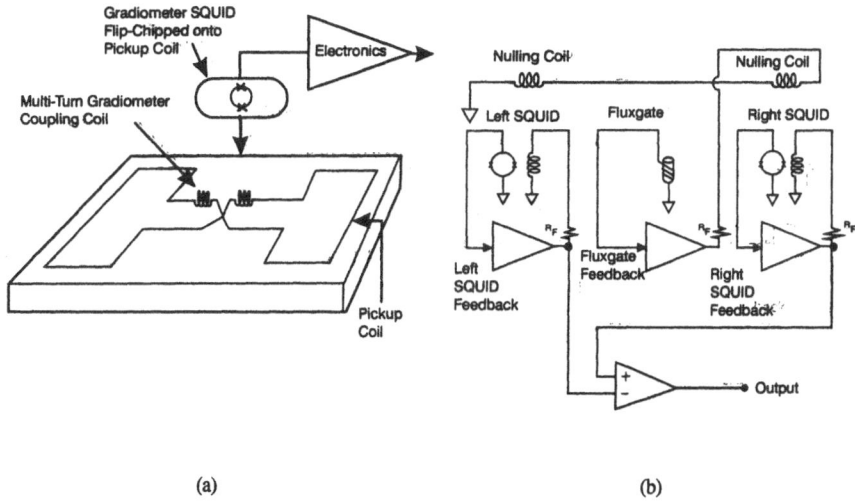

Figure 6.4. Two gradiometer concepts using thin-film SQUID circuitry: (a) an intrinsic gradiometer circuit with counterwound loop structure (requiring multi-layer processing) for signal subtraction prior to the SQUID stage and (b) the three-sensor gradiometer (TSG) using two SQUID-based magnetometers in differential modes with output signal subtraction and a third magnetometer (in this case a fluxgate) to null out the mean field on the two SQUID magnetometers (in order to extend the range of the gradiometer).

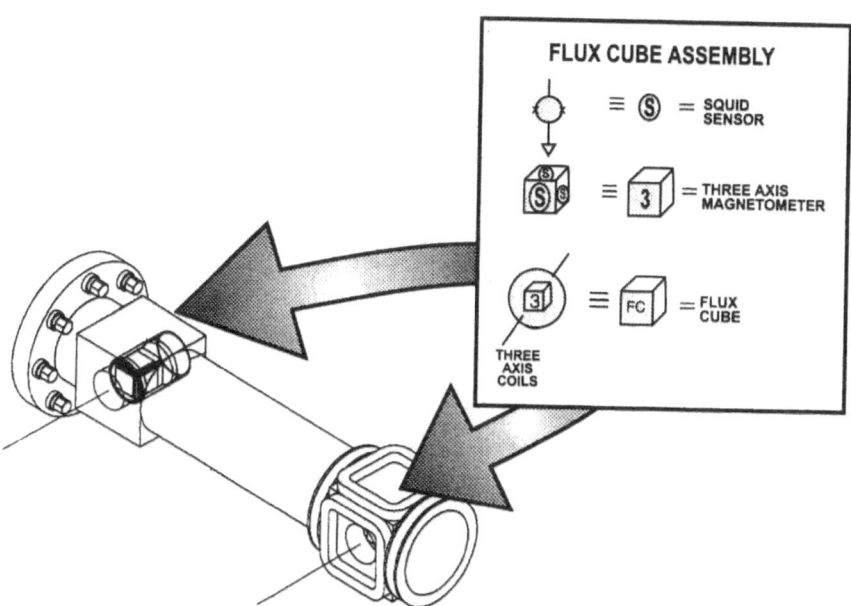

Figure 6.5. An artist's conception of the test article to evaluate high-T_c TSG circuits.

7. Conclusions and Summary

Magnetic sensors provide one tool potentially valuable for mobile search operations and surveys for targets with a significant magnetic signature. Superconducting SQUID-based sensors theoretically represent the most sensitive of known magnetic sensors. SQUID-based magnetometers have been demonstrated with sensitivities on the order of 1 fT/$\sqrt{Hz}$ at frequencies down to 0.1 Hz, while fluxgate and total-field magnetometers are demonstrating sensitivities down to 1-10 pT/$\sqrt{Hz}$ at 0.1 Hz.

The U.S. Navy has developed the Superconducting Gradiometer/Magnetometer Sensor, a superconducting gradiometer that has provided long-range detection compared to conventional, non-superconducting magnetic sensors. This sensor has been utilized to demonstrate a capability for buried mine detection. As a result of the multi-channel approach, the sensor provides an accurate localization capability and multi-target discrimination. The magnetic detection and classification signal processing developed in conjunction with the sensor has proven to be effective, providing a fully automated, real time capability. This high-tech sensor did provide reliable, rugged performance in undersea tows conducted over a period of seven years. The sensor has been operated in the same tow vehicle adjacent to a sonar without a loss in performance.

This technology is available off-the-shelf to provide the greatest capability for magnetic detection and localization ever demonstrated. Work continues to demonstrate its utility in environmental cleanup. A large range of other applications involving the detection of buried or underwater ferrous targets may be pursued using this technology: geophysical survey, civil engineering, the detection of hidden facilities and tunnels, antisubmarine warfare, police investigations, and archeological/treasure hunting. It is expected that the superconducting technology for high-performance sensors will be utilized by a limited number of niche companies or governmental agencies specializing in surveys. Cost and logistics/support requirements will likely limit the numbers of units fielded. In particular, the supply of liquid helium will make it difficult to support operations in many remote locations of the world.

In general, this technology should only be considered when all simpler, less costly approaches are inadequate. It is important to note that the performance of competing technologies, specifically fluxgate and total-field sensors, continues to improve. However, in the quest for high sensitivity, superconducting magnetic sensors are projected to provide greater sensitivity than any other known technology and the theoretical limits are being approached, at least, for laboratory devices. The ability to fold this projected greater performance into practical devices working outside the laboratory environment is promising, based on the previous success with the U.S. Navy's SGMS.

The ideal situation for operating sensitive magnetic sensors onboard moving platforms is to provide an environment in proximity to the sensor that is ferromagnetically clean, free of other metallic eddy-current sources, and is distant from electrical and electronic equipment, power plants, electrical distribution paths and propulsion systems. In the MADOM program, the SGMS was operated successfully in an underwater fiberglass tow body far removed from the towing craft. In order to gain widespread acceptance of magnetics in general, especially for a high-performance sensor, it is very desirable to provide system concepts in which the magnetic sensor can be operated in user convenient configurations, preferably onboard the primary vehicle, be it a tractor, plane or boat. Hardware and signal-processing approaches to provide such versatile performance likely represents the greatest challenge in order to gain widespread acceptance of the technology.

It is only a matter of time before high-T_c devices will be available with performance comparable to that which has already been demonstrated with low-T_c devices. The benefits in reduced size, longer hold times, and reduced logistics and support make the high-T_c approach attractive compared to its low-T_c counterpart. The localization capability afforded by 5-channel gradiometers (and not previously afforded by conventional magnetic sensors) is expected to add impetus to the acceptance of magnetics for a number of applications. The development of a 5-channel fluxgate gradiometer is currently in progress. As a result of a much lower cost and no special support requirements, such sensors will likely be sold in larger numbers for short-range applications. Their wider usage would then enhance the opportunity to display the utility of the greater classification and localization capability afforded by the 5-channel approach. With greater acceptance of the 5-channel approach, end users will likely want to obtain sensors with greater sensitivity. It is likely that the eventual development of high-performance, reliable high-T_c gradiometers will work synergistically with an increased acceptance of the more powerful magnetic signal-processing approaches. Hopefully, the time scales for this supply and demand will be commensurate and so expedite sensor developmental efforts.

Acknowledgments

Among the many individuals who supported these developments at the CSS, the authors would like to recognize the contributions from G. I. Allen, P.J. Carroll,. D.W. Colberg, M.C. Froelich, L.H. Fry, M. Gershenson, J.F. McCormick, T.B. Pride, J.W. Purpura, T.W. Rackers, L. Vaizer, D.L. Whatley, Jr. and R.F. Wiegert. Finally the CSS would like to acknowledge the contributions from the Naval Research Laboratory and from industry by Unisys, Loral, IBM, Ball Aerospace, Quantum Magnetics, and Conductus including the more recent individual efforts from R.H. Koch, J.H. Eraker, J. Rozen, R.J. Soulen, Jr., W.J. Gallagher, J.Z. Sun, B.D. Thorson, M.J. Burns, R. Cantor and M.S. Colclough.

568

Références

1. KeKelis, G. J. and Wynn, W. M. (1994) Magnetic Field Gradiometers for Ordnance Detection, in the Proceedings for the Unexploded Ordnance Detection and Range-Remediation Conference, Walcoff & Associates, Inc.
2. Jet Propulsion Laboratory (1994) Sensor Technology Assessment for Ordnance and Explosive Waste Detection and Location, JPL Report D-11367 Revision A.
3. Lathrop, J.D. (1995) High Area Rate Reconnaissance (HARR) and Mine Reconnaissance/Hunter Exploratory Development Programs, in Dubey, A.C., Cindrich, I., Ralston, M., and Rigano K., (eds.) Detection Technologies for Mines and Minelike Targets, Proc. SPIE 2496, pp. 350-356.
4. Vrba, J.D. (1996) SQUID Gradiometers in Real Environments, this volume.
5. Wynn, W.M., Frahm, C.P., Carroll, P.J., Clark, H., Wellhoner, J., and Wynn, M. J. (1975) Advanced Superconducting Gradiometer/Magnetometer Arrays and a Novel Signal Processing Technique, IEEE Trans. on Magn., Vol. MAG-11, pp. 701-707.
6. Wynn, W.M. (1995) Magnetic Dipole Localization using the Gradient Rate Tensor Measured by a Five-Axis Gradiometer with Known Velocity, in Dubey, A.C., Cindrich, I., Ralston, M. and Rigans, K., (eds.) Detection Technologies for Mines and Minelike Targets, the International Society for Optical Engineering.
7. Superconducting Technology Division, United Scientific Corporation, Santa Clara, CA 91750.
8. The Unisys Corporation, Defense Products Group, 3333 Pilot Knob Road, St. Paul, MN 55122.
9. The Loral Federal Systems Company , 9500 Godwin Drive, Manassas, VA 22110.
10. The IBM Corporation Research Division, T. J. Watson Research Center, P.O. Box 218, Yorktown Heights, NY 10598.
11. Quantum Magnetics, 11578 Sorrento Valley Road, San Diego, CA 92121.
12. Fleming, D.L., Gershenson, M., Hastings, R., Sauter, G.F., and Sweeny, M.F. (1985) Hybrid dc SQUIDs containing all Refractory Thin Film Josephson Junctions, IEEE Trans. on Magn., Vol. MAG-21, No. 2, pp. 658-659.
13. Clarke, J. (1996) SQUID Fundamentals, this volume.
14. Matsushita, S. and Campbell, W.H. (1967) Introduction to Geomagnetic Phenomena, Academic Press (2 volumes).
15. Cladis, J.B., Davidson, G.T., and Newkirk, L.L. (1971) The Trapped Radiation Handbook, DNA 2542H.
16. Campbell, W.H., (1966) A Review of the Equatorial Studies of Rapid Fluctuation in the Earth's Magnetic Field, Ann. Geophys. 22, p. 492.
17. McLeod, M.G. (1992) A Predicted Geomagnetic Field Model for Epoch 1990.0, NRL/FR/7442-92-9414, Naval Research Laboratory, Stennis Space Center, MS 39529-5004.
18. Clem, T.R., Goldstein, M.J., Purpura, J.W., Allen, L.H., Claassen, J.H., Gubser, D.U., and Wolf, S.A. (1989) Investigation of Noise Sources in SQUID Electronics, IEEE Trans. on Magn., Vol. MAG-25, No. 2, pp. 1012-1017.
19. Wellstood, F., Heiden, C., and Clarke, J. (1984) Integrated dc SQUID Magnetometer with a High Slew Rate, Rev. Sci. Instrum 55(6), pp. 952-957.
20. Koch, R.H., Foglietti, V., Rozenk, J.R., Stawiasz, K.G., Ketchen, M.B., Lathrop, D.K., Sun, J.Z., and Gallagher, W.J. (1994) Effects of Radio Frequency Radiation on the dc SQUID, Appl. Phys. Lett. 65(1), pp. 100-102.
21. Ball Aerospace Systems Division, P.O. Box 1062, Boulder, CO 80306-1062.
22. Clarke J., and Koch, R.H. (1988) The Impact of High-Temperature Superconductivity on SQUID Magnetometers, Science, Vol. 242, pp. 217-223.
23. Koch, R., (1990) SQUIDs made from High Temperature Superconductors, Solid State Technology, pp. 255-260.
24. Koelle, D., Miklich, E., Dantsker, A.H., Ludwig, F., Nemeth, D.T., Clarke, J., Ruby W., and Char, K. (1993) High Performance dc SQUID Magnetometers with Single Layer $YBa_2Cu_3O_{7-x}$ Flux Transformers, Appl. Phys. Lett.. 63 (26), pp. 3630-3632 .
25. Cantor, R., Lee, L.P., Teepe, M., Vinetskiy, V., and Longo, J. Low Noise, Single-Layer $YBa_2Cu_3O_{7-x}$ DC SQUID Magnetometers at 77K, (1995) IEEE Trans. on Applied Superconductivity, Vol. 5, No. 2, pp. 2927-2930.
26. Heiden, C. (1996) Pulse Tube Refrigerators, this volume.
27. Conductus, Inc., 969 West Maude Ave., Sunnyvale, CA 84086.
28. Koch, R.H. (1992) Gradiometer Having a Magnetometer which Cancels Background Magnetic Field from other Magnetometers, U.S. Patent No. 5,122,744.
29. Allen, G.I., Koch, R.H., and Keefe, G. (1995) Unique Man-Portable 5 Element Fluxgate Gradiometer System, in A. C. Dubey, I. Cindrich, M. Ralston, and K. Rigano (eds.) Detection Technologies for Mines and Minelike Targets, the International Society for Optical Engineering, Proc. SPIE 2496, pp. 384-395.

SUPERCONDUCTING ACCELEROMETERS, GRAVITATIONAL-WAVE TRANSDUCERS, AND GRAVITY GRADIOMETERS

H.J. Paik
Department of Physics, University of Maryland
College Park, MD 20742, U.S.A.

Sensitive accelerometers and gravity gradiometers have been developed for precision tests of the laws of gravity, for the search for gravitational waves and new weakly-interacting particles, and for applications in inertial navigation and gravity surveying. The Meissner effect, flux quantization and SQUID magnetometers can all be utilized, along with the enhanced mechanical stability of materials at cryogenic temperatures, to obtain gravitational-wave transducers with sensitivity approaching the quantum limit and to construct gravity gradiometers with orders of magnitude improvement in sensitivity. The superconducting transducers and gravity gradiometers have achieved a displacement sensitivity for the antenna motion and a differential acceleration sensitivity of 10^{-18} m Hz$^{-1/2}$ and 10^{-12} m s^{-2} Hz$^{-1/2}$, respectively. However, the sensitivities of these devices must be improved by two to three orders of magnitude in order to detect gravitational waves coming from nearby galaxies and to be able to perform desired tests of the Equivalence Principle and the inverse-square law in space. These, in turn, require improvement of the intrinsic noise of the SQUID. We review the design, operating principles, and performance of the various superconducting inertial instruments. We also discuss application of the devices in physics, geophysics, and engineering.

1. Introduction

One of the challenges that researchers faced at Stanford University in 1970 was to develop a motion detector which could detect a fraction of the Brownian motion of a massive ($\simeq 5$ ton) aluminum cylinder cooled to 3 mK [1]. To detect expected gravitational waves of extragalactic origin, the detector was required to resolve an amplitude smaller than 10^{-20} m with an integration time of the order of 0.1 s. This led to the development of sensitive superconducting transducers based on emerging SQUID technology in the early 1970's [2, 3].

H. Weinstock (ed.), SQUID Sensors: Fundamentals, Fabrication and Applications, 569–598.
© *1996 Kluwer Academic Publishers.*

In 1975, Paik and coworkers applied basic superconducting accelerometer technology to construct the first superconducting gravity gradiometer (SGG) at Stanford [4]. This led to the development of more sophisticated SGGs at the University of Maryland in the 1980's with support from NASA and the U.S. Air Force [5, 6]. The aim of the NASA project was to develop a highly sensitive gravity gradiometer system which could be flown in a low-altitude near-polar orbit to obtain a high-resolution gravity map of the Earth [7].

Three models of SGGs with increasing complexity and sensitivity have been developed at Maryland. These instruments have been used to perform null tests of Newton's inverse-square law of gravitation in the laboratory. The Model II SGG has reached an operating sensitivity of 0.02 E Hz$^{-1/2}$ (1 E $\equiv$ 10^{-9} s^{-2}), three orders of magnitude improvement over the sensitivity achieved by room-temperature gravity gradiometers. The demanding sensitivity and platform control requirements of the SGG gravity mapping mission have prompted further innovations in technology: this has led to the development of the Model III SGG with an intrinsic sensitivity of 10^{-4} E Hz$^{-1/2}$, and a superconducting six-axis accelerometer (SSA) which detects the six degrees of freedom platform acceleration simultaneously [8].

With sensitivity improved by orders of magnitude over conventional technologies, the SGG finds useful application in geophysical measurements, mineral prospecting and inertial navigation. The great potential for practical application has inspired many groups around the world to develop similar gravity gradiometers. Variations of the Maryland SGG are now being developed by groups at the University of Western Australia, the University of Strathclyde, Scotland, the Institute for Low Temperature Physics, Ukraine and the Sandia National Laboratories, New Mexico. Recently, a consortium led by Oxford Instruments in England has started development of a spaceborne SGG with ESA (European Space Agency) support. The possibilility of constructing a high-T$_c$ SGG is also being investigated by several groups.

The exquisite sensitivities of SGGs will enable important fundamental physics experiments in space. A precision test of the inverse-square law at 100 km [9] and detection of the gravitomagnetic field of the Earth [10] have been proposed. Employing similar devices, the STEP (Satellite Test of the Equivalence Principle) mission aims at testing the Equivalence Principle to one part in 10^{17}, a five orders of magnitude improvement over the best ground experiment [11].

This paper reviews superconducting accelerometer and gradiometer technology. The principle and intrinsic noise of the basic accelerometer are presented in Section 2. In Section 3 we review the principle of resonant-mass gravitational-wave detectors and derive the conditions that the transducer must satisfy to minimize the noise of the detector. Section 4 shows how superconducting accelerometer technology is used to meet transducer requirements in gravitational-wave detectors. We then turn to gravity gradiometry. After introducing the basic principle of gravity gradiometry in Section 5, the design and performance of various types of SGGs are discussed in Section 6. Both in-line and cross-component SGGs are discussed. Section 7 describes the SSA. Finally, Section 8 discusses applications of the SGG technology in many challenging experiments in fundamental physics, as well as in geodesy and geophysics.

2. Basic Superconducting Accelerometer

2.1. PRINCIPLE OF SUPERCONDUCTING ACCELEROMETER

Figure 1 shows a superconducting accelerometer in its simplest form. The accelerometer consists of a superconducting proof mass, a superconducting sensing coil and a SQUID with input coil. A persistent current is stored in the loop formed by the sensing coil and the SQUID input coil. When the platform undergoes an acceleration, or equivalently, when a gravity signal is applied, the proof mass is displaced relative to the sensing coil, modulating its inductance through the Meissner effect. This induces a time-varying current in the loop to preserve flux quantization. Thus, the spring-mass system converts the platform acceleration or gravity signal into a displacement of the proof mass relative to the platform, the superconducting circuit converts this displacement into an electrical current through the SQUID input coil, and the SQUID converts the induced current into an output voltage signal.

If we denote the acceleration signal by $a(\omega)$ and the displacements of the proof mass and the platform by $x(\omega)$ and $X(\omega)$, respectively, we find

$$x(\omega) - X(\omega) = \frac{a(\omega)}{\omega^2 - \omega_0^2}, \tag{1}$$

where ω_0 and ω represent the (angular) resonance frequency and signal frequency, respectively, and the effect of damping has been ignored. There are three categories of operation: (1) $\omega < \omega_0$ (stiff suspension), (2) $\omega \sim \omega_0$ (resonant), and (3) $\omega > \omega_0$ (almost-free mass). In most accelerometer applications, the first condition is satisfied: i.e., the signal frequency is low compared to the suspension frequency. It is clear that, in order to obtain the best sensitivity, one needs to reduce ω_0 to about the value of ω. The resonant transducer for gravitational waves belongs to the second category. When the third condition is met, the output becomes proportional to $-X(\omega)$, and the device becomes a displacement sensor.

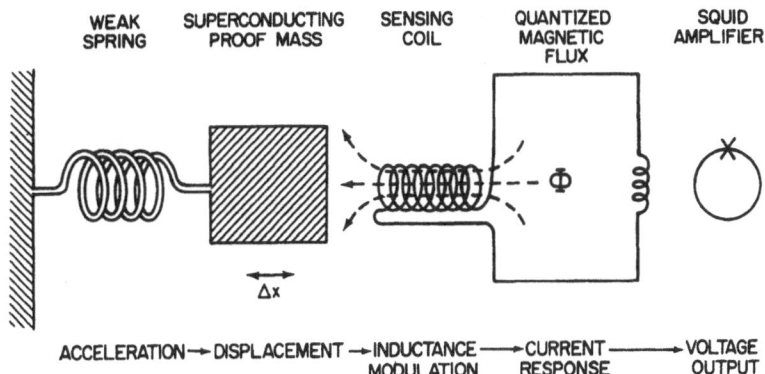

Figure 1. Schematic of a superconducting accelerometer.

2.2. INTRINSIC NOISE OF SUPERCONDUCTING ACCELEROMETER

When $\omega \ll \omega_0$, the power spectral density of the intrinsic acceleration noise of a superconducting accelerometer can be shown [12] to be

$$S_a(f) = \frac{4}{m}\left[k_B T \frac{\omega_0}{Q} + \frac{\omega_0^2}{2\beta\eta} E_A(f) \right],$$ (2)

in the limit where nonlinear modulation of the sensing-coil inductance and the SQUID back-action noise can be ignored, and where m, Q and T are the mass, quality factor and temperature of the proof mass, β is the electromechanical energy coupling constant of the transducer, η is the energy coupling efficiency from the transducer circuit to the SQUID, $E_A(f)$ is the "input energy resolution" of the SQUID, and $f = \omega/2\pi$ is the signal frequency. Equation 2 clearly exhibits the advantages of a superconducting accelerometer. As the accelerometer is cooled to liquid helium temperature ($T = 4.2$ K), T/Q is reduced typically by four orders of magnitude from its room-temperature value, and $E_A(f)$ of a SQUID amplifier is lower by a similar ratio in comparison to the corresponding value for an FET amplifier.

The energy coupling coefficient β is defined as the ratio of the induced energy in the electrical circuit to the total energy in the system. It is related to the frequencies by

$$\omega_0^2 = \omega_m^2 / (1-\beta),$$ (3)

where ω_m is the (angular) resonance frequency due to parasitic (mechanical) stiffness. When ω_m is held fixed, one finds that $\beta = \frac{1}{2}$ minimizes the amplifier noise contribution.

3. Principle of Gravitational-Wave Detection

3.1. THEORY AND OPTIMIZATION OF GRAVITATIONAL-WAVE DETECTORS

A resonant-mass gravitational-wave detector has three components: (1) a mechanical elastic solid antenna, which interacts with incoming gravitational waves, (2) an electromechanical transducer, which converts the signal energy to an electrical form, and (3) an electrical amplifier, which amplifies the electrical signal from the transducer. The antenna can be characterized by five parameters: effective mass M, effective length D, angular resonance frequency ω_a, quality factor Q_a and temperature T_a. The transducer is characterized by an impedance matrix Z_{ij} (which relates the mechanical input variables, force $f(\omega)$ and velocity $u(\omega)$, to the electrical output variables), voltage $V(\omega')$ and current $I(\omega')$. For a *passive* transducer, as is the superconducting inductive transducer discussed here, the output signal frequency ω' coincides with the input signal frequency ω, so that

$$\begin{bmatrix} f(\omega) \\ V(\omega) \end{bmatrix} = \begin{bmatrix} Z_{11} & Z_{12} \\ Z_{21} & Z_{22} \end{bmatrix} \begin{bmatrix} u(\omega) \\ I(\omega) \end{bmatrix}.$$ (4)

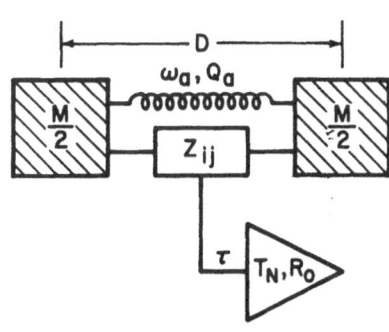

Figure 2. Schematic of a resonant-mass gravitational-wave detector.

The amplifier is characterized by three parameters: noise temperature T_N, optimum source impedance R_0 and integration time $\tau \approx (2\Delta f)^{-1}$, where Δf is the bandwidth. The detector is schematically shown in Figure 2.

A gravitational-wave pulse is characterized by a dimensionless metric perturbation h with a dominant frequency ω_s and duration τ_s. For a short pulse with $\tau_s \approx 2\pi\omega_s^{-1}$, the energy that the gravitational wave deposits into a favorably oriented, noiseless, cylindrical antenna becomes

$$E_S \approx \frac{2}{\pi^2} M\omega_s^2 (Dh)^2. \tag{5}$$

When a simple differencing algorithm is used for pulse detection, the total intrinsic noise of the detector, referred to the input of the noiseless antenna, can be shown [13] to be

$$E_N \approx k_B T_a \frac{\omega_a \tau}{Q_a} + k_B T_N \left[\frac{2(\zeta + \zeta^{-1})}{\beta_{21}\omega_s\tau} + \frac{\beta_{12}\omega_s\zeta\tau}{2} \right], \tag{6}$$

where β_{21} and β_{12} are the *forward* and *reverse* energy coupling coefficients of the transducer, respectively, $\zeta \equiv |Z_{22}|/R_0$ is the dimensionless impedance matching parameter, and the correlation between the two conjugate noise terms have been ignored. For a passive transducer, β_{21} and β_{12} are equal and are given by

$$\beta_{21} = \beta_{12} = \frac{|Z_{21}||Z_{12}|}{M\omega_s|Z_{22}|} \equiv \beta_S. \tag{7}$$

Here β_s is the *signal* coupling coefficient to the amplifier, which is equivalent to the product $\beta\eta$ introduced in Section 2.2.

The gravitational-wave pulse is detectable if $E_s > E_N$. Equation 6 shows that E_N comes from three terms: the Brownian motion noise and the two conjugate noise sources of the amplifier, the *forward* and *reverse action* noise of the amplifier. The Brownian motion noise can be made arbitrarily small, in principle, by reducing T_a/Q_a. The amplifier noise contribution can be minimized by choosing

$$\tau = \frac{2}{\beta_S \omega_s} \left(1 + \frac{1}{\zeta^2} \right)^{1/2}. \tag{8}$$

Substituting this into Equation 6 leads to

$$E_N \approx \frac{2k_B T_a}{\beta_S Q_a} \left(1 + \frac{1}{\zeta^2} \right)^{1/2} + 2k_B T_N \left(1 + \zeta^2 \right)^{1/2}. \tag{9}$$

From Equations 8 and 9, three optimization conditions follow:

$$|Z_{22}| / R_0 \leq 1,$$ (10)

$$\Delta \omega_S / \omega_S \approx \beta_S,$$ (11)

$$T_a / Q_a \ll \beta_S T_N,$$ (12)

where $\Delta \omega_S \approx \pi \tau^{-1}$ is the bandwidth of the detector. Thus a large β_S allows for a large fractional bandwidth, which reduces the Brownian motion noise of the antenna. If a near-unity β_S could be achieved *without* restricting the bandwidth of the transducer, a completely wideband detector ($\Delta \omega_S \approx \omega_S$) could be realized without compromising the signal-to-noise ratio.

When Equations 10 through 12 are satisfied, the total detector noise becomes amplifier-limited at $2k_B T_N$. A more rigorous theory with proper treatment of the correlated noise and application of the optimal filter [14] leads to the true amplifier limit of $k_B T_N$. Combining this with Equation 5, one finds

$$h_{\min} \approx \left(\frac{5k_B T_N}{M \omega_S^2 D^2} \right)^{1/2}.$$ (13)

Therefore, for a given frequency, the antenna mass (M) and size (D), and the amplifier noise temperature (T_N) are the parameters which determine the ultimate sensitivity of the resonant-mass gravitational-wave detector. For *linear* (phase-preserving) amplifiers, T_N has a quantum limit [15]:

$$T_{N,QL} = \hbar \omega_S / k_B.$$ (14)

Back-action-evasion techniques have been proposed to beat this "standard quantum limit" [16, 17]. For a cylindrical antenna with $M = 1200$ kg, $D = 3$ m and $\omega_S / 2\pi = 900$ Hz (present antennas) at the quantum limit, we obtain $T_N \approx 0.04$ μK and $h_{\min} \approx 3 \times 10^{-21}$.

3.2. PRINCIPLE OF RESONANT TRANSDUCERS

Equations 11 and 12 show that large β_S is necessary to reduce the Brownian motion noise of the detector. The large antenna mass M that appears in the denominator of Equation 7, however, tends to make β_S very small, typically $10^{-6} \sim 10^{-5}$ for $M \geq 10^3$ kg. This is equivalent to there being an impedance mismatch by five to six orders of magnitude between the antenna output and the transducer electrical input. There are two ways of overcoming this problem. One is by using the mechanical analog of a transformer, a lever, which reduces the mechanical impedance of the antenna over a wide bandwidth [18]. The other is by using the mechanical analog of a tank circuit, a mechanical resonator, which allows impedance matching over a restricted bandwidth [2]. Although the former could, in principle, achieve a wider bandwidth, a rigid lever with low stray mass and sufficient gain may be difficult to realize in practice. For this reason, the use of resonant transducers has become standard practice.

Figure 3 illustrates the principle of a resonant transducer. The antenna with mass M receives a tiny "hammer blow" from the gravitational wave, and its mode is excited. If the resonance frequency of the small mass m is tuned to that of the antenna, the antenna begins to drive the resonator, transferring its entire energy to the small mass, whose displacement then becomes $(M/m)^{1/2}$ times larger than the initial displacement of the antenna. The energy flows back and forth between the two masses with a beat period of $(2\pi/\omega_a)\,(M/m)^{1/2}$.

The resonant transducer thus improves the energy coupling β_s by the ratio of M/m, but with the bandwidth restricted by the beat frequency:

$$\Delta\omega_s \approx \omega_a(m/M)^{1/2}. \qquad (15)$$

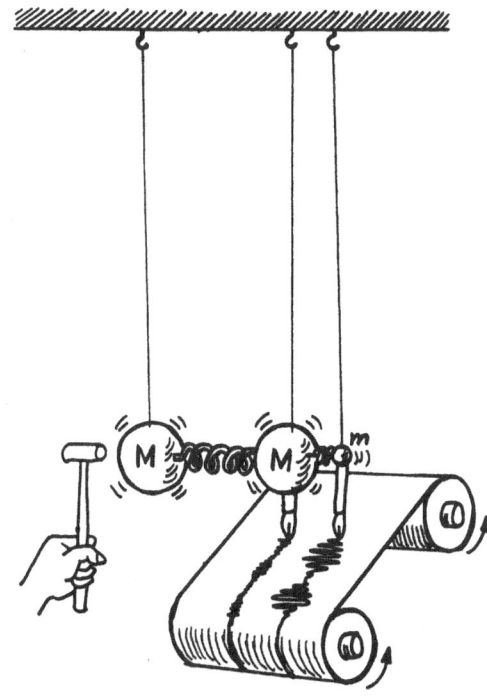

Figure 3. Principle of a resonant transducer.

One is tempted to maximize the transducer mass m; however, this generally results in a reduction of β_s and therefore limits the usable bandwidth $\Delta\omega_s$ by Equation 11. To obtain the largest $\Delta\omega_s$, one must choose a value of m_{opt} which satisfies the condition:

$$\left(\Delta\omega_s / \omega_s\right)_{\max} \approx \beta_s(m_{opt}) \approx \left(m_{opt}/M\right)^{1/2}. \qquad (16)$$

Another factor one must consider in choosing m is the electrical Q of the transducer. In order to benefit from a large β_s, one must have a sufficiently high electrical Q, since otherwise the loaded Q of the antenna, Q_a, is degraded.

The resonant transducer concept can be extended further by using a cascade of n resonators with geometrically decreasing masses [19]. If ξ is the ratio of neighboring masses, the antenna mass M and the final resonator mass m are related by $m/M = \xi^{n-1}$. Since the beat frequency is now $\omega_a\xi^{1/2}$, Equation 16 is modified to become

$$\left(\Delta\omega_s / \omega_s\right)_{\max} \approx \beta_s(m_{opt}) \approx \left(m_{opt}/M\right)^{1/(2n-2)}. \qquad (17)$$

A fractional bandwidth arbitrarily close to unity can be obtained, in principle, by increasing n. The resulting increase in $\Delta\omega_s$, however, is slow beyond $n = 3$, while the hardware quickly becomes very complex with increasing n. The practical limit for n thus appears to be 3~4 for most cases of interest [14].

4. Superconducting Transducer for Advanced Gravitational-Wave Detectors

4.1. SUPERCONDUCTING RESONANT TRANSDUCER

The superconducting resonant transducer for gravitational-wave experiments is identical to the basic accelerometer shown in Figure 1 except that two sensing coils, L_1 and L_2, are used to sense the two opposite faces of the proof mass. These coils are connected in parallel to the SQUID input coil, L_i, and a persistent current I_0 is stored in the loop formed by L_1 and L_2. By simple circuit analysis [3], one can show, in the limit where the parasitic stiffness due to the nonlinear inductance modulation can be ignored, that the resonance frequency is shifted from the uncoupled value ω_m according to

$$\omega_0^2 = \omega_m^2 + \frac{2}{1+\gamma} \frac{B^2 S}{\mu_0 m d},$$ (18)

where $\gamma \equiv L_i(L_1^{-1} + L_2^{-1})$, S is the area of each pancake coil, d is the gap between each pancake coil and the proof mass, B is the magnetic field produced by I_0 and μ_0 is the permeability of vacuum. From Equations 3 and 18, we find

$$\beta = 1 - \left(\frac{\omega_m}{\omega_0}\right)^2 = \frac{2}{1+\gamma} \frac{B^2 S}{\mu_0 m \omega_0^2 d}.$$ (19)

The SQUID is characterized by three noise spectral densities: the flux noise $S_\phi(f)$, circulating current noise $S_J(f)$ and their correlation $S_{\phi J}(f)$. In the simple case when $S_{\phi J}(f)$ can be ignored, these noise spectral densities define

$$T_N \equiv \omega_S \left(S_\phi S_J\right)^{1/2} / 2k_B,$$ (20)

$$R_{opt} \equiv \omega_S M_i^2 \left(S_J / S_\phi\right)^{1/2},$$ (21)

$$\zeta \equiv \omega_S \left(L_p + L_i\right) / R_{opt},$$ (22)

where $M_i \equiv \alpha(L_i L_s)^{1/2}$ is the mutual inductance between L_i and the SQUID loop inductance L_s and $L_p \equiv (L_1^{-1} + L_2^{-1})^{-1}$ is the parallel combination of the sensing coil inductances. The optimization conditions, Equations 10 through 12, then lead to $\alpha^2 \to 1$, $\eta \to 1$ and $\gamma \approx 1$. The third condition is easy to satisfy; however, $\alpha^2 \leq 0.5$ typically for low-noise dc SQUIDs, and $\eta \leq 0.5$ in general due to the parasitic inductance that the input coil provides to the circuit. Clearly, for gravitational-wave detector application, minimizing both the flux and circulating current noise and obtaining the highest signal coupling are equally important. This imposes the most stringent requirements for SQUID design.

Figure 4 is a cross-sectional view of the superconducting transducer which has been constructed at the University of Maryland to be integrated with the 50 mK antenna AURIGA at Legnaro, Italy [20]. The resonator is a centrally-loaded diaphragm. Niobium (Nb) is chosen for the resonator material, since it has by far the highest H_{c1}, the

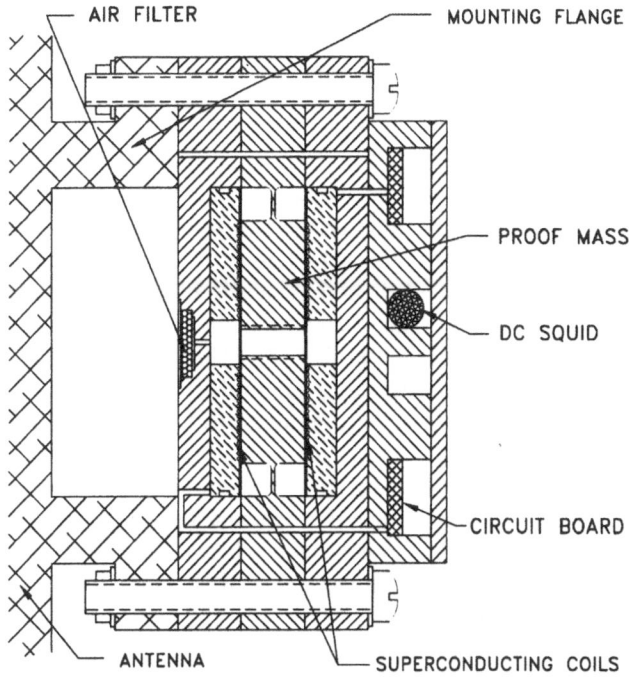

Figure 4. Cross sectional view of a superconducting transducer.

maximum field for complete flux exclusion. The entire resonator is machined from a single block of high RRR value Nb and heat-treated to obtain high mechanical and electrical Qs. Two Nb "pancake coils" are located close to the proof-mass surfaces and connected together to form a superconducting loop in which a persistent current is stored. The two sensing coils are connected in parallel to the input coil of a low-noise dc SQUID. Prior to assembly, the mating surfaces of the proof mass and the pancake coils are lapped and chemically polished to reduce d to ~25 μm and thus improve β_s.

With $m = 0.62$ kg, $S = 3.5 \times 10^{-3}$ m^2, $B = 0.1$ T (H$_{c1}$ of Nb at 4.2 K), $\omega_0/2\pi = 900$ Hz, $\eta \approx 0.4$ and $\gamma \approx 1$, this transducer can achieve $\beta \approx 0.05$ and $\beta_s \approx 0.02$. For the antenna mass of $M = 1200$ kg, this satisfies Equation 16 and allows $\Delta\omega_s/\omega_s \approx 2\%$.

The University of Maryland project to develop low-noise dc SQUIDs optimized for gravitational-wave detection has now produced a double dc-SQUID system with $S_\phi(f)/2L_s \approx 50\hbar$ in closed loop operation [21]. Although $S_I(f)$ and $S_\phi(f)$ have not yet been measured, the best estimate for the noise temperature is $T_N \approx 1.2$ μK at $f = 900$ Hz for this SQUID. The loaded Q of the antenna is expected to be $Q_a \geq 5 \times 10^6$ at $T_a = 50$ mK. Since these parameters satisfy Equation 12, the gravitational-wave sensitivity of this detector is given by Equation 13 as $h_{min} \approx 1.6 \times 10^{-20}$.

Such a sensitivity should be sufficient to detect theoretically predicted astrophysical events occurring in our galaxy. However, the event rate may be only one every several years. In order to increase the event rate to several a year, it is necessary to improve the sensitivity to $h_{min} \leq 10^{-21}$ and reach the Virgo cluster of galaxies, which is at a distance of ~15 Mpc. In the next section, we discuss an initiative being undertaken by researchers world-wide to make another quantum jump in sensitivity: the project of developing massive, ultralow-temperature, *spherical* detectors.

4.2. SPHERICAL DETECTOR OF GRAVITATIONAL WAVES

Unlike a cylinder, whose wave cross section has strong directional dependence, a sphere is omni-directional. A sphere has five degenerate quadrupole modes which interact with gravitational waves. By detecting all five modes simultaneously, four unknowns associated with each gravitational-wave event, the source direction (θ, ϕ) and the amplitudes of two polarizations (ψ_+, ψ_x), can be determined. The remaining degree of freedom can be used to discriminate against non-gravitational-wave disturbances. This multi-mode nature of the sphere was first recognized by Forward [22].

Wagoner and Paik [23] carried out a detailed mode analysis of the sphere and computed the wave cross section of the sphere. They found, in particular, that the cross section of a sphere is approximately five times as big as that of a cylinder with the same mass averaged over the source direction and polarization, due to the five modes which couple to gravitational waves. Since a sphere is at least ten times as heavy as a typical cylinder for the same resonance frequency, this means that, for a given frequency, a sphere has almost two orders of magnitude larger cross section than a typical cylindrical antenna used at present. Therefore, a massive spherical antenna operating near its quantum limit may be the ultimate solution which bridges the gap between the sensitivity expected with the new ultralow-temperature cylindrical antennas and the sensitivity required to detect events in the Virgo cluster.

Recently, Johnson and Merkowitz [24] found an arrangement of six radial-motion transducers which maintain the degeneracy of the quadrupole modes. Six resonant transducers mounted on six pentagonal faces of a hemisphere of a truncated icosahedron (or "buckey ball") split the modes evenly, thus maintaining the "spherical" symmetry. This truncated icosahedral gravitational-wave antenna (TIGA) is shown in Figure 5. The transducer locations are marked by black dots.

Here we consider design parameters which will give a total system noise of $10\hbar$ to an aluminum sphere of diameter $D = 3$ m cooled to $T_a = 50$ mK. The total mass and the lowest quadrupole-mode frequency of the antenna are 3.8×10^4 kg and 900 Hz, respectively. The effective mass of the sphere seen by each transducer is $M = 9.5 \times 10^3$ kg. We assume a three-mode ($n = 3$) transducer with a final resonator of $m = 0.17$ kg. The mass of the intermediate resonator then becomes 40 kg. This should allow $\beta_s \approx \Delta\omega_s/\omega_s \approx 0.06$. If the SQUID noise is $T_N \approx 0.2$ μK ($5\hbar$), we need $Q_a \geq 10^7$ in order to be amplifier limited. The gravitational-wave sensitivity is then given by

$$h_{\min} \approx \left(\frac{2.4 k_B T_N}{M \omega_s^2 D^2} \right)^{1/2}. \qquad (23)$$

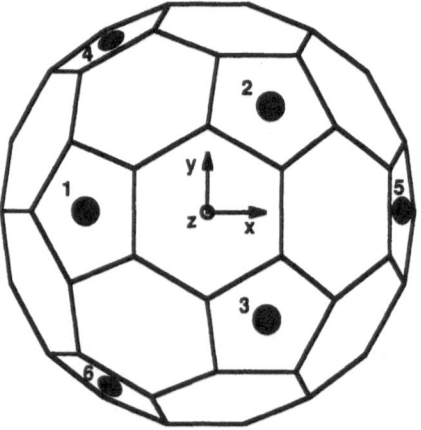

Figure 5. Truncated icosahedral gravitational-wave antenna.

This equation yields $h_{min} \approx 1.6 \times 10^{-21}$, a level that may be high enough to detect gravitational waves from neighboring galaxies. At the quantum limit, the sensitivity of the sphere corresponds to $h_{min} \approx 7 \times 10^{-22}$. In order to visualize how sensitive this is, one can imagine blowing up the sphere to the size of the Earth. Then the strain sensitivity of the quantum-limited spherical antenna corresponds to a detection of a 5-Fermi movement of the surface of the Earth! [Note: 1 Fermi $\equiv 10^{-15}$ m.]

5. Principle of Gravity Gradiometry

5.1. IN-LINE AND CROSS-COMPONENT GRADIOMETERS

The gravity field is described by a scalar potential $\phi(x_i, t)$, which is an unobservable. Its first spatial derivatives form a vector: $g_i \equiv -\partial\phi/\partial x_i$. The gravitational acceleration is indistinguishable from a platform acceleration by the Equivalence Principle. In General Relativity, gravitational acceleration can be transformed away by choosing a freely-falling frame. The intrinsic field that uniquely characterizes the gravity field is the Riemann curvature tensor. The corresponding quantity in Newtonian gravity is the gradient of the gravitational acceleration, a tensor quantity:

$$\Gamma_{ij} \equiv \partial^2\phi / \partial x_i \partial x_j . \tag{24}$$

The gravity-gradient tensor is symmetric by construction. Its trace is related to the local mass density by Poisson's equation, an expression of the inverse-square law:

$$\sum_i \Gamma_{ii} = \nabla^2\phi = 4\pi G\rho . \tag{25}$$

In free space ($\rho = 0$), this trace must vanish. This leaves only five independent components for the gravity-gradient tensor: two diagonal ("in-line") components and three off-diagonal ("cross") components.

An in-line-component gradiometer can be constructed by differencing signals between two linear accelerometers whose sensitive axes are aligned along their separation. As an example, we show in Figure 6 a schematic of the Model I SGG developed at the University of Maryland [25]. The superconducting circuit sums and differences the acceleration signals at the two proof masses to detect the diagonal-component gradient Γ_{ii} and the linear acceleration g_i simultaneously along the baseline of the gradiometer.

The power spectral density of the intrinsic gradient noise of the in-line-component SGG can be obtained from

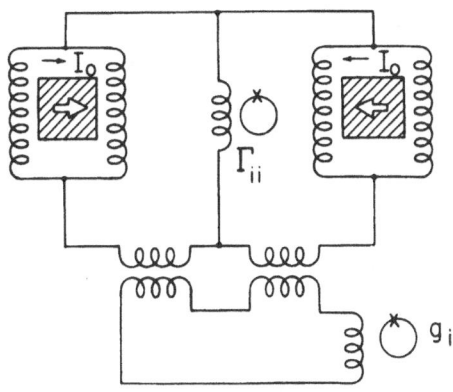

Figure 6. In-line-component SGG.

580

Equation 2 by replacing m with the reduced mass for the differential mode, $m/2$, and by dividing $S_a(f)$ by ℓ^2, where ℓ is the baseline of the gradiometer:

$$S_\Gamma(f) = \frac{8}{m\ell^2}\left[k_B T\frac{\omega_0}{Q} + \frac{\omega_0^2}{2\beta\eta}E_A(f)\right],\tag{26}$$

where all the other parameters are as defined in Equation 2. The gradient sensitivity could be improved by increasing m and ℓ. However, this may be undesirable because some errors also scale with dimension. The SGG shares with the superconducting accelerometer the benefits of operating at low temperatures: the low Brownian motion noise and the low amplifier noise by the use of a SQUID.

A cross-component gradiometer can be constructed by differencing signals between two concentric angular accelerometers whose arms are orthogonal to each other. Figure 7 shows a superconducting cross-component gradiometer [26]. The circuit sums and differences the angular acceleration signals at the two proof-mass arms to detect the off-diagonal-component gravity gradient Γ_{ij} and the common-mode angular acceleration α_k simultaneously, where k is the coordinate axis normal to the gradiometer plane. The intrinsic noise of the cross-component SGG can also be described by Equation 26, if $m\ell^2$ is replaced by the moment of inertia I of each proof-mass arm. The cross-component gravity gradiometer was pioneered by Forward [27], and a superconducting version was first developed at the University of Western Australia [28].

A *tensor* gradiometer could be constructed by combining six in-line-component gradiometers with proper relative orientation. One such configuration is the arrangement of two 45 degree orientations on each of the three orthogonal coordinate planes. This configuration of in-line-component SGGs has been suggested for detection of the relativistic gravitomagnetic field, as will be discussed in Section 8.2. Another configuration is the dodecahedral arrangement of the accelerometers, as proposed for a wideband spherical gravitational wave detector [29]. All the tensor components could then be determined from the gradiometer outputs by using the rotation property of the gradient tensor.

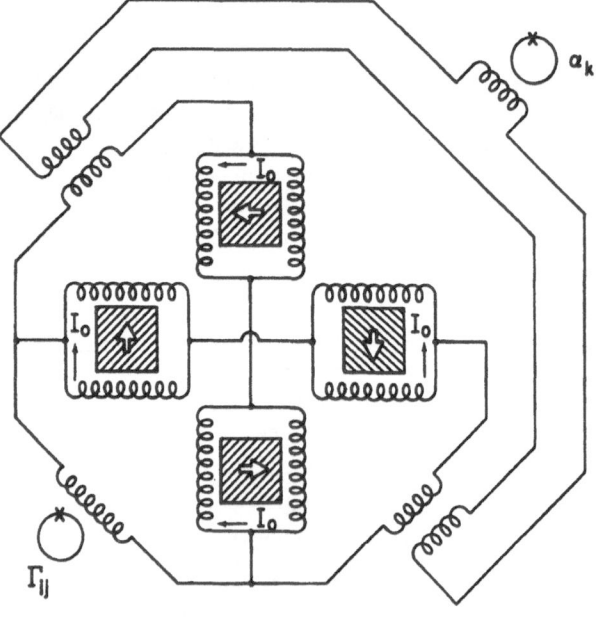

Figure 7. Cross-component SGG.

5.2. REJECTION OF DYNAMIC ERRORS

Gravity gradiometers are usually operated in a dynamically noisy environment to detect very weak differential signals. Passive vibration isolation is not possible, unlike in a gravitational-wave detector, because the signal frequencies for the gradiometry are very low in general (≤ 0.1 Hz). Active isolation, or "platform stabilization," can be applied to the angular degrees of freedom. It is difficult, however, to achieve active isolation in all six degrees of freedom. Therefore, it is important to build into the device as high common-mode-rejection ratios as possible for both linear and angular accelerations.

In order to reject the common mode along the sensitive axis, the scale factors of the component accelerometers must be matched precisely. This, in turn, requires high stablility of the scale factors. This is where a cryogenic instrument has another important advantage. An SGG has *extremely stable* scale factors due to the increased mechanical stability of materials at low temperatures and the ultrahigh stability of persistent currents.

The axial component of the common-mode acceleration can be rejected to the degree the scale factors are matched. However, the gradiometer will still be sensitive to the transverse component of the acceleration by misalignment of the sensitive axes. In an in-line-component gradiometer, linear and angular accelerations of the platform, $\vec{a}$ and $\vec{\alpha}$, couple to the differential mode through departures from parallelism and concentricity of the sensitive axes of the component accelerometers, $\delta\hat{n}$ and $\delta\hat{\ell}$, respectively [12]:

$$\delta\Gamma_a = \frac{1}{\ell}\delta\hat{n}\cdot\vec{a}, \tag{27}$$

$$\delta\Gamma_\alpha = \left(\delta\hat{\ell}\times\hat{n}\right)\cdot\vec{\alpha}. \tag{28}$$

There are corresponding error sources in a cross-component device. A departure of the rotation axes from parallelism provides a coupling mechanism for angular acceleration, whereas an asymmetric mass distribution in each moment arm (a departure of the rotation axis from the mass symmetry axis) causes linear acceleration to couple to the gradiometer.

For an SGG, the enhanced mechanical stability of materials at low temperatures guarantees that $\delta\hat{n}$ and $\delta\hat{\ell}$ are also stable. These error coefficients can be measured definitively during the initial setup, multiplied by the linear and angular accelerations of the gradiometer platform, and subtracted from the gradiometer output. By applying this "residual common-mode balance" [6], $\delta\hat{n}$ and $\delta\hat{\ell}$ can be effectively reduced to very small values.

Another important error source for a gradiometer is the centrifugal acceleration of the platform. The centrifugal acceleration is proportional to the angular velocity squared. This nonlinear nature of the centrifugal acceleration is especially troublesome because it down-converts the angular motion noise of the platform from all frequencies above the signal frequency. Therefore, the high-frequency angular velocity noise must be attenuated by applying passive vibration isolation and/or the platform must be stabilized against the angular motion noise at all frequencies.

6. Superconducting Gravity Gradiometers (SGGs)

6.1. MODEL II SGG

Figure 8 is a cross-sectional view of a component accelerometer of Model II SGG. The proof mass weighing 1.2 kg is suspended by a pair of diaphragm flexures. The diaphragms have arc-like slits cut into them, so they act like a set of folded cantilevers. The natural frequency of the spring-mass system is 4.5 Hz. This frequency was chosen as a compromise between two conflicting requirements: precise sensitive axis alignment, which favors stiff suspension, and high acceleration sensitivity, which requires soft suspension. The differential and common-mode resonance

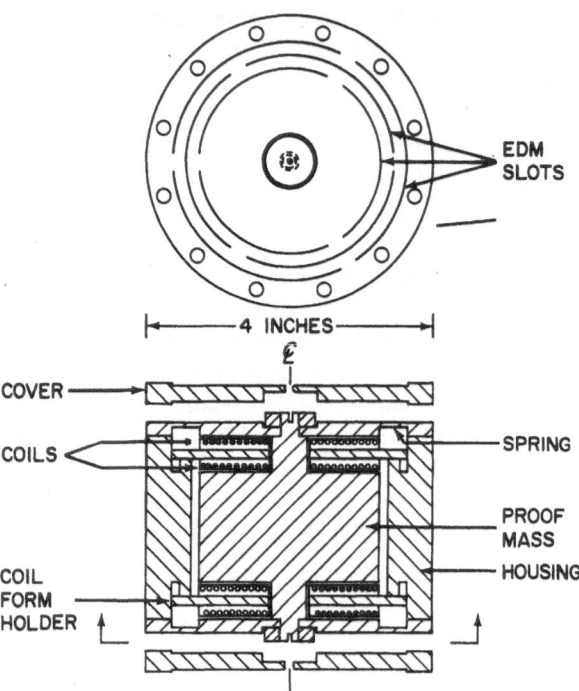

Figure 8. Component accelerometer for Model II SGG.

frequencies increase to about 9 and 50 Hz, respectively, when we trap flux in the circuits. The entire structure is machined from Nb. Six identical component accelerometers are mounted on six faces of a precision titanium (Ti) alloy cube, with the sensitive axes normal to the cube surfaces, to form a three-axis in-line-component SGG.

Figure 9 is a schematic circuit diagram for each axis of the Model II SGG. The two proof masses on opposite faces of the mounting cube are connected by a superconducting circuit to form a gradiometer. The coils L_1 through L_6 are superconducting coils wound with Nb wire in a pancake shape. The proof masses are levitated against gravity by storing a persistent current I_{c2} in the loop formed by L_3 and L_4. (Another current $I_{c1} = I_{c2}$ is stored in zero-g.)

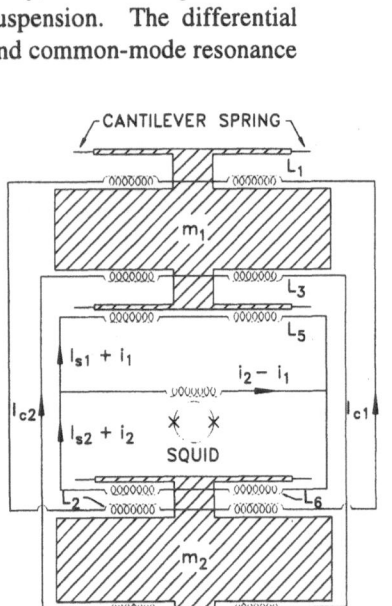

Figure 9. Circuit diagram for each axis of Model II SGG

The coils L_5 and L_6 are connected in parallel to a SQUID to form a sensing circuit. Each axis of the gradiometer has two circuits, although, for simplicity, only one is shown. In one circuit, we store the currents I_{d1} and I_{d2} with the same sense, as shown in the figure. In this case, the SQUID detects the differential acceleration, or the gravity gradient. In the other circuit, we reverse the direction of one of the currents, and the SQUID detects the common-mode motion. Signal differencing by means of stable persistent currents *before* detection is a unique feature of the SGG. This assures excellent null stability of the device, which, in turn, improves the overall common-mode rejection. Further, the SQUID sees only a small differential signal, thereby reducing the dynamic-range requirement on the amplifier and signal-processing electronics.

If we could manufacture a perfect gradiometer, one with perfectly matched spring constants, coil-to-proof-mass gaps, and coil inductances, then setting $I_{d2} = I_{d1}$ would give perfect rejection of the common-mode motion in the differential-mode output. Because of mismatches of a real device, we adjust I_{d2}/I_{d1} to maximize the common-mode rejection. We do this by applying a periodic linear acceleration to the SGG and tuning I_{d2}/I_{d1} until the signal at the shaking frequency in the differential-mode output disappears.

Although the component of the linear acceleration parallel to the sensitive axis can be rejected precisely by this current adjustment, components normal to the sensitive axis couple to the gradient output through misalignments of the sensitive axes, as described in Section 5.2. In the Model II SGG, all the misalignment angles are measured from the response of the gradiometer to accelerations applied in various directions. The results are then multiplied by the measured linear acceleration components and subtracted from the gradiometer outputs to achieve the residual common-mode balance. The misalignments were about 10^{-4} rad. The residual balance improved the common-mode rejection to 10^7.

The numerical values of the gradiometer parameters are: $m = 1.2$ kg, $\ell = 0.19$ m, $\omega_0/2\pi = 10$ Hz, $Q = 10^6$, $T = 4.2$ K, $\beta = \eta = 0.25$, $E_A(f) = [1 + (0.1\ \text{Hz})/f]\ 5 \times 10^{-31}$ J Hz^{-1} below $f \leq 0.1$ Hz (commercial dc SQUID). Equation 26 predicts a white-noise level of the instrument at 2×10^{-3} E Hz$^{-1/2}$. Below 0.1 Hz, a 1/f power noise should appear.

Figure 10 shows the actual noise spectrum of the Model II SGG obtained in the laboratory. The white-noise level corresponds to 0.02 E Hz$^{-1/2}$, dominated by uncompensated angular acceleration noise. Below 0.1 Hz, a 1/f

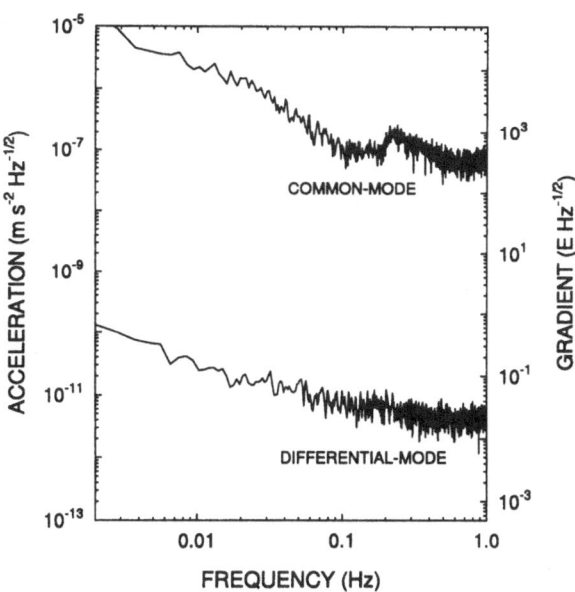

Figure 10. Noise spectrum of Model II SGG.

584

power noise appears, but with an amplitude an order of magnitude higher than expected from the SQUID noise. This excess low-frequency noise is believed to come from a combination of temperature drift of the instrument, which couples to the gradiometer through modulation of the penetration depth of the superconductor, and down-converted centrifugal acceleration noise. The demonstrated sensitivity of the SGG represents three orders of magnitude improvement over that of the conventional gravity gradiometers operated at room temperature.

Figure 11 is the response of the three-axis Model II SGG to a 95-kg mass passing by with a closest distance of 0.75 m. The three top traces are the responses of the three orthogonal axes of the gradiometer and the bottom trace represents the sum of these three outputs. The sum of the three in-line-component gradients must vanish in accordance with Equation 25. A small residual signal seen in the bottom trace is due to the finite baseline of the gradiometer, which couples to the higher-order gradients of the gravity field.

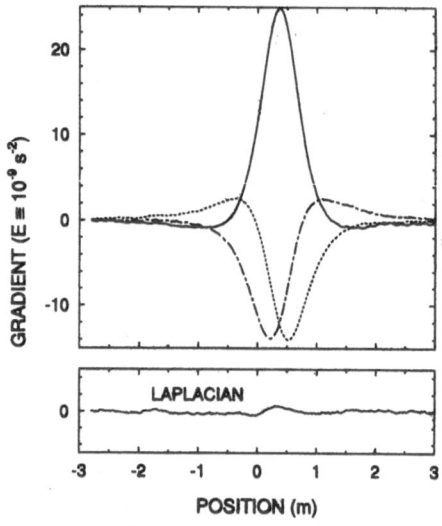

Figure 11. Response of Model II SGG to a mass swing-by.

The common-mode circuits of the SGG respond to the gravitational acceleration of the Earth, as well as to the platform acceleration. Figure 12 shows the vertical acceleration signal measured with the Model II SGG over the course of approximately two days, plotted with the calculated tidal gravity signal over the same time span. The agreement is excellent, especially considering that the calculation ignores important effects such as atmospheric loading and the elastic response of the local lithosphere. Thus, the device works also as a vector gravimeter.

The Model II SGG is being integrated with three superconducting angular accelerometers (of design similar to those shown in Figure 25) to detect and remove the angular and centrifugal acceleration errors, thus extending the

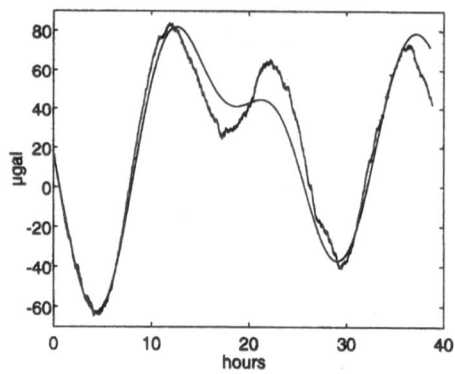

Figure 12. Tidal gravity measurement with Model SGG II.

"residual common-mode balance" to the angular degrees of freedom. Each angular accelerometer has a single proof mass supported at its center of mass by a pivot, and its rotation is sensed by a superconducting circuit, similar to the one shown in Figure 9.

6.2. MODEL III SGG

Equation 26 shows that the SQUID-limited gradient noise, $S_\Gamma^{1/2}(f)$, is proportional to the differential-mode frequency ω_0. Therefore, a factor ten improvement in sensitivity should be possible if $\omega_0/2\pi$ could be lowered to 1 Hz. With such an aim, we have developed a "superconducting negative spring," which cancels the stiffness of the mechanical springs [30].

A superconducting ball placed in the midplane of a ring-shaped superconducting coil carrying a persistent current is in unstable equilibrium, experiencing a repulsive force proportional to the displacement from its equilibrium position along the symmetry axis. The magnetic field thus produces a "negative" spring.

Figure 13 shows a cross-sectional view of a component

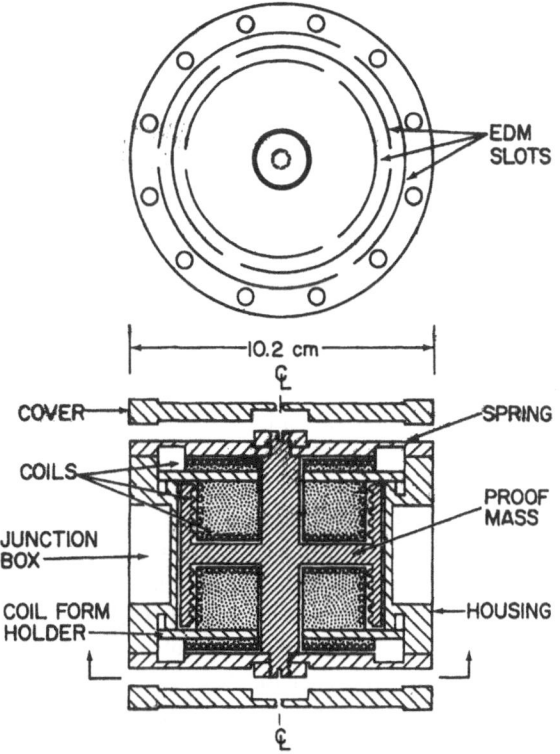

Figure 13. Component accelerometer for Model III SGG.

accelerometer of the Model III SGG, which incorporates this negative spring concept. Annular pockets have been cut out of the proof mass to make room for a coil form containing ten short superconducting solenoids. Corresponding round ridges are machined on the interior surface of the Nb proof mass. We have constructed a single-axis SGG of this design and demonstrated the negative spring.

Figure 14 shows the spring constant of the differential mode $(m\omega_0^2)$ plotted against the stored current squared. The differential-mode frequency was lowered from 7 Hz to 1 Hz, removing 98% of the stiffness. Further reduction in frequency was prevented by imprecise alignment between the ridges and the ring-shaped coils in the apparatus. With $\omega_0/2\pi = 0.3$ Hz, the intrinsic white noise of the SGG would become 2×10^{-4} E Hz$^{-1/2}$.

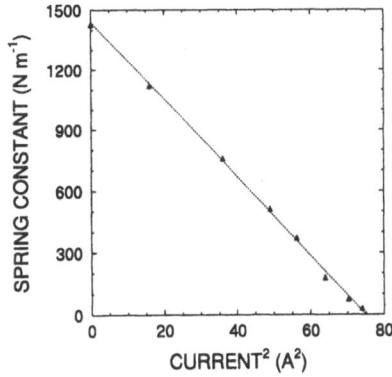

Figure 14. Stiffness of differential mode versus current squared.

586

6.3. CROSS-COMPONENT SGG

The in-line-component SGGs discussed in previous sections are under development primarily for space application. The achieved common-mode rejection of 10^7 is sufficient for a spacecraft environment, which is dynamically quiet. However, for airborne application, the linear acceleration rejection must be improved to 10^9 or higher because the SGG platform cannot be stabilized against the large translational motion experienced by aircraft, which may exceed the dimension of the cabin. The angular acceleration rejection is not as critical because the platform can be stabilized against rotational motion by using conventional techniques, as discussed in Section 5.2.

This is where a cross-component gradiometer offers an advantage. A cross-component device can be designed to be inherently insensitive to linear accelerations by employing pivoted rotating arms whose mass moments are precisely balanced prior to assembly. Linear accelerations are further discriminated by designing the pivot to be compliant only for the desired rotational acceleration, but stiff against the other degrees of freedom.

Figure 15 shows the three-axis cross-component SGG which is under development at the University of Maryland in collaboration with Sandia National Laboratories. The proof mass/pivot structure is cut from a single block of Nb by wire EDM, as in the angular accelerometers for the Model II SGG. Two component angular accelerometers on opposite faces of the mounting cube, with their rotating arms oriented orthogonal to each other, are connected together to form a single-axis SGG. Two superconducting circuits, similar to the one shown in Figure 9, are provided to detect the differential-mode (cross-component gradient) and the common-mode (angular acceleration) signals. The expected intrinsic noise of this instrument is 0.03 E $Hz^{-1/2}$, with a 1/f power noise below 0.1 Hz. We expect to achieve a common-mode rejection ratio better than 10^9 with this device.

A cross-component SGG tends to have a lower sensitivity than an in-line-component device due to the shorter baseline which can be accommodated within the same instrument volume. The cross-component SGG, however, is the preferred instrument option in situations where the common-mode rejection is a more important criterion than the intrinsic noise, as in most airborne applications.

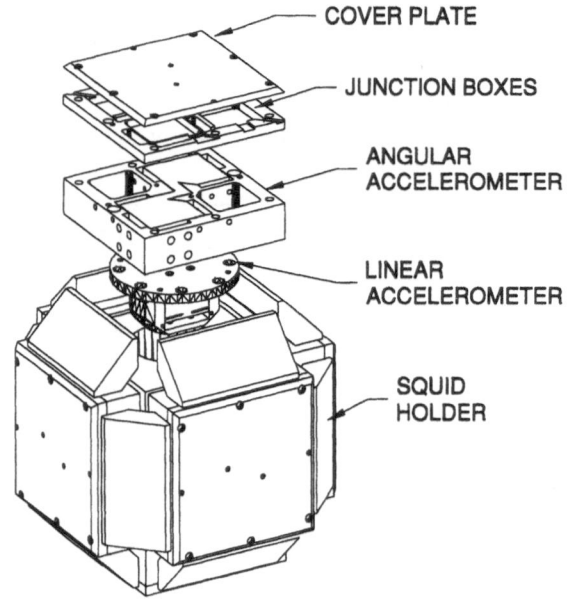

COVER PLATE

JUNCTION BOXES

ANGULAR ACCELEROMETER

LINEAR ACCELEROMETER

SQUID HOLDER

Figure 15. Exploded view of a cross-component SGG.

7. Superconducting Six-axis Accelerometer (SSA)

In order to realize the full sensitivity of the SGGs, the platform must be stabilized against motion or the acceleration errors must be compensated in all six degrees of freedom. Although the in-line and cross-component SGGs come with built-in linear and angular accelerometers, respectively, we have developed a superconducting six-axis accelerometer (SSA), a device which measures all components of linear and angular acceleration simultaneously [31, 32].

Figure 16 shows the mechanical components of the SSA. All the components are constructed of Nb except for the coil forms, which are made of a non-superconducting Ti alloy. A single Nb proof mass of inverted-cube geometry is levitated on 24 levitation coils. Its motion is detected in six degrees of freedom with six inductance bridges which are formed by 24 sensing coils. These 48 coils are mounted on eight coil forms, each carrying three levitation and three sensing coils, which, in turn, are mounted in the eight corners of the cubic cavity in the housing.

Figures 17 and 18 show the levitation and sensing circuits for a linear degree of freedom in a one-g environment. Similar circuits are used for levitation and sensing in the angular degrees of freedom. Persistent currents are stored in the levitation circuits to levitate the proof mass against gravity.

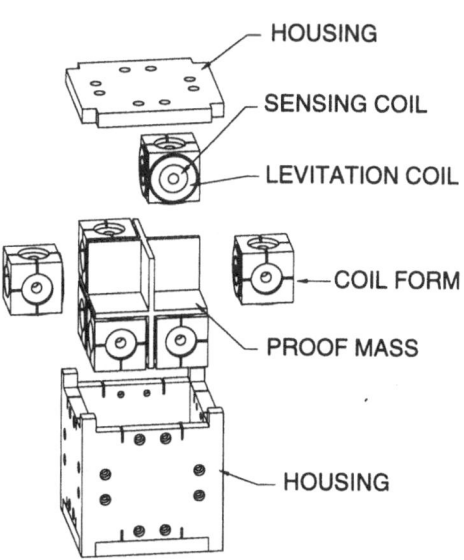

Figure 16. Mechanical components of SSA.

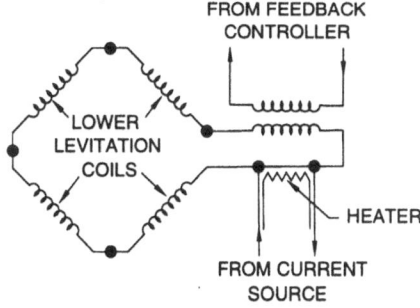

Figure 17. Levitation circuit of SSA for linear acceleration.

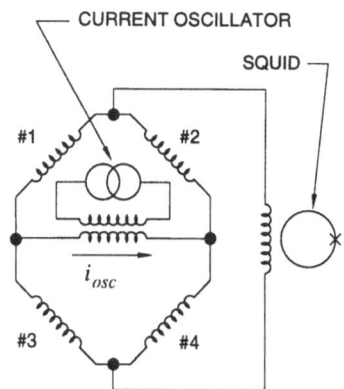

Figure 18. Sensing circuit of SSA for linear acceleration.

588

The bridges in the sensing circuits are driven with audio-frequency currents supplied externally, and their outputs are detected with SQUIDs and demodulated. Levitation currents are adjusted until the carrier signals vanish at the outputs of the bridges, in order to center the proof mass in all six degrees of freedom. The sideband outputs, which measure the platform accelerations, are fed back to the levitation circuits to operate the SSA in a force-rebalance mode. The intrinsic linear and angular acceleration noise of the SSA are 10^{-11} m s^{-2} Hz$^{-1/2}$ and 10^{-10} rad s^{-2} Hz$^{-1/2}$, respectively [8].

The SSA represents an extremely sensitive inertial navigation system, which improves the state of the art by at least three orders of magnitude. As such, it will find useful application in precision inertial guidance and inertial survey, as well as in ultrasensitive platform control. Integrated with an SGG, it represents a gradiometer-aided inertial navigation system which has the capability of removing errors due to variations in local gravity. The SSA can also be used as a vector gravimeter with angular readouts. It could be flown, integrated with a GPS receiver, in an aircraft to conduct an improved airborne gravity survey [33].

8. Applications of the SGG Technology

8.1. PRECISION TESTS OF THE GRAVITATIONAL INVERSE-SQUARE LAW

New weakly interacting particles have been proposed in order to solve remaining puzzles in supergravity and unified field theories. Many of these theories predict a departure of the gravitational inverse-square law in the form:

$$\phi(r) = -\frac{GM}{r}\left[1 + \alpha \exp\left(-\frac{r}{\lambda}\right)\right], \tag{29}$$

where α and λ are the dimensionless strength and the range of the Yukawa potential. The Poisson equation for gravitational potential, Equation 25, suggests a *null* test of the inverse-square law: If the inverse-square law is valid, the trace of the gravity-gradient tensor must vanish. Therefore, by summing the outputs of a three-axis in-line-component gradiometer in response to a locally generated time-varying gravity field, one can perform a null test of the inverse-square law [34].

We have carried out a series of laboratory tests of the inverse-square law using the SGG as a null detector [35, 36]. Figure 19 shows the source-detector configuration. The source was a 1500 kg lead (Pb)

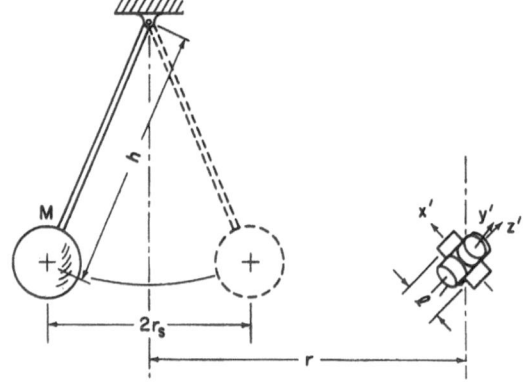

Figure 19. Source-detector configuration for the inverse-square law test.

pendulum suspended from the ceiling of the laboratory. The three-axis Model II SGG was used as the detector in the latest experiment. The gradiometer was in the "umbrella orientation," in which all three sensitive axes make the same angle, 54.7 degrees, with respect to the vertical.

In order to match the scale factors for the three orthogonal gradients, the SGG was rotated about the vertical into three azimuthal positions separated by 120 degrees, and each gradiometer output was summed over the three positions. The trace of the gradient tensor was therefore measured independently by each of the three single-axis gradiometers, allowing for a valuable cross check between the data. This method of rotating a single-axis gradiometer and summing its outputs over its three orthogonal orientations has the additional advantage of averaging out the horizontal components of all acceleration errors by symmetry [32].

Figure 20 shows the outputs of one of the SGG axes in three orientations over six pendulum cycles and their sum. The residual noise in the sum was mainly in the second and fourth harmonic of the pendulum frequency, which is due to the residual coupling of the gradiometer to the vertical acceleration of the floor driven by the pendulum motion. Data were collected and analyzed over thirty three nights to obtain the final null result of $(0.58 \pm 3.10) \times 10^{-4}$ E. This null result has been used to obtain upper limits of α as a function of λ. The strictest limit obtained with our experiment is $\alpha = (0.9 \pm 4.6) \times 10^{-4}$ for a

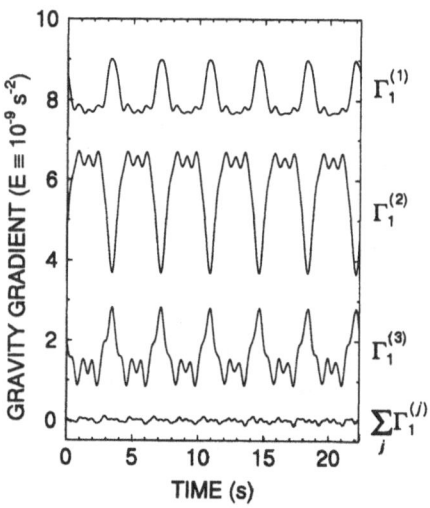

Figure 20. Inverse-square law data.

range of 1.5 m, where the error represents 2σ (two standard deviations).

Figure 21 shows the 1σ error in α plotted as a fuction of λ. The shaded region is the excluded region of the parameter space set by previous experiments. Thus, our experiment has improved the limit of the inverse-square law by more than an order of magnitude at a 1.5 m distance.

In a new laboratory experiment being planned [37], a long cylindrical shell will be used as a near-null source to reduce metrology errors and improve the limits of the inverse-square law to

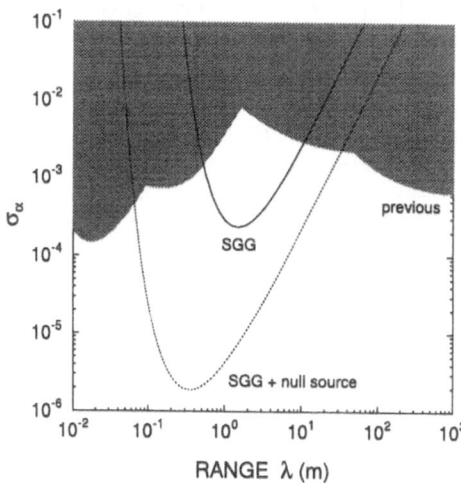

Figure 21. 1σ of the Yukawa potential obtained by experiment.

better than 1 part in 10^5 at $\lambda \approx 0.3$ m (the dotted line in Figure 21). A geophysical scale null experiment with a resolution of $\alpha \leq 10^{-5}$ at $\lambda \geq 10$ m is also envisioned in which a three-axis SGG will be moved vertically on a tower with respect to the Earth [37].

8.2. SPACE MISSIONS IN SEARCH OF NEW WEAK FORCES

Superconducting accelerometer technology enables precision fundamental physics experiments in space. An orbital test of the inverse-square law, which searches for a new force with a coupling strength eleven orders of magnitude weaker than gravity at a range of 10^2 to 10^3 km, could be performed by orbiting a three-axis SGG with a sensitivity of 10^{-4} E Hz$^{-1/2}$ in an elliptical orbit [9].

Another interesting space gravity experiment that has been proposed is detection of the gravitomagnetic field of the Earth [38]. General Relativity predicts existence of an analog of the magnetic field in gravity, the "gravitomagnetic field," which arises from a moving mass. GP-B (Gravity Probe-B) will search for the effect of such a field on a spinning body, the "frame dragging" precession of superconducting gyros in an Earth orbit [39]. The gradiometer experiment aims at detecting the gravitomagnetic Riemann tensor component directly. This field is expected to appear as an off-diagonal component in the geographic coordinate system with a strength of 8×10^{-8} E at a 650 km altitude.

The Newtonian background can be removed by flying a two-axis in-line-component SGG in the Earth-fixed frame, with one axis pointing northeast and the other northwest, and differencing the two signals [10]. With a gradiometer sensitivity of 10^{-5} E Hz$^{-1/2}$, which can be obtained by scaling up the SGG and/or reducing its differential-mode resonance frequency, one could resolve the gravitomagnetic field to one part in 10^2 in a year. This is comparable to the expected resolution of GP-B and would give a valuable cross check to the result, which may otherwise remain controversial.

In zero-g, a very soft and stable suspension can be realized for the proof masses by levitating them using magnetic fields generated by persistent currents. This eliminates the need for cumbersome mechanical springs and affords an opportunity to improve sensitive axis alignment. Superconducting differential accelerometers with magnetic levitation have been proposed for the Equivalence Principle experiment, as well as for a co-experiment searching for a mass-spin coupling force on STEP [40, 11].

In STEP, several pairs of nested cylindrical proof masses (see Figure 22), comprised of different materials, are orbited around the Earth. The centers of mass of the inner and outer proof masses are carefully matched to eliminate the sensitivity of the differential accelerometer to the gravity gradient. The shapes of the two proof masses are chosen to match their multipole moments to hexadecapole precision to reduce sensitivity to nearby mass motion such as the helium tide. If the Equivalence Principle is violated, the two masses in each pair would fall with different accelerations with respect to the Earth. When the proof masses are restrained by a restoring force, as in the STEP accelerometers, the resulting differential acceleration becomes periodic at the frequency of rotation of the differential accelerometer with respect to the Earth. This differential acceleration signal is read out by a superconducting circuit similar to the one shown in Figure 6.

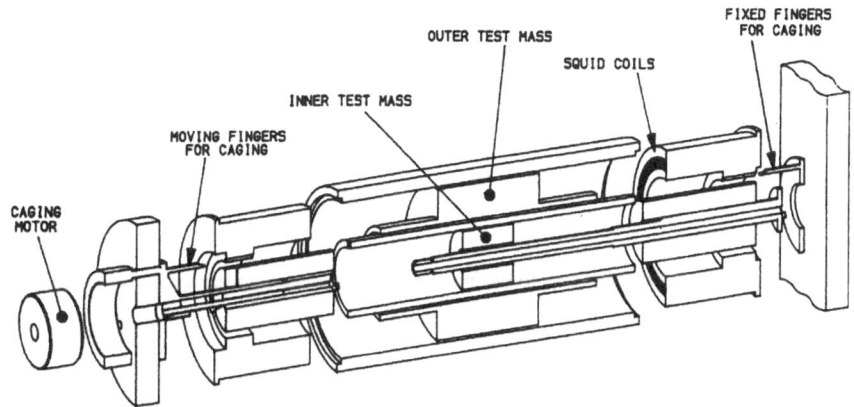

Figure 22. Equivalence Principle accelerometer on STEP.

Figure 22 is an exploded sectional view of one of the Equivalence Principle differential accelerometers for STEP. The inner and outer proof masses, constructed from different materials, are levitated radially on superconducting coils mounted on the surfaces of inner tubes. Their motions along the symmetry axis are detected and differenced by a superconducting sensing circuit coupled to a dc SQUID. With design parameters of $m = 0.2$ kg, $\omega_0/2\pi = 3 \times 10^{-3}$ Hz, $Q = 10^6$, $T = 2$ K, $\beta = \eta = 0.5$ and $E_A(f) = [1 + (0.1 \text{ Hz})/f]\ 5 \times 10^{-31}$ J Hz^{-1} (commercial dc SQUID), we find an acceleration sensitivity of 5×10^{-15} m s^{-2} Hz$^{-1/2}$ at a signal frequency of 10^{-3} Hz. With this sensitivity, the Equivalence Principle can be tested to better than one part in 10^{17} in a few months, which is a five-order-of-magnitude improvement over the best ground experiment.

8.3. GRAVITY MAPPING MISSIONS WITH SGG

Lateral variations in the strength of a planet's gravitational field provide one of the few remotely accessible measures of the state of stress and composition in the interior. Expanding Earth's gravity in spherical harmonics, our present knowledge stops at harmonic coefficients of about 40, which corresponds to a horizontal resolution of ~500 km. In the geophysics and geodesy communities, there is a pressing need to improve this spatial resolution to less than 200 km, as well as to reduce the errors in the long-wavelength components. Knowledge of gravity is equally important in understanding the interior of the planets and their evolution.

The NASA support for the SGG development has been primarily motivated by its application to gravity-mapping missions for the Earth and the planets. Mission studies have been carried out earlier by a science team led by Marshall Space Flight Center [41] and more recently by a joint team at Goddard Space Flight Center and the University of Maryland [42]. A flight test of the three-axis SGG on board the Space Shuttle has also been proposed [43].

The low-cost mission concept developed by Goddard is to put a three-axis SGG with a 10^{-3} E Hz$^{-1/2}$ sensitivity in a near-polar orbit with an initial altitude of ~350 km. The

592

spacecraft will not have reboost capability, and its orbit will be allowed to decay gradually to 200~250 km over a period of several months before it plunges into the atmosphere. With a common-mode rejection ratio of 10^7, the SGG will be able to perform at its intrinsic noise level without drag-free control. Attitude control will be provided by use of conventional reaction wheels and magnetic torquers.

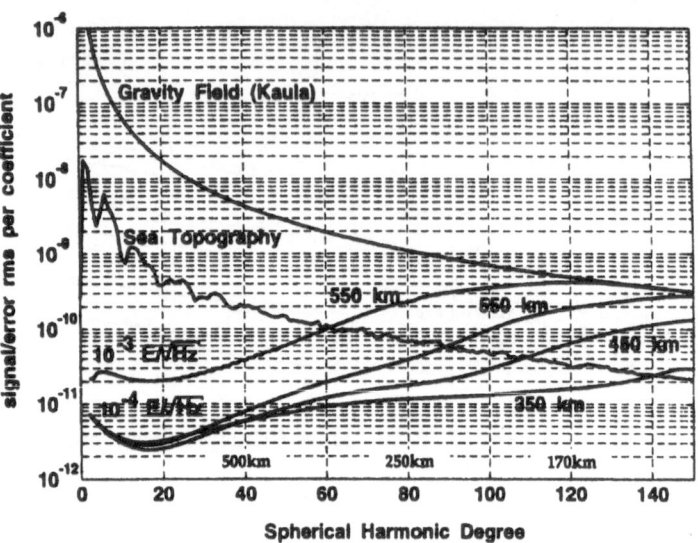

Figure 23. Expected resolution of Earth's gravity in various geodesy mission scenarios.

STEP will carry a single-axis SGG with a sensitivity of 10^{-4} E Hz$^{-1/2}$ as a geodesy co-experiment [11]. Each proof mass will be levitated magnetically, as will be discussed later in this section. The spacecraft will fly in a sun-synchronous near-polar orbit at an altitude of 400 km and will be under drag-free and attitude control to permit the SGG to operate at its full sensitivity.

Figure 23 shows the expected resolution of Earth's gravity for a number of mission scenarios: a 10^{-3} E Hz$^{-1/2}$ instrument at a 550 km altitude and a 10^{-4} E Hz$^{-1/2}$ instrument at 550, 450 and 350 km altitudes [11]. Expected signals from the solid Earth (Kaula) and ocean topography are also plotted. A 10^{-4} E Hz$^{-1/2}$ SGG at an altitude of 400 km, as envisioned for STEP, would determine the harmonic coefficients of the Earth gravity to degree 180 and would resolve gravity signals from ocean dynamics to degree 120, which correspond to resolutions of 110 and 170 km, respectively. Extrapolating the plots to 250~300 km, the effective altitude of the low-cost mission with a 10^{-3} E Hz$^{-1/2}$ sensitivity, indicates that the expected resolution of this mission is comparable to that of STEP. Hence, either mission would satisfy the scientific requirements for geophysics and geodesy, and vastly improve our understanding of Earth's dynamics and global change.

Figure 24 is an exploded view of a component accelerometer for the proposed flight instrument for the Shuttle test and the follow-up mission. Except for overall reduction in size and weight for accommodation in a small spacecraft, the instrument represents only a slight variation of our existing Model II SGG. Although the baseline was reduced by 30%, we were able to increase the dynamic mass by 20% and reduce the resonance frequency by 30% to meet the sensitivity goal of 10^{-3} E Hz$^{-1/2}$. All the accelerometer components will be made of Nb.

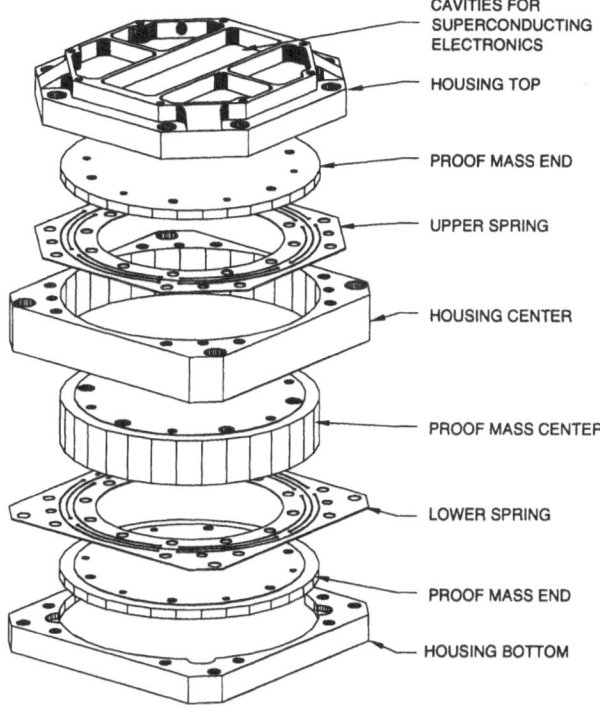

Figure 24. Exploded view of a component accelerometer for the SGG flight instrument.

Six accelerometers will be mounted on the six faces of a Ti alloy precision mounting cube, which will house three superconducting angular accelerometers of the type discussed in Section 6.1. Figure 25 shows the mounting cube and the angular accelerometers. Each accelerometer is cut from a single block of Nb by wire EDM. The mass moments with respect to the pivot will be balanced carefully before assembly. The entire SGG assembly will be soft-mounted in the dewar to provide vibration isolation at high frequencies. A cryogenic shaker will be incorporated to apply known accelerations for calibration and common-mode balance in orbit.

A geodesy mission with an SGG is also being studied by ESA. At the same time, a consortium led by Oxford Instruments is developing, under ESA support, a magnetically levitated SGG [44] with a design basically identical to that proposed for the STEP geodesy co-experiment [45]. A unique feature of this design is incorporation of superconducting coils for aligning the sensitive axes by means of persistent currents. This improves the inherent common-mode rejection of the gradiometer.

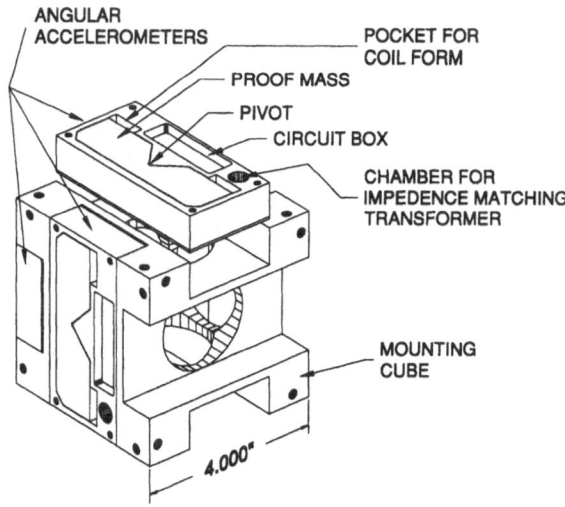

Figure 25. Superconducting angular accelerometers for the SGG flight instrument.

594

Figure 26 shows an end-on and a cross-sectional view of a magnetically levitated proof mass and associated coils. Each proof mass is made of Nb and has the shape of a belted cylindrical shell. It is levitated radially by storing persistent currents in the four elongated pancake

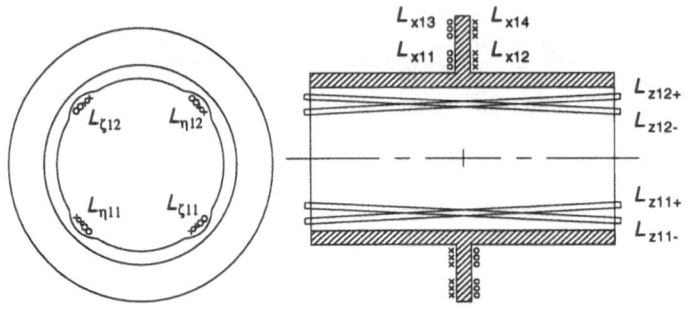

Figure 26. Proof mass and associated coils for a magnetically levitated SGG.

coils, $L_{\eta 11}$, $L_{\eta 12}$, $L_{\zeta 11}$ and $L_{\zeta 12}$, located on the four quadrants within the inner diameter of the proof mass. Four shallow circular indents are machined along the entire length of the proof mass to mate with the levitation coils. These create potential wells when they are aligned with the coils, and prevent unwanted rotation about the axis. Resistors are coupled to the levitation circuit to damp unwanted modes passively.

Two oppositely slanted meander-pattern coils, L_{z11-} and L_{z11+}, for example, interwoven with each other, are located in each quadrant, 45 degrees rotated from the radial levitation coils, near the inner surface of each proof mass. A meander-pattern coil, carrying a current, provides a repulsive force which has a strong nonlinear dependence with separation, and thus tends to align the proof mass along its axis. By varying the distribution of persistent currents in the oppositely slanted meander-pattern coils, one can "fine-tune" the alignment of the sensitive axes of the proof masses to make them truly collinear. In practice, this alignment is expected to be limited by second-order effects to $10^{-8} \sim 10^{-7}$ rad. Although the same degree of alignment could be obtained effectively by means of the "residual balance" in the mechanically suspended SGG, as discussed in Section 5.2, the direct alignment reduces second-order errors arising from the misalignment and the dynamic-range requirement of the instrument. This advantage, however, comes at the expense of increased complexity for the electrical circuit.

The differential and common-mode sensing is done by using the axial pancake coils located near the surfaces of the outer belts of the proof masses, L_{x11}, L_{x12}, L_{x13} and L_{x14}. These coils are connected in standard differencing circuits, similar to the one shown in Figure 6. With the mechanical suspension springs eliminated, the differential-mode resonance frequency, ω_0 in Equation 26, can be tuned to a very low value. The gradient signal frequency in the Earth orbit is below 0.03 Hz. Thus, $\omega_0/2\pi$ as low as 0.1 Hz could be chosen. This improves the intrinsic noise to better than 10^{-4} E Hz$^{-1/2}$, without a negative spring, for an instrument of the same size as the mechanically suspended SGG.

A magnetically levitated superconducting differential accelerometer has a potential for greater sensitivity and better common-mode rejection. For this reason, a basically identical accelerometer design has been proposed for the Equivalence Principle experiment in the STEP mission [45].

8.4. OTHER APPLICATIONS

Superconducting accelerometer technology can be applied to many other situations where measurement of a very small displacement, acceleration or pressure is called for. Basically the same measurement technique has been used to detect the quantization of dissipation in superfluid helium flowing through an orifice [46]. Viewed as a differential pressure gauge, the sensitivity of the device was 2×10^{-13} bar $Hz^{-1/2}$.

The superconducting accelerometer of Figure 1 can be turned into a sensitive thermometer by mounting the superconducting sensing coil directly onto the superconducting ground plane, thus eliminating the mechanical degree of freedom. The penetration depth of magnetic field into a superconductor is a sensitive function of temperature. Therefore, a temperature change will modulate the sensing coil inductance, as an acceleration does in the accelerometer, and the resulting current signal is detected by the SQUID. The superconducting penetration-depth thermometer has a potential sensitivity of 10^{-9} K $Hz^{-1/2}$, and a prototype thermometer has been demonstrated [47]. The device can be used as a bolometer in a particle detector or an infrared detector. A power sensitivity of 10^{-11} W $Hz^{-1/2}$ was achieved by employing thin-film coils and an inductance-bridge which was coupled to a dc SQUID [48].

9. Conclusions

In this paper, we have reviewed superconducting accelerometers, gravitational-wave transducers and gravity gradiometers. We have discussed the basic designs, the operating principles and the intrinsic noise of such devices. Where appropriate, we have also reported the performance of the devices. Reduced Brownian-motion noise and low-noise dc SQUIDs permit construction of extremely sensitive inertial instruments. The enhanced mechanical stability of materials at low temperatures and the ultimate stability of persistent currents, due to their quantum nature, result in extremely stable scale factors, allowing very high common-mode rejection ratios for superconducting differential accelerometers.

In order to improve the sensitivities of these superconducting instruments to the desired levels, the SQUID noise must be reduced further. In particular, in order to be able to open a window for observational astronomy with the new generation of gravitational-wave detectors, a sturdy, well-coupled, near-quantum-limited SQUID at frequencies around 1 kHz is required. This is a great challenge to SQUID designers but is being met successfully. On the other hand, the signal frequencies in most applications of the SGG are below 0.1 Hz. Therefore, the 1/f noise of the SQUIDs must be reduced in order to improve the sensitivities of the SGGs.

Applications of the superconducting accelerometer and gradiometer technology in fundamental physics experiments, as well as in geodesy, geophysics and other disciplines, have been discussed. In particular, the SGG technology has become the basis for several interesting space missions under study, such as STEP, the SGG follow-on mission to GP-B and SGG geodesy missions.

Acknowledgements

This work has been performed under a series of NSF, NASA, and U.S. Air Force grants and contracts. Currently, it is supported by NSF under grant PHY93-12229 and NASA under contract NAS8-38137 and grant NAG53104. I gratefully acknowledge the invaluable support of my previous colleagues at Stanford University and the entire research staff of the SGG laboratory at the University of Maryland during the course of this research. I especially thank Vol Moody, Ed Canavan, Gregg Harry and Thomas Stevenson for their help in preparing this manuscript, as well as their critical reading of the manuscript and many valuable suggestions.

References

1. Fairbank, W.M. *et al.* (1974) Search for gravitational radiation using low-temperature techniques, in B. Bertotti (ed.), *Experimental Gravitation*, Academic Press, New York, pp. 294-309.
2. Paik, H.J. (1974) Analysis and development of a very sensitive low-temperature gravitational-wave detector, Ph.D. thesis, Stanford University, Stanford, California.
3. Paik, H.J. (1976) Superconducting tunable-diaphragm transducer for sensitive acceleration measurements, *J. Applied Physics* **47**, 1168-1978.
4. Paik, H.J., Mapoles, E.R., and Wang, K.Y. (1978) Superconducting gravity gradiometers, in B.S. Deaver, Jr. *et al.* (eds.), *Future Trends in Superconductive Electronics*, AIP, New York, pp. 166-170.
5. Chan, H.A. (1982) Null test of the gravitational inverse-square law with a superconducting gravity gradiometer, Ph.D. thesis, University of Maryland, College Park, Maryland.
6. Moody, M.V., Chan, H.A., and Paik, H.J. (1986) Superconducting gravity gradiometer for space and terrestrial applications, *J. Applied Physics* **60**, 4308-4315.
7. Paik, H.J. *et al.* (1988) Global gravity survey by an orbiting gravity gradiometer, *EOS Trans.* **69**, 1601, 1610-1611.
8. Canavan, E.R., Paik, H.J., and Parke, J.W. (1991) A superconducting six-axis accelerometer, *IEEE Trans. Magnetics* **27**, 3253-3256.
9. Paik, H.J. (1989), Tests of general relativity in Earth orbit using a superconducting gravity gradiometer, *Advances in Space Research*, **9**, 41-50.
10. Mashhoon, B., Paik, H.J., and Will, C.M. (1989) Detection of the gravitomagnetic field using an orbiting superconducting gravity gradiometer. Theoretical principles, *Physical Review D* **39**, 2825-2838.
11. Blaser, J.-P. *et al.* (1993) Satellite Test of the Equivalence Principle (STEP), phase A report to the European Space Agency, SCI (93) 4.
12. Chan, H.A. and Paik, H.J. (1987) Superconducting gravity gradiometer for sensitive gravity measurements. I. Theory, *Physical Review D* **35**, 3551-3571.
13. Giffard, R.P. (1976) Ultimate sensitivity limit of a resonant gravitational wave antenna using a linear motion detector, *Physical Review D* **14**, 2478-2486.

14. Price, J.C. (1987) Optimal design of resonant-mass gravitational wave antennas, *Physical Review D* **36**, 3555-3570.

15. Heffner, H. (1962) The fundamental noise limit of linear amplifiers, *Proceedings of the IRE* **50**, 1604-1608.

16. Braginsky, V.B. and Vorontov, Y.I. (1975) Quantum-mechanical limitations in macroscopic experiments and modern experimental technique, *Soviet Physics Uspekhi* **17**, 644-650 [Usp. Fiz. Nauk. **114**, 41-53 (1974)].

17. Caves, C.M. *et al.* (1980) On the measurements of a weak classical force coupled to a quantum-limited mechanical oscillator. I. Issues of principle, *Review of Modern Physics* **52**, 341-393.

18. Press, W.H. and Thorne, K.S. (1972) Gravitational-wave astronomy, *Annual Review of Astronomy and Astrophysics* **10**, 335-374.

19. Richard, J.-P. (1984) Wide-band bar detectors of gravitational radiation, *Physical Review Letters* **52**, 165-167.

20. Cerdonio, M. *et al.* (1994) Ultracryogenic resonant antennae to detect gravitational wave bursts, *Nuclear Physics B* (Proc. Suppl.) **35**, 75-78.

21. Wellstood, F.C. (1995) Superconducting Quantum Interference Devices for gravitational wave detection, yearly report to the National Science Foundation.

22. Forward, R.L. (1971) Multidirectional and multipolarization antennas for scalar and tensor gravitational radiation, *General Relativity and Gravitation* **2**, 149-159.

23. Wagoner, R.V. and Paik, H.J. (1977) Multi-mode detection of gravitational waves by a sphere, in B. Bertotti (ed.), *Proc. International Symposium on Experimental Gravitation*, Accademia Nazionale dei Lincei, Rome, pp. 258-265.

24. Johnson, W.W. and Merkowitz, S.M. (1993) The truncated icosahedral gravitational wave antenna, *Physical Review Letters* **70**, 2367-2370.

25. Chan, H.A., Moody, M.V., and Paik, H.J. (1987) Superconducting gravity gradiometer for sensitive gravity measurements. II. Experiment, *Physical Review D* **35**, 3572-3597.

26. Paik, H.J. (1981) Superconducting tensor gravity gradiometer for satellite geodesy and inertial navigation, *J. Astronautical Sciences*, **29**, 1-18.

27. Forward, R.L. (1979) Electronic cooling of resonant gravity gradiometers, *J. Applied Physics* **50**, 1-6.

28. van Kann, F.J. *et al.* (1990) Laboratory tests of a mobile superconducting gravity gradiometer, *Physica B* **165**, 93-94.

29. Paik, H.J. (1994) Electromechanical transducers and bandwidth of gravitational-wave detectors, in E. Coccia *et al.* (eds.), *Gravitational Wave Experiments*, World Scientific, Singapore, pp. 201-219.

30. Parke, J.W. *et al.* (1984) Sensitivity enhancement of inertial instruments by means of a superconducting negative spring, in D.H. Collan *et al.* (eds.), *Proc. 10th International Cryogenic Engineering Conference*, Butterworth, Surrey, pp. 361-364.

31. Chan, H.A. *et al.* (1985) Superconducting techniques for gravity survey and inertial navigation, *IEEE Trans. Magnetics* **MAG-21**, 411-414.

32. Parke, J.W. (1990) Null test of the gravitational inverse square law and the development of a superconducting six-axis accelerometer, Ph. D. thesis, University of Maryland, College Park, Maryland.

33. Colombo, O.L. (1992) Air gravimetry, altimetry, and GPS navigation errors, in O.L. Colombo (ed.), *From Mars to Greenland: Charting Gravity with Space and Airborne Instruments - Fields, Tides, Methods, Results*, Springer-Verlag, New York, pp. 261-271.

34. Paik, H.J. (1979) New null-experiment to test the inverse-square law of gravitation, *Physical Review D* **19**, 2320-2324.

35. Chan, H.A., Moody, M.V., and Paik, H.J. (1982) Null test of the gravitational inverse square law, *Physical Review Letters* **49**, 1745-1748.

36. Moody, M.V. and Paik, H.J. (1993) Gauss's law test of gravity at short range, *Physical Review Letters* **70**, 1195-1198.

37. Paik, H.J. and Moody, M.V. (1994) Null test of the inverse-square law of gravity, *Classical and Quantum Gravity* **11**, A145-A152.

38. Braginsky, V.B. and Polnarev, A.G. (1980) Relativistic spin-quadrupole gravitational effect, *Soviet Physics JETP Letters* **31**, 415-418.

39. Everitt, C.W.F. (1974) The gyroscope experiment. I. General description and analysis of gyroscope performance, in B. Bertotti (ed.), *Experimental Gravitation*, Academic Press, New York, pp. 331-360.

40. Worden, Jr., P.W. and Everitt, C.W.F. (1974) The gyroscope experiment. III. Tests of the equivalence of gravitational and inertial mass based on cryogenic techniques, in B. Bertotti (ed.), *Experimental Gravitation*, Academic Press, New York, pp. 381-402.

41. Morgan, S.H. and Paik, H.J. (eds.) (1988) Superconducting Gravity Gradiometer Mission. II. Study Team technical report, NASA Technical Memorandum 4091.

42. Paik, H.J. *et al.* (1995) Mission concepts for a superconducting gravity gradiometer Earth survey to establish a baseline for global change studies, unpublished paper.

43. Canavan, E.R. *et al.* (1995) Predicted performance of the superconducting gravity gradiometer on the Space Shuttle, to be published in *Cryogenics*.

44. Paik, H.J. and Lumley, J.M. (1995) Superconducting gravity gradiometers on STEP and GEM missions, in T.J. Sumner (ed.), *Proc. Symposium on Fundamental Physics in Space*, London, England.

45. Paik, H.J. (1995) Principles of STEP accelerometer design, in T.J. Sumner (ed.), *Proc. Symposium on Fundamental Physics in Space*, London, England.

46. Avenel, O. and Varoquaux, E. (1985) Observation of singly quantized dissipation events obeying the Josephson frequency relation in the critical flow of superfluid ^{4}He through an aperture, *Physical Review Letters* **55**, 2704-2707.

47. Moody, M.V. and Paik, H.J. (1984) A superconducting penetration depth thermometer, in U. Eckern *et al.* (eds.), *Proc. 17th International Conference on Low Temperature Physics*, North Holland, Amsterdam, pp. 407-408.

48. Sauvegeau, J.E., McDonald, D.G., and Grossman, E.N. (1991) Superconducting kinetic inductance radiometer, *IEEE Trans. Magnetics* **27**, 2757-2760.

THE USE OF SQUIDS FOR NONDESTRUCTIVE EVALUATION

G. B. DONALDSON, A. COCHRAN AND D. McA. McKIRDY
Department of Physics and Applied Physics
University of Strathclyde
Glasgow G4 0NG
Scotland

Abstract

We open with an account of early work on SQUID NDE, describing remote magnetometry and galvanometry, and the inverse problem. Then we concentrate on our own progress in mapping the magnetic fields produced by eddy-currents as flaws in conducting structures distort their flow. The Finite Element Method, the Volume Integral Method and other approaches to the problem of deducing flaw locations are reviewed. Next we turn to developments with high-T_c SQUID systems, concentrating on 'electronic gradiometers' where the outputs of more than one SQUID are differenced to eliminate common mode fields. Flaws breaking test piece surfaces both near-side to the detector and far-side have been imaged. Moreover they have been detected also in the lower layers of multi-sheet specimens.

Flaw detection in aircraft and pressure-vessel structures are discussed, and we end by looking at the prospects for bringing the technology into widespread use.

1. Introduction

It is over ten years since the first reports of SQUID Nondestructive Evaluation (NDE) [1, 2] and many papers since have included reviews catering to research groups entering the field. The most recent has been the 1994 Applied Superconductivity Conference paper by Wikswo [3]. Previous NATO ASI Volumes also describe early developments by the present authors [4, 5]. In this Chapter we do not repeat that material: instead, after a brief account of early work, we concentrate on recent progress with eddy-current techniques in our own group at Strathclyde. We look first at the generation of low frequency eddy-currents and the mapping of their magnetic fields as their flow is distorted by surface-breaking and deep flaws in conducting structures. Next we discuss issues associated with the use of the Finite Element Method (FEM) and other techniques for deducing the actual flaw location and shape from the measured maps. Finally we turn to practical developments with high temperature superconductor devices and to issues involved in the promotion of SQUID NDE.

Throughout, we are concerned with the development of NDE methods that might be used in real engineering situations, rather than within carefully screened enclosures. It is therefore appropriate that we should end by reviewing other novel approaches to NDE, and by considering what the conservatism of the field implies for the introduction of SQUID technology.

H. Weinstock (ed.), SQUID Sensors: Fundamentals, Fabrication and Applications, 599–628.
© 1996 *Kluwer Academic Publishers.*

2. SQUIDs and NDE

2.1 HISTORICAL - WHY USE SQUIDS FOR NDE?

SQUIDs are usually presented as the most sensitive detectors of electromagnetic energy available, able to detect brain activity (5-10 fT-Hz$^{-\frac{1}{2}}$), or achieve quantum-limited performance (6×10^{-34} J-Hz$^{-\frac{1}{2}}$) in gravitational wave detection. It can be a surprise, therefore, to learn that they can also be used to detect cracks in plates of ferromagnetic steel and in eddy-current NDE, where detectors have traditionally been many orders of magnitude *less* sensitive than the best SQUIDs.

We emphasise, therefore, that although the sensitivity of SQUIDs is an important issue, they are usually employed in NDE for other special properties. These include (see [4] and [5]) the **vector properties of SQUID magnetometers, and the ability to make gradiometers** of first or higher order using balanced coils connected to the input transformer of a single SQUID. This allows field differencing before detection and eliminates the need for high electronic linearity if gradients are to be obtained by differencing the outputs of two separate magnetometers. Further properties exploited include those of **low frequency** operation (to DC if necessary) without the loss of sensitivity associated with inductive detectors, and the fact that because they are precisely periodic (in Φ_o) as flux detectors, SQUID magnetometers maintain their **differential field sensitivity in high bias fields.**

However, high SQUID flux sensitivity is not set aside completely. For a given *field* sensitivity the SQUID can operate with much smaller pickup coils than other sensors. It is even possible to use the bare SQUID as a direct field sensor. As explained in [4], the spatial resolution of a mapping method that employs pick-up coils of diameter D and a stand-off distance H is approximately the larger of D and H, (unless, as we shall see in §5.3.1, the *excitation* can be localised). Thus the SQUID, with its smaller coils, is in principle capable of higher spatial resolution. Another use of SQUID sensitivity can be with higher order gradiometers, which may be too insensitive for biomagnetism, but useful in the higher signal fields of NDE, permitting operation in "real environments" [6]. Finally there is the option simply to use less powerful excitation signals than in traditional NDE, which may be instrumentally more convenient, for example in swept frequency techniques.

SQUID NDE has used two principal approaches, which we have described in [4] and [5] as *remote magnetometry* and *remote galvanometry*. In the first, the sensor system, comprising a pick-up coil, the SQUID and its magnetic screen, is used to map the fields produced by magnetic anomalies in the test subject. These can be caused by premagnetised inclusions in a non-magnetic matrix, or by regions of unusual permeability in specimens temporarily magnetised by a locally polarising field.

In remote galvanometry the sensor system is used to detect or map fields produced by injected, induced or spontaneously-generated currents in test subjects. At a simple level, the very existence of a field may be sufficient to establish that a corrosion current is present in a test cell. At the other extreme, sophisticated processing of a magnetic field map may enable the determination of the actual pattern of current flows and hence the distribution of defects in a complex specimen.

2.2   EARLY REMOTE MAGNETOMETRY

The use of SQUID systems to detect anomalous magnetic features was first reported by Weinstock [1]. He showed that it is possible to observe domain changes in the remanent magnetic state of steel bars subjected to both elastic and inelastic stress-strain regimes. Barkhausen jumps could also be detected [7]. Work is in progress to develop a scanning system for the study of this phenomenon in extensive specimens [8, 9].

Polarising methods were introduced by the Strathclyde Group [2, 10]. Following an early biomagnetic technique [11], they used a 10-40 mT persistent mode superconducting coil to magnetise a mild steel plate beneath the planar gradiometer pick-up loops of a screened low-T_c SQUID. Undamaged regions of such a plate have a large permeability ($\mu_p \sim 700$), while surface-breaking slots are equivalent to regions with permeability $\mu_d = 1$. Along the line of a fatigue crack, regions of work hardening reduce the permeability to $\mu_d \sim 300$ - 500. When the plate is x-y scanned beneath the cryostat, undamaged regions produce no change in the flux linking the pickup coils.

The passage of a flaw, however, corresponds to the appearance of an anti-dipole of moment $M_d \sim -(\mu_p - \mu_d)BV$, where V is the flaw volume and B the magnetising field. This equivalent anti-dipole generates a net change of flux in the gradiometer coils, which is detected by the SQUID. Examples of the apparatus used and of a typical map produced by a flawed plate can be seen in Figure 1 and Figure 2.

In principle, SQUID sensitivity is good enough to detect the equivalent anti-dipolar field of a flaw as small as 35 μm^3 with a polarising field of 10 mT and a stand-off of 5 cm. In practice, residual magnetic effects and spatial fluctuations in plate permeabilities limit open crack detectability to 1 mm^2, corresponding to a distortion field of 10 nT at the sensor. This is adequate to detect flaws in underwater steel pipes through their surface coatings, but their edge effects and other magnetic inhomogeneities make it

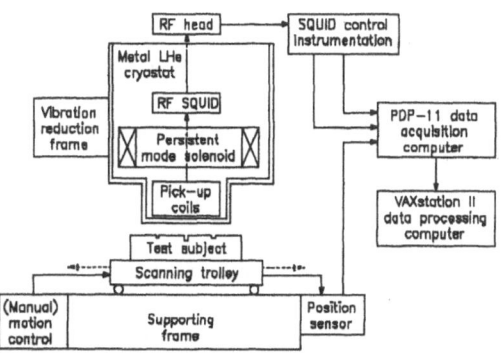

Figure 1. A SQUID NDE system for remote magnetometry.

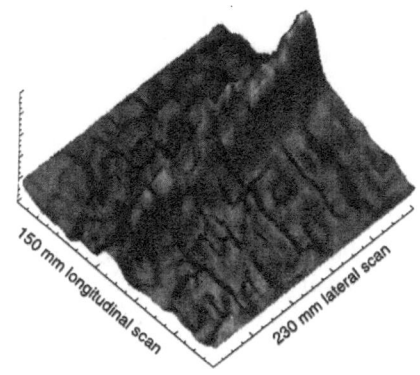

Figure 2. A fatigue crack in a mild steel plate detected by remote magnetometry.

impossible to detect cracks in welds.

A polarising method was used to observe ageing effects in non-ferromagnetic duplex stainless steel [12], by detection of the small amounts of ferritic phase that precipitate out in regions which have been thermally stressed. It was applied, too, to weakly diamagnetic materials, when holes drilled in plexiglass ($\chi \sim 10^{-5}$) were detected by Thomas *et al.* [13]; such measurements, however, can only be conducted in the magnetically quiet environment of a mu-metal enclosure.

2.3 EARLY REMOTE GALVANOMETRY

A buried metal pipe can be located quite accurately using a SQUID, by passing a small current along the pipe and detecting the separate components of the field produced outside it [1]. Indeed for pipes at depths up to 1.6 m it was also possible to detect the effects of large holes and welds as they diverted the current and distorted the field. Leakage currents into surrounding soil through damaged coatings (holidays) also have been detected [14].

Another early use for galvanometry was to detect corrosion [15, 16]. Fields produced by electrolytically-generated corrosion currents were sensed with a SQUID, and studies of their magnitudes and frequency spectra showed distinctions between various processes such as diffusion, crack propagation and etch-pit formation.

2.4   MAPPING AND THE INVERSE PROBLEM

Remote magnetometry and remote galvanometry each produce *field* maps, rather than maps of the corresponding *sources*, and it is often said that Helmholtz's theorem implies that a map $\mathbf{B}(\mathbf{r'})$ cannot be uniquely inverted to determine the distribution of magnetisation $\mathbf{m}(\mathbf{r})$ or currents $\mathbf{J}(\mathbf{r})$ producing it. This is strictly true, and is the reason that biomagnetic inversions are often iterative, having to start from heavily constrained models, which already contain some realistic physiology. We will discuss analogous iterative modelling approaches in NDE in §4.3.

$\mathbf{B}(\mathbf{r'})$ *can*, however, be inverted directly in two specific NDE situations, essentially because each has its own built-in constraints. In the first case both specimen and map are essentially two dimensional, while the second refers to 3D maps associated with low susceptibility insulators.

2.4.1 *The two-dimensional inverse problem*

Roth *et al* [17] were the first to describe an approach to calculating the current flow $j(x, y)$ in a plate or other two-dimensional structure from the field component $B_z'(x', y', z')$ perpendicular to the plate at a height z' above it. A Fourier transform of the field into a spatial-frequency spectrum is followed by convolution with an appropriate pair of Green's functions, the application of a low-pass filter to eliminate noise terms, and a reverse transform to give the current flow. The results obtainable are illustrated in Figure 3 (a) and (b), where the field measured above a printed circuit board is processed to show the current actually flowing in the tracks on the board.

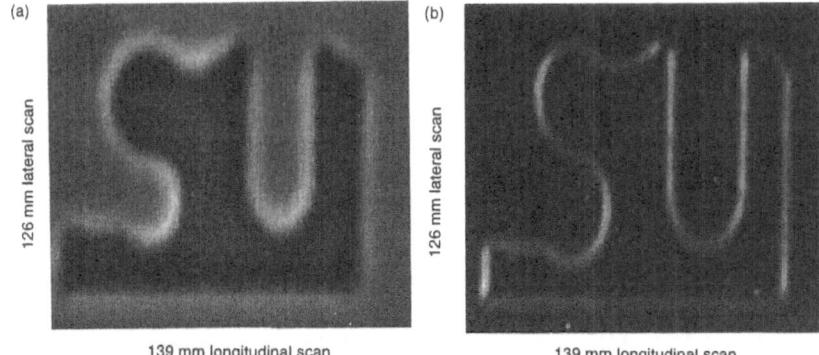

Figure 3. (a) Field measured above printed circuit board track carrying a current, (b) Current pattern computed from field map.

The solution gives a good estimate of the x and y components of the actual current distribution, provided it is confined to a thickness small compared with the stand-off distance. This will happen if the specimen is thin or if the (a.c.) current is confined to a skin depth close to the surface. Otherwise the solution is a weighted average of the current over the depth within which it flows.

2.4.2 Three-dimensional tomography

The 3D inverse problem *can* be solved exactly under the constraints that no currents flow and that the magnetisation in any volume element (voxel) of the specimen does not influence the magnetisation in any other voxel. These constraints are met in weak paramagnetic or diamagnetic insulators. The Vanderbilt group has demonstrated a formalism [18] in which SQUID measurements would be made over a number of surfaces outside the specimen, itself sequentially polarised by fields oriented in a number of directions. They can be used to generate a soluble set of linear simultaneous equations in which the unknowns are the susceptibilities χ_{ijk} of the discrete voxels (ijk). In a series of modelling studies, they have shown that holes ($\chi = 0$) in a plexiglass host ($\chi \sim 10^{-5}$) should be locatable by this tomographic technique.

2.5 IMPROVED RESOLUTION WITH LOW-T_C SQUID SYSTEMS

Much of this chapter concerns developments in applying high-T_c SQUIDS to NDE. However, because high-T_c systems are not yet mature or readily field portable, it is important also to recognise recent progress with low-T_c systems, noting especially improvements in spatial resolution. This implies reductions in the stand-off H and in the pickup-coil diameter D.

Reducing H required cryogenic developments to reduce the separation between the 4.2 K space, containing the superconductive pickup coils, and the room temperature NDE subject. Several designs [19, 20, 21] have achieved coil-to-outside-world distances of 1-1½ mm. Over the same time, coil diameters D have been reduced to about 1 mm, giving spatial resolutions of about 1-2 mm.

A typical design is our own (Figure 4), where the 4.2 K coil is pressed against the thin-walled end of a narrow liquid helium cryostat tail. The tail pierces the

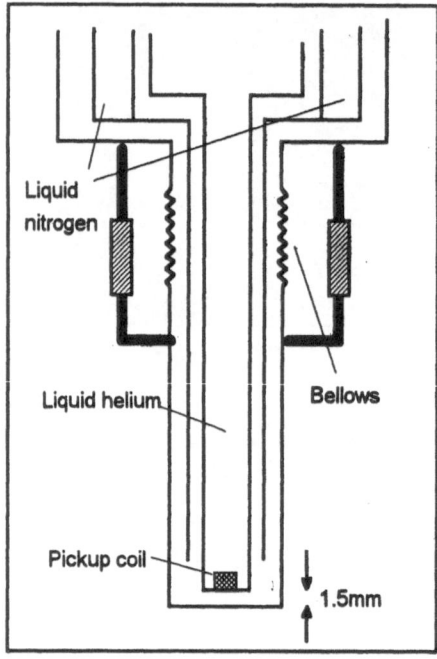

Figure 4. Schematic of Strathclyde cryostat with adjustable (1.5 mm minimum) 4.2 K to room temperature distance [19].

77 K shield, and nearly contacts the slender outside wall of the cryostat. This is achieved by a mechanical movement which compensates for differential contractions during cool-down. In Figure 4 an external bellows adjusts the outside wall position; another design [20] relies on an internal screw adjustment of the coil position.

If circumstances allow the specimen to be cooled to the same temperature as the pickup coil, much smaller stand-offs become possible. In the SQUID microscope [22], coils with $D \sim 10 \ \mu m$ can be brought within $H \sim 10 \ \mu m$ of 4.2 K samples. Resolutions on a scale of a few μm become possible. The microscope has revealed details of how flux vortices move within superconductive structures such as the thin film washers of planar SQUIDs [23], allowing new understanding of 1/f noise in these devices to emerge.

3. Recent Developments in Current Mapping

We now focus- see §1- on recent progress in our own group.

3.1 INTRODUCTION

A crack, which we take to be a thin sheet of zero conductivity, distorts the flow of any current that encounters it. The use of magnetic sensors to detect the fields produced by such distortions is not new. Within the general scope of *Eddy Current Testing* (ECT) [24], the NDE method of *AC Field Measurement* (ACFM) is well established. It usually involves an excitation coil or yoke which induces currents in the test specimen. The coil is rigidly mounted in the same module as one or more field sensors. These are usually either Faraday coils, whose output measures the rate of change of the field produced by the induced eddy-currents, or else flux gates. The outputs are made the basis of an appropriate display. Thus when the excitation/detector module is passed above a crack, the field and phase distortion produce a characteristic signature (such as a "butterfly pattern") on a screen.

Faraday systems are subject to several limitations. The induced eddy-currents and the voltages in the pickup coils both depend on rate of change of flux. Thus the magnitude of the resultant signal has an f^2 dependence on the excitation frequency f.

Table 1. The skin depth $\delta = (\rho/\pi\mu\mu_0 f)^{1/2}$ for various materials and frequencies.

	Aluminium	Stainless Steel	Mild Steel	Graphite
Resistivity $\times 10^{-8}$ Ωm	2.7	43	10	6200
Relative permeability μ	1	1	800	1
Frequency f	Skin depth mm			
1 Hz	83	330	5.6	4000
100 Hz	8.3	33	0.56	400
10 kHz	0.83	3.3	0.056	40
1 MHz	0.083	0.33	0.0056	4

This has limited operation of such systems to frequencies above about 1 kHz, when the currents tend to be confined to skin depths smaller than typical crack depths (see Table 1). Moreover in the 1-50 kHz range the need for adequate low frequency signal requires large area coils, which reduces spatial resolution.

3.2 USE OF SQUIDS IN EDDY-CURRENT TESTING

Although the currents induced in a plate by a given excitation field are frequency dependent, SQUID-based detection systems measure the *flux* produced, and not its rate of change. As a result, and also because of their high sensitivity, SQUIDs can be used to much lower frequencies than normal coil sensors. (Indeed, with current *injection*, rather than *induction*, DC operation is possible.) Moreover, with SQUID gradiometry, the effects of low frequency background fields, such as the Earth's field, its daily variations, and the laboratory 50/60 Hz sources, can be reduced significantly.

3.2.1 *Basic mapping techniques*

Figure 5 shows the basic principles of mapping by reference to low-T_c SQUID gradiometer measurements made above an artificially-cracked mild steel plate (Figure 5 (a)) [25]. Figure 5 (b) is a map of the distortion of the eddy-current fields generated in the plate by an excitation coil wound around the cryostat and carrying a 5.25 mA current at 175 Hz. The undistorted fields are about 6 μT pk-pk, but the size of the maximum distortion in Figure 5 (b) is about 10 nT pk-pk. It clearly reveals the flaw, even though it has closed up, and is invisible to the eye.

Finally, Figure 5 (c) shows the result of applying a Roth *et al.* [17] inversion to a map obtained by injecting a 190 Hz, 1.73 A current diagonally across the plate.

Similar experiments have been done by the Vanderbilt group on simulated lap joints in aircraft structures, using both injected and excited currents. They reveal defects in sub-layers, such as cracks at the edges of rivet holes and material loss due to interlayer corrosion. It is possible even to detect flaws that break the surface of the side of a plate *distant* from both excitation and SQUID coils, as we shall show in §5.3.3.

606

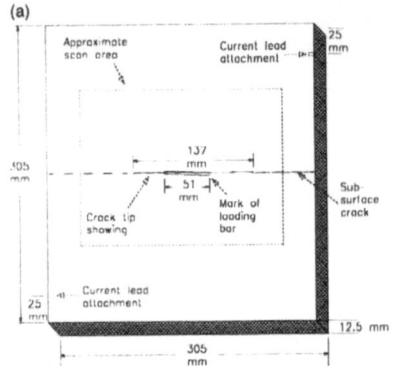

(a)

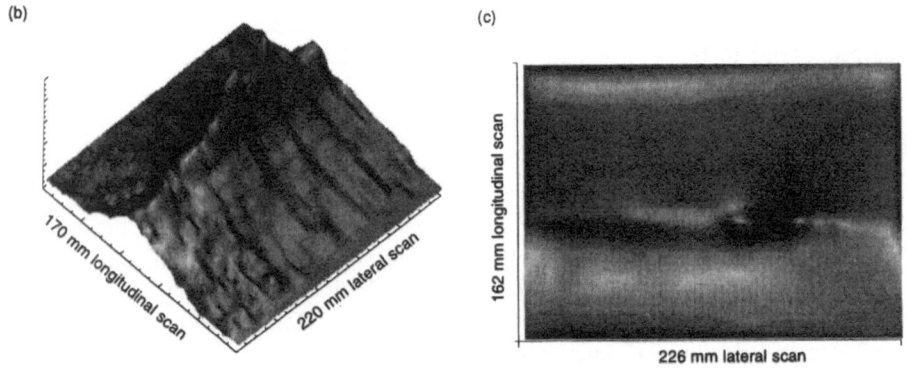

(b)

(c)

226 mm lateral scan

Figure 5. (a). Fatigue crack induced in a mild-steel plate (b) Map of distortion of 175 Hz eddy-current fields, excited by 5.25 mA coil current (c) Fourier inversion of 190 Hz injected-current map.

3.2.2 *Depth Studies*

Figure 6 shows how by varying the excitation frequency and exploiting the skin-depth magnitudes of Table 1, one can obtain information about the depth (t) of a surface breaking flaw. At low frequencies, as in Figure 6 (a), where $f \ll \rho/\pi\mu\mu_0 t^2$, so that the skin depth $\delta(f) \gg t$, induced currents flow at distances far below the bottom of the crack. It is therefore able to divert only a fraction of order t/δ of the total current, producing only a small distortion of the measured magnetic field. At high frequencies, by contrast, when $\delta(f) \ll t$, virtually all of the current will be diverted by the crack as in Figure 6 (b), and major changes will be observed in the field map $B(x', y')$ made above the specimen.

This principle was demonstrated earlier [26], when we studied a mild-steel plate with a surface-breaking semicircular crack of average depth 0.8 mm. SQUID measurements were made as a function of frequency at position C, with the sensor coil

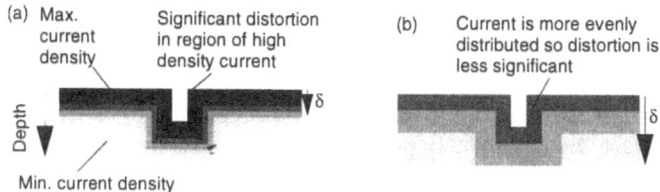

Figure 6. Schematic of eddy-current flow (a) at low frequency when the skin-depth exceeds the crack depth (b) at high frequency when the skin depth is much less than the crack depth. The crack causes significant field distortion only in the second case.

just above the crack, and at D, distant from it. The ratio $S(f) = B_C(f)/B_D(f)$ was close to unity at low frequency because the current flow was uniform to depths much greater than the crack and so was little affected by it. However, as f was increased a sharp anomaly was observed in $S(f)$ at $f = 180$ Hz (corresponding to $\delta = 0.6$ mm). As f increased still further, $S(f)$ showed more changes, because as the current became confined ever closer to the surface, the effective length of the crack increased, and the distortion pattern continued to change.

Another example is illustrated in Figure 7. Here the sample is a 13 mm-thick aluminium plate with a 6.5 mm deep, 40 mm-long slot cut in the *underside* (remote from the sensor). Measurements were made above the slot at about 5 mm from one end (C), and at a remote position (D). This time it was necessary to sense the field component parallel to the plate, because the detector was a high-T_c Mr SQUID- see §5.3.1. As Figure 7 (a) shows, the amplitude ratio $S(f) = B_C(f)/B_D(f)$ again shows three distinct frequency regions. At low frequencies, it is essentially unity, because the current is uniform throughout the plate, and any distortion produced by the (lower) surface breaking crack is on a long length scale, with little of the vertical component needed to produce a horizontal field. At about the frequency corresponding to the flaw depth ($\delta = 6.5$ mm at $f = 113$ Hz) there is once again an anomaly, though it extends over a wider range (100 Hz-3 kHz) than before. It appears because there is now significant distortion of the eddy-current into the z-direction. Finally at higher frequencies, S again becomes constant and returns to unity, because the current flows almost entirely close to the upper surface, is not affected by the crack at all. Figure 7 (b) shows that anomalies also appear in the relative phases of $B_C(f)$ and $B_D(f)$.

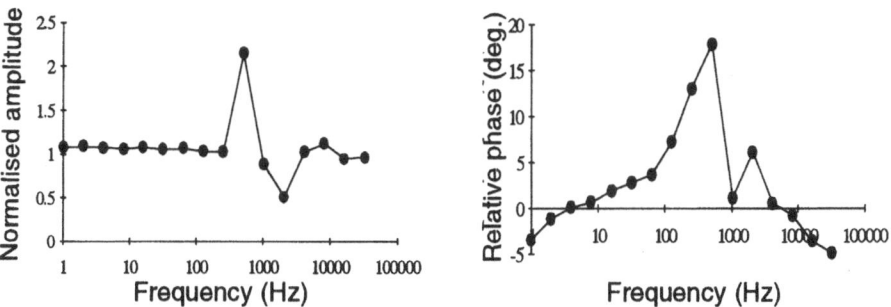

Figure 7. Swept-frequency determination of approximate depth of flaw breaking the far surface of a plate. Measurements are made above the crack (at C) and above an unflawed region (at D). (a) Frequency variation of the ratio $B_C(f)/B_D(f)$. (b) Frequency variation of the relative phases of $B_C(f)$ and $B_D(f)$.

These examples show that a crude measure of the depth of the crack can be obtained from the frequency at which $S(f)$ departs from 1, but they do not amount to a technique for determining the entire crack profile. Even the question of whether S *increases* or *decreases* at the critical frequency depends in detail on the precise dimensions of the excitation and pickup coils, on their positions and orientation relative to the crack and on the detailed shape of the crack itself. The solution requires *modelling* approaches, to which we return in §4.

3.3 EDGES

For any scheme of magnetic NDE, specimen *edges* can appear to be the biggest 'flaws'. They can cause large distortions of the currents that are to be sensed, and can mask smaller defects even when the latter are quite far from the edges. In this respect eddy-current methods are superior to injected current ones. With *injection*, current density is roughly uniform over the cross -section of the (unflawed) plate. As a result, any edge is a source of distortion and could generate a distortion field that might be detected by the SQUID at a long distance away. However *eddy-currents* are confined to a region beneath the excitation coil: when the coil is far away from any edge, the eddy-current density at that edge will be small, and the edge will produce negligible distortion field.

Injected current counter-flow schemes have been constructed in which, in the absence of any flaws, there is almost no field to detect. For example, in Figure 8 current is passed into a plate (which may or may not be flawed) and returned through an unflawed plate of matching dimensions, insulated from but just below the first. Provided the current flow patterns are the same in the upper and lower plates, and the plates are not too thick, their edge fields virtually cancel. A flaw in the upper plate will then produce a signal characteristic of the flaw itself, unmasked by geometry effects in the underlying structure. By such means, flaws have been seen in the deepest parts of so-called simulated airframe structures involving as many as seven layers of aluminium [27]. However it is important to realise that no airframe structure consists of successive layers of identical geometry, and it is very difficult to conceive of appropriate injection schemes for real situations. It follows that such approaches have few practical prospects.

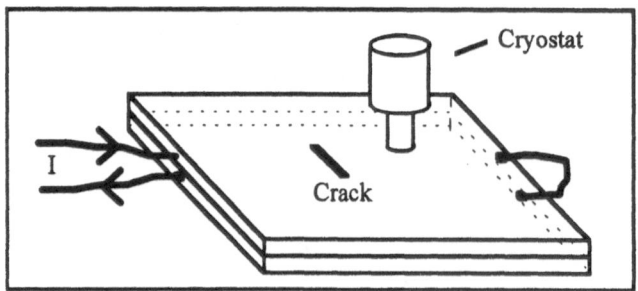

Figure 8.   Counter-current cancelling of edge effects.

4. Modelling and the Inverse Problem

4.1 INTRODUCTION

As discussed in §2.4.1, 2D inverse methods have been developed to transform a field map $B_z(x', y', z')$ into the current distribution $j(x, y)$ which produces it. Frequency scanning can provide an element of depth profiling. However, these procedures do not generate the distribution of conductivity $\sigma(\mathbf{r})$ which would properly describe a flaw. (Within a crack $\sigma = 0$, but embrittled regions have finite conductivities.) There are no analytical techniques available for taking that further step. To study the problem we have therefore turned to a number of mathematical modelling approaches- the finite element method (FEM), the volume integral method (VIM), and numerical calculation of responses to elementary sources.

There are other reasons for modelling studies. From their results, one can optimise not only the excitation system, but also the sensor design, including both the SQUID and gradiometer coils. Moreover, the simplicity of the experimental parameters of SQUID systems makes them convenient to use in the verification of new NDE techniques. For example, high-T_c SQUIDs behave as ideal point sensors with a uniform spectral response over several frequency decades and are thus easily included in any model.

A further rationale for mathematical modelling is that the forward problem is comparatively easy, especially for simple crack models such as rectangular or semicircular cuts. In principle one can compute the ideal current flow $J_i(\mathbf{r})$ and the ideal field $B_i(\mathbf{r}')$ for any excitation coil geometry and frequency. One could then tackle the inverse problem by comparing the ideal values with the measured ones, and iterating to minimise the differences by generating more realistic models that converge on the actual crack shape. The 'simple parameters' of SQUID systems could be useful here too: they should be easy to build into the automatic processing routines which must evolve if inverse problems are to be routinely tackled by iterating forward solutions.

4.2 FINITE-ELEMENT METHOD (FEM) MODELLING

FEM is attractive because of its flexibility. It is not restricted to 1D or 2D, and allows mesh arrangements that yield current and magnetic field distributions for any crack geometry. Results using the commercially available OPERA and ELEKTRA packages [28] are shown in Figure 9. These depict the currents which a circular coil induces, at 1 Hz, 256 Hz and 8192 Hz, near an air-filled rectangular slot cut in the lower surface of an aluminium plate. The model results confirm the intuitive account in §3.2.2: currents are relatively little distorted at low frequencies, become significantly perturbed when the scale length of distortions $\delta(f)$ shortens to become comparable to the flaw dimensions, and flow only above the slot at high frequencies.

The result of a crucial test of such calculations is shown in Figure 10 which is the FEM equivalent of the $S(f)$ result, shown in Figure 7 (b) for the crack created in the lower surface of an aluminium plate. A matching three-regime behaviour is observed, though the anomaly has a modelled maximum of only 1.1 as opposed to the measured value of 2.5. The discrepancy arises for two reasons. The first is the extreme sensitivity of the field component B_y to the sensor position relative to the slot, and to the stand-off value. A 1 mm error in positioning can alter B_y by a factor of 100. The second relates to tilt errors in the cryostat. The unflawed plate generates a $B_z \sim 100$ times bigger than the typical B_y produced by a flaw. A cryostat misalignment of as little as 1 degree can result in a B_z contribution that reverses the sign and magnitude of the anomaly.

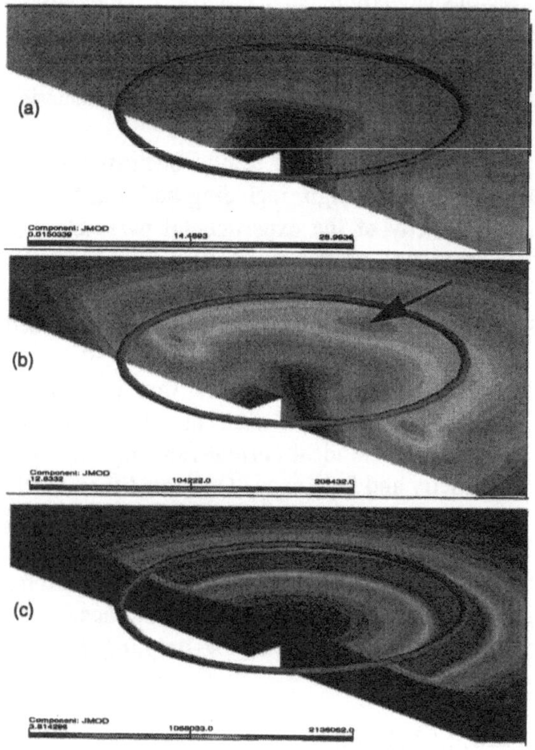

A long term solution would be to model and measure B_z rather than B_y. This should be done in the context of complete maps of the plate, rather than in ratios of values at just the two points C and D. However FEM at this level would require impracticably large computing resources. Just one OPERA/ELEKTRA map for a plate with a simple rectangular flaw involves a mesh, optimised for the two positions, with $\sim 80,000$ nodes and takes 20 hours to solve on a medium-speed work-station. It is likely that FEM will be restricted to the task of developing excitation schemes and new gradiometer structures [29] which can minimise sensitivities to the position and alignment errors explained above, until still more powerful computers become available.

Figure 9. Finite-element modelling of currents induced by a circular induction coil in a 13mm thick aluminium plate with a slot in its lower surface: (a) $f = 1$ Hz ($\delta = 83$ mm) (b) $f = 256$ Hz ($\delta = 5.2$ mm) (c) $f = 8192$ Hz ($\delta = 0.91$ mm).

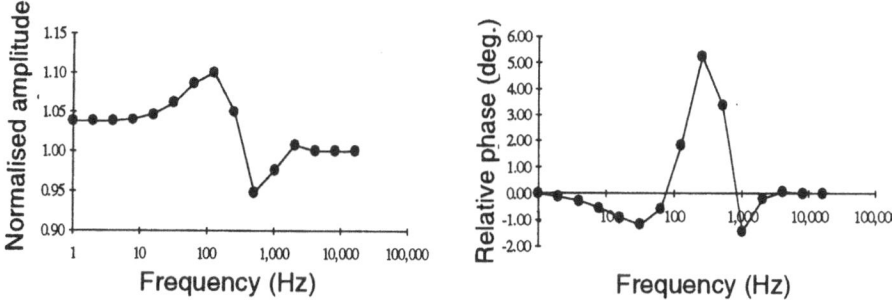

Figure 10. $S(f)$ obtained from FEM modelling of the plate for which experimental data was shown in Figure 7.

4.3 VOLUME-INTEGRAL-METHOD (VIM)

The VIM approach to eddy-current NDE, looking at isolated flaws in simple plate structures, was developed by McKirdy [30] among others. Its main advantage is that modelling can be confined to the flaw region itself. By comparison with FEM, this reduces computational needs by up to three orders of magnitude.

4.3.1 *Theoretical Principles*

Full details of the application of VIM to SQUID NDE can be found elsewhere [31]. Here we give only a summary. The integral equation for the electric field $E(r)$ is

$$E(r) - (\sigma_f - \sigma_p) \int_{V'} G(r,r').E(r')dV' = E^0(r) \tag{1}$$

Here $E(r)$ and $E^0(r)$ are the total and source electric fields in the flaw, while σ_f and σ_p are the conductivities in the flaw and test-plate respectively. $G(r, r')$ is the electric Green's tensor for the plate, but the integration is restricted to the flaw volume V. $E^0(r)$ can be found analytically or numerically, but the Green's functions can only be calculated for a few special geometries.

Taking the curl of Equation 1, because $-i\omega B = \nabla \times E$, we have

$$B(r) + \frac{(\sigma_f - \sigma_p)}{i\omega} \int_{V'} \nabla \times G(r,r').E(r')dV' = B^0(r) \tag{2}$$

and from this using the reciprocity theorem to calculate the change ΔZ_{12} which the crack causes in the transfer impedance between the excitation coil (denoted 2) and the SQUID sensor coil (1), one can obtain the excess field ΔB_i at the sensor in the ith direction due to the crack volume, as

$$\Delta B_i = \left(\frac{\sigma_p - \sigma_f}{-i\omega} \right) \int_V E_1(r).E_2(r)dV \tag{3}$$

where E_1 and E_2 are the fields induced in the crack by unit magnetic dipole (at frequency ω) in the sensor and excitation coils respectively.

Equation 2 can be discretised in the form

$$\mathbf{Ae} = \mathbf{e}^0 \qquad (4)$$

where $\mathbf{A}$ denotes the discrete form of the integral equation operator.

We now consider a hypothetical flaw as defined by a set of M model parameters

$$\mathbf{m} = \{m_k,\ k=1,....,M\} \qquad (5)$$

which for a circular cut might be the ($M = 3$) positions of the two ends and the maximum depth. We can differentiate Equation 5 with respect to these parameters, obtaining

$$\mathbf{A}\frac{\partial \mathbf{e}}{\partial m_k} = -\frac{\partial \mathbf{A}}{\partial m_k}\mathbf{e} \qquad (6)$$

Since this has the same form as Equation 4, then provided that has been solved by a direct method such as LU (Lower-Upper) decomposition, then the set of M equations in Equation 6 can also be solved with comparatively little extra effort.

The model parameters are used to produce a set of model field data $\mathbf{d}^* = \{ d_k^*,\ k=1,...,N\}$ which are compared with the *measured* set $\mathbf{d} = \{ d_k,\ k = 1,...,N\}$. A least-squares solution to the problem of minimising $\delta \mathbf{d} = \|\mathbf{d}-\mathbf{d}^*\|$, which amounts to a multivariate version of the iterative Gauss-Newton process for solving a set of overdetermined simultaneous equations, is to make changes $\delta \mathbf{m}$ in the model parameters, given by

$$\delta \mathbf{m} = (\mathbf{G}^T\mathbf{G})^{-1}\mathbf{G}^T\delta \mathbf{d} \qquad (7)$$

where $G_{ij} = \partial d_i/\partial m_j$ is a *sensitivity matrix*.

Solving the inverse problem thus becomes the repeated application of *forward modelling*. An initial choice of model parameters $\mathbf{m}$ is used to calculate $\mathbf{d}^*$, which after comparison with the measured data $\mathbf{d}$ is used through Equation 7 to calculate changes in $\mathbf{m}$. Iteration should converge on the set $\mathbf{m}_{solution}$ which best predicts the actual data, though it is important to realise that a poor starting model could ultimately diverge. For NDE, however, a simple model with only a few parameters (as in a rectangular, circular or low order polynomial crack) usually will converge to a final shape which is adequate for any normal test criteria.

4.3.2 Practical Considerations

We have developed Volume-Integral-Method code using synthetic rather than real data. Our model corresponded to a 6.5 mm-deep circular flaw 0.15 mm wide, cut with a 44.45 mm diameter rotary saw in a 12.7 mm-thick plate, so that the flaw length at the surface was 31.41 mm. The assumed material was aluminium, with a conductivity

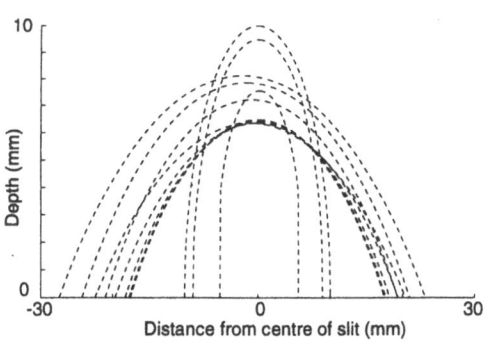

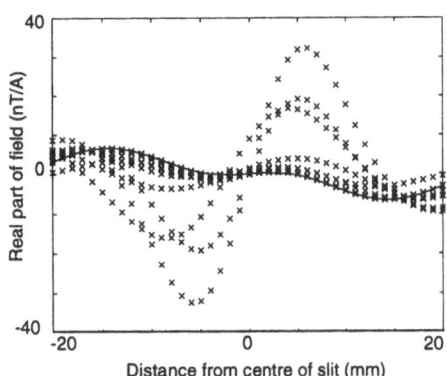

Figure 11. Convergence of successive VIM iterations to true crack shape.

Figure 12. Convergence of successive VIM iterations to true field profile.

corresponding to a skin depth $\delta(f)$ of 7.18 mm at the chosen modelling frequency of 270 Hz. Model excitation was provided by a double-D coil (see §5.3.1) of diameter 63 mm, with a stand-off of 1.5 mm, while detection was assumed to sense $\partial B_y/\partial y$ with a baseline of 25 mm, symmetrically positioned with respect to the cryostat axis. We calculated 41 data values ($\mathbf{d} = \{\, d_k,\ k = 1,\dots\dots41\}$) using Equation 3.

Our trial model was based on a semicircular crack (depth = 10 mm, ends located at $\pm$ 10 mm). Figure 11 and Figure 12 show how only 10 iterations of this model converge on the correct shape for the crack and the field it produces, yielding dimensions of 6.49 mm depth and 32.04 mm length. For NDE, such accuracies would be more than acceptable.

For a much shallower cut (1.8 mm-depth, 17.6 mm-length, made with a circular saw of diameter 44.5 mm), the basic trial model defined above converged to an incorrect length. However a model closer to the true crack (3.6 mm-depth, 24 mm-length) converged after 6 iterations to 1.93 mm depth and 18.26 mm length.

We now plan to test the method with Gaussian noise in the synthetic field data. Use with real experimental data must wait until we have much better precision in the measurement of the stand-off distance of the inducing coil, but this is common in all electromagnetic NDE. Until now our experiments have been concentrated on demonstrating the performance of SQUIDs rather than on producing data so well characterised as to be suitable for inversion.

4.4 ELEMENTARY SOURCE MODELLING

Our third modelling technique is intended to facilitate sensor design, and in particular, complex designs for gradiometric SQUID sensors [29, 32, 33, 34]. It involves the calculation of responses to elementary sources such as the magnetic dipole, the infinite current-carrying wire and the current dipole. The last of these is the most general, but the others are useful because special numerical solutions are available, and corresponding experimental sources are easy to set up.

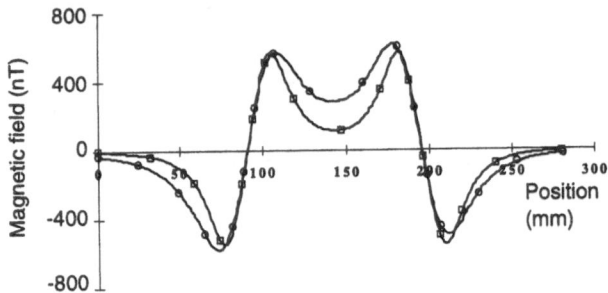

Figure 13. Comparison between experimental data (O)
and modelling results (□) for a Mr. SQUID gradiometer
(§5.3.1) scanned above a V- shaped current-carrying wire.

As an illustration of the modelling results we discuss data related to the high-T$_c$ Mr SQUID electronic gradiometer discussed later (§5.3.1). It was scanned in 2D above a horizontal wire carrying 13.9 mA at 270 Hz arranged in a V-shape. In Figure 13 we compare experimental data and modelling predictions (based on calculations on a series of current dipoles) along a line where the two parts of the V are 101 mm apart, showing that the absolute amplitude calibration of the system is accurate and that ac signals can be measured with high signal-to-noise ratios (SNRs) even when they are below the environmental noise level (typically 200 nT for the 50 Hz field in our laboratory).

The value of this technique to eddy-current NDE is not seen as lying in precise determination of flaw characteristics. Rather, it is likely to be a basic tool for the design and comparison of excitation systems and also gradiometers, including higher-order ones. Both can be developed through multi-coil structures to give improved localisation of excited current and of positional sensitivity.

5. Developments with High-T$_c$ SQUIDs

5.1 INTRODUCTION- HIGH-T$_C$ SQUIDS AND NDE

Earlier discussions of SQUID NDE [4, 5] were guarded on the prospects for using high-T$_c$ devices, because of issues of noise, the difficulty of producing gradiometers in the absence of fine high-T$_c$ wire, and the problems of moving unscreened SQUIDs around in the Earth's field.

In spite of this, we made some early NDE demonstrations [26] with an RF step-edge junction SQUID [35], and subsequently with a simple commercial high-T$_c$ SQUID [36]. Though the results were not as good as those from low-T$_c$ systems, it was clear that the convenience of the much smaller liquid nitrogen cryostats pointed towards high-T$_c$ systems and away from low-T$_c$.

Recently, high-T$_c$ SQUID white noise at 77 K has been brought close to the theoretical limit, and 1/f corners have been moved to well below 1 Hz. Some integrated thin-film gradiometers have been produced, and though they are not yet suitable for NDE, they show promise. Also, the electronic gradiometer has been introduced by Tavrin *et al.* [37]: here two SQUIDS (for first order) or three SQUIDs (for second order), mounted in appropriate, but adjustable, axial positions, have their outputs electronically differenced to produce the necessary signals. For small field amplitudes (as in cardiography), the electronics are adequately linear and have enough dynamic range to produce good gradient signals, rejecting uniform fields such as the 50/60 Hz mains.

Finally, the work of Koch on three-SQUID gradiometers [38, 39] has allowed high-T_c systems to become movable during operation. Two SQUIDs have their outputs electronically differenced to provide a gradient signal, while a third SQUID (or even a room temperature sensor such as a flux-gate) measures the changes in the background field and its low-frequency variations. The third output is used to null any change in the uniform component of the DC field seen by all the SQUIDs. This eliminates the noise and hysteresis associated with vortex motion when a high-T_c SQUID moves in the 0.1 mT Earth's field.

To compete on size and cost with existing NDE products, SQUID systems would have to use simple cryogenics and the smallest possible cryogenic-to-room-temperature distances to allow for small coils and high resolutions. One way forward, therefore, is to build on high-T_c achievements, such as those described above, to develop movable systems that can be used above test specimens in the open laboratory. As we now show, such systems are already a rudimentary possibility, with high-T_c SQUIDs performing many of the NDE functions previously feasible only with low-T_c devices.

5.2 SYSTEM REQUIREMENTS

In SQUID biomagnetism, background noise dictates extreme design strategies. Low SQUID noise at frequencies down to 1 Hz and below is needed, and even for such situations, large-area pickup coils (several cm^2) and higher order gradiometers are normally needed, together with enclosures such as mu-metal or eddy-current screened rooms. Performance above 1 kHz is of little interest since little physiological activity occurs on time-scales below 1 ms.

SQUID systems for eddy-current NDE require quite different capabilities. They need perform well only down to about 10 Hz, since operation at lower frequencies will probably result in inspection rates too slow to be acceptable. However, for skin depths to be small enough to encompass the smallest cracks, they must perform up to several tens of kHz. For good spatial resolution, pick-up coils must be as small (of the order of a few mm^2) as is consistent with the minimum coil stand-off that is cryogenically possible (about 1 mm). However, the fields to be detected can be several orders of magnitude above those of biomagnetism, so that small pickup coils are not necessarily a problem. Indeed, if sensor sensitivity appeared to be inadequate, as it might in a very high order gradiometer intended to provide good depth sensitivity, there is the option to increase the defect signal merely by increasing the eddy-current excitation level - something not available in biology.

Most important is the requirement for NDE systems to be movable, and usable, in an unscreened workshop or factory environment, where the Earth's field and other interference are ever-present. This is vital for two of the main candidate areas for SQUID NDE applications - see §6.1 - *Ageing Aircraft* (N. America) and *Nuclear and Chemical Plant Life Extension* (Japan). These specific requirements have long been met in the low-T_c context, with the SQUID screened by a superconducting Meissner enclosure and only balanced gradiometric coils exposed to the environment. It is well demonstrated in a 7-channel Conductus low-T_c system at the Hitachi Mechanical Research Laboratory. This is mounted on a robot arm programmed to scan above irradiated steel samples [40]; moreover, it operates within about 20 m of the site of the vibration tests of the persistent-mode levitation coils to be used in the new MAGLEV railway project. Moves to high-T_c technology, predicated on the need for smaller systems, better spatial resolution, and lower costs, must additionally ensure that they match these low-T_c capabilities.

5.3 ELECTRONIC GRADIOMETRY

The Conductus Mr. SQUID, launched for the educational market, can be seen now as a breakthrough for SQUID NDE. It has no pickup-coil or flux transformer and so its field sensitivity is inadequate for biomagnetism. However the sensitivity is quite satisfactory for NDE. Mr. SQUID has all the other desiderata- a small sensing area (only 70 µm square) formed by the SQUID itself, suitable frequency response and reliability. Unfortunately, in the current model the SQUID is encapsulated under plastic about 7 mm from the end of its circuit board, and cannot come closer to the outside world than 11 mm. Moreover it can measure fields only parallel to the cryostat base (B_x or B_y). However, these problems are not fundamental: the board could be redesigned so as to measure B_z with an effective stand-off of only 2 mm.

We used two Mr. SQUIDs to make a gradiometer similar to that of the Julich group [37] in which the fields are differenced after electronic processing rather than in pick-up coils. The aim was to reduce the effect of background fields, with the ultimate aim of moving the system, rather than the test piece, in an unshielded environment.

5.3.1 *Equipment*
The two SQUIDs, on their individual circuit boards, were rigidly mounted 25 mm apart on a single block, without any mechanical adjustability - see Figure 14. The block is spring-loaded against the cryostat base so that the sensor stand-off (minimum 11 mm) is accurately known.

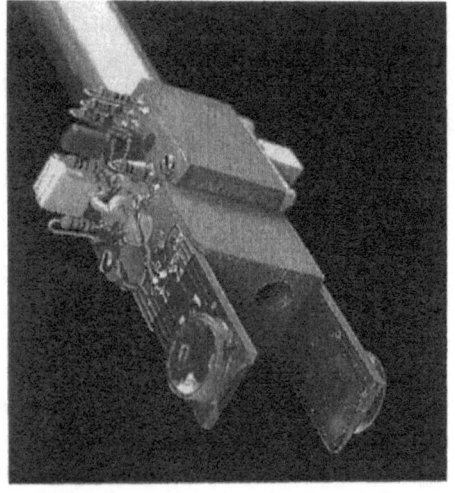

Figure 14. Arrangement of two high-T_c Mr. SQUIDs as an electronic gradiometer.

Separate SQUID electronics units were required for each channel. They were designed to realise the full SQUID-noise-limited potential of Mr. SQUID, which is $10^{-4} \Phi_0$-$Hz^{-\frac{1}{2}}$, rather than the $10^{-3} \Phi_0$-$Hz^{-\frac{1}{2}}$ performance of the commercial package. With the improved electronics, the field resolution of each SQUID was ~10 pT-$Hz^{-\frac{1}{2}}$. Each channel had a bandwidth from DC to 33 kHz, and an independently available output. The electronics package also differenced these outputs with enhanced SNR, thus providing a quasi-first-order gradient signal.

Each SQUID operated in its own flux-locked loop so that its electronics had to be capable of the usual reset if it experienced an out-of-range signal. Since the differencing circuit had a ±10 V output, it too was provided with a reset capability. All resets took place within 1 ms, so that tracking capability was not lost.

The liquid nitrogen cryostat, about 60× less in weight and volume than its liquid helium predecessor, was insulated with a vacuum space, charcoal getter and superinsulation. To avoid field distortion or noise effects, it was made from cotton-epoxy composite, the only metal parts being a vacuum valve and some top-plate aluminium components. The base plate was 4 mm thick and the cryogen hold time was 24 hours. The size and weight compare reasonably with other NDE packages, including those in which the NDE system, rather than the test-piece, is moved during scanning.

The arrangements for scanning our test-pieces underneath a cryostat [36, 41] involve a single computer controlled stepper motor, which drives a wooden table carrying the specimen in an x-y raster scan, beneath the sensor system, which is mounted in a wooden gantry.

Apart from the occasional use of direct current injection, we have normally employed eddy-current methods with the high-T_c gradiometer. Two excitation-coil arrangements have been used. They are illustrated in Figure 15.

The first is a 5 turn 4.5 mm diameter spiral copper coil on an epoxy glass printed circuit board, which formed a cantilever attached to the underside of the cryostat: thus the coil was in light contact with the test piece during scanning. In typical use the coil was excited by a 77 mA, 2.7 kHz current. This arrangement proved to be most useful for detecting and locating surface-breaking flaws, because the excited currents are confined to a small area and depth, so that when they encounter the flaw there is the maximum proportional distortion in the field they produce.

The second coil was a 63 mm diameter double-D structure, which is a type used

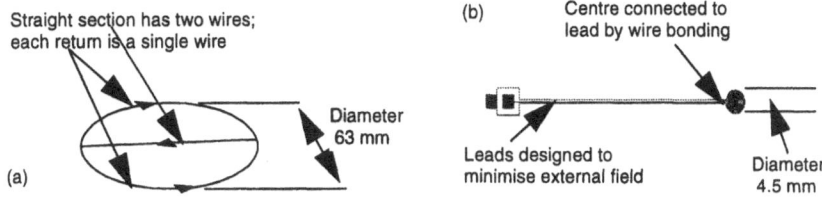

Figure 15. Eddy-current excitation coils used with Mr. SQUID gradiometer (a) Double-D induction coil (b) Spiral coil.

618

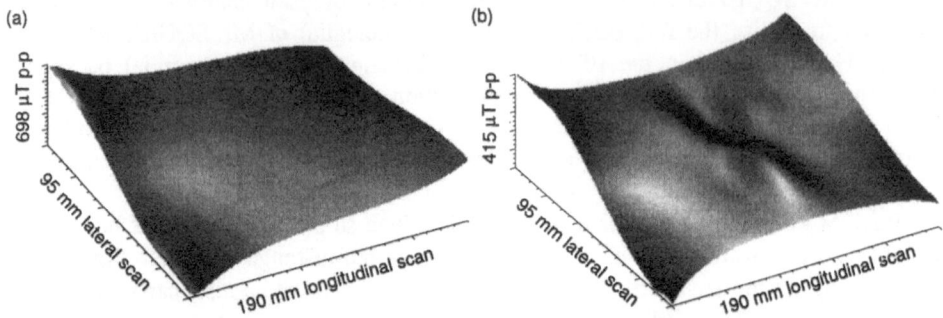

Figure 16. (a) Output of electronic gradiometer. (b) Output of one of the two SQUIDs in the gradiometer assembly. The excitation is 140 mA, 530 Hz directly injected, in each case.

elsewhere in electromagnetic NDE [42]. It too was mounted on the underside of the cryostat and excited at 77 mA, but at low frequencies- typically 250 Hz. The excitation covers a wide area, and reaches to a depth corresponding to the far side of a 10-15 mm thick aluminium specimen. The straight part of the double-D was positioned perpendicular to the planes of the two SQUIDs, minimising direct detection of the induction field and maximising the sensitivity to distortion fields produced by a flaw.

5.3.2 *Basic Performance of Gradiometer*

Figure 16 shows high-T_c measurements made above a 300×300×13 mm aluminium plate, with a central slot 6.5 mm deep × 6 mm wide × 40 mm long. Figure 16 (a) shows the gradiometer output, and Figure 16 (b) that of a single SQUID of the pair. In this case the excitation was a directly-injected 140 mA, 530 Hz current. Neither the flaw nor the excitation would be realistic in practice, but the results show the potential for the gradiometer as a device. The single SQUID result shows no indication of the slot because large signals from the current spreading out around the sides of the plate dominate and because the small signals from the slot are buried in environmental interference, itself too small to be seen on the same scale as the large signals. However, the gradiometer result clearly shows the expected peak at one end of the slot and the trough at the other. There are two benefits of using a gradiometer here: the large but distant signals from spreading current are reduced, and interference is reduced.

The resolution achievable using the gradiometer with the spiral induction coil, now excited by 47 mA at 2.7 kHz, is exhibited in Figure 17. The

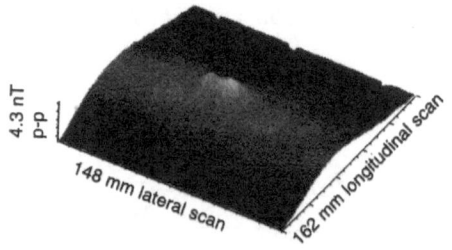

Figure 17. Gradiometer signal obtained with 47 mA, 2.7 kHz current in small spiral excitation coil.

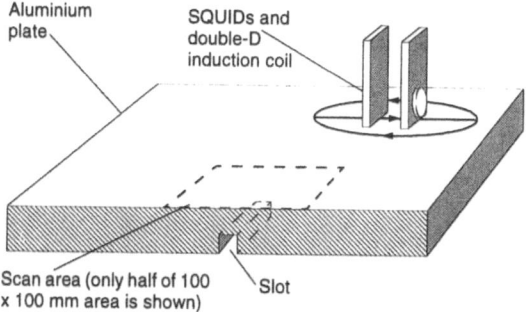

Figure 18. High-T$_c$ gradiometer arrangement for studying a slot breaking the far surface of a 13 mm aluminium plate. Excitation is by a double D coil with 140 mA at 270 Hz.

specimen was another 13 mm aluminium plate, now with a much smaller arc slit in its upper surface. This had a surface length of about 20 mm, a width of 150 mm, and penetrated 1.8 mm at its deepest. By comparison with Figure 16, we see in Figure 17 that a high spatial resolution can be achieved using a small source, even with relatively distant SQUIDs. Detection of such a small feature, at stand-offs corresponding to thick layers of surface protection, is noteworthy in itself for NDE.

5.3.3 Slit Breaking on Distant Surface - Use of Double-D Coil

The 13 mm thick plate with the 6.5 mm- deep slot- see §5.3.2- was used again. Now, however, it was turned over so that the nearest point of the slot was 6.5 mm beneath the scanned surface (Figure 18). In this case, although the slot is large, the 6.5 mm subsurface-distance is large in NDE terms also.

The double-D coil of Figure 15 was used because it provides a field decay into the plate that is much slower than with a smaller coil. The excitation was 140 mA at 270 Hz. The result (from the gradiometer only) is shown in Figure 19 (a) and compared in Figure 19 (b) with the result of a VIM (§4.3) prediction of the response based on the known test sample and detection system geometries. The resolution and elegant symmetry of the result both reflect the matching dipolar nature of both the double-D excitation coil and the double SQUID detector. The low background noise

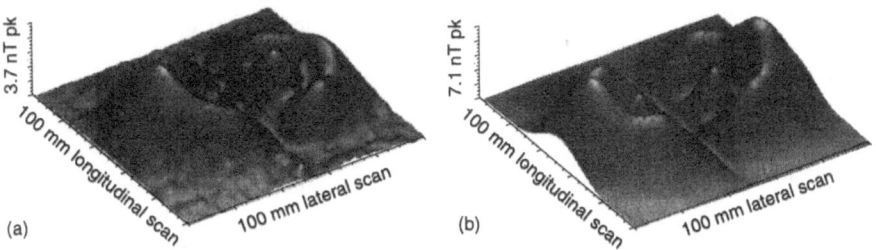

Figure 19. Gradiometer eddy-current image of slot cut in underside of 13 mm plate- see Figure 18. (a) Measured image (b) Predicted result obtained from VIM calculation. The complexity of the result is due to the geometry of the double-D induction coil.

reflects the uniform field rejection capabilities of the gradiometer.

Note that apparently similar cases involving distinct metal layers, one of which contains a through-slot [27], are electromagnetically quite different from this one. In the multilayer case, current flows only around the sides of the slot. By contrast, in our more realistic situation, current also flows over the slot, making the distortion more diffuse and difficult to detect.

5.3.4 Scanning the Gradiometer rather than the Test Piece

The balance of the two Mr. SQUIDs is good enough to scan our unshielded cryostat above our test pieces [43]. The arrangement is shown in Figure 20 and is based on a study of an aluminium plate cut twice to depths of 6.5 and 3 mm, respectively. Excitation was provided by the double-D coil with a stand-off of 11 mm and a scan speed of 6 mm s^{-1}. The results (Figure 21) show no significant difference between the measurements whether it is the test piece or the SQUID gradiometer which is moved.

We repeated the measurements with the plate covered by a thin paper insulating sheet and an additional 3.2 mm thick aluminium sheet. The signals were about 4× smaller because of the screening effect of the upper plate, but still very clear.

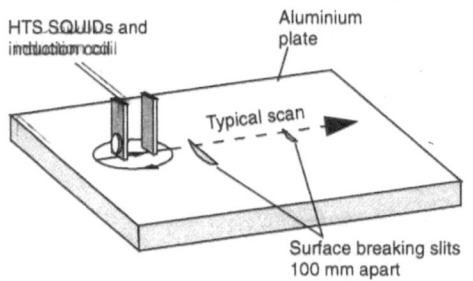

Figure 20. Arrangement for scanning Mr SQUID-based electronic gradiometer above two saw cuts, of depth 6 mm and 3 mm receptively. Eddy-currents were induced by the double-D coils (140 mA, 270 Hz).

These results suggest that unscreened scanning of high-T_c systems over realistic structures *will* be practicable. However, at this point we can scan the gradiometer only in straight lines, maintaining the component of the static field seen by the two SQUIDs roughly constant. A twisting motion of the SQUIDs will cause such severe flux motion in their washers that flux lock is lost. However, static-field cancellation, using the techniques of Koch (§5.1) should remove this limitation.

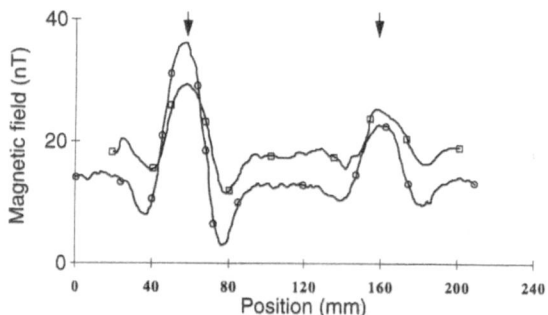

Figure 21. Electronic gradiometer scans for the two cuts illustrated in Figure 20, of depths 6 mm and 3 mm respectively. Upper trace: Scan made with SQUIDs in motion Lower trace: Scan made with test plate in motion.

5.3.5 Practical Performance - a Crack in a Pressure Vessel

We now report a high-T_c SQUID NDE demonstration on the specimen supplied by British Gas and shown in Figure 22. It was a section from a fibreglass-clad aluminium pressure vessel which had been cycled 40,000 times until it failed because of a crack through the metal wall. This has been an intractable problem for NDE because poor acoustic propagation through the fibreglass into the aluminium makes ultrasonic inspection difficult, while conventional ECT gives poor results at the high stand-off imposed by the cladding.

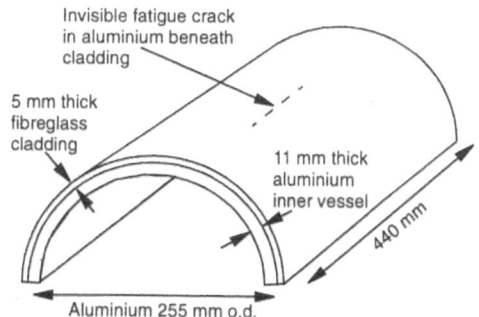

Figure 22. Section of fibreglass-clad aluminium pressure vessel with a fatigue through-crack in its metal wall (supplied by British Gas).

We raster-scanned the specimen longitudinally, with a 3° rotation after each traverse. The straight section of the double-D induction coil, carrying a 270 Hz, 77 mA current, was oriented perpendicular to the length of the specimen at a 2 mm radial stand-off from the fibreglass. We first recorded 2D maps of the in-phase and quadrature components of the gradiometer output.

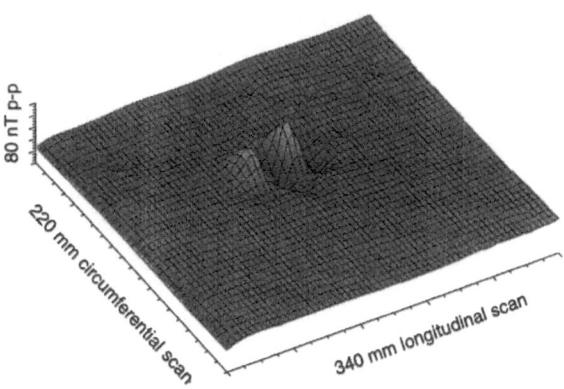

Figure 23. Field magnitude map for flawed pressure vessel of Figure 22.

Then, in a separate test at a smaller stand-off, we repeated the scan directly above the crack, this time recording single SQUID and gradiometer outputs.

The 2D field map in Figure 23 shows that the crack has been detected clearly. The circumferential length is very close to the 63 mm diameter of the induction coil, as expected from a combination of the coil diameter and the negligible width of the crack. The shape of the circumferential signature is a simple peak, again qualitatively predictable from the coil and crack configurations.

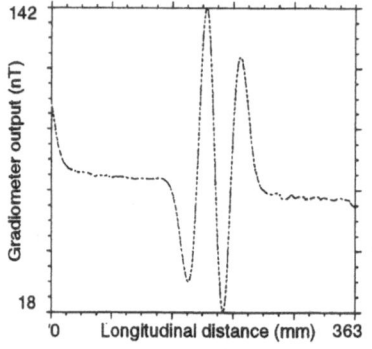

Figure 24. Gradiometer output from a single scan along the vessel above the region of the crack- see Figure 22.

The longitudinal characteristics of the 2D signal are most easily seen in Figure 24, which shows the real part of the gradiometer output from a single scan. The pairs of peaks and troughs are characteristic of our modelling studies of the double-D coil configuration (§5.3.1), and there is a significant signal over a length of about 121 mm. In this simple through-crack example, we would expect this length to be the sum of the double-D diameter (63 mm) and the true crack length, suggesting that the latter is 58 mm. We verified this independently, by removing the fibreglass cladding and using the corner echoes obtained by an ultrasonic angle probe to determine a length of 60 mm.

6. Problems and Prospects

6.1 PRACTICAL PROBLEMS

Although there are many problems to be solved in NDE, there are two particular topics of present international interest on which SQUIDs could focus. The first is of particular interest in the USA and Europe and focuses on ageing aircraft, while the second attracts more attention in Japan and relates to ageing, embrittlement and crack development in chemical and nuclear reactor pressure vessels.

6.1.1 *Ageing Aircraft*

'Ageing aircraft' refers to the need to extend the lives of civilian and military aircraft well beyond those for which they were originally designed. Lap joints, in which three or more layers of aluminium are riveted together, can develop interlayer corrosion and cracking at rivet holes. There has been at least one spectacular failure, with part of the outer skin of an elderly passenger plane tearing away in service, and there is a need to be able to detect the relevant flaws reliably and in time to effect necessary repairs.

When the flaws are in the lower layers of a lap joint, the corrosion or cracking cannot be detected with ultrasound. Nor are conventional eddy-current methods usable, because skin-depth limitations restrict detection to the upper layers.

The Vanderbilt group's work with the low-T_c microSQUID [44] has shown that lower-layer flaws can be detected, though only within a screened room and at the cost of substantial computing time. Now, however, our own work on a flaw in a gas pressure vessel (§5.3.5) shows that for some flaw types, unscreened high-T_c SQUIDs may be capable of similar performance and that VIM methods may allow very rapid computing convergence to one of a few simple realistic models.

It is important to emphasise the need for system scanning and rapid signal processing, because the industry has standards for minimum acceptable area inspection rates. These are of the order of a few linear feet per hour along joints typically made with rows of 3 or 4 rivets side by side. For magnetic scanning these rates can already be achieved, for top-layer corrosion and flaws, by a hand-held planar eddy-current magneto-optical inspection method [45], which produces images in real time. However it does not have the two-component capability (amplitude and phase) of other techniques involving SQUIDs, and so is less successful at detecting defects in second or deeper layers.

6.1.2 *Reactor Vessels*

As with aircraft there is an emerging need to extend the lives of nuclear power stations, and other comparable structures, beyond their design lives. To achieve this demands techniques that can detect not only cracks and flaws, but also other ageing phenomena such as embrittlement caused by thermal stress or neutron irradiation. The materials of particular interest are duplex stainless steel and mild steel, sometimes in layer arrangements. The issue has found particular attention in Japan, partly as a result of the failure, some years ago, of a steam pipe in the primary stage of a pressurised water reactor. Programmes have been set up by JAPEIC (the National Agency for Power Station Inspection) among which is one involving SQUIDs, conducted by Hitachi. This involves simple extensions of early remote magnetometry methods (§2.2), in which a 7-channel low-T_c system is scanned above test pieces using a robot. With this technique it has been shown that embrittlement, including effects due to irradiation [40], can indeed be detected and mapped. However, no quantitative information is yet available.

It seems clear that some of these problems merit exploration by eddy-current techniques as well as by static magnetisation methods. For example, embrittlement leads to changes in electrical conductivity, which though not as extreme as in an open crack, can certainly be included in VIM modelling (§4.3). Unfortunately, no such studies have been made. Experimental work would, of course, be impossible for NDE groups unequipped to deal with specimens which have been subjected to prolonged irradiation.

Since the number of installations involved is quite small, but represents a huge capital investment, the need for an NDE system to be inexpensive (§5.1) is partly relaxed. Moreover, these installations are subject to the most detailed record keeping. NDE techniques which detected *changes* in properties, such as resistivity, would probably be acceptable in such situations, by contrast with aircraft, where absolute detection of flaws on a one-time basis is critical.

6.2 ARE HIGH-T_C SQUIDS VITAL?

In §5.1 we said that one way to meet fundamental NDE system needs would be to develop the high-T_c SQUID systems in appropriate ways, and we demonstrated some progress in later sections. Nevertheless low-T_c systems already possess some of the properties of portability and reliable use in unscreened environments that are still being sought or developed in high-T_c systems. Moreover, low-T_c gradiometers of high order and high dynamic range are easily produced using superconducting wire or films. The only serious defect of these systems lies in the need for liquid helium cryogenics. This is not because of the much-cited cost of the cryogen, but because of the complexity and size of the cryostat if the 4.2 K-to-room temperature distance is to be as small as possible and is to be used in many orientations, and because of the consequential rapid boil-off and short hold times.

With the appearance of cryogen-free superconducting magnets, made possible by recent progress in cryocooler technology, one must ask if analogous low-T_c SQUID systems, in which the SQUID and its superconductive screen are cooled by a mechanical system such as a Stirling cycle or pulse-tube cooler, might be pursued for NDE. Podney, indeed, has developed a low-T_c NDE system [46] in which a SQUID is cooled by a combined Gifford-McMahon Joule-Thomson cryocooler (Heliplex HS-4). The SQUID is connected to excitation and pick-up coils through an cryogenic umbilical within which cooling is by copper braid connected to the cold pad of the cryocooler. The umbilical has a sapphire window at its end, and the coils can be scanned at a distance of about 3 mm above a test surface. The system gave convincing results on a standard corroded wing root test-piece.

Nevertheless the future probably does not lie with such systems. The first is in the power requirements and physical size of sub-10 K coolers, and the second in the vibrations they produce, which show up as magnetic signature fields that the pickup coils will detect unless they and the cryocooler are well separated, as they are in Podney's system. Together these will probably render low-T_c NDE systems too heavy and unwieldy to compete with other technologies, such as magneto-optic flaw visualisation systems.

We conclude that it is essential to continue to develop capability in the high-T_c area. Koch's 3-SQUID sensor [38] (§5.1) points the way. Maintaining the SQUIDs in low field eliminates the 'Lock Jumping' and 'Reset' problems that are a feature of simple movable electronic gradiometry. Coupled to an increase in the bandwidth of the SQUID electronics that allows flux cancellation of local broadcast RF signals, this approach has made it possible to deploy high-T_c systems in hostile environments such as are found within vehicles towed behind aircraft.

Compact cryostats are likely to be cooled by liquid nitrogen that is either free boiling (77 K) or sub-cooled (to 65 K) by simple drip-feed evaporative chambers. Such cryostats can have hold times of at least one working day, and be usable in all orientations (including upside down, for inspection of the undersides of aircraft wings).

Perversely, the 3-SQUID system could generate challenges to *any* role for SQUIDs in NDE. Other magnetic sensors, such as flux-gate magnetometers and thin film magnetoresistors, have recently achieved sensitivities in the 10-100 $pT\text{-}Hz^{-\frac{1}{2}}$ range, typical of what is used by SQUID NDE. While these sensors cannot be used to measure small changes in large fields, or in true gradiometers, the use of a third sensor to provide cancellation of local fields and of techniques such as flux locked loops has opened the way to highly-linearised pseudo-gradiometers involving only room temperature sensors [47, 48]. These may prove to have potential for environmental and NDE magnetic sensing, though there are bandwidth and other limitations on them at present.

The future for NDE with *SQUIDs* may after all lie in the more extreme properties of the devices, notwithstanding our opening comments (§2.1). These include their exceptional flux sensitivity. This could be exploited in the NDE situation by using sensor coils very much smaller than they currently are: only in this way could true multichannel capability, with many detectors on a single substrate, be produced. Likewise, only *true* higher-order gradiometers (where flux is differenced before detection by a single detector- a SQUID), can realise the full potential of swept-frequency depth profiling: this is because the spatial sensitivity of gradiometers can be tailored to have peaks and zeroes at chosen positions [29].

7. Summary and Conclusions

We have emphasised progress in one particular technique of SQUID NDE, namely eddy-current testing, and have shown that, although there are improvements still to be made in the material science, design and fabrication of high-T_c SQUIDs, present devices are already adequate to demonstrate their application to real problems. Further progress depends as much on *circuits and systems* as on *SQUIDs*. Screening and wiring for gradiometers of significant baseline lengths require attention. Cryostats and multichannel systems must be developed.

However these are *not* the largest hurdles. A typical Applications of Superconductivity meeting still reveals the complacent view that if the best effort is put into better films, junctions and SQUIDs, industrial utilisation can be taken as a foregone conclusion. In fact, issues of developing the technology specifically for the target technologists, and promoting it to them, are probably more important than anything else. They will be time consuming and expensive, particularly in areas such as ageing aircraft, where speed, flexibility and reliability, with no false positives or negatives, will be crucial.

A recent review of market opportunities for advanced NDE sensors [49] points out that the world NDE market is rather small (totalling about $1 billion per year). Moreover, it is extremely diverse: the review identifies 22 different NDE techniques from *Acoustic* to *X-Radiography*, into only 3 of which SQUIDs might fit (*Eddy Current Testing, Electromagnetic Techniques* and *Magnetic*), so that the population of prospective SQUID users is small. Finally, it identifies the need to detect corrosion underneath pipe-lagging, flaws in pre-tensioned concrete, corrosion in aircraft skins, and flaws in glass reinforced plastic as important issues for the future. These are all areas in which SQUIDs have been shown to produce results, at least in the laboratory. Yet the review does not mention SQUIDs at all. In fact, in an entire

626

book on recent developments in *electromagnetic* NDE [24], SQUIDs merit only two paragraphs.

To ensure that resources continue to flow towards the realisation of the full potential of SQUID NDE, it is vital for the superconducting device community to reach existing NDE practitioners *now*, and to provide them with convincing demonstrations, such as locating sub-surface flaws in real aircraft structures. Several groups have begun this task, but there is still much work to do.

8. Acknowledgements

The Strathclyde work reported here has been supported by the UK Engineering and Physical Sciences Research Council and Defence Research Agency, by a BP/Royal Society of Edinburgh Research Fellowship, and by Oxford Instruments Ltd, Quantum Magnetics, and the US Air Force Office of Scientific Research. We thank British Gas for supplying the pressure vessel specimen and Katherine Kirk for help with the ultrasonic testing mentioned in §5.3.5. We are grateful to Chris Carr, Udo Klein, Jan Kuznik and Morag Walker, who, as will be apparent from the References, actually carried out many of the experiments we have discussed. Finally, we thank Harold Weinstock for the opportunity to contribute to the ASI and to these Proceedings.

References

1. Weinstock, H. and Nisenoff, M. (1985) Nondestructive evaluation of metallic structures using a SQUID gradiometer, in H.D.Hahlbohm and H.Lubbig (eds), *SQUID'85*, de Gruyter, Berlin, pp. 853-858.
2. Bain, R.J.P., Donaldson, G.B., Evanson, S. and Hayward, G. (1985) SQUID gradiometric detection of flaws in ferromagnetic structures, in H.D.Hahlbohm and H.Lubbig (eds), *SQUID'85*, de Gruyter, Berlin, pp. 841-846.
3. Wikswo, J.P. (1995) SQUID magnetometers for biomagnetism and nondestructive testing: important questions and initial answers, *IEEE Trans. Appl. Superconductivity* **5**, 74-120.
4. Donaldson, G.B. (1989) SQUIDs for Everything Else, in H Weinstock and M Nisenoff (eds.), *Superconducting Electronics*, Springer Verlag, pp. 175-207.
5. Donaldson, G.B., Cochran, A and Bowman, R.B. (1993) More SQUID applications, in H. Weinstock and R. Ralston (eds.), *The New Superconducting Electronics*, Kluwer Academic Publishers, Dordrecht, pp. 181-220.
6. Vrba, J (1996) Squid gradiometers in real environments- *This Volume*.
7. Weinstock, H, Erber, T and Nisenoff, M (1985) Threshold of Barkhausen emission and onset of hysteresis in iron, *Phys. Rev. B* **31**, 1535-1553.
8. Weinstock, H, Mignogna, R.B., Schechter, R.S. and Simmonds, K.E. (1992) An improved system for the nondestructive evaluation of steel in H. Koch and H. Lubbig (eds.) *Superconducting devices and their applications*, Proc SQUID'91, Springer Proceedings in Physics **64**, 572-575.
9. Weinstock, H (1995) private communication.
10. Cochran, A, Donaldson, G.B, Evanson, S, and Bain, R.J.P (1993) First Generation SQUID-based Nondestructive Testing System, *IEE Proc. A* **140**, 113 - 120.
11. Wikswo, J.P., Opfer, J.E. and Fairbank, W.M. (1974), *Am. Inst. of Phys. Conf. Proc.* **18**, 1335.
12. Donaldson, G.B., Evanson, S, Otaka, M., Hasegawa K., Shimizu, T and Takaku K. (1990) Use of a SQUID magnetic sensor to detect ageing effects in duplex stainless steels, *British Journal of NDT* **32**, 238-240.
13. Thomas, I.M., Ma, Y.P., Tan, S. and Wikswo, J.P. (1993) Spatial resolution and sensitivity of magnetic susceptibility imaging, *IEEE Trans. Appl. Superconductivity* **3**, 1949-1952.
14. Murphy, J.C. (1988), *Private communication*.
15. Bellingham, J.G., Macvicar, M.L.A. and Nisenoff M (1987) SQUID technology applied to the study of electrochemical corrosion *IEEE Trans. Magnetics* **MAG-23**, 477-479.
16. Hibbs, A.D. (1993) Measurement of electrochemical corrosion currents using a multichannel SQUID magnetometer, *J. Electrochem Soc.* ******, ***"To appear"- noted in More SQUID Apps Ref [45]***, Hibbs, A.D., Chung, R. and Pence, J.S. (1993) Corrosion Current Measurements with a High Resolution Scanning Magnetometer in D.O. Thompson and D.E. Chimenti (eds.) Review of Progress in Quantitative NDE, Vol. 13A, Plenum Press, New York, pp. 1955-1962.
17. Roth, B.J, Sepulveda, N.G. and Wikswo, J.P. (1989) Using a magnetometer to image a two-dimensional current distribution *J. Appl. Phys.* **65**, 361-372.
18. Wikswo, J.P., Ma, Y.P., Sepulveda, N.G., Thomas, I.M. and Lauder, A. (1993) Magnetic susceptibility imaging for nondestructive evaluation, *IEEE Trans. Appl. Superconductivity* **3**, 1995-2001.
19. Cochran, A., Donaldson, G.B., Morgan, L.N.C., Bowman, R.M. and Kirk, K.J. (1993) SQUIDs for NDT: the technology and its capabilities, Brit. J. of NDT **35**, 173-181.
20. Wikswo, J.P., Friedman, R.N., Kilroy, A.W., van Egeraat, J.M. and Buchanan, D.S., (1990) Preliminary measurements with MicroSQUID, in S.J. Williamson, M. Hoke, G. Stroink and M. Kotani (eds.) *Advances in Biomagnetism*, Plenum Press (New York), pp. 681-694
21. Hibbs, A.D., Sager, R.E., Cox, D.W., Aukerman, T.H., Sage, T.A. and Landis, R.S. (1992) A High Resolution Magnetic Imaging System based on a SQUID Magnetometer, *Rev. Sci. Inst.* **63**, 3652-3658.
22. Vu, L.N. and van Harlingen, D.J. (1993) Design and implementation of a scanning SQUID microscope, *IEEE Trans. Appl. Superconductivity* **3**, 1918-1921.
23. Kirtley, J.R., Ketchen, M.B., Stawiasz, K.G., Sun, S.Z., Gallagher, W.J., Blanton, S.H. *et al.* High resolution scanning SQUID microscope, *Applied Phys. Lett.* **66**, 1138.
24. A useful account of ECT can be found in Blitz, J. (1991) Electrical and Magnetic Methods of Nondestructive Testing, Adam Hilger, Bristol.
25. Cochran, A., Donaldson, G.B., Morgan, L.N.C., Bowman, R.M. and Kirk, K. J. (1993) SQUIDs for NDT: the Technology and its Capabilities, *Brit. J. NDT* **35**, 173-182.
26. Cochran, A., Morgan L.N.C., Bowman R.M., Kirk K.J. and Donaldson. G.B. (1993) SQUID systems for nondestructive testing by AC field mapping, *IEEE Trans. Appl. Superconductivity* **3**, 1926-1929.

References (continued)

24. A useful account of ECT can be found in Blitz, J. (1991) Electrical and Magnetic Methods of Nondestructive Testing, Adam Hilger, Bristol.
25. Cochran, A., Donaldson, G.B., Morgan, L.N.C., Bowman, R.M. and Kirk, K. J. (1993) SQUIDs for NDT: the Technology and its Capabilities, *Brit. J. NDT* **35**, 173-182.
26. Cochran, A., Morgan L.N.C., Bowman R.M., Kirk K.J. and Donaldson. G.B. (1993) SQUID systems for nondestructive testing by AC field mapping, *IEEE Trans. Appl. Superconductivity* **3**, 1926-1929.
27. Alzayed, N.S., Fan, C., Lu, D.F., Wong, K.W., Chester, M. and Knapp,D.C. (1994) Deep nondestructive testing using a bulk high-T_c RF SQUID, *IEEE Trans. Appl. Superconductivity* **4**, 81-86.
28. Vector Fields Ltd, OPERA and ELEKTRA FEM Software, Kidlington, Oxford, OX5 1JE, UK.
29. Walker, M.E., Cochran, A., Klein, U., Bain, R.J.P. and Donaldson, G.B. (1995) Modelled Response of Planar Asymmetric Gradiometers, in D. Dew-Hughes (ed.) *Applied Superconductivity (Edinburgh) 1995*, IOP Conference Series 148, IOP Publishing (Bristol), pp. 1597-1600.
30. McKirdy, D.McA. (1989) Recent improvements to the application of the Volume Integral Method of eddy current modelling, *J. Nondestr. Eval.* **8**, 45-52.
31. McKirdy, D.McA., Cochran, A., McNab, A. and Donaldson, G.B., (1995) Theoretical Consideration of Fatigue Crack Detection and Characterisation Using SQUID Sensors, to be published in *Proc. Conference on Quantitative NDE (Seattle), 1995*.
32. Donaldson, G.B. and Bain, R.J.P. (1984) An improved design of high order superconducting gradiometer coils for magnetic monopole detection, *App. Phys. Letters*, **45**, 990-992.
33. Donaldson, G.B., Pegrum C.M. and Bain R.J.P. (1986) Integrated thin-film SQUID instruments, in H.D. Hahlbohm and H. Lubbig (eds.), *SQUID'85*, Walter de Gruyter, Berlin, pp. 729-755.
34. Jones, A.E. and Bain, R.J.P. (1995) A generalisation of planar magnetic gradiometer design *via* orthogonal polynomials, *J. Computational Phys.* **118**, 191-197.
35. Kuznik, J., Hatrick, D., Meek, A., Macfarlane, J.C. and Donaldson, G.B. (1993) Noise and flux creep in bi-epitaxial DC SQUID, in *Extended Abstracts of 4th International Superconductive Electronics Conference* (Centennial Conferences, Boulder), 141-145.
36. Cochran, A., Macfarlane, J.C., Morgan, L.N.C., Kuznik, J., Weston, R., Hao, L., Bowman, R.M. and Donaldson, G.B. (1994) Using a 77 K SQUID to Measure Magnetic Fields for NDE, *IEEE Trans. Appl. Supercond.* **4**, 128-135.
37. Tavrin,Y., Zhang, Y., Wolf, W. and Braginski, A.I. (1994), A second-order gradiometer operating at 77K, *Superconductor Science and Technology* **7**, 265-268.
38. Koch, R.H., Rozen, J.R., Sun, J.Z. and Gallagher, W.J. (1993) Three SQUID gradiometer, *Appl. Phys. Lett.* **63**, 403-405.
39. Koch, R.H., (unpublished).
40. Otaka, M., Hitachi Mechanical Engineering Research Laboratory, private communication.
41. Cochran, A. and Donaldson, G.B. (1992) Improved techniques for structural NDT using SQUIDs, in H. Koch and H. Lubbig (eds.), Superconducting Devices and their Applications', *SQUID'91*, Springer-Verlag, Berlin, pp.576-580.
42. Beissner, R.E. and Temple, J.A.G. (1990) Calculation of Eddy Current Fields for Coils of Arbitrary Shape, *Rev. Prog. Quantitative Nondestr. Eval.* **9**, 257-264.
43. Carr, C., Cochran, A, Kuznik, J., McKirdy, D.McA and Donaldson, G.B. Electronic Gradiometry for NDE in an Unshielded Environment with Stationary and Moving HTS SQUIDs, to be published in *Cryogenics*.
44. Wikswo, J.P., Ma, Y.P., Sepulveda, D.J., Staton, D.J., Tan, S. and Thomas, I.M. (1993) Superconducting magnetometry: a possible technique for aircraft NDE, in M.T Valley, N.K. Grande and A.S. Kobayashi (eds.), *Nondestructive Inspection of Aging Aircraft*, Proc. SPIE, **2001**, 191-199.
45. Physical Research Inc.- see Reference 3.
46. Podney, W.N., Eddy Current Evaluation of Airframes Using Refrigerated SQUIDs, *IEEE Trans. Appl. Superconductivity* **5**, 2490-2492 (1995).
47. Avrin, W., to be published in *Proc. Conference on Quantitative NDE (Seattle), 1995*.
48. Koch, R.H., Keefe, G.A. and Allen, G., Room Temperature Three Sensor Magnetic Field Gradiometer, to be published.
49. Dixon, T.E., Silk, M.G. and MacKeith, D.J. (1994) *Insight* **36**, 256 and 342.

THE MAGNETIC INVERSE PROBLEM FOR NDE

J.P. WIKSWO, JR.
Vanderbilt University
Department of Physics and Astronomy
Box 1807 Station B
Nashville, TN 37235, U.S.A.

Abstract. A major motivation for the use of SQUID magnetometers for nondestructive evaluation (NDE) is the need to increase the sensitivity or reliability of detection of particular types of flaws, such as hidden cracks or corrosion in the aluminum structure of aircraft. SQUID NDE holds promise over existing NDE techniques because of the combination of sensitivity, spatial resolution, accuracy, noise rejection and frequency response offered by SQUID magnetometers. However, in order to compete commercially with other NDE techniques, it is important that SQUID NDE provide quantitative images that are visually comparable to those provided by ultrasound, x-ray or thermal imaging. This chapter describes how maps of magnetic fields can be converted through a solution to the inverse problem into images of current or magnetic susceptibility. Inverse imaging approaches that have been examined for SQUID NDE include Fourier-transform spatial filtering, lead field analysis, alternating projections and blind deconvolution. More advanced tomographic techniques also are under development: the application of the current from two or more directions, with appropriate analysis algorithms, can provide maps of the conductivity distribution throughout the sample. Similarly, magnetic susceptibility tomography may allow three-dimensional magnetic imaging of diamagnetic and paramagnetic objects.

1. Introduction

Part of the appeal of radiographic and ultrasonic NDE techniques is that they provide a direct image of the flaw, whereas SQUIDs, which measure the magnetic field through a small pickup coil, do not. Laboratory experiments with scanning SQUID magnetometers and the direct injection of current

H. Weinstock (ed.), SQUID Sensors: Fundamentals, Fabrication and Applications, 629–695.
© *1996 Kluwer Academic Publishers.*

into test plates, combined with image deconvolution techniques that can provide a unique solution to the two-dimensional inverse problem, have demonstrated that maps of the magnetic field above the plate can be used to obtain images of the current distribution in the plate. However, the direct injection of current may not be practical in many cases, most importantly in the NDE of painted aircraft. Conventional eddy-current techniques do not require electrical contact to the test object, and provide signals whose wave shape, phase shift and amplitude are determined by the flaw, the probe and the sample, but interpretation of these data requires a skilled operator. Scanning eddy-current techniques, using either conventional probes or SQUIDs, can produce image-like displays of the spatial variation of the signal, but the quantitative relationship between the signal and the flaw may be difficult to ascertain. Eddy-current techniques using sheet inducers, such as the magneto-optic imager (MOI)[17] and some SQUID measurements offer the advantage of producing nearly-uniform applied current distributions so that inverse imaging techniques can be applied to obtain meaningful results. In order to understand how each of these inverse techniques works, we shall review both the measurements that create the field maps, and the variety of inverse techniques used to convert these maps into images more directly related to the distribution of field sources that produced the field. The purpose of this chapter is not to review SQUID NDE, but to provide for the first time an overview of the inverse approaches used in SQUID NDE and in the measurement of cellular magnetic fields. The examples used will be drawn largely from the extensive work at Vanderbilt on the magnetic imaging problem; a broader review of SQUID NDE instrument development and measurements, along with more complete citations, is presented elsewhere[71, 16, 70, 76, 25]. In this chapter, I do not address the numerous techniques that have been developed for the forward problem of calculating the magnetic field from current and magnetization distributions for a variety of flaw and sample geometries[51, 52, 35, 36, 37, 38].

As shown in Fig. 1, SQUIDs can be used in a number of different NDE modes[80]. The easiest SQUID NDE measurements involve intrinsic currents, for example in a printed or integrated circuit, as is shown in Fig. 1a. Remanent magnetization, perturbations in applied currents, or Johnson noise in metals can be detected as shown in Fig. 1b-d. If you apply an ac field, you can image the eddy currents (Fig. 1e), or if you use either cyclic stress or simultaneously applied ac and dc magnetic fields, you can study the hysteretic magnetization in steel, shown in Fig. 1f. If you apply a dc magnetic field, as in Fig. 1g, SQUIDs can be used to image the magnetization of diamagnetic and paramagnetic materials. The generation of the magnetic field above the sample in each of these modes is governed by one of two equations: the law of Biot and Savart relating current and magnetic

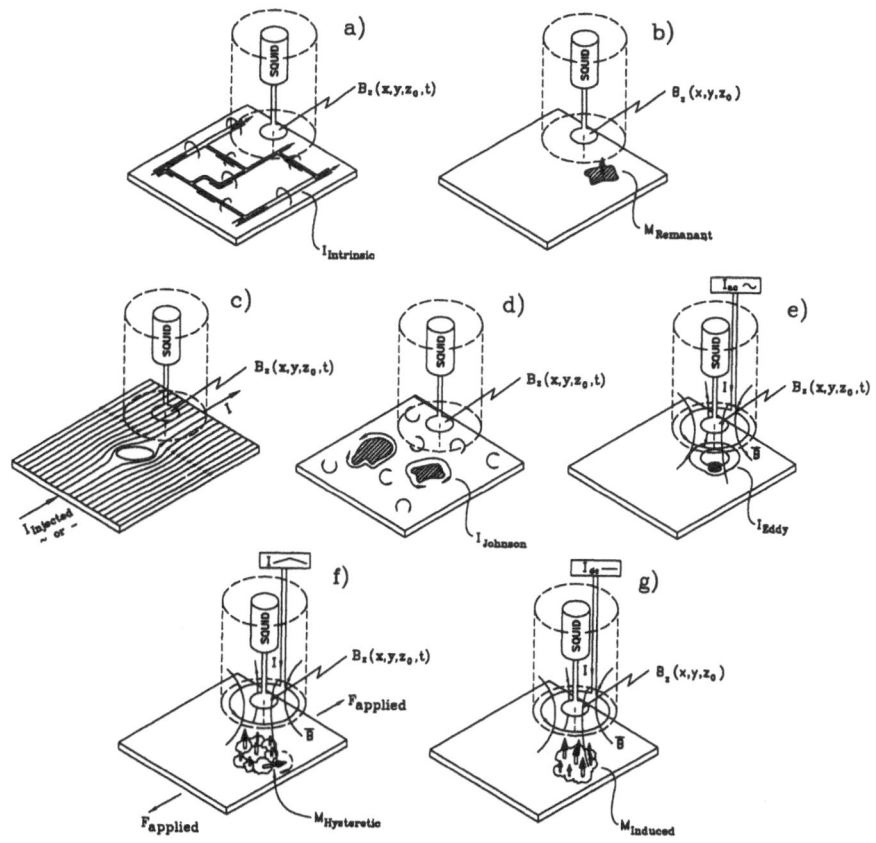

Figure 1. Modes of SQUID NDE. By scanning the sample beneath the SQUID, it is possible to image (a) intrinsic currents, (b) remanent magnetization, (c) flaw-induced perturbations in applied currents, (d) Johnson noise in conductors, (e) eddy currents and their perturbations by flaws, (f) hysteretic magnetization in ferromagnetic materials in the presence of an applied stress, and (g) diamagnetic and paramagnetic materials in an applied field. (From [80], with permission)

field, or the equation for the magnetic field produced by a point magnetic dipole. We shall examine each equation and how it can be inverted.

2. Imaging Current Distributions

2.1. THE LAW OF BIOT AND SAVART

Let us start with the simplest case of determining the distribution of currents in a planar circuit, as shown in Fig. 1a. In general, the magnetic field

$\vec{B}(\vec{r})$ at the point $\vec{r}$ is given by the law of Biot and Savart

$$\vec{B}(\vec{r}) = \frac{\mu_o}{4\pi} \int_v \frac{\vec{J}(\vec{r}') \times (\vec{r} - \vec{r}')}{|\vec{r} - \vec{r}'|^3} d^3 r' \, , \tag{1}$$

where $\vec{J}$ is the current density at point $\vec{r}'$.

It is also instructive to rewrite Eq. 1 in terms of the curl of the current distribution[72]

$$\vec{B}(\vec{r}) = \frac{\mu_o}{4\pi} \int_s \frac{\vec{J}(\vec{r}') \times \hat{n}}{|\vec{r} - \vec{r}'|} d^2 r' + \frac{\mu_o}{4\pi} \int_v \frac{\nabla' \times \vec{J}(\vec{r}')}{|\vec{r} - \vec{r}'|} d^3 r' \, , \tag{2}$$

where $\hat{n}$ is the normal to the surface s that bounds the object. The first integral represents the magnetic field due to the discontinuity of the tangential component of the current at the boundary of the object, and the second is that produced by any curl within the object. Note that within a homogeneous conductor, any current distribution that has zero curl will not contribute to the external magnetic field and, hence, will be magnetically silent[74, 78, 48]. This implies that in homogeneous, three-dimensional conductors, currents that obey Ohm's law,

$$\vec{J}(\vec{r}) = \sigma \vec{E}(\vec{r}) = -\sigma \nabla \phi(\vec{r}) \, , \tag{3}$$

do not contribute to the magnetic field because the curl of a gradient is identically zero, and the conductivity σ is a constant within the volume v. Batteries, such as a dry cell dropped into a conducting medium, or the microscopic equivalent, a current dipole, are non-Ohmic and hence have curl. The first term in Eq. 2 would indicate that whenever Ohmic currents encounter a boundary or a discontinuity in conductivity, a magnetic field could be produced, since a discontinuity in the component of the current density tangential to a boundary is equivalent to a curl.[1] In a thin sheet that would approximate a two-dimensional conductor, the sample has two parallel surfaces in proximity that have curls of opposite sign which cancel each other far from the sheet; the curl that contributes most strongly to the magnetic field is that from any edges in the conductor. A wire bent into a pattern has a curl all along its surface. We shall address the role of spatial variations in the conductivity σ in two-dimensional conductors in a later section. For now, we shall concentrate on sheet conductors with a constant (homogeneous) conductivity σ.

---

[1]The insulating boundary on the side of a cylindrical battery produces a discontinuity of tangential current, and this provides an alternate description for the magnetic field produced by a battery in a homogeneous conducting medium.

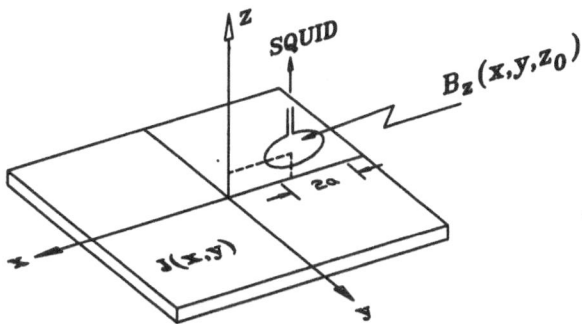

Figure 2. The geometric arrangement and coordinates for a SQUID that is scanned over a current-carrying sample. (From [47], with permission)

In typical measurements, the component of the magnetic field normal to the sample, B_z, is mapped by scanning the SQUID pickup coil of radius a over the sample, at a fixed height z_o, as shown in Fig. 2. In this case, we can expand the cross product in Eq. 1 and rewrite the law of Biot and Savart as a pair of integrals

$$B_z(\vec{r}) = \frac{\mu_0}{4\pi} \int_v \frac{J_x(\vec{r}\,')(y - y')}{|\vec{r} - \vec{r}\,'|^3} d^3 r' - \frac{\mu_0}{4\pi} \int_v \frac{J_y(\vec{r}\,')(x - x')}{|\vec{r} - \vec{r}\,'|^3} d^3 r' . \quad (4)$$

Equations 1 and 4 are convolution integrals. The source of the field is the current density $\vec{J}(\vec{r}\,')$; the remaining terms of the integrand are a function of both $\vec{r}$ and $\vec{r}\,'$, and form the Green's function $G(\vec{r}, \vec{r}\,')$. To calculate the magnetic field, we integrate the product of $\vec{J}$ and G over the entire region where $\vec{J}$ is non-zero, *i.e.*, we convolve $\vec{J}$ and G to determine $\vec{B}$. Note that in Eq. 1, G is a vector function that contains the cross product, but in Eq. 4, it is a pair of scalar functions. In this light, the inverse problem simply reduces to the deconvolving of these convolution integrals. Therein lies the challenge.

2.2. INVERTING THE LAW OF BIOT AND SAVART

The fundamental difficulty in magnetic imaging of currents is that there is no unique solution to the inverse problem of determining the three-dimensional current distribution in Eq. 1 from the measurements of the magnetic field outside of the object. The easiest proof of non-uniqueness is to note that it is possible to superimpose on any three-dimensional current distribution another source distribution that is magnetically silent, such as when the radial internal currents of a spherically-symmetric battery are superimposed upon an infinite, homogeneous conductor. This is exactly analogous to the electrostatic where there is no electric field outside a spherical capacitor formed by a pair of concentric spheres carrying opposite charge. In an electrostatics inverse problem, spherical capacitors of arbitrary radius, charge and sphere separation can be added or subtracted from the source distribution without changing the potentials measured on the surface that bounds all sources. Hence, any attempt to solve such an inverse problem will be unable to determine which spherical capacitor configurations exist, and which do not. Constraints might be used to prevent the inverse algorithm from creating spherical capacitors *de novo*, but there is a whole class of higher-order silent sources. The spherical harmonic multipole solution to Laplace's equation is designed specifically to eliminate those silent sources[84, 85]. The usual approach to this problem for either the electric or magnetic inverse is to restrict the possible sources and invert the resulting set of constrained equations. We shall not dwell on the three-dimensional current imaging problem, which has been the subject of great interest in magnetoencephalography and is treated in detail elsewhere in this book.

In contrast to the three-dimensional current-imaging problem, the two-dimensional problem does have a unique inverse. As we shall show in this chapter, there is a wide variety of inverse algorithms for magnetic imaging of two-dimensional current distributions, including spatial filtering, dipole fitting, the Hosaka-Cohen transformation, alternating projections, lead-field analysis, the finite-element method and blind deconvolution. While there may be a unique inverse solution, even the two-dimensional inverse problem can be ill-conditioned, in that the ability to determine the inverse solution with the desired accuracy or spatial resolution can be strongly dependent upon measurement noise, source-detector distance and the exact nature of the source current distribution. We shall show that this can be overcome in part by applying constraints to limit the effects of noise or to most fully utilize *a priori* knowledge of the current distribution.

2.2.1. *The Spatial Filtering Approach*

The most elegant method to obtain a two-dimensional current image from a magnetic field map is to use Fourier-transform deconvolution (inversion) of the law of Biot and Savart[15, 26, 47, 1, 54]. Following the notation of Roth *et al.*[47], let us assume that we are imaging the magnetic field produced by a two-dimensional current $\vec{J}(x, y)$ distributed through a slab of conducting material of thickness d that extends to infinity in the xy plane. We shall assume that we measure the component B_x at a height $z >> d$, so that we can integrate Eq. 1 over z' and immediately obtain[2]

$$B_x(x, y, z) = \frac{\mu_o d}{4\pi} z \int_{-\infty}^{\infty} \int_{-\infty}^{\infty} \frac{J_y(x', y')}{[(x - x')^2 + (y - y')^2 + z^2]^{3/2}} dx' dy' . \quad (5)$$

If we compute the spatial Fourier transform (FT) of B_x using the fast Fourier transform (FFT) or more accurate techniques[57], so that

$$b_x(k_x, k_y, z) = FT\{B_x(x, y, z)\} , \quad (6)$$

we can use the convolution theorem to write the law of Biot and Savart as a simple multiplication in the spatial frequency domain:

$$b_x(k_x, k_y, z) = g(k_x, k_y, z) j_y(k_x, k_y) , \quad (7)$$

where the Green's function

$$G((x - x'), (y - y'), z) = \frac{\mu_o d}{4\pi} \frac{z}{[v^2 + w^2 + z^2]^{3/2}} \quad (8)$$

has the Fourier transform

$$g(k_x, k_y, z) = \frac{\mu_o d}{2} e^{-\sqrt{k_x^2 + k_y^2} z} . \quad (9)$$

Note that g falls exponentially with both k and z, so that the high spatial-frequency contributions to the current distribution are attenuated in the magnetic field, $i.e.$, the Green's function acts as a spatial low-pass filter. The further the magnetometer is from the current distribution, the harsher is the filtering.

Inverse Spatial Filtering.   A supposedly simple division allows us to solve Eq. 7 for $j_y(k_x, k_y)$

$$j_y(k_x, k_y) = \frac{b_x(k_x, k_y, z)}{g(k_x, k_y, z)} . \quad (10)$$

The desired image of $J_y(x', y')$ is obtained with the inverse Fourier transform

$$J_y = FT^{-1}\{j_y(k_x, k_y)\} . \quad (11)$$

---

[2]Hereafter, when we are discussing the two-dimensional current image, we shall assume that the sample has a thickness d that is negligible compared to z, so that $\vec{J}$ has neither a z-component nor z-dependence. Thus, we shall be able to use the standard current density $\vec{J}$, with units of amperes/meter^2, rather than a two-dimensional equivalent. The thickness d will be written explicitly in the equations. $\vec{J}d$ is simply a two-dimensional current density with units of amperes/meter.

A similar set of equations would allow us to determine J_x from a map of B_y. This inverse solution is unique[47].

As we shall see in more detail shortly, the problem with this inverse process, in general, is in the division in Eq. 10. If b_x is small, but non-zero, at spatial frequencies for which $1/g$ is large, j_y tends toward infinity. Unfortunately, g falls exponentially with k, so that $1/g$ diverges exponentially. Since the magnetometer will have an upper limit to the spatial frequencies that it can detect (determined by the diameter of the pickup coil), Eq. 10 is destined to blow up! Before we see how this can be avoided, it is useful to look at the z component of the field.

From Eq. 4, it follows that the Fourier transform of B_z is[47]

$$b_z(k_x, k_y, z) = i\frac{\mu_o d}{2}e^{-\sqrt{k_x^2+k_y^2}z}\left(\frac{k_y j_x(k_x, k_y)}{\sqrt{k_x^2 + k_y^2}} - \frac{k_x j_y(k_x, k_y)}{\sqrt{k_x^2 + k_y^2}}\right). \qquad (12)$$

This shows us, in general, that a single image of B_z would be inadequate to determine both J_x and J_y, and would provide only a linear combination of the two. However, if we assume that the current distribution is continuous, it must have zero divergence in the quasistatic limit, i.e.,

$$\nabla \cdot \vec{J}(x,y) = 0 . \qquad (13)$$

For two-dimensional current with no source or sinks, this reduces to

$$\frac{\partial J_x}{\partial x} + \frac{\partial J_y}{\partial y} = 0 \qquad (14)$$

If we compute the Fourier transform of this equation, we then have the added constraint

$$-ik_x j_x(k_x, k_y) - ik_y j_y(k_x, k_y) = 0 , \qquad (15)$$

which allows us to solve Eq. 12 for either J_x or J_y and then use the continuity equation to obtain the other. While this procedure works well for planar current distributions, it may not be applicable for mapping effective surface currents on the upper boundary of three-dimensional current distributions in thick objects, since current flowing on the surface can disappear as J_z into the bulk conductor of the object without affecting B_z.

Noise and Windowing. We now can address briefly the issue of noise and windowing, which is treated in greater detail elsewhere for two-dimensional imaging[47] and for cylindrical models of the head[82, 61, 7] and abdomen[7]. The value of j_y in Eq. 10 will diverge if there is noise in the absence of signal, *i.e.*, if the denominator goes to zero faster than does the numerator. However, because the noise in a SQUID measurement is white at high temporal frequencies, a scanned SQUID image will have white noise at high spatial frequencies. The low-pass characteristic of the law of Biot and Savart is manifest as a Green's function that tends towards zero exponentially at high spatial frequencies, so that an unstable inverse is almost guaranteed except in cases where the magnetometer is very close to the current distribution. The solution to this dilemma is to filter spatially the magnetic field data by eliminating all contributions at high spatial frequencies, so that the numerator is identically zero above a frequency k_{max}. The process of low-pass filtering in the frequency domain (or windowing in the temporal domain), will ensure a stable inverse, but at the cost of losing high spatial frequency information. The steps in this process, and how it affects the image quality, are shown in Figs. 3 and 4 in which we simulate the forward problem of conversion of a current source distribution into a magnetic field map (down the left column) and, after adding noise, the inverse problem of converting the field map into a reconstructed source image (up the right column). Figure 5 shows a comparison of the results of the spatial filtering/Fourier transform inverse approach for equivalent theoretical and experimental images for a current test pattern. Figure 6 shows a magnetic field map obtained with MicroSQUID and the corresponding current image for a slice of isolated cardiac tissue[56, 55]. These measurements provided some of the first proof that cardiac tissue is best described by bidomain models with unequal anisotropy ratios in the intracellular and extracellular spaces[75].

The trade-off between stability of the inverse and the cutoff of high spatial frequencies is unfortunate, in that the presence of noise limits the ability to enhance the spatial resolution of a current image obtained with a distant magnetometer. Figure 7 shows how the quality of an image is degraded as the magnetometer is moved away from the source distribution. The mean square deviation (MSD) of the image from the actual source is 0.03 at 0.1 mm, 0.07 at 0.3 mm, 0.44 at 1.0 mm, and 0.95 at 3 mm.

The primary strengths of the technique are the speed with which the field map can be deconvolved and the ease with which it can be low-pass filtered: a fast Fourier transform of the data is computed, the data are multiplied in the frequency domain by the low-pass filtered Green's function, and the inverse FFT is computed. For a particular measurement geometry, the Green's function needs to be computed only once.

Apodizing. The discussion so far has assumed that the magnetic field is measured at a point. In practice, SQUID magnetometers have pickup coils whose diameter is comparable to the spacing between the coil and the

638

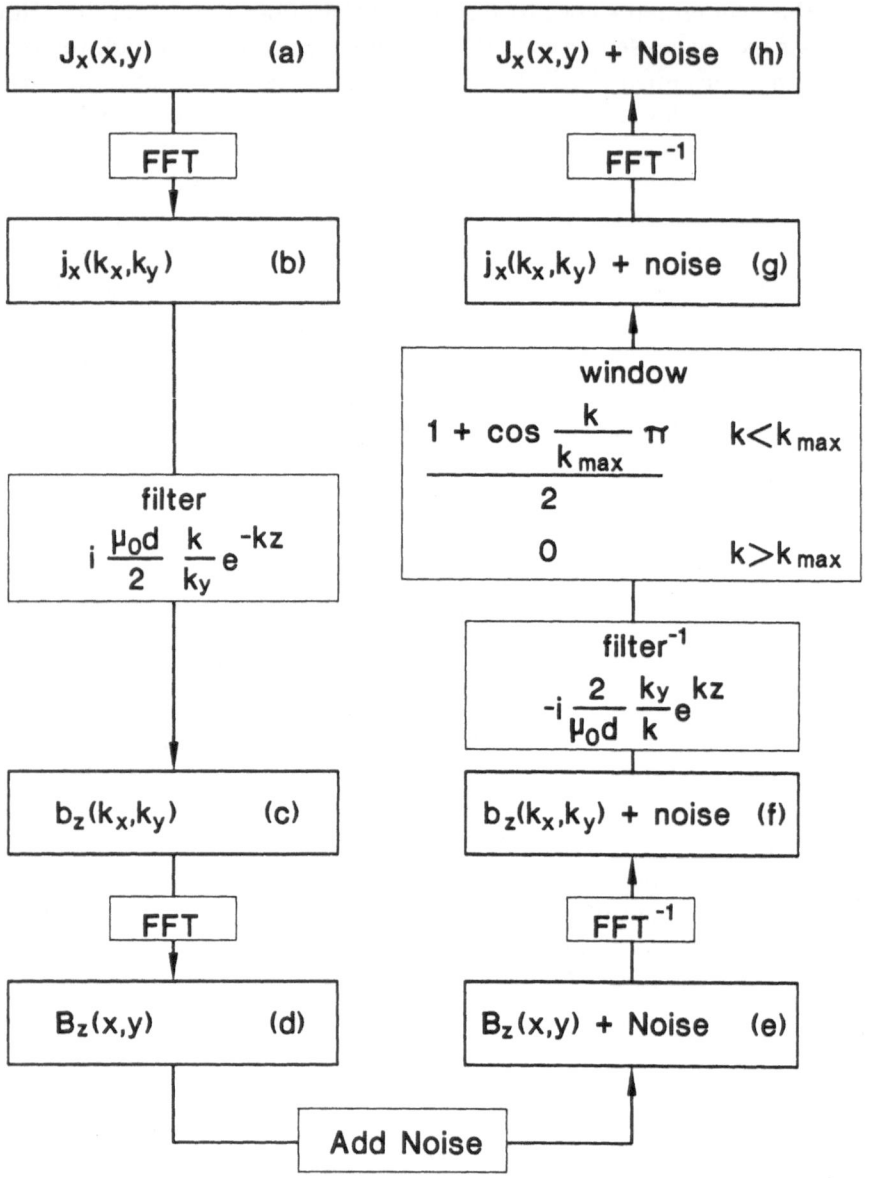

Figure 3. Schematic diagram summarizing the forward and inverse problems using the spatial filtering/Fourier transform approach and the assumption that the current in the plane is conserved. Starting with a known x component of the current density, $J_x(x, y)$, we calculate the z component of the magnetic field, $B_z(x, y, z)$, then add noise, and calculate an image of the current density from the noisy magnetic field. Windowing is usually required to provide a stable inverse solution. (From [47], with permission)

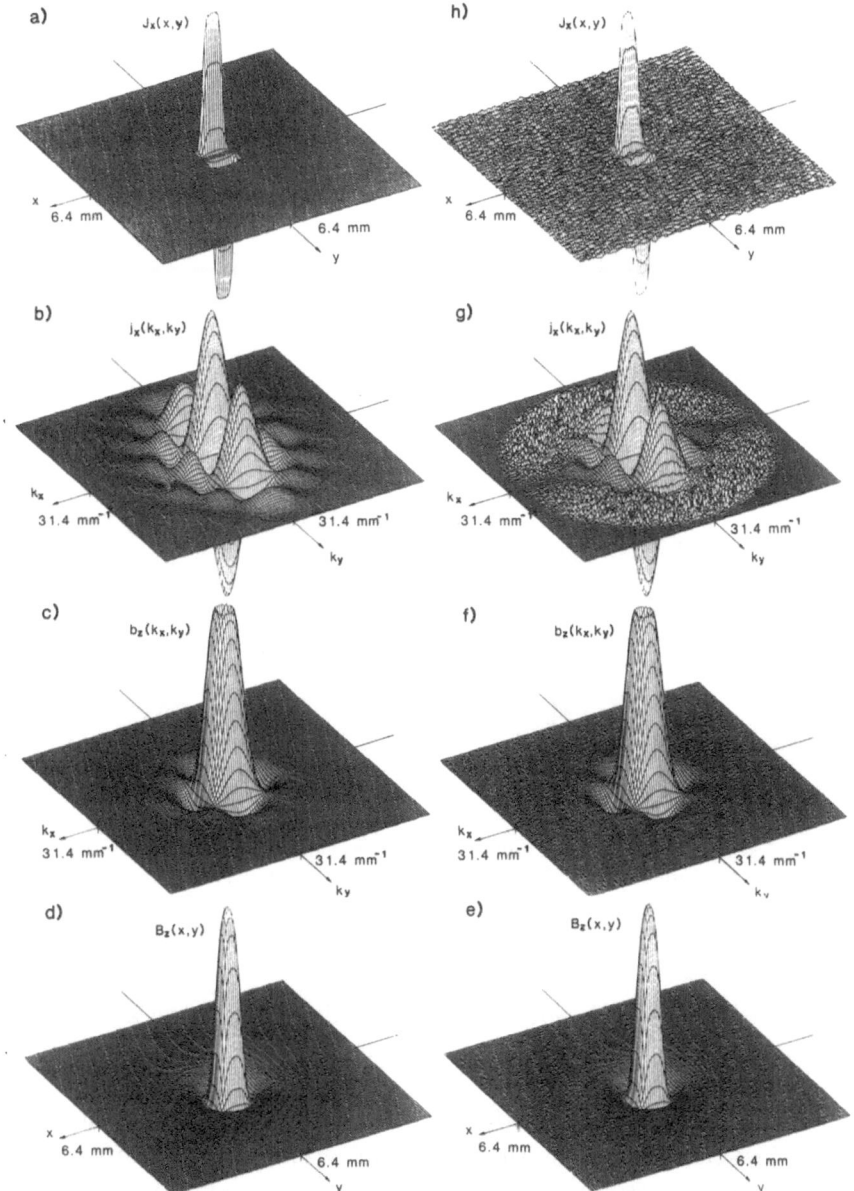

Figure 4. Forward and inverse spatial filtering/Fourier transform calculations for $z = 0.3$ mm. (a) The x component of the current density $J_x(x, y)$, for a square loop of current, (b) the imaginary part of the Fourier transform of J_x, $j_x(k_x, k_y)$, (c) the real part of the Fourier transform of B_z, $b_z(k_x, k_y)$, at $z = 0.3$ mm, and (d) the z component of the magnetic field, $B_z(x, y)$, with peak amplitude of 756 pT. (e) The same B_z as in (d), but with 0.5-pT amplitude white noise added, (f) the real part of the Fourier transform of B_z with added noise, (g) the imaginary part of the Fourier transform of the J_x image, multiplied by a Hanning window with $k_{max} = 30$ mm^{-1}, and (h) J_x of the current-density image. (From [47], with permission)

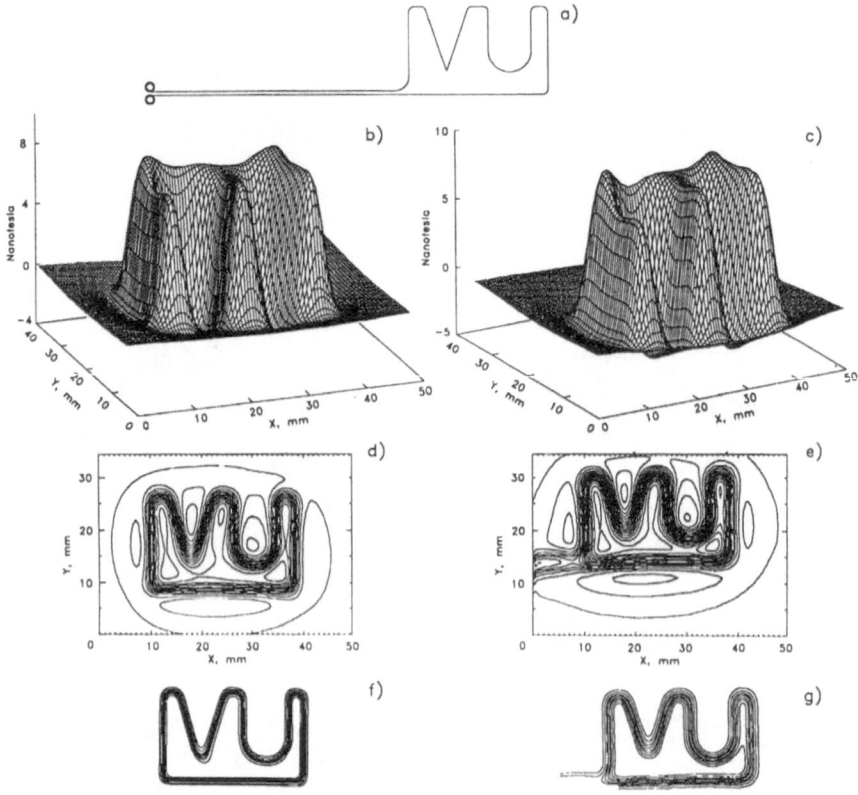

Figure 5. Magnetic imaging of a current pattern. (left) Theoretical and (right) experimental images obtained with MicroSQUID from a printed circuit pattern (a) that spells "VU". (b) and (c) Surface plots of the magnetic field; (d) and (e) isocontour plots; and (f) and (g) the deconvolved current images. (From [83], with permission)

sample, consistent with optimizing the trade-offs between sensitivity and spatial resolution [73]. Since the SQUID measures flux, the pickup coil integrates the magnetic field that threads the coil, and thus the shape of the coil is convolved with the field distribution in a manner that results in further low-pass spatial filtering of the field. The relationship between coil diameter, spatial resolution and sensitivity is non-trivial. A spatial-filtering analysis shows, as will be discussed in more detail in a later chapter[77], that a noisier, smaller pickup coil placed close to a sample can in certain circumstances provide better images than a larger, quieter SQUID further away.

One would suspect that a deconvolution approach could be used to eliminate the effect of the finite coil area. The image processing approach can

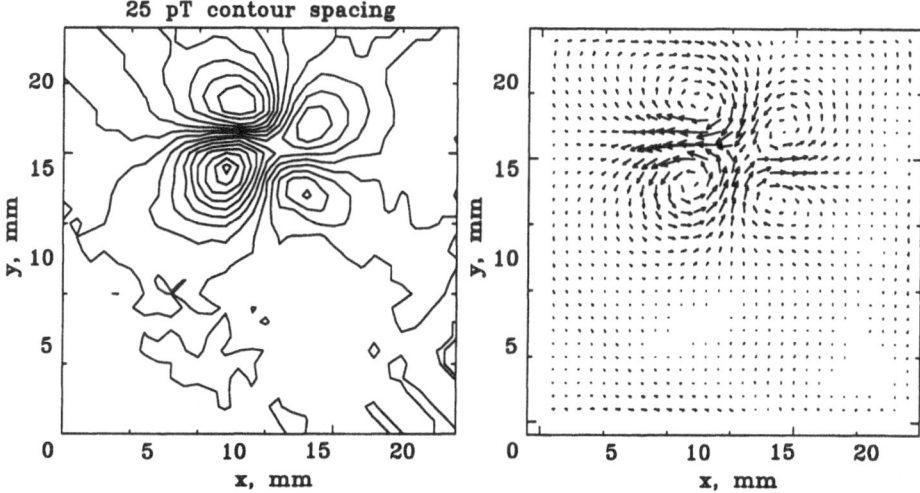

Figure 6. Left: The magnetic-field image recorded above an isolated slice of cardiac tissue 6 ms after stimulation. Right: The current image obtained by inverse spatial filtering. The length of each arrow describes the strength of the current at that location. (From [56], with permission)

be extended to examine and correct for the effects of finite-diameter pickup coils[49]. For a SQUID with a finite-diameter pickup coil, the magnetic field B_z will be convolved with the spatial sampling (or turns) function H of the coil to give the detected flux Φ. Figure 8 shows four different pickup coils[49]. In Fourier space, the magnetic-field distribution from the sample $b(k_x, k_y)$ will be multiplied by the turns function $h(k_x, k_y)$ to give the flux $\phi(k_x, k_y)$. Ideally, the effect of the coil could be corrected by dividing $\phi(k_x, k_y)$ by $h(k_x, k_y)$ to obtain $b(k_x, k_y)$, so that an inverse Fourier transform would give the coil-corrected B_z. Unfortunately, for typical coils, the edges of the coil introduce zeros in their spatial-frequency transfer function, $h(k_x, k_y)$, which complicate deconvolution of the images at high spatial frequencies: the zero in the forward transfer function produces an infinity in the inverse function, making it difficult to obtain any information from the field at or even near that spatial frequency and its higher harmonics. This could be ameliorated with windowing, as before, but this leads once again to the loss of high-frequency information instead of to the desired extent of resolution enhancement. As shown in Fig. 9, even with windowing, the contribution near the zeros produces spikes that appear in the frequency domain at the edge of the window, and these spikes produce noise in the spatial images.

A solution to this problem that has been demonstrated theoretically[49]

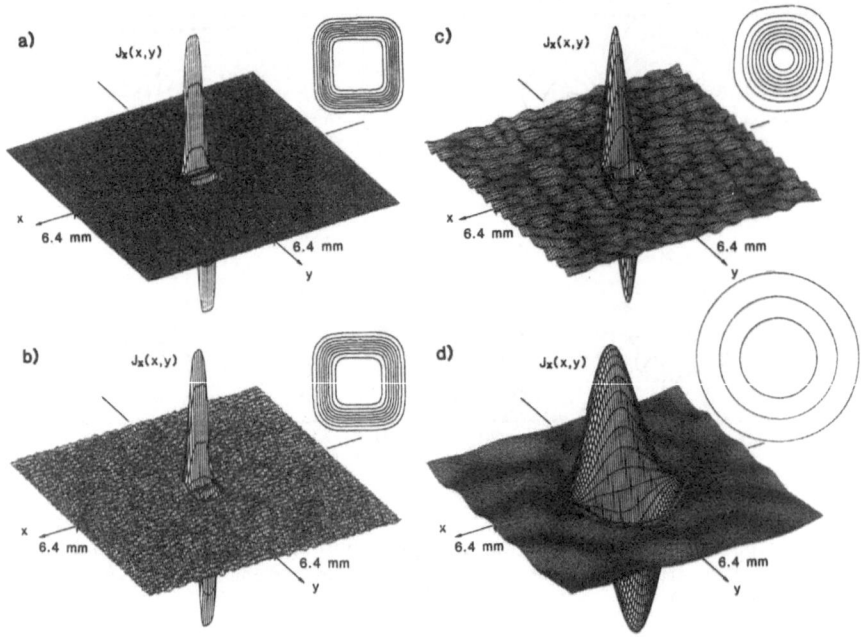

Figure 7. Spatial filtering/Fourier transform inverse image of the current density for four different values of z, calculated from the z component of the magnetic field produced by the square current loop used in Fig. 4 after magnetometer noise of 0.5 pT has been added to the magnetic field. Plots of both $J_x(x,y)$ and the current lines (upper right inset) are shown. Each line corresponds to 0.1 μA, except in (d), where each line is 0.05 μA. (a) $z = 0.1$ mm, $k_{max} = 100$ mm^{-1}, MSD = 0.03; (b) $z = 0.3$ mm, $k_{max} = 30$ mm^{-1}, MSD = 0.07; (c) $z = 1.0$ mm, $k_{max} = 10$ mm^{-1}, MSD = 0.44; (d) $z = 3.0$ mm, $k_{max} = 3$ mm^{-1}, MSD = 0.95. (From [47], with permission)

but not yet implemented, is to adjust the spacing of individual turns in a planar coil, as shown in Fig. 8, so that the zeros in the transfer function $h(k_x, k_y)$ shown in Fig. 9, are either eliminated or forced to very high spatial frequencies. This process, termed apodizing, could provide a significant enhancement in spatial resolution for certain SQUID imaging applications, particularly SQUID microscopy. The current image in Fig. 9d from the apodized coil has lower noise yet provides excellent spatial resolution, despite the fact that the apodized coil has a larger outer diameter than the other three coils.

Inward Continuation. Because there is no unique solution to the three-dimensional magnetic inverse problem, most solutions to this inverse problem use heavily-constrained models for the sources. Alternatively, if the

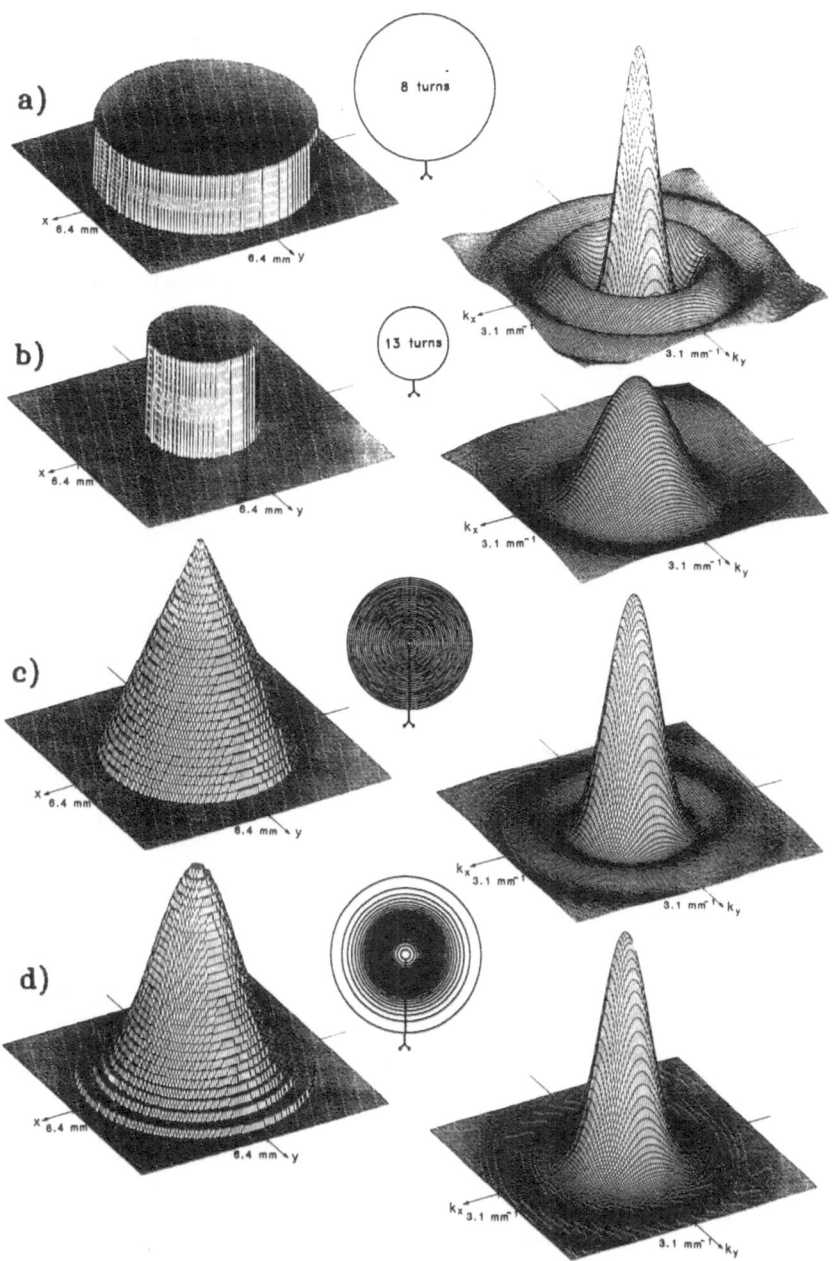

Figure 8. Four SQUID pickup coils. Left: the turns function $H(r)$; middle: the coil; right: the Fourier transform of the turns function $h(k_x, k_y)$. The coils are (a) 8 turns with a 5-mm diameter, (b) 13 turns with a 2.5-mm diameter, (c) a pancake coil with 29 turns, and (d) an apodized coil with 29 turns and a turns functions given by $e^{-a^2 r^2}$ (Adapted from [49], with permission)

644

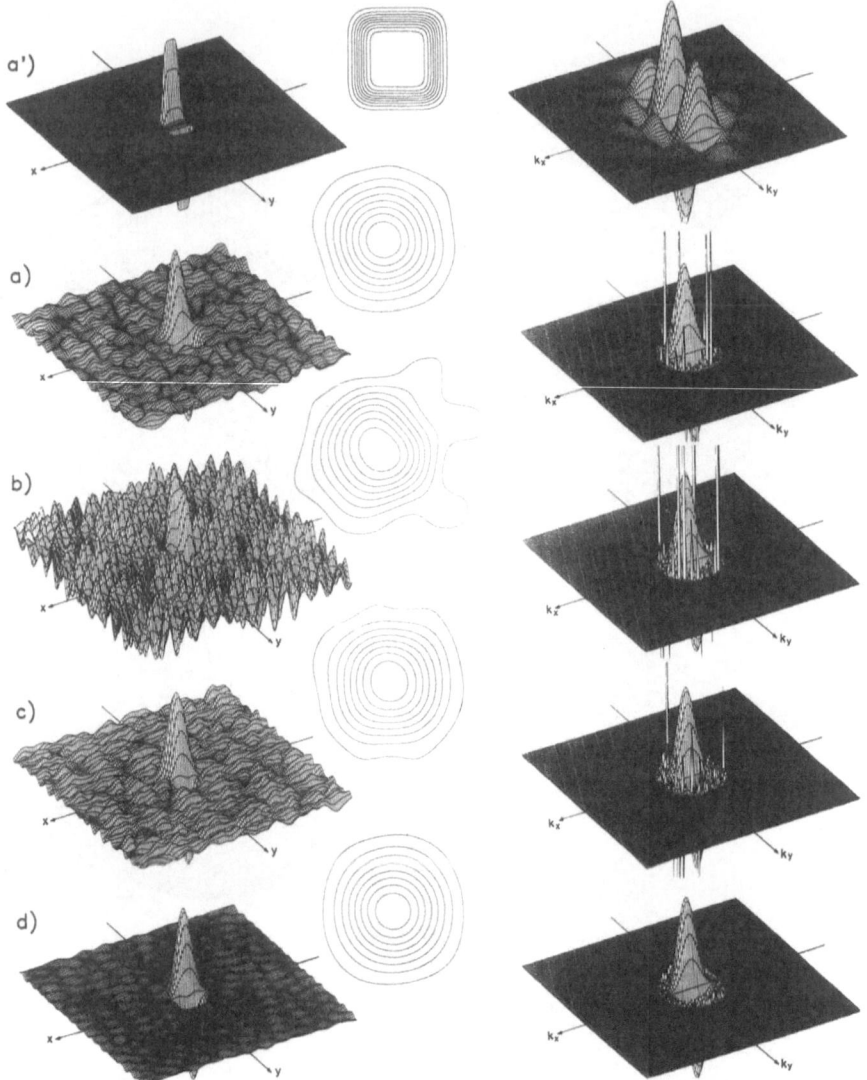

Figure 9. A comparison of the imaging ability of the four different SQUID pickup coils in Fig. 8. (a'): (left) The x-component of the current density for a square current loop with sides of length 5 mm and a Gaussian width with length constant 0.75 mm. (center) The current lines that describe the square current loop–the current density is in the direction tangent to the lines, and the amplitude is proportional to the spacing of the lines. (right) The imaginary part of the Fourier transform of the x component of the current density. This current was used as the source distribution for computing the subsequent images, which were calculated using $z=5$ mm, $I_{noise}=15$ pA, and $k_{max}=1.8$ mm^{-1}. (a-d): (left) The x-component; (center) current lines of the image of the current density from the four coils; (right) The imaginary part of the Fourier transform of the current-density images. (Adapted from [49], with permission)

sources are restricted to a well-defined, bounded volume, Laplace's equation can be used to continue fields observed at some distance inward to the bounding surface, thereby sharpening or deblurring the field. As discussed in more detail elsewhere[61, 7], this approach can be used to convert, in a model-independent manner, a map of the magnetic field measured a centimeter or two from the surface of the scalp into a sharper map of the magnetic field on the surface of the cortex without having to utilize any physical or physiological constraints. Inward continuation deeper into the cortex could be invalidated by active current sources near the cortical surface.

The inward continuation is unique, but the stability of the calculation can be affected by measurement geometry, source configuration and noise. While inward continuation has not been utilized heavily in biomagnetism, it is a common technique in geophysics to sharpen geomagnetic features deep beneath the soil surface. In the spatial frequency domain, the inward continuation of the field measured at z_1 to the point z_2 is[47]

$$b_z(k_x, k_y, z_2) = e^{\sqrt{k_x^2+k_y^2}(z_1-z_2)} b_z(k_x, k_y, z_1) . \tag{16}$$

When z_2 is less than z_1, the exponential term acts as a high-pass amplifier, amplifying the high spatial-frequency components of the image. The advantage of this approach is that it allows sharpening of the image without a specific model for the source. The disadvantages are that it works only between two values of z for which there are no current sources, and that it is subject to noise-related instabilities.

Imaging Discontinuous Currents. In the spatial frequency domain, the z component of the magnetic field from a current-carrying plate of thickness d is given by Eq. 12, which can be written in a more compact form as

$$b_z = \frac{i\mu_0 d}{2} \left(\frac{k_y}{k} j_x - \frac{k_x}{k} j_y \right) e^{-kz} \quad z \gg d , \tag{17}$$

where

$$k = \sqrt{k_x^2 + k_y^2} . \tag{18}$$

If current is injected into a planar sample through vertical wires, as shown in Fig. 10, the magnetic field from the vertical wires is horizontal and does not contribute to the vertical magnetic field B_z. If we know the location of the wires that inject the current, we can modify the two-dimensional continuity condition (Eq. 13) to include this information[58] as follows:

$$\nabla \cdot \vec{J}(x,y) = F(x,y) , \tag{19}$$

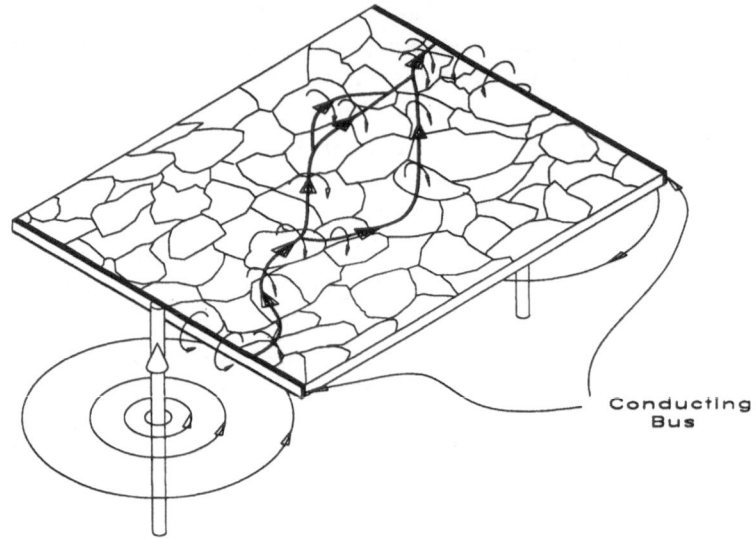

Figure 10. Current injection into a granular substrate.

where $F(x, y)$ describes the current source and sink distribution in the sheet, *i.e.*, I_z in the wires. In the spatial frequency domain, this becomes

$$-ik_x j_x - ik_y j_y = f(k_x, k_y) . \tag{20}$$

We then obtain

$$\begin{aligned}
J_x &= FT^{-1}\{j_x\} &(21)\\
&= FT^{-1}\left\{-i\frac{2k_y}{\mu_o k d}e^{kz}b_z + i\frac{k_x f}{k^2}\right\} &(22)\\
J_y &= FT^{-1}j_y &(23)\\
&= FT^{-1}\left\{i\frac{2k_x}{\mu_o k d}e^{kz}b_z + i\frac{k_y f}{k^2}\right\} . &(24)
\end{aligned}$$

Figure 11 shows a comparison of the original current distribution and the distribution reconstructed by this approach. Note that the source and the image are in good agreement in the immediate vicinity of the current-injection electrodes. As we shall see later, the inability of the spatial filtering technique to constrain currents to flow within known boundaries, limits this technique to situations where the magnetometer is very close to the current distribution.

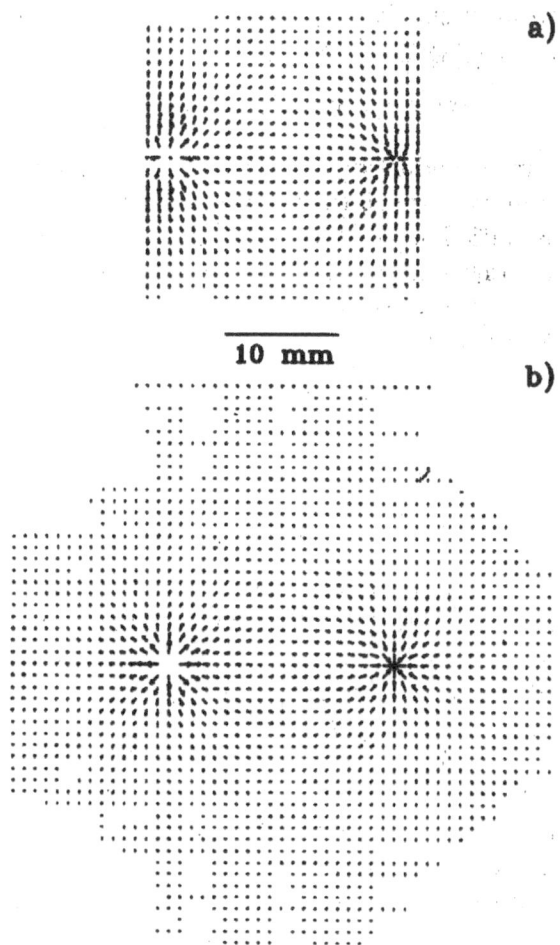

Figure 11. The results of using an unconstrained filtering technique to solve a bounded current source. (a) The original current distribution; (b) the reconstruction using inverse spatial filtering. (From [62], with permission)

Current-Injection Tomography. Magnetic imaging and the Fourier transform/spatial-filtering approach can be used to determine the conductivity distribution within a planar conducting object[57]. We start with Ohm's law, which gives us the current density $\vec{J}(\vec{r})$ in an inhomogeneous, isotropic conductor with conductivity distribution $\sigma(\vec{r})$

$$\vec{J}(\vec{r}) = \sigma(\vec{r})\vec{E}(\vec{r}) . \tag{25}$$

As we saw in Eq. 2, the curl of the current density can be thought of as the "source" of the magnetic field. In inhomogeneous conductors, $\nabla \times \vec{J}$

has two contributions: one from spatial variations ($\nabla\sigma$) and another from electromotive forces (EMFs) ($\nabla \times \vec{E}$),

$$\nabla \times \vec{J} = \nabla\sigma \times \vec{E} + \sigma\nabla \times \vec{E} \ . \qquad (26)$$

A boundary between conducting and insulating regions is simply a sudden spatial variation in σ, at which the tangential current produces a magnetic field. In the quasistatic limit with no sources of EMF within the conductor, *i.e.*, with currents injected and removed only on the surface of the object, the curl of $\vec{E}$ must be zero, since it is derivable from a scalar potential. In this case, $\nabla \times \vec{J}$ becomes

$$\nabla \times \vec{J}(\vec{r}) = \frac{\nabla\sigma(\vec{r})}{\sigma}(\vec{r}) \times \vec{J}(\vec{r}) \ , \qquad (27)$$

which can be written as

$$\nabla \times \vec{J}(\vec{r}) = (\nabla \ln \sigma(\vec{r})) \times \vec{J}(\vec{r}) \ . \qquad (28)$$

As pointed out by Staton *et al.*[57], the strategy of conductivity imaging via current-injection tomography is to reconstruct $\sigma(\vec{r})$ from a known $\vec{J}(\vec{r})$ using Eq. 28. However, it is not possible to reconstruct the conductivity distribution using a single distribution of injected current. One problem with a single configuration of injected current is that it is possible to have a nonzero $\nabla \ln \sigma$ and nonzero $\vec{J}$, but have zero $\nabla \times \vec{J}$. This occurs if both $\nabla \ln \sigma$ and $\vec{J}$ are parallel, for example in a rectangular conducting bar in which the conductivity varies only in one direction, but the current flows uniformly in the bar along that direction. In this case, the currents do not indicate the presence of gradients in the conductivity distribution, and the current distribution could be the result of uniform current injection into a isotropic homogeneous bar. It is thus important that tomography be employed by injecting current into the sample from multiple directions, with magnetic imaging of the current distribution for each configuration of current injection. At the very least, current should be injected in orthogonal directions: two directions for a thin conducting plate, three directions for a rectangular conducting slab, as illustrated in Fig. 12.

If the current density $\vec{J}$ were known everywhere in a three-dimensional conducting sample, it would in principle be possible to solve Eq. 27. However, there is no unique solution to the general three-dimensional magnetic inverse problem for currents, and hence determination of a three-dimensional conductivity distribution from magnetic measurements alone will, in general, be impossible. However, the two-dimensional current inverse problem does have a unique solution[47], and we can use current tomography with magnetic measurements of $J(\vec{r})$ to determine the conductivity distribution in a planar sample.

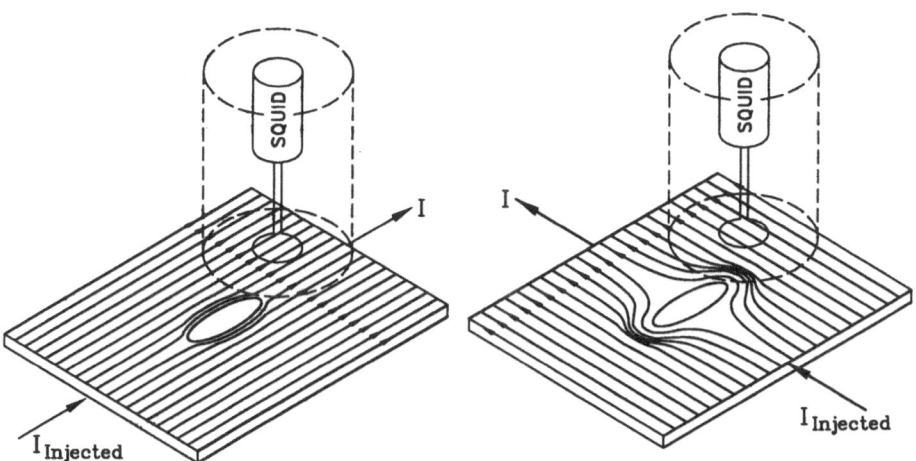

Figure 12. Current-injection tomography in a thin, conducting plate. When current is injected (left) along the major axis of an elliptical flaw (x direction), the current is perturbed less than when injected (right) orthogonally along the minor axis (y direction). The combined magnetic imaging of both current distributions contains information on the length and width of the flaw. (From [57], with permission)

To illustrate this approach, let us examine the minimal case of using only two orthogonal distributions of injected current for a planar sample. We can use a magnetometer and spatial-filtering inverse techniques described previously to obtain images of the two resulting current distributions

$$\vec{J_1}(x, y) = J_{1x}(x, y)\hat{i} + J_{1y}(x, y)\hat{j} \tag{29}$$

$$\vec{J_2}(x, y) = J_{2x}(x, y)\hat{i} + J_{2y}(x, y)\hat{j}. \tag{30}$$

For our two-dimensional case, the normal component of the curl of each current-density distribution obeys Eq. 28

$$\nabla \times \vec{J_1}(x, y) \cdot \hat{n} = J_{1y}\left(\frac{\partial \ln \sigma(x, y)}{\partial x}\right) - J_{1x}\left(\frac{\partial \ln \sigma(x, y)}{\partial y}\right)$$

$$\equiv c_1(x, y) \tag{31}$$

$$\nabla \times \vec{J_2}(x, y) \cdot \hat{n} = J_{2y}\left(\frac{\partial \ln \sigma(x, y)}{\partial x}\right) - J_{2x}\left(\frac{\partial \ln \sigma(x, y)}{\partial y}\right)$$

$$\equiv c_2(x, y). \tag{32}$$

The unknown quantities in this pair of equations are $\partial \ln \sigma(x, y)/\partial x$ and $\partial \ln \sigma(x, y)/\partial y$. In tomography, with a large number of directions of injected current, there will be a highly overdetermined set of equations for these two unknown distributions that would be solved by singular-value decomposition or another numerical technique. In our simple two-current example, we can reduce the pair of equations into a more tractable form analytically

$$\frac{\partial \ln \sigma(x,y)}{\partial x} = (J_{x1}c_2 - J_{x2}c_1)/(J_{x1}J_{y2} - J_{x2}J_{y1}), \equiv f_x(x,y) \quad (33)$$

$$\frac{\partial \ln \sigma(x,y)}{\partial y} = (J_{y1}c_2 - J_{y2}c_1)/(J_{x1}J_{y2} - J_{x2}J_{y1}), \equiv f_y(x,y) . \quad (34)$$

To obtain the conductivity distribution, we simply numerically integrate each of these equations

$$\ln \sigma(x,y) = \ln \sigma(x_o,y_o) + \int_{x_o}^{x} f_x(x',y)dx' + \int_{y_o}^{y} f_y(x_o,y')dy' . \quad (35)$$

The conductivity distribution is thus

$$\sigma(x,y) = \sigma(x_o,y_o)e^{\left(\int_{x_o}^{x} f_x(x',y)dx' + \int_{y_o}^{y} f_y(x_o,y')dy'\right)} . \quad (36)$$

Note that in the integration, we include explicitly as a constant of integration the conductivity $\sigma(x_o,y_o)$ at the single point where we began the integration. If that conductivity is not known, we can determine only normalized conductivity $\sigma(x,y)/\sigma(x_o,y_o)$. Alternatively, if we measure the resistance between the current injection electrodes, we should be able to obtain a value for $\sigma(x_o,y_o)$. Figure 13 shows the results of a numerical simulation of this process for an elliptically-shaped, Gaussian flaw in a copper plate, given by

$$\sigma(x,y) = \sigma_o \left(1 - 0.8e^{\frac{1}{2}\left[\left(\frac{x}{a}\right)^2 + \left(\frac{y}{b}\right)^2\right]}\right), \quad (37)$$

where $\sigma_o = 5.8 \times 10^7$ S/m, $a = 5$ mm and $b = 2$ mm. The thin, conducting plate is of length $l = 10$ cm, width $w = 10$ cm and thickness 100 μm. For the simulation, a total of 10 mA of current was injected uniformly into the plate along the major axis of the elliptical flaw (x-direction), and also in a separate current injection along the minor axis of the elliptical flaw (y-direction) as illustrated in Fig. 12. The forward problem consisted of calculating the electrical potentials, current densities, curl of the current densities, and magnetic fields. The inverse problem consisted of reconstructing the conductivity distribution from the magnetic fields. As discussed in more detail by Staton et al.[57], the accuracy of this method is determined by the ability to image current densities and curl distributions from the magnetic fields. The error in the inverse procedure depends on a variety of factors: the distance between the current distribution and detector, the signal-to-noise ratio of the measurement, the distance between magnetic measurements,

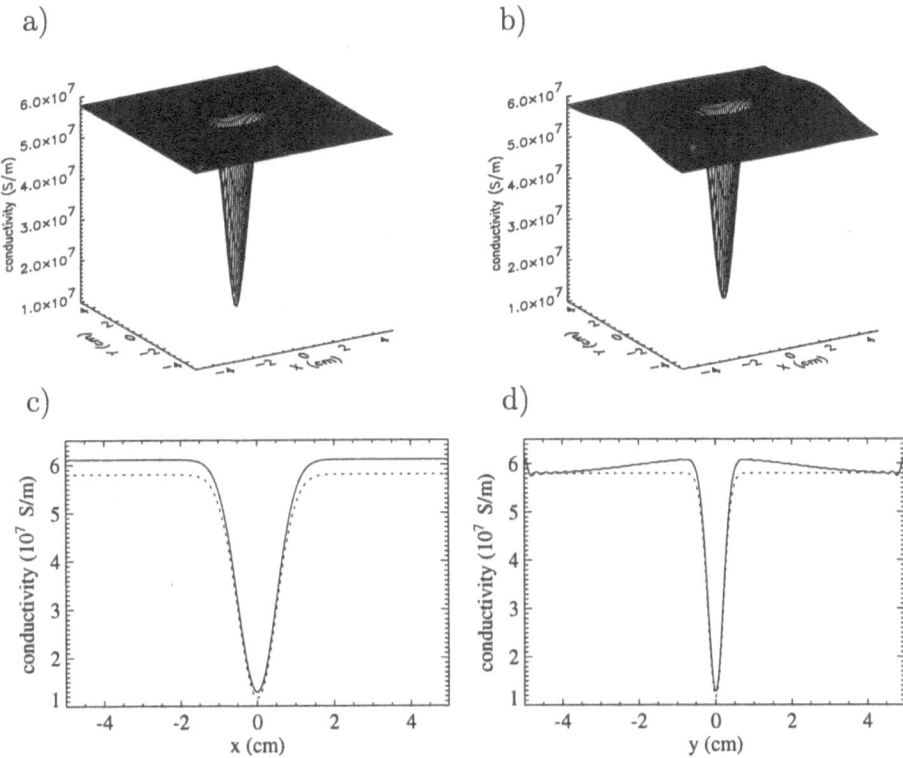

Figure 13. A comparison of (a) original conductivity distribution and (b) magnetically reconstructed solution using current-injection tomography, and cross-section comparison of original conductivity distribution (dotted line) and reconstructed solution (solid line) along the (c) major axis and (d) minor axis of the elliptically-shaped, Gaussian flaw. (From [57], with permission)

and the spatial extent of the measurement. This calculation demonstrates the feasibility of the technique; further studies will require adding noise and varying the sampling parameters, and should also address the possible advantages of applying current to the sample from multiple directions (more than the minimally required number of two or three).

Other Applications of Spatial Filtering. As can be seen from the breadth of applications we have just outlined, spatial filtering is a powerful approach. Other applications of the technique relevant to magnetic imaging include the interpretation of gradiometers as spatial filters[8, 9, 10, 46, 11], and the use of filters to determine multipole moments from measurement of the normal component of the field above the surface of a sphere or plane[2].

Summary of Findings for Spatial Filtering. From this brief summary and the more detailed analyses in the literature, we find from spatial filtering that the coil-to-source spacing limits spatial resolution, that coil diameter should equal coil-to-source spacing, that inverse spatial filtering is unique in one and two dimensions, that inward continuation separates the problems of uniqueness from those of instability, that noise limits stability, that apodizing can improve spatial resolution, and that windowing to filter out noise and improve stability compromises spatial resolution. The spatial filtering approach is powerful and easy to implement with commercial software packages, but it is limited primarily by the difficulty with this approach in applying specialized boundary conditions and utilizing other *a priori* knowledge of the sources. Current tomography offers promise for the characterization of the conductivity distribution within a planar conducting object, but further algorithm development is required.

2.2.2. *The Hosaka-Cohen Transformation*

A "current pattern" in the xy-plane can be obtained from B_z by the Hosaka-Cohen (HC) transformation[19]

$$\mathbf{HC} \equiv \frac{\partial B_z}{\partial y}\hat{i} - \frac{\partial B_z}{\partial x}\hat{j} , \tag{38}$$

where the field is measured a distance z above the currents. The components of **HC** resemble the currents, *i.e.*,

$$J_x \quad \sim \quad \frac{\partial B_z}{\partial y} \equiv HC_x \tag{39}$$

$$J_y \quad \sim \quad -\frac{\partial B_z}{\partial x} \equiv HC_y . \tag{40}$$

The HC transformation is believed valid for two-dimensional current distributions, but has been applied to data from three-dimensional sources[19, 64]. The uncertainty lies with the derivation of the HC transformation, which is based upon the curl of $\vec{B}$. Ampere's law states that

$$\vec{J} = \frac{1}{\mu_o}\nabla \times \vec{B}$$

$$= \frac{1}{\mu_o}\left\{\left(\frac{\partial B_z}{\partial y} - \frac{\partial B_y}{\partial z}\right)\hat{i} - \left(\frac{\partial B_z}{\partial x} - \frac{\partial B_x}{\partial z}\right)\hat{j} + \left(\frac{\partial B_y}{\partial x} - \frac{\partial B_x}{\partial y}\right)\hat{k}\right\} .$$

Within a current distribution of infinite extent with no z-dependence,

$$\frac{\partial B_y}{\partial z} \simeq \frac{\partial B_x}{\partial z} \simeq 0 . \tag{41}$$

If we add the tighter constraint that $J_z = 0$ everywhere, then

$$\vec{J} \simeq \frac{1}{\mu_o}\left\{\frac{\partial B_z}{\partial y}\hat{i} - \frac{\partial B_z}{\partial x}\hat{j}\right\} . \tag{42}$$

Within a constant, this is the same as Eq. 38. However, this equation, as derived, applies only within the conductor, since outside a conductor both $\vec{J}$ and $\nabla \times \vec{B}$ are everywhere zero in the quasistatic limit. In practice, one cannot readily measure $\vec{B}$ inside a conductor. However, the symmetry for a two-dimensional planar current distribution is the same as for a three-dimensional current distribution with no J_z and no z-dependence of either J_x or J_y. The validity of the Hosaka-Cohen transformation is unclear when it is possible only to measure $\vec{B}$ at height z above the xy-plane. We have recently examined the Hosaka-Cohen transformation using Fourier spatial filtering[58, 63].

In the spatial-frequency domain, we had the exact inverse solution for two-dimensional current distributions

$$j_x(k_x, k_y) = i\frac{2}{\mu_o d}\frac{k_y}{\sqrt{k_x^2 + k_y^2}} e^{\sqrt{k_x^2+k_y^2}z} b_z(k_x, k_y) \tag{43}$$

$$j_y(k_x, k_y) = -i\frac{2}{\mu_o d}\frac{k_x}{\sqrt{k_x^2 + k_y^2}} e^{\sqrt{k_x^2+k_y^2}z} b_z(k_x, k_y) . \tag{44}$$

This can be expanded in a Taylor's series

$$j_x(k_x, k_y) = i\frac{2}{\mu_o d}\frac{k_y b_z(k_x, k_y)}{k}(1 + kz + \frac{1}{2!}k^2 z^2 + \cdots) , \tag{45}$$

where

$$k = \sqrt{k_x^2 + k_y^2} . \tag{46}$$

If k and z are sufficiently small, then

$$j_x(k_x, k_y) \simeq i\frac{2}{\mu_o d}k_y b_z(k_x, k_y)f , \tag{47}$$

where

$$f = 1/k + z + kz^2/2 . \tag{48}$$

Similarly,

$$j_y(k_x, k_y) \simeq -i(\frac{2}{\mu_o d})k_x b_z(k_x, k_y)f . \tag{49}$$

For comparison, the two-dimensional Fourier transform of **HC** is

$$j_x(k_x, k_y) \propto ik_y b_z(k_x, k_y) \tag{50}$$

$$j_y(k_x, k_y) \propto -ik_x b_z(k_x, k_y) . \tag{51}$$

By fitting a straight line to the exact expression, we obtain[58, 63]

$$j_x \simeq i\frac{2\pi z}{\mu_o d}k_y b_z(k_x, k_y) \tag{52}$$

$$j_y \simeq -i\frac{2\pi z}{\mu_o d}k_x b_z(k_x, k_y) . \tag{53}$$

By taking the inverse Fourier transform, this becomes

$$J_x \simeq \frac{2\pi z}{\mu_o d}\frac{\partial B_z}{\partial y} \tag{54}$$

$$J_y \simeq -\frac{2\pi z}{\mu_o d}\frac{\partial B_z}{\partial x} , \tag{55}$$

and we see that the Hosaka-Cohen transformation is the first-order approximation to the exact inverse result. The approximation is best for small z and k. Figure 14 compares the results of the Hosaka-Cohen transformation and the exact result for data similar to those in Fig. 5.

2.2.3. Dipole Fitting

The conceptually easiest inverse solution is to divide the two-dimensional object into N elements, and to then assign an unknown current dipole to each element, i.e., $p_{x,i}$ and $p_{y,i}$, where the magnetic field from the ith elemental current dipole is given by

$$B_{z,i}(\vec{r}) = \frac{\mu_o}{4\pi}\frac{p_{x,i}(\vec{r'})(y - y') - p_{y,i}(\vec{r'})(x - x')}{|\vec{r} - \vec{r'}|^3} . \tag{56}$$

A least-squares solution, singular-value decomposition, or an iterative approach can be used to determine the components of each elemental dipole[6]. The solution will be most stable when the magnetometer is close to the sample. While this approach can be used in three dimensions, troublesome instabilities can occur if more than a few dipoles are used. Even in two dimensions, there are several problems with this approach. For a large number of dipoles, the approach is computationally demanding. If there are no constraints between the adjacent dipoles, current will not be conserved. While a stable solution to the equation may be obtained, interpretation of the results in such a situation may be unclear. Furthermore, the presence of noise can lead to instabilities in the solution, particularly if a very fine discretization is used, and it can become necessary to devise constraints to avoid the appearance of large opposing dipoles in adjacent elements. Minimum norm techniques may improve the performance of this approach[43].

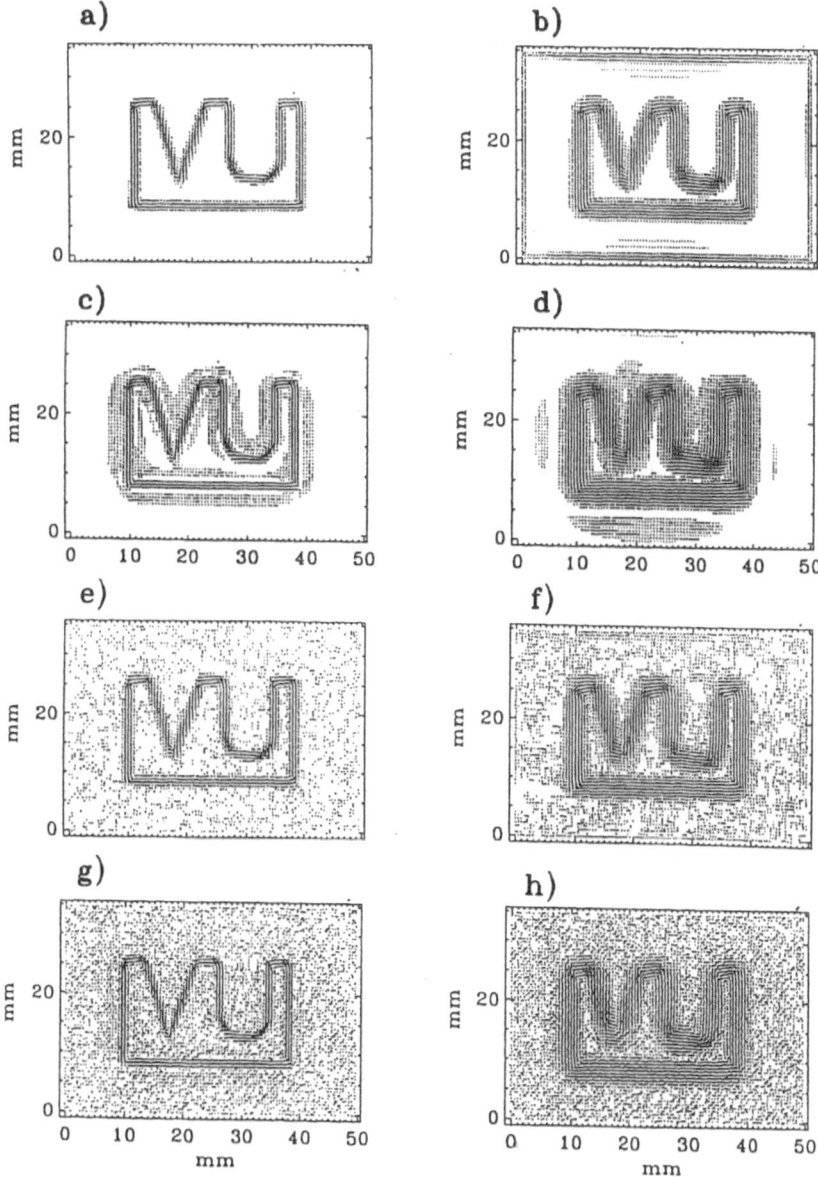

Figure 14. A comparison of the exact and **HC** reconstructions for a 'VU' current source similar to that in Fig. 5. Left: $z = 1.5$ mm; Right: $z = 3.0$ mm. Top half (a,b,c,d): no noise; bottom half (e,f,g,h): SNR = 10. (a,b,e,f) Exact reconstruction; (c,d,g,h) **HC** reconstruction. (Adapted from [58, 63], with permission).

2.2.4. *Lead Field Analysis*

The intrinsic limitation of the three-dimensional magnetic inverse problem is the presence of degrees of freedom in the hypothetical current source distribution that are magnetically silent, such as spherical batteries or radially symmetric arrays of current dipoles. One particularly useful method to prevent the mathematical instabilities associated with the potential existence of silent sources is to use lead-field expansions to constrain the inverse problem so as to limit the set of sources only to those that produce measurable magnetic fields in the particular magnetometer arrangement used.

Lead Fields. To understand this approach, we first need to define lead fields[69, 81, 3, 42]. Suppose that a magnetometer measures B_z at a specific point $\vec{r}$. As shown in Eq. 1, this measurement represents a convolution of the source current distribution $\vec{J}(\vec{r}')$ and the Green's function $G(\vec{r}, \vec{r}')$ over the points $\vec{r}'$ where both $\vec{J}$ and G are non-zero. This Green's function is equivalent to a mutual inductance, M_{pq}, between a particular source element, $d\vec{p}$, of a current dipole distribution and a single, specific pick-up coil, q. The principle of reciprocity states that M_{pq} equals M_{qp}, which is equivalent to saying that the Green's function is the same whether the current source is at $\vec{r}$ and the coil is at $\vec{r}'$ or visa versa, *i.e.*, $G(\vec{r}, \vec{r}') = G(\vec{r}', \vec{r})$. The lead field $\vec{L}_i^J(\vec{r}')$ for the ith magnetometer at $\vec{r}_i$ is simply the current distribution $\vec{J}$ that would be produced were the magnetometer pick-up loop replaced by a coil carrying a low-frequency ac current. In this approach, the output of the ith SQUID magnetometer, a scalar variable, is simply the convolution of the source currents and the lead field for that magnetometer

$$B_i(\vec{r}) = \int_v \vec{L}_i^J(\vec{r}') \cdot \vec{J}(\vec{r}') d^3 r' . \tag{57}$$

Note the index i in $\vec{L}_i^J$ contains within it the location $\vec{r}_i$, and hence we need not carry the dual-position notation of $G(\vec{r}, \vec{r}')$. If a current source at a particular point $\vec{r}'$ is orthogonal to $\vec{L}_i^J(\vec{r}')$ at that same point, then that source cannot contribute to the field detected by the SQUID at $\vec{r}_i$. Reciprocally, a measurement of that component of the magnetic field can not determine the strength of that component of the current source at $\vec{r}'$, although it may be able to detect another component at that location[3].

---

[3]This is an easy explanation for the inability of a SQUID to measure magnetic fields from radial dipoles in a conducting sphere: it is impossible, by means of external coils, to induce in the sphere eddy currents with radial components, *i.e.*, radial lead fields. For this reason, it is impossible to use external coils to create a lead field that would detect a spherical battery. Also note that were we trying to measure a magnetization $\vec{M}$ rather than a current $\vec{J}$, we would utilize the magnetic lead field $\vec{L}^B$, rather than the current lead field $\vec{L}^J$.

The requirement that the solution set is to be restricted to those currents that produce fields that can be measured by the chosen magnetometers is equivalent to stating that all of the imaged currents have to lie along the lines of the lead fields of one or more of the magnetometers.

Lead-Field Expansions. Equation 57 suggests that we use the set of lead-fields for our N magnetometers to constrain the inverse source distribution[22, 58, 20]. While lead-field expansions were originally developed for three-dimensional inverse problems, to maintain the consistency of our derivations we shall restrict ourselves to measurement of two-dimensional current distributions. We begin with the law of Biot and Savart for B_x at the ith measurement point above a two-dimensional current distribution

$$B_{x,i}(\vec{r}_i) = \frac{\mu_0 d}{4\pi} \int_s \frac{J_y(\vec{r}')(z_i - z')}{|\vec{r}_i - \vec{r}'|^3} dx' dy' , \qquad (58)$$

where S is the surface containing the sources. We can write this in terms of the lead-field for that magnetometer

$$B_{x,i}(\vec{r}_i) = \int_s \vec{L}_i^J(\vec{r}') \cdot \vec{J}(\vec{r}') d^2 r' . \qquad (59)$$

The dot product and the orientation of the lead field automatically selects the J_y component of the source current.

Let us suppose that we can expand $\vec{J}(\vec{r}')$ in terms of the non-orthogonal set of m lead-field functions $\{\vec{L}_i^J\}$

$$\vec{J}(\vec{r}') = \sum_{k=1}^{m} A_i \vec{L}_k^J(\vec{r}') w(\vec{r}') , \qquad (60)$$

where $w(\vec{r}')$ is the *a priori* probability density for the current density, and the A_k's are the expansion coefficients. If we know, for example, that the currents in a particular region are zero, then we can at the beginning of our calculation set $w(\vec{r}')$ to zero at those locations. To determine the A_k's, we substitute Eq. 60 into Eq. 59, and exchange the order of summation and integration

$$B_{x,i}(\vec{r}_i) = \sum_{k=1}^{m} A_i \int_s \vec{L}_i^J(\vec{r}') \cdot \vec{L}_k^J(\vec{r}') w(\vec{r}') d^2 r' . \qquad (61)$$

The field is thus

$$B_{x,i} = \sum_{k=1}^{m} P_{ik} A_k \qquad (i = 1, 2, \ldots, N) , \qquad (62)$$

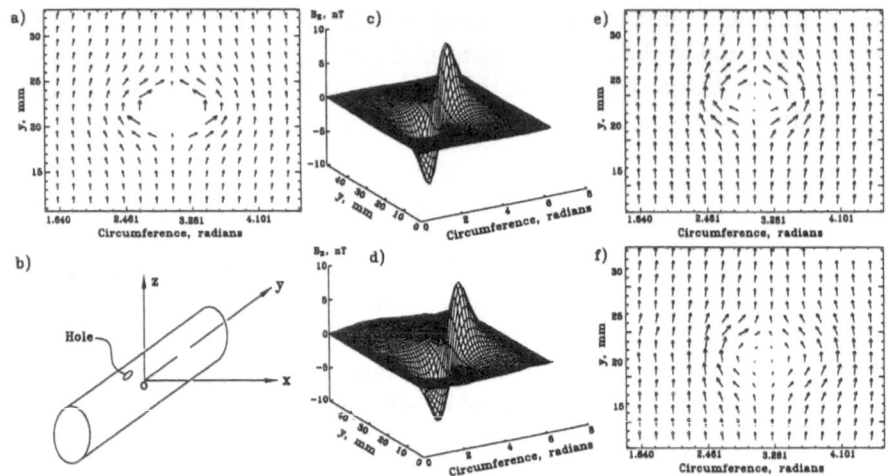

Figure 15. An example of a lead-field inverse for a current-carrying tube with a small hole. (a) Original current distribution; (b) geometry; (c) calculated magnetic field; (d) measured magnetic field; (e) currents reconstructed from theoretical magnetic field; (f) currents reconstructed from measured magnetic field. (Adapted from [59])

where

$$P_{ik} = \int_s \vec{L}_i^J(\vec{r}') \cdot \vec{L}_k^J(\vec{r}')w(\vec{r}')d^2r' . \tag{63}$$

While computation of the P_{ik} can be tedious, it needs to be done only once for a particular measurement geometry. Singular-value decomposition or some other numerical technique is used to solve this set of equations for the $\{A_k\}$, which are then substituted into Eq. 60 to find $\vec{J}(\vec{r}')$. The process can be iterated to refine the images, where $w(\vec{r}')$ is adjusted each time. Figure 15 shows how this can be used to determine the distribution of currents flowing on the surface of a current-carrying tube with a small flaw[58, 20, 21]. This approach has the advantage that constraints can be applied through both the lead fields and the weighting function $w(\vec{r}')$, but it is difficult to apply boundary conditions and other *a priori* knowledge of the source. In the case of the current-carrying tube, the lack of an analytical expression for the Fourier transform for the Green's function for currents flowing on the cylinder, precluded the easy application of the Fourier transform approach. One limitation of the approach is that if there are too many magnetic field points, or too many basis functions are chosen, then the inner products used to determine the P_{ik} are so similar to each other that the related matrix equation (Eq. 62) becomes highly singular. Thus, for this approach, the number of magnetic-field data points and basis functions must be limited[58].

As summarized by Tan[58], the lead-field analysis introduced by Ioannides[22] provides a method to constrain the current-imaging space. However, because the lead-field functions are defined over all space by using the *a priori* probability density function $w(\vec{r})$, lead-field analysis can incorporate only a constraint condition, such as the absence of current from a particular region, but not boundary conditions, such as the specification that current can flow only tangentially to an insulating boundary. Secondly, the lead-field interpolation functions are neither complete nor orthogonal, and this approach can recover only the components of the current distribution to which the pickup coil is sensitive. Therefore, the reconstructions are sensitive to the choice of the measurement locations and hence the lead-field functions.

2.2.5. *The Finite-Element Method*

One of the potentially most powerful approaches to the two-dimensional magnetic inverse problem may be the finite-element inverse, which can readily incorporate known source geometry and a wide variety of boundary conditions[58, 62]. Constrained reconstruction was originally proposed to solve the unbounded inverse Fourier transform problem in magnetic resonance imaging[18]. This method used a series of box-car functions as the interpolation functions to represent the original function, so that the solution of the inverse Fourier transform is bounded. Tan *et al.*[62] examined the applicability of this approach to the magnetic-imaging problem and then developed a more flexible approach that utilized the finite-element interpolation functions.

We can write the law of Biot-Savart for the z-component of the magnetic field above a two-dimensional current distribution $\vec{J}(x', y')$ in the x'y' plane at $z' = 0$ in the expanded form

$$B_z(x, y, z) = \frac{\mu_o d}{4\pi} \int \frac{J_x(x', y')(y - y') - J_y(x', y')(x - x')}{[(x - x')^2 + (y - y')^2 + z^2]^{3/2}} dx' dy' . \quad (64)$$

In order to reconstruct the current image $\vec{J}$ from the magnetic-field data recorded in the xy plane at a height z above the current distribution, we section the current-source space into a mesh of elements, as shown in Fig. 16, that represents our prior knowledge about the conductor geometry. We can represent the current distribution $\vec{J}(\vec{r})$ anywhere within an element (k) by using a set of two-dimensional interpolation functions and the values of the current at the nodes in the mesh

$$J_x^k = \sum_j J_{xj}^k N_j^k(x', y')$$

$$J_y^k = \sum_j J_{yj}^k N_j^k(x', y') , \quad (65)$$

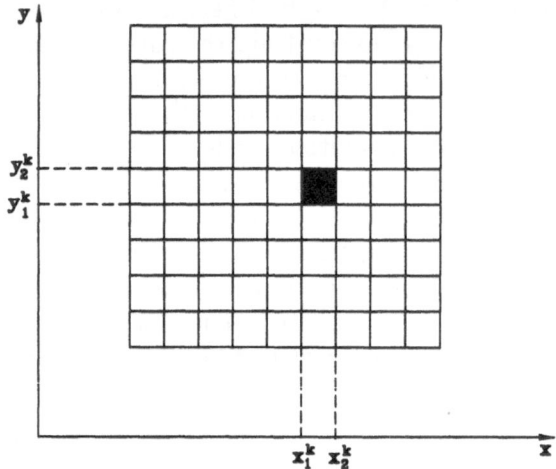

Figure 16. A finite-element mesh describing a square conducting sheet. (From [62], with permission)

where $\{N_j^k\}$ are the two-dimensional interpolation functions for the kth element, $\{J_{xj}^k, J_{yj}^k\}$ is the set of the nodal values that need to be determined, and j is the index for all interpolation functions in a single element. With this approach, we need to specify the values of the currents only at the nodes of the mesh; the interpolation functions provide us with analytical expressions for determining the current at all other points in the sample. As we did with the lead-field expansion for the current, we substitute Eq. (65) into Eq. (64), sum up all the elements, and reverse the order of summation and integration to obtain a set of linear equations that describe the magnetic field

$$[B_{zi}] = \sum_k [A_{zi,xj}^k][J_{xj}^k] - [A_{zi,yj}^k][J_{yj}^k] , \qquad (66)$$

where the subscript i stands for the ith measurement and

$$A_{zi,xj}^k = \frac{\mu_0 d}{4\pi} \int\int_{(k)} \frac{N_j^k(x',y')(y_i - y')}{[(x_i - x')^2 + (y_i - y')^2 + z^2]^{3/2}} dx' dy' \qquad (67)$$

$$A_{zi,yj}^k = \frac{\mu_0 d}{4\pi} \int\int_{(k)} \frac{N_j^k(x',y')(x_i - x')}{[(x_i - x')^2 + (y_i - y')^2 + z^2]^{3/2}} dx' dy' . \qquad (68)$$

If we measure the B_{zi}, calculate all of the A^k and invert Eq. 66, we should be able to solve for the J_{xj}^k and J_{yj}^k. However, because we are measuring only one variable (B_z) over the mesh of the field map and are trying to determine two variables (J_x and J_y) over a coarser mesh of a current map, the solution of Eq. 66 for J_x and J_y, can be difficult or impossible because the equations can be highly singular and unstable. As before, we can incorporate the current-continuity condition to improve greatly the stability of the solution from B_z. Similar equations can be derived for measurement of either B_x or B_y

$$[B_{xi}] = \sum_k [A^k_{xi,yj}][J^k_{yj}] \tag{69}$$

$$[B_{yi}] = \sum_k [A^k_{yi,xj}][J^k_{xj}] , \tag{70}$$

where

$$A^k_{xi,yj} = \frac{\mu_0 d}{4\pi} \int \int_{(k)} \frac{N^k_j(x',y')z}{[(x_i - x')^2 + (y_i - y')^2 + z^2]^{3/2}} dx'dy'$$

$$A^k_{yi,xj} = -\frac{\mu_0 d}{4\pi} \int \int_{(k)} \frac{N^k_j(x',y')z}{[(x_i - x')^2 + (y_i - y')^2 + z^2]^{3/2}} dx'dy' . \tag{71}$$

Because B_x is determined solely by J_y, and B_y by J_x, the determination of a single component by the measurement of only one tangential component is straightforward. The continuity equation can be used to determine the other component. Alternatively, the independent measurement of B_x and B_y would allow the imaging of both J_x and J_y, and hence the determination of whether or not current was conserved on the surface being mapped. This may be of great practical importance for the creation of maps of an effective surface corrosion current for three-dimensional objects, since the surface-current distribution will be determined by underlying galvanic activity.

It is time consuming to calculate the A_k and their inverses, but since they depend only upon the geometry of the finite-element mesh and the measurement arrangement, they must be calculated only once for each measurement configuration. The inversion of the matrix, on the other hand, is computationally modest.

Continuous Two-Dimensional Current Distributions. In SQUID NDE, we encounter two types of current-imaging problems: ones with a continuous current distribution, such as when a uniform current sheet that is perturbed by a localized flaw, for which the electrodes used to apply the current to the test object are distant and can be ignored; and those problems with a discontinuous current distribution, as would occur when current is injected into the planar sample by vertical wires, as shown in Fig. 10. As an example of the first case, we performed a simulation using the square current pattern shown in Fig. 17a. For this application, we used a bilinear finite element

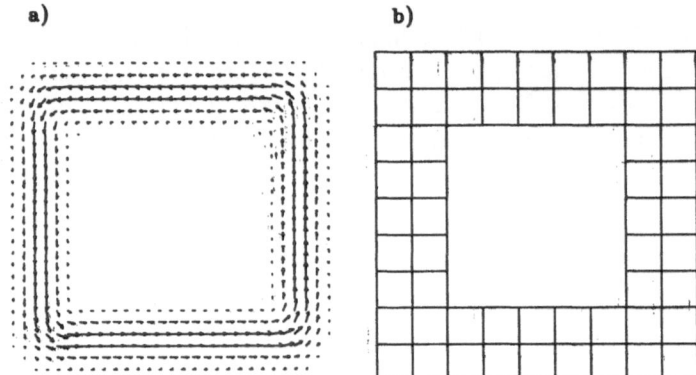

Figure 17. (a) The simulated current source used to test the continuous-current reconstruction algorithm. (b) The grid used to section the current space corresponding to (a). (Adapted from [62], with permission)

with an interpolation function of the form[62]

$$N_j^k(x, y) = a_j^k + b_{j'}^k x + c_j^k y + d_j^k xy ,\tag{72}$$

where a_j^k, b_j^k, c_j^k and d_j^k are the parameters specific to the kth element. Inside the kth element, the current distribution $\vec{J}(x, y)$ can be expressed in terms of the vector current density $\vec{J}_n$ at the four corners

$$\begin{aligned} J_x^k &= \mathcal{J}_{x1}^k N_1^k + \mathcal{J}_{x2}^k N_2^k + \mathcal{J}_{x3}^k N_3^k + \mathcal{J}_{x4}^k N_4^k \\ J_y^k &= \mathcal{J}_{y1}^k N_1^k + \mathcal{J}_{y2}^k N_2^k + \mathcal{J}_{y3}^k N_3^k + \mathcal{J}_{y4}^k N_4^k . \end{aligned}\tag{73}$$

We can utilize within each element a set of normalized, local coordinates, termed natural coordinates η and ξ, that range from $+1$ to -1, so that we can write the continuity condition (Eq. (13)) as

$$\frac{\partial J_x}{\partial \xi}\frac{\partial \xi}{\partial x} + \frac{\partial J_y}{\partial \eta}\frac{\partial \eta}{\partial y} = 0 .\tag{74}$$

Substituting Eq. (73) into the above equation and sorting the coefficients by the order of the polynomial, we obtain a polynomial expression for the continuity equation[62]

$$\frac{1}{a^k}(-\mathcal{J}_{x1}^k + \mathcal{J}_{x2}^k - \mathcal{J}_{x3}^k + \mathcal{J}_{x4}^k) + \frac{1}{b^k}(-\mathcal{J}_{y1}^k - \mathcal{J}_{y2}^k + \mathcal{J}_{y3}^k + \mathcal{J}_{y4}^k) +$$
$$\frac{1}{a^k}(\mathcal{J}_{x1}^k - \mathcal{J}_{x2}^k - \mathcal{J}_{x3}^k + \mathcal{J}_{x4}^k)\eta + \frac{1}{b^k}(\mathcal{J}_{y1}^k - \mathcal{J}_{y2}^k - \mathcal{J}_{y3}^k + \mathcal{J}_{y4}^k)\xi = 0 .\tag{75}$$

Since the coordinates ξ and η are independent of each other, each term of the polynomial must be individually zero to ensure that the continuity condition is satisfied in every element. Thus, we obtain a set of equations governing the coefficients J_{xj}^k and J_{yj}^k

$$\frac{1}{a^k}(-J_{x1}^k + J_{x2}^k - J_{x3}^k + J_{x4}^k) \ + \ \frac{1}{b^k}(-J_{y1}^k - J_{y2}^k + J_{y3}^k + J_{y4}^k) = 0$$
$$(J_{x1}^k - J_{x2}^k \ - \ J_{x3}^k + J_{x4}^k) = 0$$
$$(J_{y1}^k - J_{y2}^k \ - \ J_{y3}^k + J_{y4}^k) = 0 \,. \tag{76}$$

By incorporating the continuity condition (Eq. (76)) into one of the reconstruction equations (Eqs. (66), (69), or (70)), we can obtain the images of the current density using only one component of the magnetic field[58, 62].

Since the finite-element method can deal with each individual element, any kind of boundary condition is easy to incorporate into the solution. For instance, a bounded current source usually will not allow current to flow out of the edge, which corresponds to the boundary condition

$$\vec{J} \cdot \hat{n} = 0 \,. \tag{77}$$

Because the interpolation functions in the finite-element method are designed so that the nodal values are simply the current densities at the nodes, if the nodal values of the current component normal to the edge are zero, then this boundary condition is satisfied along that edge.

To demonstrate this approach, we consider a 11.7 mm × 12.6 mm current loop as shown in Fig. 17a. The current distribution does not have sharp edges to avoid problems with spatial aliasing that would occur with the spatial-filtering inverse. We assumed that we know the shape of the conductor, and hence can create the finite-element mesh in Fig. 17b. In an ideal case, when no noise is present in the data and the magnetic field is recorded very close to the current source, the filtering technique, which involves only a fast Fourier transform (FFT), a two-dimensional multiplication and an inverse FFT, has the advantages of dealing with a large amount of data quickly and provides an excellent result. However, even a small amount of noise will reduce severely the quality of the image produced by the filtering technique, producing current noise over the entire image plane, while the finite-element method controls the effects of the noise in the magnetic field by restricting the current to within the correct boundary. To demonstrate this, we calculated the z-component of the magnetic field as would be measured at $z = 1.5$ mm and at $z = 3$ mm, and then added spatially white noise so that the signal-to-noise ratios (SNR) were 20-to-1 and 5-to-1, respectively. For the finite-element inverse, we sampled the magnetic field over a 25 mm × 25 mm area with 1 mm spacing, $i.e.$, 26 × 26 points. Since the spatial-filtering inverse can readily use more data points without requiring the inversion of a giant matrix, we sampled the field over the

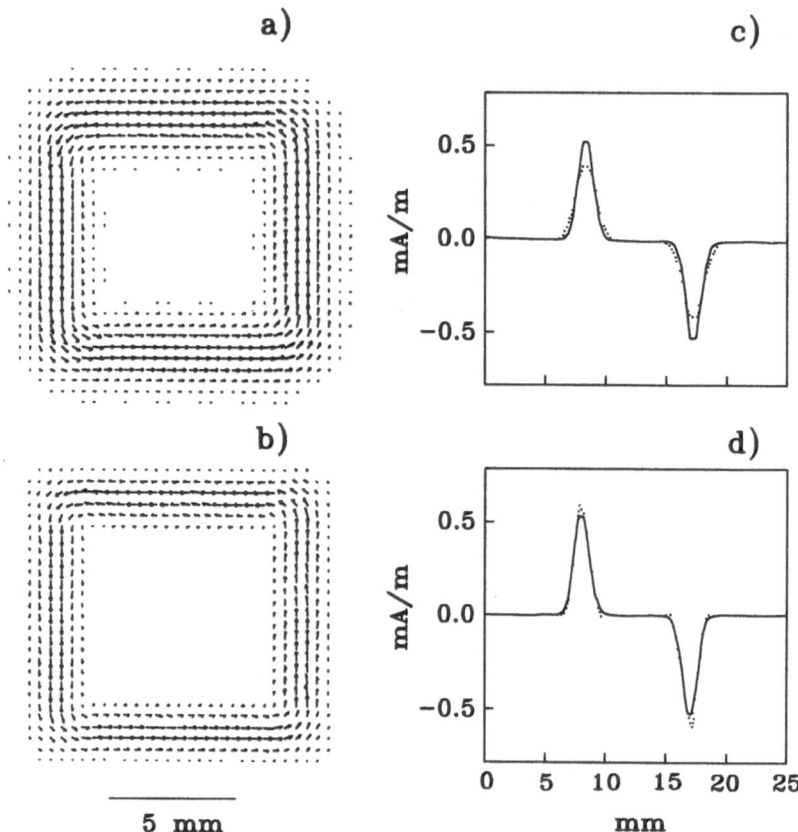

Figure 18. The reconstructions by the filtering technique (a,c) and the finite element method (b,d) at $z = 1.5$ mm with SNR = 20. (a,b) Reconstructed images of the current; (c,d) cross section of J_x. The solid lines in (c) and (d) are the original current distribution while the dotted lines are the reconstruction. (From [62], with permission)

same region with a 60×60 mesh. Figure 18(a,b) shows the results of the filtering technique and the finite-element method, respectively, for $z = 1.5$ mm and an SNR of 20. The mean-square deviation (MSD) for the result from the filtering technique (Fig. 18a) increases from 0.002 to 0.133 for the noise-free case (not shown), which means that even a 5% noise level will degrade the quality by a factor of sixty over the reconstruction from the noise-free data. In contrast, the MSD for the finite-element method (Fig. 18b) increases only from 0.026 to 0.040. While the finite-element approach with 26×26 noise-free field points at 1.5 mm provides images that are of a lower quality than obtained by the 60×60 data used with the filtering approach, the image quality is degraded less quickly by noise for the finite-element approach, so that in the presence of only a small amount of noise,

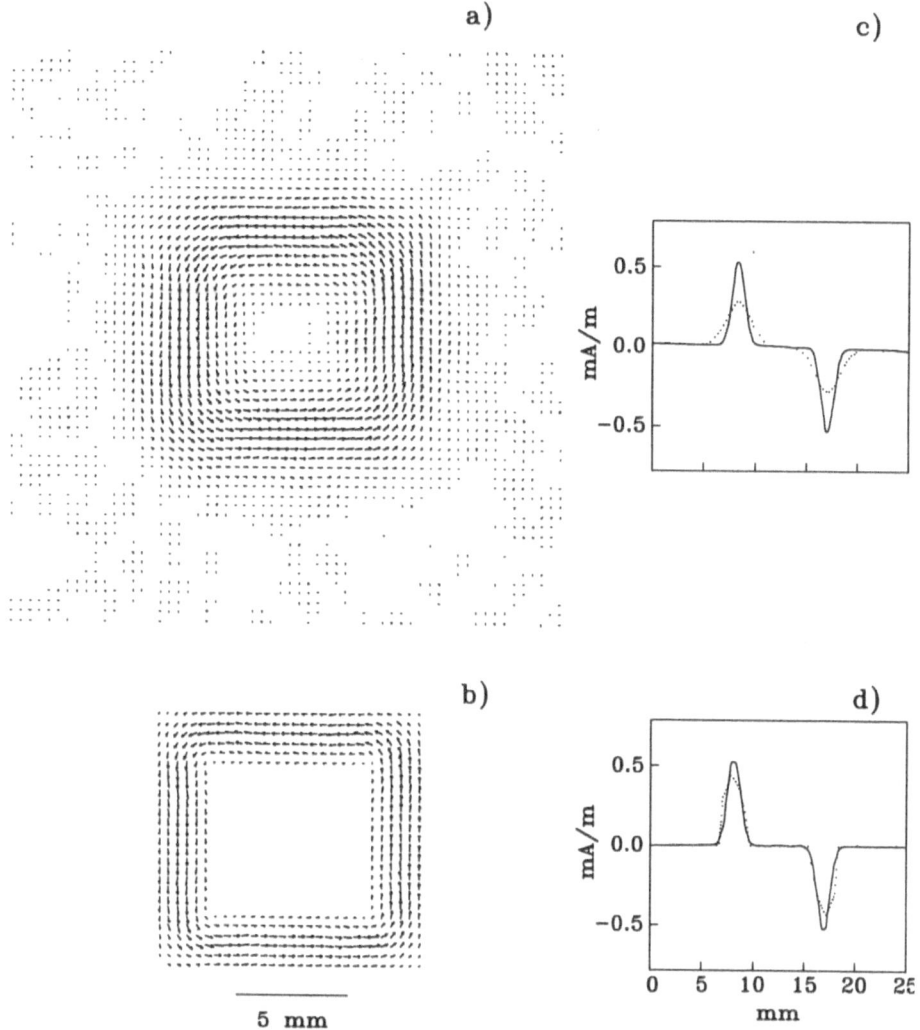

Figure 19. The reconstructions by the filtering technique (a,c) and the finite-element method (b,d) at $z = 3.0$ mm with SNR = 5 (a,b) Reconstructed images of the current; (c,d) cross section of J_x. The solid lines in (c) and (d) are the original current distribution, while the dotted lines are the reconstruction. (From [62], with permission)

the finite-element approach provides superior results.

The advantage of the finite-element approach is even more apparent for B_z measured at 3 mm with a SNR of 5, shown in Fig. 19. The filtering result in Fig. 19a has an MSD of 0.42, whereas the finite-element method, shown in Fig. 19b, has an MSD of 0.12.

Discontinuous Two-dimensional Current Sources. In measurements when current is injected into a conductor, as in Fig. 10, there is a discontinuity in the current in the plane at the location of the two electrodes. In the filtering technique, we addressed this with source/sink terms in the equation of continuity; in the finite-element approach, the continuity equation (Eq. 76) can be modified for the elements containing the sources[62]. However, the accuracy of the reconstruction in the immediate vicinity of the electrode may be unsatisfactory. The usual approach would be to modify the mesh to have a very fine discretization in the immediate vicinity of the electrodes. However, with the finite-element inverse approach, such mesh refinement would drastically increase the size of the matrix that had to be inverted by singular-value decomposition. An alternate approach is to superimpose two currents: a divergent current $\vec{J}_d$ that correctly represents the current in the immediate vicinity of the two electrodes, and a continuous component $\vec{J}_c$, such that

$$\vec{J} = \vec{J}_d + \vec{J}_c . \tag{78}$$

For the divergent part, we assume that the current corresponds to that associated with the potential ϕ produced by a pair of point electrodes at voltages $\pm V$ in an unbounded homogeneous conducting sheet, where

$$\phi = V \log \sqrt{(x - x_1)^2 + (y - y_1)^2} - V \log \sqrt{(x - x_2)^2 + (y - y_2)^2} , \tag{79}$$

so that

$$
\begin{aligned}
J_{dx} &= \frac{\partial \phi}{\partial x} \\
&= \frac{V\,(x - x_1)}{(x - x_1)^2 + (y - y_1)^2} - \frac{V\,(x - x_2)}{(x - x_2)^2 + (y - y_2)^2} \tag{80} \\
J_{dy} &= \frac{\partial \phi}{\partial y} \\
&= \frac{V\,(y - y_1)}{(x - x_1)^2 + (y - y_1)^2} - \frac{V\,(y - y_2)}{(x - x_2)^2 + (y - y_2)^2} . \tag{81}
\end{aligned}
$$

Then the deconvolution problem reduces to trying to find the divergence-free component $\vec{J}_c$ such that the total current $\vec{J}$ will produce the correct magnetic field while satisfying both the boundary condition

$$(\vec{J}_d + \vec{J}_c) \cdot \hat{n} = 0 , \tag{82}$$

and also the continuity condition

$$
\begin{aligned}
\nabla \cdot \vec{J}_d &= V\delta(\vec{r} - \vec{r}_1) - V\delta(\vec{r} - \vec{r}_2) \tag{83} \\
\nabla \cdot \vec{J}_c &= 0 . \tag{84}
\end{aligned}
$$

Examples of this approach are presented in[62].

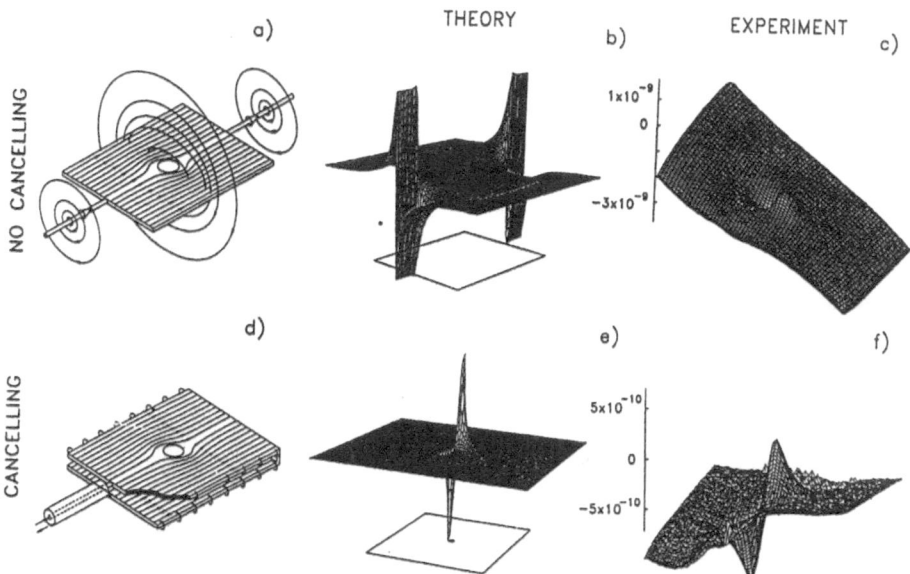

Figure 20. The effects of conductor edges and the use of a cancelling plate. (a) The injection of current into a planar conductor with a hole. (b) The predicted magnetic field. (c) The measured magnetic field (in tesla) in the vicinity of the hole. (d) The use of a cancelling plate and a coaxial cable to deliver the current to the sample. (e) The predicted magnetic field. (f) The measured magnetic field (in tesla). (From [79], with permission)

Eliminating Edge Fields. When current is passed through a conducting object of finite dimension, the discontinuity of the current at the edges of the object can produce large magnetic fields. We first recognized this when we were trying to determine the smallest possible hole in a plate that we could detect by SQUID imaging of injected currents. Figure 20 shows how this can be accomplished by using a cancelling plate that has a uniform current distribution passing beneath the sample, but in the opposite direction[79]. This approach may be particularly useful for SQUID NDE measurements on long samples, such as at the end of a continuous aluminum extruder or rolling mill.

An alternate approach is to use spatial filters, which can take the form of actual or synthetic gradiometers, that are matched to the magnetic signature of the flaw of interest[53]. Figure 21 shows how the edge effects of the internal structure of the lower wing splice of an F-15 aircraft can be eliminated by digital filtering. More importantly, this study also demonstrates that planar gradiometers can be designed to detect small flaws in large background signals.

The finite-element technique also can be used to eliminate edge fields, thereby enhancing the signals from any interior structure[62]. Equation 2

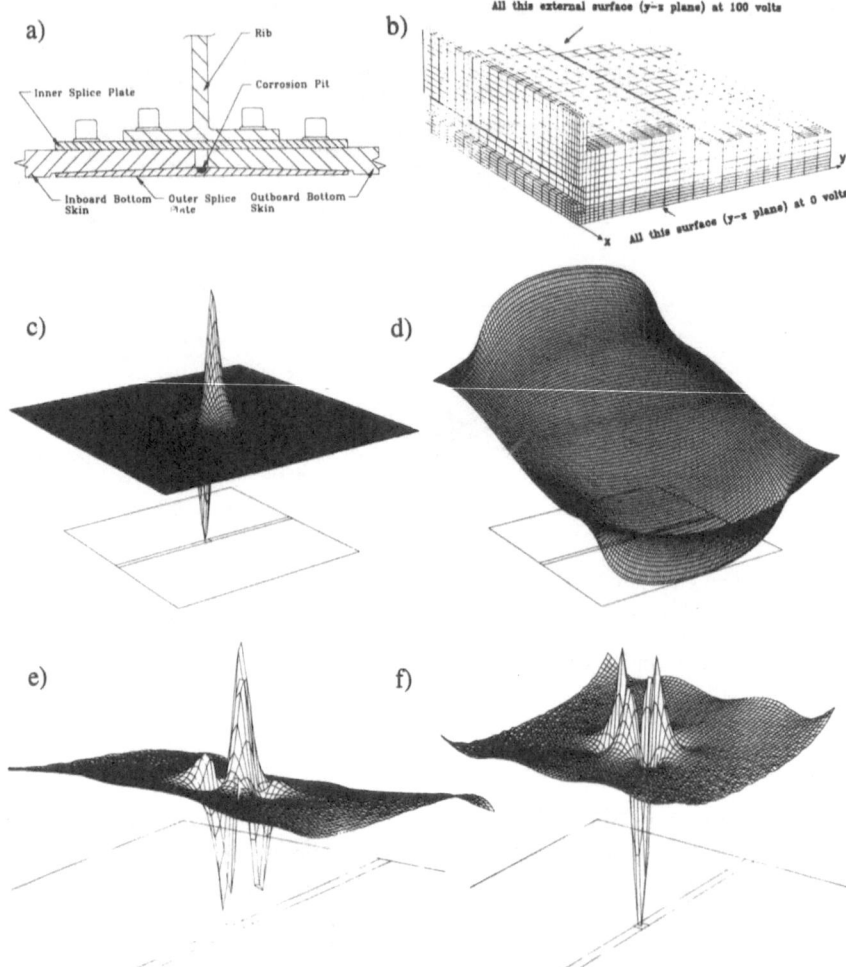

Figure 21. Simulations of the magnetic signature of a small region of hidden corrosion in an F-15 lower wing splice. (a) Schematic cross section of the splice. (b) The finite-element mesh describing one-quarter of a section of wing splice. A uniform current is applied parallel to the x-axis. (c) The magnetic signature of the flaw alone. (d) The magnetic field from the wing structure and the flaw. The peak-to-peak signal arising from the edges of the various plates in the splice is 400 times larger than that of the flaw. (e) and (f) The simulated output of a SQUID gradiometer configured to reject large-scale spatial variations such as from the wing structure and selectively detect fields from localized flaws. (Adapted from [53])

showed that we can write the law of Biot and Savart in terms of a surface integral of the discontinuity in the tangential component of the current density $\vec{J}$, and the volume integral of the curl of $\vec{J}$. In two dimensions, the surface integral reduces to a line integral over the boundary b of the sample, and the volume integral to an integral over the two-dimensional surface s, so that

$$B_z(\vec{r}) = \frac{\mu_o d}{4\pi} \oint_b \frac{\vec{J}(\vec{r}') \cdot \vec{dl}'}{|\vec{r} - \vec{r}'|} + \frac{\mu_o d}{4\pi} \int_s \frac{[\nabla' \times \vec{J}(\vec{r}')]_z}{|\vec{r} - \vec{r}'|} dx' dy' . \quad (85)$$

Following the approach in[62], we can use Ohm's law

$$\vec{J} = -\sigma \nabla \phi \quad (86)$$

and

$$\begin{aligned}
\nabla \times \vec{J} &= -\nabla \times (\sigma \nabla \phi) \\
&= -\nabla \sigma \times \nabla \phi \\
&= \nabla \times (\phi \nabla \sigma)
\end{aligned} \quad (87)$$

to rewrite Eq. (85) as

$$B_z(\vec{r}) = \frac{\mu_o d}{4\pi} \oint_b \frac{\vec{J}(\vec{r}') \cdot dl'}{|\vec{r} - \vec{r}'|} + \frac{\mu_o d}{4\pi} \int_s \frac{[\phi(\vec{r}')\nabla'\sigma(\vec{r}') \times \vec{r}']_z}{|\vec{r} - \vec{r}'|} dx' dy' . \quad (88)$$

If the medium is homogeneous and isotropic, $\nabla'\sigma = 0$, and the second integral is zero, which means that the magnetic field depends only on the current tangential to the edge, $i.e.$, the first integral. If the test object has a non-uniform or anisotropic conductivity, the second term will contribute. The finite-element technique provides us with a powerful tool for separating these two contributions: The magnetic field is used with the finite-element inverse to determine the current distribution in a homogeneous, isotropic sample of the correct shape. This current is then used with the first integral of Eq. 88 to compute the magnetic field produced by the edges. If this is identical to the measured field, the sample is homogeneous and isotropic. If the two fields differ, then there are internal inhomogeneities or anisotropies. This approach may provide a means for enhancing the sensitivity of SQUID imaging to detect internal flaws in metallic structures.

2.2.6. Alternating Projections to Enhance High Frequency Information
A variety of general-purpose techniques common to image processing can be used to enhance the magnetic images obtained with scanning magnetometers. Examples include high- and low-pass spatial filtering[4], and background subtraction using polynomial fits. Alternating projections is a useful iterative technique to apply either constraints or *a priori* knowledge about the source[39, 28]. As an example of this approach, suppose that we have an unknown current distribution, $\vec{J}^u(x', y')$ that corresponds to a wire of unknown shape. This current produces a magnetic field $B_z(x, y)$ by a known

Green's function $G(x, y, x', y')$. In the spatial frequency domain, we have that

$$b_z(k_x, k_y) = FT\{B_z(x, y)\} \tag{89}$$

$$g(k_x, k_y) = FT\{G(x, y, x', y')\} . \tag{90}$$

The inverse problem is simply

$$j_x(k_x, k_y) = \frac{b_z(k_x, k_y)}{g(k_x, k_y)} . \tag{91}$$

However, as we discussed before, this results in numerical instabilities when $g(k_x, k_y)$ is small at large k. We avoid this problem by low-pass filtering the inverse process to obtain an approximate current image $\hat{j}_x(k_x, k_y)$, where

$$\hat{j}_x(k_x, k_y) = LPF \left\{ \frac{b_z(k_x, k_y)}{g(k_x, k_y)} \right\} , \tag{92}$$

with $LPF = 0$ when $k_{LPF} < \sqrt{k_x^2 + k_y^2} < \infty$, and $LPF = 1$ otherwise. But a spatial low-pass filter blurs the image and reduces the spatial resolution. The challenge is to return as much of the high-frequency information as possible while avoiding the instabilities that plague the Fourier inverse approach. Because we know that the current is contained in a wire, we can sharpen the image by assuming that signals below a certain amplitude in the spatial domain are noise and eliminate them by a thresholding operation. We then use the added high-frequency information in the spatial-frequency domain to replace the high-frequency information that was lost in the low-pass filtering operation. Iteration between the spatial and spatial-frequency domains is why this approach is termed alternating projections.

We prepare for the iteration by assuming that our zeroth current distribution is the one that was obtained by the low-pass filtered Fourier inverse

$$j_x^o(k_x, k_y) = \hat{j}_x(k_x, k_y) . \tag{93}$$

We select a threshold T and let $n = 1$. As step 1, we compute

$$J_x^n(x, y) = FT^{-1}\{j_x^{n-1}(k_x, k_y)\} . \tag{94}$$

The second step is to form the thresholded image $\tilde{J}_x^n(x, y)$ by setting to zero all the values of $J_x^n(x, y)$ whose amplitudes are less than T, which sharpens the current distribution and adds high-frequency information. Step 3 converts this image into the spatial-frequency domain,

$$j_x^n(k_x, k_y) = FT\{\tilde{J}_x^n(x, y)\} . \tag{95}$$

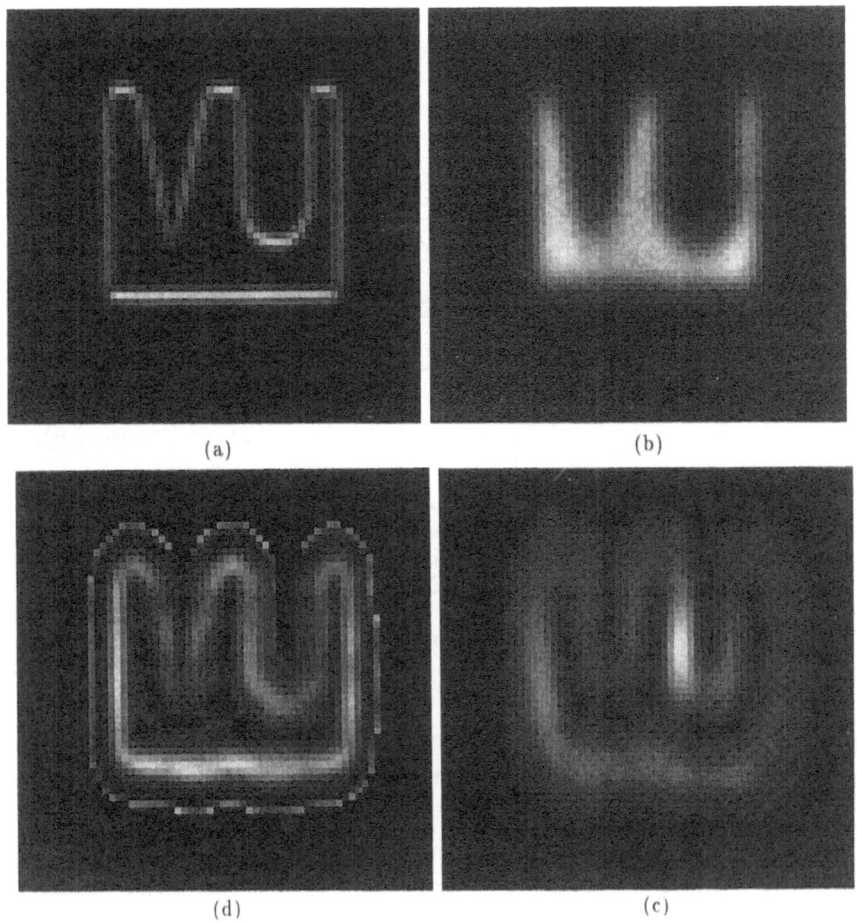

Figure 22. Alternating-projection enhancement of magnetic field maps. (a) The original current image; (b) the magnetic field at 1.5 mm with noise; (c) the recovered image after the first iteration; (d) after ten iterations. (From [28], with permission)

The difficulty is that the low-frequency information in this image is not as accurate as that in the original $\hat{j}_x(k_x, k_y)$. We correct for this with step 4: we substitute the more accurate $\hat{j}_x(k_x, k_y)$ for the less accurate $j_x^n(k_x, k_y)$ in the region $0 < \sqrt{k_x^2 + k_y^2} < k_{LPF}$. We keep the sharpened $j_x^n(k_x, k_y)$ in the region $\sqrt{k_x^2 + k_y^2} > k_{LPF}$. Finally, we decide whether or not to terminate the iteration, and if not, we let $n = n + 1$ and return to step 1. The results of this process are shown in Fig. 22. This is but one example of alternating projections. The technique can be further improved by adding a current-continuity constraint as a third projection[28]. Other criteria can be used to sharpen or modify the image, such as the selective enhancement of line-like structures in the image[39]. As with other sharpening algorithms, it

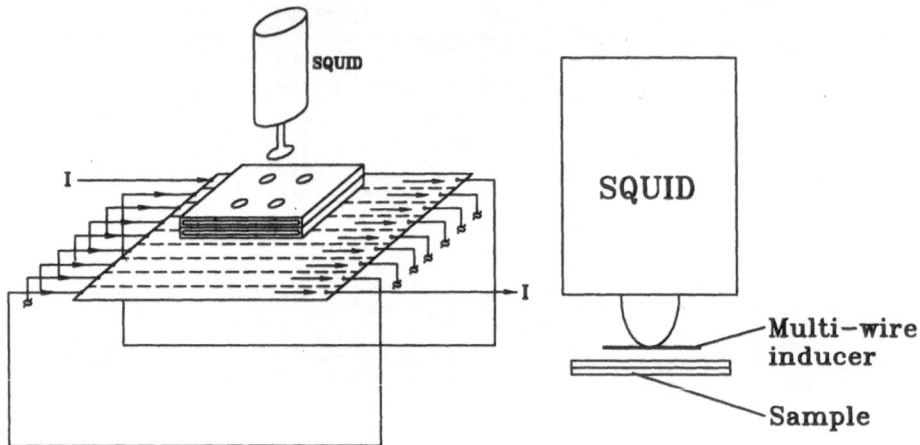

Figure 23. Schematic representations of two sheet inducers used for SQUID NDE. (a) A large multi-strip inducer below the sample. The sample and the inducer are scanned beneath the SQUID. (b) A small multi-wire inducer attached to the SQUID so that the sample can be scanned beneath the SQUID/inducer system, or the system can be scanned above the sample. In either case, frequencies can be as low as 100 Hz, with inducer currents less than a 100 mA. (From [34], with permission)

is important to realize that the alternating-projection approach does not reconstruct the original current distribution in a quantitative manner as do the other techniques we have reviewed, but instead processes the observed field pattern in a manner that can make it resemble more closely the original currents, with no guarantee of quantitative accuracy.

2.3. PHASE-SENSITIVE EDDY-CURRENT ANALYSIS

So far, we have discussed only the injection of current into conducting samples. While the injected-current technique is useful for high-precision measurements on test samples, the need to make good electrical contact with the sample would make it difficult to use on painted structures such as airplanes. We have adapted a standard eddy-current technique, also used in the MagnetoOptic Imager (MOI)[17], in which the ac magnetic field is applied tangential to the surface of the test object by a sheet conductor parallel to the test surface. This induces a large-extent sheet current in the test specimen, and thus produces flaw perturbation fields quite similar to those obtained with direct-current injection. Figure 23 shows schematic representations of the two types of sheet inducers that we have developed. While we have only just begun to apply deconvolution techniques to these images, we present this approach to point out that SQUID data from this technique produce images suitable for deconvolution, and to demonstrate

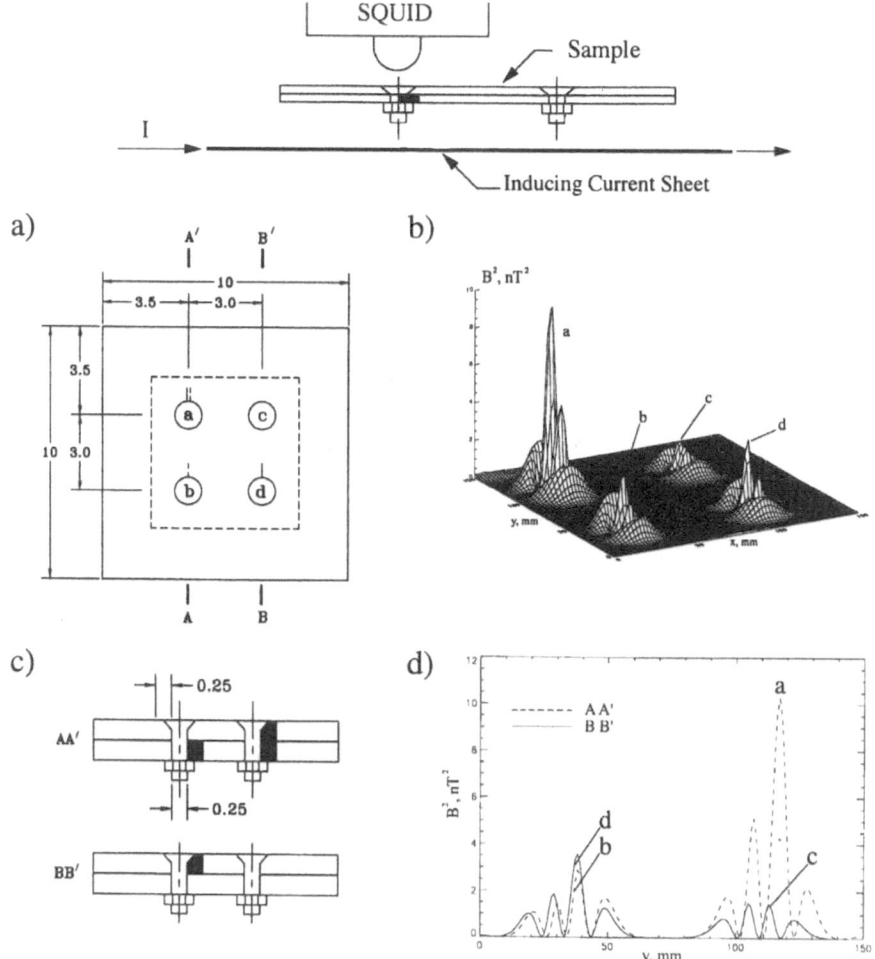

Figure 24. SQUID eddy-current detection of second-layer cracks. Top: The experimental setup (not to scale). Bottom: SQUID images of the cracks beneath rivets obtained at Vanderbilt with the MicroSQUID magnetometer and a sheet inducer. (a) The 25 cm × 25 cm test sample. The dashed region was scanned. (b) A surface plot of the data. (c) A cross-section of the sample. (d) Cross-sections of the data through the peaks. See text for additional details. (From [34], with permission)

that phase-sensitive analysis techniques can provide an additional degree of feature discrimination based upon flaw depth. The potential applications for SQUID eddy-current imaging are vast, and include detecting hidden corrosion damage, closed fatigue cracks, and second-layer cracks at rivets.

Figure 24 illustrates a series of measurements[32, 34, 33] made using the large inducer in Fig. 23a to examine a section of simulated aircraft wing that has cracks adjacent to fasteners. The test sample is made of two layers of 7075-T6 aluminum panels bolted together by four 6 mm diameter aluminum flat-head bolts and nuts. Each panel is 25 cm × 25 cm and 3 mm thick. The crack defects beneath the surface are simulated by electric discharge machined (EDM) slots that are through the entire layer and 6 mm into the metal. Adjacent to bolt a are slots in both the top and bottom layers. Bolt b has a slot in the bottom layer and bolt d has a slot in the top layer (see cross section AA' and BB' in (c)). Bolt c has no slots and provides the signature of an unflawed hole and fastener. The dashed line indicates the 150×150 mm² mapping area. A sheet inducer, which was a set of strips carrying a current of 15 mA and 1017 Hz in the x direction, was placed below the sample. The induced eddy current is disturbed by both the bolts and slots.

In order to increase the signal-to-noise ratio of the measured magnetic field and to allow discrimination of the magnetic field produced by the eddy current from that produced by the inducer current, the output of the SQUID magnetometer is connected to a two-phase lock-in amplifier that determines the components of the signal that are in phase and in quadrature to the inducer current. These two signals can be combined to obtain the signal at an arbitrary phase angle[34, 33]. Figure 24b shows the surface plot of the magnetic field obtained at a phase angle of 50°. The field has been squared for better visualization. The signal is largest for bolt a (with slots in both layers), and smallest for bolt c (without slots). In (d), the cross sections are taken from the peak of the signal. The dashed line was taken from section AA', which shows the signal from bolt b (left) and bolt a (right). The solid line was taken from section BB' which shows the signal from d (left) and c (right). Each signal has four peaks: the sharper peaks reflect the contribution of the current densities near the surface, while the broader peaks reflect the contribution of the current densities below the surface, as predicted by theory[33]. The signal for bolt c, which is without the crack, shows small and symmetrical peaks. The peak signal from bolt d, which has a top-layer crack, is slightly larger than the peak signal from bolt b, which has a second-layer crack. All three signals from the bolts with cracks show the asymmetric peaks, with the larger peak on the side for which there is a crack. While the signatures of the holes with cracks are larger than the one without the crack, it is difficult to distinguish the first layer cracks from the second-layer ones. The challenge is to devise a data-analysis technique that provides depth discrimination that enhances the signals from second- and third-layer flaws.

In the geometry shown in Fig. 23, the phase of the eddy current induced in a planar sample is a function of depth[33]. At the surface of the sample, the eddy currents lead the magnetic field by approximately 90°. At low frequencies, the eddy-current phase is reasonably constant within the sample.

At frequencies such that the skin depth is one-tenth of the thickness of the plate, there are large changes in the phase shift between the surface and the center of the sample. By using a vector lock-in amplifier and software phase-rotation techniques, we can image the component of the magnetic signal from the eddy currents at any desired phase relative to the applied field[33]. Figure 25 shows data acquired with the small inducer (Fig. 23b) and the sample used in Fig. 24, with current perpendicular and parallel to the flaw. When the induced current is parallel to the flaw (right column), we obtain signatures from the four holes that resemble each other, but have a shape that changes slightly with phase. When the induced current is perpendicular to the flaw (left column), we see a very large signature from the hole with cracks on both layers, a small signature from the hole with no cracks and intermediate signatures from the holes with a crack in either the first or second layer. The most important observation is that as the phase of the image is adjusted, the signature of the holes changes in a depth-dependent manner. At 95° phase, the largest signal comes from the hole with cracks in both layers, and the smallest from the hole with no cracks. The hole with the first-layer crack is not very different from the hole without cracks, whereas the hole with the second-layer crack gives a very large signature. The second-layer hole provides a larger signal than does the first layer hole! Thus the phase sensitive analysis can provide depth selectivity. We are currently working towards coupling this analysis with deconvolution techniques. Because of a frequency-dependent reversal of the sign of the eddy currents with depth, this technique also offers the possibility of three-dimensional current tomography[33].

2.4. BLIND DECONVOLUTION

In all of the analyses examined so far, we have utilized a detailed knowledge of the Green's function to guide the inverse process. Suppose instead that the Green's function relating source $\vec{J}$ to the field component F is not known. James Cadzow and his group at Vanderbilt are examining a technique, called blind deconvolution[12], that uses an iterative approach to produce an image, derived from F, that maximizes the absolute value of the kurtosis K of the image. Kurtosis is a statistical measure of a random variable; if the variable, $i.e.$, the field $F(x)$, is assumed to be a random function of the coordinate x, then the kurtosis is given by

$$K(x) = \frac{E\{x^4\} - 3E\{x^2\}^2}{E\{x^2\}^2} \tag{96}$$

$$= \frac{E\{x^4\}}{E\{x^2\}^2} - 3 , \tag{97}$$

676

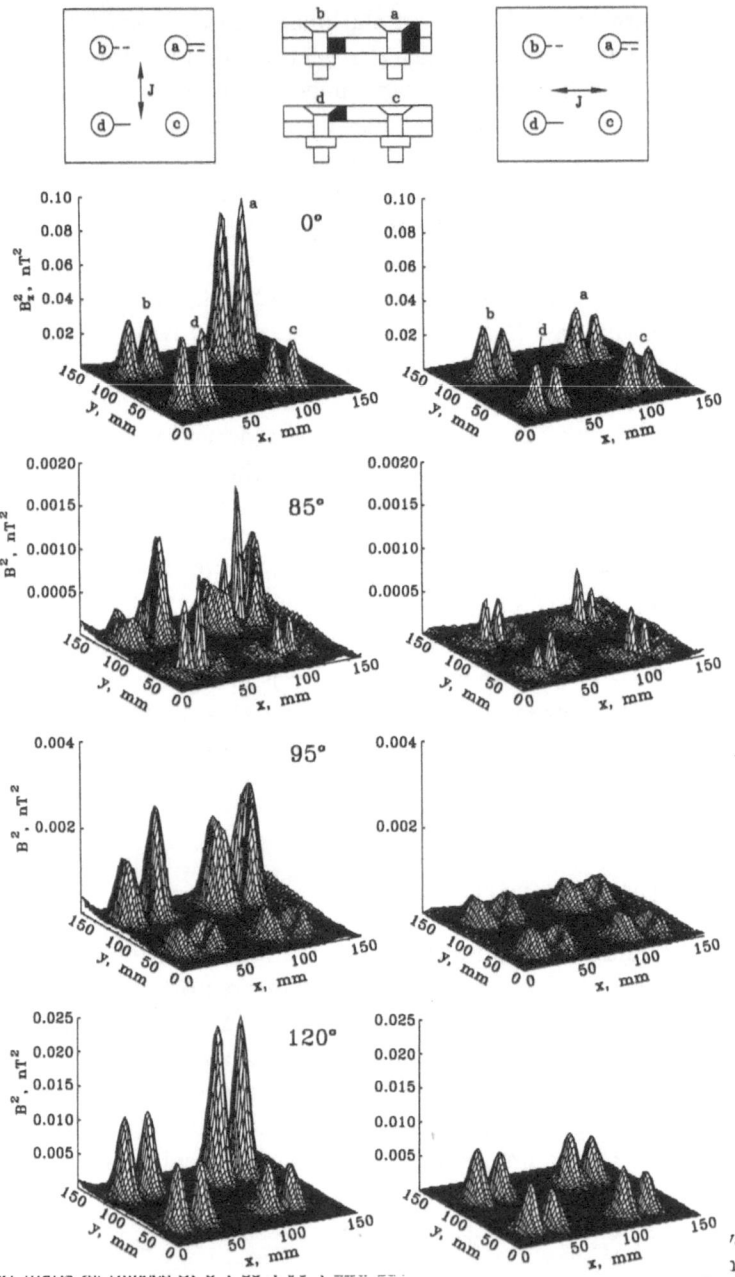

Figure . ~~~~~~ ~~ ~~~~~ ~~ ~ , ~~ , ~~ , ~~~ ~~ ~~~ ~~~~~sis.
The field maps at phases of 0°, 85°, 95°, and 120° ~~~~ ~~~~ ~~~~ sheet
inducer (Fig. 23b) operating at 500 Hz for the sample in Fig. 24. The maps in the left
column were obtained with the excitation current across the cracks, while the maps in the
right column were obtained when the excitation current was parallel to the cracks. Note
that in the left-hand figure at 95°, the signature from the second-layer flaw (left back) is
larger than that from the first-layer flaw (left front). (From [34], with permission)

where it is assumed that the signal has zero mean

$$E\{x\} \equiv 0 \qquad (98)$$

and the various expectation values, or moments, of the field distribution are given by

$$E\{x\} = \int x F(x) dx \qquad (99)$$

$$E\{x^2\} = \int x^2 F(x) dx \qquad (100)$$

$$E\{x^4\} = \int x^4 F(x) dx . \qquad (101)$$

If the original signal is uncorrelated, it has been proven that the maximization of the kurtosis will restore $\vec{J}$ from F. This can then be used to determine the Green's function that relates the two. Figure 26 demonstrates that this approach can improve the quality of two-dimensional images obtained with a scanning SQUID magnetometer, although neither the exact nature of the Green's function recovered by blind deconvolution nor the effects of spatial correlations within the image are as yet well understood. This approach is currently being extended to a more general approach that utilizes other cumulants in addition to kurtosis[14].

As can be seen from the preceding analyses, the deconvolution of magnetic field maps to obtain images of current distributions can be addressed with a wide variety of mathematical techniques. There are others, such as an iterative perturbative approach using cubic splines[23] and a volume-integral approach[38], that have not been discussed here, and undoubtedly more techniques will be developed. A likely candidate for further development as an inverse technique is the boundary-integral method[13]. The geophysics literature is rich in sophisticated techniques[29, 40, 41]. A major limitation of the entire deconvolution process is that it is still more of an art than a science: there is no simple recipe to determine which approach is preferable for a particular combination of measurement geometry, noise and current distribution. Until such a recipe is developed, it may be necessary to determine empirically the optimal technique for a particular application.

3. Imaging Magnetization Distributions

3.1. THE DIPOLE FIELD EQUATION

Virtually all materials are magnetic, *i.e.*, they perturb to some extent an applied magnetic field. The perturbation is large if the material is iron, and very small if it is water or plastic. The perturbations of this field can

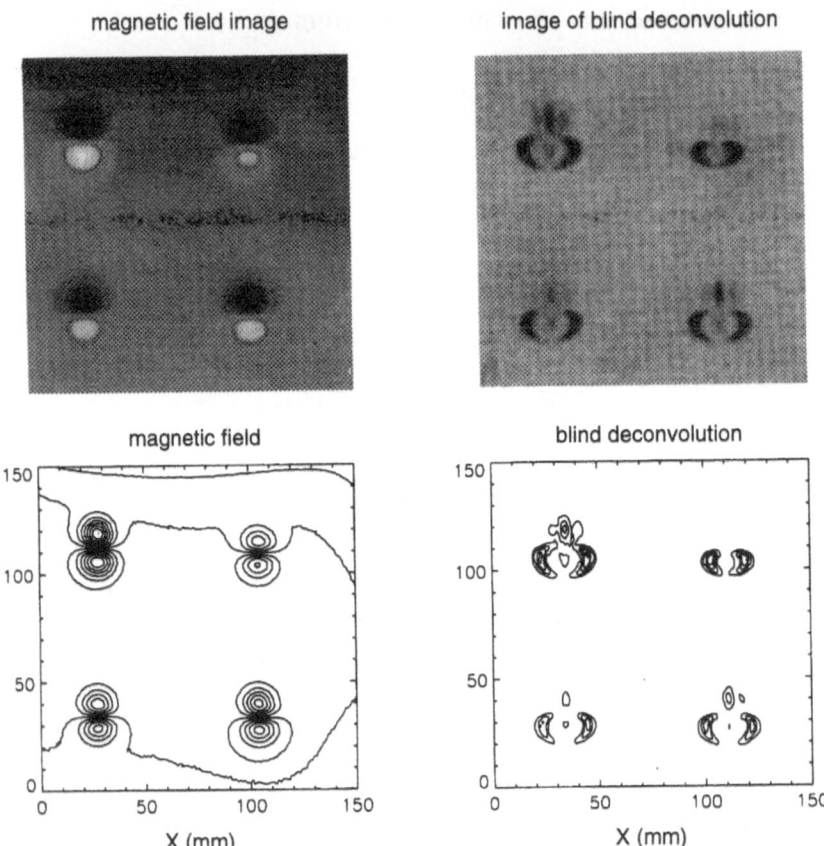

Figure 26. An example of blind deconvolution. Left: The magnetic-field image (top) and contour map (bottom) from the data in Fig. 24 obtained from a two-layer sample simulating an aircraft lap joint. Right: The results of the blind-deconvolution algorithm operating on the data on the left. In each image, the signal in the upper right has no cracks, while that at the upper left has a 6-mm crack in both layers. The signal at the lower left has a crack in the upper layer alone, while that at the right has a crack in the lower layer only. (Courtesy of James Cadzow and Yu Pei Ma).

be imaged, as shown in Fig. 1g, and can be used to determine the nature of the object that produced the perturbation, generally in terms of the magnetization or magnetic susceptibility of the object. To see how this is accomplished, we need to start with a single magnetic dipole. The magnetic field $\vec{B}$ at the point $\vec{r}$ produced by a point magnetic dipole $\vec{m}$ at the point $\vec{r}'$ is given by

$$\vec{B}(\vec{r}) = \frac{\mu_o}{4\pi} \left\{ \frac{3\vec{m}(\vec{r}') \cdot (\vec{r} - \vec{r}')}{|\vec{r} - \vec{r}'|^5} (\vec{r} - \vec{r}') - \frac{\vec{m}(\vec{r}')}{|\vec{r} - \vec{r}'|^3} \right\} . \qquad (102)$$

Since this equation is linear in the dipole moment $\vec{m}$, if the location of the dipole is known, this equation can be readily inverted[27] to obtain the three components of $\vec{m}$ from measurement of the three components of $\vec{B}$ at a single point $\vec{r}$.

$$\vec{m}(\vec{r}') = \frac{4\pi}{\mu_o} \mid \vec{r} - \vec{r}' \mid^3 \left\{ \frac{3}{2} \frac{\vec{B}(\vec{r}') \cdot (\vec{r} - \vec{r}')}{\mid \vec{r} - \vec{r}' \mid^2} (\vec{r} - \vec{r}') - \vec{B}(\vec{r}') \right\} . \quad (103)$$

This equation is particularly useful when using a three-axis, vector SQUID magnetometer to find the components of a magnetic dipole[81]. However, if the location of the dipole is unknown, the nonlinearity of Eq. 102 in $\vec{r}$ and $\vec{r}'$ makes the inverse process much harder; in general, there is no closed-form analytical solution for determining both $\vec{m}$ and $\vec{r}'$ from measurements of $\vec{B}$ at multiple locations. This should give an indication of the difficulties that will be encountered in attempting to use magnetic-field maps to determine the distribution of magnetization in a three-dimensional object, which is equivalent to determining the location and strength of the many dipoles that represent different small regions of the object.

We have to address two points: the generation of the magnetization within the object, and the relationship between this magnetization and the external magnetic field. The latter is a straightforward extension of the dipole-field equation: each differential volume element d^3r' in the object is assigned a dipole moment $d\vec{m}(\vec{r}')$ that is equal to $\vec{M}(\vec{r}')d^3r'$, so that we can simply integrate Eq. 102

$$\vec{B}(\vec{r}) = \frac{\mu_o}{4\pi} \int_v \left\{ \frac{3\vec{M}(\vec{r}') \cdot (\vec{r} - \vec{r}')}{\mid \vec{r} - \vec{r}' \mid^5} (\vec{r} - \vec{r}') - \frac{\vec{M}(\vec{r}')}{\mid \vec{r} - \vec{r}' \mid^3} \right\} d^3r' . \quad (104)$$

3.2. FIELDS FROM MAGNETIC MEDIA

Let us suppose that an object made of magnetically-linear material is placed in a magnetic field produced by a distant electromagnet. The magnetization $\vec{M}(\vec{r}')$ at a source point $\vec{r}'$ is determined by the product of the magnetic susceptibility $\chi(\vec{r}')$ and the applied magnetic field intensity $\vec{H}(\vec{r}')$

$$\vec{M}(\vec{r}') = \chi(\vec{r}')\vec{H}(\vec{r}') . \quad (105)$$

The magnetic induction field $\vec{B}$, hereafter referred to as the "magnetic field," at the same source point $\vec{r}'$ is given by

$$\vec{B}(\vec{r}') = \mu_o\{\vec{H}(\vec{r}') + \vec{M}(\vec{r}')\} , \quad (106)$$

where μ_o is the permeability of free space. We can express this in terms of the susceptibility by substituting Eq. 105 into Eq. 106 to obtain

$$\vec{B}(\vec{r}') = \mu_o\{1 + \chi(\vec{r}')\}\vec{H}(\vec{r}') \tag{107}$$

$$= \mu_o\mu_r(\vec{r}')\vec{H}(\vec{r}') \tag{108}$$

$$= \mu(\vec{r}')\vec{H}(\vec{r}') , \tag{109}$$

where the relative permeability μ_r is given by

$$\mu_r(\vec{r}') = 1 + \chi(\vec{r}') \tag{110}$$

and the absolute permeability μ is

$$\mu(\vec{r}') = \mu_o\mu_r(\vec{r}') . \tag{111}$$

As we shall see, the difficulty with susceptibility and magnetization imaging is that the field measured by the SQUID is not the local field within the object, but the field in the source-free region outside of the object.

3.2.1. Ferromagnetic Materials

Ferromagnetic materials have high permeabilities, in the range $\approx 10^3 \leq \mu_r \leq \approx 10^5$, so that

$$\mu_r = 1 + \chi \approx \chi . \tag{112}$$

If the materials are "hard," they exhibit significant hysteresis; if they are "soft," they do not. In either case, the magnetic field within ferromagnetic materials can be from 10^3 to 10^5 times the applied field. The determination of the magnetization within a ferromagnetic material must be made in the strong-field limit: $\vec{B} = \mu_o(\vec{H} + \vec{M}) = \mu\vec{H}$ at any point in the material, so that $\vec{M}$ at one point is affected by $\vec{M}$ at other points in the material. Self-consistency requires simultaneous solution of $\vec{H}$ and $\vec{M}$ everywhere, since $\vec{M}(\vec{r})$ is determined by both $\vec{H}(\vec{r})$ and $\chi(\vec{r})$, even if the applied field $\vec{H}$ was initially uniform before the object was placed in the field. In this strong-field case, the magnetic inverse problem, i.e., the inversion of Eq. 104, is difficult to impossible, particularly if there is a remanent (hard) magnetization superimposed upon the induced (soft) magnetization. While it is difficult to induce a soft, spherically-symmetric, magnetically-silent magnet with external fields, it is in principle possible to have such a distribution in a hard component of magnetization, and this leads to the previously discussed nonuniqueness problem.

3.2.2. Paramagnetic and Diamagnetic Materials

The situation is much friendlier for the magnetic imaging of paramagnetic $(0 \leq \chi \leq 10^{-3})$ and diamagnetic $(-10^{-6} \leq \chi \leq 0)$ materials, in that

$$\mu_r = 1 + \chi \approx 1 . \tag{113}$$

As a result, the induced magnetic field $\vec{B}$ is 10^{-6} to 10^{-3} times the applied field, and is proportional to the applied field, since paramagnetic and dia- magnetic materials are linear and non-hysteretic at practical applied fields. The most significant feature of the low susceptibility of these materials is that Eq. 104 can be evaluated in the weak-field limit, also known as the Born approximation: at any point in the material we can **ignore** the con- tributions to the applied field at $\vec{r}'$ from the magnetization elsewhere in the object. In the Born approximation, the magnetization is independent of the magnetization elsewhere in the sample, and hence is a local phenomenon, in contrast to ferromagnetism. Because $\vec{M}$ is so weak for diamagnetic and paramagnetic materials, if we know $\vec{H}_o$ everywhere, we shall then know $\vec{B}_o$ to at least one part in 10^3 for a paramagnetic material with $\chi = 10^{-3}$, and to 1 part in 10^6 for a diamagnetic one with $\chi = 10^{-6}$. Thus, we have eliminated a major problem in obtaining a self-consistent, macroscopic so- lution that is based upon the microscopic constitutive equation given by Eq. 106. Because of their periodic flux-voltage characteristic, and the ability to thermally release magnetic flux trapped in pickup coils, SQUID magne- tometers readily can measure only the very small perturbation $\vec{B}_p(\vec{r})$ in the applied magnetic field[30, 31]. We thereby can eliminate $\vec{B}_o$ and $\vec{H}_o$ from the imaging problem, and need them only to determine the magnetization. The measured magnetic field, $\vec{B}_p(\vec{r})$, thus is given by Eq. 104 where

$$\vec{M}(\vec{r}') = \frac{\chi(\vec{r}')}{\mu_o}\vec{B}_o(\vec{r}') = \chi(\vec{r}')\vec{H}_o(\vec{r}') \, . \tag{114}$$

If $\vec{H}_o$ is uniform, then the spatial variation of $\vec{M}(\vec{r}')$ is determined only by $\chi(\vec{r}')$. For isotropic materials, χ is a scalar, and the direction of $\vec{M}$ is the same as that of $\vec{B}_o$; otherwise, a tensor susceptibility is required.

3.3. INVERTING THE DIPOLE FIELD EQUATION FOR DIAMAGNETIC AND PARAMAGNETIC MATERIALS

The general inverse problem for magnetic media involves solving for the vector magnetization $\vec{M}(\vec{r}')$ in Eq. 104, or, after dividing by the applied field, the susceptibility $\chi(\vec{r}')$. This inverse problem has no unique solution. In the quasistatic limit, the curl of $\vec{B}$ is zero in a current-free region, indi- cating that the field can be expressed as the gradient of the magnetic scalar potential ϕ_m, where $\vec{B} = -\mu_o\nabla\phi_m$. The magnetic potential outside of the surface S that bounds the sample of volume v is given by[44]

$$\phi_m(\vec{r}) = \frac{1}{4\pi}\int_s \frac{\vec{M}(\vec{r}') \cdot \hat{n}}{|\vec{r}-\vec{r}'|}d^2r' + \frac{1}{4\pi}\int_v \frac{\nabla' \cdot \vec{M}(\vec{r}')}{|\vec{r}-\vec{r}'|}d^3r' \, . \tag{115}$$

Hence, we see that measurements of quasistatic magnetic fields in free space outside a magnetized body provide information about the divergence of the magnetization distribution, rather than the magnetization itself, just as measurements of the magnetic field outside of a current distribution provide information only about the curl of the currents (Eq. 2). Thus, any magnetization distribution that is divergence-free will be magnetically silent and cannot affect the external magnetic field. Beardsley[5] points out that in special cases, such as in a film with a magnetization that does not vary with thickness, the combined measurement of the external magnetic field and the angular deflection of an electron beam passing through the sample (differential phase contrast Lorentz microscopy) can provide the requisite information required to determine unambiguously the magnetization within the film. In this chapter, we restrict ourselves to SQUID measurements alone, and hence are faced, once again, with a potentially insolvable inverse problem.

For magnetically-soft materials, the magnetization is provided by external magnetic fields, and hence, divergence-free magnetizations are avoided. This provides an important constraint to the problem. If we apply only a uniform B_z field, and if we know that for our sample $\vec{M}(\vec{r}') = \chi \vec{H}(\vec{r}')$, we need to solve for the scalar magnetization $M_z(\vec{r}')\hat{z}$ in the slightly simpler equation

$$\vec{B}(\vec{r}) = \frac{\mu_o}{4\pi} \int_v \left\{ \frac{3M_z(\vec{r}')(z-z')}{|\vec{r}-\vec{r}'|^5}(\vec{r}-\vec{r}') - \frac{M_z(\vec{r}')\hat{z}}{|\vec{r}-\vec{r}'|^3} \right\} d^3r' . \qquad (116)$$

In two dimensions, there is a unique inverse solution to this problem, to be presented in the next section. After that discussion, we shall show that in three dimensions, however, there is still a problem with non-uniqueness.

3.3.1. Two-Dimensional Magnetization Imaging

For two-dimensional samples, we can apply our inverse spatial filtering approach to determine the magnetization or susceptibility distributions from the magnetic field[58, 59, 60]. If the source is restricted to two-dimensions, such as a thin sheet of diamagnetic or paramagnetic material, Eq. 116 reduces to a two-dimensional surface integral. We shall for now assume that we are applying only a z-component field $H_o\hat{z}$, and are measuring only the z-component of the sample-induced magnetic field $\vec{B}$ at a height $(z-z')$ above the two-dimensional sample, so that we have

$$B_z(\vec{r}) = \frac{\mu_o}{4\pi} \int_{x'=-\infty}^{\infty} \int_{y'=-\infty}^{\infty} \left\{ \frac{3M_z(\vec{r}')(z-z')^2}{|\vec{r}-\vec{r}'|^5} - \frac{M_z(\vec{r}')}{|\vec{r}-\vec{r}'|^3} \right\} dx'dy' . \qquad (117)$$

In practice, the integrals need not extend beyond the boundary of the source object, outside of which $\vec{M} \equiv 0$. In order to solve this equation for $M_z(\vec{r}')$, we define a Green's function,

$$G_z(\vec{r} - \vec{r}') = \frac{\mu_o}{4\pi} \left\{ \frac{3(z - z')^2}{|\vec{r} - \vec{r}'|^5} - \frac{1}{|\vec{r} - \vec{r}'|^3} \right\},\tag{118}$$

so that Eq. 117 becomes

$$B_z(\vec{r}) = \int_{-\infty}^{\infty} \int_{-\infty}^{\infty} M_z(\vec{r}')G_z(\vec{r} - \vec{r}')dx'dy' .\tag{119}$$

We compute the two-dimensional spatial Fourier transform of the magnetic field

$$b_z(k_x, k_y, z) = FT\{B_z(x, y, z)\}\tag{120}$$

so that we can use the convolution theorem to express Eq. 117 in the spatial frequency domain as

$$b_z(k_x, k_y, z) = g_z(k_x, k_y, z - z')m_z(k_x, k_y) ,\tag{121}$$

where $g_z(k_x, k_y, z - z')$ is the spatial Fourier transform of the Green's function, i.e.,

$$g_z(k_x, k_y, z - z') = \frac{\mu_o}{4\pi}\{2\pi k e^{-k(z-z')}\},\tag{122}$$

with

$$k = (k_x^2 + k_y^2)^{\frac{1}{2}} ,\tag{123}$$

and $m_z(k_x, k_y)$ is the Fourier transform of the magnetization $M_z(x', y')$. The inverse problem then reduces to a division in the spatial frequency domain

$$m_z(k_x, k_y) = \frac{b_z(k_x, k_y, z)}{g_z(k_x, k_y, z)} .\tag{124}$$

As we have seen before, it may be necessary to use windowing techniques to prevent this equation from blowing up because of zeros in the Green's function occurring at spatial frequencies for which there is a contribution to the magnetic field from either the sample or from noise. Typically, the window $w(k_x, k_y)$ is a low-pass filter which attenuates high-frequency noise in the vicinity of the zeros of g_z, so that Eq. 124 becomes

$$m_y(k_x, k_y) = \frac{b_z(k_x, k_y, z)}{g_z(k_x, k_y, z)}w(k_x, k_y) .\tag{125}$$

As the final step, we use the inverse Fourier Transform (FT^{-1}) to obtain an image of the magnetization distribution

$$M_z(x', y') = FT^{-1}\{m_z(k_x, k_y)\} ,\tag{126}$$

which can then be used to obtain the desired susceptibility image

$$\chi(x', y') = \frac{M_z(x', y')}{H_o(\vec{r}')} .$$ (127)

This outlines the basic approach to two-dimensional magnetic susceptibility imaging; many of the techniques demonstrated for current-density imaging also will be applicable. It is important to note that once the Green's function (and its inverse) and the window have been specified, it also is possible to proceed directly from $B_z(x, y)$ to $\chi(x', y')$ by evaluating the appropriate convolution integral in xy-space.

Applications of this technique include localizing dilute paramagnetic tracers[67] (Fig. 27), and the imaging of plastic[66] (Fig. 28), rock[68], and even water[80].

3.3.2. Magnetic Susceptibility Tomography

If the source is three-dimensional, a somewhat more general approach must be taken. We can start with Eq. 102, the dipole-field equation for the magnetic field $d\vec{B}(\vec{r})$ produced by a single magnetic dipole $d\vec{m}(\vec{r}')$. If the dipole moment arises from the magnetization of an incremental volume dv in an applied field $\vec{H}(\vec{r}')$, we have that

$$d\vec{m}(\vec{r}') = \chi(\vec{r}')\vec{H}(\vec{r}')dv' .$$ (128)

The dipole-field equation then becomes

$$d\vec{B}(\vec{r}) = \frac{\mu_o \chi(\vec{r}')}{4\pi} \left\{ \frac{3\vec{H}(\vec{r}') \cdot (\vec{r} - \vec{r}')}{|\vec{r} - \vec{r}'|^5}(\vec{r} - \vec{r}') - \frac{\vec{H}(\vec{r}')}{|\vec{r} - \vec{r}'|^3} \right\} dv' .$$ (129)

This equation can be written as

$$d\vec{B}(\vec{r}) = \vec{G}(\vec{r}, \vec{r}', \vec{H})\chi(\vec{r}')dv' ,$$ (130)

where we introduce a vector Green's function

$$\vec{G}(\vec{r}, \vec{r}', \vec{H}) = \frac{\mu_o}{4\pi} \left\{ \frac{3\vec{H}(\vec{r}') \cdot (\vec{r} - \vec{r}')}{|\vec{r} - \vec{r}'|^5}(\vec{r} - \vec{r}') - \frac{\vec{H}(\vec{r}')}{|\vec{r} - \vec{r}'|^3} \right\} .$$ (131)

The components of $\vec{G}$ are simply

$$\vec{G} = G_x\hat{x} + G_y\hat{y} + G_z\hat{z} ,$$ (132)

where

$$G_x(\vec{r}, \vec{r}', \vec{H}) = \frac{\mu_o}{4\pi} \left\{ \frac{3\vec{H} \cdot (\vec{r} - \vec{r}')}{|\vec{r} - \vec{r}'|^5}(x - x') - \frac{H_x}{|\vec{r} - \vec{r}'|^3} \right\}$$ (133)

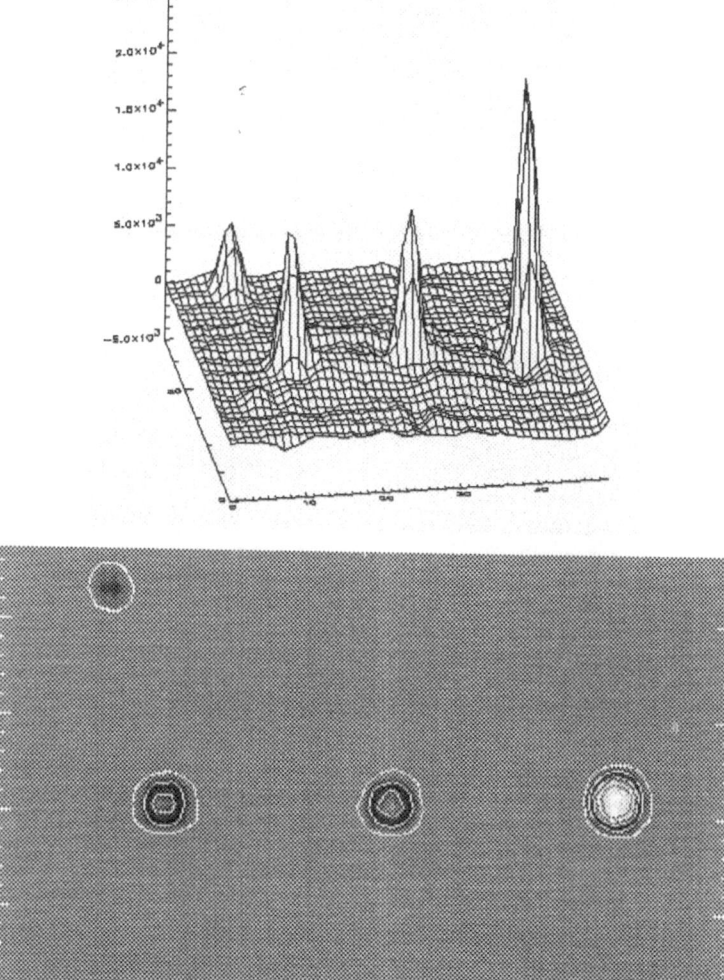

Figure 27. Magnetic decoration of surface defects. A nickel NDE sample had electric discharge machined (EDM) slots with dimensions of 100 μm, that were filled with a superparamagnetic tracer. The magnetic field was recorded 2.0 mm from the sample with a 174-μT applied field. These susceptibility images display the location and size of surface defects, including one (upper left) that was the result of a previously undetected scratch. MicroSQUID can find flaws as small as 2×10^{-12} m^3. (Adapted from [67], with permission)

$$G_y(\vec{r}, \vec{r}', \vec{H}) = \frac{\mu_o}{4\pi} \left\{ \frac{3\vec{H} \cdot (\vec{r} - \vec{r}')}{|\vec{r} - \vec{r}'|^5}(y - y') - \frac{H_y}{|\vec{r} - \vec{r}'|^3} \right\} \qquad (134)$$

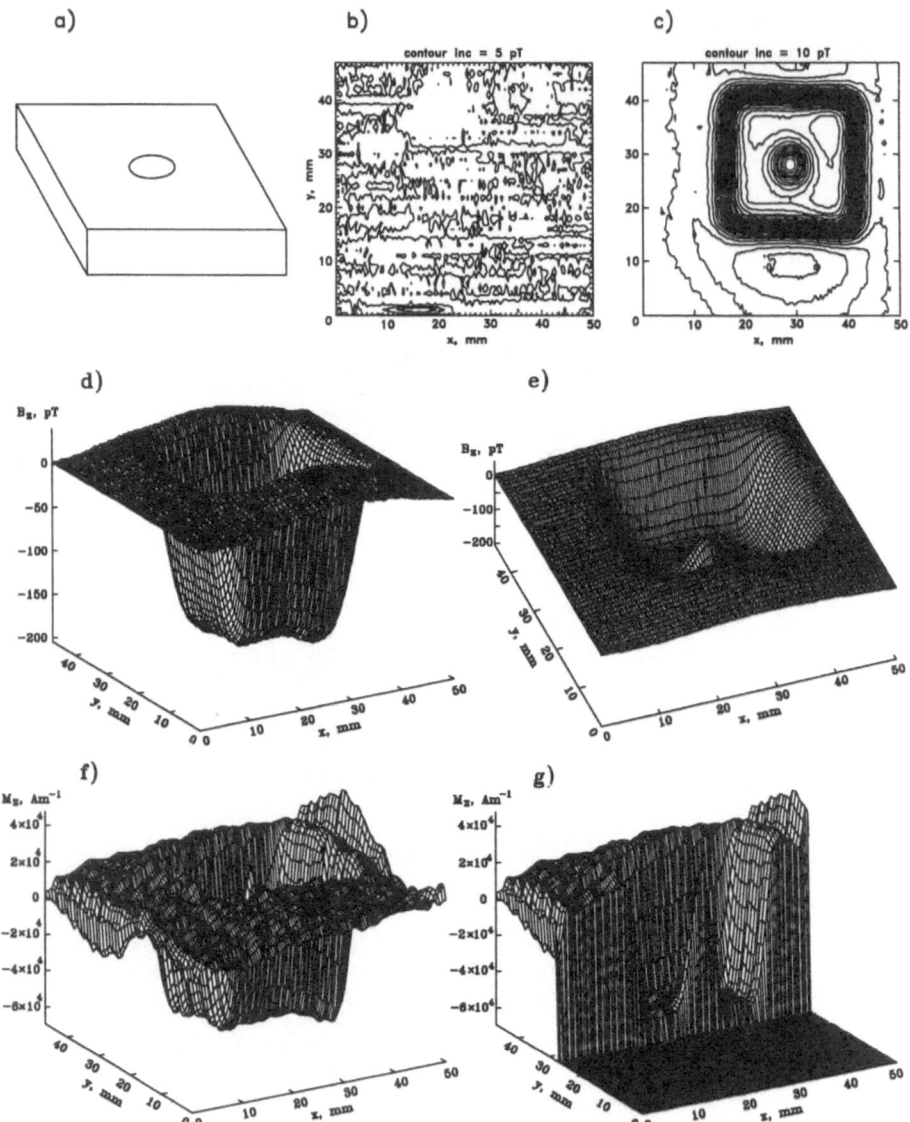

Figure 28. Susceptibility images of plexiglas. A 25.4-mm square sample of plexiglas containing a one 4.5-mm diameter hole was magnetized in a 110-μT applied field and scanned at a distance of 2.0 mm. (a) The sample, (b,c) contour maps of the recorded field B_z with (b) $B_{app} = 0\,\mu$T and (c) $B_{app} = 100\mu$T. (d,e) Surface maps of the recorded field ($B_{app} = 100\,\mu$T) from two perspectives. (f,g) Magnetization distributions computed from the measured field for (f) the entire sample and (g) a section of the sample. (Adapted from [66], with permission)

$$G_z(\vec{r}, \vec{r}', \vec{H}) = \frac{\mu_0}{4\pi} \left\{ \frac{3\vec{H} \cdot (\vec{r} - \vec{r}')}{|\vec{r} - \vec{r}'|^5}(z - z') - \frac{H_z}{|\vec{r} - \vec{r}'|^3} \right\}. \tag{135}$$

The three components of the magnetic field in Eq. 130 now can be written as

$$dB_x(\vec{r}) = \chi(\vec{r}')G_x(\vec{r}, \vec{r}', \vec{H}) \tag{136}$$

$$dB_y(\vec{r}) = \chi(\vec{r}')G_y(\vec{r}, \vec{r}', \vec{H}) \tag{137}$$

$$dB_z(\vec{r}) = \chi(\vec{r}')G_z(\vec{r}, \vec{r}', \vec{H}). \tag{138}$$

Note that $\vec{H}$ may in turn be a function of x', y', and z'. In contrast, the Green's function in Eq. 118 did not contain $\vec{H}$. However, $\vec{H}$ is assumed to be known and only adds a geometrically-variable scale factor into the Green's function. The increased complexity of Eqs. 133 through 135 arises from our desire to include $\vec{H}$ as a vector field with three independently-specified components.

If we know both the location $\vec{r}'$ of a source that is only a single dipole, and also the strength and direction of $\vec{H}$ at that point, we can make a single measurement of the magnetic field at $\vec{r}$ to determine $\chi(\vec{r}')$. It is adequate to measure only a single component of $\vec{B}(\vec{r}')$ as long as that component is non-zero. The problem becomes somewhat more complex when there are either multiple dipoles or a continuous distribution of dipoles. In that case, we need to sum or integrate Eq. 130 over the entire source object

$$\vec{B}(\vec{r}) = \int_{x'} \int_{y'} \int_{z'} \vec{G}(\vec{r}, \vec{r}', \vec{H})\chi(\vec{r}')dv'. \tag{139}$$

To proceed numerically, we shall assume that we can discretize the source object into m elements of volume v_j, where $1 \leq j \leq m$. The field from this object is then

$$\vec{B}(\vec{r}) = \sum_{j=1}^{m} \vec{G}(\vec{r}, \vec{r}_j', \vec{H})\chi(\vec{r}_j')v_j. \tag{140}$$

A single measurement of $\vec{B}$ will be inadequate to determine the susceptibility values for the m elements. Thus, we must make our measurements at n measurement points $\vec{r}_i$, where $1 \leq i \leq n$ identifies each such measurement. The three field components measured at a single point would constitute, in this notation, three independent scalar measurements which happen to have the same value for $\vec{r}$. Equation 140 becomes

$$\vec{B}_i(\vec{r}_i) = \sum_{j=1}^{m} \vec{G}(\vec{r}_i, \vec{r}_j', \vec{H})\chi(\vec{r}_j')v_j. \tag{141}$$

To simplify analysis, we can convert to matrix notation. In this case, the vector Green's function $\vec{G}$ becomes an $n \times m$ matrix $\tilde{\tilde{G}}$ that contains as each of its rows the Green's functions that relate a single measurement to every source element, i.e., Eq. 138. The n field measurements can be written as the n elements of an $n \times 1$ column matrix $\tilde{B}$. The magnetic susceptibility of each of the m source elements can be described by an $m \times 1$ column matrix $\tilde{\chi}$. The volume of each source element can be incorporated into either the $\tilde{\tilde{G}}$ or $\tilde{\chi}$ matrices. The measurements are related to the sources by

$$\tilde{B} = \tilde{\tilde{G}}\tilde{\chi}. \tag{142}$$

If $n = m$, the system of equations will be exactly determined, but it may not be possible to obtain a solution because of measurement noise or non-orthogonality, i.e., linear dependence of the n equations. The alternative is to choose $n > m$, so that the system becomes over-determined, and a least-squares solution can be attempted. Ideally, only those measurements needed to increase the orthogonality of the equations will be added. While there are several ways to proceed, we shall consider only the general approach of multiplying both sides of Eq. 142 by $\tilde{\tilde{G}}^T$, the transpose of $\tilde{\tilde{G}}$,

$$\tilde{\tilde{G}}^T \tilde{B} = \tilde{\tilde{G}}^T \tilde{\tilde{G}}\tilde{\chi}. \tag{143}$$

$\tilde{\tilde{G}}^T \tilde{\tilde{G}}$ is now an $m \times m$ matrix that in principle can be inverted. This allows us to solve for $\tilde{\chi}$

$$\left[\tilde{\tilde{G}}^T \tilde{\tilde{G}}\right]^{-1} \tilde{\tilde{G}}^T \tilde{B} = \tilde{\chi}. \tag{144}$$

The ability to compute the inverse of the $\tilde{\tilde{G}}^T \tilde{\tilde{G}}$ matrix is determined by the measurement noise, by how well the measurements span the source space, and by the orthogonality of the G matrix. Typically, if this inversion process is attempted for measurements made in a single plane over a complex source, the near elements of the source will dominate, and the matrix will be ill-conditioned.

Magnetic susceptibility tomography[50] can be used to avoid the ill-conditioned nature of Eq. 144 by applying the magnetic field from a number of different directions and by measuring the magnetic field at multiple locations all around the object. Since $n \gg m$, the system of equations becomes highly overdetermined, and standard techniques, such as singular value decomposition, can be used to determine the susceptibility of each voxel. Since the direction of the magnetization of each voxel of the material is known for every measurement, and this direction is varied, the domination of the $\overset{\approx}{G}^T \overset{\approx}{G}$ matrix by a small set of measurements can be avoided. The stability of the inversion of Eq. 144 can be enhanced by using non-uniform magnetizing fields[45].

The first demonstration of this technique, shown in Fig. 29, used a uniform magnetizing field and a 64-voxel cube[50]. The reconstruction was reasonably accurate in the absence of noise, but the matrix was sufficiently ill-conditioned that stable inverse solutions were difficult to obtain with modest amounts of noise. The matrix will be more readily inverted if non-uniform magnetizing-fields are utilized, since there are only three independent uniform fields that can be applied to an object[45], and since a non-uniform magnetizing field eliminates the uncertainty of equivalent spherical sources with identical dipole moments[4].

Two- and three-dimensional magnetic susceptibility imaging offers potential advantages for imaging composites with dilute magnetic tracers[24], for magnetic imaging of materials such as plastic, titanium or aluminum that are normally considered nonmagnetic, monitoring macrophage activity[65], and, with susceptibility tomography, three-dimensional magnetic images of metabolism and iron storage in the liver and possibly regional oxygenation of the brain.

4. Conclusions: Magnetic Imaging

SQUID NDE is in competition with a wide variety of NDE techniques, such as magnetic-particle inspection, magneto-optical inspection (MOI), X-ray, ultrasound, thermal imaging, optical/laser interferometry, neutron radiography, and scanned eddy-currents or eddy current arrays. Each of these techniques produces images that aid in the detection and discrimination of flaws. I firmly believe that the acceptance of a new NDE technique, such as SQUID NDE, by those individuals that use the standard NDE techniques in the field is determined by two things: does the new technique provide

---

[4]Suppose that we consider two uniformly magnetized spheres, centered at the origin, that have different radii but with magnetizations that are scaled such that the two spheres have the same magnetic dipole moment. Because the magnetic field outside of a uniformly magnetized sphere is perfectly dipolar, it is impossible to distinguish between these two sources by means of magnetic measurements made beyond the radius of the larger sphere. The existence of two different source distributions with identical fields in the region outside of all possible sources is the death knell for the inverse problem: without additional constraints, an inverse algorithm will be unable to control the degree of freedom within the source that corresponds to the radius of a uniformly magnetized sphere whose susceptibility scales inversely with the sphere volume. A multipole analysis of the external fields can give the dipole moment of the source to an arbitrary accuracy, but cannot determine both the radius of a uniformly magnetized sphere and its susceptibility if they are scaled for a constant dipole moment.

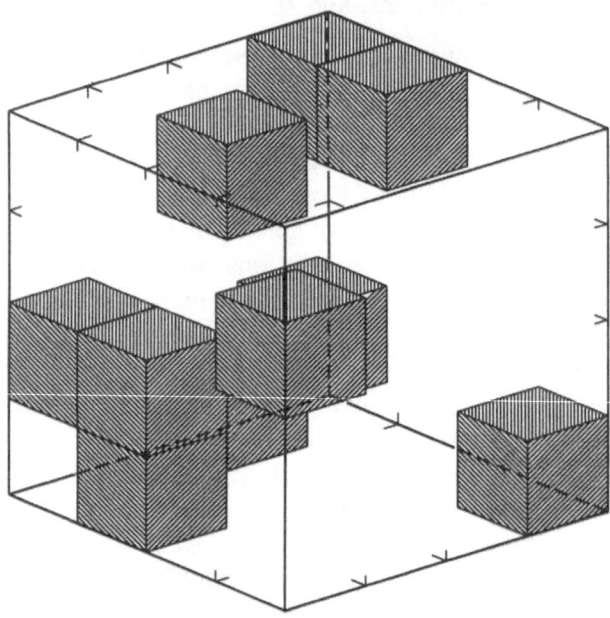

Figure 29. Susceptibility tomography to reconstruct a simulated 64-element cube. Top: The ten shaded elements have a susceptibility of $+2 \times 10^{-5}$; all others have a susceptibility of $+2.5 \times 10^{-5}$. Bottom: (Upper) gray-scale representation of the actual susceptibility distribution; (lower) the distribution determined by a simulated susceptibility tomography measurement using a fixed magnetizing field and maps of the normal component of the perturbation field over the six faces of the cube. (From [50], with permission)

improved sensitivity and specificity over existing techniques, and is it easy and cost effective to use? For just these reasons, the trend in NDE is toward techniques that provide images. Studies at Vanderbilt and elsewhere show that SQUIDs provide an unparalleled sensitivity at low frequencies that should help meet the first criterion for acceptance. The techniques outlined in this chapter, even in their present rudimentary state of development, demonstrate that scanned SQUIDs can produce field images, and that deconvolution techniques can produce source images. Since the stability of the image deconvolution process is determined by coil-to-source distance, noise, and the current or magnetizing-field distribution, it is clear that sensitive, high-resolution, closely-spaced SQUIDs will be advantageous. Given the promise of these and other advanced imaging techniques, it is enticing to consider the future use of SQUID cameras based upon digital SQUID techniques to provide a thousand integrated digital SQUIDs on a single chip[76]. Meanwhile, there is much work that can be done to explore even more fully the mathematics and experimental techniques of magnetic imaging with SQUIDs.

5. Acknowledgements

The preparation of this manuscript and much of the research described within it was funded by grants from the Air Force Office of Scientific Research and the Electric Power Research Institute. I am indebted to James Cadzow, William Jenks, Xangkang Li, Yu Pei Ma, Bradley Roth, Nestor Sepulveda, Daniel Staton, Shaofen Tan and Ian Thomas for their major contributions to the Vanderbilt magnetic-imaging effort, and their papers from which I have drawn heavily in preparing this chapter. I thank William Jenks, Margaret Khayat, Eduardo Parente Ribeiro, Daniel Staton and Leonora Wikswo for their comments on this manuscript, and Licheng Li for her care in preparing the illustrations.

References

1. Alvarez, R.E. (1990) "Biomagnetic Fourier imaging," *IEEE Trans. Med. Imaging*, **Vol. 9, no. 3**, pp. 299–304.
2. Alvarez, R.E. (1991) "Filter functions for computing multipole moments from the magnetic field normal to a plane," *IEEE Trans. Med. Imaging*, **Vol. 10, no. 3**, pp. 375–381.
3. Arzbaecher, R.C., and Brody, D.R., "The lead field: vector and tensor properties," in *The Theoretical Basis of Electrocardiology*, C.V. Nelson and D.B. Geselowitz, Eds., (Clarendon Press, Oxford, 1976), pp. 175–201.
4. Barbosa, C.H., Bruno, A.C., Scavarda, Lima, E.A., Ribeiro, P.C., and Kelver, C. (1995) "Image processing techniques for NDE SQUID," *IEEE Trans. Applied Supercond.*, **Vol. 5, no. 2**, pp. 2478–2485.
5. Beardsley, I.A. (1989) "Reconstruction of the magnetization in a thin film by a combination of Lorentz microscopy and external field measurements," *IEEE Trans. Mag.*,

Vol. 35, no. 1, pp. 671–677.

6. Benzing, W., Scherer, T., and Jutzi, W. (1993) "Inversion calculation of two dimensional current distributions from their magnetic field," *IEEE Trans. Applied Supercond.*, **Vol. 3, no. 1**, pp. 1902–1905.

7. Bradshaw, L.A. (1995) "Measurement and modeling of gastrointestinal bioelectric and biomagnetic fields," Ph.D. Dissertation, Department of Physics and Astronomy, Vanderbilt University.

8. Bruno, A.C., Ribeiro, P.C., von der Weid, J.P., and Eghrari, I.R. (1985) "Spatial discrimination: An alternative approach," in *Biomagnetism: Applications and Theory*, H. Weinberg, G. Stroink, and K. Katila, Eds., Pergamon Press, pp. 67–72.

9. Bruno, A.C., Ribeiro, P.C., and von der Weid, J.P. (1986) "Discrete spatial filtering with SQUID gradiometers in biomagnetism," *J. Appl. Phys.*, **Vol. 59, no. 7**, pp. 2584–2589.

10. Bruno, A.C., Guida, A.V., and Ribeiro, P.C. (1988) "Planar gradiometer input signal recovery using a Fourier technique," in *Biomagnetism '87*, K. Atsumi, S. Ueno, T. Katila, and S.J. Williamson, Eds., Tokyo Denki Univ. Press, Tokyo, pp. 454–457.

11. Bruno, A.C. and Ribeiro, P.C. (1989) "Spatial deconvolution algorithm for superconducting planar gradiometer arrays," *IEEE Trans. Mag.*, **Vol. 25, no. 2**, pp. 1219–1222.

12. Cadzow, J.A., and Li, X. (1995) "Blind deconvolution," *Digital Signal Processing*, **Vol. 5, no. 1**, pp. 3–20.

13. Cruse, T.A. (1978) "Two-dimensional BIE fracture mechanics analysis," *Applied Mathematical Modeling*, **Vol. 2**, pp. 287–293.

14. Cadzow, J.A. "Blind deconvolution via cumulant extrema," in preparation.

15. Dallas, W.J. (1985) "Fourier space solution to the magnetostatic imaging problem," *Applied Optics*, **Vol. 24, no. 24**, pp. 4543–4546.

16. Donaldson, G.B. (1989) "SQUIDs for everything else," in *Superconducting Electronics*, H. Weinstock and M. Nisenoff, Eds., New York: Springer Verlag, pp. 175–207.

17. Fitzpatrick, G.L., Thome, D.K., Skaugset, R.L., and Shih, E.Y.C. (1993) "The present status of magneto-optic eddy current imaging technology," *Review of Progress in QNDE*, **Vol. 12**, pp. 617–624.

18. Haacke, E.M., Liang, Z. and Izen, S.H. (1989) "Constrained reconstruction: A superresolution, optimal signal-to-noise alternative to the Fourier transform in magnetic resonance imaging," *Med. Phys.*, **Vol. 16**, pp. 388–397.

19. Hosaka, H. and Cohen, D. (1976) "Visual determination of generators of the magnetocardiogram", *J. Electrocardiology*, **Vol. 9**, pp. 426–432.

20. Hurley, D.C., Ma, Y.P., Tan, S., and Wikswo, Jr., J.P. (1993) "Imaging of small defects in nonmagnetic tubing using a SQUID magnetometer," *Res. Nondestr. Eval.*, **Vol. 5**, pp. 1–29.

21. Hurley, D.C., Ma, Y.P., Tan, S., and Wikswo, Jr., J.P. (1993) "A comparison of SQUID imaging techniques for small defects in nonmagnetic tubes," *Review of Progress in QNDE*, **Vol. 12**, pp. 633–640.

22. Ioannides, A.A., Bolton, J.P.R., and Clarke, C.J.S. (1990) "Continuous probabilistic solutions to the biomagnetic inverse problem," *Inverse Problems*, **Vol. 6**: pp. 1–20.

23. Ioannides, A.A., and Grimes, D.I.F., (1986) "A method for reconstructing line current sources from magnetic field data," *Inverse Problems*, **Vol. 2**: pp. 331–352.

24. Jenks, W.G., Ma, Y.P., Parente Ribeiro, E., and Wikswo, Jr., J.P., "SQUID NDE of composite materials with magnetic tracers," in preparation.

25. Jenks, W.G., Sadeghi, S.S.H., and Wikswo, Jr., J.P. "A review of SQUID magnetometers for non-destructive testing," *Journal of Physics D*, in press.

26. Kullmann, W., and Dallas, W.J. (1987) "Fourier imaging of electrical currents in the human brain from their magnetic fields," *IEEE Trans. Biomed. Eng.*, **Vol. BME-34, no. 11**, pp. 837–842.

27. Mark Leifer, personal communication, 1974.

28. Li, X., Cadzow, J.A., and Wikswo, Jr., J.P. "Reconstruction of electric current

density distribution from magnetometer measurements," in preparation.

29. Lines, L.R. and Treitel, S. (1984) "Tutorial: A review of least-squares inversion and its application to geophysical problems," *Geophys. Prosp.*, **Vol. 32**, pp. 159–186. .

30. Ma, Y.P., Thomas, I.M., Lauder, A., and Wikswo, Jr., J.P. (1993) "A high resolution imaging susceptometer," *IEEE Trans. Applied Supercond.*, **Vol. 3, no. 1**, pp. 1941–1944.

31. Ma, Y.P., Thomas, I.M., and Wikswo, Jr., J.P. "Magnetic susceptibility imaging systems," in preparation.

32. Ma, Y.P. and Wikswo, Jr., J.P. (1994) "SQUID eddy current techniques for detection of second layer flaws," *Review of Progress in QNDE*, **Vol. 13**, pp. 303–309.

33. Ma, Y.P. and Wikswo, Jr., J.P. (1995) "Techniques for depth-selective, low-frequency eddy current analysis for SQUID-based non-destructive testing," *J. Nondestr. Eval.*, **Vol. 14, no. 3**, pp. 149–167.

34. Ma, Y.P. and Wikswo, Jr., J.P. (1996) "Depth-selective SQUID eddy current techniques for second layer flaw detection," *Review of Progress in QNDE*, **Vol. 15**, pp. 401–408.

35. Ma, Y.P. and Wikswo, Jr., J.P. "The magnetic field produced by a cylindrical flaw with finite depth in a thick current carrying plate," in preparation.

36. Ma, Y.P. and Wikswo, Jr., J.P. "The magnetic field produced by an elliptical flaw in a current carrying plate," in preparation.

37. Ma, Y.P. and Wikswo, Jr., J.P. "Magnetic field of a subsurface spherical flaw inside a current-carrying conductor, " in preparation.

38. McKirdy, D. McA., Cochran, A., Donaldson, G.B., and McNab, A. (1996) "Forward and inverse processing in electromagnetic NDE using SQUIDs," *Review of Progress in QNDE*, **Vol. 15**, pp. 347–354.

39. Oh, S., Ramon, C., Marks II, R.J., Nelson, A.C., and Meyer, M.G. (1993) "Resolution enhancement of biomagnetic images using the method of alternating projections," *IEEE Trans. Biomed. Eng.*, **Vol. 40, no. 4**, pp. 323–328.

40. Oldenburg, D. (1990) "Inversion of electromagnetic data: An overview of new techniques," *Surveys in Geophysics*, **11**, pp. 231–270.

41. Marcuello-Pascual, A., Kaikkonen, P., and Pous, J. (1992) "2-D inversion of MT data with a variable model geometry," *Geophys. J. Int.*, **110**, pp. 297–304.

42. Plonsey, R., *Bioelectric Phenomena*, (McGraw Hill, New York, 1969),

43. Ramon, R., Meyer, M.G., Nelson, A.C., Spelman, F.A., and Lamping, J. (1993) "Simulation studies of biomagnetic computed tomography," *IEEE Trans. Biomed. Eng.*, **Vol. 40**, pp. 317–322.

44. Reitz, J.R. and Milford, F.J. (1967) *Foundations of Electromagnetic Theory*, Addison Wesley, p. 189.

45. Parente Ribeiro, E., personal communication.

46. Ribeiro, P.C., Bruno, A.C., Paulsen, C.C., and Symko, O.G. (1987) "Spatial Fourier transform method for evaluating SQUID gradiometers," *Rev. Sci. Instrum.*, **Vol. 58, no. 8**, pp. 1510–1513.

47. Roth, B.J., Sepulveda, N.G., and Wikswo, Jr., J.P. (1989) "Using a magnetometer to image a two-dimensional current distribution," *J. Appl. Phys.*, **Vol. 65**, pp. 361–372.

48. Roth, B.J. and Wikswo, Jr., J.P. (1986) "Electrically-silent magnetic fields," *Biophys. J.*, **Vol. 50**, pp. 739–745.

49. Roth, B.J. and Wikswo, Jr., J.P. (1990) "Apodized pickup coils for improved spatial resolution of SQUID magnetometers," *Rev. Sci. Instrum.*, **Vol. 61**, pp. 2439–2448.

50. Sepulveda, N.G., Thomas, I.M., and Wikswo, Jr., J.P. (1994) "Magnetic susceptibility tomography for three-dimensional imaging of diamagnetic and paramagnetic objects," *IEEE Trans. Mag.*, **Vol. 30, no. 6**, pp. 5062–5069.

51. Sepulveda, N.G., Staton, D.J., and Wikswo, Jr., J.P. (1992) "A mathematical analysis of the magnetic field produced by flaws in two-dimensional current-carrying conductors," *J. Nondestruc. Eval.*, **Vol. 11, no. 2**, pp. 89–101.

52. Sepulveda, N.G. and Wikswo, Jr., J.P. (1996) "A numerical study of the use of

SQUID magnetometers to detect hidden flaws in conducting objects," *J. Appl. Phys.*, **Vol. 79, no. 4**, pp. 2122–2135.

53. Sepulveda, N.G. and Wikswo, Jr., J.P. "Differential operators and their applications to magnetic measurements using SQUID magnetometers," unpublished.

54. Smith, W.E., Dallas, W.J., Kullmann, W.H., and Schlitt, H.A. (1990) "Linear estimation theory applied to the reconstruction of a 3-D vector current distribution," *Applied Optics*, **Vol. 29, no. 5**, pp. 658–667.

55. Staton, D.J. (1994) "Magnetic imaging of applied and propagating action current in cardiac tissue slices: Determination of anisotropic electrical conductivities in a two dimensional bidomain," Ph.D. Dissertation, Department of Physics and Astronomy, Vanderbilt University.

56. Staton, D.J., Friedman, R.N., and Wikswo, Jr., J.P. (1993) "High resolution SQUID imaging of octupolar currents in anisotropic cardiac tissue," *IEEE Trans. Appl. Supercond.*, **Vol. 3, no. 1**, pp. 1934–1936.

57. Staton, D.J., Rousakov, S.V. and Wikswo, Jr., J.P. (1996) "Conductivity imaging in plates using current injection tomography," *Review of Progress in QNDE*, **Vol. 15**, pp. 845–851.

58. Tan, S. (1992) "Linear system imaging and its applications to magnetic measurements by SQUID magnetometers," Ph.D. Dissertation, Department of Physics and Astronomy, Vanderbilt University.

59. Tan, S., Ma, Y.P., Thomas, I.M., and Wikswo, Jr., J.P. (1993) "High resolution SQUID imaging of current and magnetization distributions," *IEEE Trans. Applied Supercond.*, **Vol. 3, no. 1**, pp. 1945–1948.

60. Tan, S., Ma, Y.P., Thomas, I.M., and Wikswo, Jr., J.P. (1996) "Reconstruction of a two-dimensional susceptibility distribution from the magnetic field of non-ferromagnetic materials," *IEEE Trans. Mag.*, **Vol. 32, no. 1**, pp. 230–234.

61. Tan, S., Roth, B.J., and Wikswo, Jr., J.P. (1990) "The magnetic field of cortical current sources: The application of a spatial filtering model to the forward and inverse problems," *Electroencep. and Clin. Neurophys.*, **Vol. 76**, pp. 73–85 .

62. Tan, S., Sepulveda, N.G., and Wikswo, Jr., J.P. (1995) "A new finite-element approach to reconstruct a bounded and discontinuous two-dimensional current image from a magnetic field map," *J. Comput. Phys.*, **Vol. 122**, pp. 150–164.

63. Tan, S. and Wikswo, Jr., J.P. "A comparison of the Hosaka-Cohen transformation and the exact imaging algorithm for determining current distributions from magnetic field maps," in preparation.

64. Thomas, I.M., Freake, S.M., Swithenby, S.J., and Wikswo, Jr., J.P. (1993) "A distributed quasi-static ionic current source in the 3-4 day old chicken embryo," *Phys. Med. Biol.*, **Vol. 38**, pp. 1311–1328.

65. Thomas, I.M. and Friedman, R.N. (1995) "Study of macrophage activity in rat liver using intravenous superparamagnetic tracers," in *Proc. 9th Inter. Conf. on Biomagnetism, Vienna*, C. Baumgartner, L. Deecke, G. Stroink, and S.J. Williamson, Eds., IOS Press, Amsterdam, Netherlands, pp. 809–813.

66. Thomas, I.M., Ma, Y.P., Tan, S., and Wikswo, Jr., J.P. (1993) "Spatial resolution and sensitivity of magnetic susceptibility imaging," *IEEE Trans. Appl. Supercond.*, **Vol. 3, no. 1**, pp. 1937–1940.

67. Thomas, I.M., Ma, Y.P., and Wikswo, Jr., J.P. (1993) "SQUID NDE: Detection of surface flaws by magnetic decoration," *IEEE Trans. Appl. Supercond.*, **Vol. 3, no. 1**, pp. 1949–1952.

68. Thomas, I.M., Moyer, T.C., and Wikswo, Jr., J.P. (1992) "High resolution magnetic susceptibility imaging of geological thin sections: Pilot study of a pyroclastic sample from the Bishop tuff," *Geophys. Res. Lett.*, **Vol. 19, no. 21**, pp. 2139–2142.

69. Tripp, J.H. (1983) "Physical concepts and mathematical models," in *Biomagnetism: An Interdisciplinary Approach*, S.J. Williamson, G.L. Romani, L. Kaufman and I. Modena, Eds., Plenum, New York, pp. 101–139.

70. Weinstock, H. (1991) "A review of SQUID magnetometry applied to nondestructive

evaluation," *IEEE Trans. Mag.*, **Vol. 27, no. 2**, pp. 3231–3236.

71. Weinstock, H. and Nisenoff, M (1985) "Non-destructive evaluation of metallic structures using a SQUID gradiometer," in *SQUID '85, Proc. 3rd International Conference on Superconducting Quantum Devices*, H.D. Hahlbohm and H. Lubbig, Eds., Berlin: de Gruyter, pp. 843–847.

72. Wikswo, Jr., J.P. (1978) "The calculation of the magnetic field from a current distribution: Application to finite element techniques," *IEEE Trans. Mag.*, **Vol. MAG-14**, pp. 1076–1077.

73. Wikswo, Jr., J.P. (1978) "Optimization of SQUID differential magnetometers," *AIP Conference Proceedings*, **Vol. 44**, pp. 145–149.

74. Wikswo, Jr., J.P. (1983) "Theoretical aspects of the ECG-MCG relationship," in *Biomagnetism: An Interdisciplinary Approach*, S.J. Williamson, G.L. Romani, L. Kaufman and I. Modena, Eds., Plenum, New York, pp. 311–326.

75. Wikswo, Jr., J.P. (1994) "The complexities of cardiac cables: Virtual electrode effects," *Biophys. J.*, **vol. 66**, pp. 551–553.

76. Wikswo, Jr., J.P. (1995) "SQUID magnetometers for biomagnetism and nondestructive testing: Important questions and initial answers," *IEEE Trans. Applied Supercond.*, **Vol. 5, no. 2**, pp. 74–120.

77. Wikswo, Jr., J.P. "Magnetic imaging of cellular action currents," these proceedings.

78. Wikswo, Jr., J.P. and Barach, J.P. (1982) "Possible sources of new information in the magnetocardiogram," *J. Theoretical Biol.*, **Vol. 95**, pp. 721–729.

79. Wikswo, Jr., J.P., Crum, D.B., Henry, W.P., Ma, Y.P., Sepulveda, N.G., and Staton, D.J. (1993) "An improved method for magnetic identification and localization of cracks in conductors," *J. Nondestr. Eval.*, **Vol. 12, no. 2**, pp. 109–119.

80. Wikswo, Jr., J.P., Ma, Y.P., Sepulveda, N.G., Tan, S., Thomas, I.M., and Lauder, A. (1993) "Magnetic susceptibility imaging for nondestructive evaluation," *IEEE Trans. Applied Supercond.*, **Vol. 3, no. 1**, pp. 1995–2002.

81. Wikswo, Jr., J.P., Malmivuo, J.A.V., Barry, W.H., Leifer, M.C., and Fairbank, W.M. (1979) "Theory and application of magnetocardiography," in *Cardiovascular Physics*, D.N. Ghista, E. Van Vollenhoven and W. Yang, Eds., Karger, Basil, pp. 1–67.

82. Wikswo, Jr., J.P. and Roth, B.J., (1988) "Magnetic determination of the spatial extent of a single cortical current source: A theoretical analysis," *Electroenceph. and Clin. Neurophys.*, **Vol. 69**, pp. 266–276.

83. Wikswo, Jr., J.P., van Egeraat, J.M., Ma, Y.P., Sepulveda, N.G., Staton, D.J., Tan, S., and Wijesinghe, R.S. (1990) "Instrumentation and techniques for high-resolution magnetic imaging," *Digital Image Synthesis and Inverse Optics*, A.F. Gmitro, P.S. Idell, and I.J. LaHaie, Eds., *SPIE Proceedings*, **Vol. 1351**, pp. 438–470.

84. Wikswo Jr., J.P. and Swinney, K.R. (1984) "A comparison of scalar multipole expansions," *J. Appl. Phys.*, **Vol. 56**, pp. 3039–3049.

85. Wikswo, Jr., J.P. and Swinney, K.R. (1985) "Scalar multipole expansions and their dipole equivalents," *J. Appl. Phys.*, **Vol. 57**, pp. 4301–4308.

INDEX

Accelerometer, 571-572, 587-588
Adaptive method
  coefficients, 164, 167-171, 173
  frequency-dependent, 118, 165
  frequency-independent, 167
  noise cancellation, 117-118, 164-171,
    173
  procedure, 166-168, 173
A/D conversion, 473
Additional positive feedback, 17-18,
  88-100
Aging aircraft, 622-623, 673-675
Alternating projections, 669-672
Amplifier, 84-87
Arrhythmia, 407-409, 432-438

Balancing, 118, 127, 133-136, 167,
  169, 174
Bandwidth limit, 72-76
Bias mode, 79-82
Biomagnetic system
  dewar, 467-468
  multichannel, 469-471
  shielded room, 467-468
  single channel, 466-467
Biot-Savart law, 631-633
Blind deconvolution, 675-677
Boundary element method, 364,
  383-385

Cardiac current, 308-312, 317-327
  bidomain model, 312-313
  electrophysiology, 313-314
Cerebral plasticity, 481-484
Clinical studies, 479-481
Cochlear implant, 479
Common mode
  signal, 121-122, 126, 132, 134,
    143-144, 146, 153
  term, 132, 140, 148

CT, 491, 494, 496, 500-501, 504,
  507, 509-510
Current dipole, 368-369
Current imaging, 347-350, 631-677
Cut-off
  distance, 146
  frequency, 146

Data acquisition, 471-474
DC SQUID, 1-2, 5-18, 30-33, 44-51
  64-67, 179-233 257-266,
  268-269
Digital signal processing, 473
Digital SQUID, 105-111
Dipole field
  biological, 368-369, 379-382
  equation, 677-679, 681-689
  mobile operation, 521
Dipole fitting, 654-655
Dynamic range, 124-125, 174

ECG, 314-317, 388-390, 399-400, 403,
  406-408, 413-414
EEG, 363-364, 391-392, 491- 497, 501,
  507
Eddy current
  NDE, 604-608, 672-675
  room, 120
  shielding, 428-429
  signal, 132-134
  term, 132, 135
  vector, 117, 129-130, 132-133, 135
EEG, 149, 151
Electronics
  digital EEG, 149
  digital SQUID, 125, 149, 151
  feedback loop, 134, 151
Electrophysiology
  cells, 447-452
  heart, 398-399

Environment, 117-118, 120, 124, 126,
 147, 149, 151-157, 160-169
Epilepsy, 479, 482

Fabrication
 HTS, 45-48, 235-288
 LTS, 194-195
Feedback
 configuration, 82-84
 loop, 67-70
Fetal magnetocardiography, 409-410
Finite element
 analysis, 659-669
 method, 385-386
Flux focuser, 263, 266-269
Flux pinning, 236
Flux transformer, 127, 135, 137-138,
 149, 266-274
fMRI, 491-492, 498-501, 507
Forward solution, 414-420, 452-460

Geophysical application, 38-40, 53
Gradient
 1st, 125, 128-129, 140-141, 148-149
  158-160, 165, 172
 2nd, 126, 141, 143, 172
 3rd, 118, 143, 158, 172
 coefficients, 148
 spatial, 152, 172
 tensor, 128, 140-141, 143, 165
Gradiometer
 1st-order, 118, 122, 127-130,
  139-141, 148-150, 153-174
 2nd-order, 119, 127, 140-142, 148,
  152, 155, 161, 164-168, 173
 3rd-order, 118, 127, 136-138,
  142-149, 152-174
 asymmetric, 138
 axial, 128, 135-137, 140, 142
 gravity, 38-40, 579-586, 588-595
 hardware, 121-123, 127-128, 136,
  149
 ideal, 141-142, 147

Gradiometer (continued)
 k-th-order, 127-128, 142
 mobile operation, 525-527
 optimization, 340-346
 planar, 140
 software, 127-128, 139-141, 147-148,
  162-165, 174
 symmetric, 128
Gravity gradiometer. *See* Gradiometer
Gravity-wave detection, 35-38, 572-579

Harmonic distortion, 70-72
Hemispheric asymmetry, 481
Hosaka-Cohen transformation, 652-654
HTS SQUID, 44-52, 21-222, 235-288,
 296-300, 485, 614-622,
 623-625

Image processing, 552-554, 677-689
Integrator, 76-79
Inverse
 problem, 602-603, 609-613,
  629-695
 solution, 420-424, 435-436,
  460-466

Josephson junction, 1-5, 40-43
 bicrystal, 242-244
 biepitaxial, 250
 RSJ model, 2-5, 241-242, 250
 step-edge, 244-249, 251-257

Lead field analysis, 656-659

MADOM, 551-557
Magnetic noise, 119-121
 car traffic, 121, 123, 174
 environmental, 118-121, 127, 136,
  145, 149, 164, 168, 174
 low frequency, 118, 122, 161-163
 moving dipole, 146
 rotating dipole, 147-148, 153-157
Magnetocardiogram, 53, 409-410

Magnetometer, 25-28, 44-45, 118,
        124-127, 130-131, 134-141,
        147, 155-157, 161-162, 165,
        170-171, 198-210, 213-222,
        235-288, 524, 601-602
Mapping
  analysis, 431-434
  magnetization, 682-684
  NDE, 602-603, 609-613
  sensory areas, 478-479
Maxwell equations, 370-371, 445-447
MCG, 149, , 314-315, 388-390,
        395-412, 413-416
MEG, 117-118, 124, 147-160, 163-167,
        173-174, 363-364, 391-392,
        479, 491-492, 500-501,
        506-507, 509, 511
Microscope (SQUID), 331-339
MicroSQUID, 308, 328-333, 350
Microwave resonator, 51-52
Mobile operation, 520, 529-532
MRI, 363, 386-387, 445, 479, 491,
        496, 501-511
Multilayer fabrication, 45-48
Multiloop SQUID, 274-276
Multiple SQUID circuit, 84-88
Myocardischemia, 401-402
NDE, 599-628, 629-695

Neuromagnetic measurement, 475-478
Noise
  accelerometer, 572
  brain, 164
  cancellation, 117-119, 148-152,
        155-157,164-174
  environmental, 118-121, 136, 145,
        149, 164, 168, 174
  f^1, 13-17, 44
  flux, 6, 25, 27, 52-53
  low frequency, 118, 122, 161-163
  mobile operation, 539-558
  Nyquist, 3
  pulse tube refrigerator, 300-302

Noise (continued)
  spectrum, 161
  thermal, 7-8, 12-13
  white, 43, 144, 160-164
NQR, 33-35

Packaging, 211
PET, 363, 491-498, 500-502, 504-511
Phase shift, 134-135
PLD, 236, 245, 248, 252, 272
Poisson equations, 367, 372-374
Power line
  field, 123-124, 174
  frequency, 120, 123
  signal, 161-162

Read-out electronics, 17-18, 195-198
Refrigeration, 289-305
  general, 290-291
  pulse tube, 291-305
    basics, 291-296
    HTS SQUID, 296-300
    noise, 300-302
Relaxation oscillator, 100-105
Resonator, 51-52, 277-279
RF SQUID, 1-2, 18-25, 51-52,
        259-269, 276-279

SGMS, 532-539
Shielded room, 117-121, 126-127, 135,
        144-145, 149, 152-162,
        171-174, 428-431, 467-468,
        470
Slew rate, 123-124, 174
Source modeling, 417-420
Spatial filter, 143, 174, 634-652
SPECT, 363
SPET, 491-492, 495-498, 506, 508-511
Steel, 601-602, 605-608, 623
Susceptometer, 29-30
Susceptibility tomography, 684-689

Target identification, 555-557

700

Time delay, 134

Vibration, 117-118, 125-127, 146,
        158-160, 162, 170, 174
Voltmeter, 30, 53

Washer SQUID, 257-266, 276-279

GLOSSARY OF ABBREVIATIONS AND ACRONYMS

A/D: Analog-to-digital
A/DC: Analog-to-digital converter
AEF: Auditory evoked field
AP: Accessory pathway
APF: Additional positive feedback
ASW: Antisubmarine warfare
AV: Atrio-ventricular

BCF: Bias current feedback
BEM: Boundary element method
BPTR: Basic pulse tube refrigerator
BSPM: Body surface potential map

CT: Computed tomography

DIPTR: Double inlet pulse tube refrigerator
DLC: Diamond-like carbon
DROS: Double relaxation oscillation SQUID
DSP: Digital signal processor

ECD: Equivalent current dipole
ECG: Electrocardiogram
ECT: Eddy-current testing
EDM: Electric discharge machined
EEG: Electroencephalogram
EPI: Echoplanar imaging
EPS: Electrophysiological studies

FEM: Finite-element method
FET: Field-effect transistor
[^{18}F]FDG: ^{18}F-labelled 2-fluoro-deoxy-D-glucose
FFT: Fast Fourier transform
FLL: Flux-locked loop
fMRI: Functional MRI
FT: Fourier transform
FTA: Flux transformer assembly

GP-B: Gravity probe - B

HC: Hosaka-Cohen (transformation)
HTS: High-temperature-superconducting

IBE: Ion-beam etching

LPF: Low-pass filter
LTS: Low-temperature-superconducting

MADOM: Magnetic and acoustic detection of mines
MCG: Magnetocardiogram
MCM: Mine countermeasures
MEG: Magnetoencephalogram
MFM: Magnetic field map
MI: Myocardial infarction
MOI: Magneto-optic imager or magneto-optic inspection
MRI: Magnetic resonance imaging
MSD: Mean-square deviation
MUDSS: Mobile underwater debris survey system

NDE: Nondestructive evaluation
NMR: Nuclear magnetic resonance
NQR: Nuclear quadrupole resonance

OPTR: Orifice pulse tube refrigerator

PARSIPS: Parallel signal and image processing system
PBCO: Praseodymium barium copper oxide
PET: Positron emission tomography
PLD: Pulsed laser deposition
PTR: Pulse tube refrigerator

RIE: Reactive ion etching
ROS: Relaxation oscillation SQUID
RRR: Residual resistance ratio
RSJ: Resistively-shunted-junction (model)

SAS: Synthetic aperture sonar
SEJ: Step-edge junction
SGG: Superconducting gravity gradiometer
SGMS: Superconducting gradiometer/magnetometer sensor
SHAD: Second harmonic detection
SNAP: Selective niobium anodization process
SNR: Signal-to-noise ratio
SNS: Superconducting-normal-superconducting (junction)
SPECT: Single positron emission computerized tomography

SPET: Single photon emission tomography
SQUID: Superconducting quantum interference device
SSA: Superconducting six-axis accelerometer
STEP: Satellite test of the equivalence principle
STO: Strontium titanate

THD: Total harmonic distortion

UXO: Unexploded ordnance

VF: Ventricular fibrillation
VIM: Volume-integral method
VT: Ventricular tachycardia

YBCO: Yttrium barium copper oxide